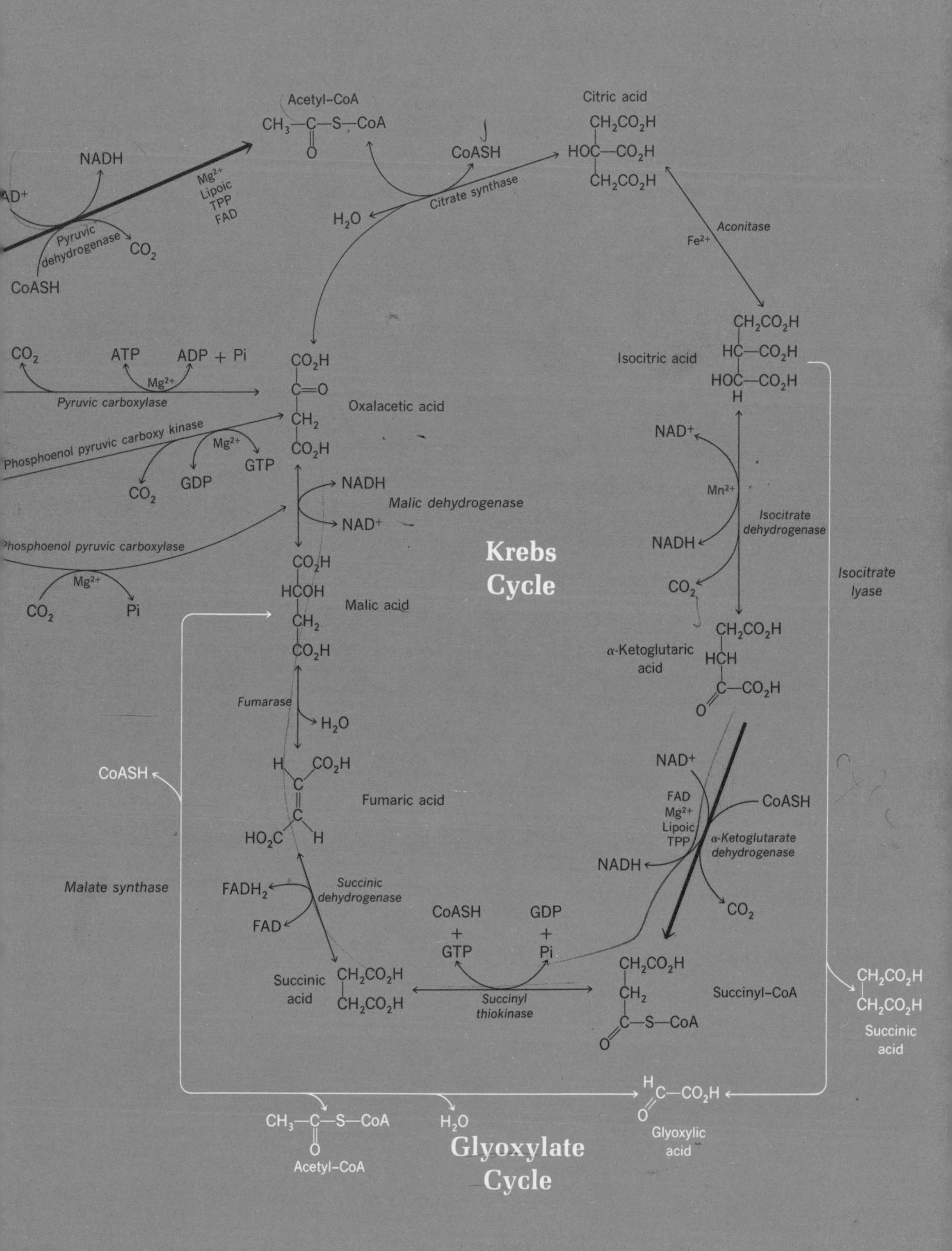

Acetyl–CoA
CH_3—C—S—CoA
O
NADH
NAD^+
Mg^{2+}
Lipoic
TPP
FAD
Pyruvic dehydrogenase
CO_2
CoASH
CoASH
Citrate synthase
H_2O
Citric acid
CH_2CO_2H
HOC—CO_2H
CH_2CO_2H
Aconitase
Fe^{2+}
CO_2
ATP
ADP + Pi
Mg^{2+}
Pyruvic carboxylase
Phosphoenol pyruvic carboxy kinase
Mg^{2+}
GTP
GDP
CO_2
CO_2H
C=O
CH_2
CO_2H
Oxalacetic acid
NADH
NAD^+
Malic dehydrogenase
Phosphoenol pyruvic carboxylase
Mg^{2+}
CO_2
Pi
CO_2H
HCOH
CH_2
CO_2H
Malic acid
Krebs Cycle
Isocitric acid
CH_2CO_2H
HC—CO_2H
HOC—CO_2H
H
NAD^+
Mn^{2+}
Isocitrate dehydrogenase
NADH
CO_2
Isocitrate lyase
α-Ketoglutaric acid
CH_2CO_2H
HCH
C—CO_2H
O
Fumarase
H_2O
H
CO_2H
C
C
HO_2C
H
Fumaric acid
CoASH
Malate synthase
NAD^+
FAD
Mg^{2+}
Lipoic
TPP
CoASH
α-Ketoglutarate dehydrogenase
NADH
CO_2
$FADH_2$
FAD
Succinic dehydrogenase
CoASH
+
GTP
GDP
+
Pi
Succinic acid
CH_2CO_2H
CH_2CO_2H
Succinyl thiokinase
CH_2CO_2H
CH_2
C—S—CoA
O
Succinyl–CoA
CH_2CO_2H
CH_2CO_2H
Succinic acid
H
C—CO_2H
O
Glyoxylic acid
CH_3—C—S—CoA
O
Acetyl–CoA
H_2O
Glyoxylate Cycle

Outlines of Biochemistry

Outlines of Biochemistry

THIRD EDITION

ERIC E. CONN
and
P. K. STUMPF
Department of Biochemistry
and Biophysics
University of California at Davis

JOHN WILEY & SONS, INC.
New York London Sydney Toronto

Library of Congress Cataloging in Publication Data

Conn, Eric E.

Outlines of biochemistry.

Includes bibliographical references.
1. Biological chemistry. 2. Metabolism.
I. Stumpf, Paul Karl, 1919– joint author.
II. Title.

QP514.2.C65 1972 574.1'92 72-325
ISBN 0-471-16844-0

Printed in the United States of America

10 9 8 7 6 5 4 3

Preface

This revision of *Outlines* is long overdue. Indeed, the rapid advances in the area of protein and nucleic acid metabolism made those sections in the previous edition out of date shortly after its publication. Research in other areas such as metabolic regulation, photosynthesis, and nitrogen fixation has produced new information that should be contained in an introductory textbook of biochemistry. Those sections of the book dealing with these topics have been completely rewritten. A new chapter comparing the features of procaryotic and eucaryotic cells has been prepared and is located at the beginning of Part II. The remainder of the book has been updated and, in many instances, the material has been rephrased, hopefully to be more meaningful to the beginning student.

Continuing experience with the book in our large introductory course in biochemistry at Davis has led to some rearrangement of the material. The first part continues to treat the chemistry of biological compounds. This is followed by the largest part of the book, which deals with the metabolism of the energy-yielding compounds. These are the biochemical monomers, the carbohydrates, the lipids, and the small nitrogen compounds—amino acids, purines, and pyrimidines and their nucleotides. Because the basic life processes of photosynthesis and the nitrogen cycle are primarily concerned with monomers, chapters on this topic are included in this part as well. Part II is introduced with the chapter entitled "The Cell—Its Biochemical Organization," and this is followed by chapters on biochemical energetics, enzymes, and vitamins and coenzymes—Chapters 6, 7, and 8 in the previous edition.

The final part of the book is concerned with the metabolism of informational molecules, the biopolymeric nucleic acids and proteins. This material, together with the chapter on metabolic regulation, has been completely redone and hopefully contains facts and principles which will not so rapidly become outdated.

Three appendices are provided, the one reviewing modern concepts in organic chemistry being replaced by one dealing with the biochemical literature.

Our efforts in preparing this third edition have been aided by those who have used the previous editions and were kind enough to forward their suggestions and criticisms. The students enrolled in Biochemistry 101AB at Davis are particularly to be acknowledged for their skill in finding errors and their suggestions concerning the organization of the text. We wish to acknowledge the assistance of our colleagues, G. E. Bruening, M. E. Dahmus, and I. H. Segel, who read certain chapters. We are particularly grateful to J. L. Wilson and P. V. Benko, who read the entire text and made many valuable suggestions. In addition to the earlier translations into Japanese, Italian, and Spanish, the book has now been translated into Portuguese.

Davis, California
May, 1972

E. E. C.
P. K. S.

Preface to the Second Edition

At first thought one might assume that an introductory textbook of biochemistry should not require revision after only three years. A wellknown textbook of thermodynamics published in 1923 went unrevised for 38 years. However, the authors of introductory biochemistry texts soon find themselves facing the same problems as the lecturer in an introductory biochemistry course. While a majority of the material presented in the course or text is neither new nor changes from year to year, still a significant portion of the subject matter will deal with some of the most recent developments in the field. At the same time, each year brings new findings which make the older, apparently established knowledge even more meaningful and, therefore, deserve their place in the appropriate lecture or chapters. Thus both the lecturer and the author experience the need to revise their material.

The preparation of this second edition is our attempt to meet that need. It also provides us with an opportunity to reorganize certain portions of the book with which students have had difficulty. Finally, as with all second editions, we are given the chance to correct errors—factual and typographical—which existed in the first edition.

Examination of the second edition will find its size increased by 75 pages although the number, titles, and sequence of the chapters remain the same. Chapters 3, 7, 13, and 14 have been extensively revised to present this familiar material in a different way and to include new findings which are relevant to these chapters. Recent developments in metabolic regulation and biochemical genetics have required the rewriting of Chapters 20 and 21. Similarly, because much has been learned about the control of cellular metabolism during glycolysis, the tricarboxylic acid cycle, and synthesis of lipids, references to these recent advances are included in Chapters 9, 11, 13, and elsewhere. Finally, new research observations in the rapidly developing areas of protein synthesis, photosynthesis, and nitrogen fixation have been incorporated in the chapters in which these topics are discussed. As with the previous edition, one of us is responsible for each chapter, this assignment being determined by our familiarity with the subject either through our research or teaching.

One important change in terminology was made in the second edition. Based on our teaching experience, we made the decision to convert the terms

NAD^+ and NADH to DPN^+ and DPNH, and $NADP^+$ and NADPH to TPN^+ and TPNH. It is difficult for both the instructor and the student to avoid confusion in the usage of the rather awkward terminology recently adopted by European biochemists. The more traditional DPN^+, TPN^+, DPNH, and TPNH are terms that are easily recognized in the lecture hall and are thus used throughout this textbook. Some might complain that these terms are inaccurate. We do not consider this argument valid if the terms are properly defined. Certainly CoASH is not a technically accurate abbreviation but it is an international symbol for this cofactor and is so recognized.

In preparing this edition we are indebted to our friends and colleagues who have kindly offered suggestions for improvement and have tactfully pointed out errors in the first edition. Special acknowledgment, however, should be paid to the students enrolled in Biochemistry 101 on the Davis Campus during the past three years. Their suggestions and those of Dr. I. H. Segel, who has instructed in this course with one of us, have been especially helpful and are much appreciated.

Finally, the authors wish to emphasize again that this book is designed to introduce the subject of biochemistry to students in a one-semester course. As a result of attempting to restrict the size of the book, important areas of modern biochemistry may not be covered. If our concept that this text develops the basic knowledge of biochemistry is correct, then we hope that our colleagues who frown on the lack of coverage of some areas will seize the opportunity and cover these deficiencies by their own lectures. While we would admit that the introduction can be accomplished through the use of different topics, we prefer to do it through the theme of intermediary metabolism. We are grateful that our opinion is shared by some number of teachers who adopted the first edition as a text. The book has been translated into Japanese, Italian, and Spanish.

Davis, California
May, 1966

E. E. C.
P. K. S.

Preface to the First Edition

This book is the outgrowth of experience gained from teaching a one-semester course in general biochemistry on both the Berkeley and the Davis campuses of the University of California.

It is our opinion that the subject of biochemistry can be introduced to upper-division undergraduates and first-year graduate students, majors and nonmajors, in one semester. In writing our book we have attempted to acquaint the students with a skeleton—a rather substantial skeleton to be sure—of intermediary metabolism. This has required a brief review of the chemical properties of compounds of biological interest; it appears in Part I. Part II opens with a plunge into energetics and enzymology to prepare the student for the backbone of the course. This is found in the chapters which describe the metabolism of carbohydrates, lipids, amino acids, and proteins. Not even an introductory course in biochemistry is complete without an attempt to integrate this knowledge of the metabolism of the major cell constituents through a consideration of the interrelations among the carbon, nitrogen, and energy cycles. This attempt is made in Part III.

Since every student in a biological science feels compelled to take an introductory biochemistry course, the instructors of that course are obliged to present some of the important areas of modern biochemistry. The problem is one of careful selection of topics and limited discussion of experimental data. This procedure requires generalization, but the student learns when he is made aware of the significant exceptions.

The appendix is employed to review pH and buffer problems, common weaknesses of most students entering biochemistry. This is followed by a brief discussion of some of the concepts of modern organic chemistry that can be applied to biochemical reactions. Finally, a description of methods commonly employed in biochemical research is included to familiarize the student with the terms employed. The student is advised to make use of the appendix as he proceeds through the book. Key references are given at the end of each chapter for the student to explore the area under discussion further. From these references he can be easily guided into the much larger literature of the particular area.

Although the theme of our book is intermediary metabolism, we do not claim to have covered all areas of metabolism. In addition, some of our readers may consider the chemical treatment of the major cellular components too brief. Finally, those accustomed to the discussion of the classical subjects usually found only in medical biochemistry texts will find these topics either missing or drastically abbreviated. We ask that our colleagues again consider the audience of upper-division and graduate students, majors and nonmajors, for whom this book was written.

Davis, California
January, 1963

E. E. C.
P. K. S.

Contents

PART I

Chemistry of Biological Compounds

pH and Buffers

Introduction

Living cells contain carbohydrates, lipids, amino acids, proteins, nucleic acids, nucleotides and related compounds in varying amounts. Although these compounds have an almost infinite number of chemical structures, the mass of these compounds is accounted for almost entirely by only six elements—carbon (C), hydrogen (H), oxygen (O), nitrogen (N), phosphorus (P), and sulfur (S). Moreover, two of the elements, hydrogen and oxygen, combine to make the most abundant cellular component, H_2O, which does not fall into any of the categories listed above. Over 90% of blood plasma is H_2O; muscle contains about 80% H_2O, and H_2O constitutes more than half of most other plant or animal tissues.

While H_2O is the most abundant cell component, it is also an indispensible compound for life. The nutrients which a cell consumes, the oxygen it uses in oxidation of those nutrients, and the waste products it produces are all transported by H_2O. It is useful therefore to note that this familiar, important chemical has a number of exceptional properties that make it peculiarly well-suited for its job as the solvent of life.

Some Important Properties of Water

Many of the physical properties of H_2O are uniquely different. Consider, for example, the group of compounds listed in Table 1-1. These compounds may be compared with H_2O either because they have good solvent properties or because they have the same number of electrons (isoelectronic). As can be seen, H_2O has the highest boiling point, the highest specific heat of vaporization, and by far the highest melting point of all these compounds. Pauling has expressed the anomolous behavior of H_2O in another way by comparing it with the hydrides of other elements in Group VI of the periodic table—H_2S, H_2Se, and H_2Te. When this is done, we would predict that H_2O should have a boiling point of $-100°C$ instead of $+100°C$ which it possesses.

Table 1-1

Some Physical Properties of Water and Other Compounds

Substance	Melting point (°C)	Boiling point (°C)	Heat of vaporization (cal/g)	Heat capacity (cal/g)	Heat of fusion (cal/g)
H_2O	0	100	540	1.000	80
Ethanol	−114	78	204	0.581	24.9
Methanol	−98	65	263	0.600	22
Acetone	−95	56	125	0.528	23
Ethyl acetate	−84	77	102	0.459	—
Chloroform	−63	61	59	0.226	—
NH_3	−78	−33	327	1.120	84
H_2S	−83	−60	132	—	16.7
HF	−92	19	360	—	54.7

The water molecule is highly polarized because the electronegative oxygen atom tends to draw electrons away from the hydrogen atoms, leaving a net positive charge surrounding the proton. Because of this polarization, water molecules behave like dipoles since they can be oriented toward both positive and negative ions. This property in turn accounts for the unusual ability of water to act as a solvent. Positive or negative ions in a crystal lattice can be approached by dipolar water molecules and brought into solution. Once in solution, ions of both positive and negative charge will be surrounded by protective layers of water molecules and further interaction between those ions of opposite charge will be subsequently decreased.

The high boiling and melting points of H_2O and its high heat of vaporization are the result of an interaction between adjacent water molecules known as hydrogen bonding. Briefly put, the term *hydrogen bond* refers to the interaction of a hydrogen atom that is covalently bonded to one electronegative atom with a second electronegative atom. There is a tendency for the hydrogen atom to associate with the second electronegative atom by sharing its electron pair, and a weak bond of approximately 4.5 cal/mole can exist. (In biological material the two atoms most commonly involved in hydrogen bonding are nitrogen (N) and oxygen (O).) In liquid water small transient chains of water molecules will occur due to this interaction.

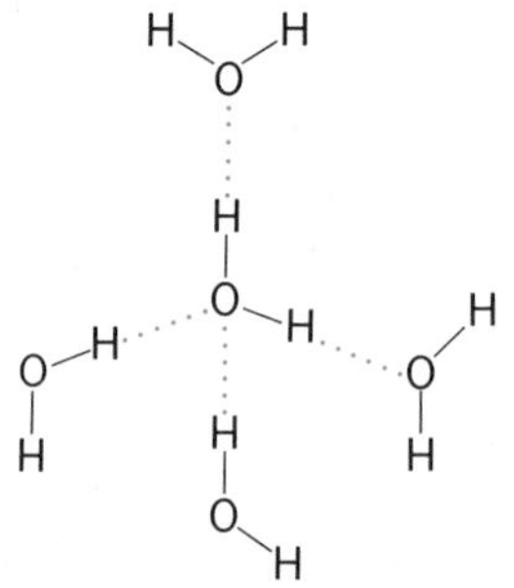

The energy necessary to disrupt the hydrogen bond (4–10 kcal/mole) is much less than that required to break an O—H covalent bond, and in solution hydrogen bonds are broken and formed readily. The additive effect with hydrogen bonding of water is a major factor in explaining many of the unusual properties of H_2O. Thus, the extra energy required to boil water and melt ice may be attributed largely to extensive hydrogen bonding.

Other unusual properties of water make it an ideal medium for living organisms. Thus, the specific heat capacity of H_2O—the number of calories required to raise the temperature of 1 g of water from 15 to 16°C—is 1.0 and is unusually high among several of the solvents just considered (ethanol, 0.58; methanol, 0.6; acetone, 0.53; chloroform, 0.23; ethyl acetate, 0.46). Only liquid ammonia is higher at 1.12. The higher the specific heat of a substance the less the change in temperature which results when a given amount of heat is absorbed by that substance. Thus, H_2O is well-designed for keeping the temperature of a living organism relatively constant. It is this property of water that also made the oceans of the earth an ideal environment for the origin of life and evolution of the primeval forms.

The heat of vaporization of H_2O, as already mentioned, is unusually high. Expressed as the specific heat of vaporization (calories absorbed per gram vaporized) the value for water is 540 at its boiling point and even higher at lower temperatures. This high value is very useful in helping the living organism keep its temperature constant, since a large amount of heat can be dissipated by vaporization of H_2O.

The high heat of fusion of H_2O (80 cal/g compared with 25 for ethanol, 22 for methanol, 17 for H_2S, 23 for acetone) is also of significance in stabilizing the biological environment. While cellular water rarely freezes in higher living forms, the heat released by H_2O on freezing is a major factor in decreasing the actual lowering of the temperature of a body of water during the winter. Thus, a gram of H_2O must give up eighty times as much heat in freezing at 0°C as it does in being lowered from 1 to 0°C just before freezing.

One final example of a property of H_2O that is of biological significance may be cited. This is the fact that H_2O passes through its maximum density at 4°C. That is, H_2O expands on solidifying and ice is less dense. This phenomenon is rare, but its importance for biology has long been recognized. If ice were heavier than liquid H_2O, it would sink to the bottom of the container on freezing. This would mean that oceans, lakes and streams would freeze from the bottom to the top and once frozen would be extremely difficult to melt. Such a situation would obviously be incompatible with those bodies of H_2O serving as the habitat of many living forms as they do. As it is, however, the warmer, liquid H_2O falls to the bottom of any lake and the ice floats on top where heat from the external environment can reach it and melt it.

Additional properties of water such as high surface tension and a high dielectric constant have significance in biology. However, we refer the student to the classical publication by L. J. Henderson, *The Fitness of the Environment*, in which this subject is discussed in more detail and instead consider the mechanism by which the concentration of hydrogen ion (H^+) in aqueous solutions is controlled. To do this, we review the *law of mass action* and the *ion product of water*.

The Law of Mass Action

For the reaction

$$A + B \rightleftharpoons C + D \tag{1-1}$$

in which two reactants A and B interact to form two products C and D, we may write the expression

$$K_{eq} = \frac{C_C \cdot C_D}{C_A \cdot C_B} \tag{1-2}$$

This is an expression of the law of mass action, applied to reaction 1-1, which states that *at equilibrium, the product of the concentrations of the substances formed in a chemical reaction divided by the product of the concentrations of the reactants in that reaction is a constant known as the equilibrium constant,* K_{eq}. This constant is fixed for any given temperature. If the concentration of any single component of the reaction is varied, it follows that the concentration of at least one other component must also change in order to meet the conditions of the equilibrium as defined by K_{eq}.

To be precise we should distinguish between the concentration of the reactants and products in this reaction and the *activity* or *effective concentration* of these reactants. It was early recognized that the concentration of a substance did not always accurately describe its reactivity in a chemical reaction. Moreover, these discrepancies in behavior were appreciable when the concentration of reactant was large. Under these conditions the individual particles of the reactant may exert a mutual attraction on each other or exhibit interactions with the solvent in which the reaction occurs. On the other hand, in dilute solution or low concentration, the interactions are considerably less if not negligible. In order to correct for the difference between concentration and effective concentration, the activity coefficient γ was introduced. Thus,

$$a_A = C_A \times \gamma \tag{1-3}$$

where *a* refers to the activity and C_A to the concentration of the substance. The activity coefficient is not a fixed quantity but varies in value depending on the situation under consideration. In very dilute concentrations the activity coefficient approaches unity, because there is little if any solute–solute interaction. At infinite dilution the activity and the concentration are the same. For the purpose of this book, we do not usually distinguish between activities and concentrations; rather, we use the latter term. This is not a serious deviation from accuracy, since the reactants in many biochemical reactions are quite low in concentration. In addition, the H^+ concentration in most biological tissues is approximately 10^{-7} mole/liter, at which concentration the activity coefficient would be unity.

Dissociation of Water and Its Ion Product, K_w

Water is a weak electrolyte which dissociates only slightly to form H^+ and OH^- ions:

$$H_2O \rightleftharpoons H^+ + OH^- \quad (1\text{-}4)$$

The equilibrium constant for this dissociation reaction has been accurately measured, and at 25°C it has the value 1.8×10^{-16} mole/liter. That is,

$$K_{eq} = \frac{C_{H^+}C_{OH^-}}{C_{H_2O}} = 1.8 \times 10^{-16}$$

The concentration of H_2O (C_{H_2O}) in pure water may be calculated to be $\frac{1000}{18}$ or 55.5 moles/liter. Since the concentration of H_2O in dilute aqueous solutions is essentially unchanged from that in pure H_2O, this figure may be taken as a constant. It is, in fact, usually incorporated into the expression for the dissociation of water, to give

$$\begin{aligned} C_{H^+}C_{OH^-} &= 1.8 \times 10^{-16} \times 55.5 = 1.01 \times 10^{-14} \\ &= K_w = 1.01 \times 10^{-14} \end{aligned} \quad (1\text{-}5)$$

at 25°C.

This new constant K_w, termed the *ion product* of water, expresses the relation between the concentration of H^+ and OH^- ions in aqueous solutions; for example, this relation may be used to calculate the concentration of H^+ in pure water. To do this, let x equal the concentration of H^+. Since in pure water one OH^- is produced for every H^+ formed on dissociation of a molecule of H_2O, x must also equal the concentration of OH^-. Substituting in equation 1-5, we have

$$\begin{aligned} x \cdot x &= 1.01 \times 10^{-14} \\ x^2 &= 1.01 \times 10^{-14} \\ x &= C_{H^+} = C_{OH^-} = 1.0 \times 10^{-7} \text{ mole/liter} \end{aligned}$$

pH

In 1909, Sörensen introduced the term pH as a convenient manner of expressing the concentration of H^+ ion by means of a logarithmic function; pH may be defined as

$$\text{pH} = \log \frac{1}{a_{H^+}} = -\log a_{H^+} \quad (1\text{-}6)$$

where a_{H^+} is defined as the activity of H^+. In this text no distinction is made between activities and concentrations, and so

$$\text{pH} = \log \frac{1}{[H^+]} = -\log [H^+] \quad (1\text{-}7)$$

Moreover, to indicate that we are dealing with concentrations, we use brackets [ ] to indicate them. Thus, the concentration of H^+ (C_{H^+}) is represented as $[H^+]$. We may point out the difference between activities and concentrations by the following example: The pH of 0.1M HCl when measured with a pH meter is 1.09. This value can be substituted in equation 1-6, as the pH meter measures activities and not concentrations (see Appendix 2):

$$1.09 = \log \frac{1}{a_{H^+}}$$
$$a_{H^+} = 10^{-1.09}$$
$$a_{H^+} = \text{antilog } \bar{2}.91$$
$$a_{H^+} = 8.1 \times 10^{-2} \text{ mole/liter}$$

Since the concentration of H^+ in $0.1M$ HCl is 0.1 mole/liter, the activity coefficient γ may be calculated:

$$\gamma = \frac{a_{H^+}}{[H^+]}$$
$$= \frac{0.081}{0.1}$$
$$= 0.81$$

It is important to stress that the pH is a logarithmic function; thus, when the pH of a solution is decreased one unit from 5 to 4, the H^+ concentration has increased tenfold from $10^{-5}M$ to $10^{-4}M$. When the pH has increased three-tenths of a unit from 6 to 6.3, the H^+ concentration has decreased from $10^{-6}M$ to $5 \times 10^{-7}M$.

If we now apply the term of pH to the ion product expression for pure water, we obtain another useful expression:

$$[H^+] \times [OH^-] = 1.0 \times 10^{-14}$$

We take the logarithms of this equation:

$$\log [H^+] + \log [OH^-] = \log (1.0 \times 10^{-14})$$
$$= -14$$

and multiply by -1:

$$-\log [H^+] - \log [OH^-] = 14$$

If we now define $-\log [OH^-]$ as pOH, a definition similar to that of pH, we have an expression relating the pH and pOH in any aqueous solution:

$$\text{pH} + \text{pOH} = 14 \tag{1-8}$$

Brönsted Acids

A most useful definition of acids and bases in biochemistry is that proposed by Brönsted. He defined *an acid as any substance that can donate a proton,* and *a base as a substance that can accept a proton*. Although other definitions of acids, notably one proposed by G. N. Lewis, are even more general, the Brönsted concept should be thoroughly understood by students of biochemistry.

The following substances shown in red are examples of Brönsted acids:

$$HCl \longrightarrow H^+ + Cl^-$$
$$CH_3COOH \longrightarrow H^+ + CH_3COO^-$$
$$NH_4^+ \longrightarrow NH_3 + H^+$$

and the generalized expression would be

$$HA \longrightarrow H^+ + A^-$$

The corresponding bases are

$$Cl^- + H^+ \longrightarrow HCl$$
$$CH_3COO^- + H^+ \longrightarrow CH_3COOH$$
$$NH_3 + H^+ \longrightarrow NH_4^+$$

The corresponding base for the generalized weak acid HA is

$$A^- + H^+ \longrightarrow HA$$

It is customary to refer to the acid–base pair as follows: HA is the *Brönsted acid* because it can furnish a proton; the anion A^- is called the *conjugate base* because it can accept the proton to form the acid HA.

Dissociation of Strong Electrolytes

Strong electrolytes are substances that, in aqueous solution, are dissociated almost completely into charged particles known as ions. Sodium chloride, even in its solid, crystalline form exists as Na^+ ions and Cl^- ions. We may represent the dissociation of NaCl as being complete:

$$Na^+Cl^- \longrightarrow Na^+ + Cl^-$$

Strong acids and bases are electrolytes that are almost completely dissociated into their corresponding ions in aqueous solution. Thus, hydrochloric acid (HCl), a familiar mineral acid, is completely dissociated in H_2O:

$$HCl \longrightarrow H^+ + Cl^- \qquad (1\text{-}9)$$

We should, however, represent the reaction of HCl in H_2O more accurately as an *ionization*,

$$HCl + H_2O \longrightarrow H_3O^+ + Cl^-$$

in which the electrically neutral HCl has reacted with H_2O to form Cl^- anion and the hydrated proton or hydronium ion, H_3O^+. In the terminology of Brönsted, the Brönsted acid HCl, on ionization, has contributed a proton to the conjugate base H_2O to form a new Brönsted acid (H_3O^+) and the conjugate base of HCl, the chloride ion, Cl^-:

$$\underset{(\text{Conjugate acid})_1}{HCl} + \underset{(\text{Conjugate base})_2}{H_2O} \longrightarrow \underset{(\text{Conjugate acid})_2}{H_3O^+} + \underset{(\text{Conjugate base})_1}{Cl^-}$$

It should also be remembered that not only is the proton contributed by HCl hydrated to form the hydronium ion (H_3O^+), but also that the Cl^- is hydrated. It is common practice to omit the water of hydration in chemical reactions and to represent the ionization of a strong acid like HCl as a simple dissociation according to reaction 1-9.

Ionization of Weak Acids

A weak acid, in contrast to a strong acid, is only partially ionized in aqueous solution. Consider the ionization of the generalized weak acid, HA:

$$\underset{(\text{Conjugate acid})_1}{HA} + \underset{(\text{Conjugate base})_2}{H_2O} \rightleftharpoons \underset{(\text{Conjugate acid})_2}{H_3O^+} + \underset{(\text{Conjugate base})_1}{A^-} \qquad (1\text{-}10)$$

The proton donated by HA is accepted by H_2O to form the hydronium ion, H_3O^+. The equilibrium constant for this ionization reaction is known as an ionization constant, K_{ion}:

$$K_{eq} = K_{ion} = \frac{[H_3O^+][A^-]}{[HA][H_2O]} \tag{1-11}$$

Because, as we have already seen, the concentration of H_2O in aqueous solution is itself a constant, 55.5 moles/liter, we can combine K_{ion} and $[H_2O]$ to obtain a new constant, K_a:

$$K_a = K_{ion}[H_2O] = \frac{[H_3O^+][A^-]}{[HA]} \tag{1-12}$$

Moreover, because $[H_3O^+]$ is the same as the hydrogen ion concentration, we see that K_a becomes

$$K_a = \frac{[H^+][A^-]}{[HA]} \tag{1-13}$$

This expression in turn is identical with the equilibrium constant we would have written if HA is considered as a weak acid that partially *dissociates* to yield protons and A^- anions:

$$HA \rightleftharpoons H^+ + A^-$$

$$K_{eq} = \frac{[H^+][A^-]}{[HA]} \tag{1-14}$$

Ionization of Weak Bases

The ionization of a weak base, defined in the chemical sense as a substance that furnishes OH^- ions on dissociation, can be represented as

$$BOH \rightleftharpoons B^+ + OH^- \tag{1-15}$$

$$K_{eq} = K_b = \frac{[B^+][OH^-]}{[BOH]}$$

For ammonium hydroxide (NH_4OH), the K_b is given in chemical handbooks as 1.8×10^{-5}. It is therefore important to realize that the extent of dissociation of NH_4OH is identical with that of acetic acid (CH_3COOH). The important difference, of course, is that NH_4OH dissociates to form hydroxyl ions (OH^-) whereas CH_3COOH dissociates to form protons (H^+), and that the pH of 0.1M solutions of these two substances is by no means similar.

One of the most common types of weak base encountered in biochemistry is the group called organic amines (e.g., the amino groups of amino acids). Such compounds, when represented with the general formula $R{-}NH_2$, do not contain hydroxyl groups that can dissociate as in reaction 1-15. On the other hand, such compounds can ionize in H_2O to produce hydroxyl ions:

$$\underset{(\text{Conjugate base})_1}{RNH_2} + \underset{(\text{Conjugate acid})_2}{H_2O} \rightleftharpoons \underset{(\text{Conjugate acid})_1}{RNH_3^+} + \underset{(\text{Conjugate base})_2}{OH^-} \tag{1-16}$$

In this reaction, H_2O serves as an acid to contribute a proton to the base RNH_2.

Using the Brönsted definition of a base as a substance (A^-) that accepts a proton, we can write the general expression

$$\underset{(\text{Conjugate base})_1}{A^-} + \underset{(\text{Conjugate acid})_2}{H_2O} \rightleftharpoons \underset{(\text{Conjugate acid})_1}{HA} + \underset{(\text{Conjugate base})_2}{OH^-} \quad (1\text{-}17)$$

The equilibrium constant K_{ion} for this ionization may be written in analogy with equation 1-11 as

$$K_{\text{ion}} = \frac{[HA][OH^-]}{[A^-][H_2O]} \quad (1\text{-}18)$$

Combining K_{ion} and $[H_2O]$ as previously, we have, in analogy with equation 1-12,

$$K_b = \frac{[HA][OH^-]}{[A^-]} \quad (1\text{-}19)$$

Equation 1-19 can be used for calculating the $[OH^-]$ of a solution of a weak base; the chemical handbooks list values for the K_b of such substances. The pOH in turn can be calculated and from this the pH may be obtained (equation 1-8). However, there is a direct relationship between the K_b of a weak base and the K_a of its conjugate acid that is useful in obtaining directly the pH of mixtures of weak bases and their salts.

Solving equation 1-19 for $[OH^-]$, we have

$$[OH^-] = \frac{K_b[A^-]}{[HA]} \quad (1\text{-}20)$$

Similarly, solving equation 1-13 for $[H^+]$, we have

$$[H^+] = \frac{K_a[HA]}{[A^-]} \quad (1\text{-}21)$$

Then, substituting for $[H^+]$ and $[OH^-]$ in the following expression, which was defined earlier (equation 1-5):

$$[H^+][OH^-] = K_w$$

$$\frac{K_a[HA]}{[A^-]} \cdot \frac{K_b[A^-]}{[HA]} = K_w \quad (1\text{-}22)$$

which simplifies to

$$K_a \cdot K_b = K_w \quad (1\text{-}23)$$

Substituting the value of K_w at 25°C, we have

$$K_a \cdot K_b = 10^{-14} \quad (1\text{-}24)$$

Taking logarithms and multiplying by -1, we have

$$\log K_a + \log K_b = \log K_w$$

$$-\log K_a - \log K_b = -\log K_w \quad (1\text{-}25)$$

Then, just as pH has been defined as $-\log [H^+]$, we can define pK_a and pK_b as $-\log K_a$ and $-\log K_b$, respectively. Equation 1-25 then becomes

$$pK_a + pK_b = -\log K_w = 14 \quad (1\text{-}26)$$

The Henderson-Hasselbalch Equation

Henderson and Hasselbalch have rearranged the *mass law* as it applies to the ionization of weak acids into a useful expression known as the Henderson–Hasselbalch equation. If we consider the ionization of a generalized weak acid HA:

$$HA \rightleftharpoons H^+ + A^-$$

$$K_{\text{ion}} = K_a = \frac{[H^+][A^-]}{[HA]}$$

Rearranging terms, we have

$$[H^+] = K_a \frac{[HA]}{[A^-]}$$

Taking logarithms, we find

$$\log [H^+] = \log K_a + \log \frac{[HA]}{[A^-]}$$

and multiplying by -1,

$$-\log [H^+] = -\log K_a - \log \frac{[HA]}{[A^-]}$$

If $-\log K_a$ is defined as pK_a and $\log [A^-]/[HA]$ is substituted for $-\log [HA]/[A^-]$, we obtain

$$pH = pK_a + \log \frac{[A^-]}{[HA]} \tag{1-27}$$

This form of the Henderson–Hasselbalch equation can be written in a more general expression in which we replace $[A^-]$ with the term "conjugate base" and [HA] with "conjugate acid":

$$pH = pK_a + \log \frac{[\text{Conjugate base}]}{[\text{Conjugate acid}]} \tag{1-28}$$

This expression may then be applied not only to weak acids such as acetic acid, but also to the ionization of ammonium ions and those substituted amino groups found in amino acids. In this case, NH_4^+ ions or the protonated amino groups RNH_3^+ are the conjugate acids which dissociate to form protons and the conjugate bases NH_3 and RNH_2, respectively:

$$NH_4^+ \rightleftharpoons NH_3 + H^+$$
$$RNH_3^+ \rightleftharpoons RNH_2 + H^+$$

Applying equation 1-28 to the protonated amine, we have

$$pH = pK_a + \log \frac{[RNH_2]}{[RNH_3^+]} \tag{1-29}$$

Biochemistry handbooks usually list the K_a (or pK_a) for the conjugate acids of substances we normally consider as bases (e.g., NH_4OH, amino acids, organic amines). If they do not, the K_b (or pK_b; see equation 1-19) for the ionization of the weak base will certainly be listed, and the K_a (or pK_a) must first be calculated before employing the generalized Henderson–Hasselbalch equation. Although more care must be taken to identify correctly the conjugate acid–base pairs in that expression, its usage leads directly to the pH of mixtures of weak bases and their salts.

Some Representative Problems

Let us calculate the concentration of H^+ in 1.0*M* acetic acid (CH_3COOH) and then determine the degree of ionization of a solution of acetic acid of this concentration. Before starting, it should be realized that, were acetic acid completely ionized like the *strong* mineral acid HCl, the $[H^+]$ would be 1.0 mole/liter. However, since acetic acid is a weakly ionized acid, equation 1-13 must be employed. If we let x equal the concentration of H^+ formed by the ionization of the acetic acid, then x will also be the concentration of CH_3COO^-, because these two ions are formed in equal amounts when acetic acid ionizes. The amount of CH_3COOH remaining after the ionization equilibrium has been established will then be $1 - x$. Therefore,

Initial concentration (mole/liter)	Concentration at equilibrium after ionization (mole/liter)
$[CH_3COOH] = 1.00$	$[CH_3COOH] = (1.00 - x)$
$[H^+] = 0.00$	$[H^+] = x$
$[CH_3COO^-] = 0.00$	$[CH_3COO^-] = x$

Substituting the values that exist at equilibrium in the expression for the ionization of acetic acid, we have

$$\frac{x^2}{1 - x} = 1.8 \times 10^{-5} \qquad (1\text{-}30)$$

This quadratic equation, when solved for x (see Appendix 1), is found to equal 0.0042*M*. Thus, $[H^+] = [CH_3COO^-] = 0.0042M$. The concentration of the undissociated CH_3COOH will therefore be $1.00 - 0.0042$ or 0.9958*M*. At 25°C, a 1*M* solution of acetic acid is dissociated or ionized to the extent of only 0.4%. The pH of this solution, which is 0.0042*M* in hydrogen ion, may be calculated from equation 1-7:

$$\begin{aligned} \text{pH} &= -\log 0.0042 = -\log (4.2 \times 10^{-3}) \\ &= -\log 4.2 - \log 10^{-3} \\ &= -0.62 + 3 \\ &= 2.38 \end{aligned}$$

It is possible to simplify the solution of equation 1-30 above. In considering the ionization of relatively concentrated solutions of weak electrolytes, the term in the denominator $(1 - x)$ may be simplified by not correcting for the amount of acid (x) which dissociated, provided x is small. In the example just given, the amount that dissociated was negligible (only 0.4%) and may be ignored. When this approximation is made,

$$\begin{aligned} x^2 &= 1.8 \times 10^{-5} \\ x &= \sqrt{18 \times 10^{-6}} \\ x &= 4.2 \times 10^{-3} \\ [H^+] = [CH_3COO^-] &= 0.0042M \end{aligned}$$

Another important relation is emphasized by calculating the H^+ concentration when the concentration of anion is equal to the concentration of unionized weak acid. Such a relation would exist in a solution prepared by mixing 0.1 mole of sodium acetate (8.2 g) with 0.1 mole of acetic acid (6 g) in sufficient water to make a liter of solution. Under these conditions, $[CH_3COO^-] = [CH_3COOH] = 0.1M$, and when these are substituted in equation 1-13,

$$\frac{[H^+][CH_3COO^-]}{[CH_3COOH]} = 1.8 \times 10^{-5}$$

$$\frac{[H^+][0.1]}{[0.1]} = 1.8 \times 10^{-5}$$

$$[H^+] = 1.8 \times 10^{-5}$$

$$pH = 5 - \log 1.8$$

$$pH = 4.74$$

Thus, $H^+ = K_a$ when the concentration of the anion of the acid is equal to the concentration of the unionized acid. Since different weak acids have different K_a's, equimolar mixtures of these acids and their corresponding salts will each have a different pH.

Note that the Henderson–Hasselbalch equation (equation 1-28) would not be useful in calculating the $[H^+]$ of a solution containing only the weak acid. However, in the problem in which we calculated the $[H^+]$ of a mixture containing 0.1 mole of sodium acetate and 0.1 mole of acetic acid, the Henderson–Hasselbalch equation is particularly useful. In this case, equation 1-28 becomes

$$pH = pK_{a_{HAc}} + \log \frac{[CH_3COO^-]}{[CH_3COOH]} \qquad (1\text{-}31)$$

The acetic acid concentration $[CH_3COOH]$ in this mixture will be $0.1M$ *minus* the small amount a of CH_3COOH that dissociates, and the acetate ion concentration $[CH_3COO^-]$ will be $0.1M$ *plus* the small amount a of acetate ion produced in the dissociation just mentioned. Equation 1-31 therefore becomes

$$pH = pK_{a_{HAc}} + \log \frac{[0.1 + a]}{[0.1 - a]} \qquad (1\text{-}32)$$

Although it is possible to calculate a, the quantity is usually negligible and can be ignored. This approximation, together with the approximation previously introduced by substituting concentrations for activities (equation 1-3), can be incorporated into the expression for the equilibrium constant by the use of a corrected *equilibrium constant*, K'_{eq}. This constant will vary, of course, with the concentration of the reactants, and the conditions for its use must be specified, usually in terms of the ionic strength.

The K_a for acetic acid is 1.8×10^{-5} mole/liter; pK_a is therefore $-\log (1.8 \times 10^{-5})$ or 4.74. Substituting this value and neglecting a, we find that equation (1-32) becomes

$$pH = 4.74 + \log \frac{0.1}{0.1}$$
$$= 4.74$$

The student should be thoroughly familiar with calculations involving the Henderson–Hasselbalch equation. Appendix 1 contains problems illustrating the use of the equation.

Titration Curves

The titration curve obtained when 100 ml of 0.1N CH_3COOH is titrated with 0.1N NaOH is shown in Figure 1-1. This curve can be obtained experimentally in the laboratory by measuring the pH of 0.1N CH_3COOH before and after addition of different aliquots of 0.1N NaOH. The curve may also be calculated by the Henderson–Hasselbalch equation for all the points except the first, where no NaOH has been added, and the last, where a stoichiometric amount (100 ml) of 0.1N NaOH has been added. Clearly, the Henderson–Hasselbalch equation cannot be used to determine the pH at the limits of the titration where the ratio of salt to acid is either zero or infinite.

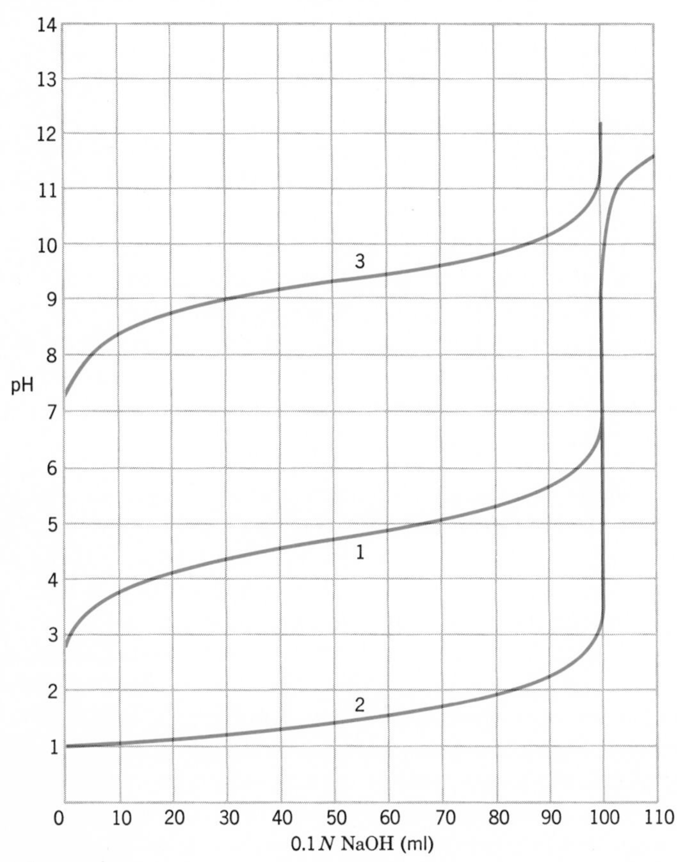

Figure 1-1

Titration curve of 100 ml of 0.1N CH_3COOH (1), 100 ml of 0.1N HCl (2), and 100 ml of 0.1N NH_4Cl (3) with 0.1N NaOH.

In considering the grosser aspects of the titration curve of acetic acid, we see visually that the change in pH for unit of alkali added is greatest at the beginning and end of the titration, whereas the smallest change in pH for unit of alkali added is obtained when the titration is half complete. In other words, an equimolar mixture of sodium acetate and acetic acid shows less change in pH initially when acid or alkali is added than a solution consisting mainly of either acetic acid or sodium acetate. We refer to the ability of a solution to resist a change in pH as its *buffer action,* and it can be shown that a buffer exhibits its *maximum* action when the titration is half complete or when the pH is equal to the pK_a (equation 1-31). In Figure 1-1 the point of maximum buffer action is at the pH of 4.74.

Another way of representing the condition which exists when the pH of a mixture of acetic acid and sodium acetate is at the pK_a is to state that the acid at this pH is half ionized. That is, half the "total acetate" species is present as undissociated CH_3COOH, while the other half is in the form of acetate ion, CH_3COO^-. Since at its pK_a any weak acid will be half ionized, this is one of the most useful ways of distinguishing between individual weak acids. The pK_a is also a characteristic property of each acid, because the ionization constant is a function of the inherent properties of the weak acid.

The titration curve of 0.1N HCl is also represented in Figure 1-1. The Henderson–Hasselbalch equation is of no use in calculating the curve for HCl, since it applies only for weak electrolytes. However, the pH at any point on the HCl curve can be calculated by determining the milliequivalents of HCl remaining and correcting for the volume. Thus, when 30 ml of 0.1N NaOH have been added, 7.0 meq of HCl will remain in a volume of 130 ml. The concentration of H^+ will therefore be 7.0/130 or 0.054M. If the activity coefficient is neglected, the pH may be calculated from equation 1-7 as 1.27.

Curve 3 in Figure 1-1 is the titration curve obtained when 100 ml of 0.1N NH_4Cl is titrated with 0.1N NaOH. In this titration the protons contributed by NH_4^+ are neutralized by the OH^- ions provided by the NaOH:

$$NH_4^+ + OH^- \longrightarrow NH_3 + H_2O$$

Again, the Henderson–Hasselbalch equation is of no value when calculating the pH of the solution of NH_4Cl before any NaOH has been added. The pH of this solution may be calculated by using equation 1-19 to first obtain the $[OH^-]$. The $[H^+]$ or pH may then be calculated from equation 1-5 or equation 1-8. However, the Henderson–Hasselbalch equation can be employed to determine any point on the curve when some of the NH_4Cl has been neutralized.

Up to this point we have considered only the monobasic acid, acetic acid. Polybasic or *polyprotic* acids, commonly encountered in biochemistry, are acids capable of ionizing to yield more than one proton per molecule of acid. In each case the extent of dissociation of the individual protons may be described by a K_{ion} or K_a. In the case of phosphoric acid (H_3PO_4) three protons may be furnished on complete ionization of a mole of this acid:

$$H_3PO_4 \rightleftharpoons H^+ + H_2PO_4^- \qquad K_{a_1} = 7.5 \times 10^{-3} \qquad pK_{a_1} = 2.12$$
$$H_2PO_4^- \rightleftharpoons H^+ + HPO_4^{2-} \qquad K_{a_2} = 6.23 \times 10^{-8} \qquad pK_{a_2} = 7.21$$
$$HPO_4^{2-} \rightleftharpoons H^+ + PO_4^{3-} \qquad K_{a_3} = 4.8 \times 10^{-13} \qquad pK_{a_3} = 12.32$$

This means that at the pH of 2.12 the first ionization of H_3PO_4 is half complete; the pH must be 12.32, however, before the third and final ionization of H_3PO_4 is 50% complete. At the pH of 7.0, which is frequently encountered in the cell, the second proton of phosphoric acid ($pK_{a2} = 7.21$) will be about half dissociated. At this pH both the mono- and dianions of phosphoric acid or phosphate esters will be present in approximately equal concentrations. For phosphoric acid the two predominant ionic species will be $H_2PO_4^-$ and HPO_4^{2-}. In the case of α-glycerol phosphate the two following ions will be present in about equal concentration at pH 7.0:

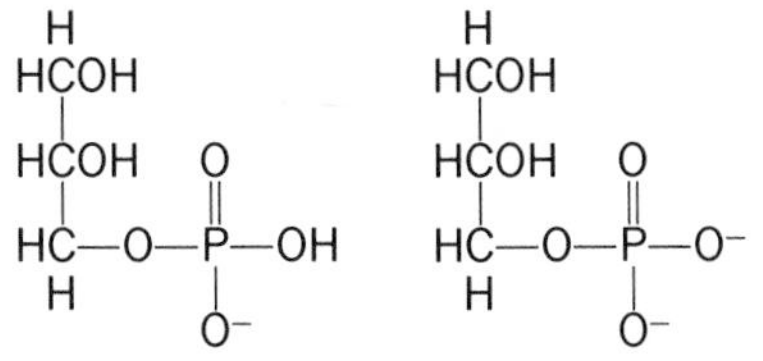

```
  H                  H
 HCOH               HCOH
  |                  |
 HCOH   O           HCOH   O
  |     ||           |     ||
 HC—O—P—OH          HC—O—P—O⁻
  H     |            H     |
        O⁻                 O⁻
```

Many of the common organic acids encountered in intermediary metabolism are polyprotic; for example, succinic acid ionizes according to the following scheme:

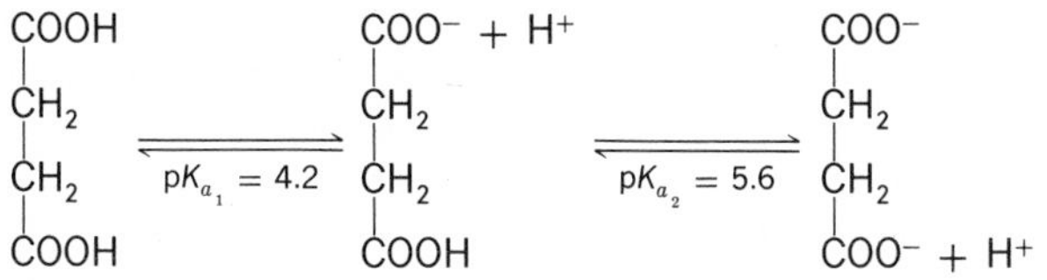

```
COOH                    COO⁻ + H⁺                 COO⁻
|                       |                         |
CH2                     CH2                       CH2
|     ⇌ (pKa1 = 4.2)    |     ⇌ (pKa2 = 5.6)      |
CH2                     CH2                       CH2
|                       |                         |
COOH                    COOH                      COO⁻ + H⁺
```

At pH 7.0 in the cell, succinic acid will exist predominantly as the dianion $^-OOC—CH_2—CH_2—COO^-$. Furthermore, most of the organic acids which serve as metabolites (palmitic, lactic, and pyruvic acids, for example) will be present as their anions (palmitate, lactate, and pyruvate). This has led to the use of the names of the *ions* when these compounds are discussed in biochemistry. In writing chemical reactions, however, it will be the practice in this text to use the formulas for the undissociated acid.

Table 1-2 lists the pK_a's for several of the organic acids commonly encountered in intermediary metabolism.

If he has not already done so, the student is urged to review his knowledge of chemical stoichiometry. The meanings of gram molecular weight and gram equivalent weight (mole and equivalent, respectively) and the significance of molarity, molality, and normality must be thoroughly understood. Biochemistry is a quantitative science, and the student must recognize immediately such terms as millimole and micromole. In connection with titrations it is also important to remind the student that the H^+ concentrations of $0.1N$ H_2SO_4 and $0.1N$ CH_3COOH are by no means similar but that 1 liter of each of these solutions contains the same amount of total titratable acid.

Determination of pK_a

The ability to ionize is a valuable property of many biological compounds. Organic acids, amino acids, proteins, purines, pyrimidines, and phosphate esters are examples of biochemicals which are ionized to varying degrees

Table 1-2

The pK_a of Some Organic Acids

	pK_{a_1}	pK_{a_2}	pK_{a_3}
Acetic acid (CH_3COOH)	4.74		
Acetoacetic acid (CH_3COCH_2COOH)	3.58		
Citric acid ($HOOCCH_2C(OH)(COOH)CH_2COOH$)	3.09	4.75	5.41
Formic acid ($HCOOH$)	3.62		
Fumaric acid ($HOOCCH{=}CHCOOH$)	3.03	4.54	
DL-Glyceric acid ($CH_2OHCHOHCOOH$)	3.55		
DL-Lactic acid ($CH_3CHOHCOOH$)	3.86		
DL-Malic acid ($HOOCCH_2CHOHCOOH$)	3.40	5.26	
Pyruvic acid ($CH_3COCOOH$)	2.50		
Succinic acid ($HOOCCH_2CH_2COOH$)	4.18	5.56	

in biological systems. Since the pH of most biological fluids is near 7, the extent of dissociation of some of these compounds may be complete there. The first ionization of H_3PO_4 will likewise be complete; the second ionization (pK_{a_2} = 7.2) will be approximately half complete.

One of the characteristic qualitative properties of a molecule is the pK_a of any dissociable group it may possess. The experimental determination of the pK_a of dissociable groups is therefore an important procedure in describing properties of an unknown substance. The pK_a may be determined in the laboratory by measuring the titration curve experimentally with a pH meter. As known amounts of alkali or acid are added to a solution of the unknown, the pH is determined, and the titration curve can be plotted. From this curve the inflection point (pK_a) may be determined by suitable procedures.

Buffers

With a thorough understanding of the ionization of weak electrolytes it is possible to discuss buffer solutions. A *buffer solution is one that resists a change in pH on the addition of acid or alkali.* Most commonly, the buffer solution consists of a mixture of a weak Brönsted acid and its conjugate base; for example, mixtures of acetic acid and sodium acetate or of ammonium hydroxide and ammonium chloride are buffer solutions.

There are many examples of the significance of buffers in biology; the ability to prevent large changes in pH is an important property of most intact biological organisms. The cytoplasmic fluids which contain dissolved proteins, organic substrates, and inorganic salts resist excessive changes in pH. The blood plasma is a highly effective buffer solution almost ideally designed to keep the range of the pH of blood within 0.2 pH unit of 7.2–7.3; values outside this range are not compatible with life. Further appreciation of the buffered nature of the living cell results from recognizing that many of the metabolites constantly being produced and utilized in the cell are weak Brönsted acids. In addition, enzymes responsible for the catalysis of reactions in which these metabolites participate exhibit their maximum catalytic action at some definite pH (Chapter 8).

In the laboratory the biochemist also wishes to examine reactions *in vitro* under conditions where the change in pH is minimal. He obtains these conditions by using efficient buffers, preferably inert ones, in the reactions under investigation. The buffers may include weak acids such as phosphoric, acetic, glutaric, and tartaric acids or weak bases such as ammonia, pyridine, and tris-(hydroxymethyl)amino methane.

Let us consider the mechanism by which a buffer solution exerts control over large pH changes. When alkali (for instance, NaOH) is added to a mixture of acetic acid (CH_3COOH) and potassium acetate (CH_3COOK), the following reaction occurs:

$$OH^- + CH_3COOH \longrightarrow CH_3COO^- + H_2O$$

This reaction states that OH^- ion reacted with protons furnished by the dissociation of the weak acid and formed H_2O:

$$CH_3COOH \rightleftharpoons CH_3COO^- + H^+ \xrightarrow{OH^-} H_2O$$

On the addition of alkali there is a further dissociation of the available CH_3COOH to furnish additional protons and thus to keep the H^+ concentration or pH unchanged.

When acid is added to an acetate buffer the following reaction occurs:

$$H^+ + CH_3COO^- \longrightarrow CH_3COOH$$

The protons added (in the form of HCl, for example) combine instantly with the CH_3COO^- anion present in the buffer mixture (as potassium acetate) to form the undissociated weak acid CH_3COOH. Consequently the resulting pH change is much less than would occur if the conjugate base were absent.

In discussing the quantitative aspects of buffer action we should point out that two factors determine the effectiveness or *capacity* of a buffer solution. Clearly, the molar concentration of the buffer components is one of them. The buffer capacity is directly proportional to the concentration of the buffer components. Here we encounter the convention used in referring to the concentration of buffers. The concentration of a buffer refers to the *sum* of the concentration of the weak acid and its conjugate base. Thus, a $0.1M$ acetate buffer could contain 0.05 mole of acetic acid and 0.05 mole of sodium acetate in 1 liter of H_2O. It could also contain 0.065 mole of acetic acid and 0.035 mole of sodium acetate in 1 liter of H_2O.

The second factor influencing the effectiveness of a buffer solution is the *ratio* of the concentration of the conjugate base to the concentration of the weak acid. Quantitatively it should seem evident that the most effective buffer would be one with *equal concentrations* of basic and acidic components, since such a mixture could furnish *equal quantities* of basic or acidic components to react, respectively, with acid or alkali. An inspection of the titration curve for acetic acid (Figure 1-1) similarly shows that the minimum change in pH resulting from the addition of a unit of alkali (or acid) occurs at the pK_a for acetic acid. At this pH we have already seen that the ratio of CH_3COO^- to CH_3COOH is 1. On the other hand, at values of pH far removed from the pK_a (and therefore at ratios of conjugate base to acid greatly differing from unity), the change in pH for unit of acid or alkali added is much larger.

Having stated the two factors that influence the buffer capacity, we may

(mol wt, 121), add 0.20 mole of HCl (200 ml of 1*N* HCl) to it, and dissolve in a final volume of 500 ml.

Additional buffer problems may be found in Appendix 1.

References

1. I. H. Segel, *Biochemical Calculations*. New York: Wiley, 1968.
   This inexpensive paperback thoroughly treats the subject of acid–base equilibria in biochemistry, and many typical calculations are discussed.
2. C. Long, ed., *Biochemist's Handbook*. New York: Van Nostrand, 1961; R. M. C. Dawson, D. C. Elliott, W. H. Elliott, and K. M. Jones, eds., *Data for Biochemical Research*. 2nd ed. New York and Oxford: Oxford University Press, 1969.
   These two handbooks are particularly useful as sources of information on biochemicals, including dissociation constants for the buffers commonly encountered.
3. L. J. Henderson, *The Fitness of the Environment*. Boston: Beacon Press, 1958.
   This classic, first published in 1913, has been reissued in a paperback edition. The modern introduction by George Wald of Harvard places the book in proper perspective for contemporary biologists.

2

Carbohydrates

Introduction

In the next four chapters, we shall describe the more important building blocks of the biosphere—the simple sugars, fatty acids, amino acids, and mononucleotides—which are assembled into the biopolymers of the cell—the polysaccharides, lipids, proteins, and nucleic acids. In the present chapter we shall examine the simple sugars, the storage carbohydrates, and the structural polysaccharides. Complex carbohydrates play not only a structural role in the cell but may serve as a reservoir of chemical energy to be enlarged and depleted as the organism wishes. As examples of structural carbohydrates we may cite cellulose, the major structural component of plant cell walls, and the peptidoglycans of bacterial cell walls. The storage carbohydrates include the more familiar starch and glycogen, polysaccharides that may be produced and consumed in line with the energy needs of the cell.

Carbohydrates may be defined as polyhydroxy aldehydes or ketones, or as substances that yield one of these compounds on hydrolysis. Many carbohydrates have the empirical formula $(CH_2O)_n$ where n is 3 or larger. This formula obviously contributed to the original belief that this group of compounds could be represented as *hydrates of carbon*. It became clear that this definition was not suitable when other compounds were encountered that had the general properties of carbohydrates but contained nitrogen or sulfur in addition to carbon, hydrogen, and oxygen. Moreover, the important simple sugar deoxyribose, found in every cell as a component of deoxyribonucleic acid, has the molecular formula $C_5H_{10}O_4$ rather than $C_5H_{10}O_5$.

Carbohydrates can be classified into three main groups: monosaccharides, oligosaccharides, and polysaccharides. Monosaccharides are simple sugars that cannot be hydrolyzed into smaller units under reasonably mild conditions. The simplest monosaccharides fitting our definition and empirical formula are the *aldose* glyceraldehyde and its isomer, the *ketose* dihydroxy acetone. Both of these sugars are *trioses* because they contain three carbon

```
                    H
                   /
CH2OH—CHOH—C             CH2OH—C—CH2OH
                   \\                ||
                    O               O
```

Glyceraldehyde Dihydroxy acetone

atoms. Note that the monosaccharides may be described not only by the type of functional group, but also by the number of carbon atoms they possess.

Oligosaccharides are hydrolyzable polymers of monosaccharides that contain from two to six molecules of simple sugars. Thus, disaccharides are oligosaccharides which on hydrolysis yield two molecules of monosaccharides. For the most part, monosaccharides and oligosaccharides are crystalline compounds which are soluble in water and frequently have a sweet taste.

Polysaccharides are very long chains, or polymers, of monosaccharides that may be either linear or branched in structure. If the polymer is made up from a single monosaccharide, the polysaccharide is called a *homo*polysaccharide. If two or more different monosaccharides are found in the polymer, it is called a *hetero*polysaccharide. Some of the monosaccharides that are bound together by glycosidic bonds to form polysaccharides are glucose, xylose, and arabinose. Polysaccharides are usually tasteless, insoluble compounds with high molecular weights.

Stereoisomerism

The study of carbohydrates and their chemistry immediately introduces the topic of stereoisomerism. It is desirable therefore to review the subject of isomerism as it is treated in organic chemistry.

The subject of isomerism may be divided into *structural isomerism* and *stereoisomerism.* Structural isomers have the same molecular formula but differ from each other by having different structures; stereoisomers have the

```
   H  H  H  H           H     H     H
   |  |  |  |           |     |     |
H—C—C—C—C—H        H—C—— C —— C—H
   |  |  |  |           |     |     |
   H  H  H  H           H    CH3    H
```

n-Butane Isobutane

same molecular formula and the same structure, but they differ in *configuration*, that is, in the arrangement of their atoms in space. Structural isomers, in turn, can be of three types. One type is that of the *chain isomers*, in which the isomers have different arrangements of the carbon atoms. As an example, *n*-butane is a chain isomer of isobutane. Another type of structural isomers is that of the *positional isomers; n*-propyl chloride and isopropyl chloride, in which the two compounds involved have the same carbon chain but differ in the position of a substituent group, are positional isomers. The

```
   H  H  H              H  H  H
   |  |  |              |  |  |
H—C—C—C—Cl         H—C—C—C—H
   |  |  |              |  |  |
   H  H  H              H  Cl H
```

n-Propyl chloride Isopropyl chloride

third type of structural isomer is that of the *functional group* isomers, in which the compounds have different functional groups. Examples are *n*-propanol and methylethyl ether.

$H_3C—CH_2—CH_2OH$ $H_3C—CH_2—O—CH_3$

n-Propanol Methylethyl ether

The subject of stereoisomerism can be divided into the smaller areas of *optical isomerism* and *geometrical* (or *cis-trans*) *isomerism*. The latter type of isomerism is illustrated by the *cis-trans* pair, fumaric and maleic acids.

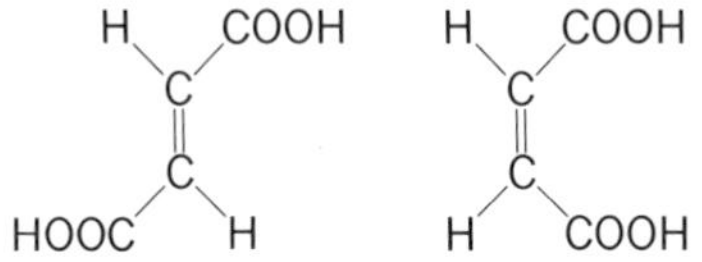

```
    H     COOH          H     COOH
     \   /               \   /
      C                    C
      ||                   ||
      C                    C
     /  \                 /  \
HOOC     H               H    COOH
  Fumaric acid          Maleic acid
    (trans)                (cis)
```

Optical Isomerism. This is the type of isomerism commonly found in carbohydrates; it is usually encountered when a molecule contains one or more *asymmetric* carbon atoms. The subject of stereoisomerism was extensively developed after van't Hoff and LeBel introduced the concept of the *tetrahedral carbon atom*. Today it is recognized that in many compounds the carbon atom has the shape of a tetrahedron in which the carbon nucleus sits in the center of the tetrahedron and the four covalent bonds or bond axes extend out to the corners of the tetrahedron (Structure 2-1).

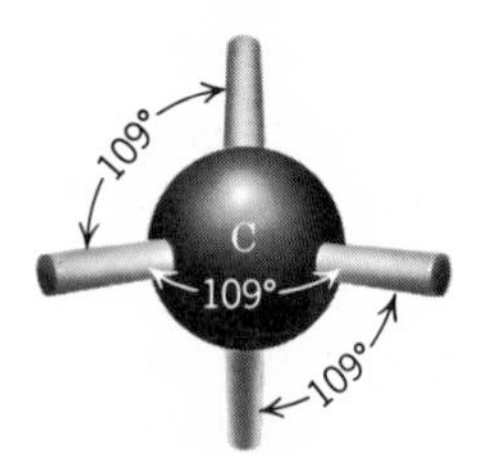

Structure 2-1

When four different groups are attached to those bonds, the carbon atom in the center of the molecule is said to be an *asymmetric carbon atom*. This is indicated in Structure 2-2 where the compound C(ABDE), containing a

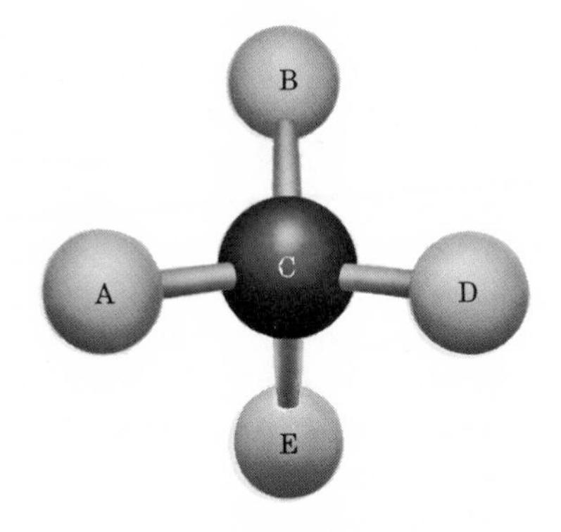

Structure 2-2

single asymmetric carbon atom, is represented as having the four groups A, B, D, and E attached. These groups may be arranged in space in two different ways so that two different compounds are formed. These compounds are obviously different; that is, they cannot be superimposed on each other. Instead, one compound is related to the other as a right hand is related to a left hand. Such chiral (Greek cheir = hand) molecules are said to possess "handedness" and are therefore mirror images of each other; if one molecule is held before a mirror, the image in the mirror corresponds to the other molecule (Structure 2-3).

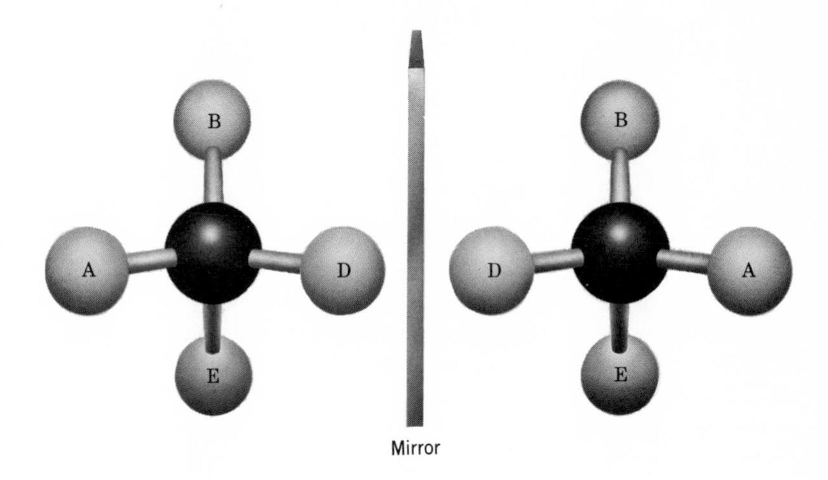

Structure 2-3

These mirror image isomers constitute an *enantiomeric pair*; one member of the pair is said to be the *enantiomer* of the other.

Optical Activity. Almost all the properties of the two members of an enantiomeric pair are identical—they have the same boiling point, the same melting point, the same solubility in various solvents. They also exhibit optical activity; in this property they differ in one important manner. One member of the enantiomeric pair will rotate a plane of polarized light in a clockwise direction and is therefore said to be *dextro*rotatory. Its mirror image isomer or enantiomer will rotate the plane of polarized light to the same extent but in the opposite or counterclockwise direction. This isomer is said to be *levo*rotatory. It must be noted, however, that not all compounds possessing asymmetric carbon atoms are chiral and exhibit optical activity. On the other hand, a molecule may possess chirality, exhibit optical activity, and not contain an asymmetric carbon atom.

The subject of optical activity and the ability of optically active compounds to rotate plane-polarized light is dealt with in introductory organic chemistry. The student should review the principles of light refraction that make it possible to construct a Nicol prism which can polarize light into two planes. The student should also review the construction of the polarimeter, the device that measures quantitatively the extent to which plane-polarized light is rotated when it passes through optically active materials. Finally, the student should review the meaning of *specific rotation* $[\alpha]_D^T$ which is given by the formula

$$[\alpha]_D^T = \frac{\text{Observed rotation } (^\circ)}{\text{Length of tube (dm)} \times \text{Concentration (g/ml)}}$$

Projection and Perspective Formulas. In the study of carbohydrates many examples of optical isomerism are encountered, and it is necessary to have a means for representing the different possible isomers. One way of representing them is to use the *projection formula* introduced in the nineteenth century by the illustrious German organic chemist, Emil Fischer. The projection formula represents the four groups attached to the carbon atom as being projected onto a plane. This projection can be represented for the asymmetric molecule depicted previously as shown in Structure 2-4. In the Fischer projection

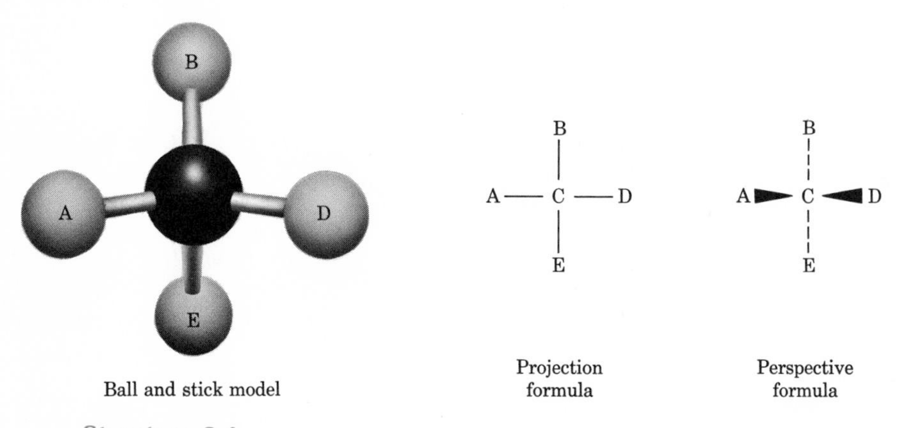

Ball and stick model

Projection formula

Perspective formula

Structure 2-4

formula, the horizontal bonds are understood to be in front of the plane of the paper while the vertical bonds are behind. This relationship is seen more clearly in the *perspective formula*. Here dashed lines indicate bonds extending behind the plane of the page while solid wedges identify bonds standing in front of the plane of the page. The projection and perspective formulas can be used to distinguish between the compound just shown and its mirror image isomer below. These two pairs of formulas represent all the enantiomers:

Projection formula (B above, D—C—A, E below) Perspective formula (B above, D▶C◀A, E below)

The perspective formula can be rotated in all planes without fear of confusing the two enantiomers. Caution is required in the use of the Fischer formulas; while they may be rotated a full 180° in the plane of the paper, rotation of only 90° results in the enantiomer because of the convention that the horizontal bonds are represented as being in front of the plane of the paper. The Fischer formula cannot be removed from the plane of the paper.

D-Glyceraldehyde As a Reference Compound. With the existence of a large number of optical isomers in carbohydrates it is also necessary to have a reference compound. The simplest monosaccharide that possesses an asymmetric carbon atom has been chosen as the reference standard; this compound is the triose *glycerose* or *glyceraldehyde*. Since this compound has a chiral center, it can exist in two optically active forms. These may be represented by their Fischer formulas or by a simplified version of the ball and stick models:

Fischer formula

CHO | H—C—OH | CH_2OH CHO | HO—C—H | CH_2OH

Ball and stick model

CHO, H, OH, CH_2OH CHO, HO, H, CH_2OH

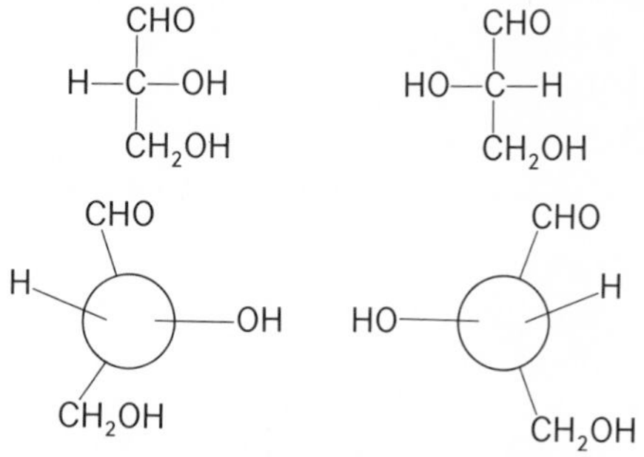

It should be clear that these two forms are related to each other as mirror image isomers. Although they will have the same melting point, boiling point, and solubility in H_2O, they will differ in the direction in which they rotate plane-polarized light. The isomer that rotates light in the clockwise direction is identified with the symbol (+) to indicate that it is the dextrorotatory enantiomer. At the turn of this century that isomer was also assigned the Fischer formula in which the hydroxyl group is on the right when the aldehyde group is at the top. Moreover, it was agreed that this form should be designated as D(+)-glyceraldehyde. For clarification, both the projection formula and the frequently seen ball and stick representations are given:

D(+)-Glyceraldehyde

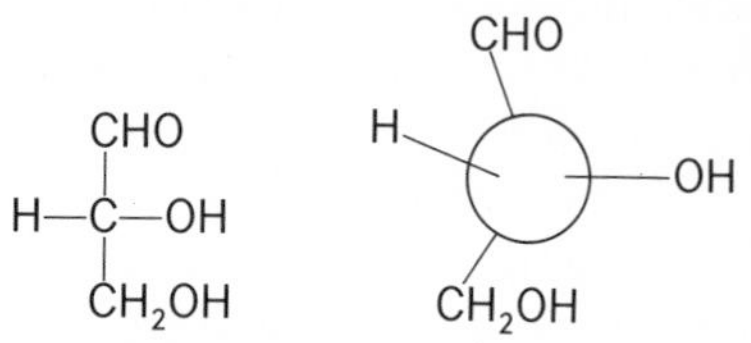

CHO | H—C—OH | CH_2OH CHO, H, OH, CH_2OH

Fischer projection Ball and stick model

This particular assignment had a 50:50 chance of being correct, and it was not until 1949 that Bijvoet, using x-ray diffraction, determined the actual configuration of the atoms in (+)-tartaric acid and thereby showed the choice to have been the correct one. Some 30 years earlier, D(+)-glyceraldehyde was shown by a series of chemical reactions to be related to levorotatory (−)-tartaric acid; that is, D(+)-glyceraldehyde and (−)-tartaric acid had the same configuration on the reference carbon atom. Bijvoet's study therefore established the absolute configuration of all the compounds which in previous decades had been shown to be related in configuration to either D(+)-glyceraldehyde or its enantiomer, L(−)-glyceraldehyde.

While D(+)-glyceraldehyde serves an important role as a reference compound for optical isomers of sugars, the hydroxy and amino acids, and related compounds, there are many compounds for which it cannot serve this function. Instead, a procedure proposed by three distinguished European organic chemists, R. S. Cahn, C. Ingold, and V. Prelog serves this purpose, and compounds can be designated **R** (for *rectus*) and **S** (for *sinister*), depending on the arrangement of atoms on the asymmetric carbon atom. According to the Cahn–Ingold–Prelog rule, D(+)-glyceraldehyde would be termed **R**(+)-glyceraldehyde. While this rule is employed in many areas of organic chemistry, biochemists do not yet appear to be using this terminology with reference to the carbohydrates or amino acids, and this will also be our practice.

Cyanohydrin Synthesis. As an illustration of the use of D-glyceraldehyde as a reference compound, consider the formation of tetrose sugars from a triose by the *Kiliani–Fischer* synthesis. This synthesis is a process by which the chain length of an aldose (aldotriose, aldotetrose, aldohexose, etc.) may be increased. In the initial reaction, the sugar is allowed to react with HCN. Addition of cyanide to the aldehyde group generates a new asymmetric carbon atom, and *two* diastereomeric cyanohydrins are produced. These cyanohydrins may be hydrolyzed to carboxylic acids, dehydrated to form the corresponding γ-lactones, and finally reduced by sodium amalgam to yield two diastereomeric aldoses containing one more carbon atom than the starting sugar.

The application of the Kiliani–Fischer synthesis to the D and L forms of glyceraldehyde is shown in Figure 2-1. In the initial reaction with D-glyceraldehyde, two cyanohydrins are formed in which the configuration at the carbon atom adjacent to the nitrile group is reversed. When hydrolysis, lactonization, and reduction are complete, two new sugars, the tetroses D-erythrose and D-threose, are formed. Note that these tetroses differ only in the position of the hydroxyl group on carbon atom 2, the carbon atom adjacent to the aldehyde group. They do not differ in the configuration on the asymmetric carbon atom 3, the asymmetric carbon atom which was present initially in D-glyceraldehyde. Since they have the same configuration on this *reference carbon atom,* the asymmetric carbon atom having the highest number (the farthest removed from the functional aldehyde group), these two tetroses are known as D-sugars. Similarly, two new L sugars, L-threose and L-erythrose, are formed in the cyanohydrin synthesis from L-glyceraldehyde (Figure 2-1).

At this point it is profitable to consider the stereochemical relations that exist between these four tetroses and the reference compound D-glyceraldehyde. First, these four tetroses have the same structural formula, $CH_2OH—CHOH—CHOH—CHO$, and they are therefore stereoisomers rather than structural isomers. Second, with regard to their stereoisomerism, they obviously belong to the class of optical isomers rather than that of geometric isomers. Third, there are two pairs of enantiomers among the four tetroses:

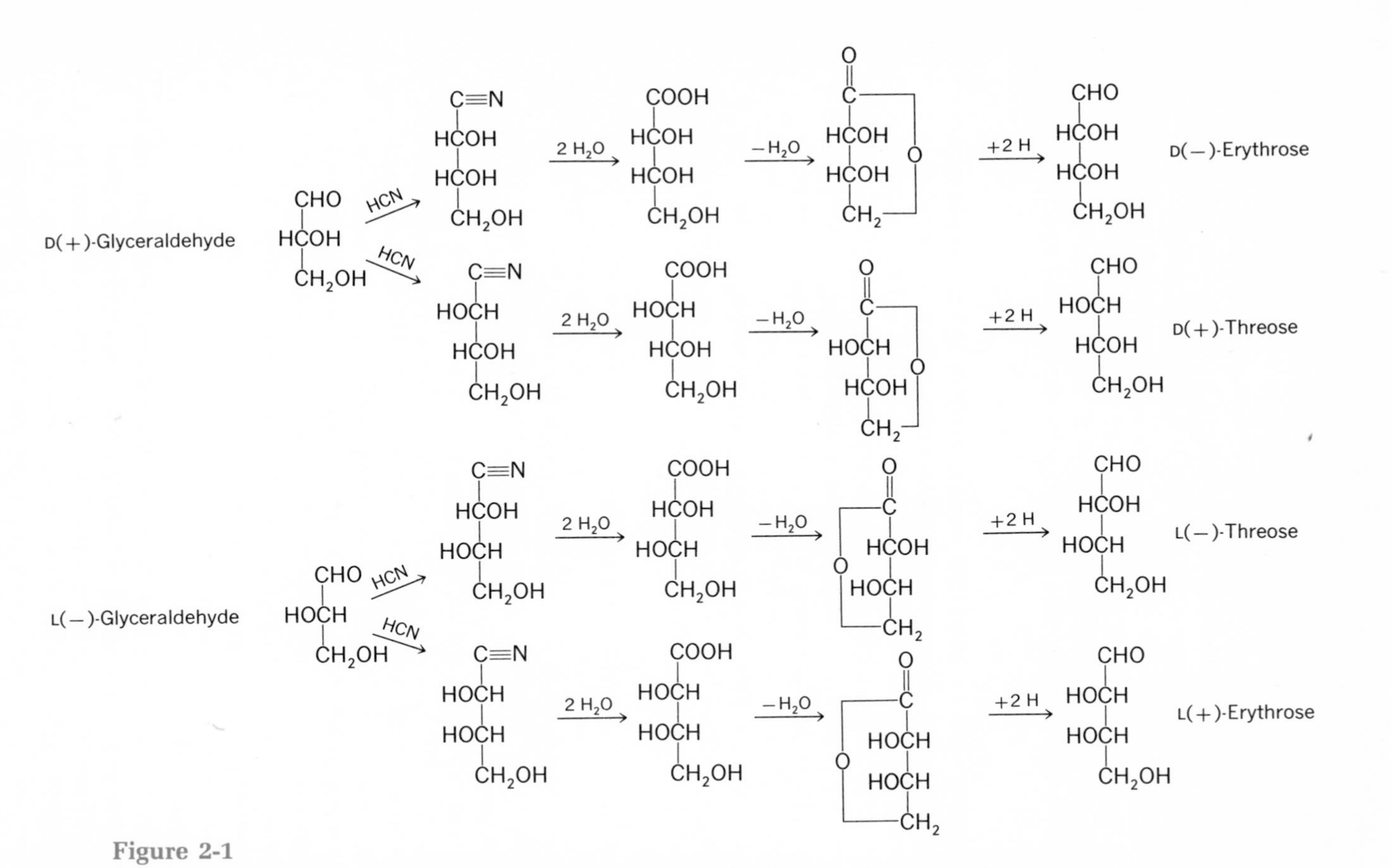

Figure 2-1
Application of the cyanohydrin synthesis to D-glyceraldehyde and L-glyceraldehyde.

D-erythrose is the mirror image isomer of L-erythrose. The same relation exists between D-threose and L-threose. Fourth, the two D sugars are related structurally to D-glyceraldehyde because they have the same configuration about the reference carbon atom. In most of the common sugars the reference carbon atom is the *penultimate* carbon atom, the carbon atom next to the last carbon from the functional or aldehyde group. Fifth, note that the symbols D and L *bear no relation* to whether the tetrose is dextrorotatory or levorotatory. D-Erythrose is levorotatory, whereas D-threose is dextrorotatory. The direction in which light is rotated is a specific property of the molecule under consideration and is not related directly to the configuration about the penultimate carbon atom (except in the glyceraldehydes).

As the number of asymmetric carbon atoms increases in a carbohydrate molecule the number of optical isomers also increases. In the trioses, where there is one asymmetric carbon atom, there are two optical isomers; in the tetroses, where there are two asymmetric carbon atoms, there are four optical isomers; in the aldohexoses, where there are four asymmetric carbon atoms, there are sixteen optical isomers. van't Hoff established that 2^n represents the number of possible optical isomers, where *n* is the number of asymmetric carbon atoms. In the ketohexoses, where *n* is 3, there are eight possible optical isomers.

Consider the four common hexoses whose projection formulas are

D(+)-Glucose	D(+)-Mannose	D(+)-Galactose	D(−)-Fructose
CHO	CHO	CHO	CH_2OH
HCOH	HOCH	HCOH	C=O
HOCH	HOCH	HOCH	HOCH
HCOH	HCOH	HOCH	HCOH
HCOH	HCOH	HCOH	HCOH
CH_2OH	CH_2OH	CH_2OH	CH_2OH

The following statements can be made about their isomerism: All four sugars are D sugars, because they have the same configuration as D-glyceraldehyde on the penultimate carbon atom; the use of the term D has no bearing on whether these sugars are dextro- or levorotatory.

D-Fructose is a structural isomer of the other three hexoses. Although it has the same molecular formula ($C_6H_{12}O_6$), it has a different functional group; it is a ketose rather than an aldose.

The three aldohexoses are stereoisomers, more specifically, optical isomers. Because no one of the three is an enantiomer of either of the other two, they are related as *diastereomers*. If two optical isomers are not enantiomers, they are related as diastereomers. Since these isomers are diastereomers they have different melting points, different boiling points, different solubilities, different specific rotations, and, in general, different chemical properties. Clearly, the three aldoses are only three of the possible *sixteen* optical isomers; there are eight pairs of enantiomorphs in the sixteen aldohexoses.

D(+)-Glucose may be said to be an *epimer* of D(+)-mannose because these compounds differ from each other by their configuration on a single asymmetric carbon atom. Similarly, D(+)-glucose is an epimer of D(+)-galactose. On the other hand, there is no epimeric relationship between D(+)-mannose and D(+)-galactose.

The Structure of Glucose

Emil Fischer received the Nobel Prize in Chemistry for his studies on the structure of glucose, more specifically for establishing the configuration of the four asymmetric carbon atoms in that aldohexose relative to D(+)-glyceraldehyde. From Fischer's work, chemists were able to write the projection and ball and stick formulas for D- and L-glucose (Structure 2-5).

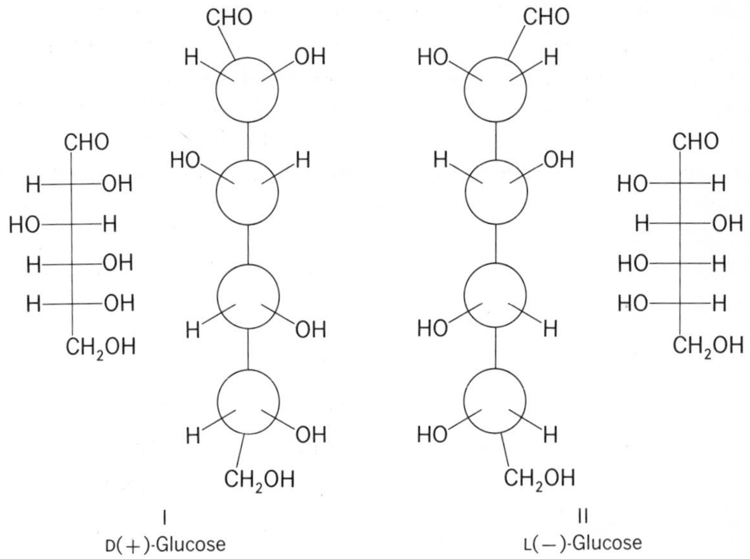

I
D(+)-Glucose

II
L(−)-Glucose

Structure 2-5

If a ball and stick model represented above is actually constructed and the —CHO and —CH_2OH are held so that they extend away from the holder (behind the plane of the paper), the remainder of the carbon atoms will form a ring extending toward the holder, and the H and OH groups will project out even more toward the holder.

Although the sugars have been considered as polyhydroxy aldehydes or ketones to this point, there is abundant evidence to indicate that other forms (of glucose, for example) exist and indeed predominate both in the solid phase and in solution. For instance, aldohexoses undergo the Kiliani–Fischer synthesis with difficulty, although cyanohydrin formation with simple aldehydes is usually rapid. Glucose and other aldoses fail to give the Schiff test for aldehydes. Solid glucose is quite inert to oxygen, and yet aldehydes are notoriously autoxidizable. Finally, it is possible to show that two crystalline forms of D-glucose exist. When D-glucose is dissolved in water and allowed to crystallize out by evaporation of the water, a form designated as α-D-glucose is obtained. If glucose is crystallized from acetic acid or pyridine, another form, β-D-glucose, is obtained. These two forms of D-glucose show the phenomenon of mutarotation. A freshly prepared aqueous solution of α-D-glucose has a specific rotation $[\alpha]_D^{20}$ of +113°; when the solution is left standing it changes to +52.5°. A fresh solution of β-D-glucose, on the other hand, has an $[\alpha]_D^{20}$ of +19°; on standing it also changes to the same value, +52.5°.

The explanation of the existence of the two forms of glucose, as well as the other anomalous properties described, is found in the fact that the aldohexoses and other sugars react internally to form cyclic hemiacetals. Hemiacetal formation is a characteristic reaction between aldehydes and alcohols:

$$R{-}C(H){=}O + R'OH \longrightarrow R{-}C(H)(OH){-}OR'$$

This reaction occurs with glucose because of the proximity of the alcoholic hydroxyl group on carbon atom 5 to the aldehyde group of carbon atom 1. As noted above, the angles of the tetrahedral carbon atom can bend the glucose molecule into a ring; the C-5 hydroxyl group therefore reacts to form a six-membered ring. When the C-4 hydroxyl reacts, a five-membered ring results; a seven-membered ring is too strained to permit the C-6 hydroxyl of an aldohexose to form a hemiacetal.

When the process of ring formation is depicted as in Figure 2-2, it is easier to understand why the cyclic hemiacetal has the structure shown. Simple rotation of the bond between carbon atoms 4 and 5 in a counterclockwise direction moves the C-5 hydroxyl into position for reaction with the aldehyde group. In so doing, the —CH_2OH group now occupies the position formerly occupied by the hydrogen at C-5. As the hemiacetal ring is formed, note that the C-1 becomes an asymmetric carbon atom. Therefore, two diastereomeric molecules are possible. These isomers are the α and β forms of glucose; they are diastereomers, however, rather than enantiomers, for the α form differs from the β form only in the configuration around the hemiacetal carbon atom. More specifically, they are known as *anomers*, because they differ only in the configuration at the hemiacetal carbon. Since the cyclic forms of the aldohexoses have five asymmetric carbon atoms, there are thirty-two optical isomers of the cyclic aldohexoses consisting of sixteen pairs of enantiomers.

The English chemist, W. H. Haworth proposed that the first five carbon atoms of the aldohexoses and the oxygen atom of the ring might better be represented as a hexagonal ring in a plane perpendicular to the plane of the paper. The side of the hexagon that is nearer to the reader would then be indicated by a thickened line. When this is done, the substituents on the carbon atom then will extend above or below the plane of the six-membered ring. Carbon atom 6, a substituent on C-5, will therefore be above the plane of the ring. The Haworth formulas for α-D(+)-glucose and β-D(+)-glucose may then be compared with the Fischer projection formulas for these diastereomers (Structure 2-6).

With respect to assigning structures to the α and β anomers, Fischer originally suggested that, in the D series, the more dextrorotatory compound

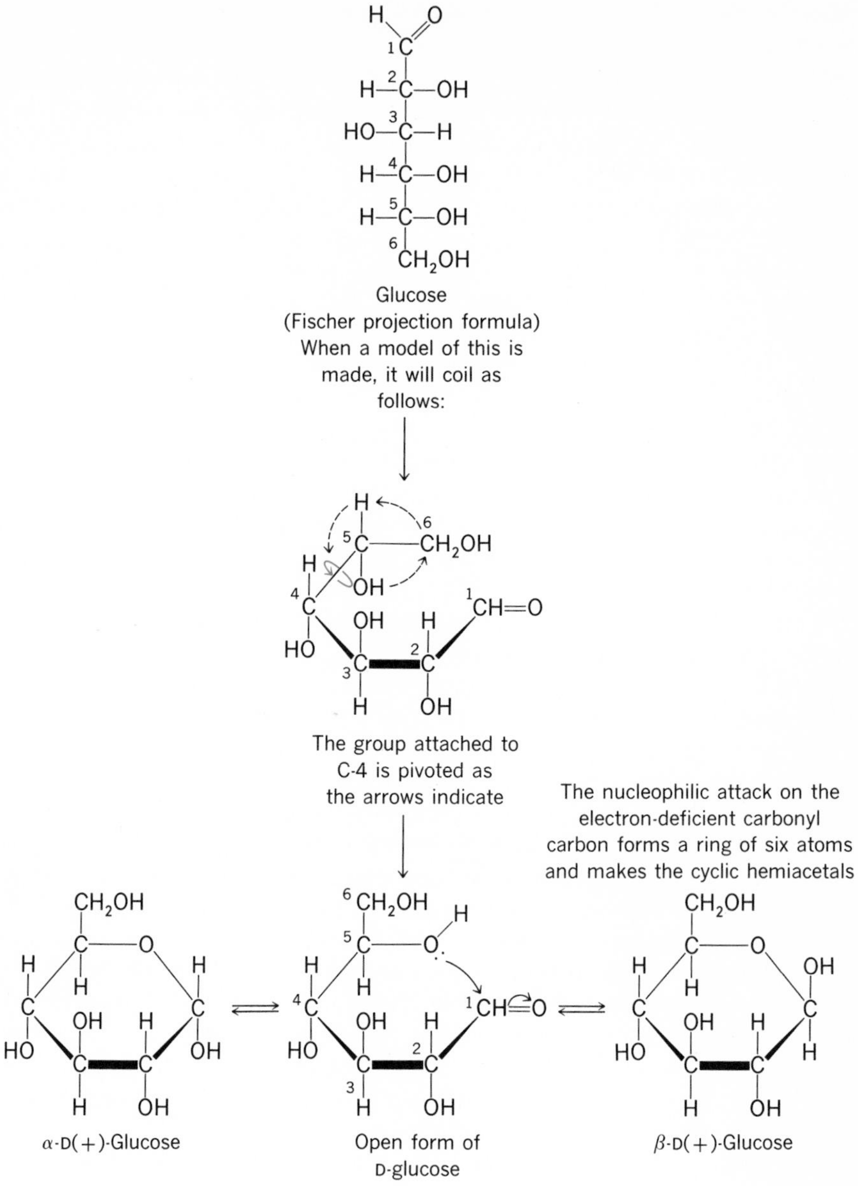

Glucose
(Fischer projection formula)
When a model of this is made, it will coil as follows:

The group attached to C-4 is pivoted as the arrows indicate

The nucleophilic attack on the electron-deficient carbonyl carbon forms a ring of six atoms and makes the cyclic hemiacetals

α-D(+)-Glucose

Open form of D-glucose

β-D(+)-Glucose

Figure 2-2
Scheme depicting the formation of the hemiacetal forms of D-glucose. Note that an equilibrium exists between the α and β forms and the open-chain form.

be called the α anomer while, in the L series, the α anomer would be the more levorotatory substance. Later, Freudenberg proposed that the α and β anomers be classified with respect to their configuration rather than sign or magnitude of rotation. He suggested that those anomers having the same configuration at both the anomeric and reference carbon atoms be called α while the β anomer would have different configurations at these two carbon atoms. The latter convention is used today, and the α and β forms of D-glucose are those shown in Structure 2-6.

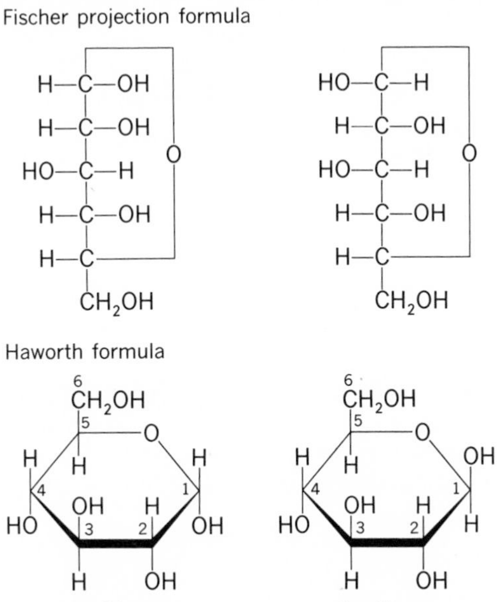

Fischer projection formula

Haworth formula

α-D-Glucose

β-D-Glucose

Structure 2-6

Note that, with respect to the Haworth formula, this means that the anomeric hydroxyl is written below the plane of the ring (or the same as the hydroxyl at C-2 and C-4) in the case of α-D-glucose. The β anomer then has the anomeric hydroxyl above the plane of the ring. The six-membered ring sugars may be considered derivatives of pyran, while the five-membered rings are considered related to furan. Hence, it is customary to refer to the

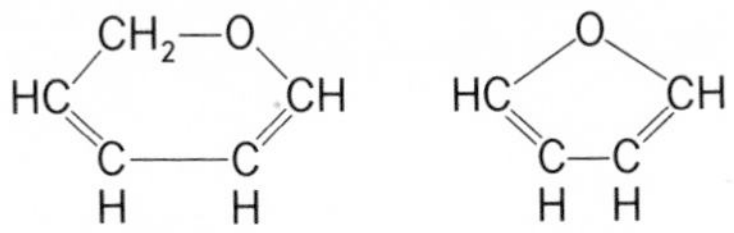

α-Pyran

Furan

pyranose or *furanose* form of the monosaccharide. Furanose forms of glucose are less stable than the pyranose forms in solution; combined forms of furanose sugars (as in the fructose unit of sucrose) are found in nature, however.

There is still one final aspect of the structure of glucose to be mentioned; this is its *conformation*. Because the C—O—C bond angle of the hemiacetal ring (111°) is similar to that of the C—C—C ring angles (109°) of the cyclohexane ring, the pyranose ring of glucose, rather than forming a true plane, is puckered in much the same way as cyclohexane. Review of the structure of cyclohexane will recall that the ring can exist in two conformations, the *chair* and *boat* forms. The chair conformation of glucose minimizes torsional strain and further, the conformational structure in which a maximum number of bulky groups (—OH and —CH_2OH) are *equatorial* rather than *axial* to an axis passing through the ring is preferred. The diagram

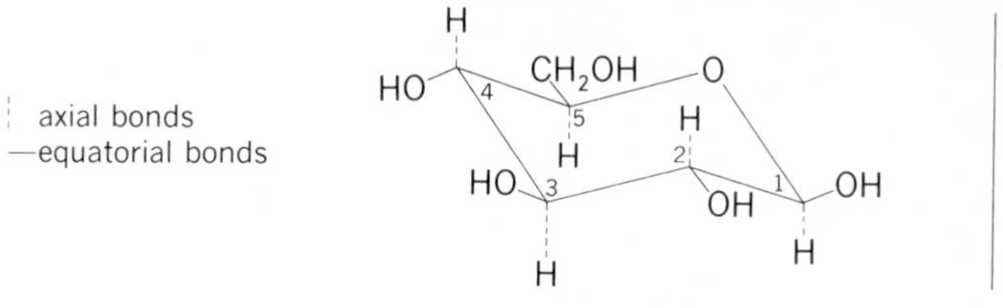

¦ axial bonds
—equatorial bonds

β-D(+)-Glucopyranose

axis

here shows that β-D(+)-glucopyranose can achieve a conformation in which all bulky groups are equatorial (or perpendicular) to an axis passing through the plane of the ring. This conformation is thermodynamically more stable than that in which the hydroxyls and the —CH_2OH are axial (parallel to the axis shown). α-D-Glucopyranose can have a conformation in which all bulky groups *except* the anomeric hydroxyl are equatorial, and the preferred structure for this form may be represented as

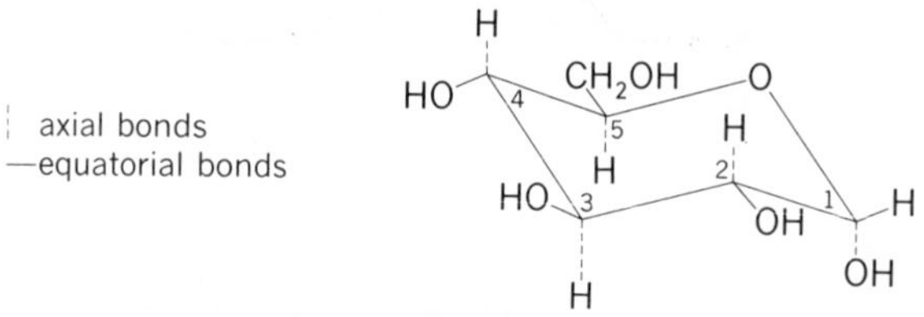

¦ axial bonds
—equatorial bonds

α-D(+)-Glucopyranose

One of the two anomers, therefore, the β anomer with *all* bulky groups equatorial, should predominate in solution over the α isomer with one axial group, the anomeric hydroxyl. Thus, in aqueous solution, β-D(+)-glucopyranose is present to the extent of about 63% after mutarotation, while α-D(+)-glucopyranose comprises about 36%. The linear polyhydroxy aldehyde form accounts for less than 1% of the total carbon present as glucose (see Figure 2-2).

Structures of Other Monosaccharides

Pyranose forms for the other aldohexoses mentioned on page 31 may be written by proper arrangement of the hydroxyl group on C-2, C-3, and C-4.

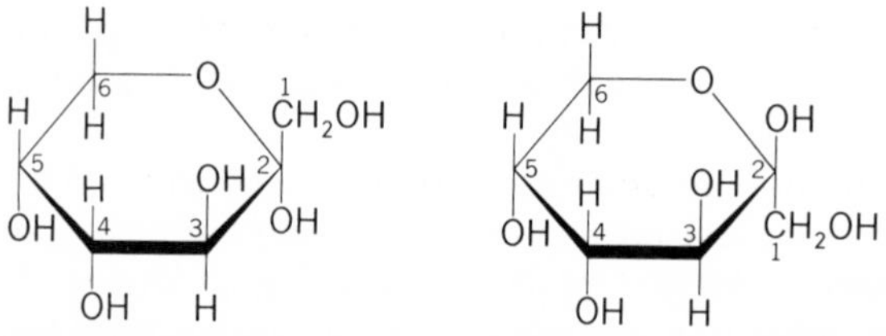

α-D(−)-Fructopyranose

β-D(−)-Fructopyranose

Structure 2-7

Similarly, the Haworth formulas for α-D-fructopyranose and β-D-fructopyranose may be written as shown in Structure 2-7. Note, however, that the five-member furanose structure is the one encountered for fructose when the hemiketal (from the ketone group of the ketohexoses) group is substituted as in sucrose (see below) and fructosans (Structure 2-8).

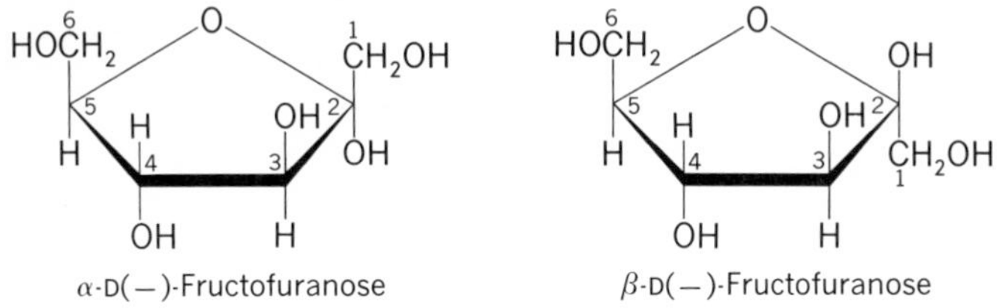

α-D(−)-Fructofuranose β-D(−)-Fructofuranose

Structure 2-8

The ubiquitous pentose, D-ribose, a component of ribonucleic acid, exists as a furanose; 2-deoxy-D-ribose, a component of 2-deoxyribonucleic acid is also a furanose sugar. Both α and β isomers can exist in solution, but in the nucleic acids, the β isomer is the one which is found (Structure 2-9).

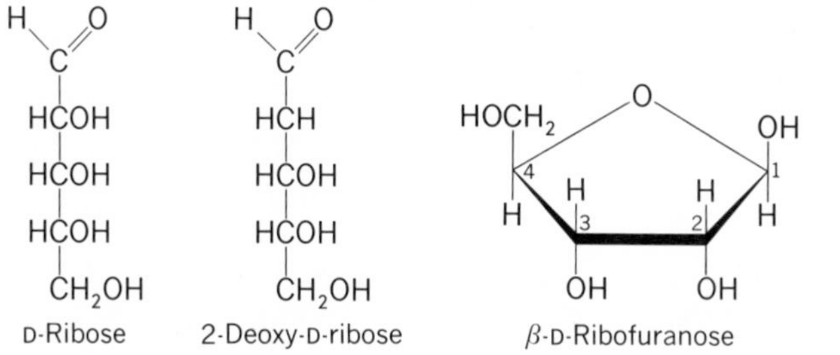

D-Ribose 2-Deoxy-D-ribose β-D-Ribofuranose

Structure 2-9

Four other monosaccharides that play important roles in the metabolism of carbohydrates during photosynthesis are the aldotetrose, D-erythrose, the ketopentoses, D-xylulose and D-ribulose, and the ketoheptose, D-sedoheptulose.

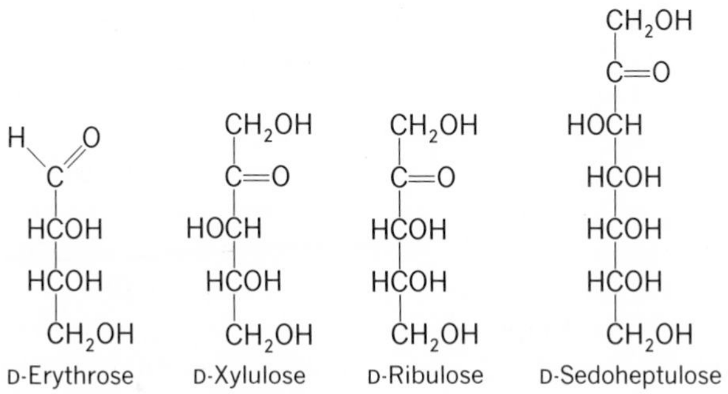

D-Erythrose D-Xylulose D-Ribulose D-Sedoheptulose

While five-membered hemiacetal (erythrose) or hemiketal (xylulose, ribulose) structures of these monosaccharides may be written, the metabolically active forms are the phosphate esters in which the primary alcohol (—CH_2OH)

group has been esterified with H_3PO_4 thereby preventing its participation in a furanose ring.

Two other deoxy sugars are found in nature as components of cell walls. These are L-rhamnose (6-deoxy-L-mannose) and L-fucose (6-deoxy-L-galactose).

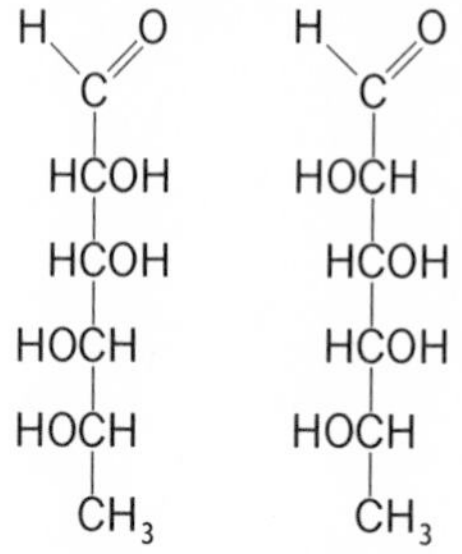

L-Rhamnose	L-Fucose
H–C=O	H–C=O
HCOH	HOCH
HCOH	HCOH
HOCH	HCOH
HOCH	HOCH
CH_3	CH_3

L-Rhamnose L-Fucose

Two amino sugars, D-glucosamine and D-galactosamine, exist in which the hydroxyl group at C-2 is replaced by an amino group. The former is a major component of chitin, a structural polysaccharide found in insects and crustaceans. D-Galactosamine is a major component of the polysaccharide of cartilage. Their hemiacetal forms are shown here. The derived forms of the amino sugars will be described later.

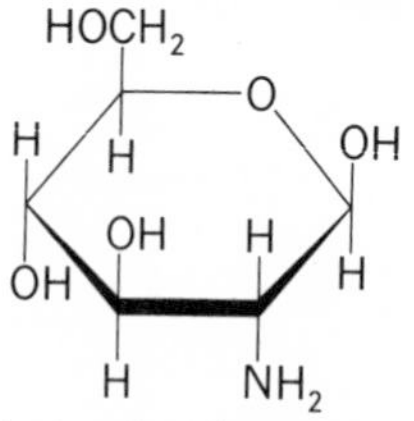

2-Amino-β-D-glucopyranose (β-D-Glucosamine)

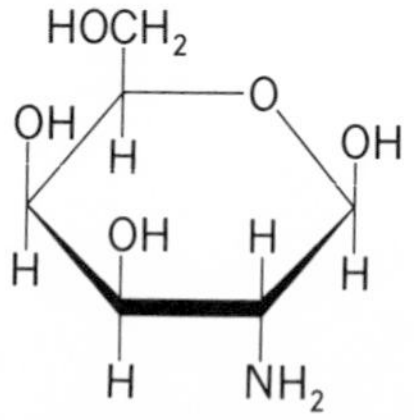

2-Amino-β-D-galactopyranose (β-D-Galactosamine)

Properties of Monosaccharides

Mutarotation. We have already referred to the phenomenon of mutarotation exhibited by the anomeric forms of D-glucopyranose. Mutarotation is a property exhibited by the hemiacetal and ketal forms of sugars that are free to form the open-chain sugar. As pointed out in Figure 2-2, the open-chain polyhydroxy aldehyde or ketone is an intermediate in the interconversion of the α and β forms during mutarotation.

When glucose is exposed to dilute alkali for several hours, the resulting mixture contains both fructose and mannose. If either of these sugars is treated with dilute alkali, the equilibrium mixture will contain the other sugar as well as glucose. This reaction, known as the Lobry de Bruyn–von Ekenstein transformation, is due to the enolization of these sugars in the presence of alkali. *Enediol* intermediates that are common to all three sugars are responsible for the establishment of the equilibrium. At higher concentrations of alkali, the monosaccharides are generally unstable and undergo oxidation, degradation, and polymerization.

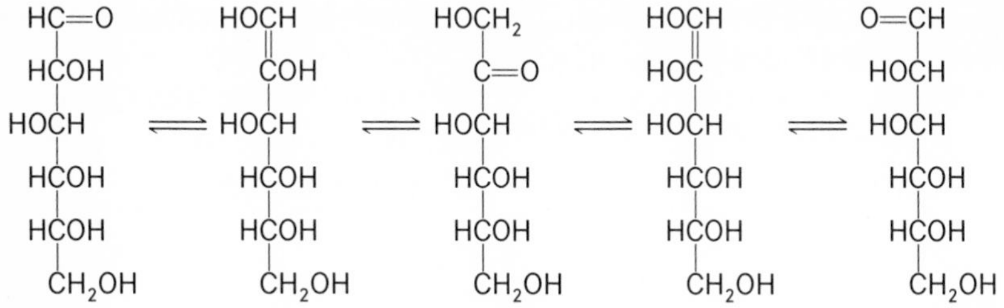

D-Glucose		*trans*-Enediol		D-Fructose		*cis*-Enediol		D-Mannose
HC=O		HOCH		$HOCH_2$		HOCH		O=CH
HCOH		‖ COH		C=O		‖ HOC		HOCH
HOCH	⇌	HOCH	⇌	HOCH	⇌	HOCH	⇌	HOCH
HCOH		HCOH		HCOH		HCOH		HCOH
HCOH		HCOH		HCOH		HCOH		HCOH
CH_2OH		CH_2OH		CH_2OH		CH_2OH		CH_2OH

Isomerization in dilute alkali

By contrast, monosaccharides are generally stable in dilute mineral acids, even on heating. When aldohexoses are heated with strong mineral acids, however, they are dehydrated, and hydroxymethyl furfural is formed:

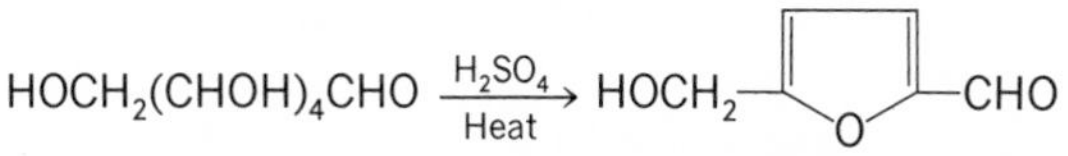

$$HOCH_2(CHOH)_4CHO \xrightarrow[\text{Heat}]{H_2SO_4} HOCH_2\text{—(furan ring, O)—}CHO$$

Hydroxymethyl furfural

Under the same condition, pentoses yield furfural:

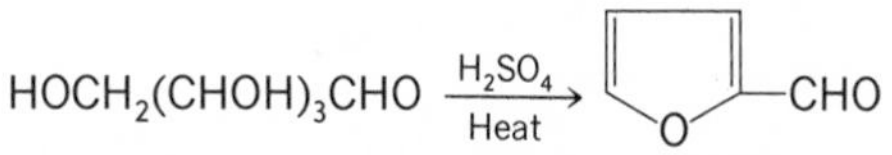

$$HOCH_2(CHOH)_3CHO \xrightarrow[\text{Heat}]{H_2SO_4} \text{(furan ring, O)—}CHO$$

Furfural

This dehydration reaction is the basis of certain qualitative tests for sugars, since the furfurals can be reacted with α-naphthol and other aromatic compounds to form characteristic colored products.

Reducing Sugars. Carbohydrates may be classified as either reducing or nonreducing sugars. The reducing sugars, which are the more common, are able to function as reducing agents because free or potentially free aldehyde and ketone groups are present in the molecule. The reducing properties of these carbohydrates are usually observed by their ability to reduce metal ions, notably copper or silver, in alkaline solution. Benedict's solution is a common reagent for detecting reducing sugars; in this reagent Cu^{2+} is maintained in solution as its alkaline citrate complex. When the Cu^{2+} is reduced, the resulting Cu^{+} ion is less soluble and Cu_2O precipitates out of the alkaline solution as a yellow or red solid. The reducing sugar in turn is oxidized, fragmented, and polymerized in the strongly alkaline Benedict's solution.

The aldehyde group of aldohexoses is readily oxidized (as shown by its oxidation by Cu^{2+}) to the carboxylic acid at neutral pH by mild oxidizing agents or by enzymes. The monocarboxylic acid that is formed is known as an aldonic acid (e.g., galactonic acid from galactose). In the presence of a strong oxidizing agent like HNO_3, both the aldehyde and the primary alcoholic function will be oxidized to yield the corresponding dicarboxylic or aldaric acid (e.g., galactaric acid). One of the more important oxidation products of monosaccharides is the monocarboxylic acid obtained by the oxidation of only the primary alcoholic function, usually by specific enzymes, to yield the corresponding uronic acid (e.g., galacturonic acid). Such acids are components of many polysaccharides.

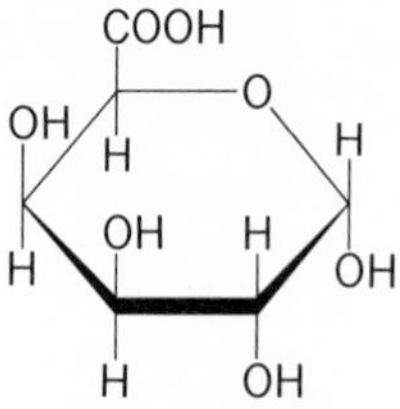

α-D-Galacturonic acid

Much use is made of the oxidizing agent, periodic acid, in carbohydrate analysis. This reagent will cleave carbon–carbon bonds if both carbons have hydroxyl groups or if a hydroxy and an amino group are on adjacent carbon atoms. Thus, the glucoside α-methyl-D-glucose will react as shown below. The carbon atoms whose bonds are severed are converted to aldehydes (R—CHO). If there happen to be three hydroxyl groups on adjacent carbon atoms, as in this case, the central carbon atom is released as formic acid:

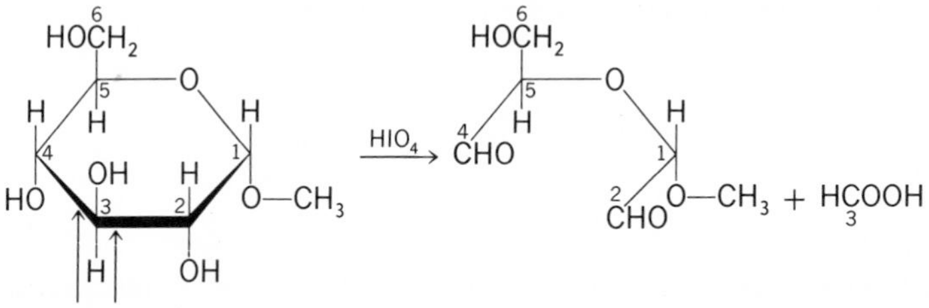

$$\xrightarrow{HIO_4} \quad \ldots + HCOOH$$

The aldehyde and ketone functions of monosaccharides may be reduced chemically (with hydrogen or $NaBH_4$) or with enzymes to yield the corresponding sugar alcohols. Thus, D-glucose when reduced yields D-sorbitol, and D-mannose produces D-mannitol. Sorbitol is found in the berries of many

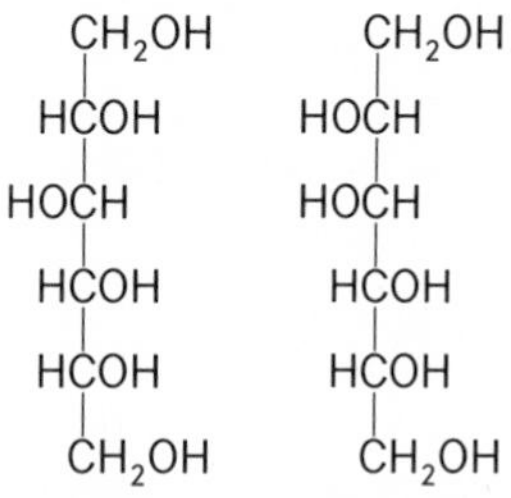

D-Sorbitol	D-Mannitol
CH_2OH	CH_2OH
HCOH	HOCH
HOCH	HOCH
HCOH	HCOH
HCOH	HCOH
CH_2OH	CH_2OH

higher plants, especially in the *Rosaceae*; it is a crystalline solid at room temperature but has a low melting point. D-Mannitol is found in algae and fungi. Both compounds are soluble in H_2O and have a sweet taste.

Glycoside Formation. One of the most important properties of monosaccharides is their ability to form glycosides or acetals. Consider as an example the formation of the methyl glycoside of glucose. When D-glucose in solution is treated with methanol and HCl, two compounds are formed. Determination of their structure has shown that these two compounds are the diastereomeric α- and β-methyl-D-glucosides. These glucosides, and glycosides in general, are acid labile but are relatively stable at alkaline pH. Since the formation of the methyl glycoside converts the aldehydic group to an acetal group, the glycoside is not a reducing sugar and does not show the phenomenon of mutarotation.

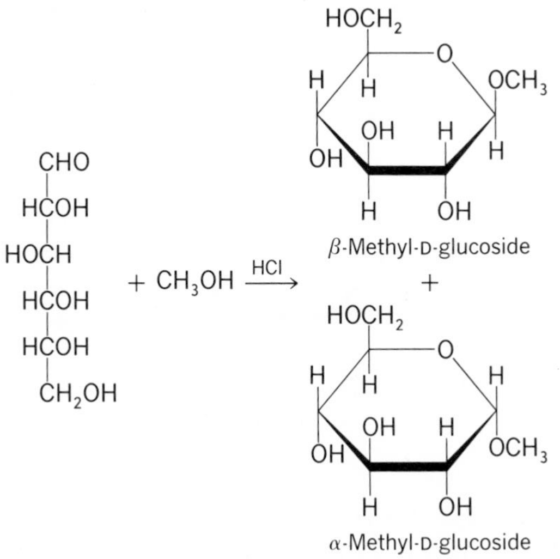

CHO
HCOH
HOCH
HCOH
HCOH
CH_2OH

$+ CH_3OH \xrightarrow{HCl}$

$HOCH_2$ O H H OCH_3 OH H OH H H OH

β-Methyl-D-glucoside

+

$HOCH_2$ O H H H OH H OH OCH_3 H OH

α-Methyl-D-glucoside

When the alcoholic hydroxyl group on a second sugar molecule reacts with the hemiacetal (or hemiketal) hydroxyl of another monosaccharide, the resulting glycoside is a disaccharide. The bond between the two sugars is known as a *glycosidic bond*. Polysaccharides are formed by linking together a large number of monosaccharide units with *glycosidic bonds*.

While the anomeric hydroxy group of sugars may be methylated with ease, as in the formation of methyl glycosides just described, methylation of the remaining hydroxyl functions requires much stronger methylating agents. Nevertheless, the remaining four hydroxyl groups of methyl-α-D-glucopyranoside can be reacted with methyl iodide or dimethyl sulfate to yield the pentamethyl derivative. Such compounds, in turn, are useful in determining the ring structure of the parent sugar as in the following example:

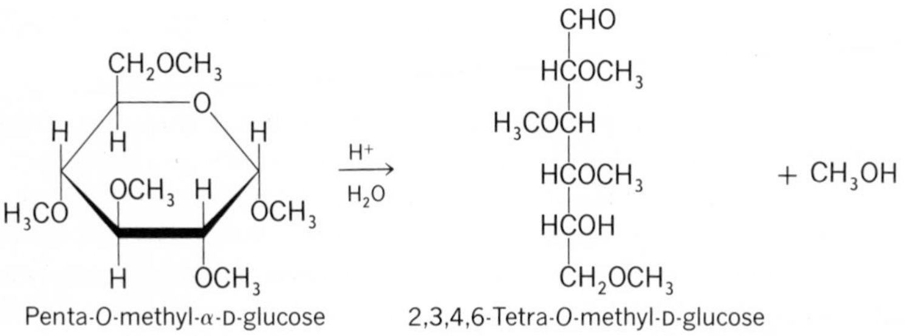

CH_2OCH_3 O H H H OCH_3 H H_3CO OCH_3 H OCH_3

Penta-O-methyl-α-D-glucose

$\xrightarrow[H_2O]{H^+}$

CHO
$HCOCH_3$
H_3COCH
$HCOCH_3$
HCOH
CH_2OCH_3

2,3,4,6-Tetra-O-methyl-D-glucose

$+ CH_3OH$

The methyl group on the hemiacetal carbon, being a glycosidic methyl is readily hydrolyzed by acid. The remaining methyl groups, being methyl ethers, are not. Therefore, treatment of the pentamethylglucose derivative pictured here with dilute acid at 100°C will yield the 2,3,4,6-tetra-O-methyl-D-glucose. Treatment of the pentamethyl derivative in which the sugar is in a furanose ring yields 2,3,5,6-tetra-O-methyl-D-glucose instead.

Ester Formation. Another derivative useful in structural determination is the acetyl derivative of the carbohydrates. Thus, when α-D-glucopyranoside is treated with acetic anhydride, all the hydroxyl functions are acetylated to yield the penta-O-acetyl glucose pictured here. These acetyl groups, being esters, can be hydrolyzed either in acid or alkali.

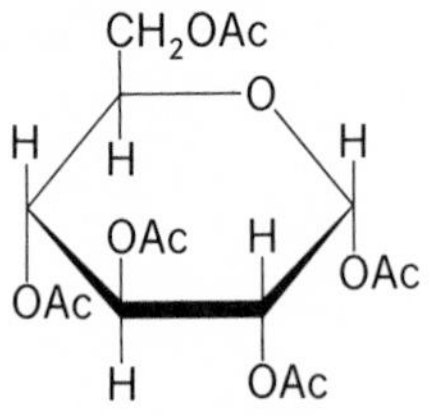

Penta-O-acetyl-α-D-glucose
(Ac = CH_3—C(=O)—)

An important type of carbohydrate derivative encountered in intermediary metabolism is the phosphate ester. Such compounds are frequently formed by the reaction of the carbohydrate with adenosine triphosphate (ATP) in the presence of an appropriate enzyme. An example is fructose-1,6-diphosphate.

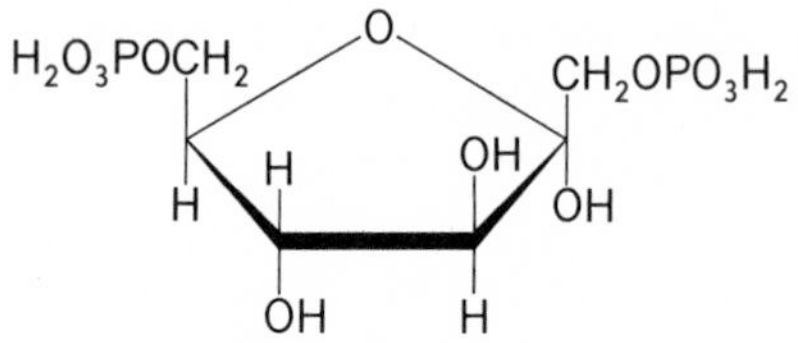

α-D-Fructose-1,6-diphosphoric acid

The correct name of the nonionized form of this compound is α-D-fructofuranose-1,6-diphosphoric acid. As noted in Chapter 1, such phosphate esters are relatively strong acids with values of approximately 2.1 and 7.2 for pK_{a_1} and pK_{a_2}. Thus, at neutral pH, the sugar phosphates are anions and are normally referred to by the name of the anion, i.e., fructose-1,6-diphosphate.

Oligosaccharides

Disaccharides. The oligosaccharides (see page 24 for definition) most frequently encountered in nature are disaccharides which on hydrolysis yield 2 moles of monosaccharides. Among the disaccharides encountered is the sugar maltose; this sugar is obtained as an intermediate in the hydrolysis of starch by enzymes known as amylases. In maltose one molecule of glucose is linked through the hydroxyl group on the C-1 carbon atom in a glycosidic bond to the hydroxyl group on the C-4 of a second molecule of glucose.

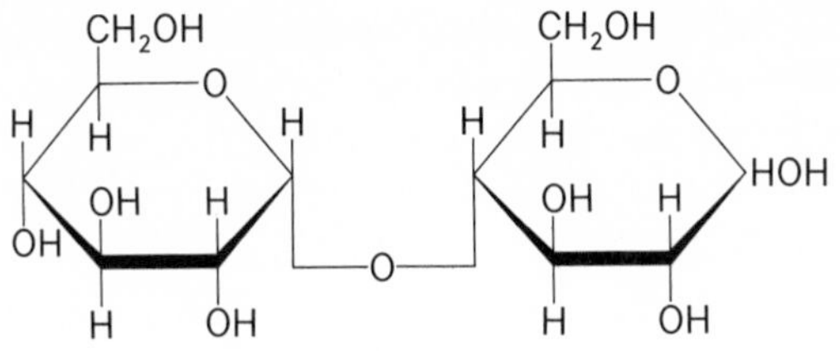

Maltose

Because the configuration on the hemiacetal carbon atom involved in the bonding is of the α form and because it is linked to the 4 position on the second glucose unit, this linkage is designated as α-1-4. The second glucose molecule possesses a free anomeric hydroxyl that can exist in either the α or β configuration; this free anomeric hydroxyl thus confers the property of mutarotation on maltose, and the disaccharide is a reducing sugar. That maltose has the structure shown was determined by analyzing the two products obtained on acid hydrolysis of its *octa*methyl derivative. The latter was prepared by treating maltose with dimethyl sulfate. The fully methylated

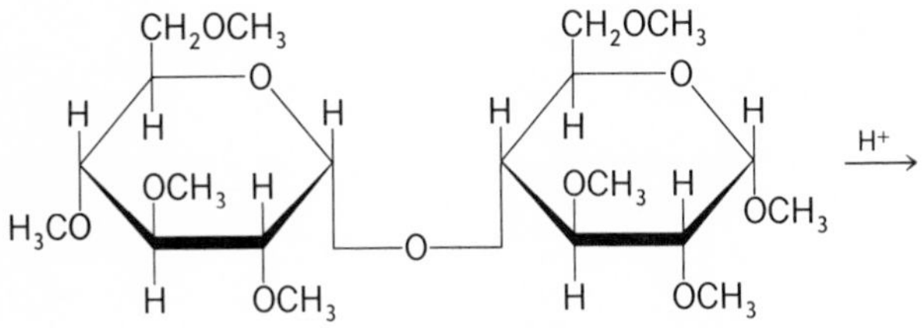

Octamethyl-D-maltose $\xrightarrow{H^+}$

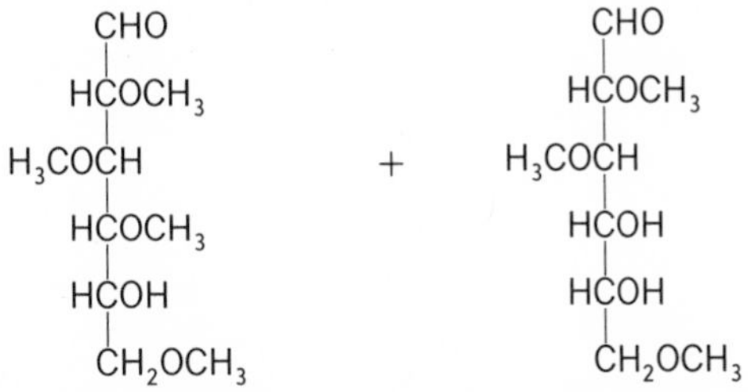

2,3,4,6-Tetra-O-methyl-D-glucose + 2,3,6-Tri-O-methyl-D-glucose

maltose yields 2,3,4,6-tetra-O-methyl-D-glucose and 2,3,6-tri-O-methyl-D-glucose. While the anomeric carbon of maltose is methylated on treatment of the disaccharide, this O-methyl glycosidic bond as well as the glycosidic bond between the two glucose units of the disaccharide is acid labile and both are cleaved on hydrolysis with acid.

The disaccharide cellobiose is identical with maltose except that the former compound has a β-1-4 glycosidic linkage. Cellobiose is a disaccharide formed during the hydrolysis of cellulose. It is a reducing sugar and undergoes mutarotation. Treatment of cellobiose with dimethyl sulfate would also yield an octamethylated sugar, and acid hydrolysis would yield the same products that were obtained from octamethyl maltose.

Isomaltose, another disaccharide obtained during the hydrolysis of certain polysaccharides, is similar to maltose except that it has an α-1-6 glucoside

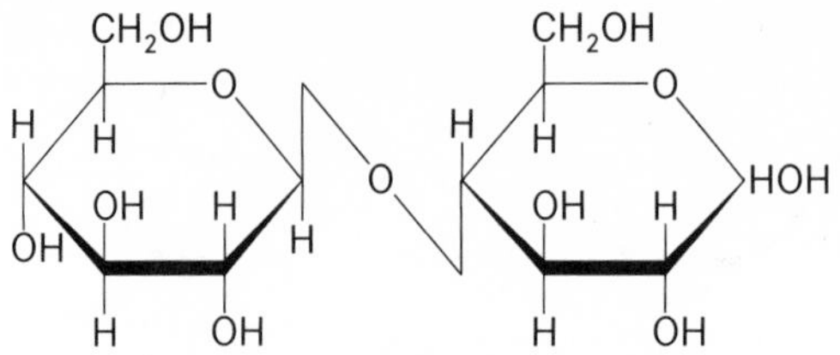

Cellobiose

linkage. Exhaustive methylation and acid hydrolysis of octamethyl isomaltose would yield 2,3,4,6-tetra-*O*-methyl-D-glucose and 2,3,4-tri-*O*-methyl-D-glucose.

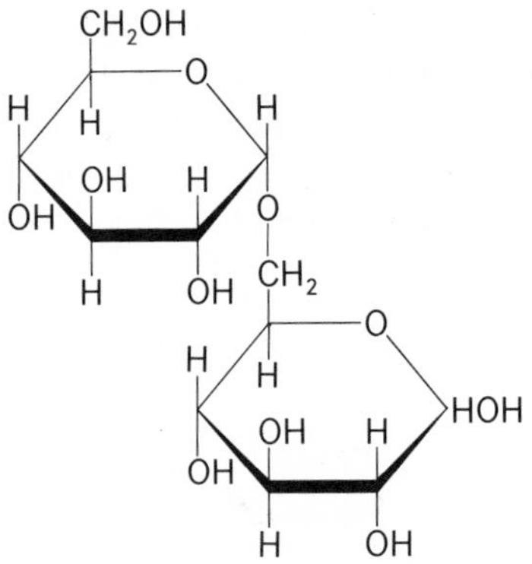

Isomaltose

Lactose is a disaccharide found in milk; on hydrolysis it yields 1 mole each of D-galactose and D-glucose. It possesses a β-1-4 linkage, is a reducing sugar, and can undergo mutarotation.

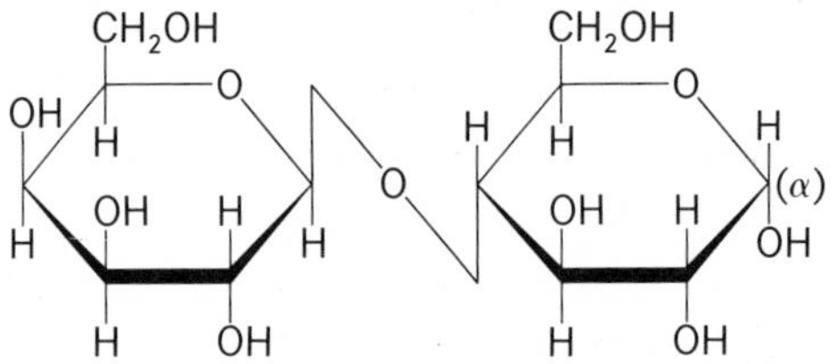

Lactose

Sucrose, the sugar of commerce, is widely distributed in higher plants. On hydrolysis it yields 1 mole each of D-glucose and D-fructose. Sugar cane and sugar beets are the sole commercial sources. In contrast to all the other mono- and disaccharides described previously, it is not a reducing sugar. This fact means that the reducing groups in both of the monosaccharide constituents must be involved in the linkage between the two sugar units. That is, the C-1 and C-2 carbon atoms, respectively, of the glucose and fructose moieties must participate in glycoside formation. Further, acid hydrolysis of the octamethyl derivative of sucrose is known to yield 2,3,4,6-tetra-*O*-methyl-D-glucose and 1,3,4,6-tetra-*O*-methyl-D-fructose. The fact that

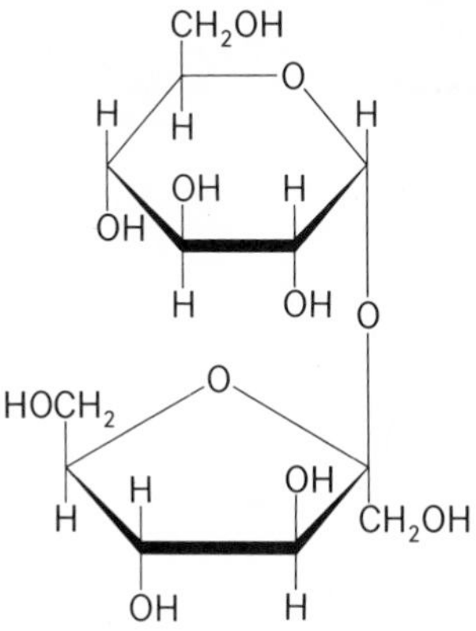

Sucrose

the configuration in the fructose is β, while that in the glucose is α, is known from x-ray studies and work with enzymes that specifically hydrolyze α or β linkages.

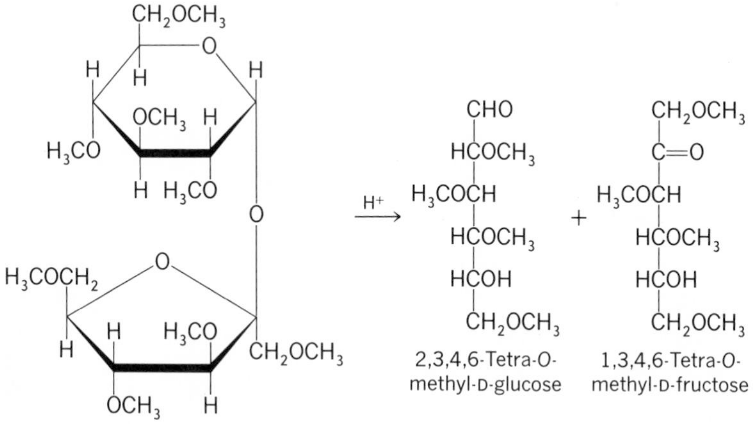

Octamethyl sucrose $\xrightarrow{H^+}$ 2,3,4,6-Tetra-O-methyl-D-glucose + 1,3,4,6-Tetra-O-methyl-D-fructose

Polysaccharides

The polysaccharides found in nature either serve a structural function or play an important role as a stored form of energy. All polysaccharides can be hydrolyzed with acid or enzymes to yield monosaccharides and/or monosaccharide derivatives. D-Glucose, the monomeric unit of starch, glycogen, and cellulose, is the most abundant carbohydrate building block in the biosphere.

Storage Polysaccharides. Starch, the storage polysaccharide of higher plants, consists of two components, amylose and amylopectin, which are present in varying amounts. The amylose component consists of D-glucose units linked in a linear fraction by α-1-4 linkages; it has a nonreducing end and a reducing end (Structure 2-10). Its molecule weight can vary from a few thousand to 150,000. Amylose gives a characteristic blue color with iodine due to the ability of the halide to occupy a position in the interior of a helical

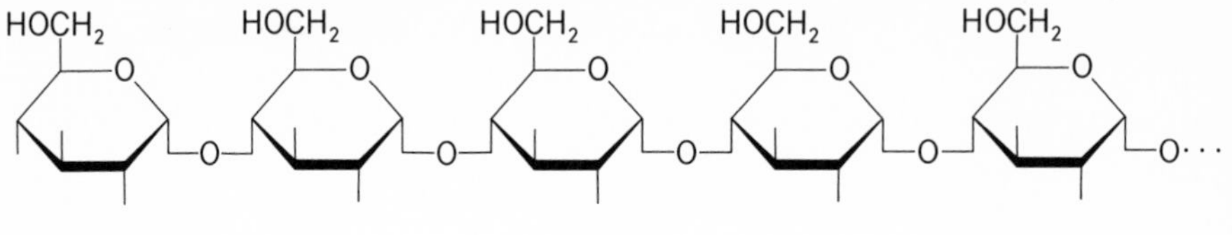

Amylose

Structure 2-10

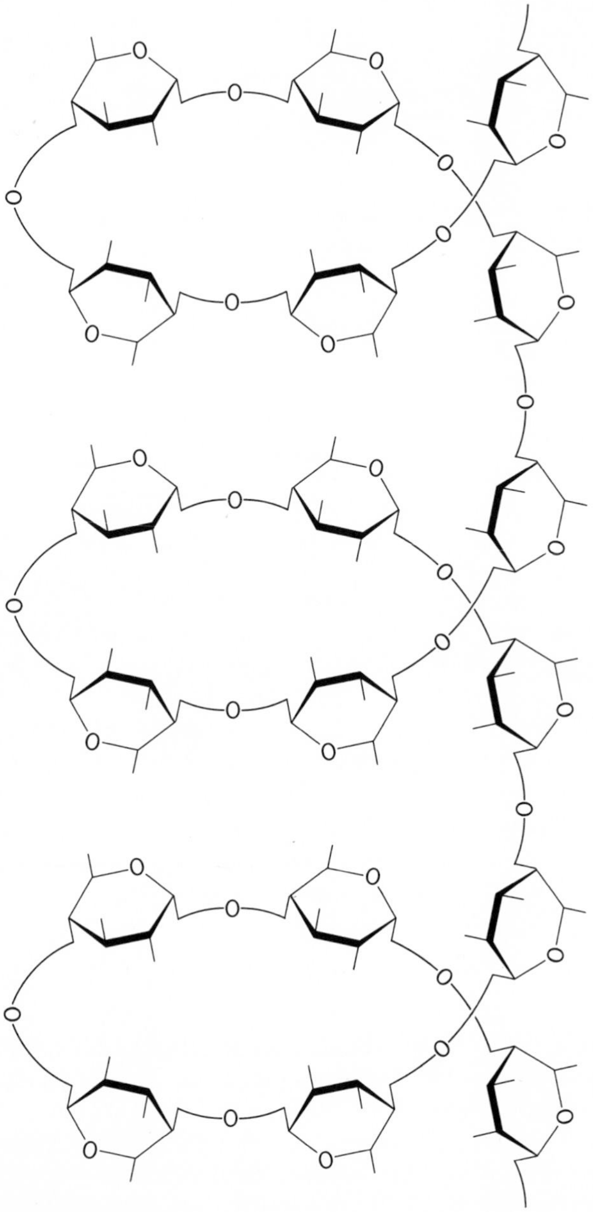

Structure 2-11

coil of glucose units that is formed when amylose is suspended in water (Structure 2-11).

Amylopectin is a branched polysaccharide; in this molecule shorter chains (about 30 units) of glucose units linked α-1-4 are also joined to each other by α-1-6 linkings (from which isomaltose can be obtained). (See Structure 2-12.) The molecular weight of potato amylopectin varies greatly and may be 500,000 or larger. Amylopectin produces a purple to red color with iodine.

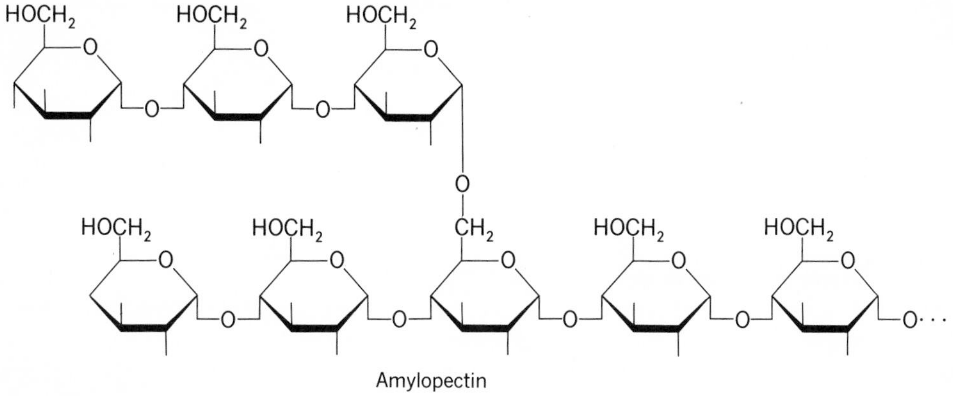

Amylopectin

Structure 2-12

Much has been learned about the structure of starch not only from studies with exhaustive methylation and oxidizing agents, but also by the action of enzymes on the polysaccharide. One enzyme, α-amylase, found in the digestive tract of animals (in saliva and the pancreatic juice), hydrolyzes the linear amylose chain by attacking α-1-4 linkages at random throughout the chain to produce a mixture of maltose and glucose. β-Amylase, an enzyme found in plants, attacks the nonreducing end of amylose to yield successive units of maltose.

Amylopectin can also be attacked by α- and β-amylase, but the α-1-4 glycosidic bonds near the branching point in amylopectin and the α-1-6 bond itself are not hydrolyzed by these enzymes. Therefore, a highly branched core—called a limit dextrin—of the original amylopectin is the product of these enzymes. A separate "debranching" enzyme, an α-1-6 glucosidase, can hydrolyze the bond at the branch point. Therefore, the combined action of α-amylase and the α-1-6 glucosidase will hydrolyze amylopectin ultimately to a mixture of glucose and maltose.

The storage polysaccharide of animal tissues is glycogen; it is similar in structure to amylopectin in that it is a branched, homopolysaccharide composed of glucose units. It is more highly branched than amylopectin, however, having branch points about every 8–10 glucose units. Glycogen is hydrolyzed by α- and β-amylases to form glucose, maltose, and a limit dextrin.

A final example of a nutrient polysaccharide will suffice. This is inulin, a storage carbohydrate found in the bulbs of many plants (dahlias, Jerusalem artichokes). Inulin consists chiefly of fructofuranose units joined together by β-2-1 glycosidic linkages.

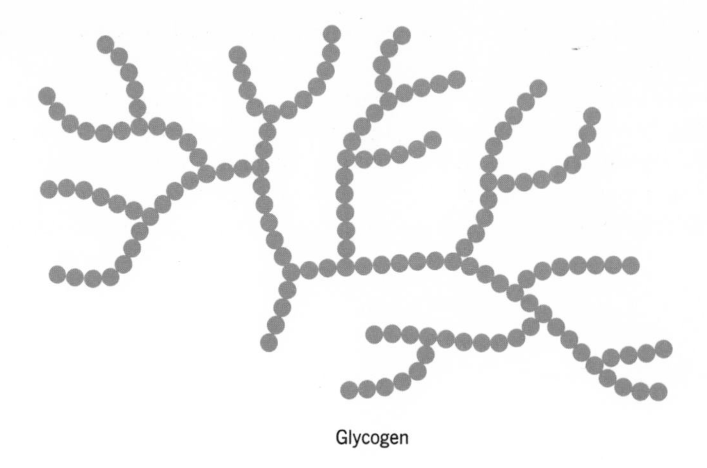

Glycogen

Structural Polysaccharides. The most abundant structural polysaccharide is cellulose, a linear, homopolysaccharide composed of D-glucopyranoside units linked β-1-4. Cellulose is found in the cell walls of plants where it contributes in a major way to the structure of the organism (see page 140). Lacking a skeleton of bone onto which organs and specialized tissues may be organized, the higher plant relies on its cell walls to bear its own weight whether it is a sunflower or a sequoia. The wood of trees is composed primarily of cellulose and another polymer called *lignin*.

In contrast to starch, the β-1-4 linkages of cellulose are highly resistant to acid hydrolysis; strong mineral acid is required to produce D-glucose; partial hydrolysis yields the reducing disaccharide, cellobiose. The β-1-4 linkages of cellulose are not hydrolyzed by glycosidases found in the digestive tracts of humans or other higher animals. However, snails secrete a *cellulase* that hydrolyzes the polymer; termites contain a similar enzyme. Moreover, the rumen bacteria that reside within the intestinal tract of cattle and other ruminants can hydrolyze cellulose and further metabolize the D-glucose produced.

Other examples of structural polysaccharides in plants are known. Plants contain pectins and hemicelluloses. The latter are not cellulose derivatives, but rather are homopolymers of D-xylose linked β-1-4. Pectins contain arabinose, galactose, and galacturonic acid. Pectic acid is a homopolymer of the methyl ester of D-galacturonic acid.

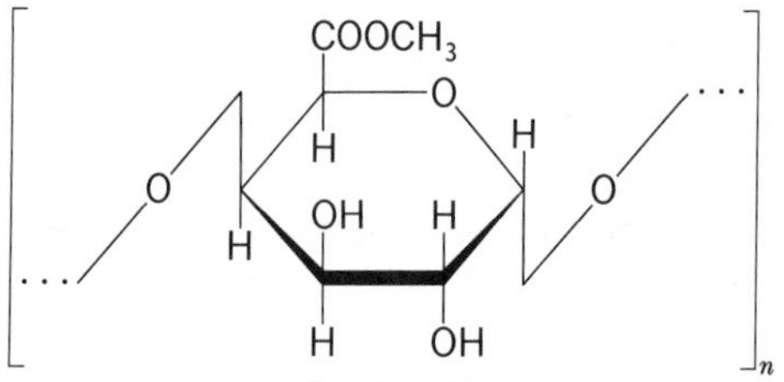

Pectic acid

Chitin, a homopolymer of *N*-acetyl-D-glucosamine, is the structural polysaccharide that constitutes the shell of crustaceans and the scales of insects.

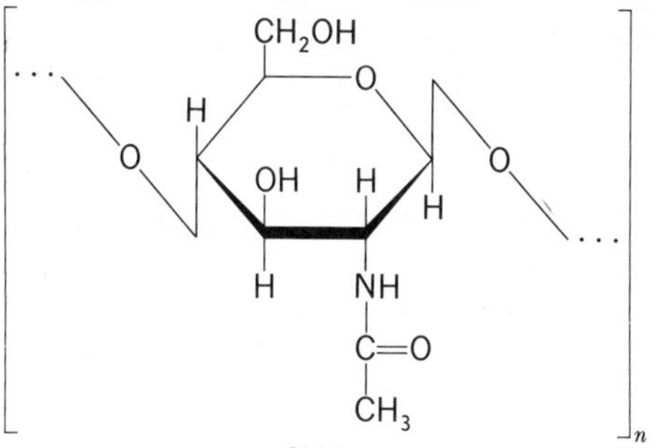

Chitin

In recent years much effort has been expended on identifying the chemical nature of animal and bacterial cell walls and related structures. Animal cells frequently do not possess a well-defined cell wall but have a *cell coat*, visible in the electron microscope, that plays an important role in the interaction with adjacent cells. These cell coats contain glycolipids, glycoproteins, and mucopolysaccharides. The chemical nature of the first two will be discussed later; the mucopolysaccharides are gelatinous substances of high molecular weights (up to 5×10^6) that both lubricate and serve as a sticky cement. One common mucopolysaccharide is hyaluronic acid, a heteropolysaccharide composed of alternating units of D-glucuronic acid and *N*-acetyl-D-glucosamine. The two different monosaccharides are linked by a β-1-3 unit to form a disaccharide which is linked (β-1-4) to the next repeating unit. Hyaluronic acid, found in the vitreous humor of the eye and in the umbilical cord, is water soluble but forms viscous solutions.

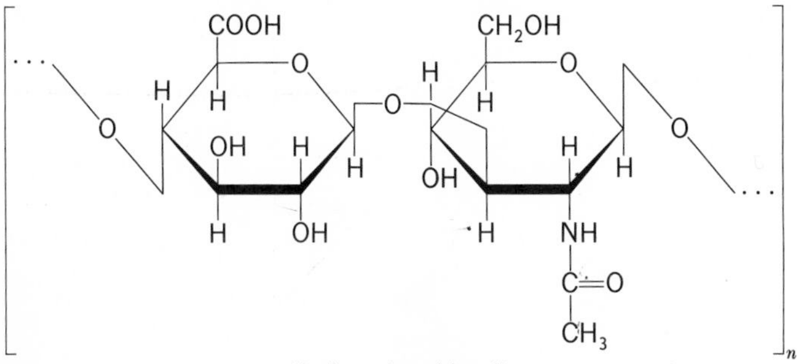

Hyaluronic acid unit

Chondroitin, similar in structure to hyaluronic acid except that the amino sugar is *N*-acetyl-D-galactosamine, is also a component of cell coats. Sulfate esters (at the C-4 and C-6 positions of the amino sugar) of chondroitin are major structural components of cartilage, tendons, and bones.

Bacterial cell walls, which determine many of the physiological characteristics of the organism they enclose, contain complex polymers of polysaccharide linked to chains of amino acids (see Chapter 6). Since the individual chains of amino acid are not as long as in proteins, such polymers have been termed peptidoglycans rather than glycoproteins. The repeating unit of peptidoglycans is a disaccharide composed of *N*-acetyl-D-glucosamine (NAG) and *N*-acetylmuramic acid (NAMA) joined by a β-1,4-glycosidic bond. *N*-Acetylmuramic acid consists of a *N*-acetyl glucosamine unit which has its C-3 hydroxyl group joined to the α-hydroxyl function of lactic acid by an ether linkage. In the peptidoglycan the carboxyl group of each lactic acid moiety is, in turn, linked to a tetrapeptide (see Chapter 4) consisting of L-alanine, D-glutamic acid, L-lysine, and D-alanine. While the peptidoglycan could then be represented as a linear polysaccharide chain with a tetrapeptide branching out at every second hexose amine unit, there is considerable evidence of cross-linking between adjacent, parallel polysaccharide chains. In the cross-

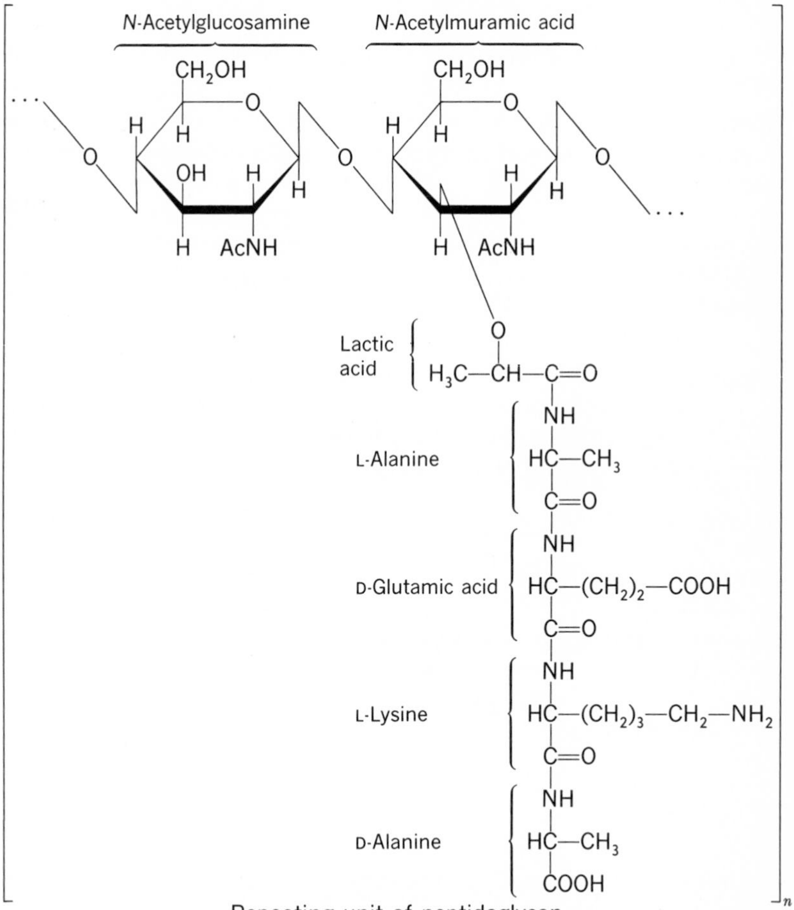

N-Acetylglucosamine — *N*-Acetylmuramic acid

Lactic acid: $H_3C—CH—C{=}O$

L-Alanine: $NH—HC(CH_3)—C{=}O$

D-Glutamic acid: $NH—HC((CH_2)_2—COOH)—C{=}O$

L-Lysine: $NH—HC((CH_2)_3—CH_2—NH_2)—C{=}O$

D-Alanine: $NH—HC(CH_3)—COOH$

Repeating unit of peptidoglycan

linking, the carboxyl group of the terminal D-alanine moiety is attached to a pentaglycine residue which in turn is attached to the ε-amino group of lysine in the next adjacent glycan unit.

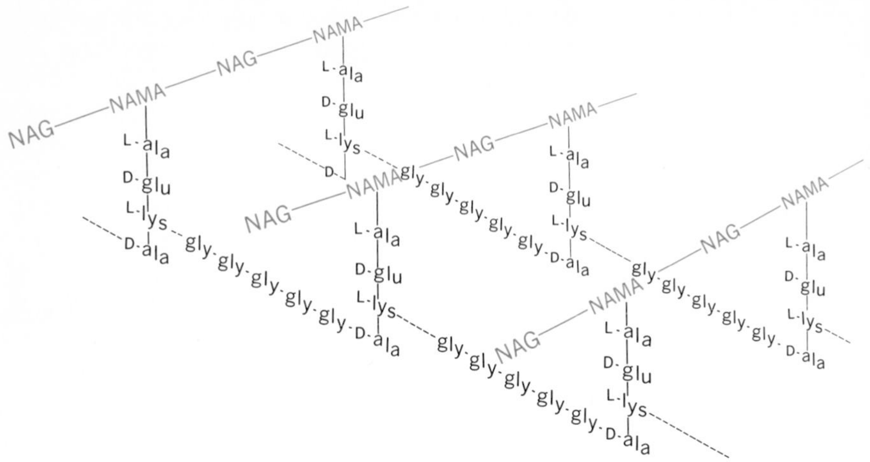

Figure 2-3

Scheme showing the cross-linking of glycine pentapeptide between tetrapeptides located on adjacent glycan backbones. The latter are shown in color.

Much remains to be learned of the structure of cell walls before we completely understand such important phenomena as the immune response and cellular growth and differentiation.

References

1. R. T. Morrison and R. N. Boyd, *Organic Chemistry*. 2nd ed. Boston: Allyn and Bacon, 1966.
   An excellent textbook for review of elementary organic chemistry.
2. W. Pigman and D. Horton, eds., *The Carbohydrates*. 2nd ed. New York: Academic Press, 1970.
   A valuable source of detailed information on the chemistry and biochemistry of carbohydrates.
3. R. Barker, *Organic Chemistry of Biological Compounds*. Englewood Cliffs, N.J.: Prentice-Hall, 1971.
   The chapter on carbohydrates is especially well done and reflects the research interests of the author.

points and are liquids at room temperature since they contain a large proportion of unsaturated fatty acids such as oleic, linoleic, or linolenic acids. In contrast, animal triacyl glycerols contain a higher proportion of saturated fatty acids, such as palmitic and stearic acids, resulting in higher melting points, and thus at room temperature they are semisolid or solid. Table 3-1 lists some of the naturally occurring fatty acids, their structures, and their melting points.

Waxes

Equally widespread are the waxes which serve as protective coatings on fruits and leaves, or which are secreted by insects (for example, beeswax). In general, waxes are a complicated mixture of long-chain alkanes, with an odd number of carbon atoms ranging from C_{25} to C_{35}; and oxygenated derivatives such as secondary alcohols and ketones as well as esters of long-chain fatty acids and long-chain monohydroxy alcohols. Being highly insoluble in water and having fully reduced hydrocarbon chains, these waxes are chemically inert. They serve admirably on leaf surfaces to protect plants from water loss and from abrasive damage. Waxes play an important role in providing a water barrier for insects, birds, and animals such as sheep. This property has been dramatically demonstrated in recent years. When extensive oil spills have occurred in the ocean, detergents have frequently been used to solubilize the oil. Under these conditions birds have great difficulty in maintaining their buoyancy, since the waxy layers covering their feathers were removed by both the oil and the detergent.

Reactivities

The chemical reactivities of triacyl glycerols reflect the reactivity of the ester linkage and the degree of unsaturation in the hydrocarbon chain. Since free fatty acids occur only to a very limited extent in the cell, the major proportion is found bound as esters (triacyl glycerols and phospholipids).

Ester bonds are susceptible to both acid and base hydrolysis. Acid hydrolysis differs from base hydrolysis in that the former is reversible and the latter is irreversible. The last step in base hydrolysis is irreversible because in the presence of excess base the acid exists as the fully dissociated anion which has no tendency to react with alcohols. In acid hydrolysis, however, the system is essentially reversible in all its steps and reaches an equilibrium rather than going to completion. Thus, strong bases are used for saponification to hydrolyze the ester bonds in the simple and complex lipids.

Ester linkages

$$\begin{array}{ll} \alpha & CH_2OCOR^1 \\ \beta & R^2COOCH \\ \alpha' & CH_2OCOR^3 \end{array} \xrightarrow[\text{In alkali called saponification}]{OH^-} \begin{array}{l} CH_2OH \\ CHOH \\ CH_2OH \end{array} + R^1COO^- + R^2COO^- + R^3COO^-$$

Triacyl glycerol → Glycerol + Fatty acids

Table 3-1

Structures of Common Fatty Acids

Acid	Structure	Melting point (°C)
Saturated fatty acids		
Lauric acid	$CH_3CH_2CH_2CH_2CH_2CH_2CH_2CH_2CH_2CH_2CH_2COOH$	44
Myristic acid	$CH_3(CH_2)_{12}COOH$	54
Palmitic acid	$CH_3(CH_2)_{14}COOH$	63
Stearic acid	$CH_3(CH_2)_{16}COOH$	70
Arachidic acid	$CH_3(CH_2)_{18}COOH$	75
Behenic acid	$CH_3(CH_2)_{20}COOH$	80
Lignoceric acid	$CH_3(CH_2)_{22}COOH$	84
Monoenoic fatty acids		
Oleic acid	$CH_3(CH_2)_7CH\overset{cis}{=}CH(CH_2)_7COOH$	13
Vaccenic acid	$CH_3(CH_2)_5CH\overset{cis}{=}CH(CH_2)_9COOH$	44
Dienoic fatty acid		
Linoleic acid	$CH_3(CH_2)_4(CH\overset{cis}{=}CHCH_2)_2(CH_2)_6COOH$	−5
Trienoic fatty acids		
α-Linolenic acid	$CH_3CH_2(CH\overset{cis}{=}CHCH_2)_3(CH_2)_6COOH$	−10
γ-Linolenic acid	$CH_3(CH_2)_4(CH\overset{cis}{=}CHCH_2)_3(CH_2)_3COOH$	—
Tetraenoic fatty acid		
Arachidonic acid	$CH_3(CH_2)_4(CH\overset{cis}{=}CHCH_2)_4(CH_2)_2COOH$	−50
Unusual fatty acids		
α-Elaeostearic acid	$CH_3(CH_2)_3CH\overset{trans}{=}CHCH\overset{trans}{=}CHCH\overset{cis}{=}CH(CH_2)_7COOH$ (conjugated)	48
Tariric acid	$CH_3(CH_2)_{10}C{\equiv}C(CH_2)_4COOH$	51
Isanic acid	$CH_2{=}CH(CH_2)_4C{\equiv}C{-}C{\equiv}C(CH_2)_7COOH$	39
Lactobacillic acid	$CH_3(CH_2)_5\underset{\diagdown CH_2 \diagup}{CH{-}CH}(CH_2)_9COOH$	28
Vernolic acid	$CH_3(CH_2)_4\underset{\diagdown O \diagup}{CH\overset{cis}{-}CH}CH_2CH{=}CH(CH_2)_7COOH$	—
Prostaglandin (PGE_2)	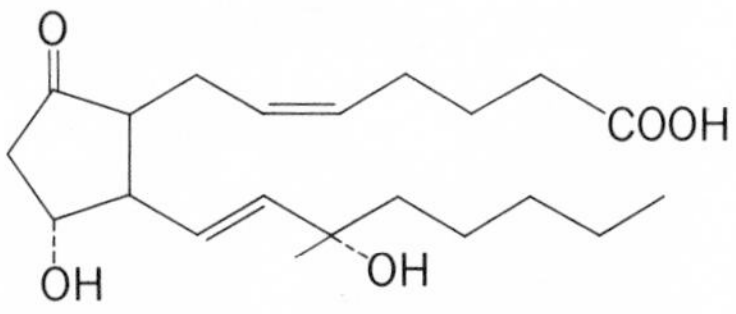	—

Free fatty acids undergo dissociation in water:

$$RCOOH \rightleftharpoons RCOO^- + H^+$$

$$K_a = \frac{[H^+][RCOO^-]}{[RCOOH]}$$

Since pK_a = $-\log K_a$, the acid strength is determined by the dissociation of the proton. Thus the pK_a of most fatty acids is about 4.76–5.0. Stronger acids have lower pK_a values and weaker acids have higher pK_a values. The effective concentration of an acid is also an important factor. Since acetic acid is very soluble in water, its acid properties are readily measured. On the other hand, palmitic acid, with its long, hydrophobic, hydrocarbon side chain, is highly insoluble in water; consequently, its acid properties are not readily measurable. See Chapter 1 for a discussion of acid dissociation.

Other properties of fatty acids reflect the nature of their hydrocarbon chains. Naturally occurring saturated fatty acids that have from one to eight carbon atoms are liquid, whereas those with more carbon atoms are solids. Stearic acid has a melting point of 70°C, but, with the introduction of one double bond, as in oleic acid, the melting point drops to 14°C, and the addition of more double bonds further lowers the melting point. When a double bond is found in the hydrocarbon chain of a fatty acid, geometric isomerism occurs. Most unsaturated fatty acids are found as the less stable

H, $(CH_2)_7COOH$ / C=C (*cis*) / H, $(CH_2)_7CH_3$
Oleic acid

H, $(CH_2)_7COOH$ / C=C (*trans*) / $H_3C(CH_2)_7$, H
Elaidic acid

H, $(CH_2)_7COOH$ / C=C (*cis*) / H, CH_2 — H, C=C (*cis*) / $H_3C(CH_2)_4$, H
Linoleic acid

cis isomers rather than as the more stable *trans* isomers. Structurally, the hydrocarbon chain of a saturated fatty acid has a zigzag configuration, as indicated in Structure 3-1, with the carbon–carbon bond forming a 109° bond angle.

23.2Å

109°

CH_3 CH_2 CH_2 CH_2 CH_2 CH_2 CH_2 CH_2 CH_2 O

4.8Å

CH_2 CH_2 CH_2 CH_2 CH_2 CH_2 CH_2 CH_2 C

OH

21.3Å

Structure 3-1

When a *cis*-9,10 double bond is introduced, as in oleic acid, the combination of the *cis* configuration and the σ and π bonds of the double bond produce the bent molecule indicated in Structure 3-2.

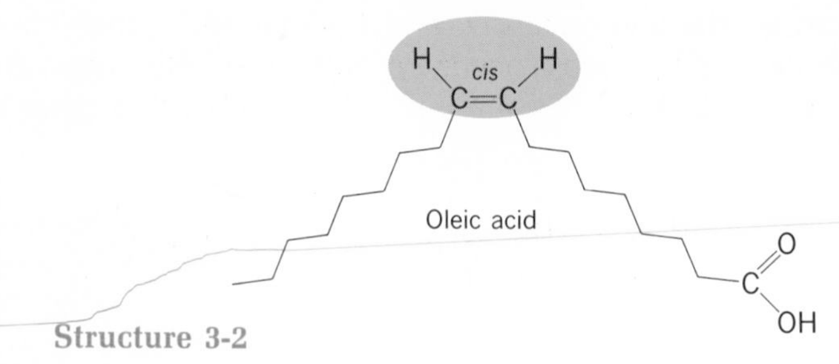

Structure 3-2

Linoleic acid, with two double bonds in the hydrocarbon chain, has its alkene chain even more severely bent (Structure 3-3). Therefore, when we

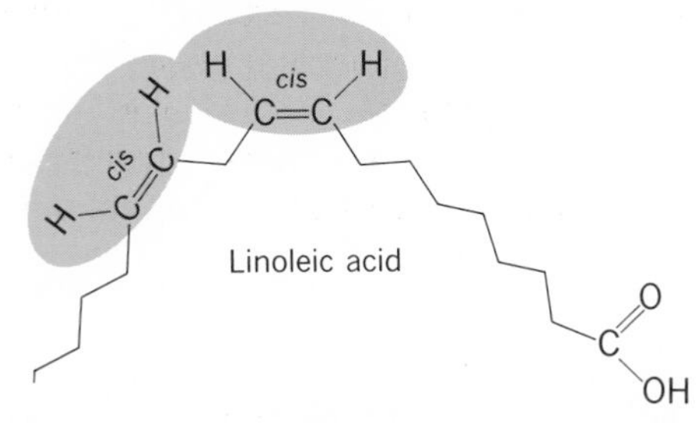

Structure 3-3

examine compounds containing double bonds in hydrocarbon chains, we should picture these not as straight chains occupying a minimum of space, but as large, bulky groups which are considerably bent if they are unsaturated. It is of interest that membranes in animal and plant cells are rich in polyunsaturated fatty acids. On the other hand, bacteria do not contain polyunsaturated fatty acids. Their principal monoenoic acid is *cis*-vaccenic acid, $CH_3—(CH_2)_5—CH=CH(CH_2)_9—COOH$.

In addition to geometric isomerism, another aspect of structural component found in naturally occurring fatty acids is the *nonconjugated double bond system* of the polyunsaturated fatty acids. Linoleic acid is an example of the nonconjugated type.

$$—CH_2—CH=CH—CH_2—CH=CH—CH_2—$$

Nonconjugated double bond system

However, an important polyunsaturated fatty acid, α-elaeostearic acid, the principal acid in tung oil, is isomeric with α-linolenic acid, but differs from it by having a conjugated triene system. Its structure is

trans *trans* *cis*

$$CH_3(CH_2)_3CH=CHCH=CHCH=CH(CH_2)_7COOH$$

and it illustrates the conjugated double bond system.

$$—CH_2—CH=CH—CH=CH—CH=CH—CH_2—$$

Conjugated double bond system

These two types of multi-double bond systems exhibit important differences in chemical reactivity. The nonconjugated or 1,4-pentadiene system has a methylene group flanked by double bonds on both sides. The methylene group may be directly attacked to form a free radical leading to a series of reactions with oxygen:

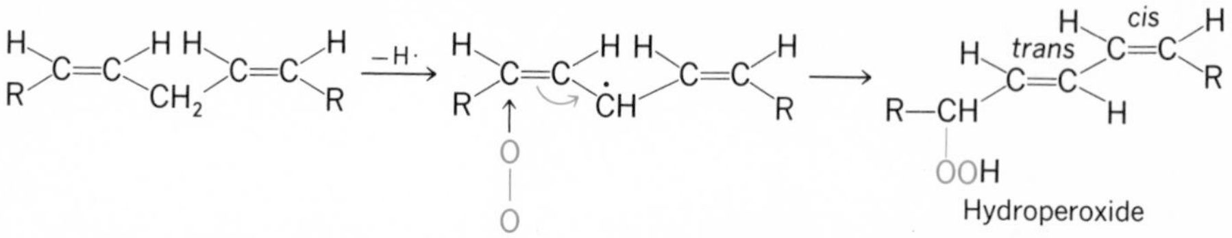

$$\text{H(R)C=CH–CH}_2\text{–CH=CH(R)} \xrightarrow{-\text{H}\cdot} \text{H(R)C=CH–}\dot{\text{C}}\text{H–CH=CH(R)} \;(+\text{O–O}) \longrightarrow \text{R–CH(OOH)–CH=CH–CH=CH–R}$$

(trans, cis) Hydroperoxide

The conjugated double bond systems are much more reactive because of extensive delocalization of π electrons. Fatty acids with these systems undergo extensive polymerization, a valuable property used by the paint industry. Both vitamin A and the carotenoids are excellent examples of important conjugative systems in biomolecules. These bond systems may play an important role in the visual processes of the retina. We will indicate other examples elsewhere in the book.

Analyses of Lipids

Within the past decade a revolution has occurred in techniques of separating and characterizing lipid classes and their components. By the judicious use of thin-layer chromatography and gas–liquid chromatography the lipid chemist can now handle easily the majority of analytical problems. Both methods are described in Appendix 2. Because these methods are rapid, adaptable to microamounts of material, and quantitative, they have displaced the older techniques involving the determination of the iodine number, saponification, and acetylation values; thus these older methods will not be described in this book.

Nomenclature

If either carbon 1 or 3 of glycerol is esterified by a fatty acid or phosphoric acid, carbon 2 becomes an asymmetric center, yielding antipodal forms. Thus, students as well as biochemists are often confused by the fact that L-3-glycerophosphate (I) is equivalent to D-1-glycerophosphate (II). To simplify

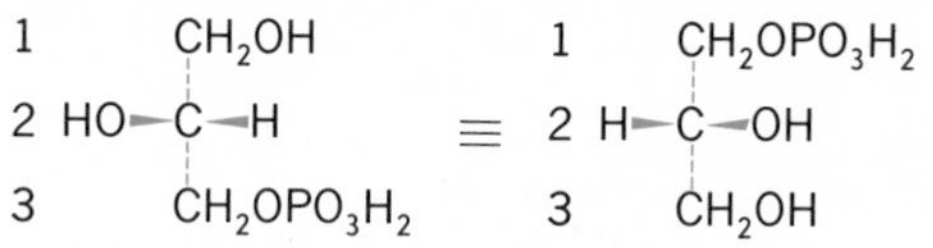

$$\begin{array}{lc} 1 & CH_2OH \\ 2 & HO–C–H \\ 3 & CH_2OPO_3H_2 \\ & \text{I} \end{array} \equiv \begin{array}{lc} 1 & CH_2OPO_3H_2 \\ 2 & H–C–OH \\ 3 & CH_2OH \\ & \text{II} \end{array}$$

this problem, in 1967 the IUPAC–IUB Commission on Biochemical Nomenclature adopted the following system for naming more clearly the derivatives of glycerol. The numbers 1 and 3 *cannot be used* interchangeably for the same primary alcohol group. The second hydroxy group of glycerol is shown to the left of C-2 in the Fischer projection, while the carbon atom above C-2 is called C-1 and the one below, C-3. This *stereospecific numbering* is indicated by the prefix *sn* before the stem name of the compound. Glycerol is thus labeled:

$$\begin{array}{ll} CH_2OH & 1 \\ HO—C—H & 2 \\ CH_2OH & 3 \end{array} \leftarrow \text{Stereospecific numbering } (sn)$$

Clearly, compound I, called *sn*-glycerol-3-phosphoric acid, is the optical antipode of *sn*-glycerol-1-phosphoric acid (III). A mixture of both would be called *rac*-glycerol phosphoric acid.

$$\begin{array}{c} CH_2OPO_3H_2 \\ HO—C—H \\ CH_2OH \\ \text{III} \end{array}$$

The stereochemistry of a phosphatidyl choline would be defined by the term 3-*sn*-phosphatidyl choline. Keeping the definition of *sn* in mind, we simply write the formula as:

$$\begin{array}{c} CH_2OCOR^1 \\ R^2COO—C—H \\ CH_2OPO_3CH_2CH_2N^+(CH_3) \end{array}$$

Lipid biochemists employ a useful shorthand notation to describe fatty acids. The general rule is to write first the number of carbon atoms, then the number of double bonds, and finally indicate the position of double bonds, counting from the carboxyl carbon. Thus, palmitic acid, a saturated C_{16} acid, is written as 16:0, oleic acid is written as 18:1(9) and arachidonic acid is written as 20:4(5,8,11,14). The *cis* configuration is assumed to be the only geometric isomer present. If the *trans* configuration occurs in the structure, it is so stated, i.e., 18:3 (6t,9t,12c).

Phospholipids

Phospholipids are so named because they contain a phosphorus atom. In addition, glycerol, fatty acids, and a nitrogenous base are key components. Several phospholipids, which are considered as derivatives of phosphatidic acid, are listed in Table 3-2. The structure of phosphatidic acid is

$$\begin{array}{c} CH_2OCOR^1 \\ R^2COO—C—H \\ CH_2—O—P(=O)(OH)_2 \end{array}$$

3-*sn*-Phosphatidic acid

Phospholipids are widespread in bacteria, animal, and plant tissues, and their general structures, regardless of their sources, are quite similar. Phospholipids, namely, phosphatidyl aminoethanol, choline, and serine, are frequently associated with membranes. They have been termed amphipathic

Table 3-2

Some Amphipathic Lipids

Phospholipid	Usual fatty acid (nonpolar component)	Base (polar component)	Common name
3-*sn*-Phosphatidyl choline 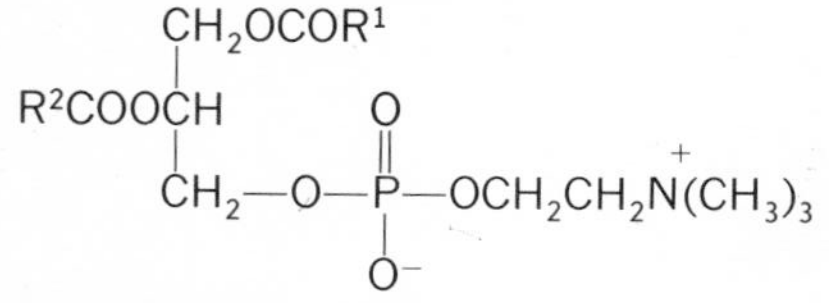 CH_2OCOR^1 R^2COOCH $CH_2—O—P(=O)(O^-)—OCH_2CH_2\overset{+}{N}(CH_3)_3$	Oleic, palmitic	Choline	Lecithin
3-*sn*-Phosphatidyl aminoethanol 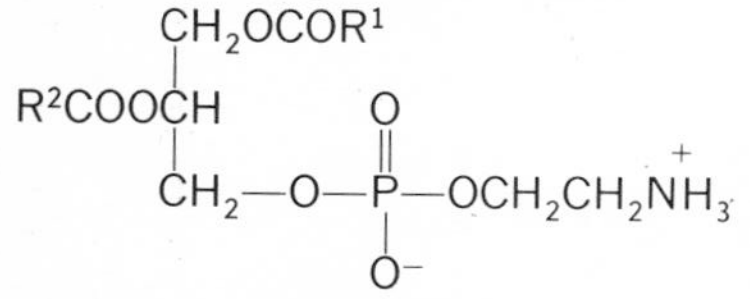 CH_2OCOR^1 R^2COOCH $CH_2—O—P(=O)(O^-)—OCH_2CH_2\overset{+}{N}H_3$	Oleic, palmitic	Aminoethanol	Cephalin
3-*sn*-Phosphatidyl serine 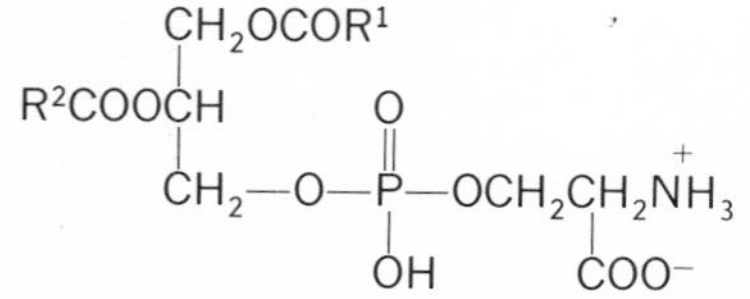 CH_2OCOR^1 R^2COOCH $CH_2—O—P(=O)(OH)—OCH_2CH(COO^-)\overset{+}{N}H_3$	Oleic, palmitic	Serine	Cephalin

3-*sn*-Phosphital aminoethanol $\alpha CH_2OCH{=}CHR^1$ $R^2COOCH\beta$ O $CH_2{-}O{-}P({=}O)(O^-){-}OCH_2CH_2\overset{+}{N}H_3$	α = Unsaturated ether β = Linoleic	Aminoethanol	Plasmalogen
1-Alkoxyl phospholipid $CH_2OCH_2R^1$ R^2COOCH $CH_2OP({=}O)(O^-)OCH_2CH_2\overset{+}{N}H_3$	R^2 probably an unsaturated fatty acid	Aminoethanol	α-Glyceryl ether
3-*sn*-Phosphatidyl inositol CH_2OCOR^1 R^2COOCH $CH_2{-}O{-}P({=}O)(OH){-}O{-}$(inositol ring: OH, OH, OH, OH, OH)	Palmitic, arachidonic	Myoinositol replaces base	Inositol phospholipid
3-*sn*-Phosphatidyl glycerol CH_2OCOR^1 CH_2OH R^2COOCH $HCOH$ $CH_2O{-}P({=}O)(OH){-}O{-}CH_2$	Polyunsaturated fatty acid	Glycerol replaces base	—

compounds since they possess both polar and nonpolar functions. We shall discuss the importance of these properties in Chapter 6.

Sphingolipids Sphingolipids include an important group of compounds closely associated with tissues and animal membranes. The central compound is called 4-sphingenine (formerly sphingosine). A variety of components can be at-

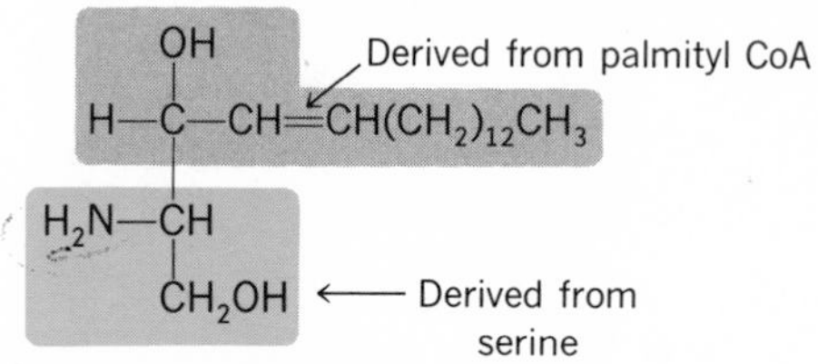

OH
H—C—CH=CH$(CH_2)_{12}CH_3$ ← Derived from palmityl CoA
H_2N—CH
CH_2OH ← Derived from serine

4-Sphingenine

tached to this structure to give important derivatives. 4-Sphingenine is formed from a rather complex series of reactions involving palmityl CoA and serine. The fully reduced compound is called sphinganine (formerly dihydrosphingosine). A list of important derivatives is presented in Table 3-3.

Glycolipids Another group of compounds are collected in the class of glycolipids since they are primarily carbohydrate–glyceride derivatives, and do not contain phosphate. These include the galactolipids and the sulfolipids, found primarily in photosynthetic tissue. Their structures are

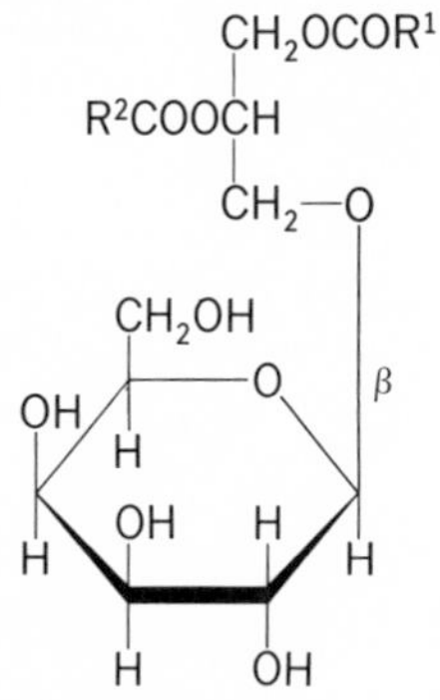

3-*sn*-Monogalactosyl diacyl glycerol

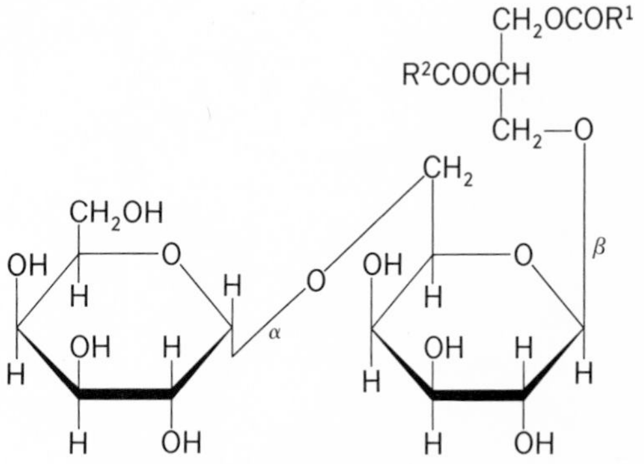

3-*sn*-Digalactosyl diacyl glycerol

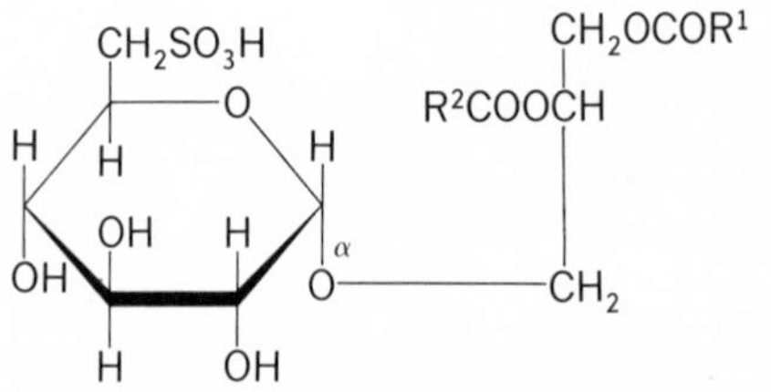

3-*sn*-Sulfonyl-6-deoxyglucosyl diacyl glycerol

Table 3-3

	Fatty acid
Sphingomyelin 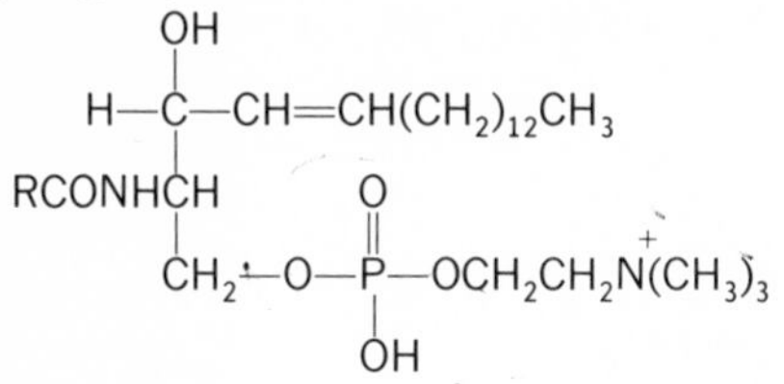 OH H—C—CH=CH$(CH_2)_{12}CH_3$ RCONHCH O CH_2—O—P—$OCH_2CH_2\overset{+}{N}(CH_3)_3$ OH	R = long-chain fatty acids
Cerebroside 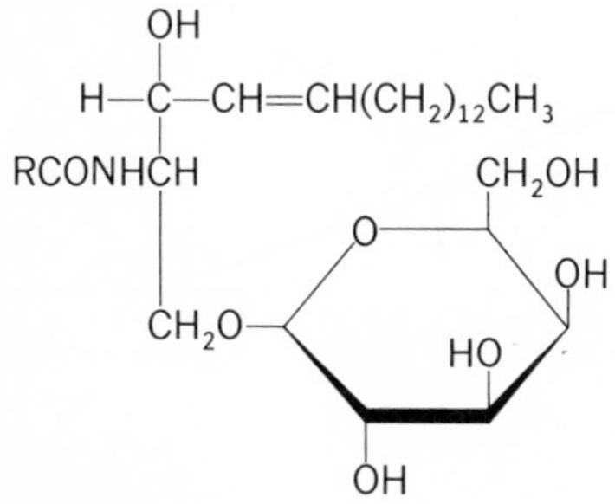 OH H—C—CH=CH$(CH_2)_{12}CH_3$ RCONHCH CH_2OH O OH CH_2O— HO OH	R = C_{24} fatty acids
Psychosine 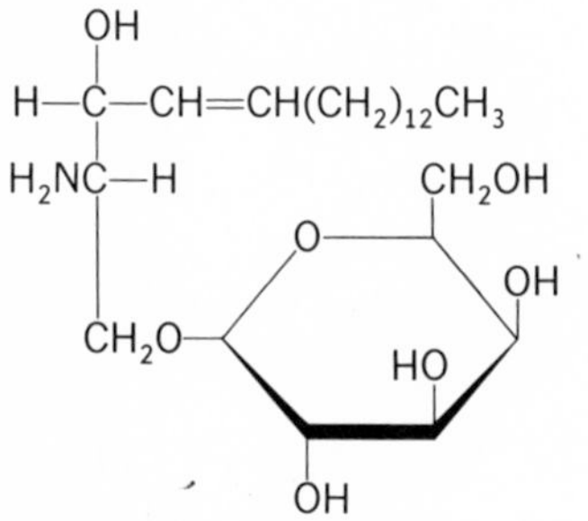 OH H—C—CH=CH$(CH_2)_{12}CH_3$ H_2NC—H CH_2OH O OH CH_2O— HO OH	

Terpenoids

The terpenoids are a very large and important group of compounds which is in actual fact made up of a simple repeating unit, the isoprenoid unit; this unit, by ingenious condensations, gives rise to such compounds as rubber, carotenoids, steroids, and many simpler terpenes. Isoprene, which does not occur in nature, has as its actual biologically active counterpart isopentenyl pyrophosphate, which is formed by a series of enzymically catalyzed steps from mevalonic acid. Isopentenyl pyrophosphate undergoes further reactions to form squalene, which in turn can condense with itself to form cholesterol. Another typical terpenoid product is β-carotene, which is cleaved in the liver to form vitamin A. The various structural relationships are indicated in the diagrams. Note the repeating isoprenoid unit in all these compounds. We shall discuss the biosynthetic reactions in Chapter 12.

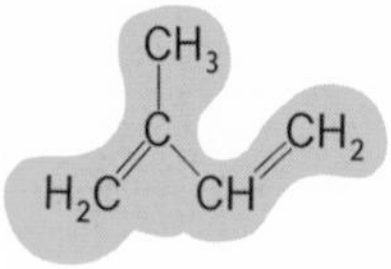

"Isoprenoid unit"

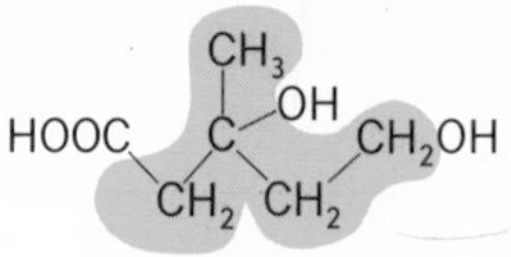

Mevalonic acid

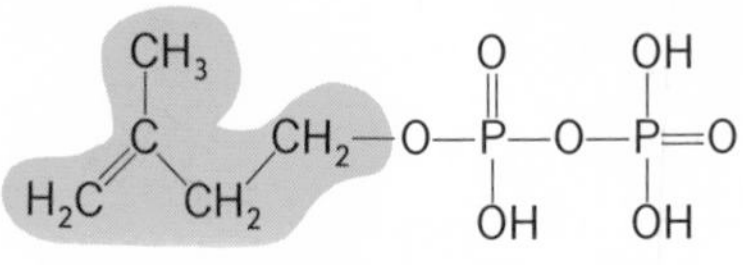

Isopentenyl pyrophosphate

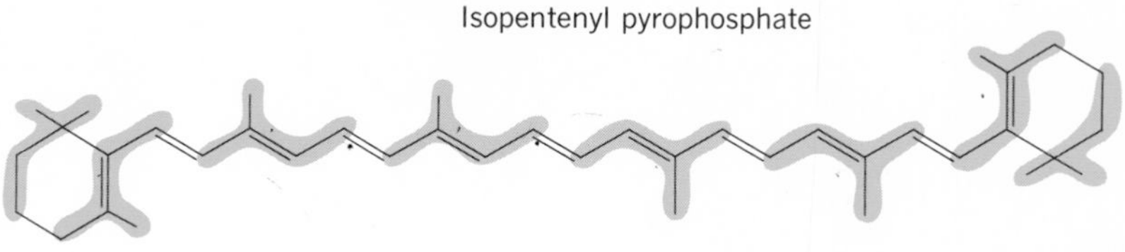

β-Carotene

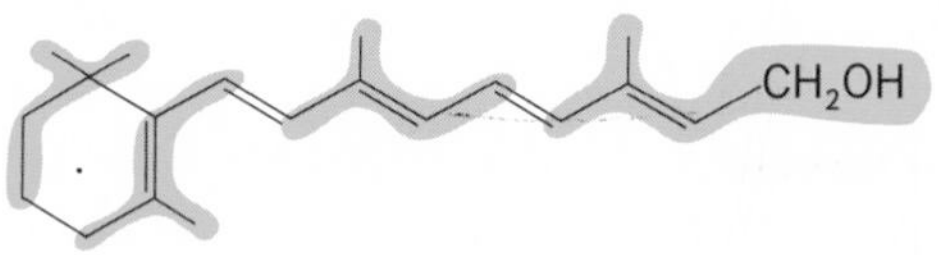

Vitamin A_1

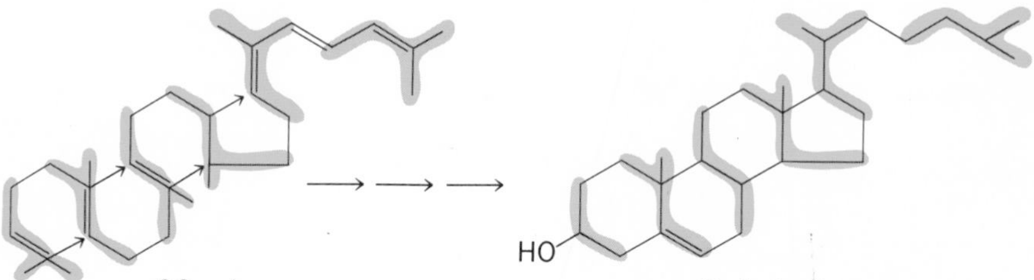

β-Squalene

Cholesterol

Function of Lipids

In recent years it has become apparent that lipids play extremely important roles in the normal function of a cell. Not only do lipids serve as highly reduced storage forms of energy, but they also play an intimate role in the structure of membranes of the cell, and of the organelles found in the cell. In Chapter 6 we shall discuss these aspects in some detail.

Lipids participate directly or indirectly in many metabolic activities as:

(1) *Activators of enzymes.* Three microsomal enzymes, namely, glucose-6-phosphatase, stearyl CoA desaturase and ω-hydroxylases, and β-hydroxy butyric dehydrogenase (a mitochondrial enzyme), require phosphatidyl choline micelles for activation. Many other examples can be cited.

(2) *Components of the electron transport system in mitochondria.* There is good evidence that the electron transport chain in the inner membranes of mitochondria is buried in a milieu of phospholipids.

(3) *A substrate.* α-Acyl-β-oleyl phosphatidyl choline specifically serves as the acceptor of a CH_3 group from *S*-adenosyl methionine, which adds across the double bond of the β-oleyl moiety to form the cyclopropane function of lactobacillic acid:

$$H_3C(CH_2)_7\overset{H}{C}{=}\overset{H}{C}{-}(CH_2)_7\overset{O}{\overset{\|}{C}}{-}O{-}\overset{CH_2OCOR}{\overset{|}{C}}{-}H \quad + \; S\text{-Adenosyl methionine}$$

$$\qquad CH_2{-}O{-}\overset{O}{\overset{\|}{P}}{-}OCH_2CH_2\overset{+}{N}(CH_3)_3 \quad (\text{P bears } O^-)$$

$$\downarrow$$

$$\underbrace{H_3C(CH_2)_7{-}\underset{H}{C}\overset{CH_2}{\diagup\diagdown}\underset{H}{C}(CH_2)_7{-}}_{\text{Lactobacillic acid}}\overset{O}{\overset{\|}{C}}O{-}\overset{CH_2OCOR}{\overset{|}{C}H} \quad + \; S\text{-Adenosyl homocysteine}$$

$$\qquad CH_2O\overset{O}{\overset{\|}{P}}OCH_2CH_2\overset{+}{N}(CH_3)_3 \quad (\text{P bears } O^-)$$

(4) *A glycosyl carrier.* The isoprenoid compound, undecaprenyl phosphate, acts as a lipophilic carrier of a glycosyl moiety in the synthesis of bacterial cell wall lipopolysaccharides and peptidoglycans.

$$\underbrace{H_3C\overset{CH_3}{\overset{|}{C}}{=}CHCH_2(CH_2\overset{CH_3}{\overset{|}{C}}{=}CHCH_2)_9CH_2\overset{CH_3}{\overset{|}{C}}{=}CHCH_2O}_{\text{Hydrophobic}}\underbrace{{-}\overset{O}{\overset{\|}{\underset{OH}{\underset{|}{P}}}}{-}O{-}\text{Sugar}}_{\text{Hydrophilic}}$$

(5) *A substrate in the indirect decarboxylation of serine to aminoethanol.* Phosphatidyl serine is decarboxylated by a specific decarboxylase to phosphatidyl aminoethanol. The direct decarboxylation of serine to aminoethanol has never been demonstrated.

$$\text{Phosphatidyl serine} \longrightarrow \text{Phosphatidyl aminoethanol} + CO_2$$

Lipoproteins

Lipids are not transported in the free form in circulating blood plasma, but move as chylomicrons, as very low density lipoproteins, or as high-density lipoproteins. In addition, stable lipoproteins occur almost exclusively as components of the fat globule membrane, and presumably serve to stabilize the emulsion. In avian yolk, the lipoproteins are somehow related to the energy requirements and lipid transport of the growing embryo.

Lipoproteins are a class of biomolecules in which the lipid components consist of triacyl glycerol, phospholipid and cholesterol (or its esters) in remarkably consistent proportions (see Table 3-4). The protein components in turn have a relatively high proportion of nonpolar amino acid residues which can participate in the binding of lipids. Studies have clearly excluded covalent and ionic bonds as being involved in the tight binding of the lipid to specific apoproteins. London–van der Waals dispersion forces, however, may play a significant role in the binding process; but the evidence is now

Table 3-4

Composition of Some Lipoproteins

			Content (%)			
Source	Lipoprotein	Molecular weight	Protein	Phospholipid	Cholesterol (free + ester)	Triacyl glycerol
Blood serum	Chylomicron	10^9–10^{10}	4	7.5	10	78
	Very low density	5–10 × 10^6	8	19	18	55
	Low density	2 × 10^6	21	28	27	10
	High density	1–4 × 10^5	58	25	12	6
Egg yolk	β-Lipovitellin	4 × 10^5	78	12	1	9
Milk	Low density	4 × 10^6	13	52	0	35

clear that the principal binding force is the hydrophobic interaction between apoproteins and lipids. Hydrophobic interaction (or bonding) is defined as the tendency of hydrocarbon components to associate with each other in an aqueous environment.

Lipoproteins are also found in the membranes of mitochondria, endoplasmic reticuli and nuclei. The electron transport system in mitochondria appears to contain large amounts of lipoproteins. Lamellar lipoprotein systems occur in the myelin sheath of nerves, photoreceptive structures, chloroplasts, and the membranes of bacteria.

Comparative Distribution of Lipids

With the advent of modern lipid techniques, much work has been directed toward an elucidation of the nature of lipids in a wide number or organisms. In general, procaryotic cells and eucaryotic cells (those without and with membrane-enclosed organelles, respectively) differ remarkably in their lipid composition. A brief survey of these differences will now be presented.

Procaryotic Cells. In general, a bacterial cell has over 95% of its total lipid complement associated with its cell membrane; the remaining 5% is distributed between its cytoplasm and the cell wall. Bacterial cells are distinctive because of the complete absence of sterols in their cells; such cells are unable to synthesize the steroid ring structure although they are capable of forming extended linear isoprenoid polymers. With the exception of the mycobacteria, triacyl glycerols are missing in bacteria, and with the exception of *Bacilli*, which do contain some 16:2(5,10) and 16:2(7,10) polyunsaturated fatty acid, bacteria do not have the capacity to synthesize the conventional polyunsaturated fatty acids. Thus, bacteria are somewhat limited in their capacity for the synthesis of a broad spectrum of fatty acids and produce only saturated, monoenoic, cyclopropane, or branched-chain fatty acids. Indeed, a number of species such as the *Mycoplasma* and mutants of *E. coli* have even lost the capacity for monoenoic fatty acid synthesis and require monoenoic fatty acids to be supplied for growth.

Eucaryotic Cells

Plants. In general, the seeds of higher plants have a rather fixed composition of fatty acids that are phenotypic expressions of their genotypes. The maturing seed synthesizes its different fatty acids at different rates and at different periods during maturation, but as the seed enters its dormant period, the composition of fatty acid is identical to that of its parent seed. The exotic fatty acids are normally found as triacyl glycerols in the mature seed and are rarely found in such tissue organelles as the chloroplast. Throughout the higher plant kingdom, chloroplasts possess a remarkably constant pattern of fatty acids and complex lipids. In particular, the polyunsaturated fatty acid α-linolenic acid is always found associated with four highly polar, complex lipids which are unique to photosynthetic tissue; monogalactosyl diacyl glycerol, digalactosyl diacyl glycerol, sulfoquinovosyl diacyl glycerol and phosphatidyl glycerol. These lipids are closely associated with the lamellar membranes of chloroplasts. Higher plants synthesize a wide range of polyunsaturated fatty acids.

Animals. The lipids of animal cells are equally complex and their composition is characteristic of a particular cell. Thus a nerve cell is rich in sphingolipids, glyceryl ethers, and plasmalogens as well as phospholipids; an adipose cell, on the other hand, consists essentially of triacyl glycerols. There is one rather remarkable feature which is unique to cells of both lower and higher forms

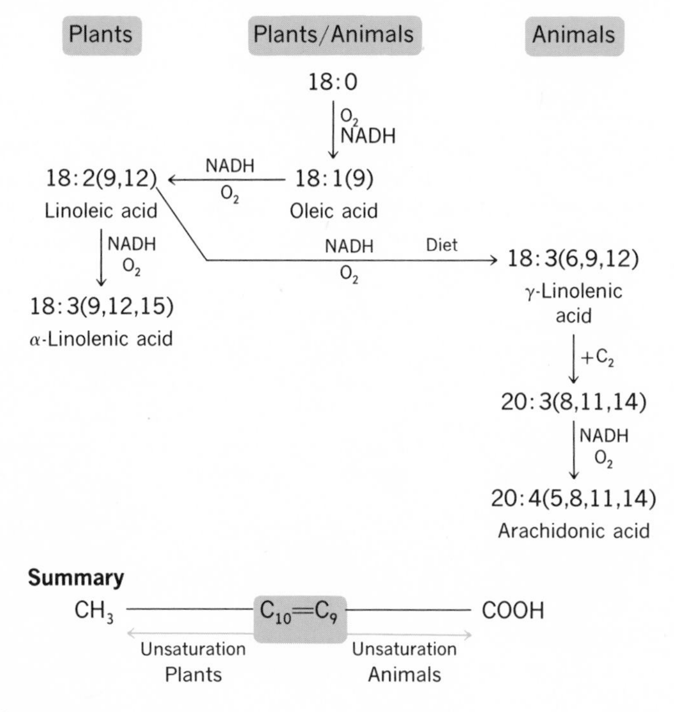

Scheme 3-1

of animal life, namely, the formation of characteristic polyunsaturated fatty acids. In general, eucaryotic cells readily synthesize oleic acid *de novo* by an aerobic mechanism in which a *cis*-9,10 position is introduced (counting from the carboxyl carbon). However, animal cells completely lack the enzyme responsible for the further desaturation of oleyl CoA to linoleyl CoA, although this specific desaturase is widespread in plant tissues. Moreover, animal cells introduce further *cis* double bonds into the hydrocarbon chain only toward the carboxyl end, whereas plant cells always introduce additional double bonds toward the methyl end, as depicted in Scheme 3-1.

In animal tissues, polyunsaturated fatty acids are necessary for an as yet unexplained nutritional requirement. A number of them can serve as precursors of a new group of hormones called *prostaglandins*. These oxygenated fatty acids (see Table 3-1) act in microgram amounts as smooth muscle stimulants, as blood pressure depressants, as abortifacients, and as antagonists for a number of hormones.

References

1. F. D. Gunstone. *An Introduction to the Chemistry and Biochemistry of Fatty Acids and Their Glycerides*. 2nd ed. London: Chapman and Hall, 1967.
   A useful if brief account of some of the basic chemistry of fatty acids.
2. R. T. Holman, ed. *Progress in the Chemistry of Fats and Other Lipids*. Oxford: Pergammon Press.
   A serial publication containing somewhat advanced but very complete reviews on a wide variety of aspects pertaining to lipids.
3. G. B. Ansell, R. M. Dawson, and J. N. Hawthorne. *Phospholipids: Chemistry, Metabolism and Function*. London: Elsevier, 1972.
   A modern and excellent treatment of all aspects of complex lipid chemistry and biochemistry.
4. A. R. Johnson and J. B. Davenport, ed. *Biochemistry and Methodology of Lipids*. New York: Wiley-Interscience, 1971.
   A very useful modern compilation of articles on a wide range of subjects pertaining to lipid chemistry and metabolism.

Amino Acids and Proteins

Introduction

In this chapter some of the more important properties of amino acids and proteins will be considered. Proteins are molecules of high molecular weight ranging from a few thousand to a million or more. These molecules contain carbon, hydrogen, oxygen, nitrogen, and usually sulfur. The elementary composition of most proteins is very similar; approximate percentages are C = 50–55, H = 6–8, O = 20–23, N = 15–18, and S = 0–4. Such figures provide little information concerning the structure of the protein molecule but are useful for making rough estimates of the protein content of biological matter and foodstuffs. Since the nitrogen content of most proteins is about 16%, and since this element is easily analyzed as NH_3 by the Kjeldahl nitrogen procedure, the protein content can be estimated by determining the nitrogen content and multiplying by 6.25 (100/16).

The fundamental structural unit of proteins is the amino acid, as may be easily demonstrated by hydrolyzing purified proteins by chemical or enzymatic procedures. For example, a protein may be hydrolyzed to its constituent amino acids in a period of 18–24 hours by the catalytic action of 6*N* HCl at 110°C in a sealed tube. Under these conditions the individual amino acids are released and may be isolated from the acid hydrolysate as their hydrochloride salts. All the naturally occurring amino acids are stable to this treatment with strong acid except tryptophan. Tryptophan may be partially or completely recovered if reducing agents are present during the hydrolysis with acid or if the hydrolysis is catalyzed by alkali (2*N* NaOH). However, the latter procedure has the disadvantage of destroying several amino acids (cysteine, serine, threonine, and arginine). In addition, treatment with alkali leads to the racemization of all of the amino acids. As we shall see, all the amino acids that occur naturally in proteins have the L configuration with respect to the reference standard, D-glyceraldehyde. On hydrolysis with alkali, the L compound will be converted to a mixture of the D and L enantiomers.

The general formula of a naturally occurring amino acid may be represented with a modified ball and stick formula or the Fischer projection formula (Structure 4-1). Because the amino group is on the carbon atom

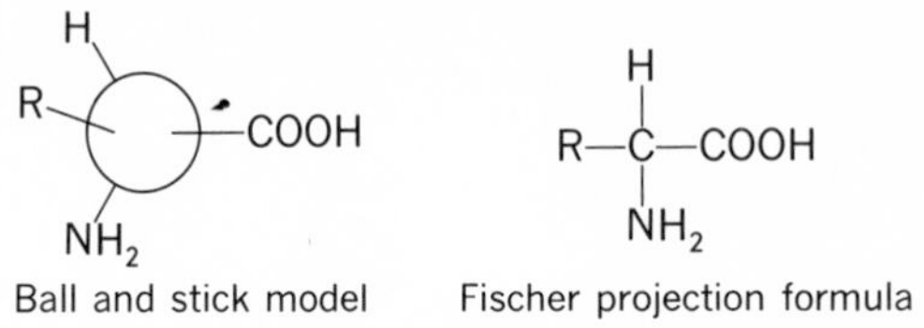

H, R, COOH, NH_2
Ball and stick model

$$\mathrm{R{-}\underset{NH_2}{\overset{H}{C}}{-}COOH}$$
Fischer projection formula

Structure 4-1

adjacent to the carboxyl group, the amino acids having this general formula are known as alpha (α) amino acids. It is also apparent that if R in this structure is not equal to H, the α carbon atom is asymmetric. Thus, two different compounds having the same chemical formula may exist; one will have the general structures shown, and the other will be the enantiomer or mirror image isomer of the first compound. It is well-known that all the naturally occurring amino acids found in proteins have the same configuration. With reference to D-glyceraldehyde as a standard, the naturally occurring amino acids have the opposite or L configuration. This relationship may be represented as shown in Structure 4-2 for the amino acid L-serine. The

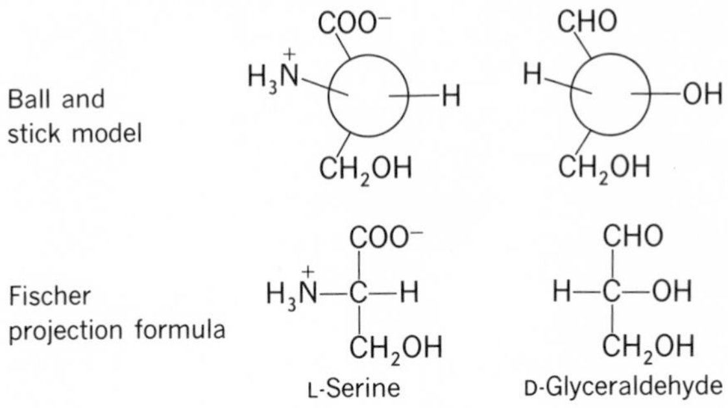

Ball and stick model: COO^-, $H_3\overset{+}{N}$, H, CH_2OH — CHO, H, OH, CH_2OH

Fischer projection formula:

$$\mathrm{H_3\overset{+}{N}{-}\underset{CH_2OH}{\overset{COO^-}{C}}{-}H}$$
L-Serine

$$\mathrm{H{-}\underset{CH_2OH}{\overset{CHO}{C}}{-}OH}$$
D-Glyceraldehyde

Structure 4-2

stereochemical relationship between D-glyceraldehyde and L-serine is apparent from these pairs of ball and stick and projection formulas, in which the aldehyde and carboxyl groups are written at the top of the structure. Furthermore, careful inspection of Structure 4-1 will disclose that the amino acid represented there has the L configuration. Note that when the carboxyl group is written to the right in the projection formula, the amino group is below the α carbon atom in an L-amino acid.

As with the carbohydrates, it is important to stress that the use of L and D conventions refers only to the relative configuration of these compounds and does not provide any information regarding the direction in which these optically active compounds rotate polarized light.

Structures of the Amino Acids Found in Proteins

The naturally occurring amino acids may be classified according to the chemical nature (aliphatic, aromatic, heterocyclic) of their R groups with appropriate subclasses. More meaningful, however, is a classification based on the polarity of the R group or residue because it emphasizes the possible functional roles which the different amino acids can play in proteins. In this classification, the twenty amino acids commonly obtained on the hydrolysis of proteins may be described as (1) nonpolar or hydrophobic; (2) polar but uncharged; (3) polar because of a negative charge at the physiological pH of 7; (4) polar because of a positive charge at physiological pH. The structures of these twenty amino acids together with certain of their distinctive features are given next:

(1) *Amino acids with nonpolar or hydrophobic R groups.* This group contains amino acids with both aliphatic (alanine, valine, leucine, isoleucine, methionine) and aromatic (phenylalanine and tryptophan) residues that are understandably hydrophobic in character. One of the compounds, proline, is unusual in that its nitrogen atom present is as a *secondary* amine rather than as a primary amine.

$H_3C{-}CH(NH_2){-}CO_2H$

L-Alanine

$(H_3C)_2CH{-}CH(NH_2){-}CO_2H$

L-Valine

$(H_3C)_2CH{-}CH_2{-}CH(NH_2){-}CO_2H$

L-Leucine

$H_3C{-}CH_2{-}CH(CH_3){-}CH(NH_2){-}CO_2H$

L-Isoleucine

$H_2C{-}CH_2{-}H_2C{-}NH{-}CH{-}CO_2H$ (ring)

L-Proline

$C_6H_5{-}CH_2{-}CH(NH_2){-}CO_2H$

L-Phenylalanine

(indolyl)$-CH_2{-}CH(NH_2){-}CO_2H$

L-Tryptophan

$H_3C{-}S{-}CH_2{-}CH_2{-}CH(NH_2){-}CO_2H$

L-Methionine

(2) *Amino acids with polar, but uncharged R groups.* Most of these amino acids contain polar R residues that can participate in hydrogen bond formation. Several possess a hydroxyl group (serine, threonine, and tyrosine) or sulfhydryl group (cysteine), while two (asparagine and glutamine) have amide groups. Glycine, which lacks an R group, is included in this grouping because of its definite polar nature, a property it possesses because its charged carboxyl and amino groups constitute such a large part of the mass of the molecule itself. Again, both aliphatic and aromatic (tyrosine) compounds are included in this group.

$H{-}CH(NH_2){-}CO_2H$

Glycine

$HOCH_2{-}CH(NH_2){-}CO_2H$

L-Serine

$H_3C{-}CH(OH){-}CH(NH_2){-}CO_2H$

L-Threonine

$HS{-}CH_2{-}CH(NH_2){-}CO_2H$

L-Cysteine

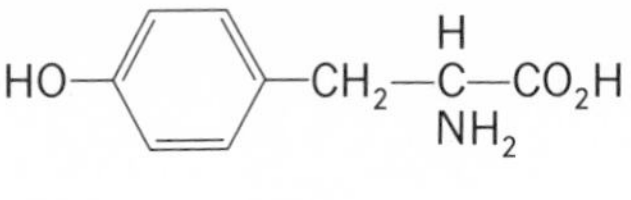

$$HO-C_6H_4-CH_2-\underset{NH_2}{\overset{H}{C}}-CO_2H$$

L-Tyrosine

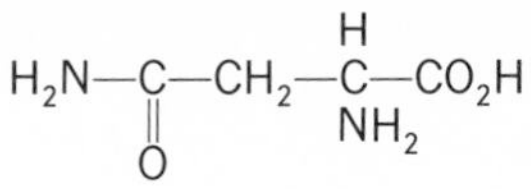

$$H_2N-\underset{O}{\overset{\|}{C}}-CH_2-\underset{NH_2}{\overset{H}{C}}-CO_2H$$

L-Asparagine

$$H_2N-\underset{O}{\overset{\|}{C}}-CH_2-CH_2-\underset{NH_2}{\overset{H}{C}}-CO_2H$$

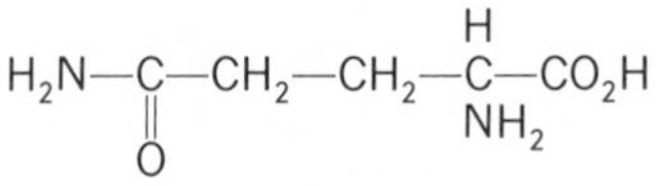

L-Glutamine

(3) *Amino acids with positively charged R groups*. Three amino acids are included in this group. Lysine, with its second (epsilon, ε) amino group ($pK = 10.5$), will be more than 50% in the positively charged state at any pH below the pK_a of that group. Arginine, containing a strongly basic guanidinium function ($pK = 12.5$), and histidine, with its weakly basic ($pK = 6.0$) imidazole group, are also included here. Note that histidine is the only amino acid which has a proton that dissociates in the neutral pH range. It is this characteristic which allows certain histidine residues to play an important role in the catalytic activities of some enzymes.

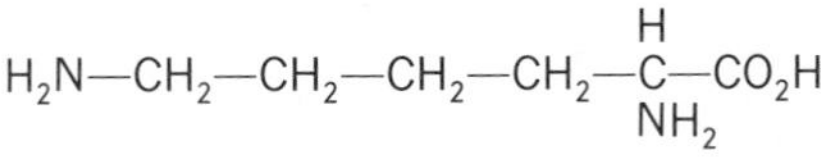

$$H_2N-CH_2-CH_2-CH_2-CH_2-\underset{NH_2}{\overset{H}{C}}-CO_2H$$

L-Lysine

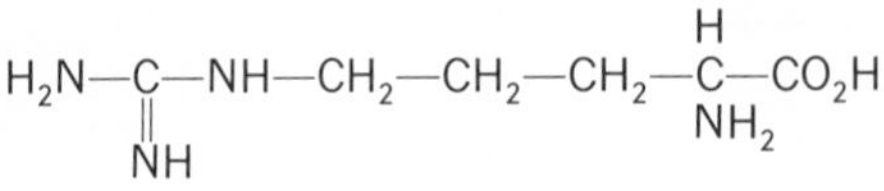

$$H_2N-\underset{NH}{\overset{\|}{C}}-NH-CH_2-CH_2-CH_2-\underset{NH_2}{\overset{H}{C}}-CO_2H$$

L-Arginine

$$HC{=}C(-CH_2-\underset{NH_2}{\overset{H}{C}}-CO_2H)\text{, ring: } HC{=}C-NH-C(H){=}N-$$

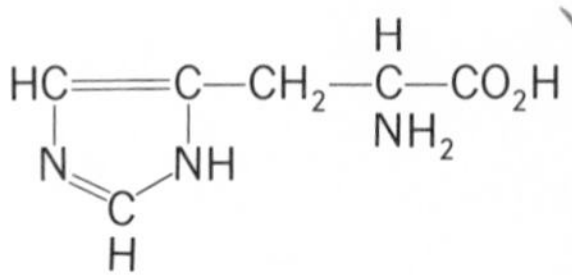

L-Histidine

(4) *Amino acids with negatively charged R groups*. This group includes the two dicarboxylic amino acids aspartic acid and glutamic acid. At neutral pH their second carboxyl groups with pK_{a_2}'s of 3.9 and 4.3, respectively, dissociate, giving a net charge of -1 to these compounds.

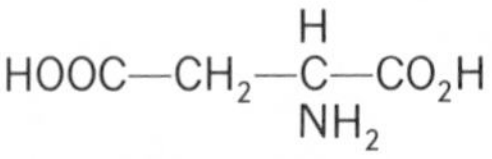

$$HOOC-CH_2-\underset{NH_2}{\overset{H}{C}}-CO_2H$$

L-Aspartic acid

$$HOOC-CH_2-CH_2-\underset{NH_2}{\overset{H}{C}}-CO_2H$$

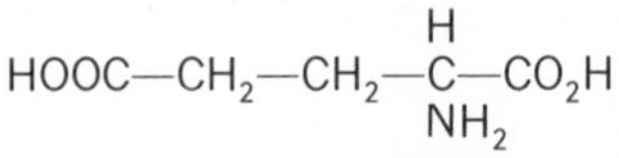

L-Glutamic acid

In addition to these twenty amino acids which are building units that have a wide distribution in all proteins, several other amino acids occur—often in high concentrations but in only a few proteins. As an example, hydroxyproline has a limited distribution in nature, but constitutes more than 12% of the structure of collagen, an important structural protein of animals. Similarly, hydroxylysine is a component of this animal protein.

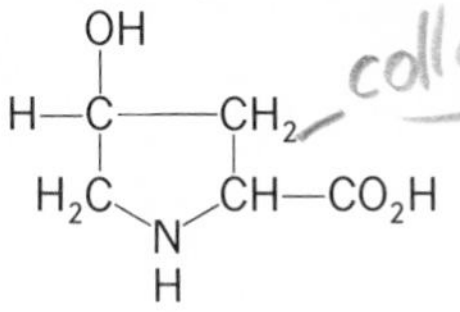

L-Hydroxyproline

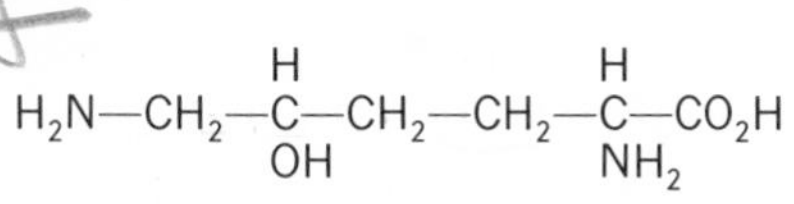

$$H_2N—CH_2—CH(OH)—CH_2—CH_2—CH(NH_2)—CO_2H$$

L-Hydroxylysine

Amino acids having the D configuration also exist in peptide linkage in nature, but not as components of large protein molecules. Their occurrence appears limited to smaller, cyclic peptides or as components of peptidoglycans of bacterial cell walls. Thus, two D-phenylalanine residues are found in the antibiotic gramicidin-S (Structure 4-3), and D-valine occurs in actinomycin-D, a potent inhibitor of RNA synthesis. D-Alanine and D-glutamic acid are found in the peptidoglycan of the cell wall of gram-positive bacteria (page 50).

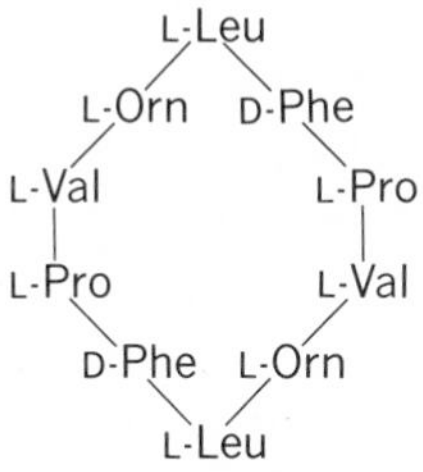

L-Leu → L-Orn, D-Phe → L-Val, L-Pro → L-Pro, L-Val → D-Phe, L-Orn → L-Leu (cyclic)

Gramicidin-S

Structure 4-3

Nonprotein Amino Acids

While the amino acids commonly found in proteins also occur as free compounds in many cells, there are a number of amino acids which are never found as constituents of proteins but which play important metabolic roles. Among these are L-ornithine and L-citrulline, which are metabolic intermediates in the urea cycle (Chapter 16) and as such participate in the biosynthesis of the amino acid arginine. An isomer of alanine, β-alanine, occurs free in nature and as a component of the vitamin pantothenic acid, coenzyme A, and acyl carrier protein (Chapter 9). The quaternary amine creatine, a derivative of glycine, plays a fundamental role in the energy storage process in vertebrates, where it is phosphorylated and converted to creatine phosphate (reaction 7-17).

$$H_2N—CH_2—CH_2—CO_2H$$

β-Alanine

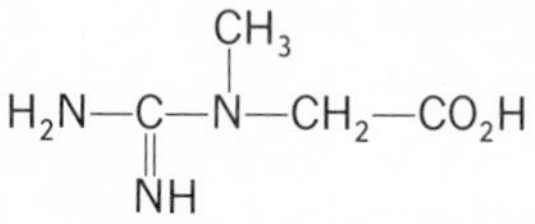

$$H_2N—C(=NH)—N(CH_3)—CH_2—CO_2H$$

Creatine

In addition to these nonprotein amino acids, for which a metabolic role may be described, more than 200 other nonprotein amino acids have been detected as natural products. Higher plants are a particularly rich source of these amino acids. In contrast to the amino acids previously described, however, these compounds do not occur widely, but may be limited to a single species or only a few species within a genus. These nonprotein amino acids are usually related to the proteinaceous ones as homologs or substituted derivatives. Thus, L-azetidine-2-carboxylic acid, a homolog of proline, may account for 50% of the nitrogen present in the rhizome of Solomon's seal (*Polygonatum multiflorum*). Orcylalanine (2,4-dihydroxy-6-methyl phenyl-L-alanine), found in the seed of the corncockle, *Agrostemma githago*, may be considered as a substituted phenylalanine.

Azetidine-2-carboxylic acid

Orcyl-L-alanine

These and the many other nonprotein amino acids that occur in nature are presently being studied in order to learn more about the conditions under which they arise and their role, if any, in the plant in which they occur.

Properties of the Amino Acids

Two readily observable properties of amino acids provide information about their structure, both in the solid state and in solution. For example, the amino acids, with certain exceptions, are generally soluble in H_2O and are quite insoluble in nonpolar organic solvents such as ether, chloroform, and acetone. This observation is not in keeping with the known properties of carboxylic acids and organic amines. Aliphatic and aromatic carboxylic acids, particularly those having several carbon atoms, have limited solubility in H_2O but are readily soluble in organic solvents. Similarly, the higher amines are usually soluble in organic solvents but not in H_2O.

Another physical property of amino acids that relates to their structure is their high melting points which often result in decomposition; the melting points of the solid carboxylic acids and amines are usually low and sharp. These two physical properties of the amino acids are not consistent with their general structural formula (Structure 4-1), which represents them as containing uncharged carboxyl and amino groups. The solubilities and melting points rather suggest structures with charged, highly polar groups.

A deeper insight into the structure of amino acids in solution is gained from considering the behavior of amino acids as electrolytes. Since alanine, for example, contains both a carboxyl and an amino group, it should react with acids and alkalis. Such compounds are referred to as *amphoteric substances*. If a solid sample of alanine is dissolved in H_2O, the pH of this solution will be approximately neutral. If electrodes are placed in solution and a difference in potential is placed across the electrodes, the amino acid will not migrate in the electric field. This result is in keeping with the representation of the amino acid as a neutral, uncharged molecule, but the same is true if alanine were represented as the zwitterion. This formula, first proposed by Bjerrum in 1923, depicts the carboxyl group as being ionized while the amino group is still protonated.

$$\underset{\text{Alanine (uncharged)}}{H_3C\text{—}\overset{\displaystyle H}{C}\text{—}COOH \atop \quad NH_2} \qquad \underset{\text{Alanine (zwitterion)}}{H_3C\text{—}\overset{\displaystyle H}{\underset{\displaystyle NH_3^+}{C}}\text{—}COO^-}$$

Titration of Amino Acids. A clear choice between these two formulas for alanine can be made when one compares the titration curve for this amino acid (Figure 4-1) with the titration curves for simple alkyl amines and carboxylic acids. If 20 ml of 0.1M alanine in solution is titrated with 0.1M NaOH, a curve with a pK_a at 9.7 is obtained when 10 ml of 0.1M NaOH have been added. This signifies that, at pH 9.7, some group capable of furnishing protons to react with the added alkali is half neutralized. Similarly, if 0.1M HCl is added to the solution of alanine, the other half of the titration curve is obtained and a pK_a of 2.3 is reached when 10 ml of 0.1M HCl have been added. The addition of alkali and acid to the zwitterion form of alanine may be represented by reaction 4-1:

$$\underset{\displaystyle NH_3^+}{H_3C\text{—}CH\text{—}COOH} \xleftarrow{H^+} \underset{\displaystyle NH_3^+}{H_3C\text{—}CH\text{—}COO^-} \xrightarrow{OH^-} \underset{\displaystyle NH_2}{H_3C\text{—}CH\text{—}COO^-} + H_2O \qquad (4\text{-}1)$$

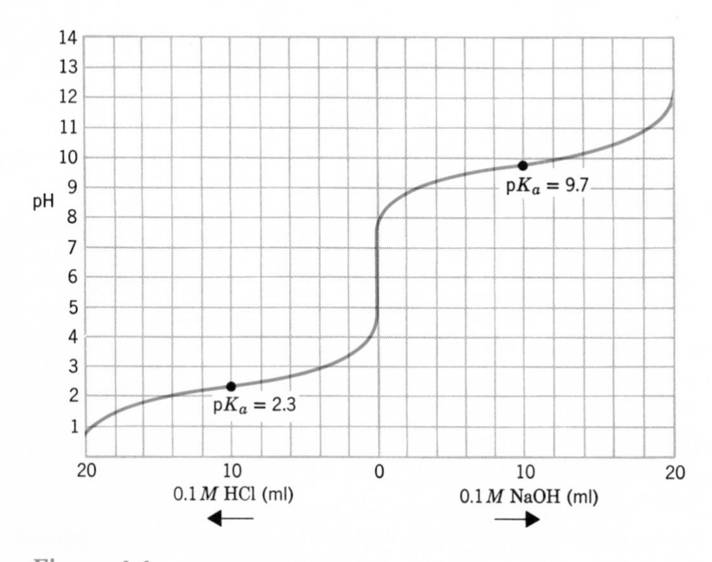

Figure 4-1

Titration curve obtained when 20 ml of 0.1M L-alanine are titrated with 0.1M NaOH and with 0.1M HCl.

Note that in the former the proton on the amino group dissociates to neutralize the hydroxyl ion added, while the ionized carboxyl group accepts the protons added during acidification. The pK_a's observed in the titration curve are consistent with this representation because, as noted in Chapter 1, the carboxyl groups of organic acids dissociate in the pH range of 3 to 5, while ammonium ions (and protonated alkyl amines) are weak acids with pK_a's in the range of 9 to 11.

If the formula with the carboxyl and amino functions uncharged (page 75) were the correct representation of alanine in neutral solution, one would write reaction 4-2 for the titration of the amino acid with alkali and acid:

$$H_3C-\underset{\underset{NH_3^+}{|}}{CH}-COOH \xleftarrow{H^+} H_3C-\underset{\underset{NH_2}{|}}{CH}-COOH \xrightarrow{OH^-} H_3C-\underset{\underset{NH_2}{|}}{CH}-COO^- + H_2O \qquad (4\text{-}2)$$

In this representation, it would be the carboxyl group of alanine which is titrated with alkali and which, according to Figure 4-1, would have to be assigned a pK_a of 9.7. Similarly, reaction 4-2 states that it is the amino group of alanine which reacts when acid is added and therefore possesses a pK_a of 2.3. It is difficult to argue that the carboxyl group of alanine has a pK_a of 9.7 when the pK_a's for acetic and propionic acid are 4.74 and 4.85, respectively. The structure of alanine does not differ sufficiently from that of propionic acid to lead us to suggest that the carboxyl group of the amino acid should be 10^5 times less acidic. As for the amino group, the pK_a of NH_4^+ is 9.26, and it is impossible to account for the fact that, as presented in reaction 4-2, the amino group of alanine should be about 10^7 times as acidic as NH_4^+.

It is important to note that the correct interpretation of the titration curve of alanine, as represented by reaction 4-1, leads to the same ionic species as would be obtained by the erroneous representation in reaction 4-2. The anion

$$H_3C-\underset{\underset{NH_2}{|}}{CH}-COO^-$$

is obtained on treatment with alkali, while the cation

$$H_3C-\underset{\underset{NH_3^+}{|}}{CH}-COOH$$

is present after acidification. These formulas do correctly represent the structure of alanine in alkaline and acid solution, respectively. Experimentally, alanine in acid solution is positively charged and migrates toward the negative cathode in an electric field. Conversely, alanine in alkaline solution is negatively charged and migrates toward the positive anode in solution.

While the pK's observed on titration constitute evidence for the zwitterionic nature of amino acids in solution, the physical properties of the solid amino acids also indicate that the amino acids are zwitterions in the solid state. Thus, they are easily soluble in H_2O and have high melting points. The zwitterion is essentially an internal salt, which should have a high melting point and be readily soluble in water. Additional evidence for the dipolar nature of the amphoteric amino acids is obtained by titrating the amino acid in formaldehyde. Formaldehyde reacts with the uncharged amino group to produce a mixture of the mono- and dihydroxymethyl derivatives:

$$R{-}\underset{\underset{H\quad H}{N}}{\overset{H}{C}}{-}COO^- \xrightarrow{HCHO} R{-}\underset{\underset{H\quad CH_2OH}{N}}{\overset{H}{C}}{-}COO^- \xrightarrow{HCHO} R{-}\underset{\underset{HOCH_2\quad CH_2OH}{N}}{\overset{H}{C}}{-}COO^-$$

These compounds are secondary and tertiary amines and, therefore, are weaker bases (or stronger acids). This is manifested in the titration curve by a lowering of the pK_a for the amino group in the presence of formaldehyde (Figure 4-2).

Other evidence of the zwitterionic nature of all amino acids is found in their spectroscopic properties, their effects on the dielectric constant of aqueous solutions, and their titrations in organic solvents.

Those amino acids having more than one carboxyl or amino group will have corresponding pK_a values for them. Thus, the pK_a for the α-carboxyl group of aspartic acid is 2.1, while the pK_a for the β-carboxyl is 3.9. The pK_a for the amino group is 9.8. The titration curve for aspartic acid is shown in Figure 4-3. Four different ionic species of aspartic acid can exist at different pH values. The student is urged to consult one of the several excellent books that discuss the titration of amino acids and practice working problems associated with these compounds.

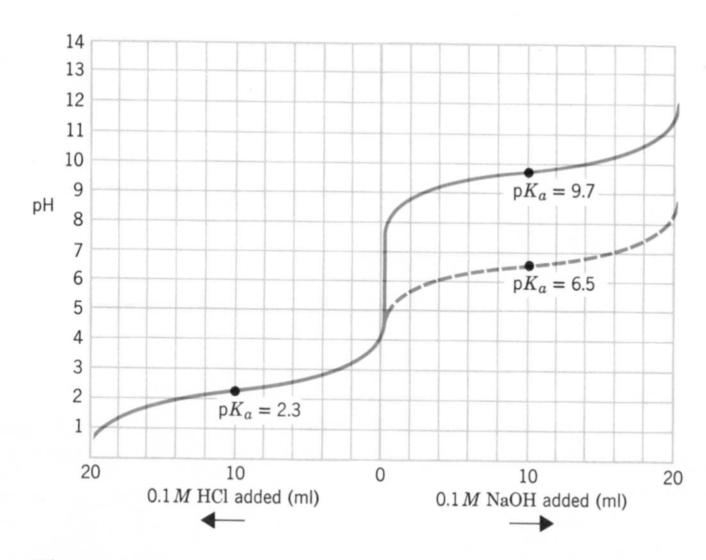

Figure 4-2

The dashed line shows the titration curve obtained when 20 ml of 0.1*M* L-alanine are titrated with NaOH in the presence of formaldehyde.

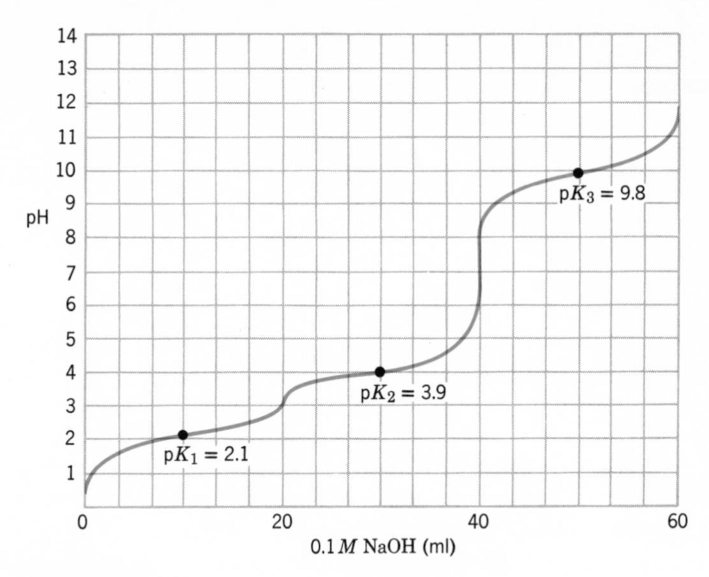

Figure 4-3

Titration curve obtained when 20 ml of 0.1*M* aspartic hydrochloride are titrated with 0.1*M* NaOH.

Some amino acids contain groups in addition to the protonated carboxyl or amino groups which are capable of dissociating protons. Thus, the sulfhydryl group of cysteine dissociates with a pK_a of 10.8:

$$H_2N-\underset{\underset{SH}{|}}{\underset{CH_2}{|}}{\overset{COO^-}{\overset{|}{C}}}-H \underset{pK_3 = 10.8}{\rightleftharpoons} H_2N-\underset{\underset{S^-}{|}}{\underset{CH_2}{|}}{\overset{COO^-}{\overset{|}{C}}}-H + H^+$$

Similarly, the guanidinium moiety of arginine dissociates to yield a proton with a pK_a of 12.5:

$$(H_2N)(R-HN)C{=}NH_2^+ \underset{pK_3 = 12.5}{\rightleftharpoons} (H_2N)(R-HN)C{=}NH + H^+$$

Other dissociable groups include the protonated nitrogen atom of the heterocyclic ring of histidine (pK_a = 6.0) and the phenolic hydroxyl (pK_a = 10.1) of tyrosine.

Reactions of Amino Acids

While the ability to act as electrolytes is an important chemical property of the amino acids, other properties dependent on the presence of carboxyl and amino groups in the molecule are equally significant. The reactions of these functional groups of the amino acids are well-known organic reactions.

Reactions of the Carboxyl Group. The carboxyl groups of amino acids may be esterified with alcohols or undergo amide formation in the presence of ammonia:

$$\underset{\displaystyle NH_3^+}{R{-}\overset{|}{C}H{-}COO^-} + C_2H_5OH \overset{H^+}{\rightleftharpoons} \underset{\displaystyle NH_3^+ \quad + OH^-}{R{-}CH{-}\overset{\displaystyle O}{\overset{\|}{C}}{-}O{-}C_2H_5} \xrightarrow[NH_3]{\text{Excess}} \underset{\displaystyle NH_2 \quad + C_2H_5OH}{R{-}CH{-}\overset{\displaystyle O}{\overset{\|}{C}}{-}NH_2}$$

When the amine involved in amide formation is not NH_3 but the α-amino group of another amino acid, a dipeptide is formed. An amide bond linking two amino acids is known as a peptide bond:

$$-\underset{\displaystyle O}{\underset{\|}{C}}-\overset{\displaystyle H}{\overset{|}{N}}-$$

The properties of this *substituted* amide linkage play a unique role in determining the structure of proteins; this subject is discussed in detail starting on page 85. A peptide containing two or more peptide bonds will react with Cu^{2+} in alkaline solution to form a violet-blue complex. This reaction, known as the Biuret reaction, is the basis for a quantitative determination of proteins.

The carboxyl group of amino acids may be decarboxylated chemically and biologically to yield the corresponding amine:

$$\underset{\displaystyle NH_2}{R{-}\underset{|}{C}H{-}CO_2H} \longrightarrow \underset{\displaystyle NH_2}{R{-}\underset{|}{C}H_2} + CO_2$$

Thus, the vasoconstrictor agent, histamine, is produced from histidine. Histamine stimulates the flow of gastric juice into the stomach and is involved in allergic responses.

Reactions of the Amino Group. The amino group of an amino acid will react with the strong oxidizing agent nitrous acid (HNO_2) to liberate N_2. This reaction, which is stoichiometric, is important in the estimation of α-amino groups in amino acids, peptides, and proteins. The amino acids proline and hydroxyproline do not react, and the ε-amino group of lysine reacts, but at a slower rate. The products are the corresponding α-hydroxy acid and N_2 gas, which can be measured manometrically.

$$\underset{\displaystyle NH_3^+}{R{-}\underset{|}{C}H{-}COOH} + HNO_2 \longrightarrow \underset{\displaystyle OH}{R{-}\underset{|}{C}H{-}COOH} + N_2 + H_2O + H^+$$

The amino group of amino acids will also undergo oxidation with the milder oxidizing agent ninhydrin to form ammonia, CO_2, and the aldehyde obtained by loss of one carbon from the original amino acid. In this reaction, one equivalent of ninhydrin serves as the oxidant for the amino acid to form the products just stated:

$$\underset{\displaystyle NH_2}{R{-}CH{-}COOH} + \text{Oxidized ninhydrin} \longrightarrow \underset{\displaystyle \| \atop O}{R{-}CH} + NH_3 + CO_2 + \text{Reduced ninhydrin} \quad (4\text{-}3)$$

A second equivalent of ninhydrin then reacts with the reduced ninhydrin and NH_3 formed in equation 4-3 to produce a highly colored product having the following structure:

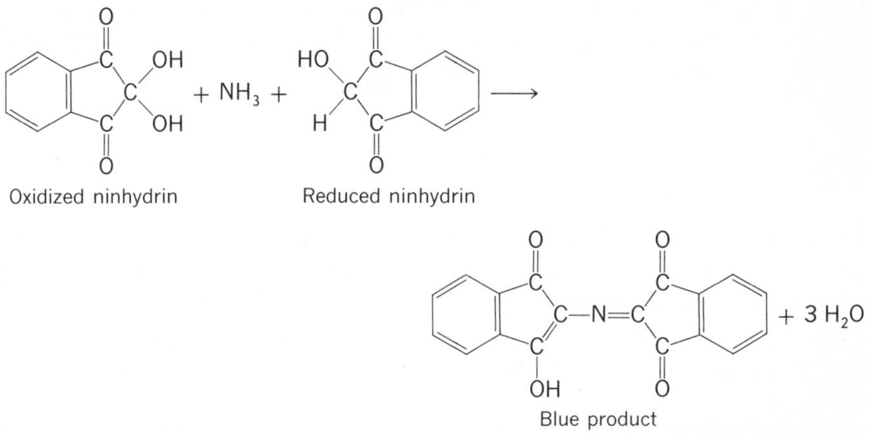

Oxidized ninhydrin + NH_3 + Reduced ninhydrin $\longrightarrow$ Blue product + $3\ H_2O$

Oxidized ninhydrin

Reduced ninhydrin

Blue product

The intense blue product is generally characteristic of those amino acids having α-amino groups. However, proline and hydroxyproline, which are secondary amines, yield yellow products, and asparagine, which has a free amide group, reacts with ninhydrin to produce a characteristic brown product. The ninhydrin reaction is extensively employed in the quantitative determination of amino acids.

Another reaction of the amino group which has found much recent use is the reaction with 1-fluoro-2,4-dinitrobenzene (abbreviated FDNB):

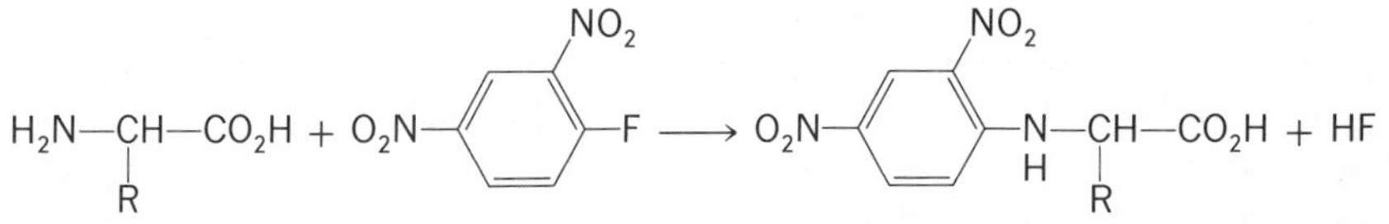

$$H_2N{-}\underset{\displaystyle R}{CH}{-}CO_2H + O_2N{-}C_6H_3(NO_2){-}F \longrightarrow O_2N{-}C_6H_3(NO_2){-}\underset{\displaystyle H}{N}{-}\underset{\displaystyle R}{CH}{-}CO_2H + HF$$

In this reaction the intensely colored dinitrobenzene nucleus is attached to the nitrogen atom of the amino acid to yield a yellow derivative, the 2,4-dinitrophenyl derivative or DNP–amino acid. The compound FDNB will react with the free amino group on the NH_2-terminal end of a polypeptide as well as the amino groups of free amino acids. Thus, by reacting a native protein or intact polypeptide with FDNB, hydrolyzing, and isolating the colored DNP–amino acid, one can identify the terminal amino acid in a polypeptide chain. The ε-amino group of lysine will also react with FDNB, but this derivative, ε-DNP–lysine, can readily be separated from the α-DNP–amino acids by an extraction procedure.

The amino group of both free amino acids and peptide chains will react with *dansyl chloride* (1-dimethylaminonaphthalene-5-sulfonyl chloride) to produce a dansyl amino acid derivative. Because the dansyl group readily fluoresces, minute amounts of amino acids may be determined by this procedure.

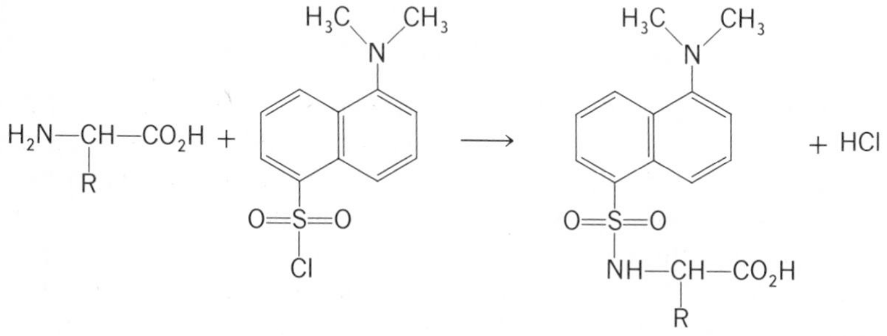

$$H_2N{-}CH(R){-}CO_2H + (CH_3)_2N{-}C_{10}H_6{-}SO_2Cl \longrightarrow (CH_3)_2N{-}C_{10}H_6{-}SO_2{-}NH{-}CH(R){-}CO_2H + HCl$$

Dansyl chloride   Dansyl amino acid derivative

The well-known reaction of isothiocyanates with amines has been ingeniously modified by Edman both to degrade a polypeptide chain and to identify the NH_2-terminal amino acid in the peptide (Appendix 2). Phenylisothiocyanate reacts with the α-amino group of an amino acid (or peptide) to form the corresponding phenylthiocarbamyl amino acid. In anhydrous acid this compound cyclizes to form a phenylthiohydantoin which is stable in acid.

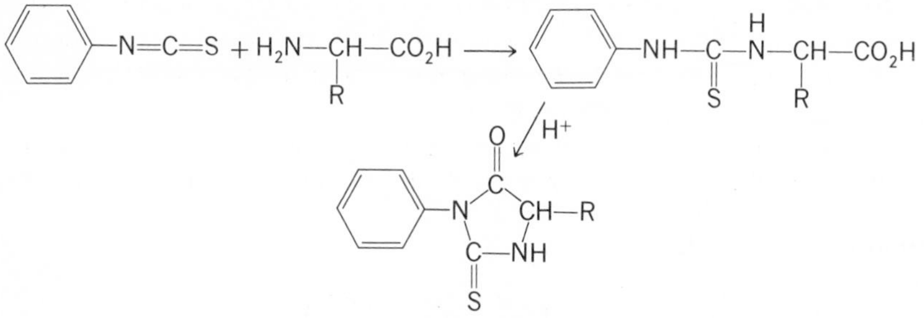

$$C_6H_5{-}N{=}C{=}S + H_2N{-}CH(R){-}CO_2H \longrightarrow C_6H_5{-}NH{-}C({=}S){-}NH{-}CH(R){-}CO_2H \xrightarrow{H^+} \text{Phenylthiohydantoin}$$

Phenylthiohydantoin

If the NH_2-terminal amino acid in a polypeptide is reacted with phenylisothiocyanate and the derivative is subsequently treated with anhydrous acid, only the phenylthiohydantoin of the NH_2-terminal amino acid is released. The remainder of the polypeptide chain is intact, hence the usefulness of this method. The NH_2-terminal group in the original peptide can of course be identified by determining the nature of the phenylthiohydantoin formed.

Reactions of the R Groups. The ionization of the R groups possessed by cysteine, tyrosine, and histidine has been referred to previously (page 78). Also of biological interest are two reactions which the R residues of serine and cysteine, respectively, undergo. The hydroxyl group of serine frequently is phosphorylated in a biologically active protein. Thus, the glycolytic enzyme phosphoglucomutase (Chapter 10) contains a serine whose hydroxyl group becomes phosphorylated during the functioning of the enzyme. The milk protein casein contains a large number of phosphorylated serine residues.

Phosphoserine moiety

The sulfhydryl group of cysteine undergoes reactions typical of the –SH group. One of these is the reversible oxidation with another molecule of cysteine to form the disulfide, cystine. Disulfide linkages between two cysteine residues in a polypeptide chain are a frequent occurrence. Insulin, for example, contains three disulfide linkages, two of which hold together two polypeptide chains (Chapter 19) in the physiologically active molecule. Ribonuclease similarly contains four disulfide bonds between four pairs of cysteine residues; if any one of these bonds is severed (by reduction, for example), the enzyme will lose its catalytic activity.

Simple Peptides

Intermediate in structural complexity between the amino acids and the proteins are the *peptides*, compounds formed by linking amino acids through peptide bonds. Several naturally occurring peptides are known and will be discussed; in addition, chemical and enzymatic hydrolysis of proteins gives rise to these compounds. Two amino acids joined by a peptide bond are known as a dipeptide; a peptide containing three amino acids is a tripeptide, etc. If a peptide contains fewer than ten amino acids, it is known as an oligopeptide; beyond that size it is known as a polypeptide.

A typical structure for an oligopeptide consisting of four amino acids is shown in Structure 4-4. Note that the molecule has a terminal amino (NH_2)

Structure 4-4

group at one end and a free carboxyl (—COOH) at the other. These ends of the polypeptide are called the amino (NH_2-) terminal end and the carboxyl (HOOC—) terminal end, respectively; the same terminology applies in the case of proteins. By convention, the NH_2-terminal amino acid in an oligopeptide or the polypeptide chain of a protein is called the first amino acid or the first residue ($a.a._1$).

Glutathione, a tripeptide that is ubiquitous in nature, illustrates the procedure for naming a simple peptide (Structure 4-5). The chemical name for glutathione is γ-glutamylcysteinyl glycine. The suffix *-yl* signifies the amino acid residue whose carboxyl group is linked in peptide linkage to the amino group of the next amino acid in the peptide. In the case of peptides containing glutamic (or aspartic) acid, the carboxyl group involved in the peptide linkage

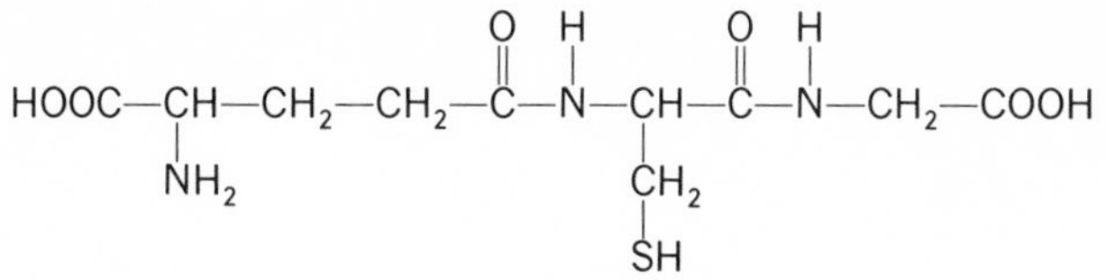

```
                            O  H     O  H
                            ‖  |     ‖  |
HOOC—CH—CH2—CH2—C—N—CH—C—N—CH2—COOH
     |                 |
     NH2               CH2
                       |
                       SH
```

γ-Glutamylcysteinyl glycine
(Glutathione)

Structure 4-5

must be identified. In glutathione, it is the γ-carboxyl that is esterified. This, however, is an unusual situation, for when glutamic acid (or aspartic acid) occurs in proteins, it is the α-carboxyl which is bound in peptide linkage. When no prefix is given, the α linkage is understood. The determination of the sequence of amino acids in a polypeptide is an essential step in the elucidation of the structure of the more complex proteins. Reactions described earlier and others not yet discussed are employed, and the Edman procedure is described in Appendix 2.

In addition to glutathione, the naturally occurring peptides include certain hormones of the pituitary gland. Vasopressin and oxytocin are nonapeptides that can assume a cyclical structure by virtue of their forming disulfide linkages between the NH_2-terminal cysteine and a cysteine in the interior of the peptide. Seven of the nine amino acids in the two peptides are identical; the COOH-terminal residue is the amide of glycine. Nevertheless, the physiological effects are quite different. Oxytocin causes the contraction of smooth muscle; vasopressin causes a rise in blood pressure by constricting the peripheral blood vessels.

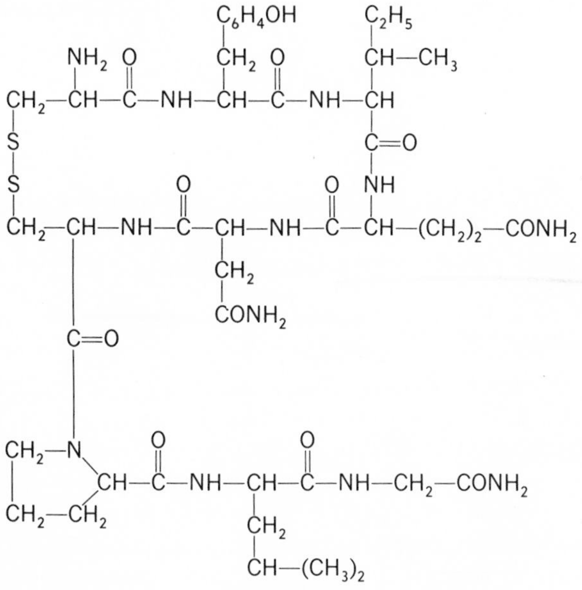

```
                     C6H4OH       C2H5
                     |            |
      NH2 O          CH2 O        CH—CH3
      |   ‖          |   ‖        |
CH2—CH—C—NH—CH—C—NH—CH
|                                 |
S                                 C=O
|                                 |
S                 O          O    NH
|                 ‖          ‖    |
CH2—CH—NH—C—CH—NH—C—CH—(CH2)2—CONH2
     |            |
     |            CH2
     |            |
     |            CONH2
     C=O
     |
     |
CH2—N          O           O
|     \        ‖           ‖
|      CH—C—NH—CH—C—NH—CH2—CONH2
|     /            |
CH2—CH2            CH2
                   |
                   CH—(CH3)2
```

Oxytocin

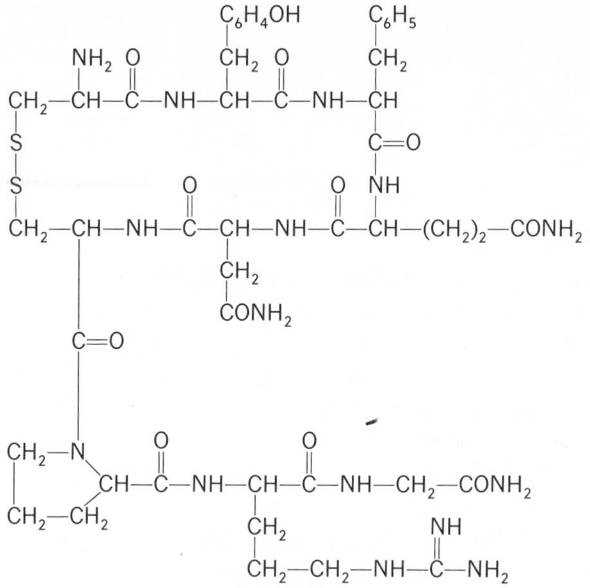

Vasopressin

Several antibiotics are polypeptides of comparatively simple structure. Gramicidin (Structure 4-3) and tyrocidin are examples of such compounds.

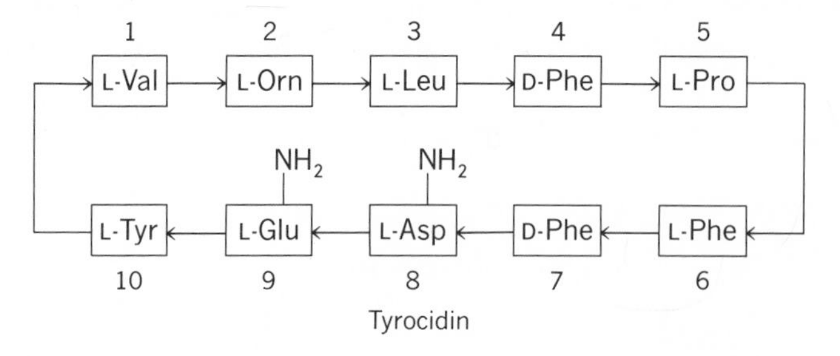

Tyrocidin

Penicillin, another antibiotic, contains the valine and cysteine residues, but these are not linked by peptide bonds. Rather, a strained four-membered ring and a sulfur-containing ring are found. The adrenocorticotrophic hor-

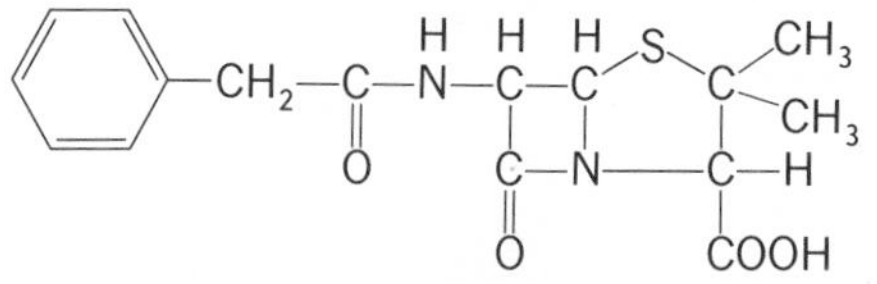

Penicillin G
(Benzyl penicillin)

mone (ACTH) contains 39 amino acid residues. Those occupying positions 4–10 are identical with residues 7–13 of the melanocyte-stimulating hormones and may be related to the ability of ACTH to stimulate the production of melanocytes, the pigment-producing cells of the skin.

Insulin (Chapter 19) produced by the pancreas, is a hormone consisting of two polypeptide chains containing a total of 51 amino acid residues.

Proteins

The polypeptide nature of proteins is shown in Figure 4-4, where a series of L-amino acids are linked by peptide bonds. In the figure, R is the side chain or residue of the amino acid and may be any one of twenty possible groups. Proteins vary in molecular weight from about 5000 to many millions. Despite this complexity, the amino acid sequence of a number of proteins have been determined. The English biochemist, Sanger, received the Nobel Prize in Chemistry for determining the precise sequence of amino acids in insulin. His methods of controlled hydrolysis of proteins to recognizable peptides and other, more recently developed techniques, have been used to elucidate the amino acid sequence of ribonuclease, the adrenal corticotrophic hormone (ACTH), and many other proteins.

A number of factors play important roles in defining the total structure of a protein:

(1) The peptide bond is the covalent bond which links amino acids together in the protein. This bond, being essentially a substituted amide, is planar in structure, since the electrons are delocalized in the amide linkage giving the C—N bond considerable double bond character, as shown by the resonance structures

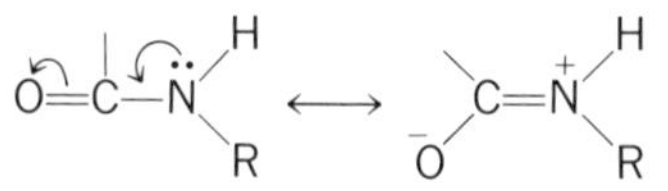

$$O{=}C{-}\ddot{N}(H)(R) \longleftrightarrow {}^{-}O{-}C{=}\overset{+}{N}(H)(R)$$

Thus, the planar peptide bond can be represented as

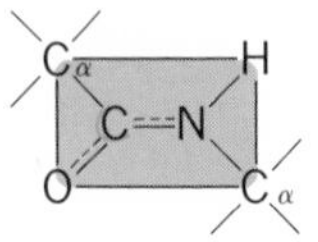

$$C_\alpha - C(=O) = N(H) - C_\alpha$$

Because of the relatively rigid plane in which the O=C and C—N atoms lie, free rotation does not occur about these axes. As depicted in Figure 4-4, a polypeptide chain consists of a series of planar peptide linkages joining α-carbons (C_α) which serve as swivel centers for the polypeptide chain.

(2) Since the peptide bond is planar, only the rotations around the C_α—N

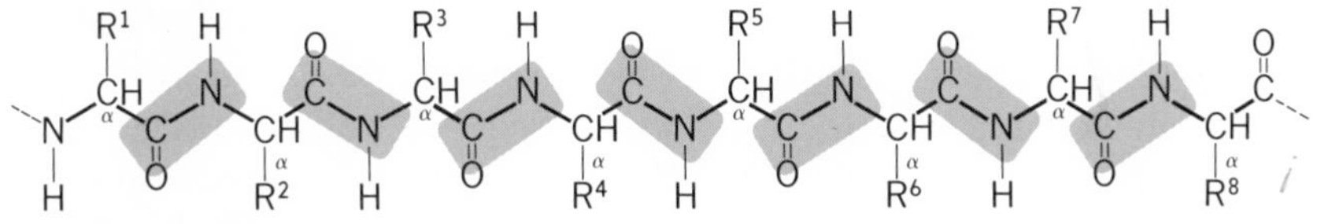

Figure 4-4

General formula of a polypeptide chain showing the linkage of adjacent amino acid residues through peptide bonds.

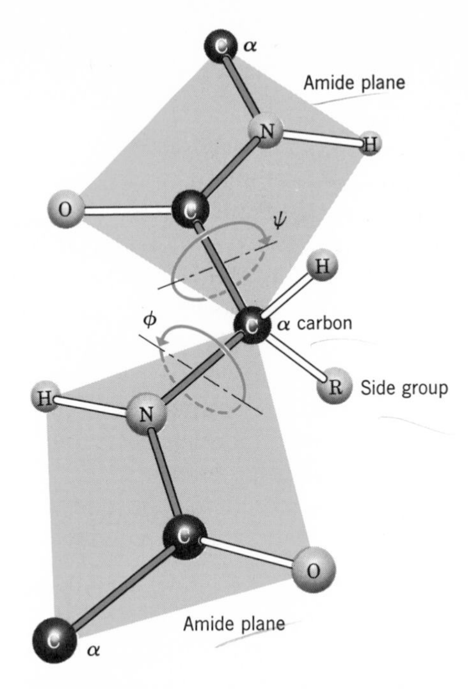

Figure 4-5

Two amide planes joined together by a carbon atom and related to each other by the rotation angles φ (C_α—N) and ψ (C_α—C). [After Dickerson and Geis (1969).]

axis (ϕ) and the C_α—C axis (ψ) are permitted (Figure 4-5). Knowledge of the ϕ and the ψ rotation values will completely define the secondary structure of a protein. From considerations of steric hindrance and the potential for hydrogen bond formation, protein chemists have deduced that a specific configuration called the right-handed α helix, α_r (Figure 4-6), is particularly favorable for a stable structure. These α helixes, which have 3.6 amino acid residues per turn, are stabilized due to hydrogen bonding between an —NH— group in the helix and the —C=O group of the fourth amino acid down the chain (Figure 4-7). Because each —NH— and —C=O can be hydrogen bonded in this manner, the α helix represents a highly favored structure. Under these conditions, the ϕ values range from 113 to 132° and the ψ values from 123 to 136°.

Since the C_α is the swivel point for the chain, the R groups associated with the C_α become extremely important. In general, if the R groups are not bulky or have no polar groups such as primary hydroxyl functions or charged $—NH_3^+$ or $—COO^-$ groups, ϕ and ψ values can be obtained for maximum α-helicity. But if lysyl (ε-NH_3^+) or aspartyl (β-COO^-) or glutamyl (γ-COO^-) are present in a cluster which would allow like charges to be positioned opposite in a helical turn and hence repelled, α-helicity is destabilized. If, however, the amino acid residues occur as

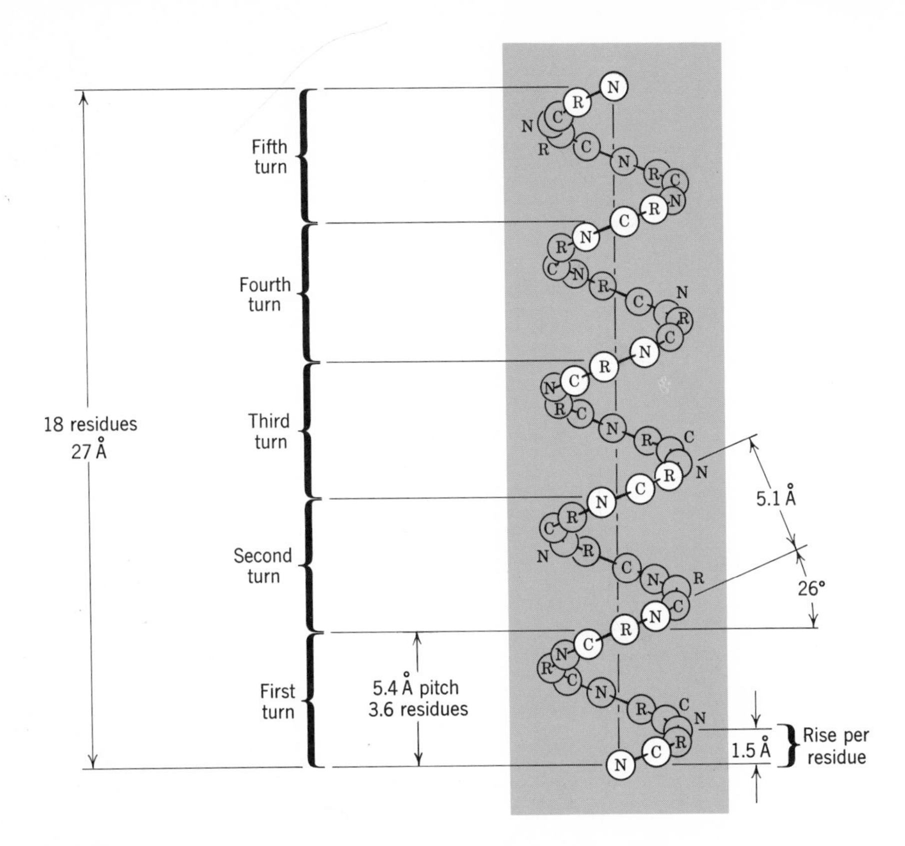

Figure 4-6

Representation of a polypeptide chain as an α-helical configuration. The shaded circles indicate atoms below the plane of the paper. The white circles represent atoms above the plane. [From L. Pauling and R. B. Corey, *Proc. Intern. Wool Textile Res. Conf., B,* 249 (1955), as redrawn in C. B. Anfinsen, *The Molecular Basis of Evolution,* John Wiley & Sons, New York, 1959, p. 101.]

isolated residues in the polypeptide chain, no helix destabilization is observed. A cluster of glycines with no R groups allows a greater degree of rotation; that is, an α-helical configuration is permitted, but since there are no R group constraints in the C_α, additional conformations are allowed—including the β-conformation. One amino acid is exceptional, namely proline. Since proline has its N_α atom in a rigid ring system, the permissible C_α–N axis value does not allow an α-helical structure but rather calls for a sharp bend. Thus, wherever a proline residue exists, an α-helical structure is disrupted. Indeed, the amino acids proline, glycine, serine, and asparagine are called helix-breaker amino acid residues.

(3) An additional role in the total structure of a protein molecule is played by the R groups. As already indicated, amino acids are divided into four general groups: nonpolar, uncharged polar, and negatively or positively

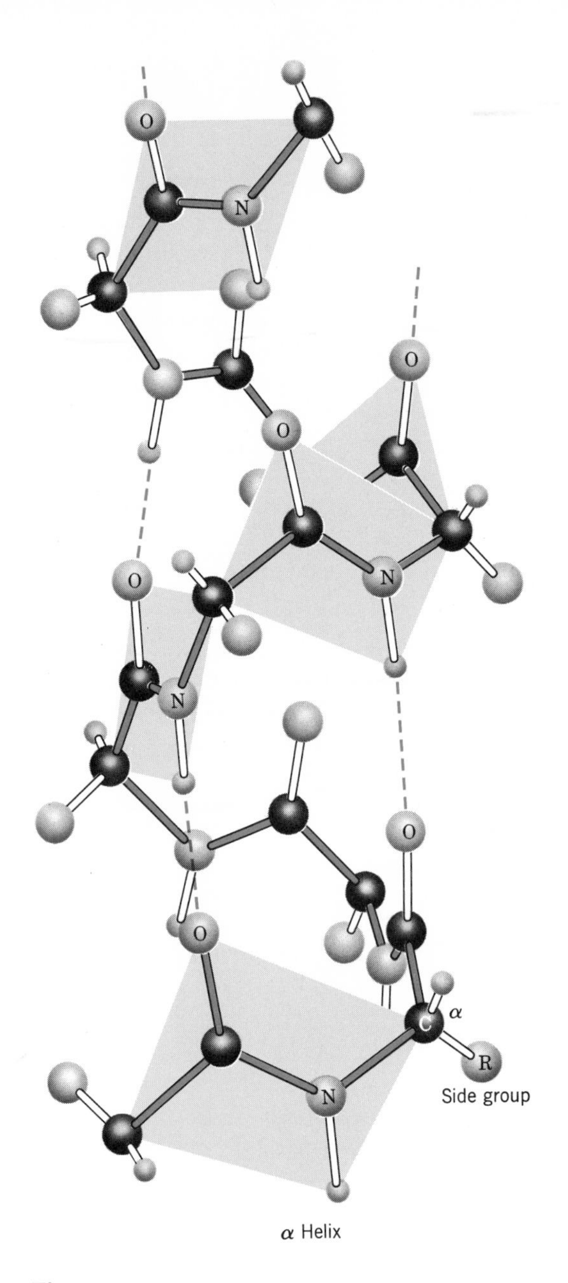

Figure 4-7

A right-handed α-helical structure of a polypeptide chain showing the unstrained hydrogen bonding stabilizing the α-helical structure and the planar configuration of the peptide bonds. [After Dickerson and Geis (1969).]

charged polar amino acids. Considerable evidence now indicates that a protein molecule, submerged in its aqueous environment, tends to expose a maximum number of its polar groups to the surrounding environment while a maximum number of its nonpolar groups are oriented internally. Orientation of these groups in this manner leads to a stabilization of protein conformation. This stabilization effect is probably related to an unfavorable decrease in entropy which would seem to accompany a transition to the reverse arrangement. If the surface of the protein molecule had a high hydrophobicity (water-repelling character), the water molecules adjacent to this surface would be forced into a more orderly, cage-like structure with lowered entropy relative to the bulk of the water. However, with polar groups on the surface, the orderly cage-like structure of water molecules may not be induced; or if it is induced, the unfavorable entropy decrease would be compensated by an enthalpy decrease due to the interactions of the polar residues with polar water molecules. The net effect would be a free-energy decrease leading to a stable state. Cytochrome c is a good example in that its tertiary structure displays hydrophilicity externally and hydrophobicity internally (page 97).

In addition, the residues of the uncharged polar amino acids are the sites of hydrogen bonding leading to potential cross-linking of chains. Charged polar groups, by responding to the pH of the surrounding medium, may markedly affect the activity of functional proteins. A number of amino acids play highly specific roles. Some of these are now described.

(1) Cysteine can cross-link with another cysteine sulfhydryl group in the same or on different polypeptide chains by oxidation to a covalent disulfide bond. The structure of insulin is an excellent example of the importance of disulfide bonds (Chapter 19). In the reduced state a cysteine residue can serve as a site for substrate attachment in a number of enzyme proteins.
(2) Histidine, with its lone electron pair in the ring nitrogen, may serve as a potential metal ligand as in the iron-containing proteins hemoglobin and cytochrome c.
(3) Lysine is intimately involved in binding pyridoxal phosphate and biotin (see Chapter 9), and like serine and histidine, may serve in making up the active site of an enzyme.
(4) Proline, because of its relatively rigid ring, forces a bend in a polypeptide chain and disrupts α-helicity.
(5) The polar amino acids, glutamic, aspartic, arginine, lysine, and histidine are ionized over wide pH ranges and can thus form ionic bonds in the protein structure. Some of the ways in which the R groups can stabilize protein structure are illustrated in Figure 4-8.

Some Definitions

To define a complicated macromolecule such as a protein in descriptive terms, biochemists have assigned four basic structural levels to proteins.

(1) *Primary structure.* This term indicates that the number and precise

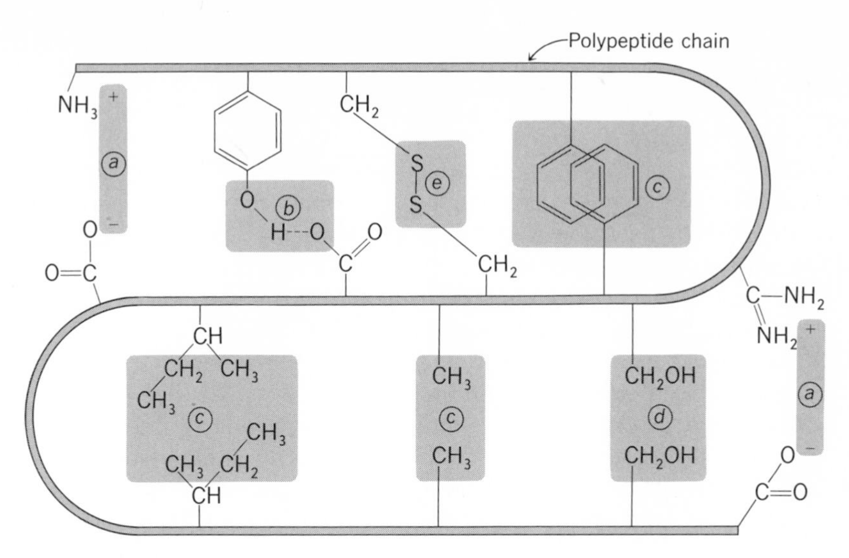

Figure 4-8

Some types of noncovalent bonds which stabilize protein structure: (*a*) electrostatic interaction; (*b*) hydrogen bonding between tyrosine residues and carboxyl groups on side chains; (*c*) hydrophobic interaction of nonpolar side chains caused by the mutual repulsion of solvent; (*d*) dipole–dipole interaction; (*e*) disulfide linkage, a covalent bond. (From C. B. Anfinsen, *The Molecular Basis of Evolution,* John Wiley & Sons, New York, 1959, p. 102.)

sequence of amino acids in the protein is known. Implied, of course, is the peptide linkage between each of the amino acids (Figure 4-4), but no other forces or bonds are indicated in this term.

(2) *Secondary structure.* The use of this term refers to the extent to which a polypeptide or protein chain may possess any helical structure. A right-handed α-helical spiral is stabilized by hydrogen bonding between the carbonyl and the imido groups of the peptide bonds which appear in a regular sequence along the chain (Figures 4-6 and 4-7).

(3) *Tertiary structure.* This term refers to the tendency for the polypeptide chain to undergo extensive coiling or folding and produce a complex, somewhat rigid structure (Figure 4-9). The stabilization of this structure is ascribed to the different reactivities associated with the R groups in the amino acid residues (Figure 4-8). Moreover, the term conformation defines the participation of the secondary and tertiary structures of the polypeptide chains in molding the total structure of the protein. The correct conformation of a protein is of prime importance in determining the fine structure of the protein and contributes greatly to the unique catalytic properties of biologically active proteins. We shall use this term extensively in our discussion of proteins with enzymic functions.

(4) *Quaternary structure.* This defines the degree of association of a protein unit. Thus the enzyme phosphorylase *a* contains two identical subunits which alone are catalytically inactive but when joined as a dimer form

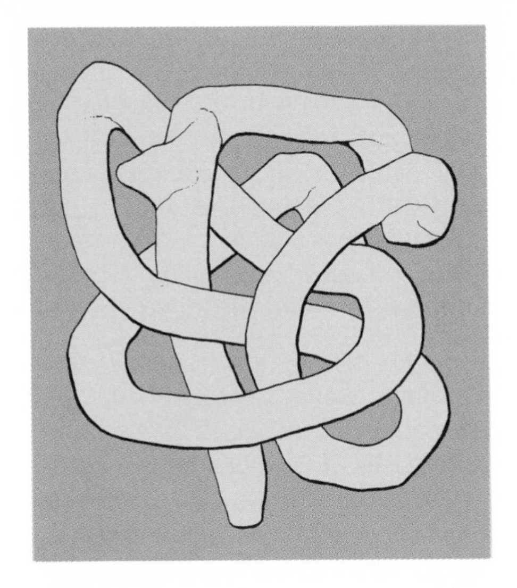

Figure 4-9

Sketch illustrating the complicated folding of a globular protein stabilized by noncovalent bonds.

the active enzyme as shown in Figure 4-10. This type of structure is called a *homogenous quaternary structure*; if there are dissimilar units, as in tobacco mosaic virus in which ribonucleic acid and protein units complex to give active virus, a *heterogenous quaternary* structure is obtained. Another term employed to describe the subunits of such a protein is *protomer*, and a protein made up of protomers would be an *oligomeric protein*. In specific terms hemoglobin is an oligomeric protein having a *heterogenous, quaternary* structure, consisting of two identical α-chain protomers and two identical β-chain protomers (i.e., $\alpha_2\beta_2$).

Figure 4-10

A protein dimer unit illustrating the quaternary structure of a complex globular protein.

Fibrous and Globular Proteins

With these basic concepts and definitions in mind, we can now examine the two broad categories of proteins, namely fibrous and globular proteins. As the name implies, fibrous proteins are composed of individual, elongated, filamentous chains which are joined laterally by several types of cross-linkages to form a fairly stable, rather insoluble structure. Important examples are keratin, myosin, and collagen. The globular proteins, on the other hand, are relatively soluble and are quite compact due to the considerable amount of folding of the long polypeptide chain. Biologically active proteins such as antigens and enzymes are of the globular type.

Fibrous Proteins. Much information is now available concerning the detailed structures of fibrous proteins. Because of their simple, repeating structure, fibrous proteins were examined by x-ray diffraction patterns. From these and other studies, three subclasses of fibrous proteins, namely the keratins, silk fibers, and collagen were identified; each class possesses a characteristic type of conformational structure (*a*) the right-handed α helix, (*b*) the antiparallel and parallel β-pleated sheet, and (*c*) the triple helix, respectively (see Figure 4-11 for general structure).

Keratins. The α-keratins, which make up the proteins of fur, claws, hooves, and feathers, consist mostly of α-helical polypeptide chains. A typical α-keratin, the wool fiber, has been thoroughly studied by a number of physical techniques including x-ray diffraction and electron microscopy. The basic unit is right-handed α helix, three of which form a left-handed coil or a protofibril, which is stabilized by cross-linking disulfide bridges. Nine protofibrils group around two additional protofibrils to form a microfibril, about 80 Å across. Each microfibril in turn is imbedded with several hundred similar fibrils in an amorphous protein matrix to form a macrofibril. A number of macrofibrils make up a cell, and these in turn are oriented in an elongated, filamentous parallel manner to form the complete wool fiber. Thus, a wool fiber consists of a very large number of polypeptide chains held together by hydrogen bonding and cross-linking disulfide bridges embedded in an insoluble protein matrix.

When α-keratin is exposed to moist heat and stretched, it is converted to a different conformational form, namely β-keratin. The hydrogen bonds stabilizing the α-helical structure are broken under these conditions and an extended parallel β-pleated sheet conformation results.

Silk. Silk, however, possesses an entirely different unique structure, namely the antiparallel β-pleated sheet. Sequence studies of silk show a repeating sequence of six residues:

$$(\text{Gly-Ser-Gly-Ala-Gly-Ala})_n$$

x-Ray crystallography, in turn, reveals extended polypeptide chains stretched parallel to the fiber axis with the neighboring chains running in the opposite direction. This arrangement assures maximum hydrogen bonding by the peptide residues of one chain to the neighboring chains. Since glycine, with its α-methylene carbon, alternates with serine and alanine, it will lie on one

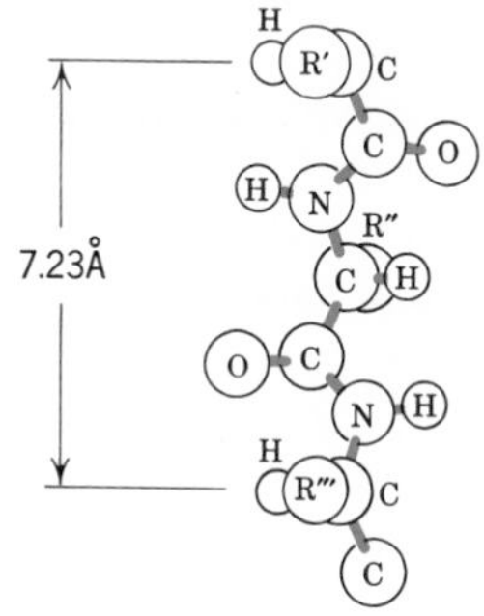

H R′ C
C O
H N R″
C H
O C
N H
H R‴ C
C

7.23Å

Extended chain

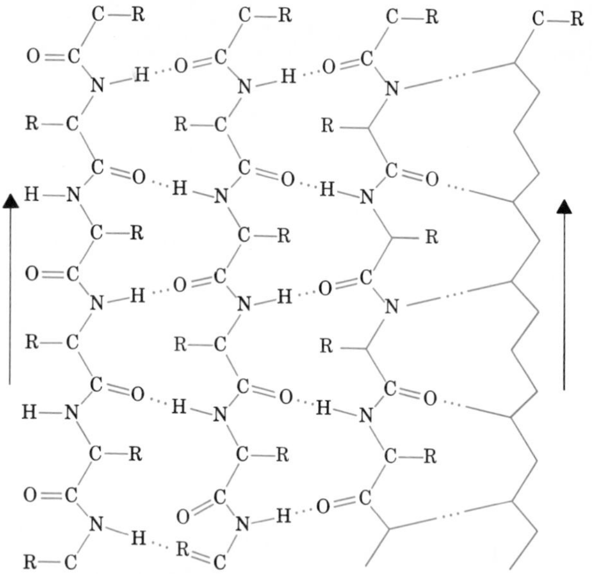

C—R C—R C—R C—R
O=C N—H···O=C N—H···O=C N—···
R—C R—C R—
C=O···H—N C=O···H—N C=O···
H—N C—R C—R —R
O=C N—H···O=C N—H···O=C N—···
R—C R—C R—
C=O···H—N C=O···H—N C=O···
H—N
C—R C—R C—R
O=C O=C N—H···O=C
N—H···R=C
R—C

Parallel chain β-pleated sheet

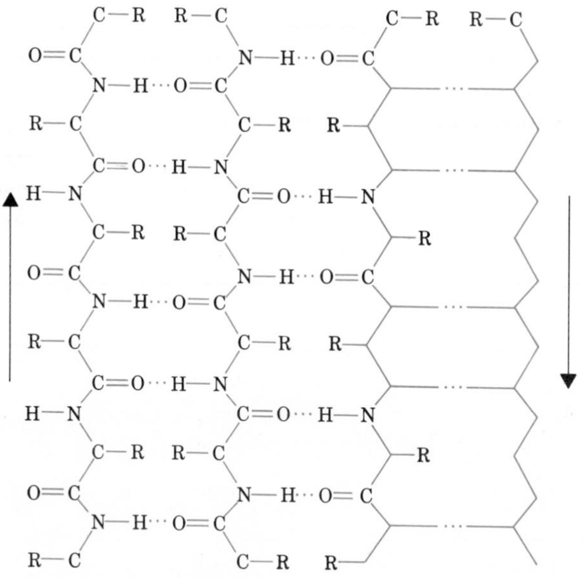

C—R R—C C—R R—C
O=C N—H···O=C
N—H···O=C
R—C C—R R—
C=O···H—N
H—N C=O···H—N
C—R R—C —R
O=C N—H···O=C
N—H···O=C
R—C C—R R—
C=O···H—N
H—N C=O···H—N
C—R R—C —R
O=C N—H···O=C
N—H···O=C
R—C C—R R—

Antiparallel β-pleated sheet

Triple helix of extended polypeptide chains typical of collagen

Figure 4-11

side of the sheet and alanine or serine will lie on the opposite side. Since silk contains no cysteinyl residues, no disulfide cross-linking occurs. The stability of this protein is therefore obtained by extensive hydrogen bonding.

Collagen. The third important structure is the triple helix structure exhibited by collagen, a protein found in skin, cartilage, and bone. Of remarkable tensile strength, collagen consists of parallel bundles of individual linear fibrils that are highly insoluble in water. The amino acid composition of collagen is rather unusual, being composed of 25% glycine and another 25% proline and hydroxyproline. Because of the high glycine and proline content, no α helix occurs. Each linear fibril is a cable consisting of three polypeptide chains. Each chain is twisted into a gentle left-handed helix, and the three chains are wrapped around each other to form an extremely strong, right-handed super helix rigidly held together by interchain hydrogen bonds (Figure 4-12).

Fibrous proteins therefore serve the *structural* needs of tissues by invoking the three basic manifestations of polypeptide chains, the α-helical, β-pleated, and triple-helix structures.

Globular Proteins. Globular proteins, the second major group of proteins, perform a multitude of functions. A typical globular protein can be described as an extensively folded and compact polypeptide chain with little if any room for molecules of water in its interior. In general, all the polar R groups of the amino acids are located on the outer surface and are hydrated, whereas nearly all the hydrophobic R groups are in the interior of the molecule. The polypeptide chain may have extensive α-helicity as in myoglobin, or it may have little as in cytochrome c and possess the β extended form of β-keratin. Figure 4-13 depicts the variation of α-helicity in a number of proteins. About ten proteins have been examined by detailed x-ray diffraction analyses.

Cytochrome c

We shall now examine cytochrome c in some detail, since much is known about the chemical structure of this important protein. Cytochrome c is a ubiquitious protein found in all aerobic organisms. Its sole function is to transport an electron from a donor of lower reduction–oxidation potential

Figure 4-12

Collagen, a triple-stranded helix, illustrates a typical fibrous protein.

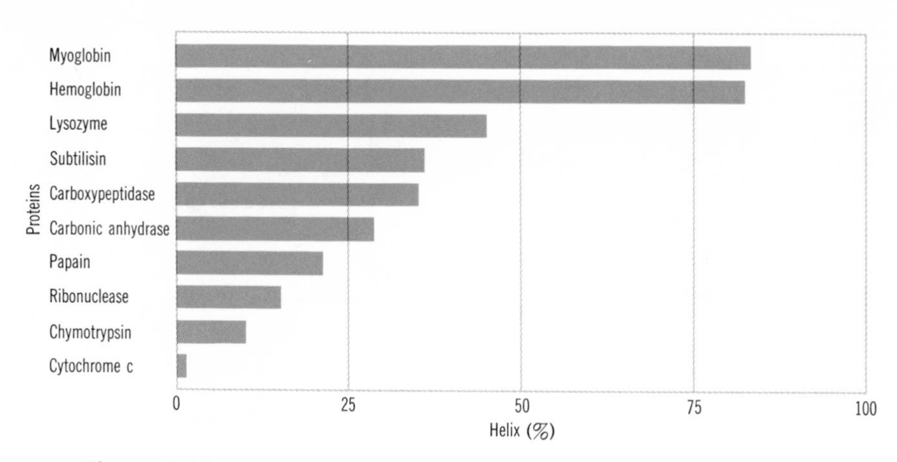

Figure 4-13

The helix content of different proteins.

to an acceptor of higher reduction–oxidation potential (see Chapter 14 for a discussion of redox pigments). Cytochrome c from over 35 species of animals, plants, and bacteria has been sequenced and found to contain only one heme and 104–115 amino acids in one continuous polypeptide chain with no disulfide bridges. The protein is basic. All cytochromes have in their structure the sequence Cys_{14}-x-x-Cys_{17}-His_{18} in which the two cysteine sulfhydryl groups are covalently linked to the heme by thioether bridges. In addition, the imidazole ring nitrogen atom of histidine-18 is coordinately bonded to one side of the heme iron as the fifth ligand. Methionine-80 extends its sulfur atom as the sixth iron ligand on the other side. These studies also reveal an important principle, namely, that widely separated amino acid residues may participate in binding a prosthetic group or even a substrate. Thus, residues 14,17,18, and 80 of cytochrome c are involved in bonding heme. We shall see later that widely separated amino acid residues in a protein can be juxtaposed by a specific conformation to form the active site of an enzyme.

The heme group is in a crevice in the center of the molecule allowing penetration of electron carriers. Surrounding the heme on all sides are closely packed hydrophobic R groups as depicted in Figure 4-14. The organizing component of the molecule is the centrally located heme group with one half of the cytochrome c molecule at the right of the crevice constructed from amino acids 1–46 while on the left the second half of the protein consists of amino acid residues 47–91. The final 13 amino acids (92–104) form the only α-helical region of the entire molecule.

Other interesting features of the molecule may be cited:

(1) Thirty-five amino acid residues are invariant in all species studied, but substitution of amino acids at other sites have occurred.
(2) The sequence of 11 amino acid residues from 70 to 80 is the longest invariant region in all cytochrome c's so far examined.
(3) Cytochrome c of vertebrates possesses *N*-acetyl glycine as the terminal

Figure 4-14(*a*)

A 4 Å low-resolution diagram of the horse heart ferricytochrome c molecule. The dark pink region depicts the heme molecule buried in the protein. (Reproduced with the permission of R. E. Dickerson.)

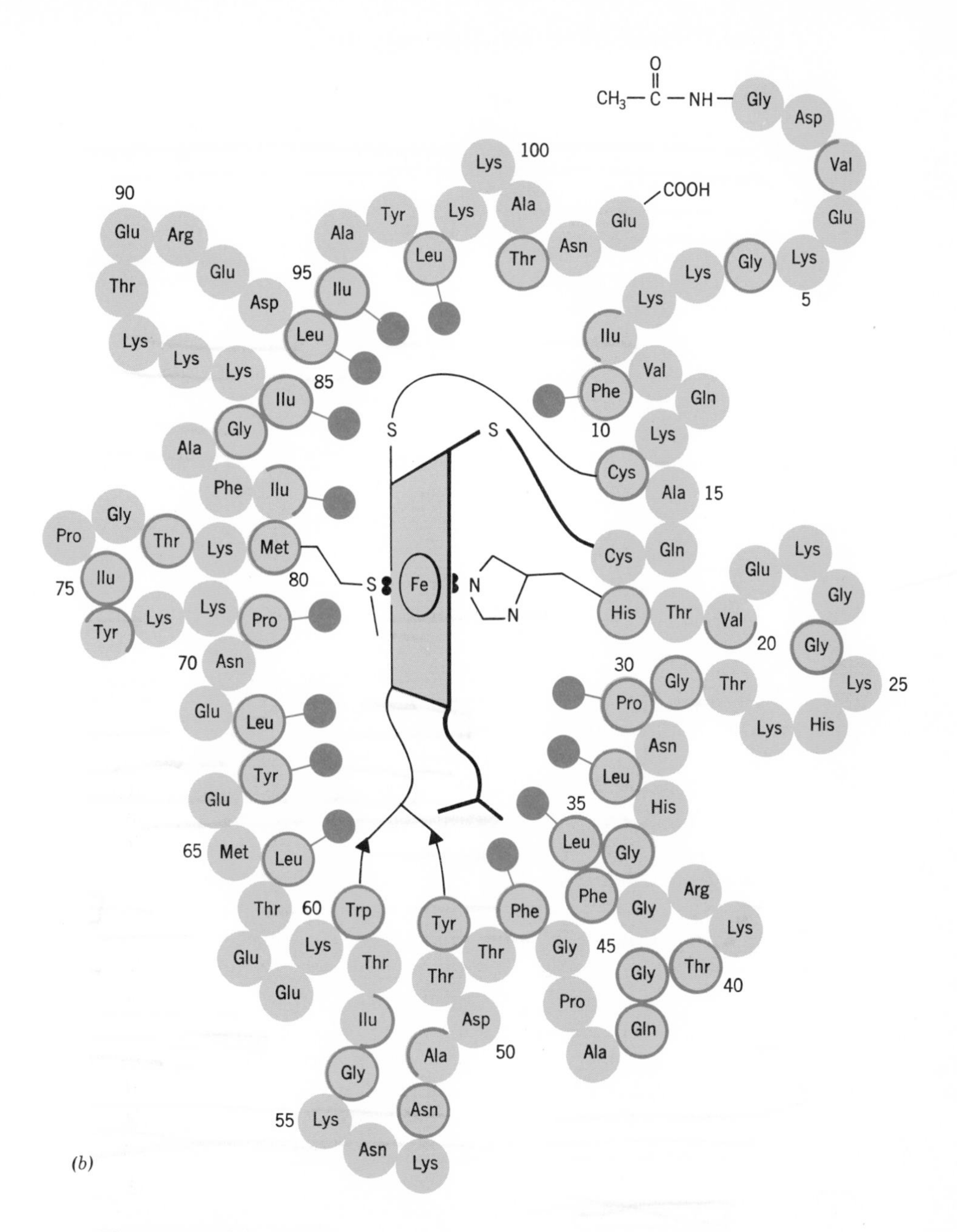

(b)

Figure 4-14(b)

A chain diagram representing the sequence of amino acids of horse heart cytochrome c. Light circles indicate amino acid residues on the outside of the molecule, and red outlined circles indicate the interior amino acid residues. Red dots mark those interior side chains of amino acids packed against the heme. Note the hydrophobicity of these side chains in contrast to the hydrophilicity of the exterior side chains. (Reproduced with the permission of R. E. Dickerson.)

group, and 103 additional amino acids. Cytochrome c from insects and fungi do not have an acetylated NH_2-terminal amino acid. Cytochrome c from higher plants has, at the amino terminus, *N*-acetyl alanine and eight additional residues.

(4) The protein undergoes considerable conformational change when the heme iron undergoes reduction and oxidation. Thus, crystals of oxidized cytochrome c shatter upon reduction, indicating very marked changes in conformation.

(5) All cytochrome c's can function equally well with any cytochrome oxidase preparation. Thus, the invariant amino acids residues completely fulfill the functional requirements of the molecule.

(6) The cytochrome c of the two primates monkey and man are almost identical; they differ by an average of 10 amino acids from other mammals such as the dog or the whale, by about 15 residues from cold blooded vertebrates, by about 30 residues from insects, and by about 50 residues from plants and procaryotes. In the eucaryotes, cytochrome c is always associated with the inner membranes of mitochondria; in procaryotes, the cytochrome c is localized in the plasma membrane.

Antibodies A large group of proteins which are classified as globulins are found in the blood plasma. Some of these proteins are produced in the spleen and lymphatic cells in response to a foreign substance called an *antigen*. The newly formed protein is called an *antibody* and specifically combines with the antigen which triggered its synthesis.

Immunoglobulins (antibodies) are classified into three major classes called IgG, IgA, and IgM. Each class has many subgroups which are produced by specific cells by normal individuals. It is almost impossible to purify these very heterogenous proteins for detailed chemical analysis. However, patients with multiple myeloma, a fatal bone cancer, produce only one type of immunoglobulin, either IgG or IgA. Since each patient makes his own unique immunoglobulin that differs in amino acid sequence from those of other patients suffering from the same disease, presumably a specific immunoglobulin-producing plasma cell—out of a population of many—propagates its unique globulin. In addition, a small protein, called a Bence–Jones protein, is eliminated into the urine of these patients. Patients with this disease therefore synthesize a specific immunoglobulin and a specific Bence–Jones protein which can be readily purified and sequenced. As a result of intensive investigation by a number of protein chemists, the following information has been obtained regarding the immunoglobulins: Each complete immunoglobulin is made up of two pairs of polypeptide chains; a pair of light (short) and a pair of heavy (long) chains. The Bence–Jones protein is identical to the light chain of the immunoglobulin. All light chains are divided into a region of variable amino acid sequence (VL) and a region of constant sequence (CL). The VL region occupies the NH_2-terminal half of the light chain (about 110 amino acid residues) and the CL region occupies the second half. The CL region is essentially identical in all human light chains. Heavy chains are also

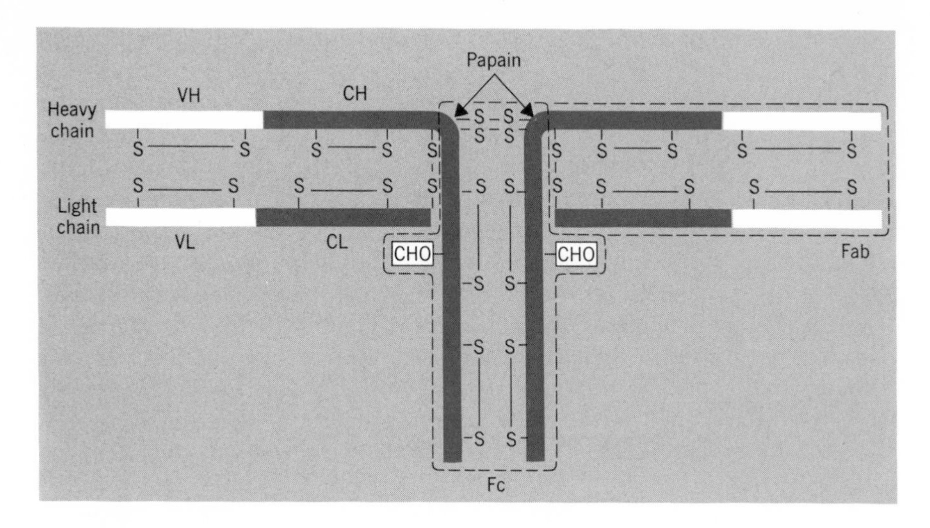

Figure 4-15

The structure of human GI immunoglobulin. The site of cleavage by papain to produce Fab and Fc fragments is shown by arrows. Interchain and intrachain disulfide bonds are indicated as are the approximate positions of carbohydrates residues (CHO) and the variable light (VL) and variable heavy (VH) and the constant light (CL) and the constant heavy (CH) regions.

divided into VH region and CH regions. The VH region is similar in size to that of light chains, but the CH region is about three times as long. A general structure can then be constructed; as shown in Figure 4-15. When the proteolytic enzyme papain is added to homogenous IgG, two fragments are obtained, a crystalline protein called Fc and a fragment called Fab, because this latter fragment combines specifically with its antigen. Since IgG has two antigen-combining sites, it is bivalent; that is, it combines with two antigen molecules. Further investigations in this highly important field will seek answers to such questions as explaining the amazing capacity of the organisms to respond in such a highly specific manner to foreign bodies, what triggers the synthesis of the antibody, and how excessive synthesis is regulated.

Blood Proteins

In addition to the immunoglobulins, three other types of proteins play extremely important roles in the maintenance of life in vertebrates. The first group are the albumins which comprise about 50% of the total plasma protein and possibly are involved in fatty acid binding and anion transport. They serve to control the osmotic pressure of the blood as well as maintain the buffering capacity of the blood pH. Serum albumin is approximately 67,000 in molecular weight and is a typical globular protein with low α-helical configuration and considerable tertiary structure.

The second important protein is fibrinogen. Accounting for about 4% of the total plasma protein, the protein plays an unusually vital role in blood clotting.

The third protein in blood with obvious implications is hemoglobin. The respiratory protein in all vertebrates, it is localized exclusively in erythrocytes. Reacting reversibly with molecular oxygen, it transports oxygen from the lungs to all parts of the body. Since hemoglobin is easily purified, a great deal of information is available concerning its structure and mechanism of action. In brief, it is a conjugated protein with four heme groups (see Chapter 16) for a discussion of porphyrins) and a protein called globin, which has a molecular weight of about 65,000. There is good evidence that the heme molecules are linked to globin via the imidazole nitrogen of the histidine residues in the globin molecule. Ferrohemoglobin, the reduced form of hemoglobin, combines reversibly with oxygen to form oxyferrohemoglobin according to the reaction

$$Hb_4 + 4\ O_2 \rightleftharpoons Hb_4(O_2)_4$$

Other conjugated proteins of considerable importance in the living cell include myoglobin, the cytochromes, and nucleoproteins (see Chapter 19 concerning abnormal hemoglobins).

Hormones These polypeptides and small proteins, found in relatively low concentrations in animal tissues, play a poorly defined but highly important role in maintaining order in a complicated complex of metabolic reactions. Included in this category are the small posterior pituitary hormones oxytocin (9 amino acids), and vasopressin (9), the larger protein glucagon (29), adrenocorticotrophic hormone (39), and the large molecule of insulin (51). The biosynthesis of insulin is discussed in some detail in Chapter 19.

Enzymes These extremely important biological catalysts are described in detail in Chapter 8 and are referred to throughout this book.

Nutrient Proteins An important function often neglected in discussing proteins is their role as a source of the essential amino acids required by man and other animals. These essential amino acids are readily synthesized by plants, but must be acquired by man, usually as proteins, in his diet (see Chapter 16).

The term protein is justifiably derived from the Greek noun *proteios* meaning "holding first place." This all too brief survey merely touches the surface of the important roles proteins play in the living cell and gives strong support to the meaning of the term protein, first suggested by Berzelius in 1838.

A brief discussion concerning the purification and characterization of proteins and the application of ultracentrifugal and electrophoretic procedures is presented in Appendix 2.

References

1. I. H. Segel. *Biochemical Calculations*. New York: Wiley, 1968.
The acid–base equilibria of amino acids and their derivatives is thoroughly and

lucidly discussed in this reference. Many typical problems are presented and their solutions discussed.

2. A. Meister. *Biochemistry of the Amino Acids*. 2nd ed., Vols. I and II. New York: Academic Press, 1965.
   This treatise is the authoritative reference on the chemistry and biochemistry of the more common amino acids found in nature.
3. L. Fowden. "The Non-Protein Amino Acids of Plants," in *Progress in Phytochemistry*, L. Reinhold and Y. Liwschitz, eds. Vol. 2. London: Interscience, 1970.
   A recent review of the nature and occurrence of these interesting compounds.
4. R. E. Dickerson and I. Geis. *The Structure and Action of Proteins*. New York: Harper & Row, 1969.
   A superb description of the structure and function of proteins, with excellent illustrations. A *must* for the inquisitive student.

Nucleic Acids and Their Components

Introduction

The nucleic acids have been the subject of biochemical investigations almost from the time they were first isolated from cell nuclei about 100 years ago. Nucleic acids occur in every living cell, where they are not only responsible for the storage and transmission of genetic information, but also translate this information for a precise synthesis of proteins characteristic of the individual cell. Like the proteins, nucleic acids are biopolymers of high molecular weight, but their repeating unit is the mononucleotide rather than an amino acid.

There are two kinds of nucleic acids, deoxyribonucleic acid (DNA) and ribonucleic acid (RNA). Their basic structures, consisting of chains of alternating phosphoric acid and sugar residues, are shown in Figure 5-1. In RNA the sugar is D-ribofuranose; in DNA, as its name implies, the sugar is 2-deoxy-D-ribofuranose:

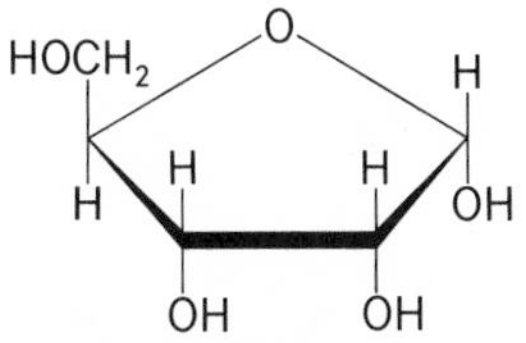

α-D-Ribose

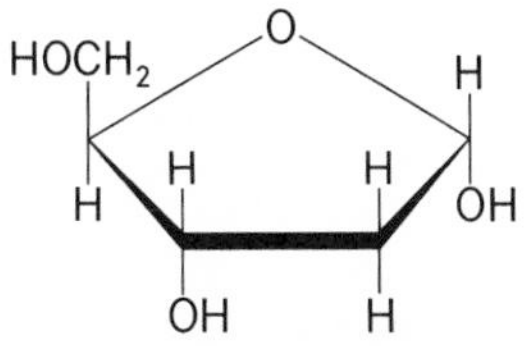

α-2-Deoxy-D-ribose

Attached to every sugar unit is the third component of the nucleic acids, a nitrogen-containing base which is either a substituted purine or a substituted pyrimidine. It is the sequence of bases on the long sugar–phosphate chains which determine the biological properties of the molecule.

Purines and Pyrimidines

Both RNA and DNA contain the two purines adenine and guanine. The general structure of a purine and the specific structures of adenine and

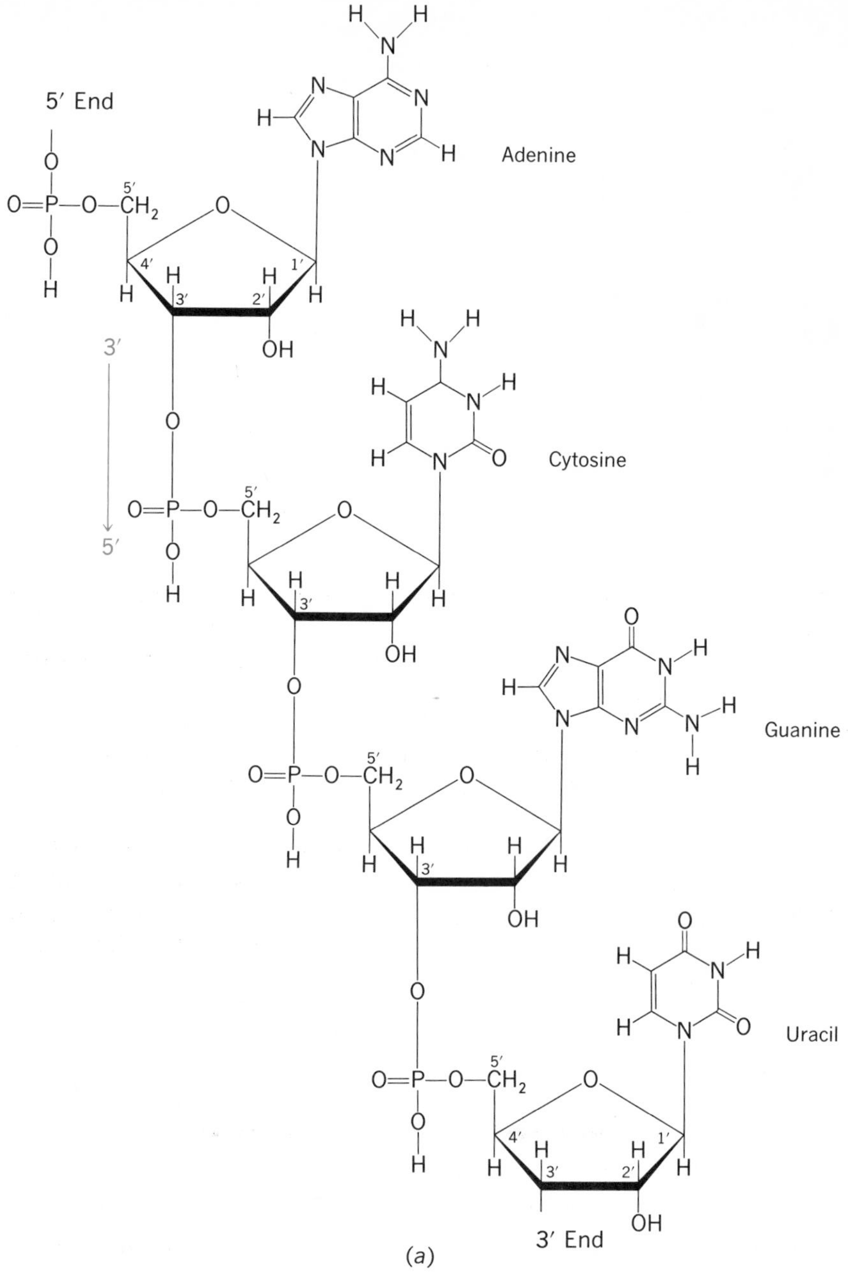

(*a*)

Figure 5-1

Structures of (*a*) a polyribonucleotide and (*b*) a polydeoxyribonucleotide, illustrating the basic skeleton of nucleosides joined by phosphodiester bridges. The terminology 3′ ⟶ 5′ phosphate diester bridging should be noted.

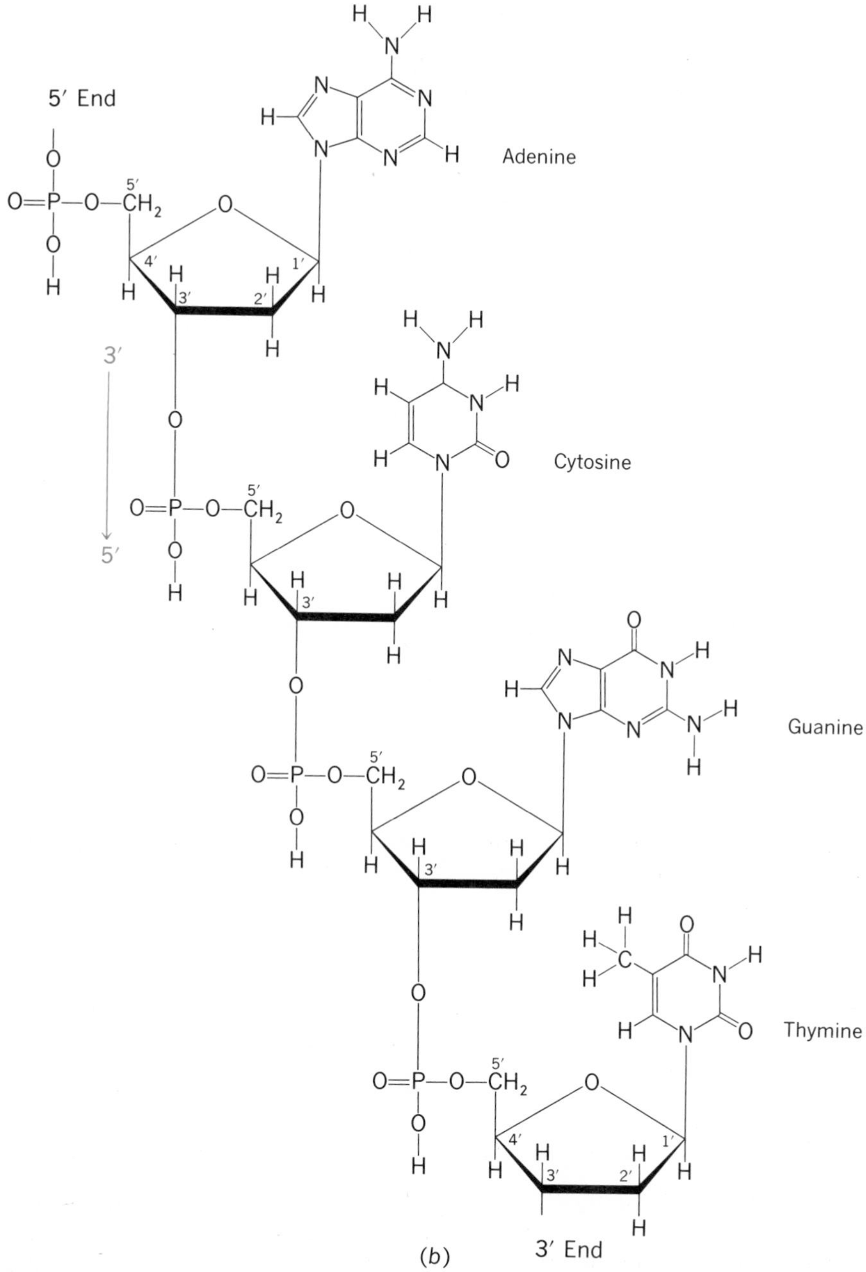

5′ End

Adenine

Cytosine

Guanine

Thymine

3′ End

(*b*)

guanine are given here, as is the numbering of the atoms in the purine. Several unusual bases have been found in the transfer RNA's. These include hypoxanthine, 1-methyl hypoxanthine, N^2-dimethyl guanine, 1-methyl guanine, N^6-(Δ^2-isopentenyl) adenine and threonylcarbamoyl adenine. Both RNA

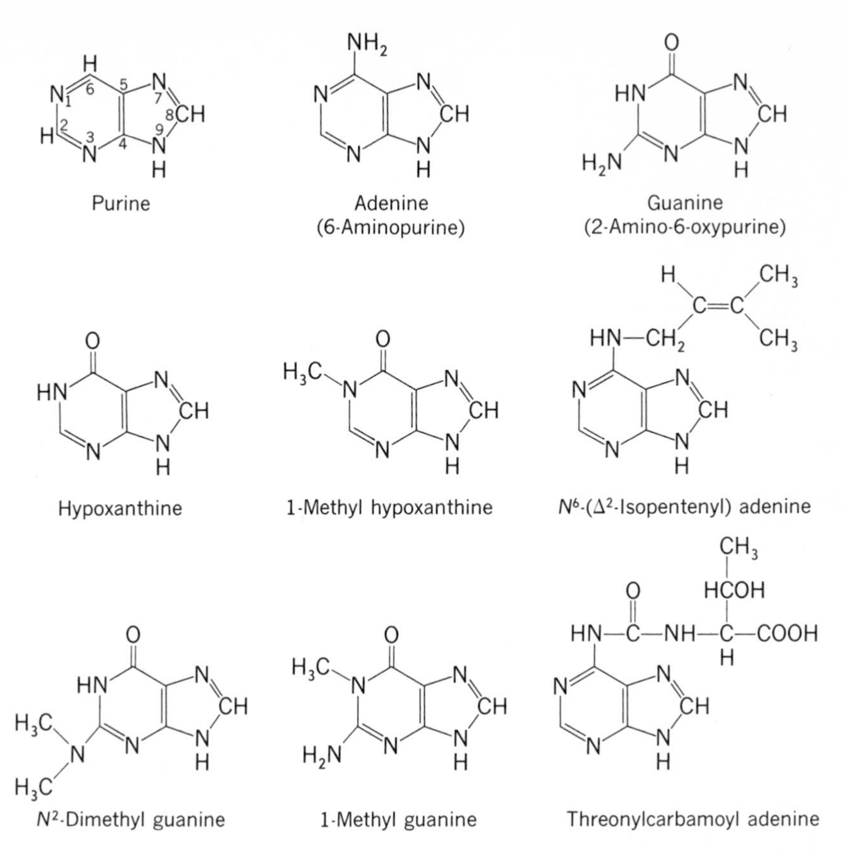

and DNA also contain the pyrimidine, cytosine, but the two kinds of nucleic acids differ in the fourth nitrogenous base: RNA contains uracil, whereas DNA contains thymine.

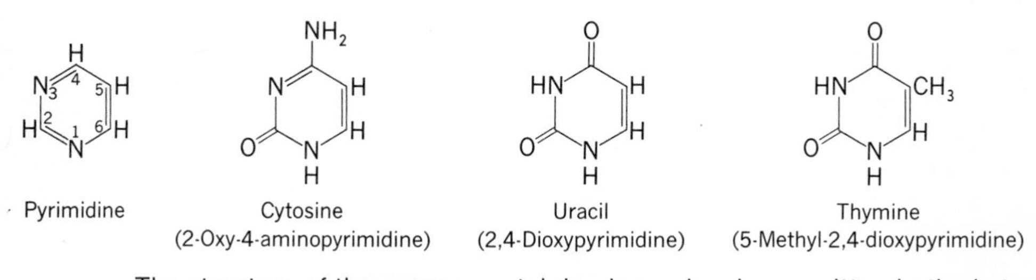

The structure of the oxygen-containing bases has been written in the keto (or lactam) form. It should be emphasized that there is an equilibrium between the keto and the enol (or lactim) forms which is dependent on the pH of the environment. It is the lactam form which predominates at physiological pH.

Lactam $\rightleftharpoons$ Lactim

In recent years other pyrimidines have been detected in purified samples of DNA; 5-methyl cytosine occurs in the DNA isolated from wheat germ and other plant sources. It has also been detected in trace amounts in the DNA from thymus gland and other mammalian sources. Cytosine is replaced by 5-hydroxymethyl cytosine in the DNA of certain bacterial viruses, namely, the T-even bacteriophages which infect *E. coli*. In tRNA are found the pyrimidine derivatives dihydrouracil, pseudouridine, and 4-thiouracil.

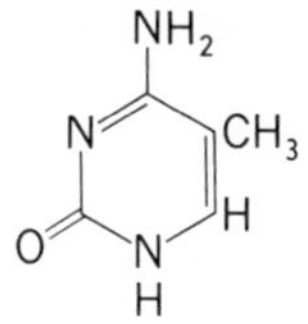

5-Methyl cytosine 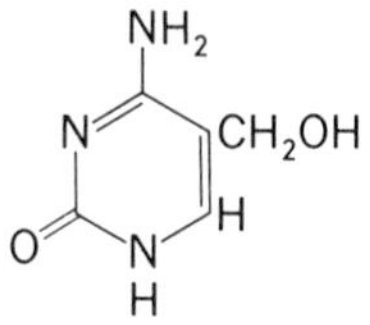 5-Hydroxymethyl cytosine 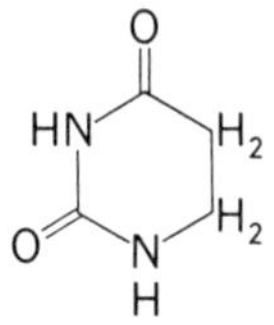 Dihydrouracil 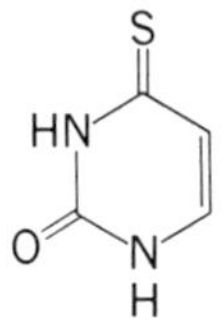 4-Thiouracil

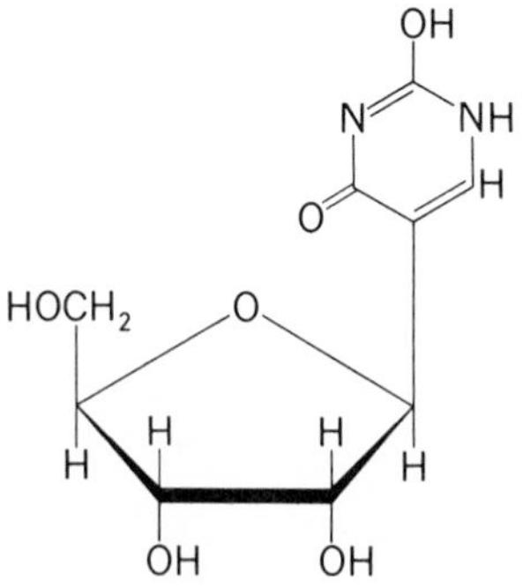

Pseudouridine

Nucleosides

The nucleosides are compounds in which purines and pyrimidines are linked to D-ribofuranose or 2-deoxy-D-ribofuranose in a *N*-β-glycosidic bond, which is the configuration in the polymeric nucleic acids. The point of attachment of the base to the sugar is the hemiacetal hydroxyl on the C-1′ carbon atom of the sugar. In the purines, it is the N-9 nitrogen atom which participates in the *N*-glycosyl bond. In the pyrimidines, the N-1 nitrogen atom is the point of attachment. Note that the carbon atoms of the sugar are designated by prime numbers (i.e., C-1′, C-5′), while the atoms in the bases lack the prime sign.

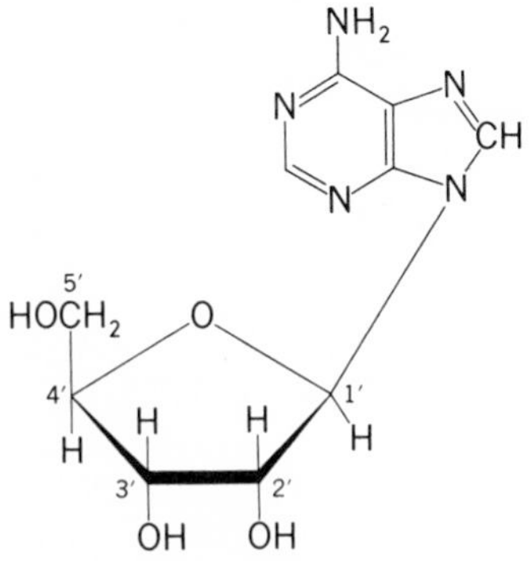

Adenosine (9-β-D-Ribofuranosyl adenine)

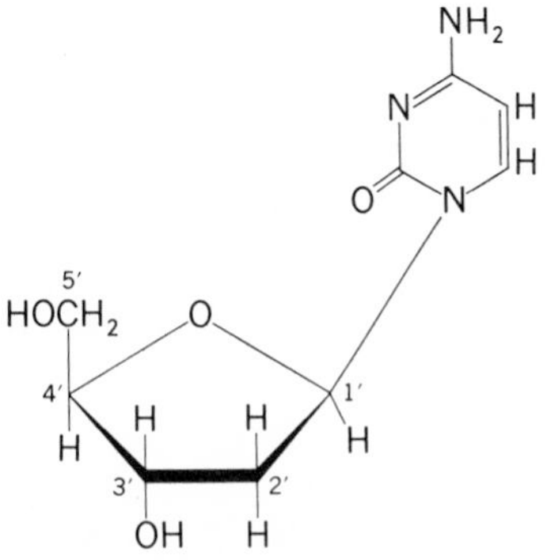

2′-Deoxycytidine (1-β-2′-Deoxy-D-ribofuranosyl cytosine)

Table 5-1
Names of Nucleosides

Base	Ribonucleoside	Deoxyribonucleoside
Adenine	Adenosine	2′-Deoxyadenosine
Guanine	Guanosine	2′-Deoxyguanosine
Uracil	Uridine	2′-Deoxyuridine
Cytosine	Cytidine	2′-Deoxycytidine
Thymine	Thymine ribonucleoside	2′-Deoxythymidine

Table 5-1 lists the trivial names of the purine and pyrimidine nucleosides which are related to the bases that occur in RNA and DNA.

Nucleotides

Nucleotides are phosphoric acid esters of the nucleosides just described. The ribose portion of a ribonucleoside has three positions (the 2′, 3′, and 5′ hydroxyl group) where the phosphate could be esterified, whereas the 2′-deoxyribonucleoside has only the 3′ and 5′ positions available. As will be seen, all these can be formed on partial hydrolysis of nucleic acids by various methods. In addition, the 5′ phosphates occur as cellular components.

One of the most important naturally occurring nucleotides is adenosine-5′-monophosphate (also called 5′-adenylic acid). This compound (AMP), together with two of its derivatives, adenosine-5′-diphosphate (ADP) and adenosine-5′-triphosphate (ATP), plays an extremely important role in the conservation and utilization of energy released during cellular metabolism. As we shall see, the physiological significance of these compounds rests in their ability to donate and accept phosphate groups in biochemical reactions.

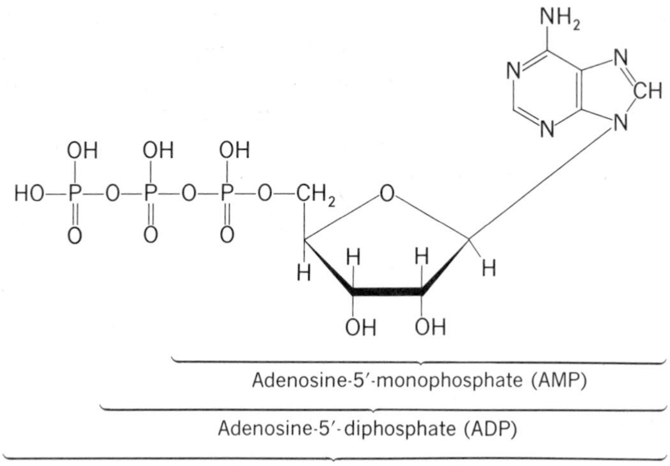

Adenosine-5′-monophosphate (AMP)

Adenosine-5′-diphosphate (ADP)

Adenosine-5′-triphosphate (ATP)

A cyclic 3′,5′-phosphate of adenosine occurs and has important regulatory properties (see Chapter 20). Mild acid hydrolysis of DNA has yielded 3′,5′-diphosphate derivatives of deoxythymidine and deoxycytidine.

Adenosine-3′,5′-monophosphate
(Cyclic adenylic acid)

Deoxythymidine-3′,5′-diphosphate

Nucleotide 5′-Diphosphates and 5′-Triphosphates. The mono-, di-, and triphosphates of adenosine have already been described. Corresponding derivatives of guanosine, cytidine, and uridine as well as deoxyadenosine, deoxyguanosine, deoxycytidine, and deoxythymidine exist and play important roles in cellular metabolism (Table 5-2). For example, the nucleoside 5′-triphosphates serve as the precursors for the synthesis of RNA and DNA (Chapter 18). Derivatives of certain nucleoside 5′-diphosphates act as coenzymes to supply sugar residues in certain reactions; other derivatives of adenosine-5′-diphosphate function in oxidation–reduction reactions. Thus, uridine-5′-diphosphate linked to glucose serves as a glucose donor (page 265). Adenosine-5′-diphosphate linked to nicotinamide forms the extremely important oxidation–reduction coenzyme, nicotinamide adenine dinucleotide, NAD^+ (page 199).

DNA and RNA

In procaryotic cells, DNA normally occurs as a highly twisted, double-stranded circle, in part associated with the inner side of the plasma membrane but

Table 5-2

The Common Ribonucleotides and 2′-Deoxyribonucleotides[a]

Ribonucleotides	2′-Deoxyribonucleotides
Adenosine-5′-monophosphate (Adenylic acid; AMP)	Deoxyadenosine-5′-monophosphate (Deoxyadenylic acid; *d*AMP)
Guanosine-5′-monophosphate (Guanylic acid; GMP)	Deoxyguanosine-5′-monophosphate (Deoxyguanylic acid; *d*GMP)
Cytidine-5′-monophosphate (Cytidylic acid; CMP)	Deoxycytidine-5′-monophosphate (Deoxycytidylic acid; *d*CMP)
Uridine-5′-monophosphate (Uridylic acid; UMP)	Deoxythymidine-5′-monophosphate (Deoxythymidylic acid; *d*TMP)

[a] Each of the 5′-monophosphates exists as the 5′-diphosphate and 5′-triphosphate. Thus, as an example, there occurs GMP, GDP, GTP, *d*GMP, *d*GDP, and *d*GTP.

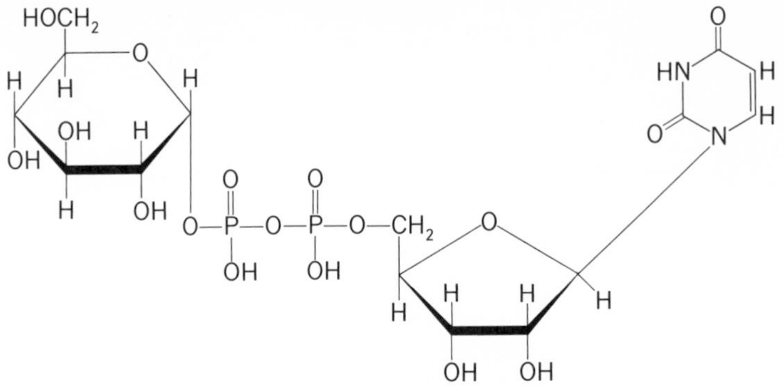

Uridine diphosphate glucose
(UDPG)

free of protein complexes. In contrast, over 98% of the total DNA in a typically differentiated eucaryotic cell is found in the nucleus as a highly twisted, double-stranded polymer bound to basic proteins called histones; the complex is known as chromatin. Much smaller amounts of DNA are always found in the matrix of eucaryotic mitochondria and in chloroplasts as small, double-stranded circles free of protein complexes.

The second nucleic acid component of the cell, RNA, occurs in multiple forms, each serving as extremely important informational links between DNA, the master carrier of information, and proteins. The smallest of these polymers is called *transfer RNA* (*tRNA*) and has a molecular weight of about 25,000. Transfer RNA consists of about 60 different molecular species. The tRNA's serve a number of functions, the most important of which is to act as specific carriers of activated amino acids to specific sites on the protein-synthesizing templates. The tRNA's comprise about 10–15% of the total RNA of the cell. A second group of RNA's include the *ribosomal RNA*'s (*rRNA*'s). These nucleic acids are always associated with a large number of proteins in a highly ordered complex called the ribosome. They make up about 75–80% of the total RNA of a cell. The third important group of RNA's are the *messenger RNA*'s (*mRNA*'s) which comprise about 5–10% of the total RNA of a cell. In bacterial cells mRNA's are highly unstable, in the sense that they are constantly being degraded and resynthesized. In eucaryotic cells the turnover rate is much lower. These nucleic acids with a base composition corresponding very closely to that of DNA are intimately involved in the transcription and translation of information programmed by DNA for the synthesis of proteins.

Chemistry of the Nucleic Acids

Isolation. In the presence of concentrated phenol and a detergent, a cell homogenate will form two liquid phases. Proteins are denatured and become insoluble in the aqueous phase, while the nucleic acids remain soluble in that phase. The aqueous phase can be readily separated from the phenol-rich phase in which some proteins have been dissolved. Addition of ethanol to the aqueous phase precipitates out the nucleic acids and many polysaccharides while the residual phenol remains in solution. The mixture of DNA and RNA can then be further treated either with a ribonuclease to degrade RNA into soluble fragments but leave the DNA intact, or alternately, the mixture can be treated with deoxyribonuclease to split the DNA, leaving RNA undegraded. After digestion of one of the nucleic acids, aqueous phenol can again be added to denature and remove any remaining protein, and the intact nucleic acid is then precipitated with ethanol. Since native DNA consists of an extremely long helical coil, addition of ethanol to a DNA solution results in the formation of long, fibrous precipitates which can be readily removed by winding the fibrous material around a stirring rod. The mass can then be dried with appropriate solvents such as acetone, and the dry DNA can be removed from the glass rod.

When the procedure is used to isolate RNA, one obtains a heterogenous mixture of tRNA, mRNA, rRNA, and degraded RNA. Either column chromatography of this mixture on columns of methylated albumin coated on Kieselguhr (MAK columns) or gradient centrifugation in sucrose solution will usually yield three fractions, the 4S (tRNA), and the 16S and 32S peaks from *E. coli* RNA which are derived in turn from 30S and 50S ribosomes or the 18–22S and 28–34S peaks from mammalian RNA (see Chapters 18–19 for description of these terms). Further purification of DNA can be carried out by chromatography on hydroxylapatite (calcium phosphate), which will yield two fractions, one containing single-stranded and the other double-stranded DNA.

Optical Properties of Nucleic Acids

The purine and pyrimidine bases found in the nucleic acids strongly absorb ultraviolet radiation of wavelength at 260 nm. This property is employed extensively in the quantitative determinations of the bases, their nucleotides, or the nucleic acids themselves (Figure 5-2). However, high-molecular-weight DNA typically has an optical density at 260 nm which is about 35–40% *less* than the optical density expected from adding up the individual absorbancies of bases in the DNA. This phenomenon is called the *hypochromic effect* and is explained by the fact that, in a helical structure (as shown in Figure 5-7, below), the bases are closely stacked one on top of the other. Interaction of the π electrons between the bases then results in a decrease in absorbancy. When this interaction is not possible, as for example in a random coil arrangement of the sugar–phosphate chain, the absorbance at 260 nm approaches the expected value. This optical property therefore is extremely useful in calculating the degree of helicity of DNA.

When highly polymerized double-stranded DNA polymers are slowly heated, the double helix "melts"; the transition from double strands to a random coil configuration occurs over a range of a few degrees. This transition from a helix to a coil results in an increase in absorbance. The midpoint temperature, T_m, of this process is the melting temperature of the helix of a specific DNA polymer (Figure 5-3). On slow cooling, an annealing occurs with an at least partial return of the coil to the helix configuration. Annealing conditions have now been developed under which the reformed helices resemble the original helices in detail. For example, the two strands have an antiparallel relationship to each other, and the bases are paired A to T and G to C (Scheme

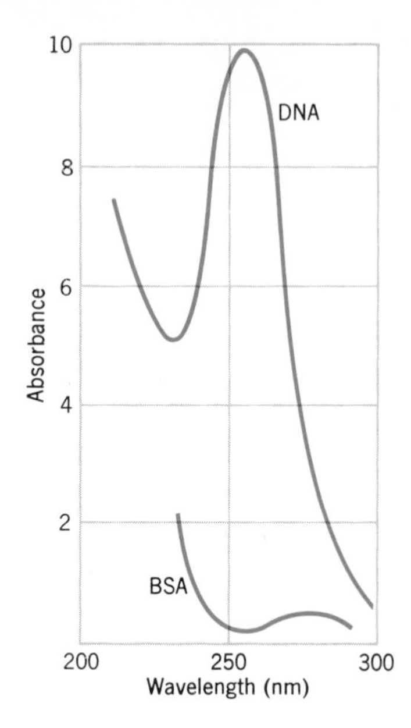

Figure 5-2
Ultraviolet absorption spectra of equal concentrations of the sodium salt of DNA and bovine serum albumin (BSA), depicting the marked difference in absorbance of the two polymers.

5-2). This specificity of pairing depends on the correct alignment of hydrogen bonds between the bases, three hydrogen bonds per G–C pair and two per A–T pair (Scheme 5-2). Since most of the nucleotide sequences in different regions of the DNA from one organism are unique, the reformed helices should have their two strands representing the same (and unique) regions of the original DNA. This has been shown to be the case and has been used to assess the relatedness of different organisms by observing the extent to which "hybrid" DNA molecules can be formed. The mere existence of the ability of DNA helices to reform (called DNA renaturation) has important implications in modern biology. Since the renaturation process is spontaneous in the test tube, the product of the renaturation reaction—the double stranded DNA—must be inherently stable and would require no additional energy or structure to maintain it in the cell. The G–C base pairs contribute more to the stability of DNA than do A–T pairs, and the T_m's of different DNA's increase linearly as a function of the percentage of G–C base pairs (Figure 5-4).

Double-stranded RNA (with G–C and A–U base pairing) is the genetic material in a few viruses, but most RNA is single-stranded, with short double-stranded regions formed by folding of the sugar–phosphate chain back

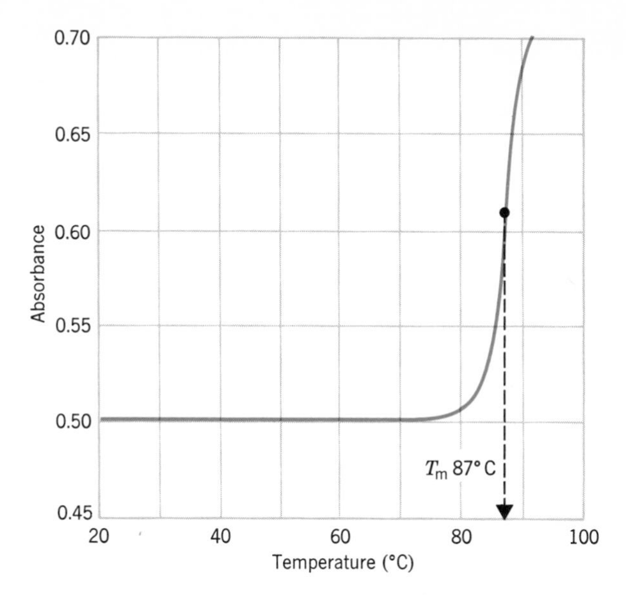

Figure 5-3

A typical thermal desaturation curve of calf thymus DNA showing disappearance of hypochromic effect with increasing temperature and the evaluation of T_m of the nucleic acid.

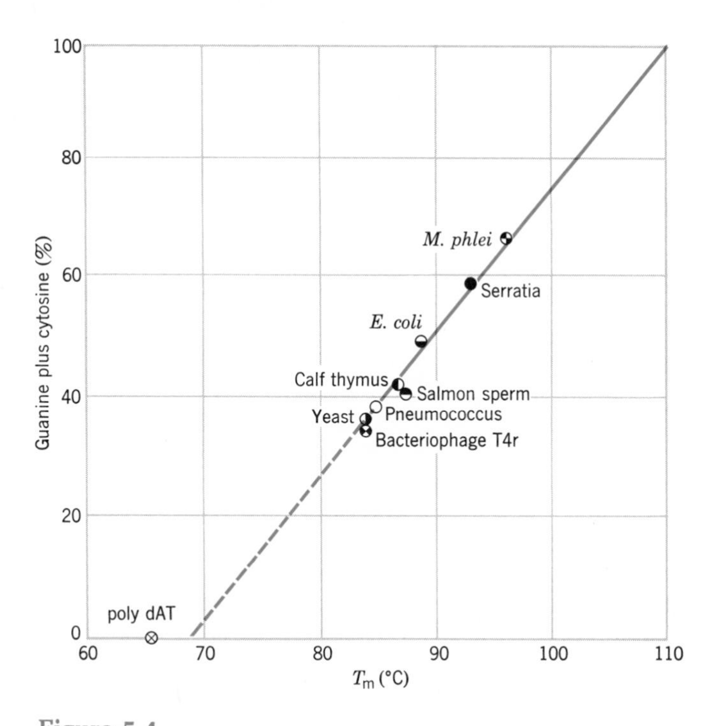

Figure 5-4

Dependence on the T_m of various samples of DNA as a function of the G–C pair content of the sample.

upon itself (see the discussion of the tRNA structure below). The melting profile of double-stranded RNA has a sharp rise, like that of double-stranded DNA. However, typical "single-stranded" RNA's show a smaller increase in absorbancy with a rise in temperature, and the melting transition is not sharp. "Single-stranded" DNA is also the genetic material of a few viruses, and it resembles "single-stranded" RNA in its properties.

"The melting" of DNA can also be accomplished with pH extremes. Nucleic acids are polyelectrolytes with one negative charge per nucleotide residue (due to the ionization of the phosphate diester) in the pH range 4–11. However, titration of a DNA solution to pH values below 4 or above 11 gradually weakens and then abruptly destroys the double-stranded structure. This behavior is explained as follows: At the low pH values, the amino groups of adenine, guanine, and cytosine are protonated with consequent disruption of the hydrogen bonding system; above pH 11, the protons of the hydroxyl groups of guanine, cytosine, and thymine (keto–enol tautomers) dissociate also with disruption of the hydrogen bonding.

Determination of Mole Ratios of Bases in Nucleic Acid

A fundamental characteristic of a nucleic acid is the composition and sequence of purine and pyrimidine bases. Compositional data are used to calculate the equivalence between A and U and between C and G in RNA; and between A and T and C and G in DNA. Similarly, the base composition and equivalence between bases can be related to the helicity of a nucleic acid as well as other physical properties of the polymers. Therefore, much effort has been spent on degrading the nucleic acid by chemical and enzymatic procedures. Dilute alkali readily hydrolyzes RNA; for example, 0.1–1*N* NaOH at room temperature for 24 hr. Under these conditions the RNA is degraded to a mixture of 2′- and 3′-nucleoside monophosphates. This important reaction is readily explained by the participation of the negatively charged 2′-alkoxide ion in a nucleophilic attack on the positively charged phosphorus atom of the phosphodiester bond thereby displacing the 5′-ribose ester and cleaving the RNA molecule (Scheme 5-1). The cyclic 2′,3′-phosphate ester is hydrolyzed to a mixture of the 2′- and 3′-monoesters by further alkaline action. These esters can then be separated quantitatively by ion exchange or paper chromatography and the separate fractions estimated spectrophotometrically for their nucleotide content. Mild conditions of hydrolysis are necessary since at higher temperatures cytidylic acid is partially converted to uridylic acid by deamination. Acid hydrolysis leads to cleavage of the *N*-glycosidic bonds and hence is not frequently employed.

The absence in DNA of the hydroxyl group at C-2′ prevents formation of the cyclic phosphate ester, and therefore DNA is not hydrolyzed by alkali. Under acid conditions the *N*-glycosidic bonds between the purine bases and deoxyribose, being the most labile bonds in the molecule, are cleaved, and apurinic acid is formed. This polymer of high molecular weight retains the sugar–phosphate backbone of the DNA but is devoid of adenine and guanine. The reaction is of some value, since by the determination of the A/G ratio, the ratio of all the four common DNA bases can be determined. However,

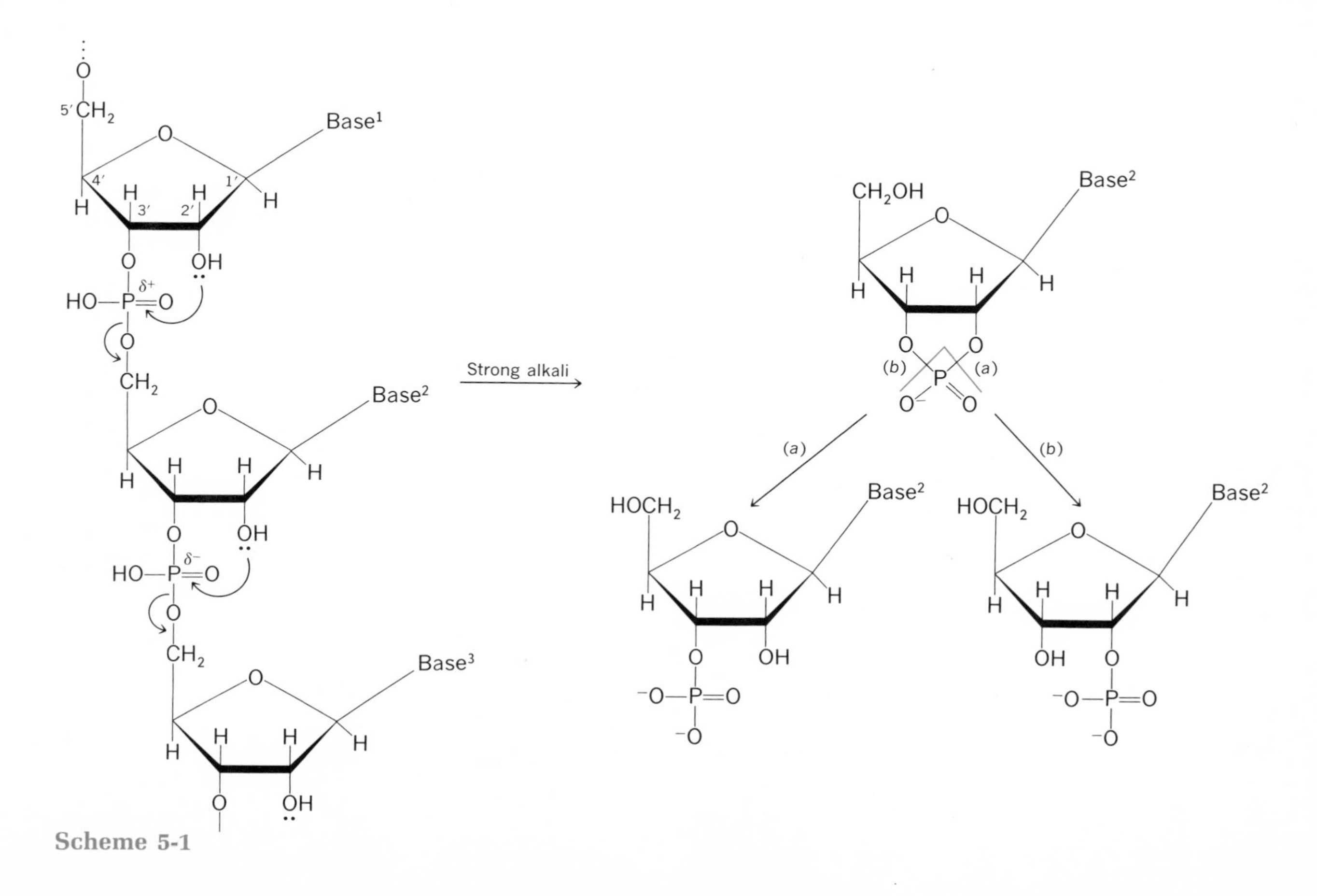

Scheme 5-1

to obtain direct evidence of the base composition, the combined action of pancreatic deoxyribonuclease (an endonuclease) and venom phosphodiesterase (a general diesterase), which splits phosphodiester bridges starting at the 3′-OH end of the chain, is employed. Deoxyribonucleoside-5′-monophosphates are formed which can then be separated by the same techniques employed in the separation of hydrolysis products of RNA. Figure 5-5*a* summarizes the action of a number of nucleases on DNA.

The action of a pure ribonuclease isolated from the pancreas on RNA is similar to that of alkali in that a 2′,3′-diester is transiently formed. In the presence of the enzyme, however, only the C-2′–phosphate bond (C-2′–P bond) of the cyclic diester is cleaved to yield the 3′-phosphates of the nucleosides. However, not every C-5′–P linkage in RNA is attacked by ribonuclease; only those diester bonds, in which the C-3′ bond is linked to a pyrimidine, are cleaved. The T_1-ribonuclease from the mold, *Aspergillus oryzae,* attacks the ribonucleic acid chain only at guanine residues, releasing the 5′-hydroxyl group of the ribose. Employing these two ribonucleases and suitable exonucleases as indicated in Figure 5-5*b*, the biochemist can degrade a ribonucleic acid under controlled conditions and elucidate its base sequence.

Structure of RNA

As indicated previously, there are three general types of RNA's; the present knowledge of each type will be briefly described.

Transfer RNA's. Since the hydrodynamic properties of all the tRNA's are very similar, it follows that molecular weights of all are in the area of 25,000 with a corresponding 4.3S value. Additional evidence suggests that 60–70% of tRNA exists as a helical structure. This and other evidence points to a "cloverleaf" structure in all tRNA's with the anticodon (i.e., the nucleotide triplet necessary for the positioning of the specific tRNA in the mRNA template during protein synthesis) located in the central petal of the cloverleaf (Figure 5-6). Pseudouridine and a number of methylated bases are characteristic minor bases always observed in tRNA's. We shall discuss in more detail the structure of tRNA in Chapter 19.

Ribosomal RNA. Several species of RNA occur in procaryotic and eucaryotic ribosomes, and these are summarized in Table 5-3. Ribosomal RNA (rRNA) has helical structure resulting from a folding back of a single-stranded polymer at areas where hydrogen bonding is possible because of short lengths of complementary structure. However, rRNA does not occur as a double-

Table 5-3
Ribosomes and Their RNA's

Procaryotic ribosomes	rRNA
30S	16S
50S	5S, 23S
Eucaryotic ribosomes	
40S	18S
60S	5S, 28S

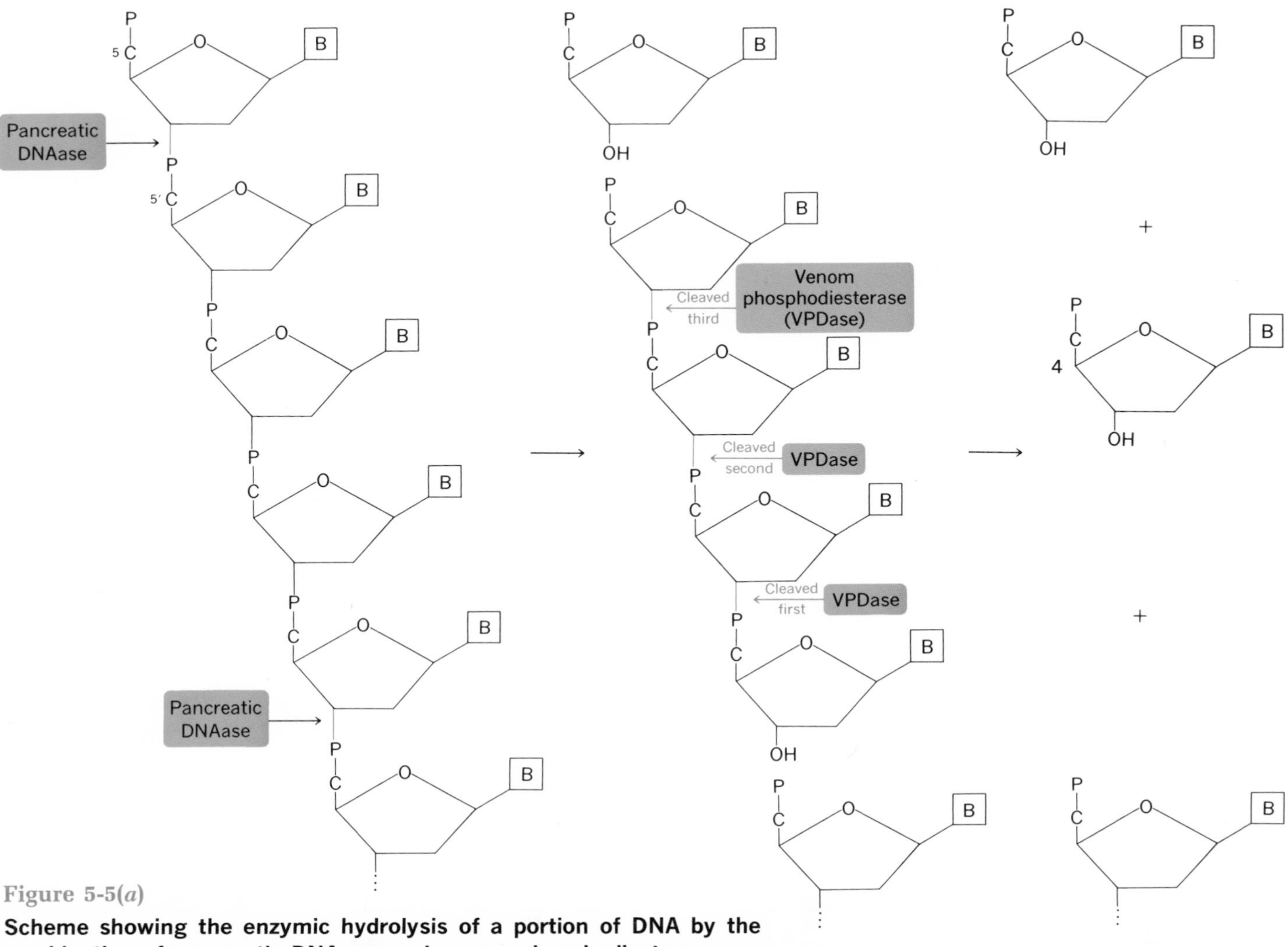

Figure 5-5(*a*)

Scheme showing the enzymic hydrolysis of a portion of DNA by the combination of pancreatic DNA-ase and venom phosphodiesterase.

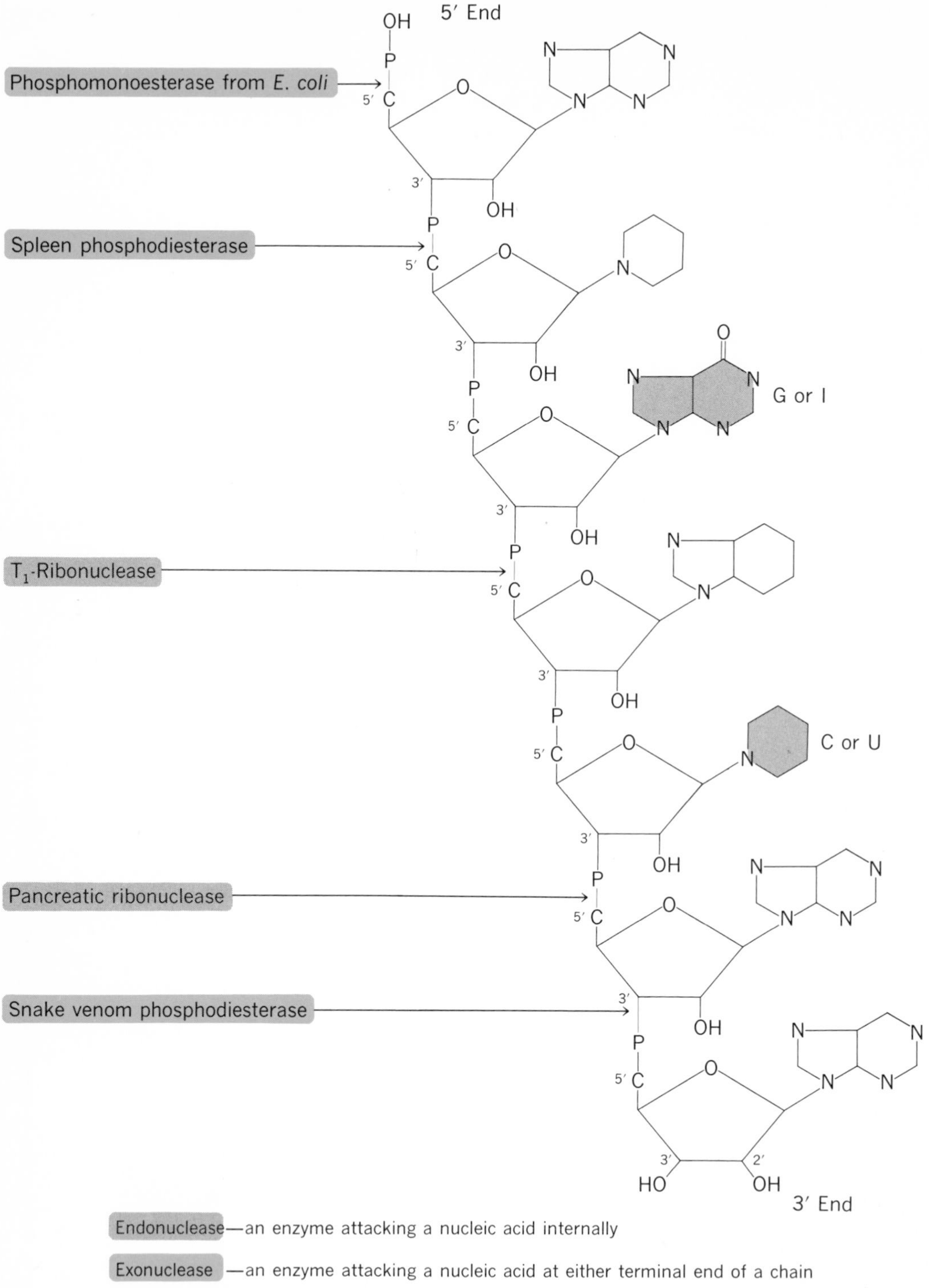

5′ End

Phosphomonoesterase from *E. coli*

Spleen phosphodiesterase

G or I

T_1-Ribonuclease

C or U

Pancreatic ribonuclease

Snake venom phosphodiesterase

3′ End

Endonuclease—an enzyme attacking a nucleic acid internally

Exonuclease —an enzyme attacking a nucleic acid at either terminal end of a chain

Figure 5-5(*b*)

Scheme showing the enzymic hydrolysis of a portion of RNA by the endonucleases T_1-ribonuclease and pancreatic ribonuclease and by the exonucleases, *E. coli* phosphomonoesterase, spleen phosphodiesterase, and snake venom phosphodiesterase.

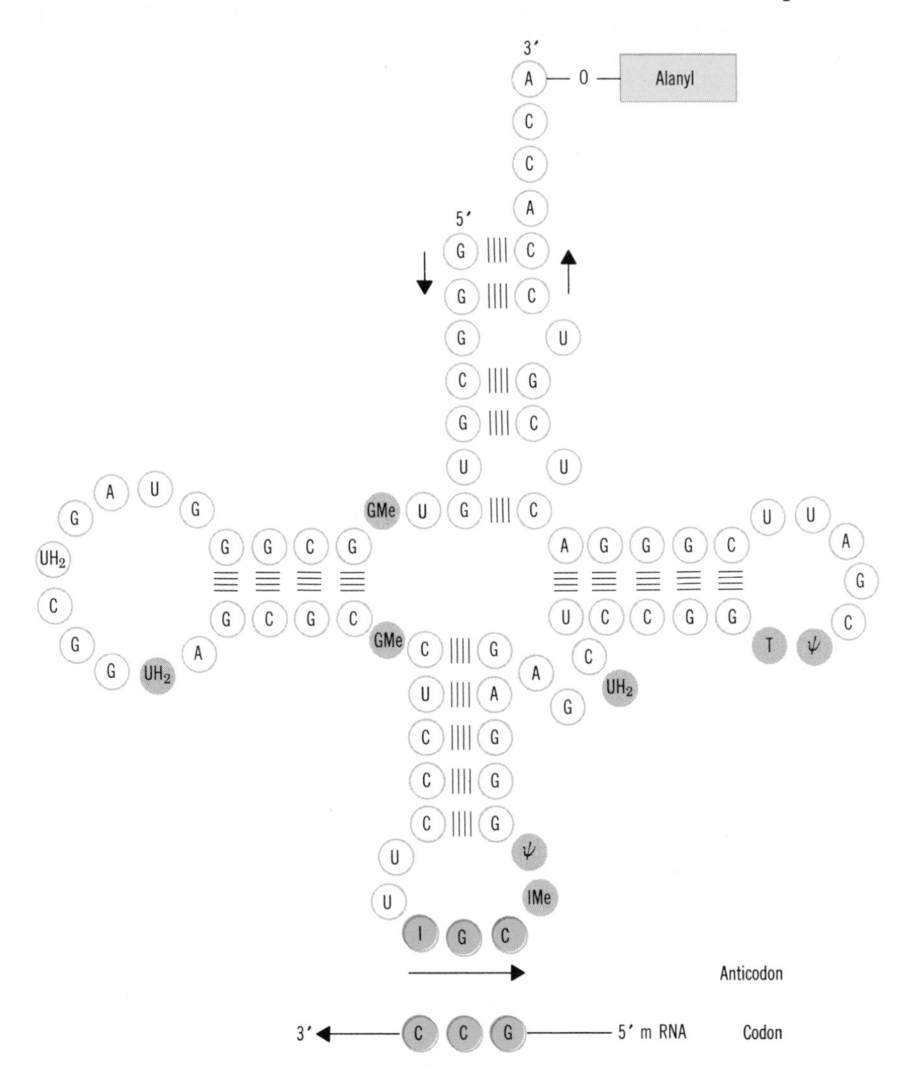

Figure 5-6

The complete sequence of alanyl tRNA illustrating the possible cloverleaf structure of the transfer ribonucleic acid and the positioning of unusual bases: ψ, pseudouridine; I, inosine; UH_2, dihydrouridine; T, ribothymidine; GMe, methyl guanosine; and IMe, methyl inosine. The dashed lines indicate hydrogen bonding.

stranded polymer. Furthermore, since rRNA does not have the extremely rigid and stable double helical structure of DNA, it may exist in several conformations. Thus, in the absence of electrolytes, or at high temperatures, a single-stranded conformation may occur. At low ionic strengths, a compact rod with regularly arranged helical regions can exist, and at high ionic strengths, a compact coil will occur. Moreover, the concentration of Mg^{2+} ion plays an important role in the macromolecule structures of RNA, presumably since the Mg^{2+} ions form coordination bonds with the phosphate groups of the nucleic acid. At a low Mg^{2+} concentration, dissociation of RNA complexes occur, whereas at high Mg^{2+} concentration, association of complexes is favored.

Messenger RNA. Because of the metabolic instability and heterogeneity of this species of RNA, careful characterization has only recently become possible. Messenger RNA appears to be principally single-stranded and complementarity with the base sequences of DNA has been demonstrated through the formation of artificial DNA–RNA double-stranded hybrid molecules.

Structure of DNA

The observation by Chargaff that the ratio of adenine to thymine and that of cytosine to guanine is very close to 1 was of basic importance in working out the structure of DNA. It was then shown that the adenine and thymine nucleotides can be so paired structurally that a maximum number of two hydrogen bonds can be drawn between these bases, whereas cytosine and guanine can be arranged spatially to permit the formation of three hydrogen bonds.

A breakthrough in the investigation of DNA structure came when Wilkins, in England, observed that DNA from different sources had remarkably similar x-ray diffraction patterns. This suggested a uniform molecular pattern of all DNA. The data also suggested that DNA consisted of two or more polynucleotide chains arranged in a helical structure. With evidence based on (*a*) the available x-ray data, (*b*) the data of Chargaff and others on base pairing and equivalence, and (*c*) titration data which suggested that the long nucleotide chains were held together through hydrogen bonding between base residues, Watson and Crick constructed their model of DNA in 1953 (see Figure 5-7).

In the Watson and Crick model of DNA two polynucleotide chains are wound into a right-handed double helix. The chains consist of deoxyribotide phosphates joined together by phosphate diesters with the bases projecting perpendicularly from the chain into the center axis. For each adenine projecting toward the central axis, one thymine must project toward adenine from the second parallel chain and be held by hydrogen bonding to adenine. Cytosine or guanine do not fit in this area and are rejected. Similarly, the specificity of hydrogen bonding between cytosine and guanine dictates their association only with each other. Thus, we have a spatial structure of two chains coiled around a common axis and held together by the specific bonding of adenine with thymine, and cytosine with guanine. Note, however, that the chains are not identical, but, because of base pairing, are precise complements of each other. Also, the chains do not run in the same direction with respect to their internucleotide linkages but rather are antiparallel. That is, if two adjacent deoxyribosides, T and C, in the same chain are linked 5′–3′, the complementary deoxyribosides A and G in the other chain will be linked 3′–5′ (Scheme 5-2).

The ability of DNA to form a double helix is of prime importance in considering its function in the cell. The double helical structure immediately suggests a mechanism for the accurate replication of genetic information.

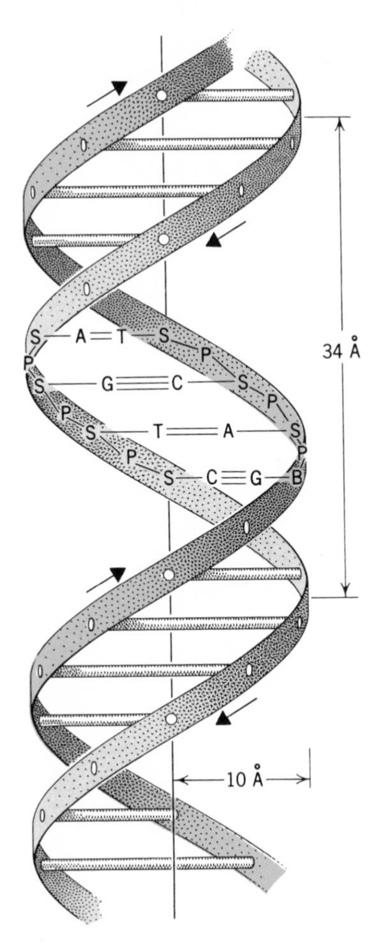

Figure 5-7

Double helix of DNA. Here, P means phosphate diester, S means deoxyribose, A═T is the adenine–thymine pairing, and G≡C is the guanosine–cytosine pairing.

Because of the complementarity of the helical structure, each strand serves as a template to specify the base sequence of a newly synthesized complementary strand. As a result, in the synthesis of two daughter molecules of DNA, each will be precisely identical to that of the parent DNA.

An important technique which tests for complementarity of regions between different nucleic acids is called *hybridization*. A mixture of two different native DNA's, one heavily labeled with ^{15}N and deuterium, is carefully heated until melting or a separation of the double strands to single strands occurs. This mixture is then very carefully cooled—this is called the annealing process—to permit a reformation of double strands wherever complementarity exists. The extent of complementarity is measured by first digesting

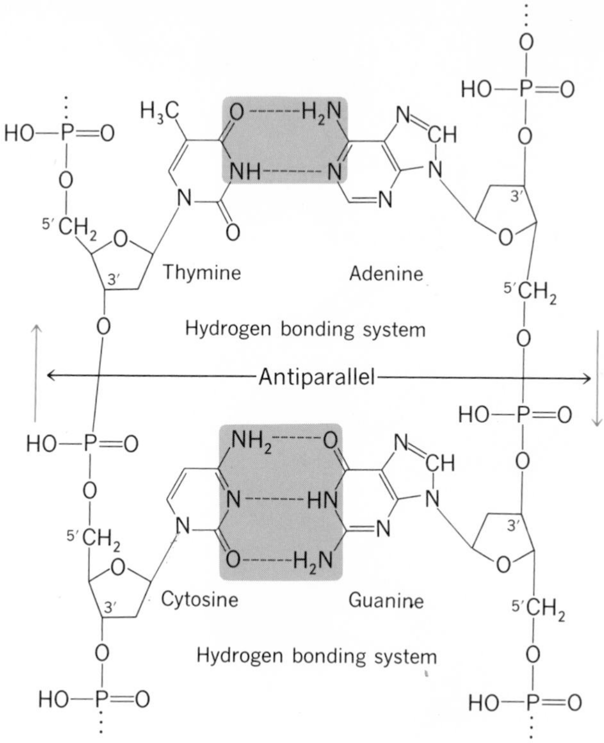

Scheme 5-2

away any single strands by the introduction of a nuclease which will hydrolyze single- but not double-stranded molecules. This treatment is followed by a density gradient centrifugation procedure which separates double-stranded heavy DNA ($^{15}N—^{15}N$) and double-stranded light DNA ($^{14}N—^{14}N$) from hybrid DNA ($^{15}N—^{14}N$) formed during the annealing process. The density of the hybrid DNA is between the densities of the heavy and light DNA's. The amount of hybrid DNA is a measure, then, of the complementarity between the two DNA preparations.

In a similar procedure, complementarity between DNA and RNA molecules can be tested. Single-stranded DNA is carefully heated with RNA labeled with a radioisotope, and the mixture is slowly cooled to permit the formation of DNA–RNA hybrids. The annealing mixture is treated with ribonuclease which digests all RNA not hybridized with DNA. This is followed by filtration of the mixture through a nitrocellulose filter which retains the DNA–RNA hybrid, but allows all free RNA and RNA fragments to pass through. The amount of DNA–RNA hybrid formed is measured by counting the radioactivity retained on the filter. We shall discuss the function of DNA in Chapter 18.

In addition to the nucleic acids found normally in the cell, an entirely unique group of nucleic acids is associated with a special class of macromolecules called *viruses*.

Because of the great variation in the structure of viruses, we shall only comment briefly on their structure. All viruses, whatever their range in particle weight, contain either DNA or RNA plus highly specific coat proteins which serve as protective shells encasing the nucleic acid core. The more complex viruses also contain lipids, carbohydrate, and functional proteins, that is, enzymes. Viruses thus far examined are either DNA-containing or RNA-containing viruses; that is, they do not contain both DNA and RNA in their structures. The specific nucleic acid each virus contains serves exclusively to carry the necessary genetic information for the successful replication of the complete virus in the host cell. The coat proteins of all viruses are not infective, since introduction of the protein into the host cell leads to no new formation of virus particles or destruction of the cell. However, introduction of highly purified nucleic acids from any of several types of viruses into the specific host cell leads to rapid replication of the complete virus particle. This indicates that indeed the nucleic acid is coding for both the replication of the virus nucleic acid and for the synthesis of its specific coat or other proteins, etc., essential for the completion of the structure of the virus particle.

Although all the more complex viruses such as smallpox and the T_2, T_4, and T_6 bacterial viruses have double-stranded DNA as the nucleic acid component, other viruses such as tobacco mosaic virus, influenza, poliomyelitis, and some bacterial viruses have single-stranded RNA. Nature always provides exceptions to these rules, since some very simple bacterial viruses are single-stranded DNA and the reoviruses have double-stranded helical RNA.

When a virus enters a host cell, the host machinery for replication, transcription, and translation may be partially or wholly diverted toward the synthesis of new complete virus particles, with the viral DNA or RNA serving as the necessary informational unit for its replication and for associative viral proteins. In addition, the viral nucleic acid will code for the synthesis of specific enzymes necessary for the formation of unique viral structures. For example, in *E. coli*, 5-hydroxymethyl cytosine is not normally present but is necessary for the successful replication of T_4-DNA virus. Viral T_4-DNA will code for the synthesis of enzymes in the *E. coli* cell which are responsible for the formation of this cytosine derivative.

The biochemistry of viruses is an exceedingly active field, and students should refer to current textbooks on the subject.

References

1. J. D. Watson. *Molecule Biology of the Gene*. 2nd ed. New York: Benjamin, 1970.
   A superb book with a good discussion of the structure of DNA and RNA.
2. J. N. Davidson. *The Biochemistry of the Nucleic Acids*. 6th Ed. New York: Wiley, 1969.
   A short but very well-written book on many aspects of nucleic acid biochemistry.
3. H. Fraenkel-Conrat, *The Chemistry and Biology of Viruses*. New York: Academic Press, 1969.
   A sound monograph covering the important aspects of virology.

PART II

Metabolism of Energy-Yielding Compounds

The Cell— Its Biochemical Organization

Introduction

In this chapter we shall attempt to define the components of cells in terms of their structure and function. We hasten to add that the descriptions will refer to general cells; to attempt to define the many specialized cells found in the animal and plant kingdom would necessarily blur the basic similarities and dissimilarities which we are attempting to emphasize.

In 1957, Dougherty first proposed the adjectives, *procaryotic* and *eucaryotic* to describe cells. These terms are now in common use. By definition, the procaryotic cell has a minimum of internal organization. It possesses no membrane-bound organelle components, its genetic material is not enclosed by a nuclear membrane, nor is its DNA complexed with histones. Indeed, histones are not found in these cells. Its sexual reproduction involves neither mitosis nor meiosis. Its respiratory system is closely associated with its plasma membrane. Typical procaryotic cells include all bacteria and the blue-green algae. All other cells are of the eucaryotic type.

A eucaryotic cell has a considerable degree of internal structure with a large number of distinctive membrane-enclosed organelles. The nucleus is the site for informational components collectively called chromatin. Reproduction involves both mitosis and meiosis; the respiratory site is the mitochondrion; and, in plant cells, the site of the conversion of radiant energy to chemical energy is the highly structured chloroplast. Figure 6-1 compares the general procaryotic and eucaryotic cells, illustrating diagramatically the differences as well as the similarities of these cells. Figure 6-2 shows freeze-etch electron micrographs of procaryotic gram-positive (G+) and gram-negative (G−) cells, and of a eucaryotic yeast cell. Figure 6-3 illustrates the general procedure for the isolation of organelle structures from eucaryotic cells.

Membranes

The semipermeable barrier between the internal and external environment of the cell is the plasma membrane. By means of a limiting membrane the

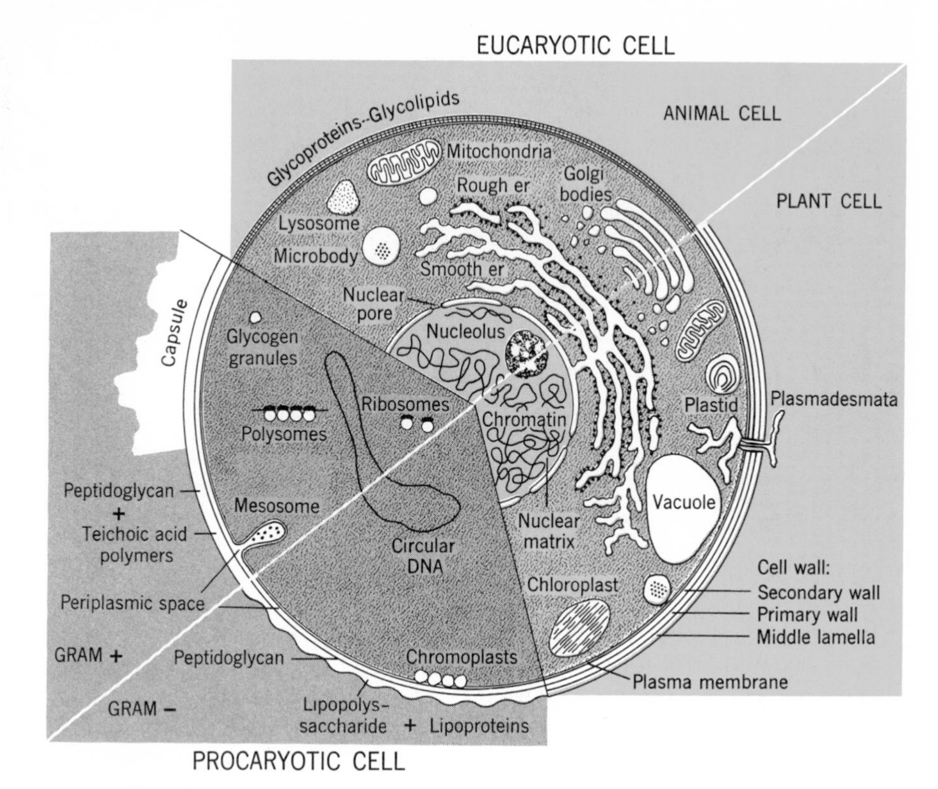

Figure 6-1

Schematic comparison of procaryotic and eucaryotic cells.

cell organizes its internal environment for specific purposes and expends energy to maintain this environment despite changes constantly occurring externally. Since the cell may also be a component of a larger unit in multicellular organisms, intercellular coordinations or interactions are necessary.

Chemical analyses of a large number of cell membranes have consistently revealed the presence of proteins and an array of complex polar lipids. Indeed, over 95% of the total phospholipid of the bacterial cell is associated with its plasma membrane.

In order to understand membrane structure, we must first explore the nature of phospholipid–protein interactions. As indicated in Chapter 3, phospholipids are called amphipathic compounds, since they contain both hydrophobic (lipophilic) and hydrophilic (lipophobic) components. All membranes always contain amphipathic lipids. Although phospholipids are insoluble in water, a suspension of phospholipid aggregates can be readily rearranged into highly water-soluble micelles by exposing the suspension to a short burst of sonication (Scheme 6-1).

Occurring in all forms and shapes, micelles can aggregate to form structures with molecular weights from a few thousand to many thousand. They are highly stable and water-soluble. In their newly ordered arrangement, the

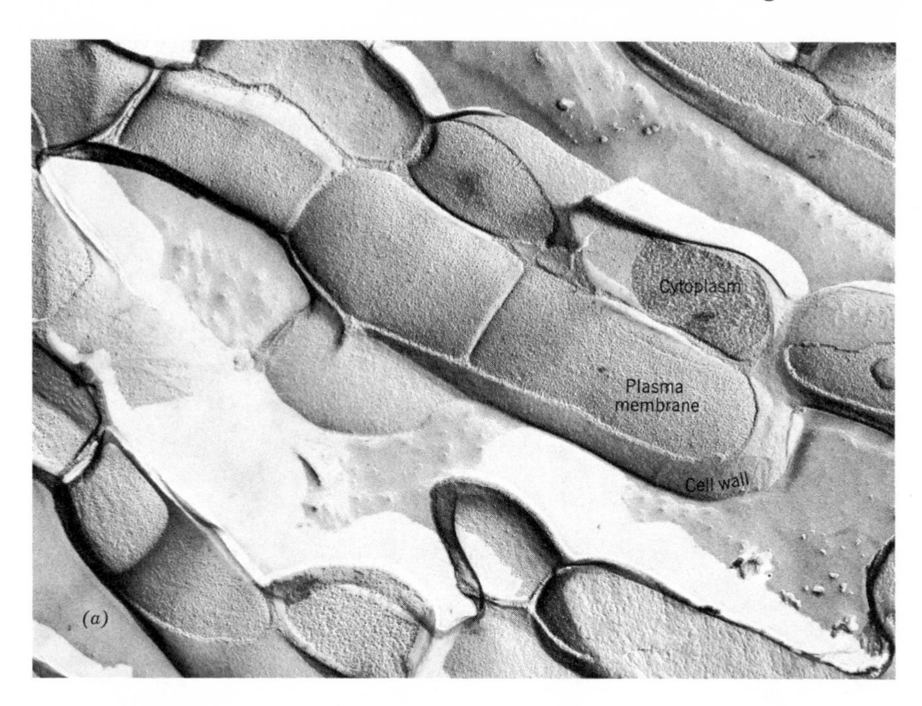

Figure 6-2(*a*)

Freeze-etch electron micrograph of a procaryotic cell, a gram-positive *Bacillus lichenformis* magnified 51,000×. (Reproduced with permission of Charles C. Remsen, Woods Hole Oceanographic Institution.)

hydrophobic functions of the amphipathic compounds, namely the hydrocarbon chains, are arranged internally to exclude water and thus are held together by hydrophobic interaction forces. The hydrophilic functions, namely the phosphoryl base moieties, are in turn highly attracted to the aqueous environment. Artificial membranes of phospholipids are readily formed by

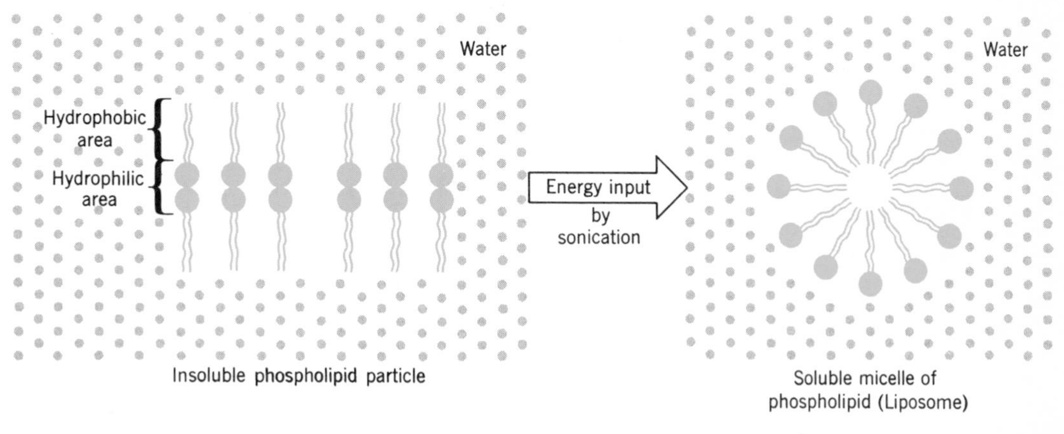

Scheme 6-1

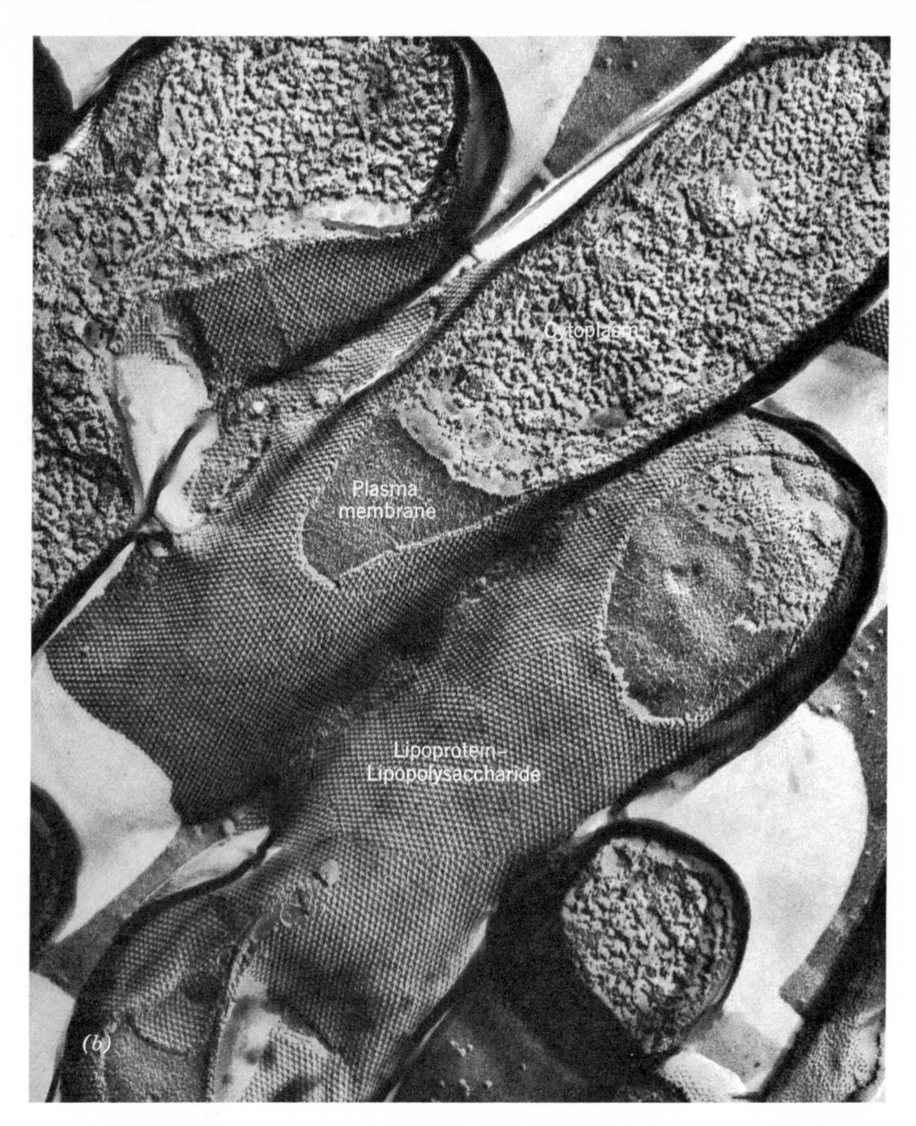

Figure 6-2(*b*)

Freeze-etch electron micrograph of a procaryotic cell, a gram-negative *Nitrosomonas* species magnified 237,000×. (Reproduced with permission of Charles C. Remsen, Woods Hole Oceanographic Institution.)

the sonication technique as well as other techniques, and are under intensive investigation as model systems.

In order to bring the protein components of membranes into the picture, Danielli and Davison in 1935 suggested that membranes were composed of a lipid bilayer, the hydrophobic ends facing each other as in a soluble micelle and held together by hydrophobic interaction while the polar ends faced outward, closely associated with a monomolecular layer of globular proteins (Scheme 6-2).

The model shown in Scheme 6-2 is the basis of the unit membrane hypothesis of Robertson, who later observed, by electron microscopy and by differential staining techniques, biomembranes with two electron-dense lines and a light central zone between them. The width of these unit membranes was 6–7 nm in eucaryotic organelles and 7–10 nm in eucaryotic surface membranes, matching the dimensions of the Danielli–Davison model.

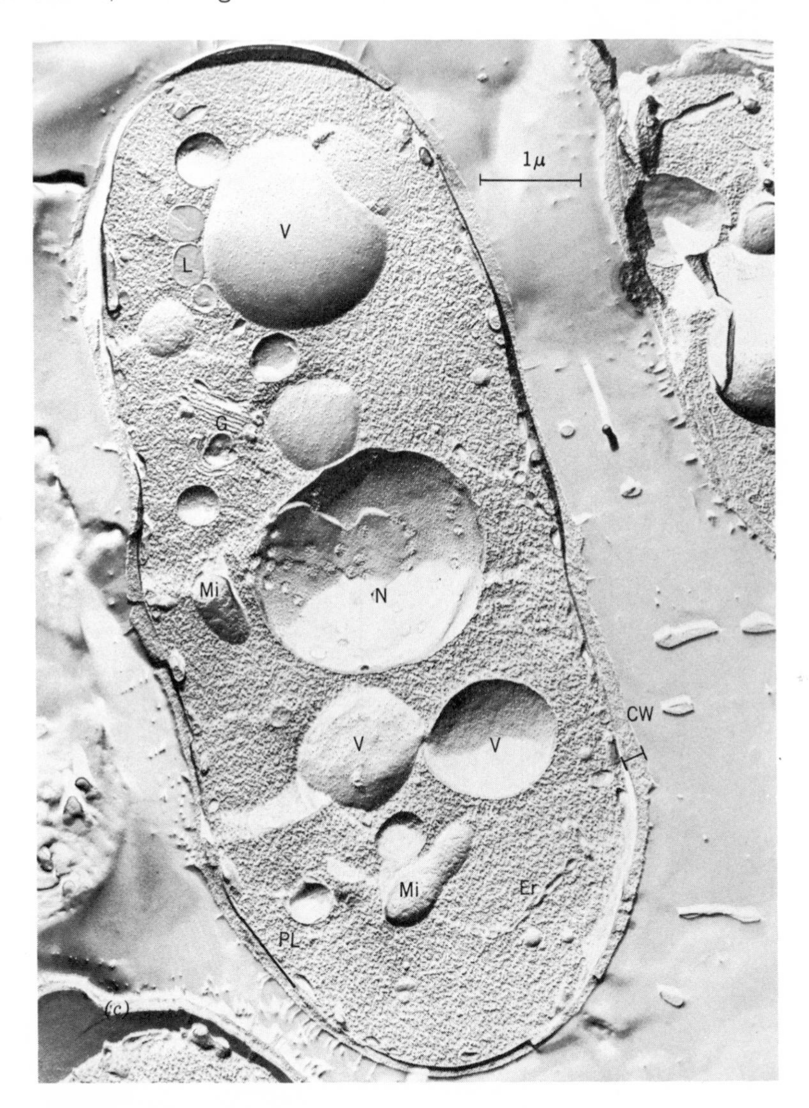

Figure 6-2(*c*)

Freeze-etch electron micrograph of an eucaryotic cell, *Schizosaccharomyces pombe* magnified 24,000×: CW, cell wall; Er, endoplasmic reticulum; G, Golgi apparatus; L, lipid body; Mi, mitochondria; N, nucleus; PL, plasmalemma; V, vacuole. (Reproduced with permission of F. Kopp and K. Mühlethaler, Swiss Federal Institute of Technology.)

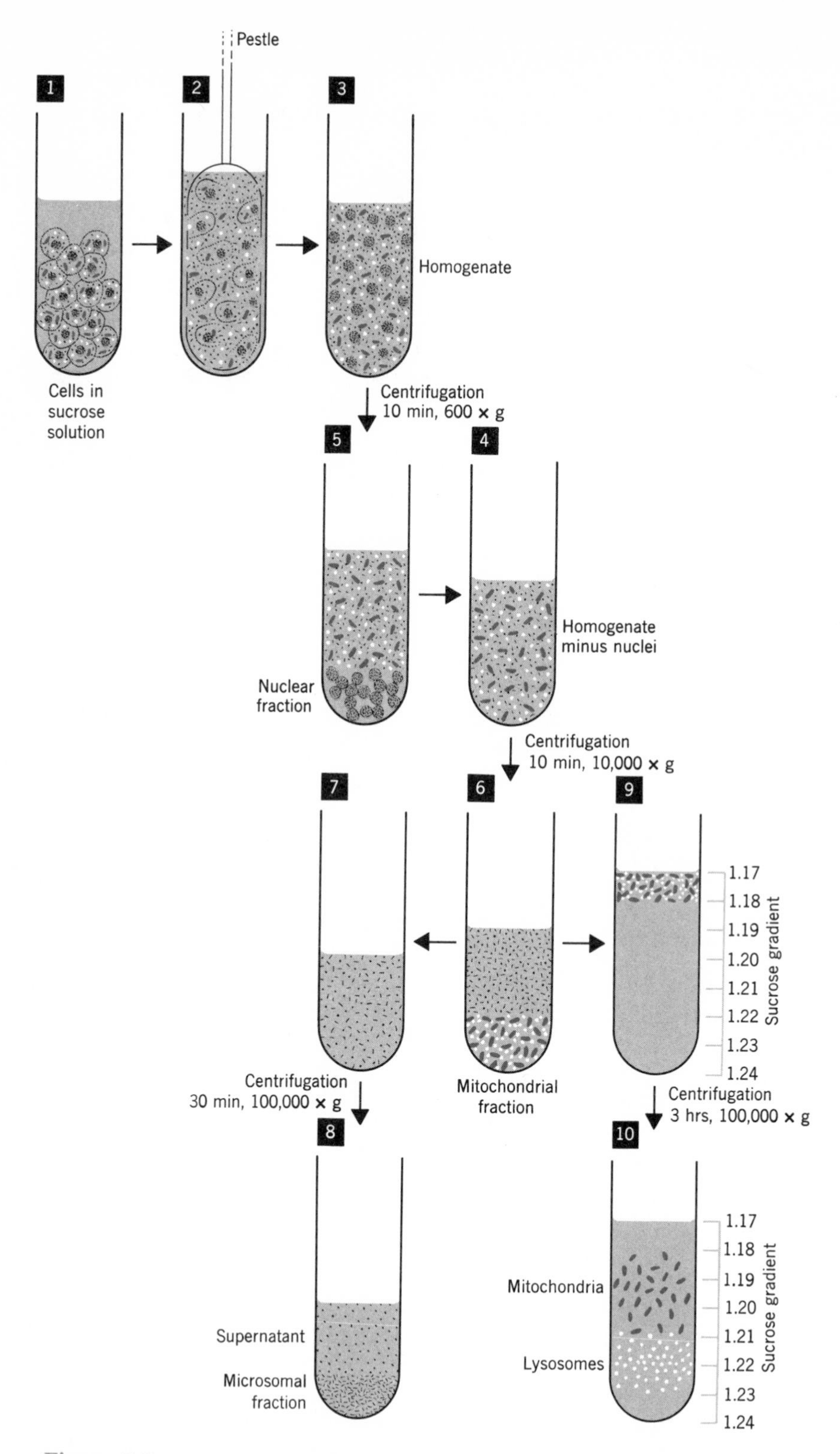

Figure 6-3

General procedure of obtaining cell organelles from eucaryotic organisms by gently disrupting the cells (2), differentially centrifuging (4–7) and then centrifuging through a gradient medium (9, 10).

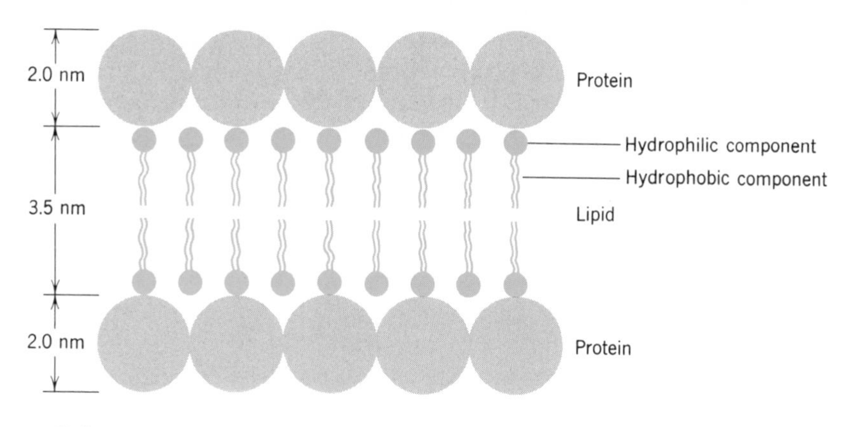

Scheme 6-2

Although the model predicts that hydrophilic bonds bind lipids to proteins, evidence is now accumulating that in many organelle membranes such as in chloroplasts, erythrocytes, and halophilic bacteria, lipid also associates with proteins hydrophobically. In fact, the recently proposed fluid mosaic model of S. J. Singer depicts the matrix of phospholipid bilayers with two types of proteins embedded in the matrix, namely the readily dissociable and soluble peripheral proteins and the difficultly dissociable and relatively insoluble integral proteins (see Scheme 6-3).

Whatever the final outcome concerning the nature of biomembranes, the semipermeable properties of these membranes must be explained in physical–chemical terms. Physical model systems clearly possess remarkable ion-selective properties which can be altered by adding well-known mitochondrial uncouplers, such as 2,4-dinitrophenol, to the membrane system. Even more impressive is the inducement of cation specificity in bilayer models by the introduction of trace amounts of macrocyclic peptides such as valinomycin. Thus, the permeability of K^+ becomes two to three hundred times larger than the Na^+ permeability in the modified bilayer containing valinomycin. This selectivity may occur since the peptide ring acts as a hydrophobic cage for the ion, permitting it to pass easily through the hydrocarbon barrier of the bilayer. Similar effects are observed with biomembranes.

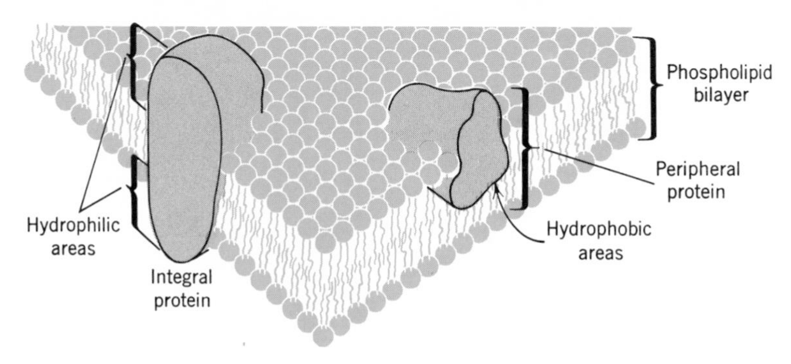

Scheme 6-3

Transport Processes

With an elementary knowledge of membranes in mind, we shall now examine transport processes which have been recently described in cells. The primary function of the biomembrane is to allow movement of all compounds necessary for the normal function of the cell across its membranal barrier. There are at least four ways in which metabolites can pass through biomembranes.

Free or Simple Diffusion. Metabolites of low molecular weights move or diffuse across the membrane. The process depends on a concentration gradient across the membrane and of course this gradient disappears as diffusion occurs. No stereospecificity is involved, i.e., D and L amino acids move across the membrane at equal rates. Simple diffusion is not believed to be an important mechanism of transport across biomembranes since the rate is too slow and no selectivity is permitted (Scheme 6-4).

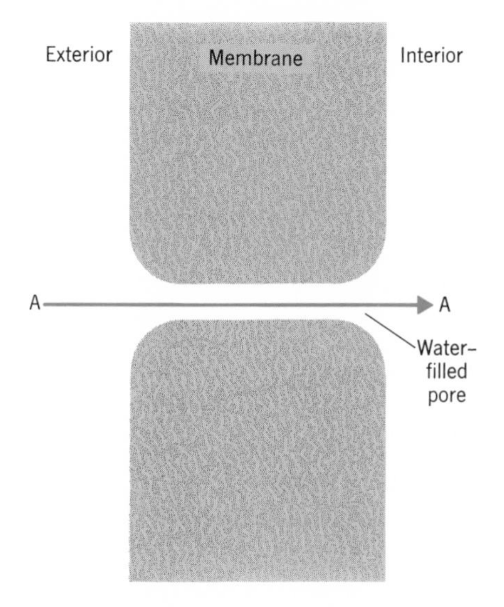

Scheme 6-4

Facilitated Diffusion. This type of diffusion is somewhat similar to simple diffusion in that a concentration gradient is required and the process does not involve an expenditure of energy. However, it differs in several respects from free diffusion. First, the membrane contains a component called a carrier or permease which catalyzes the process, that is, speeds up the rate of diffusion much more than is predicted from simple diffusion. Second, the diffusion is stereospecific. And third, the rate of penetration of the metabolite approaches a limiting value with increasing concentration on one side of the membrane. The kinetics mimic simple Michaelis–Menten enzyme kinetics (Chapter 8). Thus, K_m and V_{max} values are easily measured and characterize the carrier system.

The mechanism can be explained by a specific carrier molecule present in the membrane which forms a specific complex with the metabolite at the outer area of the membrane. The complex then, by diffusion, rotation, oscillation, or some other motion, moves to the inward area of the membrane, where it dissociates to discharge the metabolite. The only process of importance is the association, translocation, and dissociation of the complex.

Recent evidence has revealed a number of small molecular weight proteins localized in bacterial membranes which fulfill the requirements of facilitated diffusion. Arthur Pardee, in studying the transport of sulfate by *Salmonella typhimurium*, has isolated a protein with a molecular weight of 34,000. It contains no lipid, carbohydrate, phosphorus, or SH groups. One sulfate molecule is specificly bound per molecule of protein, and the binding is very strong. Binding is, however, reversible and requires no ATP. The protein exhibits no enzymic activity. Under conditions of a low sulfate source, the organism produces about 10^4 molecules of the binding protein per bacterium. Osmotic shock treatment of cells results in considerable loss of the carrier protein, suggesting that the protein is near or on the cell membrane surface (periplasmic space). An increasing number of specific membrane transport proteins have now been isolated and crystallized. These include carrier proteins for glucose, galactose, arabinose, leucine, phenylalanine, arginine, histidine, tyrosine, phenylalanine, phosphate, Ca^{2+}, Na^+, and K^+. They all have small molecular weights ranging from 9000 to 40,000.

The mechanism shown in Scheme 6-5 has been proposed.

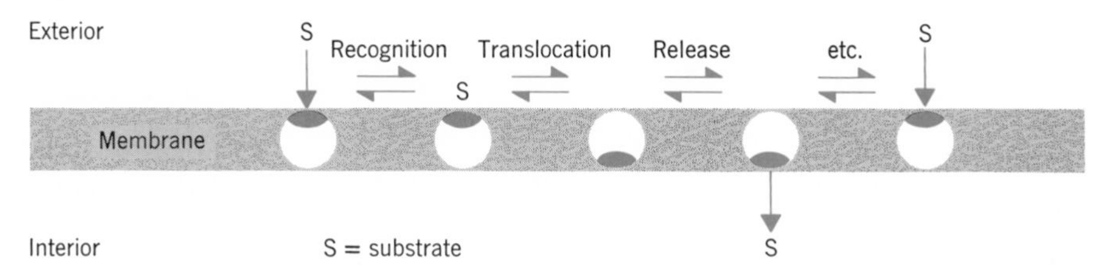

Scheme 6-5

Active Transport. In this process the transport of the *metabolite is very similar to that of facilitated diffusion*, with the important exception that the metabolite moves across the membrane against a concentration gradient which requires an expenditure of metabolic energy; that is, the metabolite moves from an area of low concentration to one of higher concentration. For example, in some cells as much as 50% of the cellular ATP is utilized for the accumulation of glycine in the cell. Thus, use of an inhibitor, such as azide or iodoacetate which markedly decreases the production of energy in the cell, greatly inhibits active transport. Either free diffusion or facilitated diffusion mechanisms would not be effected.

Two possible mechanisms might exist. One is shown in Scheme 6-6. This would suggest a change in the conformation of the carrier by an energy-utilizing system from X to X^0 which is then capable of associating with the metabolite to form AX^0. Movement to the inward direction might cause a change in the mileu leading to rapid dissociation and discharge of A to the interior of the cell with a simultaneous return of X^0 to X.

The other mechanism could be visualized as shown in Scheme 6-7. Here A undergoes a chemical change such as phosphorylation but returns to its original structure as it passes into the interior of the cell.

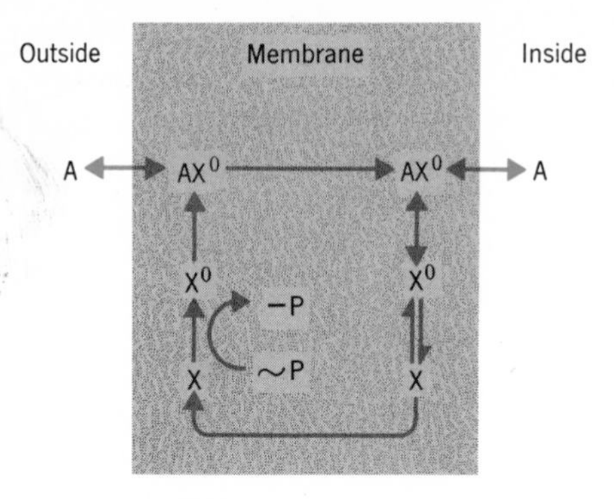

Scheme 6-6

Group Translocation. A more sophisticated transport mechanism has been proposed by S. Roseman for the transport of sugars across bacterial membranes; according to this mechanism, each of the sugars is released inside the cell as a phosphorylated derivative. Active transport is achieved since the sugar phosphate cannot escape back through the membrane.

Essentially, the transport mechanism involves the following reactions:

$$\text{Phosphoenol pyruvate} + \text{HPr} \xrightarrow{\text{Enz I, Mg}^{2+}} \text{Pyruvate} + \text{P—HPr}$$

$$\text{P—HPr} + \text{Sugar} \xrightarrow{\text{Enz II}} \text{Sugar-6-phosphate} + \text{HPr}$$

The total reaction is designated the PEP:glycose phosphotransferase system, but the system actually consists of several parts.

The histidine-containing protein, HPr, is a low-molecular-weight protein (9600) that has been obtained in a homogenous form from a number of bacteria. It is heat-stable, has no carbohydrates or phosphate, but does have 2 histidines per mole. The formation of phosphoryl—HPr with the phosphate linked to N-1 of the histidine imidazole ring is visualized as shown in Scheme 6-8.

Enzyme I and HPr are constitutive proteins; they are completely soluble; and presumably are in the cytosol of the bacterial cells. They are not in-

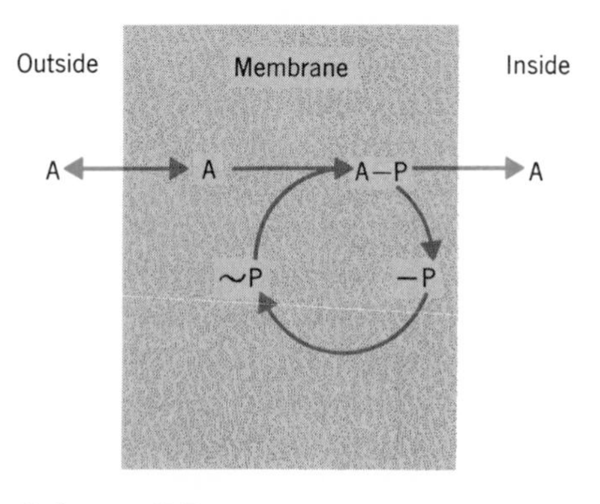

Scheme 6-7

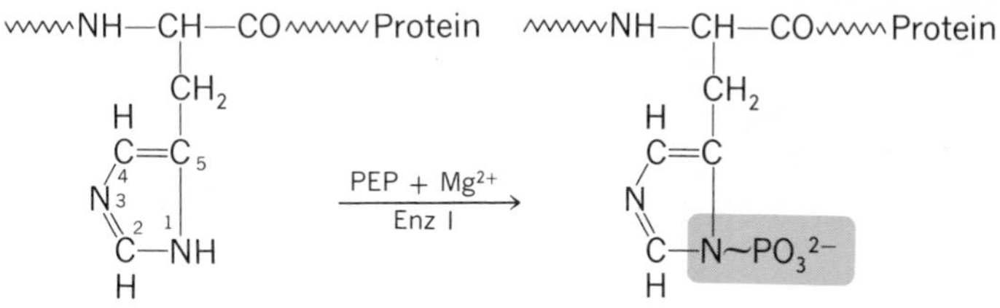

~~NH—CH—CO~~Protein (histidine residue: CH₂ attached to imidazole ring C5; ring atoms N3, C4, C2, N1—H) $\xrightarrow[\text{Enz I}]{PEP + Mg^{2+}}$ ~~NH—CH—CO~~Protein (imidazole N1—PO_3^{2-})

Scheme 6-8

creased in concentration when the cell is grown in the appropriate sugar. Neither enzyme I nor HPr bind sugars, and thus they do not serve as carriers. The second reaction catalyzed by a complex of proteins called enzyme II is, however, highly responsive to sugar specificity. Evidence suggests that the cell synthesizes a specific enzyme II protein in response to the type of sugar in which it is growing. This enzyme is thus specific for each sugar, highly inducible, and membrane-bound (probably in the periplasmic or outer region of the membrane). If grown in glucose, the organism would synthesize an enzyme II which catalyzes the transfer of a phosphoryl group from P—HPr to glucose to form glucose-6-phosphate (or mannose to form mannose-6-phosphate). If grown in ribose, the organism would form an enzyme II for the conversion of ribose to ribose-5-phosphate, etc.

Furthermore, mutants that do not exhibit transport presumably lack one of the proteins of the transferase system; that is, these mutants may possess HPr and enzyme II but lack enzyme I; some have enzyme I and enzyme II but lack HPr. This evidence adds considerable strength to the concept that group translocation is a major mechanism of sugar transport in bacterial cells.

Another type of evidence is the technique introduced by L. Heppel. If bacteria are rapidly transferred from a sucrose solution of high osmotic strength to a dilute salt solution, the cells undergo osmotic shock with the release of about 5% of their total protein into the outside medium without losing their viability. It was shown that up to 85% of the bacteria's transport capacity for certain sugars was lost by this treatment with concomitant 80% loss of HPr (Scheme 6-9). The beautiful aspect of this experiment is that the original transport capacity of the cell could be restored by incubating these shocked cells with highly purified HPr!

It is quite possible that the PEP:glycose phosphotransferase system is responsible for all sugar transport into the bacteria cell. It has been shown that one ATP (or one PEP) is required to transport one molecule of sugar against a gradient. The advantage of this system over other systems (that is, active transport, in which sugar per se enters the cell by a utilization of metabolic energy) is that the end product of the transport mechanism is a sugar-6-phosphate that is ready for metabolic activities. The energy used for transportation is thereby not wasted, but rather is efficiently conserved.

How can we depict the mechanism when Enz I and HPr are in the cytosol of the cell? Roseman has observed that Enz II, which is membrane-bound, actually consists of a complex of Enz IIa which is sugar-specific, Enz IIb

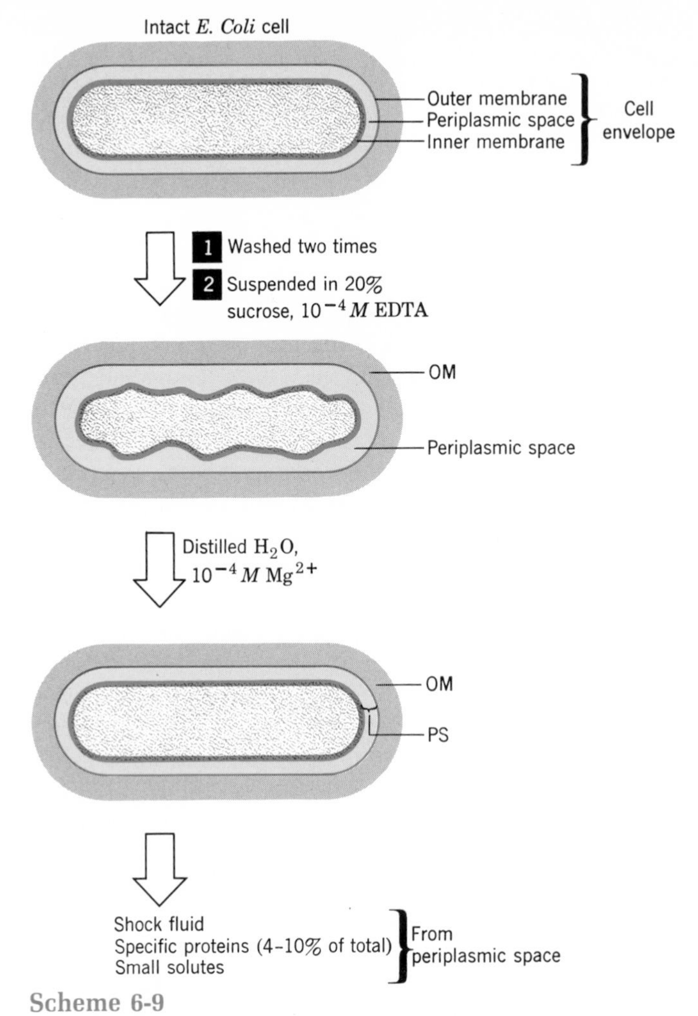

Scheme 6-9
Osmotic shock procedure.

which is present regardless of the type of sugar on which the organism is grown, a metal cation such as Mg^{2+}, and specifically phosphatidyl glycerol (Scheme 6-10).

Enzyme II is assumed to undergo a conformational change when the sugar associates with it, resulting in the positioning of the sugar in the interior of the cell but still associated with the binding site until the remaining reactions occur.

In the last few years the mysterious black box of metabolite transport has been pried open. In the near future the lid will be further opened to reveal the detailed mechanisms of transport in both the animal and plant kingdoms. Knowledge so obtained will undoubtedly find applications in medicine, agriculture, and even in industry.

Cell Walls

Unlike cell membranes, the cell walls in procaryotic cells and in most eucaryotic cells such as algae, fungi, and plants confer shape and rigidity to

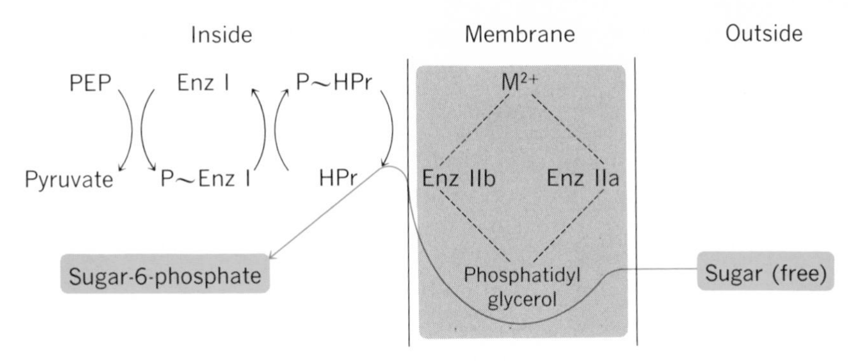

Scheme 6-10

the cell itself. Without walls, cells would most likely be spherical and extremely fragile to slight osmotic changes. As observed in Figure 6-1, cell walls of procaryotic cells are rather complicated and markedly different from those in eucaryotic cells. Bacteria are roughly divided into gram-positive and gram-negative cells based on the differential staining by a crystal violet iodine reagent. In general, gram-positive cells have thick cell walls, up to 80% of which are composed of a mesh like macropolymer called a peptidoglycan. The chemical nature of this macropolymer has been described (Chapter 2). Variations in the structure and composition of this peptidoglycan occur in a number of bacteria.

Superimposed on the peptidoglycan are the teichoic acid polymers that consist of repeating units of either glycerol or ribitol connected by internal phosphate diesters. D-Alanine is usually attached through an ester linkage to the polyhydroxyl alcohol. The teichoic acid polymers are probably intimately involved in the cell's antigenicity and susceptibility to phage infection. They definitely confer a strong negative charge on the surface of the cell wall because of their high content of ionized phosphate groups. The teichoic acids apparently are located in the region extending from the exterior of the plasma membrane to the outer regions of the peptidoglycans. Gram-positive cell walls are also characterized by the absence of any significant lipid.

Of considerable interest is the fact that the enzyme lysozyme found in tears and saliva, in bacteria and in plants, readily hydrolyzes peptidoglycans at

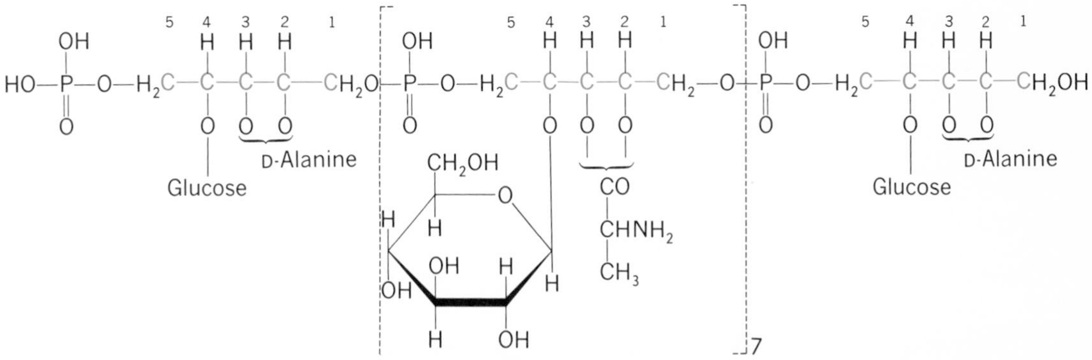

Ribitol teichoic acid from *Bacillus subtilis*

the β-1,4 linkage of *N*-acetyl muramic acid with the resulting weakening of the cell wall and subsequent rupture of the cell. Certain antibiotics such as penicillin specifically inhibit the synthesis of new cell walls in growing cells, and this leads to cell lysis. Since eucaryotic cells have entirely different cell walls or membranes when compared to procaryotic cells, penicillin has no effect on animal cells. Hence, this specificity leads to its great value in the treatment of infectious diseases caused by procaryotic cells, in particular gram-positive organisms.

As suggested in Figure 6-1, gram-negative organisms have a somewhat more complex cell wall structure. Although little if any teichoic acid is found in these organisms, and although a thin strand of peptidoglycan similar in structure to those found in gram-positive cell walls is sandwiched between the cell membrane and the outer envelope, the major component of these organisms is a giant macropolymer called a lipopolysaccharide, whose structure is known in some detail, consisting of a number of monosaccharides, α-hydroxy myristic acid, and other lipids. We do not, however, propose to discuss the details of its structure nor its biosynthesis here, because of the complexity of the subject. When lipopolysaccharides are released into the blood stream of an animal, they are very toxic, causing fever, hemorrhagic shock, and other tissue damage. They are, therefore, called endotoxins. A generalized structure is presented in Figure 6-4. The considerable knowledge concerning this heteropolysaccharide can be found within the references at the end of this chapter.

The student should not consider that the bacterial cell wall is covered with a mesh-like sheet of complex macromolecules. If this were the case, the organism would have some difficulty in obtaining metabolites for growth. The cell surface is actually punctured by a large number of pores through which biochemical compounds flow, but which prevent entry into the interior of the cell of very large molecules such as proteins or nucleic acids. One could think of a bacterial cell wall as a giant molecular sieve allowing small-molecular-weight compounds to pass through to the plasma membrane but retaining macromolecules. At the plasma membrane the transport mechanisms discussed earlier become operative.

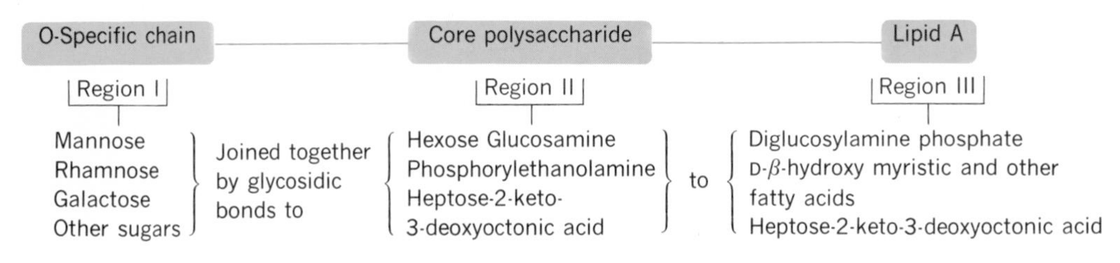

Figure 6-4
A generalized structure of a lipopolysaccharide.

Plant Cell Walls

In mature plant cells, the cell wall is composed of three distinct parts, the intercellular substance or middle lamella, the primary wall, and the secondary wall. The middle lamella is composed primarily of pectin polymers, and may also be lignified. The primary wall consists of cellulose, hemicellulose (xylans, mannans, galactans, glucans, etc.), and pectins, as well as lignin. The secondary wall, which is laid down last, contains mostly cellulose, with smaller amounts of hemicellulose and lignin.

A large number of openings called pits occur in various arrays and shapes in the secondary wall. Connecting adjacent cells are thread-like structures called *plasmodesmata,* which penetrate through the pits and the primary wall and middle lamella to the neighboring cell. It is currently thought that the endoplasmic reticulum of the cell extends through the plasmodesmata into the neighboring cell, thereby permitting a flow of material and hormones from one cell to the next.

Much work has been expended to understand the structural role of cellulose in the plant cell walls. There is considerable agreement that cellulose forms microfibrils consisting of about 2000 cellulose molecules in cross-section. These are arranged in orderly three-dimensional lattices around the cell, particularly in the secondary cell wall, to give great strength as well as plasticity to the wall.

There is now reasonably good evidence that the Golgi apparatus participates in the formation of the middle lamella and the adjacent primary cell walls as a plant cell divides during mitosis. This organelle, rich in enzymes for phospholipid and cellulose synthesis, releases small vesicles which line up and fuse in a linear fashion to form first a gel-like matrix, which then develops into the middle lamella with the deposition of hemicelluloses and pectins.

It should be noted that the plant cell wall has associated with it a significant number of hydrolases including invertase, phosphatases, nucleases, and peroxidases. The significance of these hydrolases in the cell wall is not clear.

Animal Cell Surfaces

The generalized animal cell essentially has no cell wall. However, glycoproteins are associated intimately with the plasma membrane and may also coat its surface. Membrane glycoproteins are probably involved in ion transport, cell antigenicity, and intercellular contact. Glycolipids are also localized here.

The chemistry of the membrane glycoproteins is being actively explored at present. Glycoproteins consist of oligosaccharide chains covalently linked

N-Acetylglucosamine
CH_2OH
Carbohydrate chain—O
H AcNH
NH—C=O
CH_2
Asparagine residue
NH H CO
Protein

to polypeptides. Polymers of D-galactose, D-mannose, *N*-acetyl-D-glucosamine, L-fucose, and sialic acid make up the oligosaccharide component which is in turn attached to the protein component via an asparagine-carbohydrate link.

Nucleus Although in procaryotic cells no nucleus per se is observed, a fibrillar area can be detected on the interior side of the plasma membrane. This structure, called a mesosome, is associated with an extremely involuted double-stranded circle of DNA. It has been estimated that in a single bacterial cell 2 μm long its DNA, if stretched out as a single fiber, would be over 1000 μm long, 500 times the length of its own cell body. Since histones are absent in these cells, no DNA–histone complexes exist. However, high concentrations of polyamines such as spermidine and spermine have been detected in the bacterial cell, and these compounds may participate in neutralizing the negative charges on the DNA.

In eucaryotic cells, the nucleus is a large dense body surrounded by a double membrane with numerous pores which permit passage of the products of nuclear biosynthesis into the surrounding cytoplasm. Internally, the nucleus contains chromatin or expanded chromosomes composed of DNA fibers closely associated with histones. During nuclear division, the chromosomes contract and become clearly visible in the light microscope as the DNA chains undergo their programmed changes. In addition, the nucleus contains enzymes such as DNA polymerases, RNA polymerase(s) for mRNA synthesis; and, surprisingly, the enzymes of the glycolytic sequence, citric acid cycle, and the hexose monophosphate shunt have been detected in the nucleoplasm. One to three spherical structures called the nucleolus are closely associated with the inner nuclear envelope and are presumably the sites of rRNA biosynthesis. This dense suborganelle is nonmembraneous and contains RNA polymerase, RNAase, NADP pyrophosphorylase, ATPase and *S*-adenosyl methionine–RNA–methyl transferase. There is an absence of DNA polymerase. Ribosomal RNA's are separatedly synthesized in the nucleolus and are then transported to the cytoplasm as discrete units to be assembled in the cytoplasm to form polysomes. Transfer RNA may also be synthesized in this organelle and further methylated by methyl transferase before being transferred to the cytoplasm. We shall discuss the function of these nucleic acids in Chapters 18 and 19.

Endoplasmic Reticulum Although the network of membrane-bound channels and vesicles called the endoplasmic reticulum is missing in procaryotic cells, this system is present in all eucaryotic cells (Figure 6-5).

Varying in size, shape, and number, the endoplasmic reticulum extends from the cell membrane, coats the nucleus, surrounds the mitochondria, and appears to connect directly to the Golgi bodies. There are two kinds—the rough-surfaced type known as *ergastoplasm*, which has ribosomes associated with it externally, and the smooth type, which lacks ribosomes. When cells are disrupted by homogenization and fractionated by differential centrifugation, the pellet, which sediments on centrifugation (100,000 × g, 30 min)

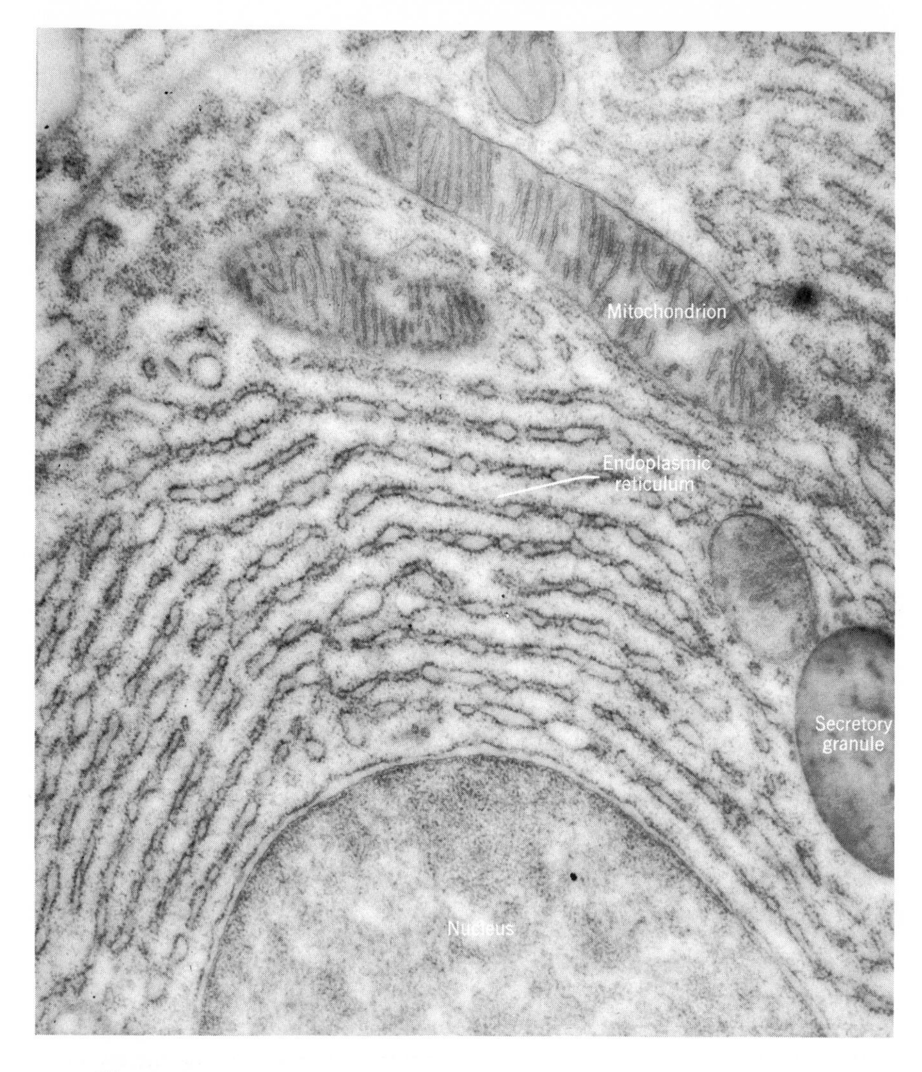

Figure 6-5

Electron micrograph of a cross-section of an exocrine cell of the guinea pig pancreas showing the common organelle structures of this eucaryotic cell. (Reproduced with the permission of G. E. Palade, Rockefeller University.)

is called the microsomal fraction (microsomes) and consists of small vesicles and fragments derived from the endoplasmic reticulum. A number of important enzymes are associated with the endoplasmic reticulum of mammalian liver cells. These include the enzymes responsible for the synthesis of sterols, triacylglycerols, and phospholipids, the detoxification of drugs by modification through methylation, hydroxylation, etc., the desaturation and elongation of fatty acids, and the hydrolysis of glucose-6-phosphate. As a word of caution, however, many of these activities are only associated with microsomes of liver cells, and are missing in microsomes from other eucaryotic tissues. Although cytochromes characteristic of mitochondria are absent, both cytochrome b_5, which may serve as a limited electron carrier system in the desaturation of fatty acids, and cytochrome P-450, which participates in hydroxylation reactions in animal cells, reside in the microsomes. Since ribosomes are associated closely with the rough-surfaced endoplasmic reticulum, protein synthesis occurs on the endoplasmic reticulum. Presumably, newly formed proteins are secreted into the vesicular system and then transferred to Golgi bodies to be used there in the formation of lysosomes and other microbodies (see Chapter 19 for a description of the role of this system in insulin biosynthesis).

A few comments are in order here concerning the ribosomal particles. In procaryotic cells ribosomes are grouped in clusters 10–20 nm in diameter, probably held together by mRNA to form polysomes. Because of the intense synthesis of proteins by growing bacterial cells, their cell matrix (sap) contains very many of these clusters. In all procaryotic cells, the intact ribosome has in general a sedimentation coefficient of 70S that dissociates reversibly in low Mg^{2+} concentrations to two subunits, a 30S and a 50S component:

$$70S \underset{Mg^{2+}\ (10^{-2}M)}{\overset{Mg^{2+}\ (10^{-4}M)}{\rightleftharpoons}} 30S + 50S$$

Ribosomes contain approximately 40% protein and 60% rRNA. In all eucaryotic cells, cytoplasmic ribosomes have an approximate 80S sedimentation behavior in the ultracentrifuge and consist of two subunits, 40S and 60S. The ribosomes found in mitochondria and chloroplasts closely resemble the bacterial ribosomes in size. We shall have much more to say about these important sites of protein synthesis in Chapter 19.

Mitochondria

Since the nineteenth century, microscopists have observed in all eucaryotic cells small, rod-shaped particles 2–3 μm long, which were called mitochondria. In 1948, A. L. Lehninger showed that in the animal cell the mitochondrion was the sole site for oxidative phosphorylation, the tricarboxylic acid cycle, and fatty acid oxidation. Because of the importance of these systems in the total economy of the cell, much research has been expended in defining the structure and function of these bodies that are found universally in eucaryotic cells but are totally missing in procaryotic cells.

All mitochondria consist of a double membrane system, an outer and an inner membrane, which by invagination extends into the matrix of the organelle as cristae. Both membranes appear to have the unit membrane structure, although this point is not completely clear. The inner surface of the inner membrane has projecting toward the matrix a cluster of knobs on stalks called inner membrane particles. E. Racker has clearly demonstrated that the inner membrane particles (85 Å diameter spheres) possess the coupling factor, F_1, which has ATPase activity and a molecular weight of 280,000. The enzyme has the interesting property of dissociating reversibly into inactive subunits at temperatures lower than 10°C. Racker visualizes

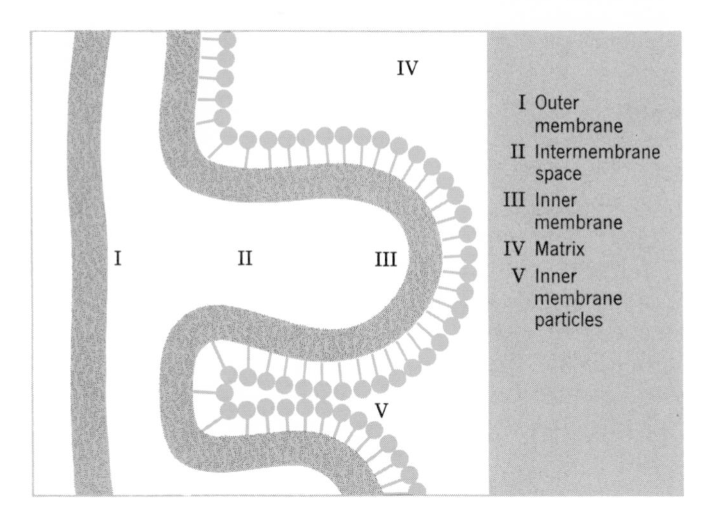

its function as catalyzing the last step of transphosphorylation resulting in ATP formation:

$$X—O \sim P + ADP \rightleftharpoons XOH + ATP$$

Considerable evidence suggests that all the enzymes of the electron transport system, namely succinic dehydrogenase, the flavoproteins, cytochromes b, c, c_1, a, and a_3, are buried in the inner membrane and are in some manner coupled to the F_1 complex, to allow oxidative phosphorylation as indicated in Figure 6-6. We shall describe the system in more detail in Chapter 14.

Although procaryotic cells have no mitochondrial bodies, their cell membranes appear to be the sites of electron transport and oxidative phosphorylation. Thus, all the cytochrome pigments and a number of dehydrogenases

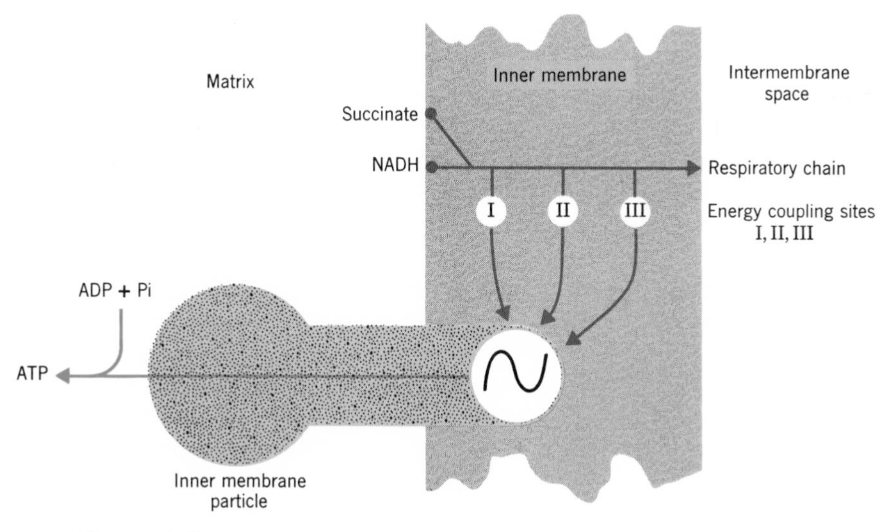

Figure 6-6

A schematic view of the relation of the inner membrane particle with the inner membrane of the mitochondrion.

associated with the tricarboxylic acid cycle, namely succinic, malic, and α-ketoglutaric dehydrogenase are localized in bacterial cell membranes. In addition, enzymes involved in phospholipid biosynthesis and cell wall biosynthesis are also found in or on the membrane structure.

The inner membrane of mitochondria possesses limited permeability, whereas the outer membrane is fully permeable to a large number of compounds with molecular weights up to 10,000. The outer membrane has a density of 1.13, and the inner membrane has a density of 1.21. The outer membrane has about three times more phospholipid than the inner membrane, and about six times more cholesterol than is found in the inner membrane; hence its slightly lower density. Phosphatidyl inositol is found exclusively in the outer membrane, whereas cardiolipin occurs almost exclusively in the inner membrane. Of the total protein in a mitochondrion, 4% is associated with the outer membrane, 21% with the inner membrane, and 67% with the matrix. Ubiquinone is present only in the inner membrane.

Although the results are still controversial, much effort has been expended in localizing the large number of enzymes associated with mitochondria. Table 6-1 lists the location of a number of enzymes in liver mitochondria.

In addition to a large number of soluble enzymes, the mitochondrial matrix contains mitochondrial DNA, a circular double-stranded molecule somewhat smaller but very similar to bacterial DNA in shape. In bacteria, mitochondria, and chloroplasts the DNA is histone-free and bound to membranes. Each mitochondrion has from 2 to 6 DNA circles amounting to about 0.2–1 μg of DNA/mg mitochondrial protein. This amount of DNA can code for about 70 polypeptide chains of 17,000 mol wt. Since DNA polymerase is found in the matrix, presumably mitochondrial DNA is independently synthesized in the mitochondria. Replication appears to be semiconservative. Of further interest is the observation that 70S ribosome particles are in the mitochondrial matrix as well as tRNA, mRNA, and protein-synthesizing enzymes which catalyze a limited type of protein synthesis. A DNA-dependent RNA polymerase has also been detected in the matrix. This complete complement of machinery for synthesizing protein may in some manner be concerned with the formation of a number of unknown mitochondrial proteins. At the present time, very little is known about the precise function of this mitochondrial protein-synthesizing machinery. Evidence suggests that the outer membrane protein is synthesized by the cytoplasmic–nuclear protein synthesizing system, and the proteins of the matrix and the inner membrane are synthesized by both the cytoplasmic nuclear system and the mitochondrial system.

A number of biochemists have submitted the provocative speculation that both mitochondria and chloroplasts resemble very closely procaryotic cells with respect to size, distribution of respiratory enzymes, and the striking similarity of their DNA and RNA components. Perhaps both mitochondria and chloroplasts originated from procaryotic endosymbionts which, over a long evolutionary period, were gradually integrated into their host.

Chloroplasts

All eucaryotic organisms with photosynthetic capabilities have chlorophyll-containing organelles called chloroplasts. Only the structure of higher

Table 6-1

Localization of Some Liver Mitochondrial Enzymes

Outer membrane	Intermembrane space	Inner membrane	Matrix
Rotenone—insensitive NADH-Cyt b_5 reductase	Adenylate kinase[a]	Cytochrome b, c, c_1, a, a_3	Malic dehydrogenase
	Nucleoside diphosphokinase	β-Hydroxybutyrate dehydrogenase	Isocitric dehydrogenase
Monoamine oxidase[a]		Ferrochelatase	Glutamic dehydrogenase
Kynurenine hydroxylase		δ-Amino levulinic synthetase	Glutamic–aspartic transaminase
ATP-dependent fatty acyl CoA synthetase		Carnitine palmityl transferase	Citrate synthase
Glycerophosphate acyl transferase		—	Aconitase Fumarase[a]
Lysophosphatidate acyl transferase		Fatty acid elongation enzymes (10)	Pyruvic carboxylase
Lysolecithin acyl transferase		Respiratory chain-linked phosphorylation enzymes	Protein synthesis enzymes
Phosphocholine transferase		Succinic dehydrogenase	Fatty acyl CoA dehydrogenase
Phosphatidate phosphatase		Cytochrome a_3 oxidase[a]	Nucleic acid polymerases
Nucleoside diphosphokinase		Mitochondrial DNA polymerase	ATP-dependent fatty acyl CoA synthetase
Fatty acid elongating system C_{14}–C_{16}		—	GTP-dependent fatty acyl CoA synthetase

[a] Marker enzymes.

plant chloroplasts will be considered here, although lower plants possess these organelles in varying size, shape, and number. In a green leaf a single palisade cell contains approximately 40 chloroplasts. Each chloroplast is about 5–10 μ in diameter and about 2–3 μ thick. About 50% of the dry weight of the chloroplast is protein, 40% lipid, and the remainder water-soluble small molecules. The lipid fraction consists of about 23% chlorophyll (a + b), 5% carotenoids, 5% plastoquinone, 11% phospholipid, 15% digalactosyl diglyceride, 36% monogalactosyl diglyceride, and 5% sulfolipid.

Chloroplasts have a bilayer membrane enclosing the outer envelope. Internally are found a large number of closely packed membraneous structures called lamellae, which contain the chlorophyll of the organelle. In one kind of chloroplast the lamellae are arranged as closely packed disks or stacks called grana (Figure 6-7*a*) which are interconnected to each other by an intergrana or stroma lamellae. The grana stacks are the sites of oxygen evolution and photosynthetic phosphorylation. In other chloroplasts the chlorophyll-containing lamellae are not arranged in stacks but rather extend the length of the organelle (Figure 6-7*b*). As the figures show, a plant species may have both kinds of chloroplasts. The matrix embedding the lamellae is called the stroma and is the site of the carbon photosynthetic enzymes involved in CO_2 fixation, ribosomes, nucleic acid synthesizing enzymes, and fatty acid synthesizing enzymes.

Chloroplasts are extremely fragile osmometers, since only a brief exposure to distilled water will result in a bursting of the outer envelope, loss of stroma protein and marked changes in the appearance of the lamellar systems.

Chloroplasts contain circular chloroplast DNA. Ribosomes are of the 70S species and are very similar to those observed in mitochondria and bacteria. A DNA-dependent RNA polymerase also occurs in intact chloroplasts.

In photosynthesizing procaryotic cells such as *Rhodosprillum rubrum,* small particles, about 60 $m\mu$ in diameter, are attached to the inner surface of the

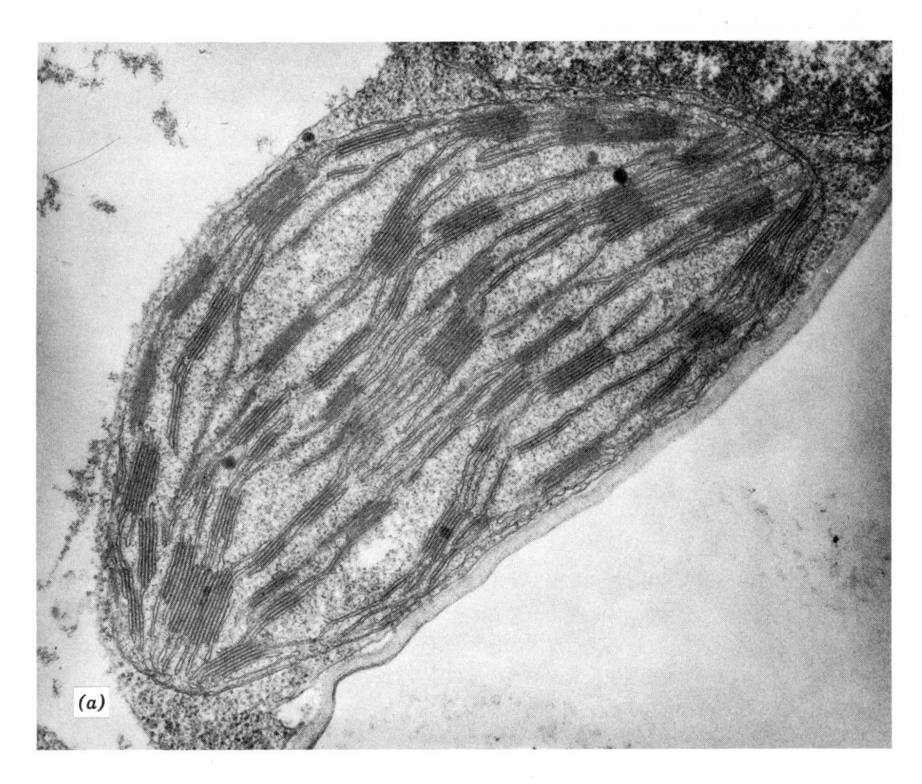

Figure 6-7(*a*)

Chloroplast of sugar cane mesophyll cell showing the grana stacks and the stroma lamellae (magnified 45,900×). (Reproduced with permission of W. M. Laetsch, University of California, Berkeley.)

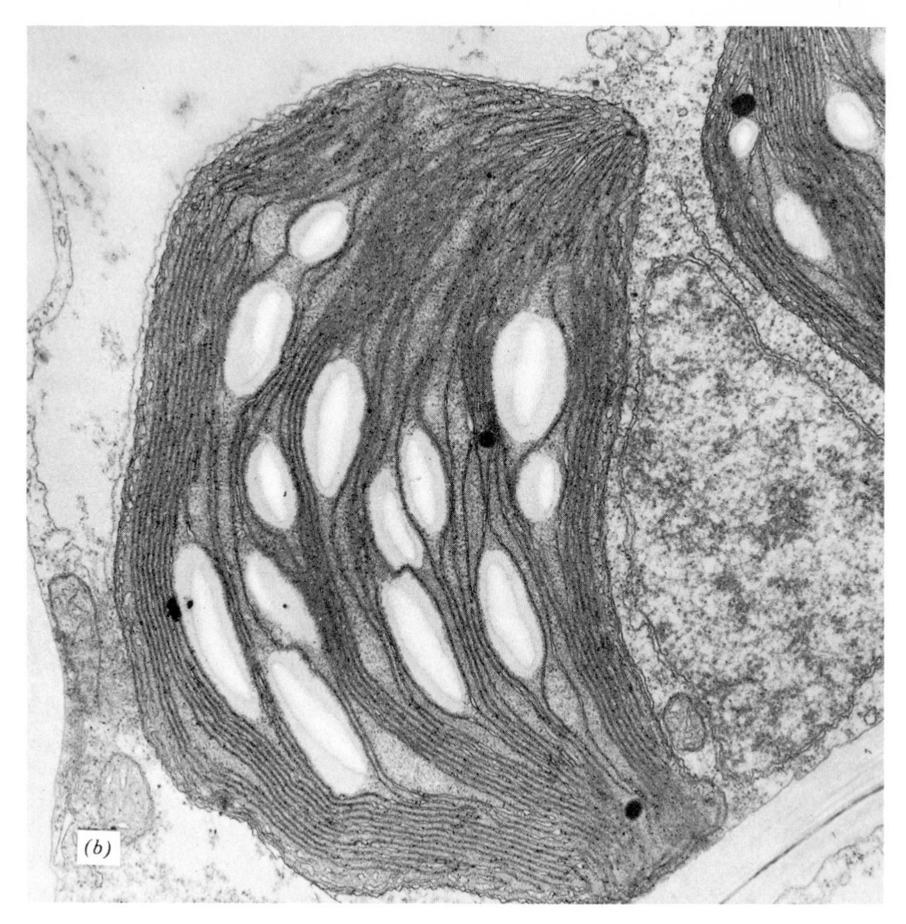

Figure 6-7(*b*)

Chloroplast of sugar cane bundle sheath cell showing the extended lamellar system with an absence of grana stacks (magnified 35,500×). White bodies are starch granules. (Reproduced with permission of W. M. Laetsch, University of California, Berkeley.)

cell membrane. These particles are called chromatophores; they have no limiting membrane and they possess all the bacteriochlorophyll. They are therefore sites of bacterial photosynthesis. In the procaryotic blue-green algae, no discrete chloroplasts are visible, but the photosynthetic lamellar membrane occupies most of the cell. These procaryotic cells therefore seem to be more advanced than the chromatophore-containing bacteria but less developed than eucaryotes such as the green algae which have membrane-limiting chloroplasts.

Lysosomes

In 1955, the Belgium biochemist de Duve discovered and described for the first time a new organelle, the lysosome. Found in all animal cells in varying numbers and types, the lysosome in general is a rather large organelle consisting of a unit membrane enclosing a matrix containing a large number of hydrolytic enzymes which are characterized by having acid pH optima.

Enzyme groups in lysosomes	
Ribonucleases	Cathepsins (6)
Deoxyribonucleases	Glycosidases (10)
Phosphatases (5)	Sulfatases

Collectively, the lysosomal enzymes act on a number of biopolymers. Thus, the proteases have a wide capacity for the hydrolysis of proteins, the acid nucleases for RNA and DNA, and the acid glycosidases for polysaccharides. A family of acid phosphatases also are present. The median value for the pH optima of these enzymes is around pH 5. Thus, the lysosomal matrix must be acidic for the enzymes to be reactive. It is attractive to consider the lysosomal membrane which has a high specific activity for NADH dehydrogenase serving as a hydrogen ion pump. All the enzymes, other than the esterases and the NADH dehydrogenase, are present as soluble proteins in the matrix of the lysosome. In autophagic processes, cellular organelles such as mitochondria and the endoplasmic reticulum undergo digestion within the lysosome.

The biogenesis and mode of action of four types of lysosomes are depicted in Figure 6-8. The biogenesis of a primary lysosome occurs at the periphery of the Golgi body with the lysosome enzymes, presumably synthesized at the ribosomal sites, being collected in Golgi vesicles and then organized as primary lysosomes. Biopolymers may move into a cell by endocytosis (engulfment by the cell membrane, and pinching off to form a phagosome). Primary lysosomes are believed to merge with the phagosomes to form a second type of lysosome, the digestive vacuole, in which under acid pH conditions the biopolymers are broken down to basic units which diffuse out of the vacuole into the cytoplasm to be incorporated again into cellular

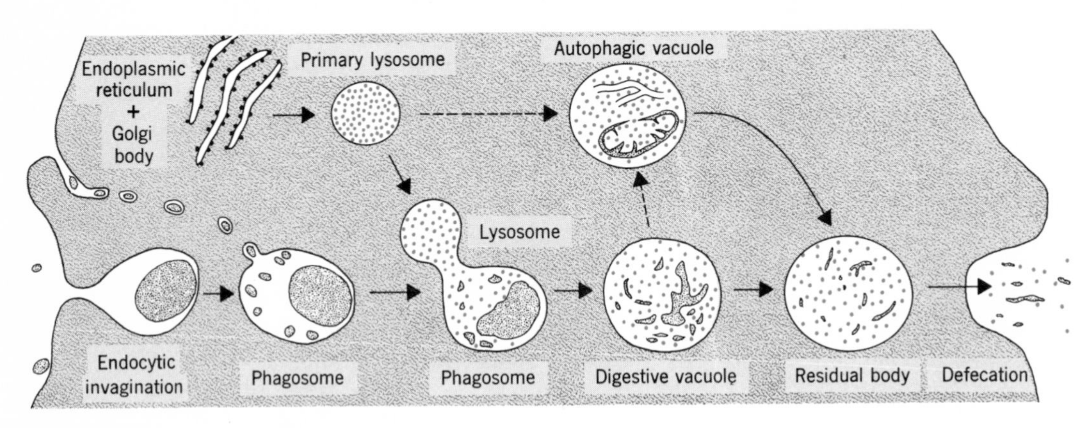

Figure 6-8

The biogenesis of lysosomes according to the investigations of C. de Duve, Rockefeller University.

components. Residual, undigested fragments in the vacuole are then expelled by the cell into the surrounding fluid via the third type, namely the residual body. At times, the primary lysosome engulfs cellular organelles such as mitochondria to form autophagic vacuoles, the fourth general type. In the death of a cell, lysosomal bodies disintegrate, releasing hydrolytic enzymes into the cytoplasm with the result that the cell undergoes autolysis. There is good evidence that in the metamorphosis of tadpoles to frogs, the regression of the tadpole's tail is accomplished by lysosomal digestion of the tail cells. Bacteria are digested by white blood cells by engulfment and lysosomal action. The acrosome, located at the head of the sperm, is a specialized lysosome and is probably involved in some manner in the penetration of the ovum by the sperm. Finally, a number of hereditary diseases involving the abnormal accumulation of complex lipids or polysaccharides in cells of the afflicted individual have now been traced to the absence of key acid hydrolases in the lysosomes of these individuals.

Microbodies

This general term covers a number of single membrane-enclosed cytoplasmic organelles which have been found in both animal and plant cells and which contain H_2O_2-producing oxidases and catalase. The particles are approximately 0.5 mμ in diameter.

In rat liver, these bodies are known as peroxisomes. Urate oxidase, D-amino acid oxidase, and L-amino acid oxidase are always localized in these organelles in addition to catalase. In the liver cell, mitochondria are about 2–5 times more numerous than peroxisomes, with lysosomes being about 1.5 times less numerous than peroxisomes.

In seeds rich in lipid, microbodies called glyoxysomes are the sites of the following enzymes: citrate synthetase, aconitase, the glyoxylate bypass enzymes isocitrate lyase and malate synthetase, malate dehydrogenase, fatty acyl CoA synthetase, crotonase, β-hydroxyacyl CoA dehydrogenase, thiolase, catalase, glutamate:oxalacetate transaminase, glycolic oxidase, and uricase. This organelle, which is present only during a short period in the germination of the lipid-rich seed, and is absent in lipid-poor seed such as the pea, is an example of a highly specialized organelle, beautifully designed to convert fatty acids to C_4 acids which can then be further converted to sucrose, etc. Since carbohydrate-rich seeds do not draw on storage lipids for carbon skeletons or energy, this organelle is appropriately not present. Since the glyoxysome contains, in addition to the glyoxylate bypass enzymes, catalase and a H_2O_2-generating glycolic oxidase, it could be considered as a highly specialized peroxisome.

In general, leaf tissues possess peroxisomes which possibly serve as sites of photorespiration in the leaf cell. This process involves the oxidation of glycolic acid (a product of photosynthetic CO_2 fixation) to CO_2 and H_2O_2. They have a single membrane and granular matrix without lamellae which houses catalase, glycolic oxidase, isozymes of malate dehydrogenase, NADP isocitrate dehydrogenase, and the transaminases glyoxylate:glutamate, hydroxypyruvate:serine, and oxaloacetate:glutamate. They are approximately 1 μ

in width, and they number from a few to one-seventh as many as mitochondria in the leaf cell.

In summary, the microbodies discussed in this section have in common very high catalase activity and one or more H_2O_2-generating oxidases spatially separated from other important sites of metabolism. They possess no sophisticated respiratory chain system nor an energy-conserving system such as occurs in other organelles. While many hypotheses can be raised concerning their function, future experimentation should allow an assessment of their importance in the total economy of the cell.

Golgi Apparatus (Dictyosomes)

Present in both plant and animal cells is a complex organization of net-like tubules or vesicles surrounded by smaller spherical vesicles. The function of this structure is not completely defined, although very recently techniques have been devised for the isolation of these organelles. Evidence strongly suggests that the Golgi bodies participate in the early stage of cell wall synthesis in higher plants as well as the organization of lysosomal structure. Intense phospholipid biosynthesis has also been observed in these organelles. Later we shall discuss the role of the Golgi bodies in the β cells of the islets of Langerhans in the synthesis of insulin (Chapter 19).

References

Cell membranes; mitochondria and chloroplasts

D. Chapman, ed., *Biological Membranes*. New York: Academic Press, 1968.

E. F. Racker, ed., *Membranes of Mitochondria and Chloroplasts*. New York: Van Nostrand Reinhold, 1970.

Transport systems

S. Roseman, *J. General Physiol.* **54,** 138 (1969).

Nucleus

G. Siebert, in *Comprehensive Biochemistry,* M. Florkin and E. H. Stotz, eds. Vol. 23. Amsterdam: Elsevier, 1968, pp. 1–37.

Nucleolus

H. Busch, in *Comprehensive Biochemistry,* M. Florkin and E. H. Stotz, eds. Vol. 23. Amsterdam: Elsevier, 1968, pp. 39–76.

Lysosomes

A. L. Tappel, in *Comprehensive Biochemistry*, M. Florkin and E. H. Stotz, eds. Vol. 23. Amsterdam: Elsevier, 1968, pp. 77–96.

Microbodies

J. F. Hogg, ed., "The Nature and Function of Peroxisomes (Microbodies, Glyoxysomes)." *Ann. N.Y. Acad. Sci.* **168,** 209–381 (1969).

7

Biochemical Energetics

Introduction

Both the procaryotic and the eucaryotic cells described in the preceding chapter exhibit many common features in their *intermediary metabolism*. By this term we mean the sum of chemical reactions which the cell's constituents undergo. In the intact cell, both synthetic and degradative processes go on simultaneously, and energy released from the degradation of some compounds may be utilized in the synthesis of other cellular components. Thus, the concept of an *energy cycle* has developed in biochemistry in which fuel molecules, representing a source of potential chemical energy, are degraded through known enzymatic reactions to produce a few different energy-rich compounds.

Playing a key role in this energy cycle is the ATP–ADP system. ADP (adenosine diphosphate) is able to accept a phosphate group from other energy-rich compounds produced during metabolism and thereby be converted into ATP (adenosine triphosphate). The ATP in turn can be utilized to drive many biosynthetic reactions and in addition serve as a primary source of energy for specific physiological activities such as movement, secretion, absorption, and conduction. In doing so it is often converted back to ADP.

To appreciate the energy changes of the ATP–ADP system as well as other energy reactions in biochemistry, it is necessary to define and understand a few fundamental terms of *thermodynamics,* a science that relates the energy changes which occur in chemical and physical processes.

The Concept of Free Energy

One thermodynamic concept particularly useful to biochemists is *free energy* (G). We may speak of the *free-energy content* of a substance A, but this quantity cannot be measured experimentally. If A is converted to B in a chemical reaction, however,

$$A \rightleftharpoons B \tag{7-1}$$

it is possible to speak of the *change* in *free energy* (ΔG). This is the *maximum amount of energy made available* as A is converted to B. If the free-energy content of the product B (G_B) is less than the free-energy content of the reactant A (G_A), the ΔG will be a negative quantity. That is,

$$\Delta G = G_B - G_A$$
$$= \text{Negative quantity when } G_A > G_B$$

For ΔG to be negative means that the reaction occurs with a decrease in free energy. Similarly, if B is converted back to A, the reaction will involve an increase in free energy, that is, ΔG will be positive. Experience has shown that reactions which occur spontaneously do so with a *decrease* in free energy ($-\Delta G$). On the other hand, if the ΔG for a reaction is known to be positive, that reaction will occur only if energy is supplied to the system in some manner to drive the reaction. Reactions having a negative ΔG are termed *exergonic;* those that have a positive ΔG are called *endergonic.*

Experience has also shown that although the ΔG for a given process is negative, this fact has no relationship whatever to the rate at which the reaction proceeds. For example, glucose can be oxidized by O_2 to CO_2 and H_2O according to equation 7-2:

$$C_6H_{12}O_6 + 6\ O_2 \longrightarrow 6\ CO_2 + 6\ H_2O \qquad (7\text{-}2)$$

The ΔG for this reaction is a very large negative quantity, approximately $-686{,}000$ cal/mole of glucose. The large $-\Delta G$ has no relationship to the rate of the reaction, however. Oxidation of glucose may occur in a matter of a few seconds in the presence of a catalyst in a bomb calorimeter. Reaction 7-2 also goes on in most living organisms at rates varying from minutes to several hours. Glucose can nevertheless be kept in a bottle on the shelf for years in the presence of air without undergoing oxidation.

The free-energy change of a reaction can be related to other thermodynamic properties of A and B by the expression

$$\Delta G = \Delta H - T\Delta S \qquad (7\text{-}3)$$

In this expression, ΔH is the *change in heat content* that occurs as reaction 7-1 proceeds at constant pressure; T is the absolute temperature at which the reaction occurs; and ΔS is the change in *entropy,* a term which expresses the degree of randomness or disorder in a system. The absolute heat H and entropy S contents of substances A and B are difficult to measure, but it is possible to measure the changes in these quantities as they are interconverted in reaction 7-1. The ΔH for a reaction may be measured in a calorimeter, a device for measuring quantitatively the heat produced at constant pressure. To describe the measurement of ΔS and the absolute entropy content of chemical substances is beyond the scope of this book. However, it follows from equation 7-3 that, as the entropy of the products increases over that of the reactants, the $T\Delta S$ term will become more positive and the ΔG will become more negative.

Determination of ΔG

For reaction 7-1 it is possible to derive the expression

$$\Delta G = \Delta G^\circ + RT \ln \frac{[B]}{[A]} \qquad (7\text{-}4)$$

where ΔG° is the *standard change in free energy*, soon to be defined; R is the universal gas constant; T is the absolute temperature; and [B] and [A] are the concentrations of A and B in moles per liter. Precisely, [B] and [A] should be replaced by the activities of A and B, a_A and a_B. As with pH, however, this correction is not usually made, because the activity coefficients are seldom known for the concentrations of compounds existing in the cell.

From equation 7-4 the ΔG for a reaction is a function of the concentrations of reactant and product as well as the standard free-energy change ΔG°. It is possible to evaluate ΔG° if we consider the ΔG at equilibrium. At equilibrium there is no net conversion of A to B, and hence the change in free energy ΔG is 0. Similarly, the ratio of [B] to [A] is the ratio at equilibrium or the equilibrium constant K_{eq}. Substituting these quantities in equation 7-4,

$$0 = \Delta G^\circ + RT \ln K_{eq}$$
$$\Delta G^\circ = -RT \ln K_{eq} \qquad (7\text{-}5)$$

When the constants are evaluated ($R = 1.987$ cal/mole/degree, $25^\circ C = 298^\circ T$, and $\ln x = 2.303 \log_{10} x$), the equation becomes (at 25°C)

$$\Delta G^\circ = -(1.987)(298)(2.303) \log_{10} K_{eq}$$
$$= -1363 \log_{10} K_{eq} \qquad (7\text{-}6)$$

This equation relating the ΔG° to K_{eq} is an extremely useful way to determine the ΔG° for a specific reaction. If the concentration of both reactants and products at equilibrium can be measured, the K_{eq} and in turn the ΔG° of the reaction can be calculated. Of course, if the K_{eq} is extremely large or extremely small, this method of measuring ΔG° is of little value, because the equilibrium concentration of the reactants and products, respectively, will be too small to measure. The ΔG° for each of a series of K_{eq} ranging from 0.001 to 10^3 is calculated in Table 7-1.

From inspection of Table 7-1 it is clear that reactions which have a K_{eq} greater than 1 proceed with a decrease in free energy. Thus, for reaction 7-1, if the $K_{eq} = 1000$ (that is, if [B]/[A] is 1000), the tendency is for the reaction to proceed in the direction of the formation of B. If we start with 1001 parts of A, equilibrium will be reached only when 1000 parts (or 99.9%) of A have been converted to B. If reaction 7-1 has a K_{eq} of 10^{-3} (that is, if [B]/[A] = 0.001), equilibrium will be attained when only 1 part or 0.1% of A has been converted to B.

Table 7-1

Relation between K_{eq} and ΔG°

K_{eq}	$\log_{10} K_{eq}$	$\Delta G^\circ = -1363 \log_{10} K_{eq}$ (cal)
0.001	−3	4089
0.01	−2	2726
0.1	−1	1363
1.0	0	0
10	1	−1363
100	2	−2726
1000	3	−4089

It is also possible to evaluate $\Delta G°$ for the situation where both the reactants and products are present at unit concentrations. When [A] = [B] = 1*M*, equation 7-4 becomes

$$\Delta G = \Delta G° + RT \ln \frac{1}{1}$$
$$= \Delta G°$$

Thus, $\Delta G°$ may be defined as the change in free energy when reactants and products are present in unit concentration, or more broadly, in their "standard state." The standard state for solutes in solution is unit molarity; for gases, 1 atm; for solvents such as water, unit activity. If water is a reactant or a product of a reaction, its concentration in the standard state is taken as unity in the expression for the ΔG (equation 7-4). If a gas is either formed or produced, its standard state concentration is taken as 1 atm. If a hydrogen ion is produced or utilized in a reaction, its concentration will be taken at 1*M* or pH = 0.

Since in the cell few if any reactions occur at pH 0 but rather at pH 7.0, the standard free-energy change $\Delta G°$ is frequently corrected for the difference in pH. Conversely, the equilibrium of a reaction may be measured at some pH other than 0. The standard free-energy change $\Delta G°$ at any pH other than 0 is designated as $\Delta G'$, and the pH for a given $\Delta G'$ should be indicated. Of course, if a proton is neither formed nor utilized in the reaction, $\Delta G'$ will be independent of pH and $\Delta G°$ will equal $\Delta G'$.

An example will demonstrate the use of these terms. In the presence of the enzyme phosphoglucomutase, glucose-1-phosphate is converted to glucose-6-phosphate. Starting with 0.020*M* glucose-1-phosphate at 25°C, it is observed that the concentration of this compound decreases to 0.001*M* while the concentration of glucose-6-phosphate increases to 0.019*M*. The K_{eq} of the reaction is 0.019 divided by 0.001, or 19. Therefore,

$$\begin{aligned} \Delta G° &= -RT \ln K_{eq} \\ &= -1363 \log_{10} K_{eq} \\ &= -1363 \log_{10} 19 \\ &= (-1363)(1.28) \\ &= -1745 \text{ cal} \end{aligned}$$

The $\Delta G°$ for this reaction will be independent of pH, since acid is neither produced nor used up in the reaction. This amount of free-energy decrease (−1745 cal) will occur when 1 mole of glucose-1-phosphate is converted to 1 mole of glucose-6-phosphate under such conditions that the *concentration of each compound is maintained at 1 M*, a situation quite different from the experimental situation just described for measuring the K_{eq}. Indeed, these conditions of *unit molarity* are difficult to maintain either in the test tube or in the cell. It should be pointed out, however, that the concentration of a particular substance (for example, glucose-6-phosphate) may frequently be maintained relatively constant at some concentration over a time interval, since it may be produced in one reaction while it is being used up in another. This condition of *steady-state* equilibrium undoubtedly exists in many biological systems and requires that thermodynamics be applied to the steady-state condition rather than to the equilibrium condition for which thermodynamics was first developed. A second complication is that the thermodynamic quantities discussed in this chapter apply only to reactions occurring in homogenous systems, whereas much metabolism occurs in heterogenous systems involving more than one phase. As a result, most of the values reported in the literature cannot be considered more than 10% accurate. Nevertheless, the concept of the standard free-energy change has found many fruitful applications in intermediary metabolism.

Energy-Rich Compounds

In all living forms, one compound repeatedly functions as a common reactant linking endergonic processes to others that are exergonic. This compound, adenosine triphosphate (ATP), is one of a group of "energy-rich" or "high-energy" compounds whose structure will now be considered. They are called "energy-rich" or "high-energy" compounds because they exhibit a large decrease in free energy when they undergo hydrolytic reactions. They are in general unstable to acid, to alkali, and to heat. In subsequent chapters their synthesis and utilization will be described in detail.

Pyrophosphate Compounds. Let us now consider the structure of ATP and its partner ADP in more detail. At pH 7.0 in aqueous solution, ATP and ADP are anions bearing a net charge of −4 and −3, respectively. This results

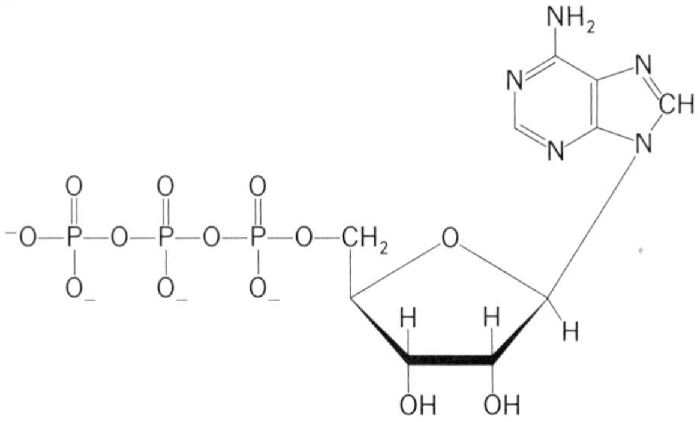

Adenosine triphosphate (ATP)

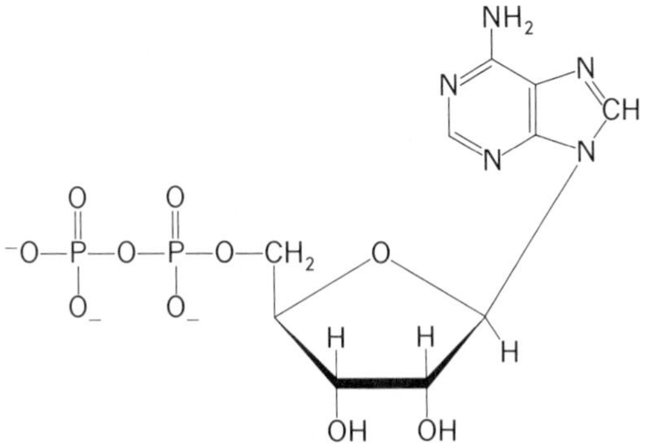

Adenosine diphosphate (ADP)

from the fact that the two dissociable protons on the interior phosphates of ATP (and the one interior phosphate of ADP) are primary hydrogens with a pK_a's in the range 2–3. The terminal phosphate of ATP (and ADP) have both a primary hydrogen with pK_a of 2–3 and a secondary hydrogen with pK_a of 6.5. Therefore, at pH 7.0, the primary hydrogen will be completely ionized and the secondary will be about 75% dissociated. In the cell, however, where a relatively high concentration of Mg^{2+} exists, both ATP and ADP will be complexed with this cation in a one-to-one ratio to form divalent and monovalent complexes, respectively.

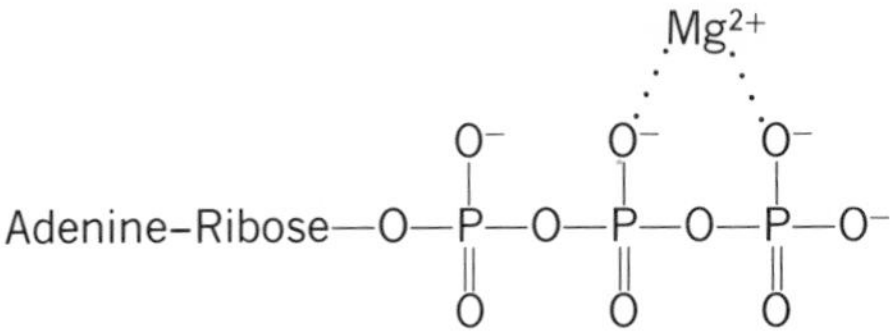

Adenine–Ribose—O—P(=O)(O⁻)—O—P(=O)(O⁻)—O—P(=O)(O⁻)—O⁻ (with Mg^{2+} coordinated to the two terminal O⁻)

$[ATP\text{-}Mg]^{2-}$ complex

Adenine–Ribose—O—P(=O)(O⁻)—O—P(=O)(O⁻)—O⁻ (with Mg^{2+} coordinated to the two O⁻)

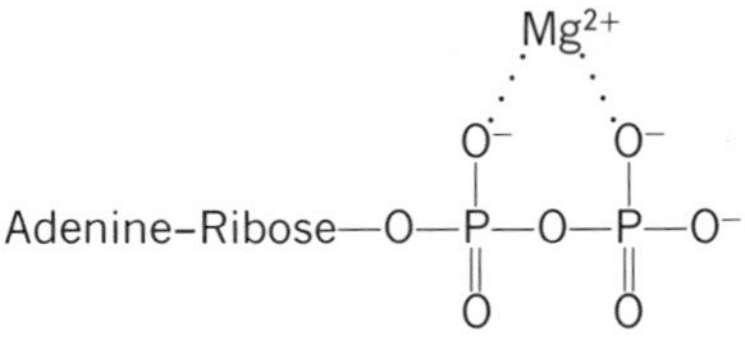

$[ADP\text{-}Mg]^{-}$ complex

It is informative to compare the $\Delta G'$ of hydrolysis of ATP with that of other phosphate compounds. The hydrolysis of the terminal phosphate of ATP may be written as in reaction 7-7:

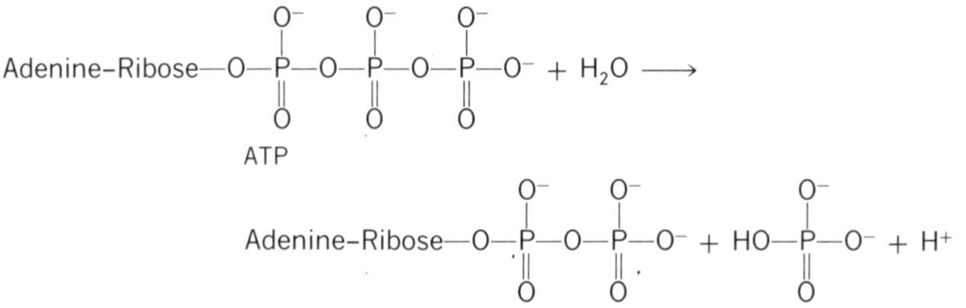

$$\text{Adenine–Ribose—O—P(=O)(O^-)—O—P(=O)(O^-)—O—P(=O)(O^-)—O^-} \;(\text{ATP}) + H_2O \longrightarrow$$

$$\text{Adenine–Ribose—O—P(=O)(O^-)—O—P(=O)(O^-)—O^-} \;(\text{ADP}) + \text{HO—P(=O)(O^-)—O^-} + H^+ \quad (7\text{-}7)$$

$\Delta G' = -7300$ cal (pH 7.0)

The $\Delta G'$ at pH 7 has been estimated to be -7300 cal/mole. This is in contrast to the hydrolysis of glucose-6-phosphate, which results in a much smaller decrease in free energy.

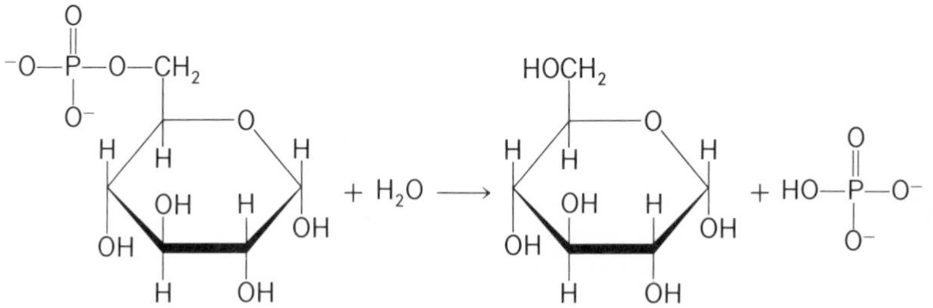

$$\text{Glucose-6-phosphate } (^-O\text{—P(=O)(O}^-\text{)—O—CH}_2\text{—glucopyranose}) + H_2O \longrightarrow \text{Glucose (HOCH}_2\text{—glucopyranose)} + \text{HO—P(=O)(O}^-\text{)—O}^- \quad (7\text{-}8)$$

$\Delta G' = -3300$ cal (pH 7.0)

We may properly ask why this large difference in the free energy of hydrolysis exists. On examining the several types of energy-rich compounds encountered in intermediary metabolism, we note several factors which are important but not all of which apply to every energy-rich compound. Regardless of the specific factors involved, it will be seen that the large decrease in free energy occurs during hydrolysis because the products are significantly *more stable* than the reactants. Important factors contributing to this stability are:

(1) Bond strain in the reactant caused by electrostatic repulsion.
(2) Stabilization of the products by ionization.
(3) Stabilization of the products by isomerization.
(4) Stabilization of the products by resonance.

In the case of ATP, the structure of importance in determining its character as an energy-rich compound is the pyrophosphate moiety which, at pH 7.0, is fully ionized:

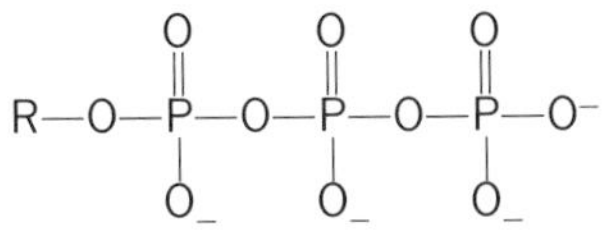

$$R{-}O{-}\overset{\overset{O}{\|}}{\underset{\underset{O^-}{|}}{P}}{-}O{-}\overset{\overset{O}{\|}}{\underset{\underset{O^-}{|}}{P}}{-}O{-}\overset{\overset{O}{\|}}{\underset{\underset{O^-}{|}}{P}}{-}O^-$$

There will be a tendency for the electrons in the P=O bond of the phosphates to be drawn closer to the *electronegative* oxygen atom, thereby producing a *partial negative charge* (δ^-) on that atom. This is compensated by a *partial positive charge* (δ^+) on the phosphorus atom resulting in a *polarization* of the phosphorus–oxygen bonding which may be indicated as:

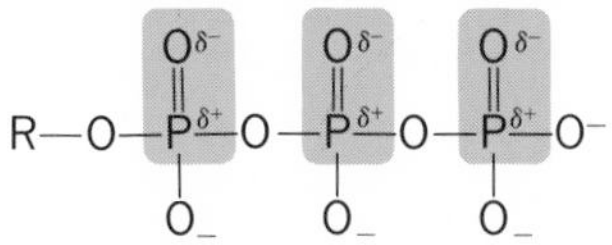

$$R{-}O{-}\overset{\overset{O^{\delta-}}{\|}}{\underset{\underset{O^-}{|}}{P^{\delta+}}}{-}O{-}\overset{\overset{O^{\delta-}}{\|}}{\underset{\underset{O^-}{|}}{P^{\delta+}}}{-}O{-}\overset{\overset{O^{\delta-}}{\|}}{\underset{\underset{O^-}{|}}{P^{\delta+}}}{-}O^-$$

The existence of residual positive charges of this nature on adjacent phosphorus atoms in the pyrophosphate structures of ATP (and ADP) means that these molecules must contain sufficient internal energy to overcome the electrostatic repulsion between the adjacent like charges. When the pyrophosphate structure is cleaved, as on hydrolysis, this energy will be released and will contribute to the total negative ΔG of the reaction. Although the P=O bond in glucose-6-phosphate can also be considered to have polar character, there is no adjacent phosphorus atom with a δ^+ charge:

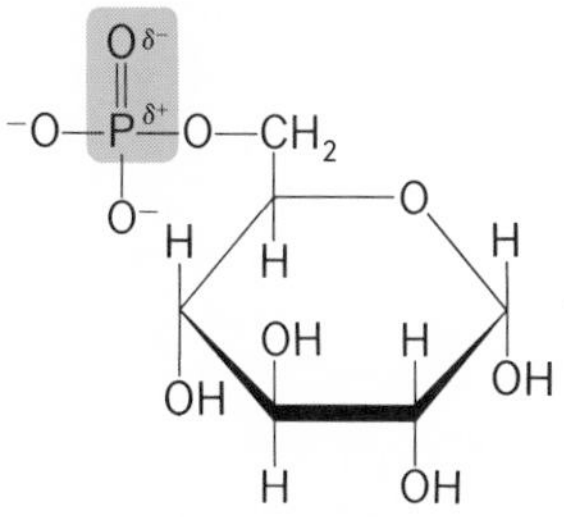

$$^-O{-}\overset{\overset{O^{\delta-}}{\|}}{\underset{\underset{O^-}{|}}{P^{\delta+}}}{-}O{-}CH_2 \text{ (glucose ring)}$$

The argument for instability due to charge repulsion does not exist with this compound, and the ΔG of hydrolysis will be less for this reason.

Obviously, the same factor applies in the hydrolysis of ADP to AMP and

$$\text{Adenine-Ribose—O—}\underset{\underset{O^{\delta-}}{\|}}{\overset{\overset{O^-}{|}}{P^{\delta+}}}\text{—O—}\underset{\underset{O^{\delta-}}{\|}}{\overset{\overset{O^-}{|}}{P^{\delta+}}}\text{—O}^- + H_2O \longrightarrow$$

ADP

$$\text{Adenine-Ribose—O—}\underset{\underset{O^{\delta-}}{\|}}{\overset{\overset{O^-}{|}}{P^{\delta+}}}\text{—O}^- + \text{HO—}\underset{\underset{O^{\delta-}}{\|}}{\overset{\overset{O^-}{|}}{P^{\delta+}}}\text{—O}^- + H^+ \quad (7\text{-}9)$$

AMP

$\Delta G' = -6500$ cal (pH 7.0)

inorganic phosphate where the observed $\Delta G'$ on hydrolysis at pH 7 is -6500 cal/mole. On the other hand, the hydrolysis of AMP to adenosine and H_3PO_4 is less ($\Delta G' = -2200$, pH 7) for lack of the same reason.

$$\text{Adenine-Ribose—O—}\underset{\underset{O^{\delta-}}{|}}{\overset{\overset{O^-}{\|}}{P^{\delta+}}}\text{—O}^- + H_2O \longrightarrow \text{Adenine-Ribose—OH} + \text{HO—}\underset{\underset{O^{\delta-}}{\|}}{\overset{\overset{O^-}{|}}{P^{\delta+}}}\text{—O}^- \quad (7\text{-}10)$$

AMP Adenosine

$\Delta G' = -2200$ cal (pH 7.0)

Although ATP is converted to ADP in many reactions of intermediary metabolism, there are a number of important reactions in which the interior pyrophosphate bond of ATP is cleaved to yield AMP and inorganic pyrophosphate:

$$\text{Adenine-Ribose—O—}\underset{\underset{O}{\|}}{\overset{\overset{O^-}{|}}{P}}\text{—O—}\underset{\underset{O}{\|}}{\overset{\overset{O^-}{|}}{P}}\text{—O—}\underset{\underset{O}{\|}}{\overset{\overset{O^-}{|}}{P}}\text{—O}^- + H_2O \longrightarrow$$

ATP

$$\text{Adenine-Ribose—O—}\underset{\underset{O}{\|}}{\overset{\overset{O^-}{|}}{P}}\text{—O}^- + {}^-\text{O—}\underset{\underset{O}{\|}}{\overset{\overset{O^-}{|}}{P}}\text{—O—}\underset{\underset{O}{\|}}{\overset{\overset{O^-}{|}}{P}}\text{—O}^- + 2\,H^+ \quad (7\text{-}11)$$

AMP

$\Delta G' = -8600$ cal (pH 7.0)

This type of cleavage is known as the *pyrophosphate cleavage* and is in contrast to the *orthophosphate cleavage* in which ADP is formed (reaction 7-7).

The ATP–ADP system is functional in nature because ADP, having been formed from ATP, can be rephosphorylated in energy-yielding reactions and be converted back to ATP. It is critical, therefore, that AMP and the pyrophosphate formed in the pyrophosphate cleavage be converted back to ATP. This is accomplished by two reactions catalyzed by enzymes widely distributed in nature. The first of these reactions, catalyzed by a pyrophosphatase, is the hydrolysis of pyrophosphate to yield 2 moles of inorganic phosphate:

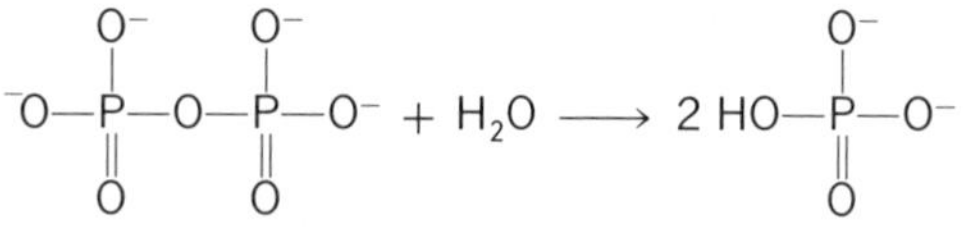

$$^{-}O-\underset{\underset{O}{\|}}{\overset{\overset{O^-}{|}}{P}}-O-\underset{\underset{O}{\|}}{\overset{\overset{O^-}{|}}{P}}-O^- + H_2O \longrightarrow 2\ HO-\underset{\underset{O}{\|}}{\overset{\overset{O^-}{|}}{P}}-O^- \qquad (7\text{-}12)$$

Pyrophosphate

$$\Delta G' = -8000 \text{ cal (pH 7.0)}$$

The second reaction is one in which ATP and AMP react to form 2 moles of ADP, which in turn can be further phosphorylated in several different energy-yielding reactions to regenerate ATP:

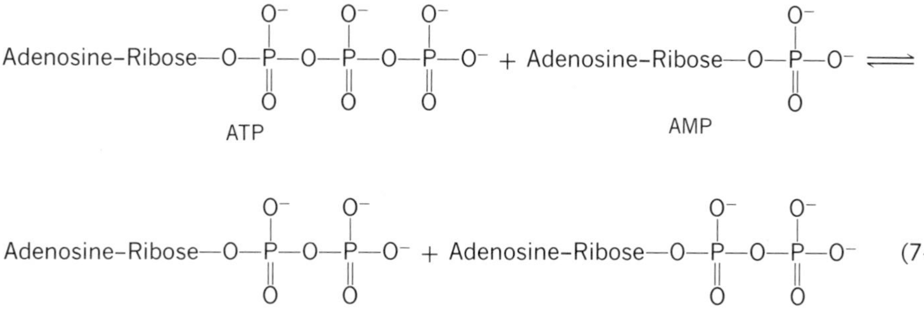

$$\underset{\text{ATP}}{\text{Adenosine–Ribose}-O-\underset{\underset{O}{\|}}{\overset{\overset{O^-}{|}}{P}}-O-\underset{\underset{O}{\|}}{\overset{\overset{O^-}{|}}{P}}-O-\underset{\underset{O}{\|}}{\overset{\overset{O^-}{|}}{P}}-O^-} + \underset{\text{AMP}}{\text{Adenosine–Ribose}-O-\underset{\underset{O}{\|}}{\overset{\overset{O^-}{|}}{P}}-O^-} \rightleftharpoons$$

$$\underset{\text{ADP}}{\text{Adenosine–Ribose}-O-\underset{\underset{O}{\|}}{\overset{\overset{O^-}{|}}{P}}-O-\underset{\underset{O}{\|}}{\overset{\overset{O^-}{|}}{P}}-O^-} + \underset{\text{ADP}}{\text{Adenosine–Ribose}-O-\underset{\underset{O}{\|}}{\overset{\overset{O^-}{|}}{P}}-O-\underset{\underset{O}{\|}}{\overset{\overset{O^-}{|}}{P}}-O^-} \qquad (7\text{-}13)$$

The $\Delta G'$ for this reaction is approximately 0 because the K_{eq} is approximately 1.0.

Other Energy-Rich Compounds

Examination of the means by which ADP can be converted back to ATP introduces two other energy-rich phosphate compounds, 1,3-diphosphoglyceric acid and phosphoenolpyruvic acid. Both of these are encountered during the conversion of glucose to pyruvic acid (see Chapter 10) and both have standard free energies of hydrolysis more negative than that of ATP.

Acyl Phosphates. 1,3-Diphosphoglyceric acid is an example of an acyl phosphate; its standard free energy of hydrolysis is -11.8 kcal/mole:

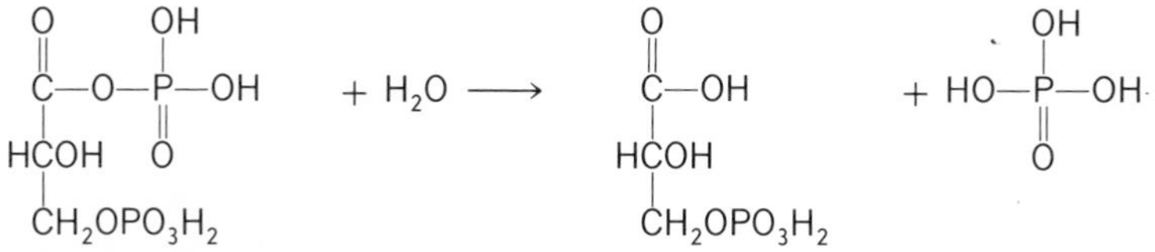

$$\begin{array}{l} \overset{\overset{O}{\|}}{C}-O-\underset{\underset{O}{\|}}{\overset{\overset{OH}{|}}{P}}-OH \\ | \\ HCOH \\ | \\ CH_2OPO_3H_2 \end{array} + H_2O \longrightarrow \begin{array}{l} \overset{\overset{O}{\|}}{C}-OH \\ | \\ HCOH \\ | \\ CH_2OPO_3H_2 \end{array} + HO-\underset{\underset{O}{\|}}{\overset{\overset{OH}{|}}{P}}-OH \qquad (7\text{-}14)$$

1,3-Diphosphoglyceric acid — 3-Phosphoglyceric acid

$$\Delta G' = -11{,}800 \text{ cal (pH 7.0)}$$

Bond strain in the acyl phosphate is a significant factor contributing to the large negative standard free energy of hydrolysis of this class of compounds. The C=O bond of the acyl phosphate group may be considered also to have considerable polar character because of the tendency for the electrons in the double bond to be drawn closer to the electronegative oxygen. Energy is required to overcome the repulsion between the partial positive charges on the carbon and phosphorus atoms, such energy being released on hydrolysis of the acyl phosphate.

The relative tendencies of reactants and products to ionize at a particular pH have an important influence on the ΔG of a reaction. This factor may also be seen in the case of 1,3-diphosphoglyceric acid. In reaction 7-14, the ionization of the reactants and products has not been indicated in the formulas. At pH 7 the reaction is more accurately represented as

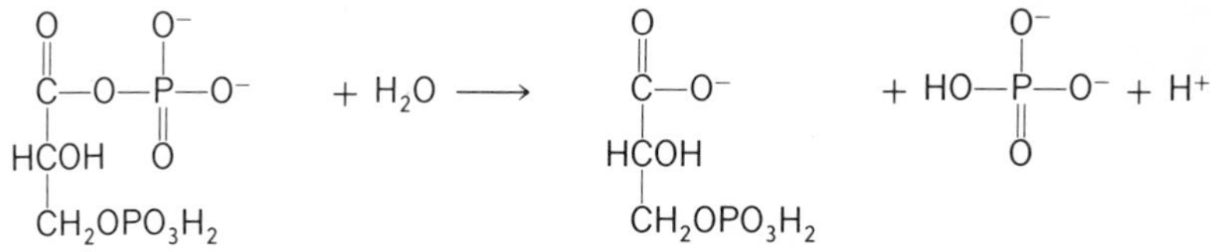

$$\underset{\text{1,3-Diphosphoglyceric acid}}{\begin{array}{l} \overset{O}{\overset{\|}{C}}-O-\overset{O^-}{\overset{|}{P}}-O^- \\ \quad\quad\quad\;\; \| \\ HCOH \quad O \\ | \\ CH_2OPO_3H_2 \end{array}} + H_2O \longrightarrow \underset{\text{3-Phosphoglyceric acid}}{\begin{array}{l} \overset{O}{\overset{\|}{C}}-O^- \\ | \\ HCOH \\ | \\ CH_2OPO_3H_2 \end{array}} + HO-\overset{O^-}{\overset{|}{\underset{\underset{O}{\|}}{P}}}-O^- + H^+ \qquad (7\text{-}15)$$

where the primary and secondary hydrogen ions are ionized, while the tertiary hydrogen (on the inorganic phosphate) is not. The carboxylic acid group ($pK = 4.8$) formed on hydrolysis will also be extensively ionized. The effect of this ionization is to reduce the concentration of the actual hydrolysis product (nonionized acid) to a low level.

It should be stressed that the extent to which ionization is a factor in the $\Delta G'$ of the reaction (i.e., the extent to which products are stabilized in a reaction) will be dependent on the *difference* in the pK_a of the newly formed ionizable group and the pH at which the reaction occurs. It may be shown that the contribution of a new group with a pK_a of 1 unit *less* than the pH of the medium is -1363 cal/mole. If reaction 7-15 were to occur at an acid pH (something less than 3) where the newly formed 3-phosphoglyceric acid is not significantly ionized, the ionization factor would contribute little to the $\Delta G'$ of hydrolysis of 1,3-diphosphoglyceric acid.

Enolic Phosphate. The second compound encountered during the conversion of glucose to pyruvate that provides for the regeneration of ATP from ADP is phosphoenolpyruvic acid (PEP). The free-energy change on hydrolysis of this energy-rich *enolic phosphate* is $-14,800$ cal at pH 7.0:

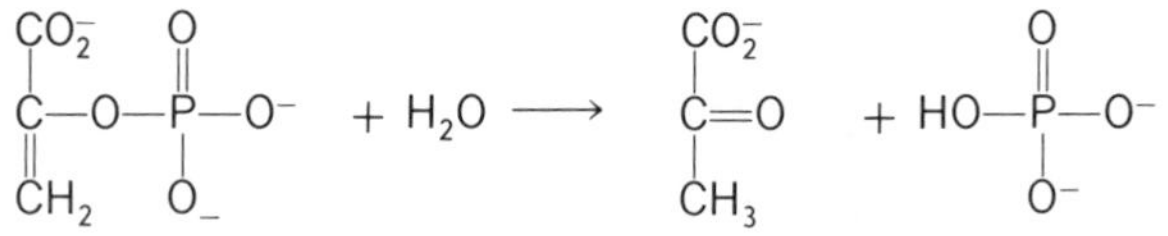

$$\underset{\text{Phosphoenolpyruvic acid}}{\begin{array}{l} CO_2^- \quad O \\ | \qquad\quad \| \\ C-O-P-O^- \\ \| \qquad\quad | \\ CH_2 \quad\; O^- \end{array}} + H_2O \longrightarrow \underset{\text{Pyruvic acid (keto form)}}{\begin{array}{l} CO_2^- \\ | \\ C{=}O \\ | \\ CH_3 \end{array}} + HO-\overset{O}{\overset{\|}{\underset{\underset{O^-}{|}}{P}}}-O^- \qquad (7\text{-}16)$$

$$\Delta G' = -14,800 \text{ cal (pH 7.0)}$$

One can appreciate the large negative ΔG observed on hydrolysis of this compound if one recognizes that the inherently unstable enolic form of pyruvic acid is stabilized in PEP by the phosphate ester group. On hydrolysis, the unstable enol may be thought of as being formed, but it will instantly isomerize to the much more stable keto structure:

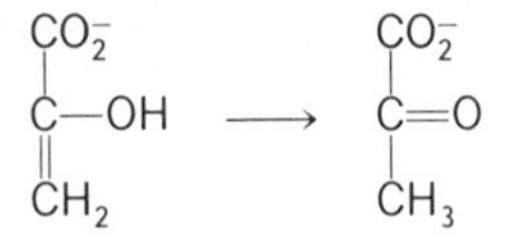

$$\underset{\substack{\text{Enol form}\\\text{(unstable)}}}{\begin{array}{c}CO_2^-\\|\\C{-}OH\\||\\CH_2\end{array}} \longrightarrow \underset{\text{Keto form}}{\begin{array}{c}CO_2^-\\|\\C{=}O\\|\\CH_3\end{array}}$$

It is estimated that the tautomerization occurs with a decrease in $\Delta G'$ of about 8000 cal/mole, therefore bringing the total $\Delta G'$ to $-14{,}800$ cal/mole. This tautomerization is of major importance in making PEP one of the most "energy-rich" phosphate compounds of biological importance.

Thiol Esters. A third type of energy-rich compound that can in turn be utilized to generate ATP from ADP (see Chapter 12) is the thioester, acetyl coenzyme A. The $\Delta G'$ of hydrolysis of this compound is approximately -7500 cal:

$$H_3C{-}\overset{\overset{\displaystyle O}{||}}{C}{-}S{-}CoA + H_2O \longrightarrow H_3C{-}\overset{\overset{\displaystyle O}{||}}{C}{-}O^- + CoA{-}SH + H^+$$
$$\Delta G' = -7500 \text{ cal (pH 7.0)}$$

An explanation for this larger $\Delta G'$ of hydrolysis is given on page 225, where the unique properties of thioesters are discussed in detail.

Guanidinium Phosphates. A fourth type of energy-rich compound that plays an important role in energy transfer and storage is the guanidinium phosphate. This type of structure is found in phosphocreatine and phosphoarginine, compounds also known as phosphagens. The phosphagens are formed by

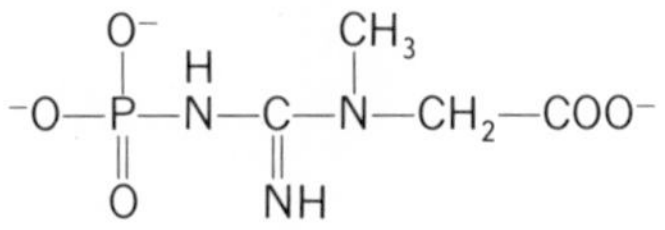

$$\underset{\text{Phosphocreatine}}{{}^-O{-}\overset{\overset{\displaystyle O^-}{|}}{\underset{\underset{\displaystyle O}{||}}{P}}{-}\overset{H}{N}{-}\underset{\underset{\displaystyle NH}{||}}{C}{-}\overset{\overset{\displaystyle CH_3}{|}}{N}{-}CH_2{-}COO^-} \qquad \underset{\text{Phosphoarginine}}{{}^-O{-}\overset{\overset{\displaystyle O^-}{|}}{\underset{\underset{\displaystyle O}{||}}{P}}{-}\overset{H}{N}{-}\underset{\underset{\displaystyle NH}{||}}{C}{-}\overset{H}{N}{-}(CH_2)_4{-}\underset{\underset{\displaystyle NH_3^+}{|}}{CH}{-}COO^-}$$

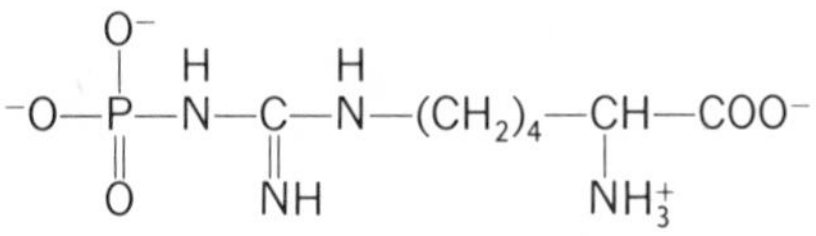

the phosphorylation of creatine or arginine with ATP in the presence of the appropriate enzyme.

$$\text{Phosphocreatine} + \text{ADP} \rightleftharpoons \text{Creatine} + \text{ATP} \qquad (7\text{-}17)$$
$$\Delta G' = -3000 \text{ cal (pH 7.0)}$$

Since, however, the standard free-energy change on hydrolyzing of these compounds is more negative by about -3000 than that of ATP, the equilibrium actually favors ATP formation. Phosphagens carry out their physiological role by furnishing a place to store energy-rich phosphate. When the concentration of ATP is high, reaction 7-17 proceeds from right to left and phosphate is stored as energy-rich phosphocreatine. Then, when the level of ATP is depleted, reaction 7-17 proceeds from left to right, and ATP concentration is increased.

The guanidinium phosphates, represented by creatine phosphate, are not inherently less stable because of bond strain as in the case of ATP and ADP. There are no obvious ionization or tautomerization processes which account for greater stability of the products over their reactants as in the case of the acyl and enolic phosphates:

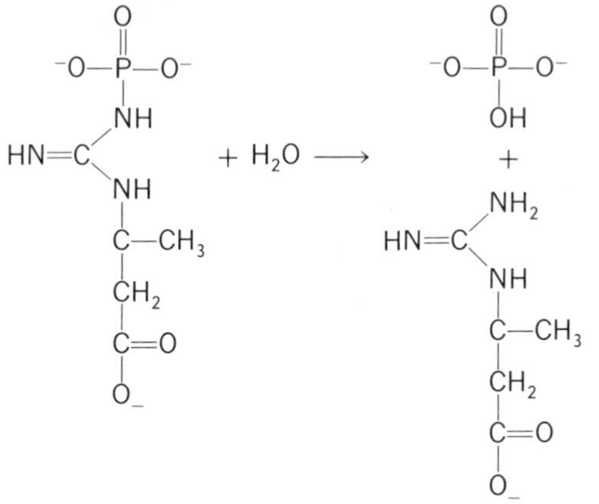

$$\text{creatine phosphate} + H_2O \longrightarrow HPO_4^{2-} + \text{creatine} \qquad (7\text{-}18)$$

$$\Delta G' = -10{,}300 \text{ cal (pH 7.0)}$$

Nevertheless, the hydrolysis products are significantly more stable than the guanidinium phosphate since one can write a greater number of *resonance forms* for the products than for the reactants. Creatine phosphate possesses twelve possible resonance forms, three of which are shown as structures I–III.

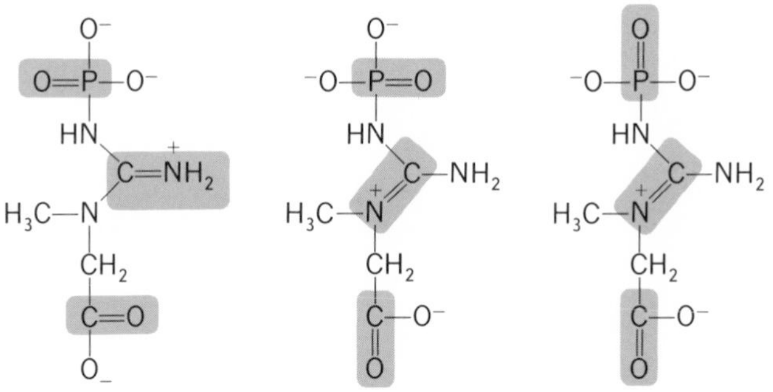

I II III

When, however, creatine lacks its phosphate group, one can write an increased number of resonance isomers which include structure IV, in which a positive charge is placed on the nitrogen atom formerly linked to the phosphate group. Since, in creatine phosphate, there is no *oxygen* atom between the P atom of the phosphate group and the ureido nitrogen, the

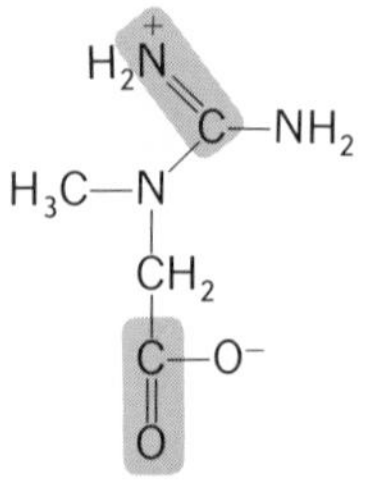

IV

partial positive charge on phosphorus would prevent a similar charge on an adjacent atom.

The five types of compounds discussed above may be contrasted with such compounds as glucose-6-phosphate or *sn*-glycerol-3-phosphate, which are phosphoric acid esters of organic alcohols and which have relatively small values for the $\Delta G'$ of hydrolysis. When all these compounds are listed in Table 7-2 one can see that there is no sharp division between "energy-rich" and "energy-poor" compounds, and that several compounds including ATP occupy intermediate positions in the table. For that matter, the unique ability of ATP to participate in so many different reactions involving energy transfer may be ascribed to its truly intermediate position between the acyl and enolic phosphates which are generated in the breakdown of fuel molecules and the numerous acceptor molecules which in the course of their metabolism are phosphorylated.

In the past it has been common practice in biochemistry to refer to high-energy and low-energy phosphate bonds. Lipmann introduced the symbol ~ph to indicate a high-energy phosphate structure. This practice has resulted in the tendency to think of the energy as concentrated in the single chemical bond. This is erroneous, because as the discussion above has stressed, the free energy change ΔG depends on the structure of the compound hydrolyzed and the products of hydrolysis. Moreover, the ΔG refers specifically to the chemical reaction involved, namely, the *hydrolysis* of the compound.

Coupling of Reactions

In the cell, the energy made available in an exergonic reaction is frequently utilized to drive a related endergonic reaction and thereby it is made to do work. This is accomplished by coupling reactions which have *common intermediates*. A specific example can best illustrate this important principle.

Table 7-2

Standard Free Energy of Hydrolysis of Some Important Metabolites

	$\Delta G'$ at pH 7.0 (cal)
Phosphoenolpyruvate	−14,800
1,3-Diphosphoglycerate	−11,800
Phosphocreatine	−10,300
Acetyl phosphate	−10,100
Pyrophosphate	−8,000
Acetyl–CoA	−7,500
ATP to ADP and Pi	−7,300
ATP to AMP and pyrophosphate	−8,600
ADP	−6,500
Glucose-1-phosphate	−5,000
Fructose-6-phosphate	−3,800
Glucose-6-phosphate	−3,300
L-Glycerol-3-phosphate	−2,200

During the conversion of glucose to lactic acid (or alcohol) the phosphorylated triose D-glyceraldehyde-3-phosphate is oxidized to 3-phosphoglyceric acid (Chapter 10). This reaction may be represented as the removal of two hydrogen atoms from the hydrated form of the aldehyde:

$$\begin{matrix} \text{H}\diagdown\text{C}{=}\text{O} \\ | \\ \text{HCOH} \\ | \\ \text{CH}_2\text{OPO}_3\text{H}_2 \end{matrix} + \text{H}_2\text{O} \longrightarrow \left[\begin{matrix} \text{OH} \\ | \\ \text{H—C—OH} \\ | \\ \text{HCOH} \\ | \\ \text{CH}_2\text{OPO}_3\text{H}_2 \end{matrix}\right] \longrightarrow \begin{matrix} \text{O} \\ \| \\ \text{C—OH} \\ | \\ \text{HCOH} \\ | \\ \text{CH}_2\text{OPO}_3\text{H}_2 \end{matrix} + 2\,\text{H}\cdot \qquad (7\text{-}19)$$

The $\Delta G'$ for reaction 7-19 is approximately $-12{,}000$ cal, indicating that the reaction is not readily reversible. However, living cells have evolved an elegant mechanism for coupling reaction 7-19 to the generation of ATP, a process which, as we have seen, has a $\Delta G'$ of about 7300 cal/mole at 37°C:

$$\text{ADP} + \text{H}_3\text{PO}_4 \longrightarrow \text{ATP} + \text{H}_2\text{O} \qquad (7\text{-}20)$$
$$\Delta G' = +7300 \text{ cal (pH 7.0)}$$

This is done through the participation of the common intermediate, 1,3-diphosphoglyceric acid, an acyl phosphate, whose formation would represent the expenditure of 11,800 cal (see reaction 7-14).

The actual reaction in which 1,3-diphosphoglyceric acid is formed is a combined oxidation–reduction and phosphorylation reaction:

$$\underset{\text{Glyceraldehyde-3-phosphate}}{\begin{matrix} \text{H}\diagdown\text{C}{=}\text{O} \\ | \\ \text{HCOH} \\ | \\ \text{CH}_2\text{OPO}_3\text{H}_2 \end{matrix}} + \text{NAD}^+ + \text{H}_3\text{PO}_4 \longrightarrow \underset{\text{1,3-Diphosphoglyceric acid}}{\begin{matrix} \text{O} \\ \| \\ \text{C—OPO}_3\text{H}_2 \\ | \\ \text{HCOH} \\ | \\ \text{CH}_2\text{OPO}_3\text{H}_2 \end{matrix}} + \text{NADH} + \text{H}^+ \qquad (7\text{-}21)$$
$$\Delta G' = 1500 \text{ cal}$$

The acyl phosphate, in a subsequent reaction, is then utilized to convert ADP to ATP:

$$\begin{matrix} \text{O} \\ \| \\ \text{C—OPO}_3\text{H}_2 \\ | \\ \text{HCOH} \\ | \\ \text{CH}_2\text{OPO}_3\text{H}_2 \end{matrix} + \text{ADP} \longrightarrow \begin{matrix} \text{O} \\ \| \\ \text{C—OH} \\ | \\ \text{HCOH} \\ | \\ \text{CH}_2\text{OPO}_3\text{H}_2 \end{matrix} + \text{ATP} \qquad (7\text{-}22)$$
$$\Delta G' = -4500 \text{ cal}$$

And the sum of the two reactions may be written as

$$\begin{matrix} \text{H}\diagdown\text{C}{=}\text{O} \\ | \\ \text{HCOH} \\ | \\ \text{CH}_2\text{OPO}_3\text{H}_2 \end{matrix} + \text{NAD}^+ + \text{H}_3\text{PO}_4 + \text{ADP} \longrightarrow \begin{matrix} \text{O} \\ \| \\ \text{C—OH} \\ | \\ \text{HCOH} \\ | \\ \text{CH}_2\text{OPO}_3\text{H}_2 \end{matrix} + \text{NADH} + \text{H}^+ + \text{ATP} \qquad (7\text{-}23)$$
$$\Delta G' = -3000 \text{ cal}$$

Moreover, the $\Delta G'$ for reaction 7-23 may be calculated by adding the $\Delta G'$ for reaction 7-21 and reaction 7-22; this amounts to -3000 cal. Note that reaction 7-23 states, in effect, that a significant amount of energy made available in the oxidation of an aldehyde to a carboxylic acid has been utilized to drive the formation of ATP rather than simply being lost to the environment as heat. Moreover, in doing so, the cell has available an overall process which it is able to utilize in converting 3-phosphoglyceric acid back to 3-phosphoglyceraldehyde, since the overall $\Delta G'$ is not as large as that for driving reaction 7-19 from right to left.

Subsequent chapters contain many examples of coupled reactions in which a common intermediate plays a key role in conserving the total energy of the system.

ΔG and Oxidation-Reduction

The ΔG of a reaction which involves an oxidation–reduction process may be related to the difference in oxidation–reduction potentials (ΔE_0) of the reactants. A detailed discussion of electromotive force is beyond the scope of this book, but some appreciation of the energetics of oxidation–reduction reactions and the term *reduction potential* is necessary.

A reducing agent may be defined as a substance that tends to furnish an electron and be oxidized:

$$Fe^{2+} \xrightarrow{\text{Oxidized}} Fe^{3+} + 1 \text{ electron}$$

Similarly, Fe^{3+} is an oxidizing agent because it can accept electrons and be reduced:

$$Fe^{3+} + 1 \text{ electron} \longrightarrow Fe^{2+}$$

Other substances such as H^+ and organic compounds such as acetaldehyde can serve as oxidizing agents and be reduced:

$$H^+ + 1 \text{ electron} \longrightarrow \tfrac{1}{2} H_2$$

$$H_3C{-}C(H){=}O + 2\,H^+ + 2 \text{ electrons} \longrightarrow H_3C{-}CH_2{-}OH$$

These reactions in which electrons are indicated as being consumed (or produced), but in which we have not indicated the donor (or acceptor), are called *half-reactions*. Clearly, the tendency or potentiality for each of these agents to accept or furnish electrons will be due to the specific properties of that compound, and hence it is necessary to have some standard for comparison. That standard is H_2, which has been arbitrarily given the *reduction potential, E_0*, of 0.000 V at pH 0 for the half-reaction

$$H^+ + 1\,e^- \longrightarrow \tfrac{1}{2} H_2 \tag{7-24}$$

Since a proton is consumed in reaction 7-24, the potential of this half-reaction will vary with pH, and at pH 7.0 the reduction potential E_0' of reaction 7-24 may be calculated to be -0.420 V. With this as a standard it is possible to determine the reduction potential of any other compound capable of oxidation–reduction with reference to hydrogen. A list of such potentials, which includes several coenzymes and substrates to be discussed in subsequent chapters, is found in Table 7-3. Note that these potentials are for the reactions written as reductions. When any two of the half-reactions in Table 7-3 are coupled, the one with the *more* positive reduction potential will go as written (i.e., as a reduction) driving the half-reaction with the *less* positive reduction potential backward (i.e., as an oxidation). Qualitatively one may observe that those compounds with the more positive reduction potentials (e.g., O_2 or Fe^{3+}) are good *oxidizing agents,* while those with the more negative reduction potentials are reducing agents (e.g., H_2 or NADH).

It is possible to derive the expression $\Delta G' = -n\mathfrak{F}\Delta E_0'$, where n is the number of electrons transferred in an oxidation–reduction reaction, $\mathfrak{F}$ is Faraday's constant (23,063 cal/V equiv.) and $\Delta E_0'$ is the difference in the reduction potential between the oxidizing and reducing agents. That is,

$$\Delta E_0' = [E_0' \text{ of half-reaction containing oxidizing agent}] - [E_0' \text{ of half-reaction containing reducing agent}]$$

For example, consider the overall reaction resulting from coupling the two half-reactions involving acetaldehyde and NAD^+.

$$\text{Acetaldehyde} + 2\,H^+ + 2\,e^- \longrightarrow \text{Ethanol} \qquad (7\text{-}25)$$

Table 7-3

Reduction Potentials of Some Oxidation–Reduction Half-Reactions of Biological Importance

Half-reaction (written as a reduction)	E_0' at pH 7.0 (V)
$\frac{1}{2}O_2 + 2\,H^+ + 2\,e^- \longrightarrow H_2O$	0.816
$Fe^{3+} + 1\,e^- \longrightarrow Fe^{2+}$	0.771
Cytochrome a–Fe^{3+} + 1 e^- $\longrightarrow$ Cytochrome a–Fe^{2+}	0.290
Cytochrome c–Fe^{3+} + 1 e^- $\longrightarrow$ Cytochrome c–Fe^{2+}	0.250
Ubiquinone + 2 H^+ + 2 e^- $\longrightarrow$ Ubihydroquinone	0.100
Dehydroascorbic acid + 2 H^+ + 2 e^- $\longrightarrow$ Ascorbic acid	0.060
Oxidized glutathione + 2 H^+ + 2 e^- $\longrightarrow$ 2 Reduced glutathione	0.040
Fumarate + 2 H^+ + 2 e^- $\longrightarrow$ Succinate	0.030
Cytochrome b–Fe^{3+} + 1 e^- $\longrightarrow$ Cytochrome b–Fe^{2+}	−0.040
Oxalacetate + 2 H^+ + 2 e^- $\longrightarrow$ Malate	−0.102
Yellow enzyme + 2 H^+ + 2 e^- $\longrightarrow$ Reduced yellow enzyme	−0.122
Acetaldehyde + 2 H^+ + 2 e^- $\longrightarrow$ Ethanol	−0.163
Pyruvate + 2 H^+ + 2 e^- $\longrightarrow$ Lactate	−0.190
Riboflavin + 2 H^+ + 2 e^- $\longrightarrow$ Riboflavin–H_2	−0.200
1,3-Diphosphoglyceric acid + 2 H^+ + 2 e^- $\longrightarrow$ Glyceraldehyde-3-phosphate + Pi	−0.290
NAD^+ + 2 H^+ + 2 e^- $\longrightarrow$ NADH + H^+	−0.320
Acetyl–CoA + 2 H^+ + 2 e^- $\longrightarrow$ Acetaldehyde + CoA–SH	−0.410
$H^+ + 1\,e^- \longrightarrow \frac{1}{2}H_2$	−0.420
Ferredoxin–Fe^{3+} + 1 e^- $\longrightarrow$ Ferredoxin–Fe^{2+}	−0.432
Acetate + 2 H^+ + 2 e^- $\longrightarrow$ Acetaldehyde + H_2O	−0.468

Half-reaction 7-25 will go as a reduction because it has the higher reduction potential. Half-reaction 7-26 then will go as an oxidation *in the opposite direction* from which it is given in Table 7-3:

$$\text{NADH} + \text{H}^+ \longrightarrow \text{NAD}^+ + 2\ \text{H}^+ + 2\ e^- \qquad (7\text{-}26)$$

The overall reaction is

$$\text{Acetaldehyde} + \text{NADH} + \text{H}^+ \longrightarrow \text{NAD}^+ + \text{Ethanol} \qquad (7\text{-}27)$$

The $\Delta E'_0$ for reaction 7-27 will be $-0.163 - (-0.320)$ or 0.157 V and the $\Delta G'$ for reaction 7-27 will be

$$\begin{aligned}\Delta G' &= (-2)(23{,}063)(0.157)\\ &= -7240 \text{ cal}\end{aligned}$$

Because this figure is a large negative quantity, the reaction is feasible thermodynamically. Whether the reaction will occur at a detectable rate is not indicated by the information at hand.

In a similar manner, the $\Delta G'$ may be calculated for the oxidation of NADH by molecular O_2, a common reaction in living tissues:

$$\text{NADH} + \text{H}^+ + \tfrac{1}{2}\,\text{O}_2 \longrightarrow \text{NAD}^+ + \text{H}_2\text{O} \qquad (7\text{-}28)$$

In this reaction, $n = 2$, and $\Delta E'_0 = 0.816 - (-0.320)$ or 1.136 V, and

$$\begin{aligned}\Delta G' &= -n\mathfrak{F}\Delta E'_0\\ &= (-2)(23{,}063)(1.136)\\ &= -52{,}400 \text{ cal}\end{aligned}$$

Although the $\Delta G'$ is a large negative quantity, this has no bearing on whether NADH is rapidly oxidized. As a matter of fact, NADH is stable in the presence of O_2 and will react only in the presence of appropriate enzymes.

The *standard* reduction potential (E_0), in analogy with the standard free-energy change ($\Delta G°$), implies some specific condition or state of the reactants in an oxidation–reduction reaction. Just as $\Delta G°$ specifies that the reactants in a hydrolytic reaction, for example, are all present in their standard state (for solutes 1 *M*), the term E_0 specifies that the ratio of the oxidant to reductant in an oxidation–reduction reaction is unity. Therefore, just as the ΔG for a reaction in which the reactants are not present at 1*M* can be related to $\Delta G°$ (equation 7-4), the E for an oxidation–reduction reaction in which the oxidized form (oxidant) and reduced form (reductant) are not present in a 1:1 ratio can be related to E_0 by the Nernst equation:

$$E = E_0 + \frac{2.303RT}{n\mathfrak{F}} \log \frac{[\text{Oxidant}]}{[\text{Reductant}]}$$

From this it may be calculated that the E will be 0.030 V more positive than E_0 (therefore more oxidizing) if the ratio of the oxidant to reductant is 10:1 and 0.060 V more positive if that ratio is 100:1. Since there is no reason that this ratio should be 1:1 in biological systems, it is clear that the actual reduction potential (E) can vary significantly from the standard reduction potential (E_0).

This is but a brief discussion of some energy relationships encountered in biochemistry. Several references follow which can be consulted for greater detail.

References

1. I. H. Segel, *Biochemical Calculations*. New York: Wiley, 1968.
   Many typical problems involving biochemical energetics are found in this book, together with their solutions.
2. L. L. Ingraham and A. H. Pardee, "Free Energy and Entropy in Metabolism," in *Metabolic Pathways*, D. M. Greeberg, ed. 3rd ed., Vol. 1. New York: Academic Press, 1967.
   This article discusses the thermodynamic relationships in metabolism in a rigorous but readable manner.
3. A. L. Leninger, *Bioenergetics*. New York: Benjamin, 1965.
4. E. Racker, *Mechanisms in Bioenergetics*. New York: Academic Press, 1965.
   Two books by authorities in the subjects that stress biochemical aspects.
5. H. M. Kalckar, *Biological Phosphorylations, Development of Concepts*. Englewood Cliffs, N.J.: Prentice-Hall, 1969.
   The author has collected the classic papers in the field and provided his own narrative of the subject.

8

Enzymes

Introduction

One of the unique characteristics of a living cell is its ability to permit complex reactions to proceed rapidly at the temperature of the surrounding environment. In the absence of the cell these reactions would proceed too slowly. The complex metabolic machinery so fundamental to a cell could not exist under such sluggish conditions. The principal agents which participate in the remarkable transformations in the cell belong to a group of proteins named enzymes.

An enzyme is a protein that is synthesized in a living cell and catalyzes or speeds up a thermodynamically possible reaction so that the rate of the reaction is compatible with the biochemical process essential for the maintenance of a cell. The enzyme in no way modifies the equilibrium constant or the ΔG of a reaction. Being a protein, an enzyme loses its catalytic properties if subjected to agents like heat, strong acids or bases, organic solvents, or other materials which denature the protein.

The high specificity of the catalytic function of an enzyme is due to its protein nature; that is, the highly complex structure of a protein can provide both the environment for a particular reaction mechanism and the template function to recognize a limited set of substrates. Since it is the primary structure that ultimately dictates the final complex (tertiary) structure, the cell can make enzymes for different functions. For example, a protein of 500 amino acid residues could have these residues arranged to be serum albumin or 10^{499} distinctly different proteins. Because of enzyme specificity, literally thousands of enzymes are required, with each enzyme catalyzing only one reaction or a reasonably closely related reaction. Thus, the study of enzyme chemistry is an essential prerequisite to an understanding of the regulation of enzyme activity and in turn the mechanisms of cellular growth and reproduction.

Let us now describe the properties of enzymes.

Effect of Enzyme Concentration and Substrate Concentration

As is true for any catalyst, the rate of an enzyme-catalyzed reaction depends directly on the concentration of the enzyme. Figure 8-1 depicts the relation between the rate of a reaction and increasing enzyme concentration in the presence of an excess of the compound which is being transformed (also called the substrate).

With a fixed concentration of enzyme and with increasing substrate concentration, a second important relationship is observed. A typical curve is shown in Figure 8-2. Let us discuss the implications of this curve in more detail.

With fixed enzyme concentration, an increase of substrate will result at first in a very rapid rise in velocity or reaction rate. As the substrate concentration continues to increase, however, the increase in the rate of reaction begins to slow down until, with a large substrate concentration, no further change in velocity is observed.

Michaelis and others in the early part of this century reasoned correctly that an enzyme-catalyzed reaction at varying substrate concentrations is diphasic; that is, at low substrate concentrations the active sites on the enzyme are not saturated by substrate and thus the enzyme rate varies with substrate concentration (phase I). As the number of substrate molecules increases, the sites are covered to a greater degree until at saturation no more sites are available and the enzyme is working at full capacity and the rate is independent of substrate concentration (phase II). This relationship is shown in Figure 8-3.

The mathematical equation that defines the quantitative relationship between the rate of an enzyme reaction and the substrate concentration and

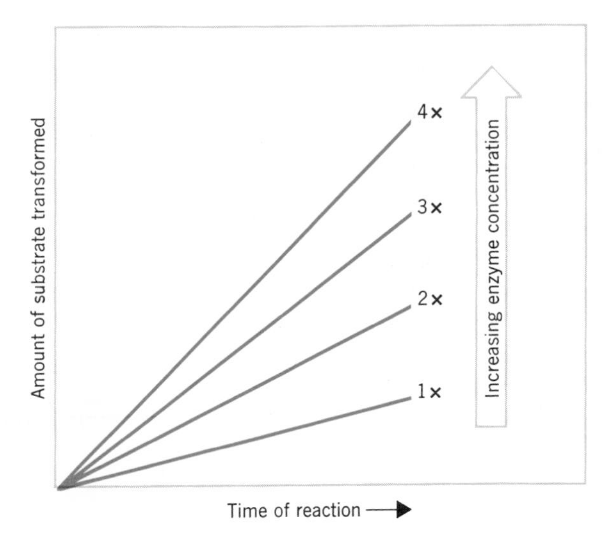

Figure 8-1
Effect of enzyme concentration on reaction rate, assuming that substrate concentration is in saturating amounts.

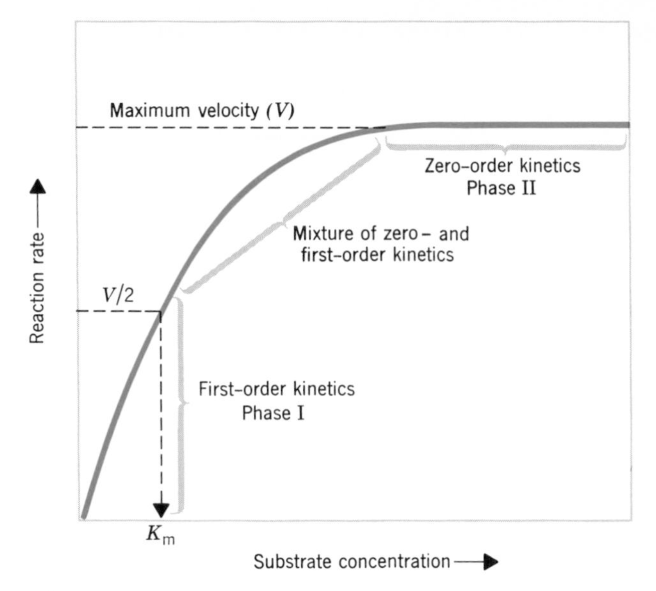

Figure 8-2

Effect of substrate concentration on reaction rate, assuming that enzyme concentration is constant.

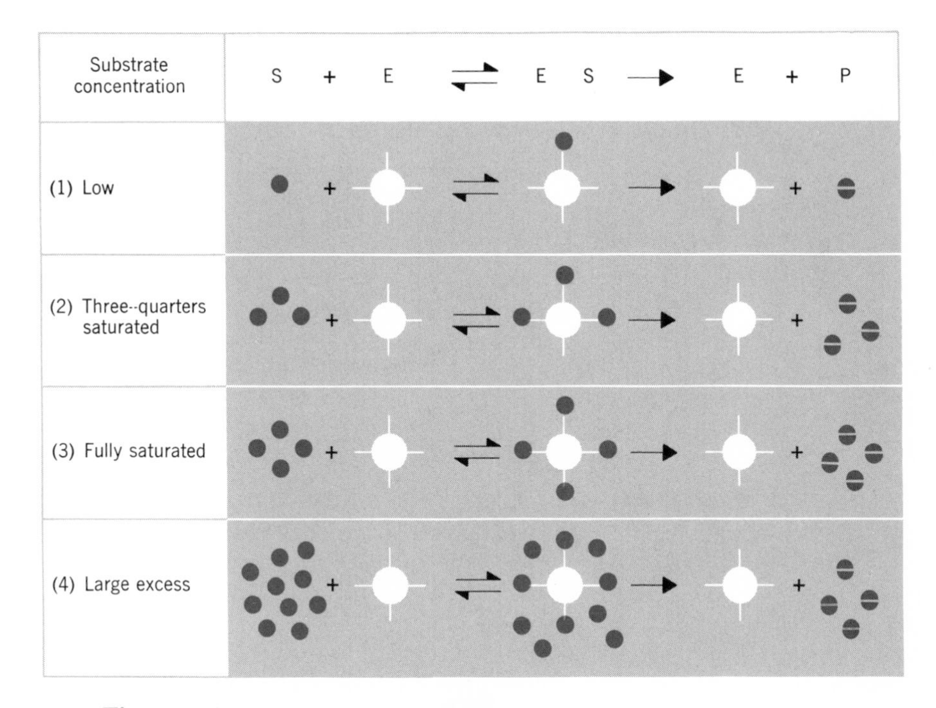

Figure 8-3

Diagrammatic demonstration of effect of substrate concentration on saturation of active sites on enzyme surface. Note that for a unit time interval, cases 3 and 4 give the same amount of P (product) despite the large excess of substrate in case 4.

thus fulfills the requirement of the hyperbolic curve (Figure 8-2) is the Michaelis–Menten equation:

$$v = \frac{V_{max}[S]}{K_m + [S]} \tag{8-1}$$

In this equation v is the observed velocity at given substrate concentration [S]; K_m is the Michaelis constant expressed in units of concentration (mole/liter); and V_{max} is the maximum velocity at saturating concentration of substrate.

Equation 8-1 is readily derived by a consideration of the following steps:

(1) A typical enzyme-catalyzed reaction involves the reversible formation of an enzyme–substrate complex (ES) which eventually breaks down to form the enzyme, E, again and the product, P. This is represented in equation 8-2:

$$E + S \underset{k_2}{\overset{k_1}{\rightleftharpoons}} ES \underset{k_4}{\overset{k_3}{\rightleftharpoons}} E + P \tag{8-2}$$

where k_1, k_2, k_3, and k_4 are the rate constants for each given reaction.

(2) A few milliseconds after the enzyme and substrate have been mixed, a concentration of ES builds up and does not change as long as S is in large excess and $k_1 \gg k_3$. This condition is called the *steady state* of the reaction, since the rate of decomposition of ES just balances the rate of formation. Recognizing that the rate of formation of ES is equal to the rate of decomposition of ES, we can write

$$\text{Rate of formation of [ES]} = \text{Rate of decomposition of [ES]}$$

$$k_1[E][S] + k_4[E][P] = k_2[ES] + k_3[ES] \tag{8-3}$$

and therefore,

$$[E](k_1[S] + k_4[P]) = [ES](k_2 + k_3) \tag{8-4}$$

$$\frac{[ES]}{[E]} = \frac{k_1[S] + k_4[P]}{k_2 + k_3}$$

$$\frac{[ES]}{[E]} = \frac{k_1[S]}{k_2 + k_3} + \frac{k_4[P]}{k_2 + k_3} \tag{8-5}$$

(3) We can simplify this equation by considering that since we are examining equation 8-2 at an early stage of the enzyme-catalyzed reaction, P will be very small and hence the rate of formation of ES by the reaction

$$E + P \xrightarrow{k_4} ES$$

will be extremely low. Thus, the term $k_4[P]/(k_2 + k_3)$ can be ignored, and equation 8-5 simplifies to

$$\frac{[ES]}{[E]} = \frac{k_1[S]}{k_2 + k_3} \tag{8-6}$$

The three constants k_1, k_2, and k_3 can be combined into a single constant,

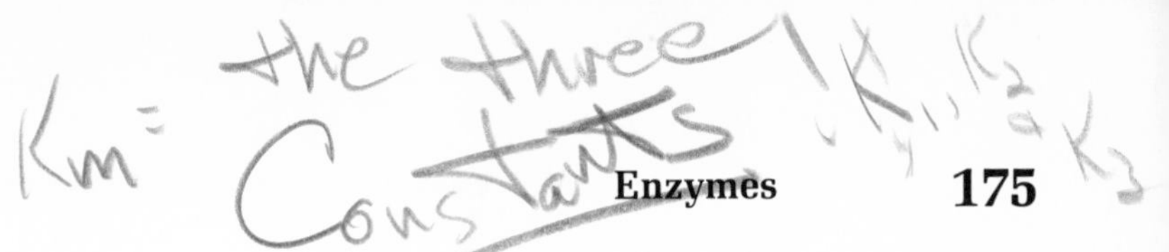

K_m, by the relationship

$$\frac{k_2 + k_3}{k_1} = K_m \tag{8-7}$$

and thus equation 8-6 can be further simplified to

$$\frac{[E]}{[ES]} = \frac{K_m}{[S]} \tag{8-8}$$

(4) We are now faced with the problem of converting [E] and [ES] into easily measurable values. We can resolve this problem if we consider that the total enzyme concentration $[E]_t$ in the reaction consists of the enzyme, [E], which is free plus that which is combined with substrate, [ES]. The free enzyme concentration [E] therefore is $[E]_t - [ES]$ and

$$\frac{[E]}{[ES]} = \frac{[E]_t - [ES]}{[ES]} = \frac{[E]_t}{[ES]} - 1$$

$$\frac{[E]_t}{[ES]} - 1 = \frac{K_m}{[S]}$$

$$\frac{[E]_t}{[ES]} = \frac{K_m}{[S]} + 1 \tag{8-9}$$

Since these terms still cannot be readily determined by the usual techniques available, we must resort to the following relationships: The maximum initial velocity (V_{max}) is attained when the total enzyme $[E]_t$ is completely complexed with saturating amounts of S or

$$V_{max} \propto [E]_t \tag{8-10}$$

Moreover, the initial velocity (v) is proportional to the concentration of enzyme present as the ES complex at a given concentration of S, or

$$v \propto [ES] \qquad \text{and thus} \qquad \frac{V_{max}}{v} \propto \frac{[E]_t}{[ES]} \tag{8-11}$$

Finally, the ratio V_{max}/v can now be substituted for $[E]_t/[ES]$ to yield

$$\frac{V_{max}}{v} = \frac{K_m}{[S]} + 1 \tag{8-12}$$

Inverting and rearranging, we obtain

$$v = \frac{V_{max}[S]}{K_m + [S]} \tag{8-1'}$$

The constant K_m is important since it provides a valuable clue to the mode of action of an enzyme-catalyzed reaction.

Thus, if we permit [S] to be very large, K_m becomes insignificant and equation 8-1 reduces to

$$v = V_{max}$$

or a zero-order reaction in which v is independent of substrate concentration.

If we select $v = \frac{1}{2} V_{max}$, equation 8-1 can be written as

$$\frac{V_{max}}{2} = \frac{V_{max}[S]}{K_m + [S]}$$

$$K_m + [S] = 2[S]$$

$$K_m = [S]$$

In agreement with the experimental curve depicted in Figure 8-2, the dimensions of K_m are expressed in moles per liter, a concentration expression.

If, however, K_m is large compared to [S], equation 8-1 becomes

$$v = \frac{V_{max}[S]}{K_m}$$

That is, v depends on S and the reaction is first-order. These conditions of first-order and zero-order kinetics are indicated in Figure 8-2, and thus the Michaelis–Menten equation fulfills the requirement of a simple enzyme-catalyzed reaction. We shall soon see, however, that enzyme kinetics can be somewhat more complex when we discuss the kinetics of regulatory enzymes later in this chapter and in Chapter 20.

Frequently, K_m has been loosely defined as the dissociation constant of an enzyme-catalyzed reaction. Since the simple reaction

$$ES \underset{k_1}{\overset{k_2}{\rightleftharpoons}} E + S$$

is defined by

$$K_s = \frac{[E][S]}{[ES]} = \frac{k_2}{k_1}$$

and since K_m is defined as $(k_2 + k_3)/k_1$, K_m will always be equal to or greater than K_s, the dissociation constant. Since $1/K_s$ is the affinity constant or k_1/k_2, $1/K_m$ will also be equal to or less than the affinity constant of the reaction.

Another important and quite practical consideration is the conclusion that the observed velocity (v) is equal to the maximum velocity (V_{max}) when $[S] \geq 100K_m$, or zero-order kinetics, and that $v = k[S]$, or first-order kinetics, when $S \leq 0.01K_m$. In setting up experimental conditions for testing enzyme activity, one attempts to operate at saturating or zero-order kinetics, since under these conditions, the enzyme activity is directly proportional to enzyme concentration and independent of substrate concentration.

Important terms such as enzyme units, specific activity, and catalytic center activity or turnover number are defined in Table 8-1.

Table 8-1

Important Terms in Enzymology

1. Enzyme unit—Amount of enzyme which will catalyze the transformation of 1 μmole of substrate per minute under defined conditions
2. Specific activity—Units of enzyme per milligram of protein
3. Catalytic center activity—Number of molecules of substrate transformed per minute per catalytic center (a newer term for turnover numbers)

Effect of Temperature

Since chemical reactions are affected by temperature, an enzyme-catalyzed reaction will also be sensitive to temperature changes. Because of the protein nature of an enzyme, however, thermal denaturation of the enzyme protein with increasing temperatures will decrease the effective concentration of an enzyme and consequently decrease the reaction rate. Up to perhaps 45°C the predominant effect will be an increase in reaction rate as predicted by chemical kinetic theory. Above 45°C an opposing factor, namely thermal denaturation, will become increasingly important, however, until at 55°C rapid denaturation will destroy the catalytic function of the enzyme protein. The dual effects of a temperature–enzyme reaction relationship are depicted in Figure 8-4.

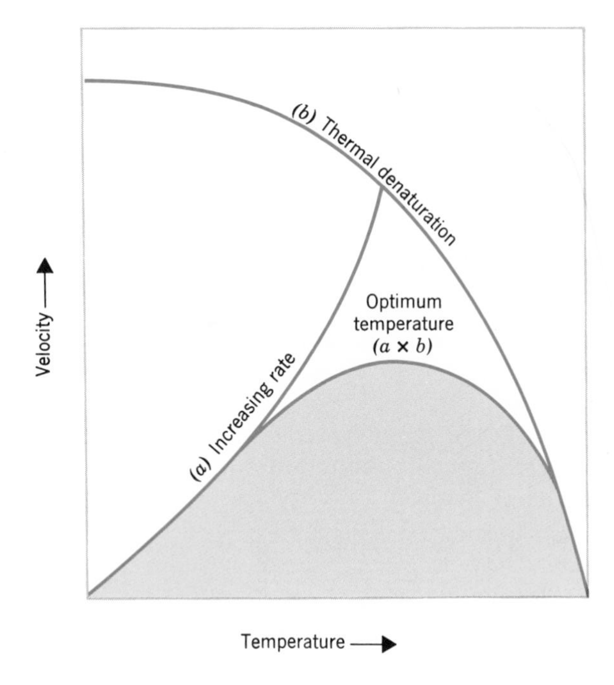

Figure 8-4

Effect of temperature on reaction rate of an enzyme-catalyzed reaction: (*a*) represents the increasing rate of a reaction as a function of temperature; (*b*) represents the decreasing rate as a function of thermal denaturation of the enzyme. The shaded area represents the combination of (*a* × *b*).

Effect of pH

Since enzymes are proteins, pH changes will profoundly affect the ionic character of the amino and carboxylic acid groups on the protein and will therefore markedly affect the catalytic site and conformation of an enzyme. In addition to the purely ionic effects, low or high pH values can cause considerable denaturation and hence inactivation of the enzyme protein.

These effects are probably the main determinants of a typical enzyme activity–pH relation. Thus a bell-shaped curve obtains with a relatively small plateau and with sharply decreasing rates on either side as indicated in Figure 8-5. The plateau is usually called the *optimal* pH point.

In enzyme studies it becomes extremely important to determine early in the investigation the optimal pH and its plateau range. The reaction mixture must then be carefully controlled with buffers of suitable buffering capacity.

In the milieu of the cell the control of the pH in various parts of the cell becomes important since a marked shift in enzyme rates will result if pH stability is not maintained. This would result in major disturbances in the closely geared catabolic and anabolic systems of the cell. Obviously, then, it would be of great value in understanding the regulation of cellular metabolism if we had better knowledge of how pH is controlled or modified in the cellular geography.

Specificity

As we have already mentioned, one important characteristic of an enzyme is its substrate specificity; that is, because of the conformation of the complex protein molecule, the uniqueness of its active site, and the structural configuration of the substrate molecule, an enzyme will select only a limited number of compounds for attack.

An enzyme will usually exhibit *group specificity;* that is, a general group of compounds may serve as substrates. Thus, a series of aldohexoses may be phosphorylated by a kinase and ATP. If the enzyme will only attack one single substrate, for example, glucose and no other monosaccharide, it is said to have an *absolute group specificity*. It may have a *relative group specificity* if it attacks a homologous series of aldohexoses.

Another important aspect of enzyme specificity is the enzyme's stereospecificity toward substrates. As has been mentioned in Chapters 2

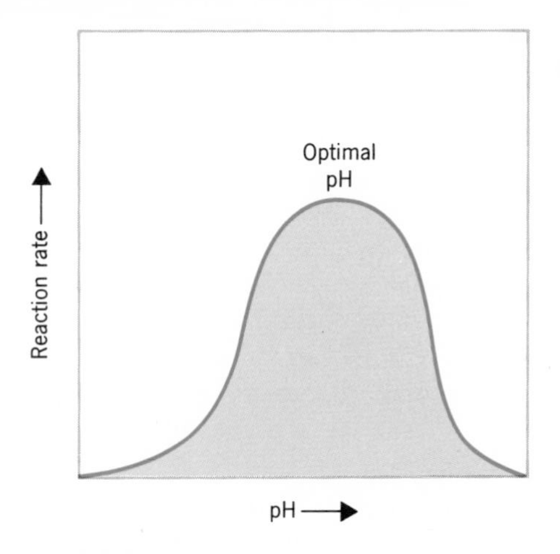

Figure 8-5

Effect of pH on an enzyme-catalyzed reaction.

and 4, an enzyme may have optical specificity for a D or L optical isomer. Thus, L-amino acid oxidase attacks only the L-amino acids, whereas D-amino acid oxidases only react with the D-amino acid isomers:

$$\text{L-Amino acids} \xrightarrow[\text{L-Amino acid oxidase}]{O_2} \alpha\text{-Keto acids} + NH_3 + H_2O_2$$

$$\text{D-Amino acids} \xrightarrow[\text{D-Amino acid oxidase}]{O_2} \alpha\text{-Keto acids} + NH_3 + H_2O_2$$

Although enzymes exhibit optical specificity, a small group of enzymes, the racemases, catalyzes an equilibrium between the L and D isomers and functions through an intermediate complex with pyridoxal phosphate. Thus, alanine racemase catalyzes the reaction

$$\text{L-Alanine} \rightleftharpoons \text{D-Alanine}$$

Still other enzymes have specificities toward geometric or *cis-trans* isomers. Fumarase will readily add water across the double bond system of the *trans* isomer fumaric acid but is completely inactive toward the *cis* isomer maleic acid.

In some enzyme-catalyzed reactions the substrate is symmetrical from the point of view of organic chemistry. Glycerol and citric acid can be considered in this category, since they have a plane of symmetry (Figure 8-6).

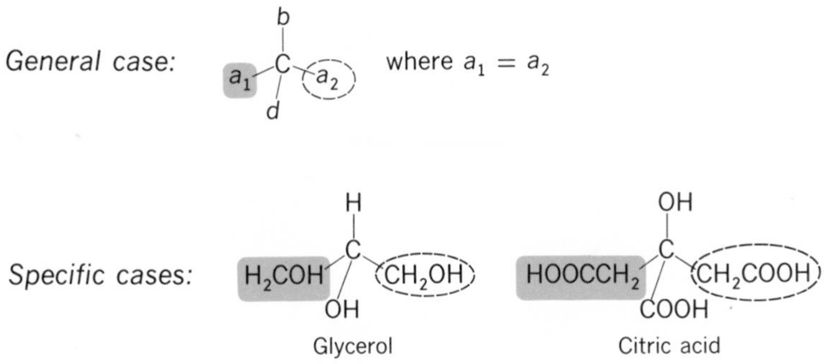

Figure 8-6

Apparently symmetrical substrates which are attacked only in the shaded area and not in the dotted area.

It has been shown, however, that these compounds behave asymmetrically when serving as substrates for enzymes. That is, $C_{a_1a_2bd}$, though symmetrical, is preferentially attacked at a_1 but not at a_2, although both groups are identical. The shaded area in glycerol and in citric acid is preferentially attacked, whereas the dotted area remains unattacked by specific enzymes. This puzzling observation was resolved when Ogston in England in 1948 made the important deduction that although a substrate may appear *symmetrical* the enzyme–substrate relationship is *asymmetrical*. The substrate will have a definite spatial relationship to the enzyme with at least three points of specific interaction between enzyme and substrate. The following specific requirements must be fulfilled:

(1) A substrate molecule must be associated with the enzyme in a specific orientation. Association between substrate and enzyme must be at not less than three sites.
(2) The reactivities of the three enzymic sites must be different or asymmetric.
(3) The compound may have two but no more identical groups (a_1 and a_2) affected by the enzyme and two dissimilar groups (b and d) all associated with a central carbon atom C.

Figure 8-7a brings out these salient features.

If the student is confused by the implications of Figure 8-7*a*, let him consider the thumb, forefinger, and middle finger of his left hand as the substrate. Let his thumb be a_1, his forefinger be a_2, and his middle finger be d. Then there is only one fit with the thumb, forefinger, and middle finger of his right hand if we specify that the forefingers, middle fingers, and thumbs of each hand must be matched (Figure 8-7*b*).

Inhibition An important number of compounds have the ability to combine with certain enzymes, but do not serve as substrates, and, therefore, block catalysis by

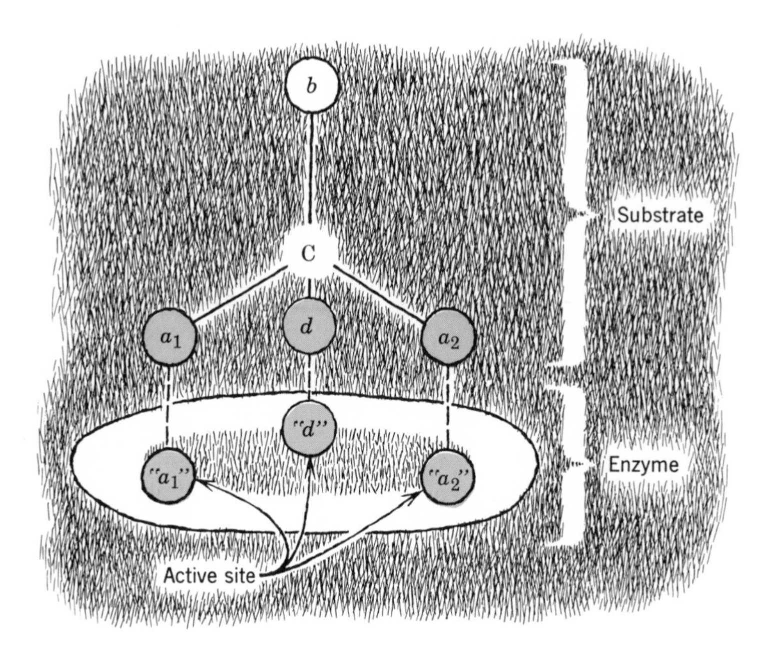

Figure 8-7(*a*)

Diagrammatic representation of the positioning of a substrate to its active site on an enzyme surface: a_1, a_2, and d are functional groups of substrate which combine with a specific site on the enzyme surface; a_1 and a_2 may be identical or they may be dissimilar. Regardless of their nature, there is only one fit on the active site of the enzyme surface.

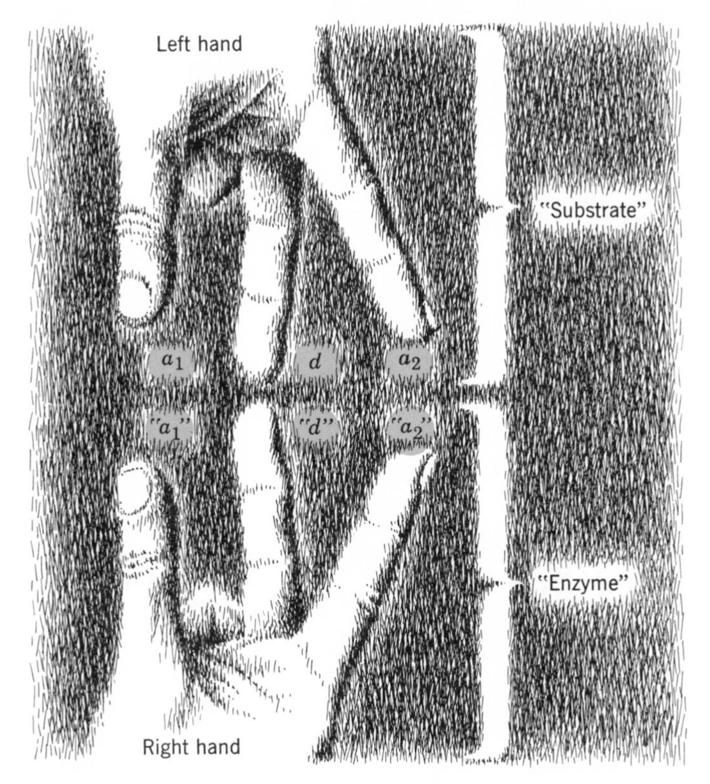

Figure 8-7(*b*)

that enzyme. These compounds are called competitive and noncompetitive *inhibitors*.

Competitive Inhibition. When a compound competes with a substrate or coenzyme for the active site on the enzyme protein and thereby reduces the catalytic activity of that enzyme, the compound is considered to be a competitive inhibitor. Thus succinic dehydrogenase readily oxidizes succinic acid to fumaric acid. If increasing concentrations of malonic acid, which closely resembles succinic acid in structure, are added, however, succinic dehydrogenase activity falls markedly. This inhibition can now be reversed by in-

$$\begin{array}{cc} \text{COOH} & \\ | & \\ \text{CH}_2 & \text{COOH} \\ | & | \\ \text{CH}_2 & \text{CH}_2 \\ | & | \\ \text{COOH} & \text{COOH} \\ \text{Succinic acid} & \text{Malonic acid} \end{array}$$

creasing in turn the concentration of the substrate succinic acid. The amount of inhibition in this type of inhibition is related to (*a*) inhibitor concentration, (*b*) substrate concentration, and (*c*) relative affinities of inhibitor and substrate. The inhibitory effect is reversible. Since the active site is directly involved, the K_m of the enzyme is altered by the inhibitor but the V_{max} is not (Figure 8-8). Figure 8-11 (page 184) shows the reciprocal plots of V and [S], and inhibitor.

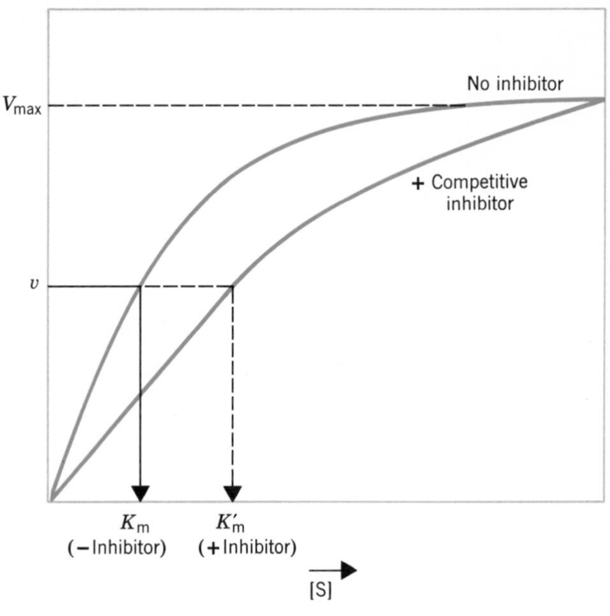

Figure 8-8

The relation between v, the rate of reaction, and substrate concentration with and without the competitive inhibitor. Note the change in K_m in the absence and presence of inhibitor with no shift in V_{max}.

Noncompetitive Inhibition. The type of inhibition that cannot be reversed by increasing substrate concentration is called noncompetitive inhibition. The inhibitor combines irreversibly with a site on the enzyme surface and *cannot be displaced* by increasing substrate concentration. The amount of inhibition in this type of inhibition is related to (*a*) inhibitor concentration and (*b*) inhibitor affinity for the enzyme. Note that substrate concentration has no effect on this system. The K_m is not altered by the inhibitor (See Figure 8-9).

A good example is the reaction of iodoacetamide on triose phosphate dehydrogenase, a sulfhydryl enzyme:

$$\text{Enzyme-SH} + \text{ICH}_2\text{CONH}_2 \longrightarrow \text{Enzyme-S—CH}_2\text{CONH}_2 + \text{HI}$$

A considerable body of information supports the rationale in medicine that many drugs function because of a specific inhibitory effect on a critical enzyme in a tissue. Thus, penicillin appears to block cell wall construction in microorganisms, and the highly dangerous nerve poison diisopropylfluorophosphate strongly inhibits acetylcholine esterase, the enzyme intimately associated with nerve function.

Specific inhibitors have played an important role in the elucidation of

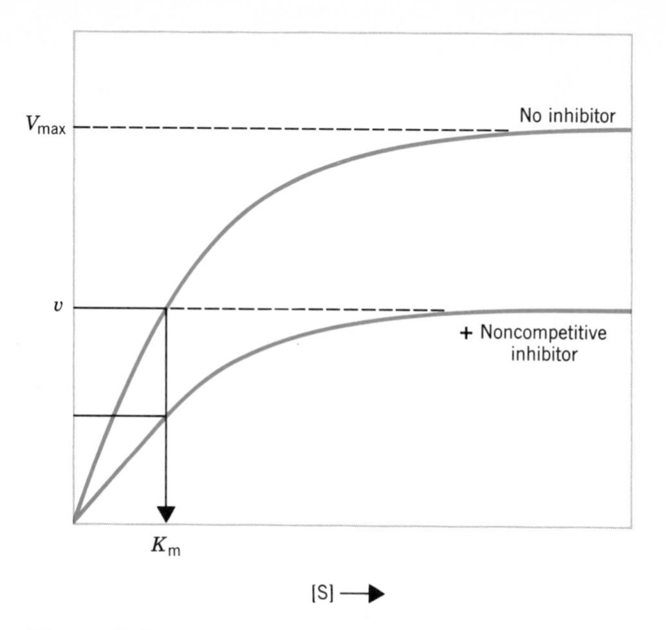

Figure 8-9

The relation between v and substrate concentration with and without the noncompetitive inhibitor. Note the shift in V_{max} but the lack of any change in K_m.

metabolic pathways in tissues. The results must be interpreted with extreme care, however, since there are very few if any inhibitors that are specific for one enzyme. For example, malonic acid for many years was thought to be metabolically inert and was therefore added in large quantities to tissue slices to inhibit succinic dehydrogenase. We now know, however, that malonic acid can be converted to malonyl–CoA which in turn can be decarboxylated to yield acetyl–CoA and CO_2. Thus, in a system containing a high malonate concentration several events take place: (*a*) succinic dehydrogenase is blocked; (*b*) ATP and CoA are redirected to the activation of malonic acid; and (*c*) acetyl–CoA, derived from the decarboxylation of malonyl–CoA, floods the various pathways which utilize this compound. The experimental results are in reality complex in interpretation rather than simple.

Graphic Analyses

Figure 8-2 depicts a very simple procedure for the determination of K_m. However, in 1934, Lineweaver and Burk showed that if the reciprocal of each side of equation 8-1 were taken, then

$$\frac{1}{v} = \frac{K_m}{V_{max}}\left(\frac{1}{[S]}\right) + \frac{1}{V_{max}} \tag{8-13}$$

which is equivalent to the straight-line equation

$$y = ax + b$$

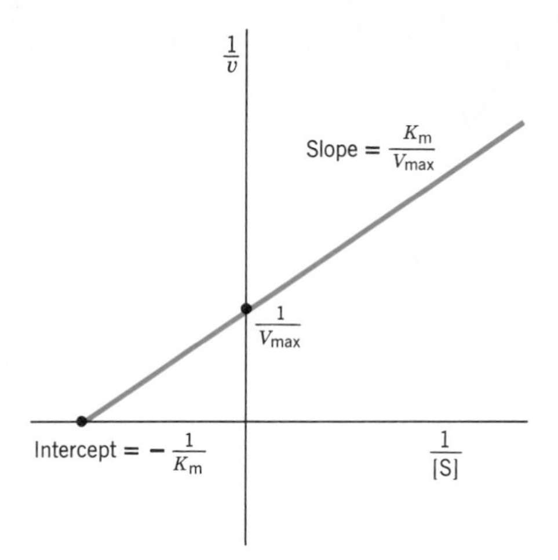

Figure 8-10

A typical Lineweaver–Burk plot of equation 8-13. Lines are extended to $1/v = 0$ to obtain greater accuracy in determining the constants.

If now a double reciprocal plot is made with $1/v$ values on the ordinate and $1/[S]$ values on the abscissa, a straight-line relation exists from which K_m can be easily evaluated (see Figure 8-10).

If a competitive inhibitor is added to the enzyme system, we know that at high concentrations of substrate, the inhibition is overcome. A reciprocal plot of $1/v$ against $1/[S]$ in the presence of this type of inhibitor will result in two nonparallel straight lines which intercept at infinitely high substrate concentration, or when $1/[S]$ is 0 (see Figure 8-11). Thus, if an enzyme

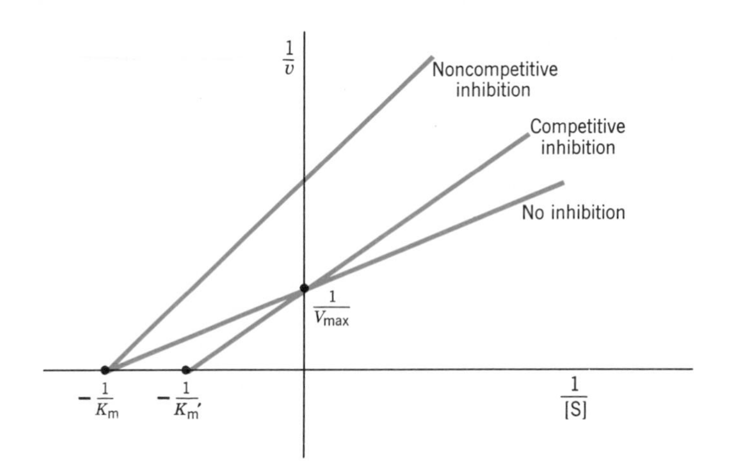

Figure 8-11

Reciprocal plots of V_{max} and [S] in the presence of a competitive and a noncompetitive inhibitor.

reaction is carried out with an inhibitor of unknown action and the data are plotted to yield this type of kinetics, we can state that the inhibition is of a competitive type.

Similarly, when an inhibitor is added to an enzyme system and *both* the slope and the intercept (on the *y* axis) are increased as indicated in Figure 8-11, but the K_m is not altered, we can state that the mode of inhibition is noncompetitive.

At high substrate concentrations either substrate activation or inhibition may occur. When these observations are treated by a Lineweaver–Burk plot, the curves illustrated below are obtained:

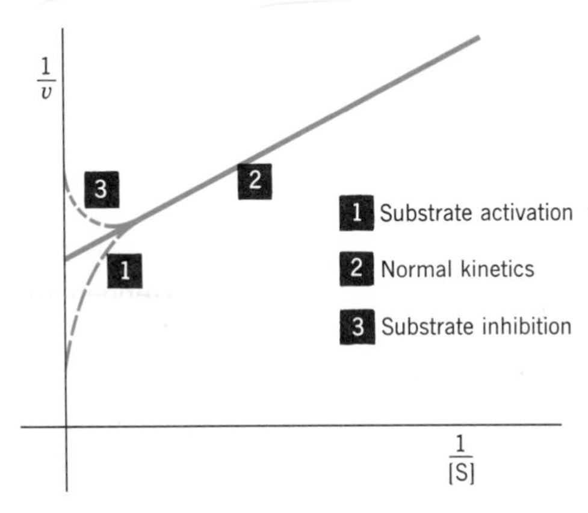

Thus, much data can be obtained by graphically analyzing rates of enzyme reactions as a function of substrate concentrations. By using the Lineweaver–Burk reciprocal plots or more sophisticated plots developed by enzyme chemists, accurate information may be evaluated to obtain important enzyme constants.

Enzymes and Activation Energy

The important feature of an enzyme-catalyzed reaction is that from the thermodynamic point of view an enzyme is a catalytic agent which speeds up a reaction by lowering the activation energy. It does so by increasing the number of molecules that are activated and therefore reactive. This can be depicted in Figure 8-12. Note here that the enzymic reaction has a lower *E* (activation energy), and therefore a larger proportion of molecules will be in the activated state susceptible to reaction. Also note that regardless of the route of reaction both the catalyzed and noncatalyzed reaction have the same ΔG of reaction. Thus we see that an enzyme does not alter the ΔG or equilibrium constant of a reaction but lowers the activation energy that molecule A must attain before it can undergo change.

Some Molecular Aspects Concerning Enzymes

Cofactors. A large number of enzymes require an additional component before the enzyme protein can carry out its catalytic functions. The general term *cofactor* encompasses this component. Cofactors may be divided rather loosely into three groups which include (*a*) prosthetic groups, (*b*) coenzymes, and (*c*) metal activators.

A prosthetic group is usually considered to be a cofactor *firmly bound* to

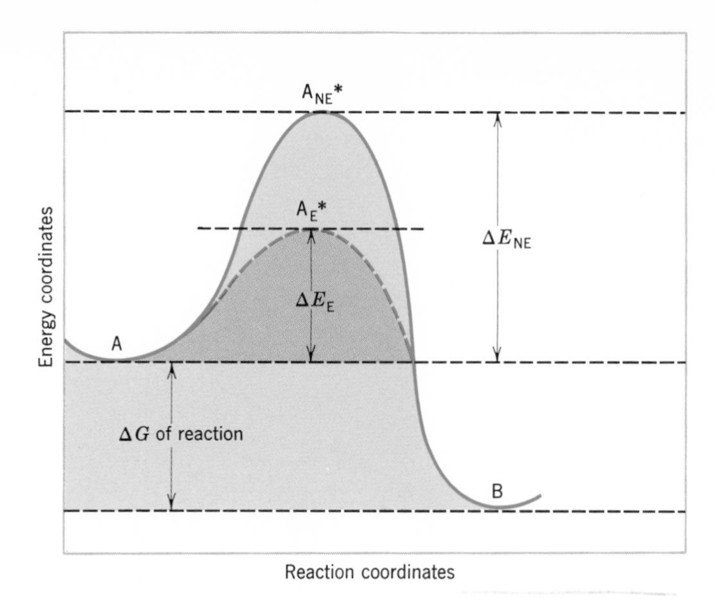

Figure 8-12

A diagram showing the energy barriers of a reaction A ⟶ B: A^*_{NE} indicates the activated complex in a nonenzymic reaction; A^*_E shows the activated complex in an enzyme-catalyzed reaction; A is the initial substrate; B is the product; ΔE_{NE} is the energy of activation for nonenzymic reaction; ΔE_E is the energy for the enzymic reaction; ΔG is the difference in free energy in A ⟶ B.

the enzyme protein. Thus, for example, the porphyrin moiety of the hemoprotein peroxidase and the firmly associated flavin–adenine dinucleotide in succinic dehydrogenase are prosthetic groups.

A coenzyme is a small, heat-stable, organic molecule which *readily dissociates* off an enzyme protein and in fact can be dialyzed away from the protein. Thus, NAD^+, $NADP^+$, tetrahydrofolic acid, and thiamin pyrophosphate are examples of coenzymes.

The metal activator group is represented by the requirement of a large number of enzymes for metallic mono- or divalent cations such as K^+, Mn^{2+}, Mg^{2+}, Ca^{2+}, or Zn^{2+}. These may be either loosely or firmly bound to an enzyme protein, presumably by chelation with phenolic, amino, phosphoryl, or carboxyl groups. On the other hand, Fe^{2+} ion bound to a porphyrin moiety and Co^{2+} bound to the vitamin B_{12} complex would be included in the group in which porphyrin and vitamin B_{12} belong.

We shall have much more to say about cofactors in Chapter 9, where we deal with the functions of vitamins and metals.

Enzymes as Catalysts

Although much is now known about the physical, chemical, and structural aspects of enzymes, the mystery of the enormous catalytic power of an enzyme remains unresolved. Once it was believed that the identification and localization of the amino acid residues associated with a catalytic site would explain the catalytic activity of any enzyme. Now biochemists realize that this approach, while still valid, is somewhat naive. In recent years enzyme chemists have designed ingenious reagents to probe and identify the active site of enzyme and in fact, at present, highly sophisticated physical techniques such as nuclear magnetic resonance spectrometry and electron spin resonance spectrometry as well as high-resolution x-ray crystallography have provided the enzyme chemist with data useful for the development of answers to the mystery of the catalytic power of a protein. As a result, the student can readily tap a large literature, where by the skillful use of physical organic principles, an enzyme chemist can "explain" the events which convert a substrate, via a Michaelis complex [ES], to the product. Whether such events actually take place is part of the problem of the explanation of the catalytic power of an enzyme. There is no question that as a substrate approaches and associates with the active site of an enzyme, a number of changes occur both to the substrate and to the protein with a reduction of the activation energy barrier to allow the conversion of substrate to product. Structurally the active site may be a crevice such as is found in papain, ribonuclease, or lysozyme, or a deep pit as in carbonic anhydrase with the catalytically essential zinc atom at its bottom. Whatever the shape of the active site, it is speculated that the correct substrate binds uniquely in the required orientation with the concomitant formation of covalent intermediates with a lower activation energy than is found in the uncatalyzed reaction. The term *productive binding* is employed here to describe the unique substrate–active site association. Another worthwhile speculation involves the binding of a favored substrate to the active site in such a way that the substrate is mechanically distorted to an energetically unfavorable conformation. The enzyme may also be in a strained conformation which is eased upon binding the substrate so that the strain energy is directed toward reducing the energy of the transition state of the substrate. These effects as well as others may all participate in catalyzing the transformation of a substrate to a product at a highly specific active site of a unique protein, namely the enzyme.

Enzymes as Proteins

Over a thousand enzymes have now been described and over one hundred have been studied in great detail. Of these about fifteen have been analyzed for their three-dimensional structure by high-resolution x-ray diffraction techniques.

Three broad groups of enzyme proteins emerge:

(1) The monomeric enzymes, that is, enzymes with only one polypeptide chain in which the active site resides.
(2) The oligomeric enzymes, that is, enzymes which contain at least 2 and as many as 60 or more subunits firmly associated to form the catalytically active enzyme protein.
(3) The multienzyme complexes in which a number of enzymes, engaged in a sequential series of reactions in the transformation of substrate(s)

Table 8-2

Monomeric Enzymes

Enzymes	Molecular weight	Amino acid residues
Lysozyme (hen egg white)	14,600	129
Ribonuclease	13,700	124
Papain	23,000	203
Trypsin	23,800	223
Carboxypeptidase A	34,600	307

to product, are tightly associated. All attempts to dissociate these enzymes lead to complete inactivation.

We shall briefly examine these three categories here and refer to them elsewhere in this book.

Monomeric Enzymes. This group of enzymes encompasses a relatively small number of enzymes, all of which participate in hydrolytic reactions. As noted in Table 8-2, their molecular weights range from 13,000 to about 35,000, and they cannot be dissociated into smaller units. A number of these proteins are highly reactive proteases and would be extremely damaging to the cell if biosynthesized in the active form. They are therefore synthesized as inactive *zymogens* by the usual ribosomal systems (see Chapters 18 and 19 for details) and subsequently are transported out of the cell into the digestive tract, where they are rapidly converted to their active form (Table 8-3).

Chymotrypsin, trypsin, and elastase, are called serine proteases, since their catalytic sites contain a highly reactive serine residue. Evidence in support of this conclusion is derived in part from the observation that the highly reactive nerve gas, diisopropylfluorophosphate reacts specifically and irreversibly with the hydroxyl function of the serine residue thereby inactivating the enzyme (Scheme 8-1).

Table 8-3

Conversion of Zymogens to Active Enzymes

Zymogen	Activating agent	Active enzyme		Inactive peptide
Pepsinogen	$\xrightarrow[\text{Pepsin}]{\text{H}^+ \text{ or}}$	Pepsin	+	Fragments
Trypsinogen	$\xrightarrow[\text{Trypsin}]{\text{Enterokinase or}}$	Trypsin	+	Hexapeptide
Chymotrypsinogen A	$\xrightarrow[\text{+ Chymotrypsin}]{\text{Trypsin}}$	α-Chymotrypsin	+	Amino acid residues
Procarboxypeptidase A	$\xrightarrow{\text{Trypsin}}$	Carboxypeptidase A	+	Fragments
Prolastase	$\xrightarrow{\text{Trypsin}}$	Elastase	+	Fragments

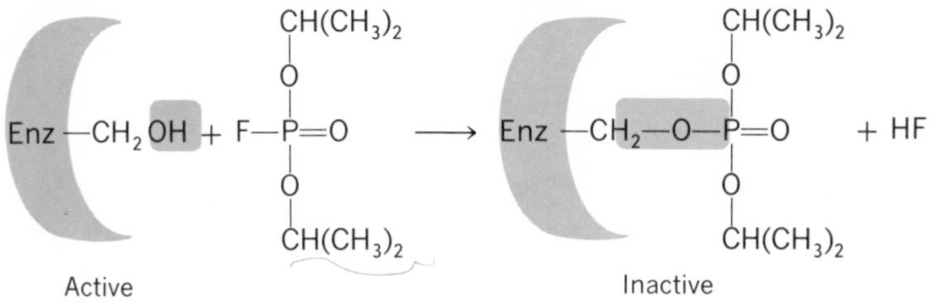

$$\text{Enz—CH}_2\text{OH} + \text{F—P(=O)[OCH(CH}_3)_2]_2 \longrightarrow \text{Enz—CH}_2\text{—O—P(=O)[OCH(CH}_3)_2]_2 + \text{HF}$$

Active   Inactive

Scheme 8-1

A number of proteolytic enzymes tend to cleave peptide bonds, depending on the nature of the R group of the C_α adjacent to the peptide bond, and this important specificity, so useful in determining the complex structure of proteins, is illustrated in Scheme 8-2.

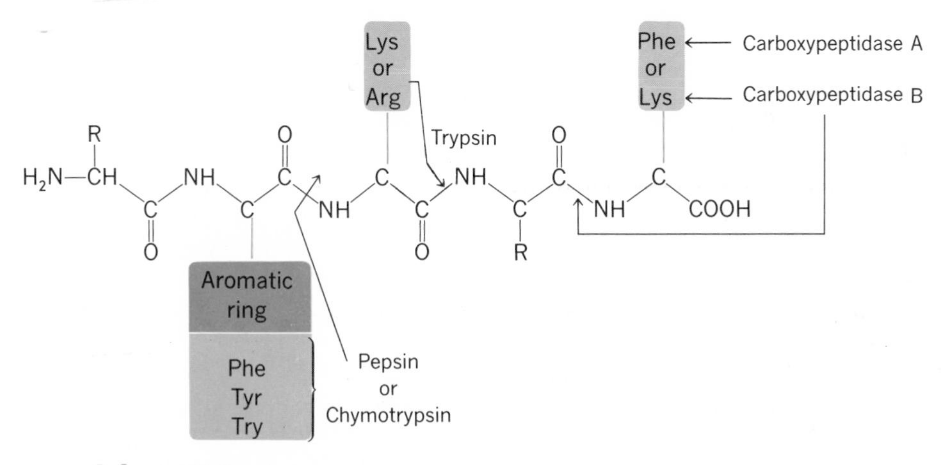

Scheme 8-2

Oligomeric Enzymes. These enzymes include proteins with molecular weights from 35,000 to more than several million and consist of a number of fascinating combinations of polypeptide units to form catalytically active enzymes. To understand fully this class of enzymes we must define a few terms:

Subunit—Any polypeptide chain in the completed functioning protein which is not covalently bound via an amide linkage to other peptide units and can thus be readily separated from other subunits

Protomer—The identical repeating unit in a protein containing a finite number of identical subunits

Oligomers—A combination of similar or different protomers to form the totally functioning enzyme protein

If we examine the glycolytic sequence enzymes (Table 8-4), one is immediately impressed by the fact that every enzyme is not a simple monomeric type of protein but oligomeric, consisting of varying numbers of subunits. Indications now would support the view that monomeric enzymes are the

Table 8-4
Glycolytic Enzymes

Enzymes	Subunits		Molecular weight
	Number	Molecular weight	
Phosphorylase *a*	4	92,500	370,000
Hexokinase	4	27,500	102,000
Phosphofructokinase	2	78,000	190,000
Fructose diphosphatase	2	29,000	130,000
	2	37,000	
Muscle aldolase	4	40,000	160,000
Glyceraldehyde-3-phosphate dehydrogenase	2	72,000	140,000
Enolase	2	41,000	82,000
Creatine kinase	2	40,000	80,000
Lactic dehydrogenase	4	35,000	150,000
Pyruvic kinase	4	57,200	237,000

exceptions and oligomeric enzymes are the rule. A few of these will now be discussed briefly to indicate simply the wide range and diversity of oligomeric enzymes and their possible functions (see also Chapter 20).

Isozymes. An enzyme which has multiple molecular forms in the same organism catalyzing the same reaction is known as an *isozyme*. The most thoroughly studied isozyme is lactic dehydrogenase (LDH) which can occur in five possible forms in organs of most vertebrates, as observed by careful starch gel electrophoretic separation. Two basically different types of LDH occur. One type, which predominates in the heart, is called heart LDH. The other type, characteristic of many skeletal muscles, is called muscle LDH. The heart enzyme consists of four identical monomers which are called H subunits. The muscle enzyme consists of four identical M subunits, each subunit of which is enzymically inactive. The two types of subunits, H and M, have the same molecular weight (35,000) but different amino acid compositions and different immunological properties. There is genetic evidence that the two subunits are produced by two separate genes. Lactic dehydrogenase can be formed from H and M units to yield a pure H tetramer and a pure M tetramer. Combinations of H and M subunits will produce three additional types of hybrid enzymes. The possible combinations of the M and subunits are therefore:

Pure M tetramer (M_4) Pure H tetramer (H_4) M_3H M_2H_2 MH_3

These various combinations have different kinetic properties, depending on the physiological roles which they perform (see Chapter 10).

Isozymes are widespread in nature, with over a hundred enzymes now known to occur in two or more molecular forms.

Allosteric enzymes. A number of extremely important enzymes called regulatory or allosteric enzymes have been intensively investigated in recent years. These enzymes are always oligomeric with a topologically distinct *regulatory* and *catalytic* site. They show sigmoid kinetics and usually can be predicted to catalyze the reaction at a branch point in a metabolic pathway. We shall have more to say about these enzymes in Chapter 20.

Bifunctional oligomeric enzymes. The typical enzyme in this category is tryptophan synthetase of *E. coli*. This enzyme consists of two proteins designated A and B. Protein A has a molecular weight of 29,500 and consists of one subunit, α. Protein B has a molecular weight of 90,000 and has two pyridoxal phosphate binding sites per mole of B. In the presence of 4*M* urea, protein B dissociates into two β subunits, each containing one pyridoxal phosphate binding site and each having a molecular weight of 45,000. The complete tryptophan synthetase consists of two A proteins and one B protein designated as $\alpha_2\beta_2$. The association of the subunits to form the fully active and associated synthetase is greatly increased by the presence of both pyridoxal phosphate and the substrate L-serine. The native synthetase $\alpha_2\beta_2$ catalyzes the reaction:

$$(1)\ \text{Indole glycerophosphate} + \text{L-Serine} \xrightarrow[\text{Pyridoxal phosphate}]{\alpha_2\beta_2} \text{L-Tryptophan} + \text{Glyceraldehyde-3-phosphate}$$

but the α subunit and the β_2 subunit catalyze the following reactions:

$$(2)\ \text{Indole glycerophosphate} \overset{\alpha}{\rightleftharpoons} \text{Indole} + \text{Glyceraldehyde-3-phosphate}$$

$$(3)\ \text{Indole} + \text{L-Serine} \xrightarrow[\text{Pyridoxal phosphate}]{\beta_2} \text{L-Tryptophan}$$

With the reconstituted $\alpha_2\beta_2$ complex, the rate of the partial reactions are 30–100-fold greater than with the individual subunits, and interestingly indole is not liberated from the enzyme complex. The coupling of reactions 2 and 3 to give reaction 1 occurs only when $\alpha_2\beta_2$ is added. This enzyme thus displays a bifunctional activity based on the presence in the complex of two separate catalytic subunits which on association yield the functionally significant reaction 1.

Multienzyme Complexes. A number of complexes have now been described that consist of an organized mosaic of enzymes in which each of the component enzymes is so located as to allow effective coupling of the individual reactions catalyzed by these enzymes. Excellent examples of this type of complex include the α-keto acid dehydrogenase complexes of bacteria and animal tissue and the fatty acid synthetase of animal and yeast cells. L. Reed in Texas and U. Henning in Germany have studied extensively the α-keto dehydrogenase complexes. The *E. coli* pyruvic acid dehydrogenase complex, for example, catalyzes the oxidation of pyruvic acid to acetyl CoA and CO_2. The mechanism of the reaction is depicted in detail in Chapters 9 and 13, but the sequence can be summarized as shown in Scheme 8-3. The total complex has a molecular weight of about 4 million and consists of three separate catalytic activities, E_I, E_{II}, and E_{III}, or pyruvic dehydrogenase, dihydrolipoyl transacetylase, and a dihydrolipoyl dehydrogenase, respectively. The complex

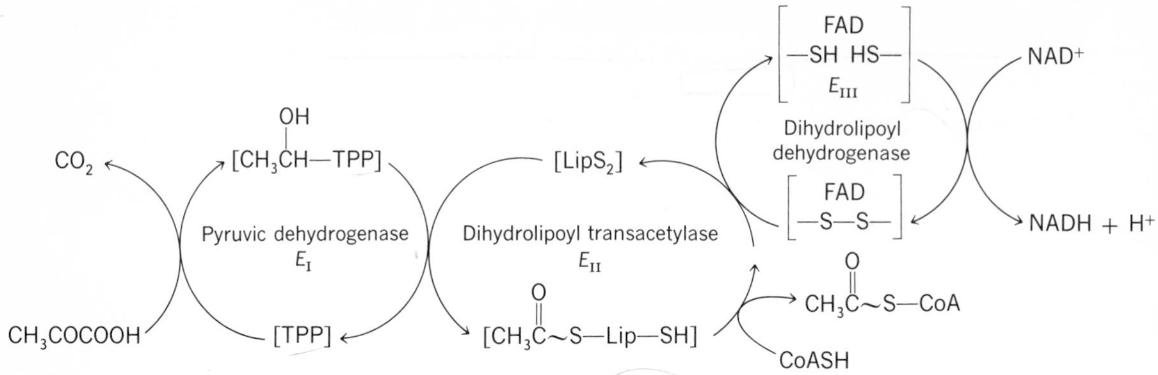

CO_2 ← $CH_3COCOOH$ → $[CH_3CH(OH)—TPP]$ (Pyruvic dehydrogenase E_I) → [TPP]

$[LipS_2]$ → $[CH_3C(=O)\sim S—Lip—SH]$ (Dihydrolipoyl transacetylase E_{II}); CoASH → $CH_3C(=O)\sim S—CoA$

[FAD —SH HS— E_{III}] ⇌ [FAD —S—S—] (Dihydrolipoyl dehydrogenase); NAD^+ → $NADH + H^+$

Scheme 8-3

is resolved by the following treatment:

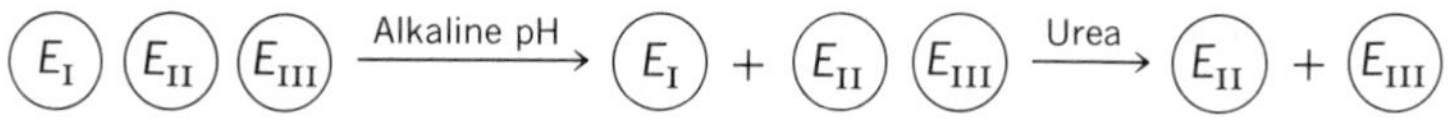

$$(E_I)(E_{II})(E_{III}) \xrightarrow{\text{Alkaline pH}} (E_I) + (E_{II})(E_{III}) \xrightarrow{\text{Urea}} (E_{II}) + (E_{III})$$

Since, in recombination studies, E_I and E_{III} will not reassociate unless E_{II} is added, E_{II} serves as the core for the reassociation process with E_I and E_{III} complexing with the core E_{II} in a definite stoichiometric manner. Beautiful electron micrographs have been taken of the complex clearly depicting the arrangements of the subunits around the central core (Figure 8-13).

An even more complex multienzyme system is the fatty acid synthetase complex which occurs as a very tightly knit group of enzymes responsible for the conversion of acetyl CoA and malonyl CoA to palmitic acid. These complexes are found in animal and yeast cells. In bacteria and in plants, these same enzymes are completely separable and readily purified. In the tight complexes of animal and yeast cells, the whole complex is an extremely efficient and effective machinery for the synthesis of fatty acid. However, it has been impossible to disaggregate the active complex to active individual units. Thus, there appear to be important noncovalent interactions between the subunits with each other so that together they are active, but separated they are inactive. We shall say more about this complex in Chapter 12.

Modification of the Specificity of an Oligomeric Enzyme by a Nonenzymic Specific Protein. In the mammary gland, the enzyme, lactose synthetase, catalyzes the synthesis of lactose by the reaction

$$\text{UDP–Galactose} + \text{Glucose} \rightleftharpoons \text{UDP} + \text{Lactose} \quad (8\text{-}14)$$

The soluble enzyme isolated from raw milk is easily separable into proteins A and B. Neither component catalyzes the above reaction. However, protein A does catalyze the reaction

$$\text{UDP–Galactose} + N\text{-Acetyl glucosamine} \rightleftharpoons N\text{-Acetyl lactosamine} + \text{UDP} \quad (8\text{-}15)$$

Addition of protein B inhibits reaction 8-15, and in the presence of glucose

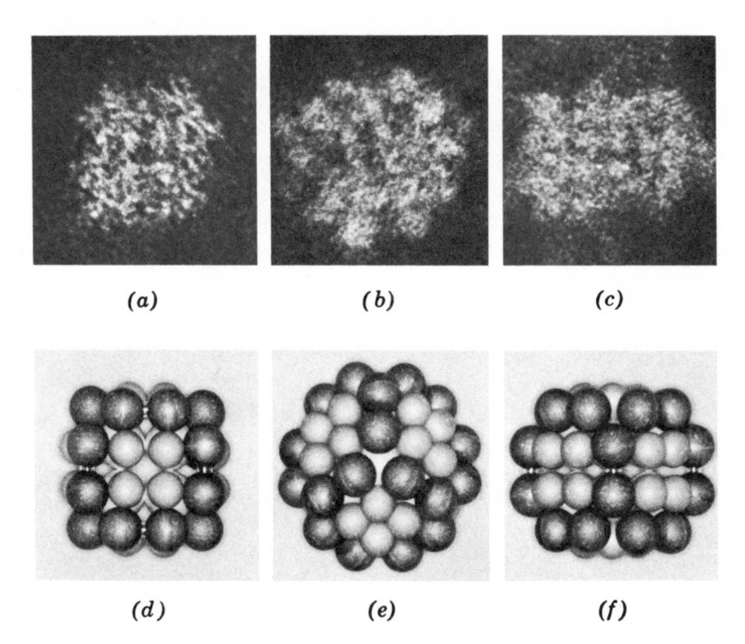

Figure 8-13

Electron micrographs of the pyruvate dehydrogenase complex ($\times$300,000) of *E. coli*. Parts *a–c* are selected images of the complex. Parts *d–f* are different views of an interpretative model that correspond to the images. The 24 small black spheres in the model represent the 24 units of the decarboxylase (90,000 mol wt). The 6 aggregates of 4 white spheres correspond to the 24 dihydrolipoyl dehydrogenase component (55,000 mol wt). A central core consisting of the dihydrolipoyl transacetylase (1.6×10^6 mol wt) is barely visible in the model. It in turn is composed of 8 spheres at the vertices of a cube. Photographs provided by L. J. Reed, University of Texas at Austin.

allows the catalysis of reaction 8-14. Thus, protein B is a specific protein which modifies the substrate specificity of protein A by a physical association to form the lactose synthetase complex. The unusual aspect of this interesting system is that protein B is α-lactalbumin, a protein found specifically only in the mammary gland but not elsewhere, whereas protein A is widely distributed in animal tissues. Hence, lactose is synthesized only in the mammary gland, since it is only in this tissue that α-lactalbumin occurs.

The role of a nonenzymic protein specifying the catalytic activity of an enzyme opens up the possibility that in oligomeric enzymes this specification may be more common then heretofore acknowledged.

Significance of Oligomeric Enzymes. Because of their multipolypeptide structures, oligomeric enzymes may exhibit properties of great importance in the proper functioning of metabolic activities. Although of a speculative nature, it is worthwhile to explore this possibility in more detail.

(1) The aggregation of specific polypeptide chains to form an oligomeric enzyme may maintain a specific conformation which would not be thermodynamically possible if such an aggregation did not occur. Indeed, the dissociation of many oligomeric enzymes to their subunits leads to complete loss of activity which is only regained by reassociation (if possible).
(2) The association of several subunits may yield an active site involving amino acid residues contributed by separate components. Thus, we know from monomeric enzyme structure that widely separated amino acid residues in ribonuclease (his 12, his 119, and lys 41) make up the active site of this enzyme (Scheme 8-4).

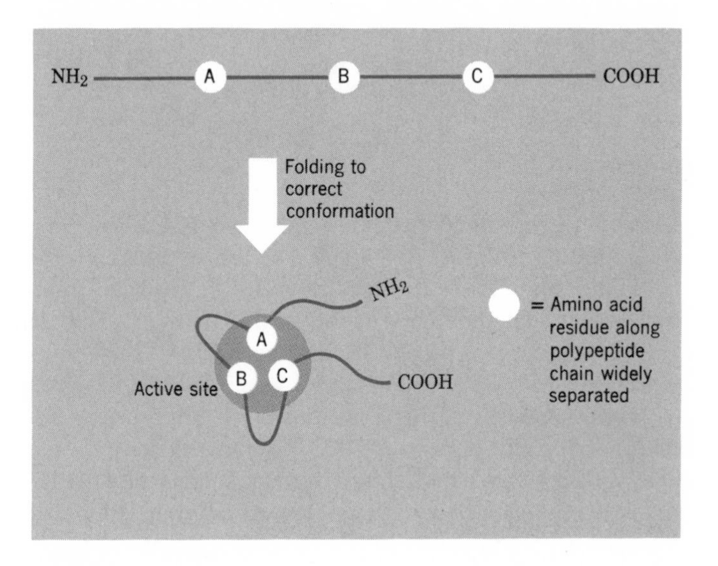

Scheme 8-4

A general case illustrating the positioning of specific, widely separated amino acids (A, B, C) in a polypeptide chain into a conformation whereby these residues are brought together to form the active site.

(3) The association of two subunits with different individual enzymic activity will yield the proper integrated enzyme reaction, as seen in the case of the previously discussed tryptophan synthetase system.
(4) The association of a nonenzymic protein with a catalytic protein such as α-lactalbumin with protein A is a fascinating example of the possible importance of this association.
(5) In a number of enzyme systems a subunit serves as a specific carrier of a substrate. For example, in acetyl CoA carboxylase of *E. coli*, the following subunits comprise the total enzymic activity.

I. Biotinyl carboxyl carrier protein (BCCP) $\underset{\text{Biotin carboxylase}}{\overset{\text{ATP + CO}_2}{\rightleftharpoons}}$ CO_2~BCCP + ADP + Pi

II. CO_2~BCCP + RH $\overset{\text{Transcarboxylase}}{\rightleftharpoons}$ BCCP + RCO_2H

Thus, acetyl CoA carboxylase consists of two catalytically active proteins, biotin carboxylase and transcarboxylase, and a specific biotinyl carboxyl carrier protein with a molecular weight of 20,000.

(6) The assembly of a series of enzymes that operate sequentially to form a product such as a fatty acid would allow highly efficient movement of intermediates from reactants to products with a minimum of competing reactions that would channel away the intermediates from the formation of the desired product, i.e., fatty acid synthesis versus β-oxidation of the same substrates.

(7) A significant number of oligomeric enzymes are regulatory enzymes with regulatory sites and catalytic sites residing on separate subunits. More will be said about these extremely important processes in Chapter 20.

In summary, oligomeric enzymes can by their very nature exhibit a number of properties of great value to the cell. The future will reveal even more the significance of the oligomeric nature of these important enzymes.

References

1. P. D. Boyer, *The Enzymes*. 3rd ed., several volumes. New York: Academic Press, 1970.
   Excellent source of information for the practicing biochemist or the advanced student on all aspects of enzyme chemistry.
2. M. Dixon and E. C. Webb, *Enzymes*. 2nd ed. New York: Academic Press, 1964.
   A substantial although older textbook for the inquiring student.
3. I. H. Segel, *Biochemical Calculations*. New York: Wiley, 1968.
   Excellent section on enzyme kinetics and the mathematical procedures for calculating kinetic data.

9

Vitamins and Coenzymes

Introduction

The term *vitamin* refers to an essential dietary factor which is required by an organism in small amounts and whose absence results in deficiency diseases. The description of the deficiency symptoms and the amounts required for their alleviation is properly the subject of nutrition. In this book the emphasis will be placed instead on the relation that exists between vitamins and coenzymes. As will be described, many coenzymes contain a vitamin as part of their structure; this relation is undoubtedly responsible for creating an "essential" role for the vitamin. The research which established this relationship is one of the most rewarding in biochemistry, because it has constituted a model for the investigation of almost all the other vitamin–coenzyme relationships.

In 1932, the German biochemist Otto Warburg published the first of a series of classic papers dealing with two important coenzymes. Warburg was investigating an enzyme system in yeast which catalyzed the oxidation of glucose-6-phosphate to 6-phosphogluconic acid. The reaction required the presence of two different proteins obtainable from yeast and a coenzyme (or coferment, as it was earlier called) which could be isolated from erythrocytes. Two separate reactions were involved; the first was the oxidation of the sugar phosphate and the simultaneous reduction of the coenzyme from red blood cells. The enzyme (a dehydrogenase) required as a catalyst for this reaction was called *Zwischenferment*. The coenzyme was subsequently known as coenzyme II because of its similarity to another coenzyme, coenzyme I, which many years earlier had been shown by Harden and Young to be involved in the anaerobic fermentation of carbohydrates. Coenzyme I was recognized to be closely related to muscle adenylic acid AMP, since the latter compound was formed on enzymatic hydrolysis of coenzyme I.

Work in Warburg's laboratory in 1935 revealed that coenzyme II contained another nitrogenous base, nicotinamide, in addition to adenine. Shortly

$$\begin{array}{l} \text{H}\diagdown\text{C}{=}\text{O} \\ \text{HCOH} \\ \text{HOCH} \\ \text{HCOH} \\ \text{HCOH} \\ \text{CH}_2\text{OPO}_3\text{H}_2 \end{array} + \text{Coenzyme II} + H_2O \xrightarrow{\textit{Zwischenferment}} \begin{array}{l} \text{COOH} \\ \text{HCOH} \\ \text{HOCH} \\ \text{HCOH} \\ \text{HCOH} \\ \text{CH}_2\text{OPO}_3\text{H}_2 \end{array} + \text{Coenzyme II–H}_2 \qquad (9\text{-}1)$$

Glucose-6-phosphate   6-Phosphogluconic acid

thereafter it was possible with this knowledge to write the structures of both coenzyme I and coenzyme II (see Structure 9-1 below).

Warburg had discovered that the reduced coenzyme II–H_2 could be reoxidized by molecular oxygen provided a second protein was present. Since this protein was yellow in color when purified extensively from brewer's yeast, Warburg called it the yellow enzyme. It provided the link for the oxidation of organic substrates to molecular oxygen, the ultimate oxidizing agent in aerobic organisms. By treatment with ammonium sulfate in acid in the cold, the protein component of the yellow enzyme was precipitated as a white solid, leaving the yellow color in solution. In Stockholm in 1934, Theorell also accomplished the separation of the yellow coenzyme from the protein component by dialysis in acid with the concomitant loss of enzymatic activity. When the coenzyme was added back to the protein component, the enzymatic activity was restored. This was the first demonstration of the reversible separation of an enzyme into its prosthetic group (coenzyme) and a pure protein component (apoenzyme).

Examining the action of this "old yellow enzyme," as it subsequently came to be known, Warburg showed that the catalyst became colorless in the presence of glucose-6-phosphate, *Zwischenferment*, and coenzyme II. The German biochemist subsequently established that this lack of color was due to the reduction of the coenzyme component of the old yellow enzyme by coenzyme II–H_2. This reaction occurred at a significant rate only when the coenzyme was firmly associated with the protein component of the old yellow enzyme.

$$\text{Coenzyme II–H}_2 + \underset{\text{Yellow}}{\text{Old yellow enzyme (oxidized)}} \longrightarrow \text{Coenzyme II} + \underset{\text{Colorless}}{\text{Old yellow enzyme (reduced)}} \qquad (9\text{-}2)$$

When exposed to air the reduced enzyme–coenzyme complex was reoxidized, and O_2 in turn was reduced to H_2O_2.

$$\text{Old yellow enzyme (reduced)} + O_2 \longrightarrow \text{Old yellow enzyme (oxidized)} + H_2O_2$$

R. Kuhn and P. Karrer had determined, simultaneously with these enzyme studies, the chemical structure of the vitamin riboflavin, which occurred as a yellow pigment in egg yolk and milk. The vitamin (page 205) became colorless on reduction with zinc in acid and regained its yellow color on reoxidation. With this information available, other properties of the coenzyme and vitamin were compared, and it was soon established that the coenzyme of the old yellow enzyme was the monophosphate of the vitamin (see Structure 9-2 below). Thus, the coenzyme role of riboflavin was established simultaneously with its description as an essential nutrient, and this was the first demonstration of the vitamin–coenzyme relation.

The overall reaction, therefore, which accounted for the oxidation of glucose-6-phosphate to phosphogluconic acid by O_2, was

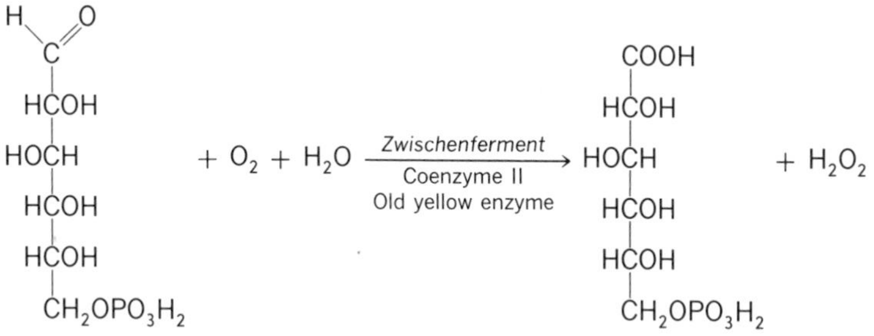

$$\begin{array}{l} HC{=}O \\ HCOH \\ HOCH \\ HCOH \\ HCOH \\ CH_2OPO_3H_2 \end{array} + O_2 + H_2O \xrightarrow[\substack{\text{Coenzyme II} \\ \text{Old yellow enzyme}}]{\textit{Zwischenferment}} \begin{array}{l} COOH \\ HCOH \\ HOCH \\ HCOH \\ HCOH \\ CH_2OPO_3H_2 \end{array} + H_2O_2$$

In this system the coenzyme II functions catalytically by being alternately reduced and oxidized. The flavin component of the old yellow enzyme functions catalytically in the same way.

Vitamins with Coenzyme Functions

Nicotinamide; Nicotinic Acid

Structure. The term *niacin* is the official name of the vitamin which is nicotinic acid or nicotinamide.

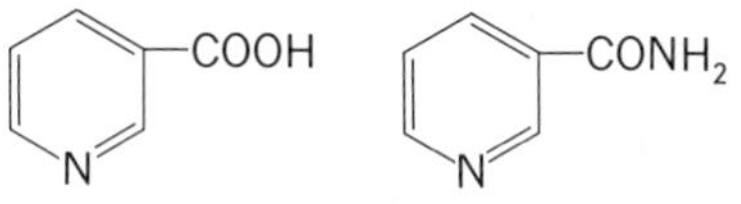

[Structures: pyridine ring with —COOH (Nicotinic acid); pyridine ring with —$CONH_2$ (Nicotinamide)]

Nicotinic acid   Nicotinamide

Occurrence. Niacin is widely distributed in plant and animal tissues; meat products are an excellent source of the vitamin. The coenzyme forms of the vitamin are the *nicotinamide nucleotide* coenzymes (Structure 9-1). The biochemical literature refers to coenzyme I as either diphosphopyridine nucleotide (DPN^+) or nicotinamide adenine dinucleotide (NAD^+). Coenzyme II is referred to as either triphosphopyridine nucleotide (TPN^+) or nicotinamide adenine dinucleotide phosphate ($NADP^+$). The names DPN^+ and TPN^+ were originally proposed by Warburg, and collectively the two coenzymes were referred to as the *pyridine nucleotide* coenzymes because of nicotinamide being a substituted pyridine. In 1964 the Commission on Enzymes of the International Union of Biochemistry proposed the names and abbreviations NAD^+ and $NADP^+$, and because of their widespread acceptance, these will be used in this text. By analogy, therefore, NAD^+ and $NADP^+$ will be referred to as the *nicotinamide nucleotide* coenzymes.

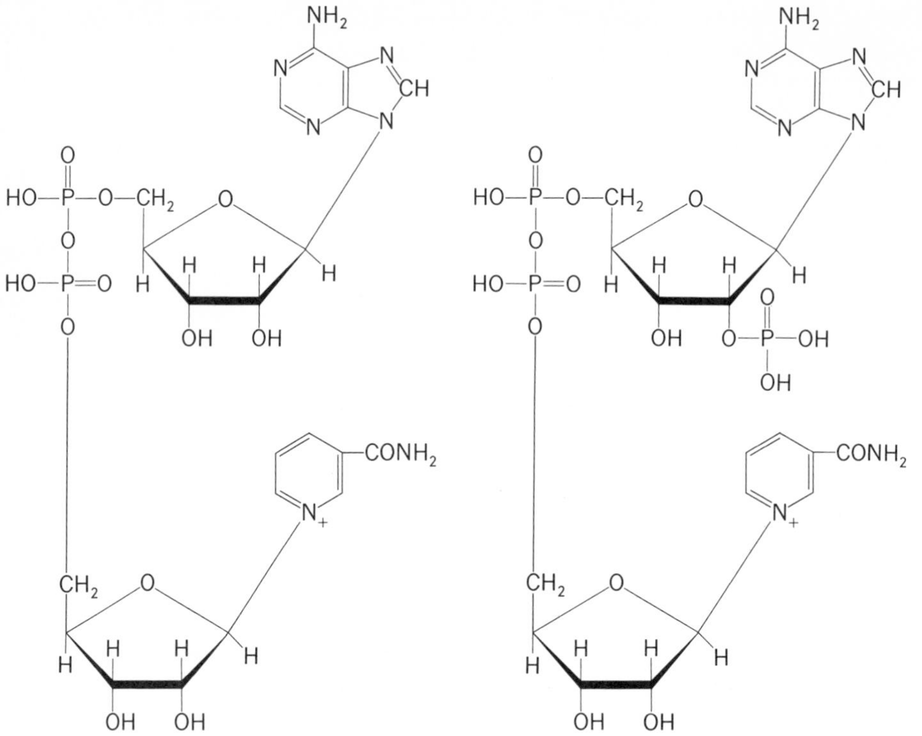

Nicotinamide adenine dinucleotide (NAD^+)
Diphosphopyridine nucleotide (DPN^+)
or Coenzyme I

Nicotinamide adenine dinucleotide phosphate ($NADP^+$)
Triphosphopyridine nucleotide (TPN^+)
or Coenzyme II

Structure 9-1

Although the structure and physiological role of these coenzymes were fairly evident by 1935, nicotinic acid was not recognized as a vitamin until 1937, when Elvehjem at the University of Wisconsin established its essential nature. A deficiency of niacin causes pellagra in man and black tongue in dogs. The coenzymes NAD^+ and $NADP^+$ are ubiquitous in nature, presumably because of their fundamental role in biological oxidations.

Biochemical function. The nicotinamide nucleotides are coenzymes for enzymes known as dehydrogenases which catalyze oxidation–reduction reactions. Thus, in the reaction catalyzed by *Zwischenferment* (9-1) the glucose-6-phosphate is oxidized and $NADP^+$ (coenzyme II) is simultaneously reduced.

Similarly, alcohol dehydrogenase, an enzyme widely distributed in nature, catalyzes the oxidation of ethanol with the concomitant reduction of NAD^+:

$$CH_3CH_2OH + NAD^+ \rightleftharpoons CH_3CHO + NADH + H^+ \qquad (9\text{-}3)$$

The apparent equilibrium constant of this reaction may be written

$$K_{app} = \frac{[CH_3CHO][NADH]}{[CH_3CH_2OH][NAD^+]}$$

When determined experimentally, K_{app} was approximately 10^{-4} at pH 7.0 and 10^{-2} at pH 9.0. The equilibrium constant is therefore obviously related to the pH; this is because a H^+ is a product of the reaction when alcohol is oxidized. Clearly, the reaction from left to right will be favored by a low H^+ concentration or high pH, while by the law of mass action the equilibrium would be displaced to the left at high H^+ concentration or low pH.

In order to understand the production of an equivalent of H^+ ion in this reaction we shall consider the reduction of NAD^+ (or $NADP^+$) in detail. An examination of the reactions catalyzed by nicotinamide nucleotide dehydrogenases shows that the reaction involves the removal of the equivalent of two hydrogen atoms from the substrate. This occurs when ethanol is oxidized to acetaldehyde. The overall process might occur by the removal of two hydrogen atoms (with their electrons), two electrons and two protons H^+ in separate steps, or a hydride ion (a hydrogen atom with an additional electron, H^-) and a proton H^+.

The oxidized and reduced forms of NAD^+ ($NADP^+$) have the formulas where R equals the remainder of the coenzyme molecule. The structure shown is

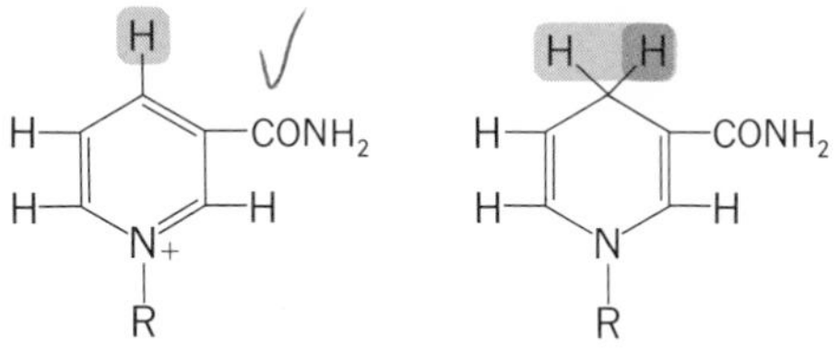

Oxidized NAD^+ or $NADP^+$ — Reduced NADH or NADPH

produced when the equivalent of one proton and two electrons have entered the nicotinamide moiety. This may occur in a single step by the addition of a hydride ion to the oxidized nucleotide at position 4, where the added hydrogen is known to enter the ring. This can be more readily pictured if we write a resonance form of oxidized NAD^+ in which the carbon at position 4 possesses the positive charge usually placed on the nitrogen atom. The proton, required to balance the reaction when a hydride ion is removed from the substrate, is released in solution.

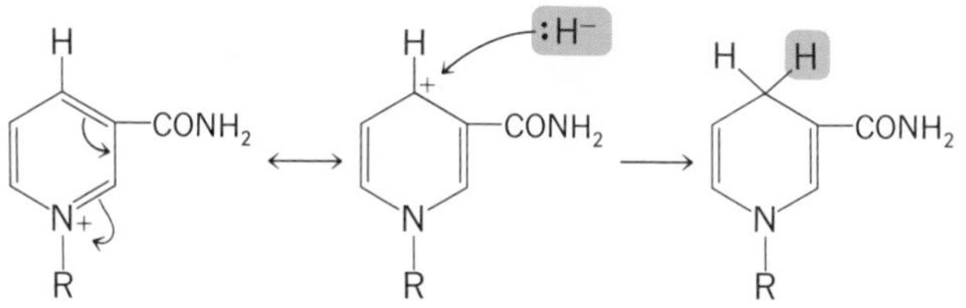

The nicotinamide nucleotide enzymes exhibit several general modes of action. The dehydrogenases that require NAD^+ and $NADP^+$ catalyze the oxidation of alcohols (primary and secondary), aldehydes, α- and β-hydroxy carboxylic acids and α-amino acids (Table 9-1). These reactions are frequently readily reversible. In other instances, the value of the equilibrium constant

Table 9-1

Some Reactions Catalyzed by Nicotinamide Nucleotide Enzymes

Enzyme	Substrate	Product	Coenzyme
Alcohol dehydrogenase	Ethanol	Acetaldehyde	NAD^+
Isocitric dehydrogenase	Isocitrate	α-Ketoglutarate + CO_2	NAD^+, $NADP^+$
Glycerolphosphate dehydrogenase	*sn*-Glycerol-3-phosphate	Dihydroxyacetone phosphate	NAD^+
Lactic dehydrogenase	Lactate	Pyruvate	NAD^+
Malic enzyme	L-Malate	Pyruvate + CO_2	$NADP^+$
Glyceraldehyde-3-phosphate dehydrogenase	Glyceraldehyde-3-phosphate + H_3PO_4	1,3-Diphosphoglyceric acid	NAD^+
Glucose-6-phosphate dehydrogenase	Glucose-6-phosphate	6-Phosphogluconic acid	$NADP^+$
Glutamic dehydrogenase	L-Glutamic acid	α-Ketoglutarate + NH_3	NAD^+, $NADP^+$
Glutathione reductase	Oxidized glutathione	Reduced glutathione	NADPH
Quinone reductase	*p*-Benzoquinone	Hydroquinone	NADH, NADPH
Nitrate reductase	Nitrate	Nitrite	NADH

may determine that under physiological conditions the reaction will proceed in only one direction. The reaction, however, may result in either the reduction or oxidation of the nicotinamide nucleotide. Because of this, the nicotinamide nucleotides may readily accept electrons directly from a reduced substrate and donate them directly to an oxidized substrate in a coupled reaction. Thus, the reduction of acetaldehyde to ethanol (in the presence of alcohol dehydrogenase) is linked to the oxidation of glyceraldehyde-3-phosphate (in the presence of triosephosphate dehydrogenase).

H–C(=O)–HCOH–$CH_2OPO_3H_2$ + H_3PO_4 → (Triosephosphate dehydrogenase) → O=C–O–PO_3H_2 / HCOH / $CH_2OPO_3H_2$

NAD^+ → NADH + H^+ (Triosephosphate dehydrogenase)

NADH + H^+ → NAD^+ (Alcohol dehydrogenase)

CH_3CHO → (Alcohol dehydrogenase) → CH_3CH_2OH

The second manner in which the nicotinamide nucleotides function is in the reduction of the flavin coenzymes. Since the flavin coenzymes are the prosthetic groups of enzymes which accomplish the oxidation or reduction of organic substrates, reduction provides a link for a reaction between nicotinamide nucleotides and these substrates. As an example we may cite the reduction of the disulfide compound oxidized glutathione by glutathione reductase. The overall reaction may be written

$$\underset{\text{Oxidized glutathione}}{\text{NADH} + \text{H}^+ + \text{GSSG}} \longrightarrow \underset{\text{Reduced glutathione}}{\text{NAD}^+ + 2\,\text{GSH}}$$

where G stands for the tripeptide moiety of the glutathione molecule (Structure 4-5). The glutathione reductases are enzymes that contain flavin adenine dinucleotide (FAD) as a prosthetic group. In the presence of NADH the flavin is first reduced, and the resulting $FAD–H_2$ in turn accomplishes the reduction of GSSG:

$$\text{NADH} + \text{H}^+ + \text{FAD} \longrightarrow \text{NAD}^+ + \text{FAD–H}_2$$
$$\text{FAD–H}_2 + \text{GSSG} \longrightarrow \text{FAD} + 2\,\text{GSH}$$

The advantage of having a flavin intermediate is that the overall reaction, which frequently has a large $\Delta G'$ is broken down into two reactions of smaller $\Delta G'$, both of which are more likely to be reversible. The flavin coenzyme, although written as a separate compound, is firmly associated with the protein of the reductase. Other compounds are reduced by the reduced nicotinamide nucleotides in the presence of enzymes containing FAD and flavin mononucleotide (FMN) as prosthetic groups; these include nitrate ion (nitrate reductase) and cytochrome c (cytochrome c reductase).

A third function which the nicotinamide nucleotides perform is as a source of electrons for the hydroxylation and desaturation of aromatic and aliphatic compounds. These important reactions are more fully discussed in Chapter 12. Finally, NAD^+ serves a unique function in the important reaction catalyzed by DNA ligase, and this is discussed in Chapter 18.

The nicotinamide nucleotides and their dehydrogenases have been a favorite subject for the study of the kinetics and the mechanisms of enzyme action. Several of the dehydrogenases are available in the form of highly purified, crystalline proteins. In addition, there is a convenient method for distinguishing the reduced nicotinamide nucleotide from its oxidized form. The method is based on the observation by Warburg that the reduced coenzymes strongly absorb light at 340 nm and that the oxidized coenzymes do not. The absorption spectra of the oxidized and reduced nicotinamide nucleotides are shown in Figure 9-1; the molar absorbancy a_m for the two coenzymes is identical. By measuring the change in the absorption of light at 340 nm during the course of a reaction, it is possible to follow the reduction or oxidation of the coenzyme. An example of such measurements is given in Figure 9-2, where the reduction of NAD^+ in the presence of ethanol and alcohol dehydrogenase is shown.

In Figure 9-2 the absorbancy at 340 nm is plotted as a function of time. After equilibrium is obtained and no further reduction of NAD^+ occurs, acetaldehyde is added. In adding a product of the reaction, the equilibrium

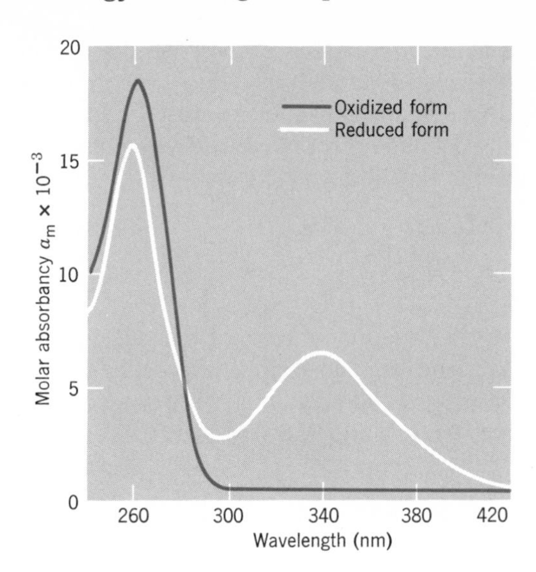

Figure 9-1

Absorption spectra of the oxidized and reduced nicotinamide nucleotides.

of reaction 9-3 is displaced to the left and some of the reduced NADH is reoxidized, as indicated by a decrease in the absorption of light at 340 nm. If additional alcohol is added now, the equilibrium is again adjusted, this time from left to right, and NAD^+ reduction results, as shown by the increase in light absorption at 340 nm (see also Appendix 2).

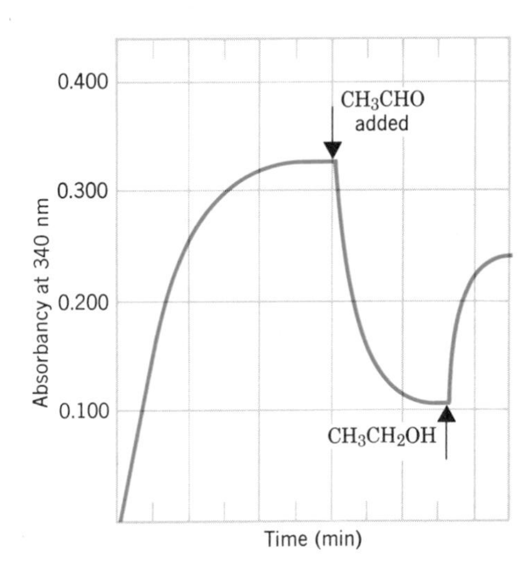

Figure 9-2

The reduction and reoxidation of NAD^+ in the presence of ethanol, acetaldehyde, and alcohol dehydrogenase.

Riboflavin

Structure. Riboflavin (vitamin B_2) consists of the sugar alcohol D-ribitol attached to a substituted isoalloxazine ring.

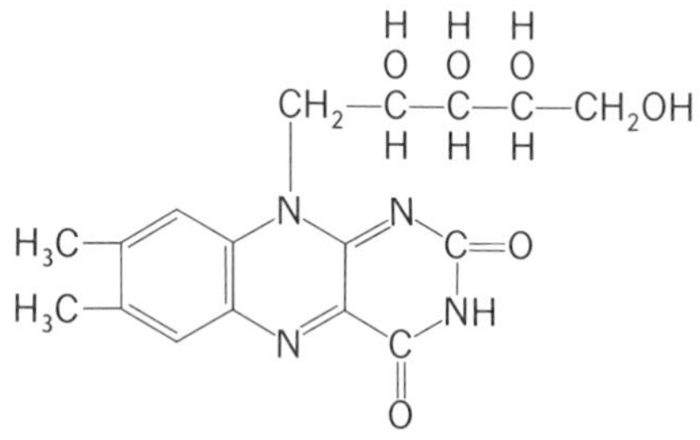

Riboflavin

Occurrence. The vitamin was detected as a growth factor for rats. It is obtained commercially from the culture medium of certain microorganisms which produce it in good yield. One of the most important chemical properties of riboflavin is its ability to undergo oxidation–reduction reactions. On reduction, the yellow color disappears since the reduced flavin is colorless. On exposure to air, the yellow color of the oxidized form reappears. As indicated in the

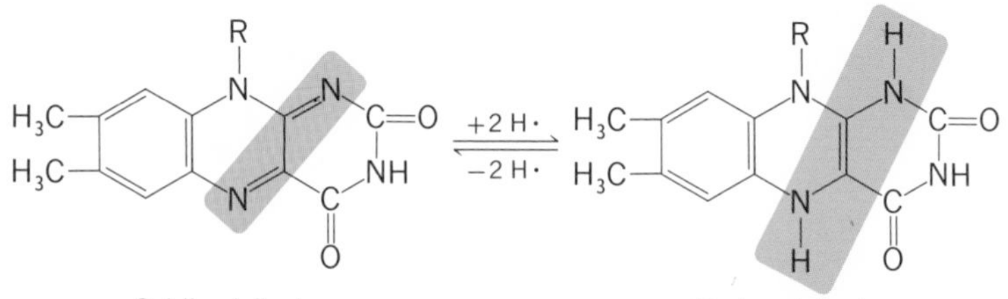

Oxidized flavin $\underset{-2\,H\cdot}{\overset{+2\,H\cdot}{\rightleftharpoons}}$ Reduced flavin

R = Remainder of FMN or FAD molecule

diagram, the reduction consists of the addition of two hydrogen atoms (two electrons + two protons) in a 1,4 addition reaction to form the reduced or leuco-riboflavin.

The vitamin occurs in nature almost exclusively as a constituent of the two flavin coenzymes flavin mononucleotide (FMN) and flavin adenine dinucleotide (FAD). Although the flavin coenzymes were given the name mono- and dinucleotide, the names are not accurate chemically speaking, since the compound attached to the flavin moiety is the sugar alcohol ribitol and not the aldose sugar ribose (Structure 9-2).

Biochemical function. The role of FMN as a coenzyme for the old yellow enzyme of Warburg has been mentioned; FAD was first demonstrated as a coenzyme for D-amino acid oxidase. These enzymes belong to a group of proteins termed *flavoproteins* which catalyze oxidation–reduction reactions (Table 9-2). In contrast to the nicotinamide nucleotide dehydrogenases, the prosthetic coenzymes FAD and FMN are firmly associated with the protein component and are carried along during the purification of the enzyme. In fact, the flavin coenzymes are usually only separated from the apoenzyme by acid treatment in the cold or, perhaps, by boiling. The latter technique destroys

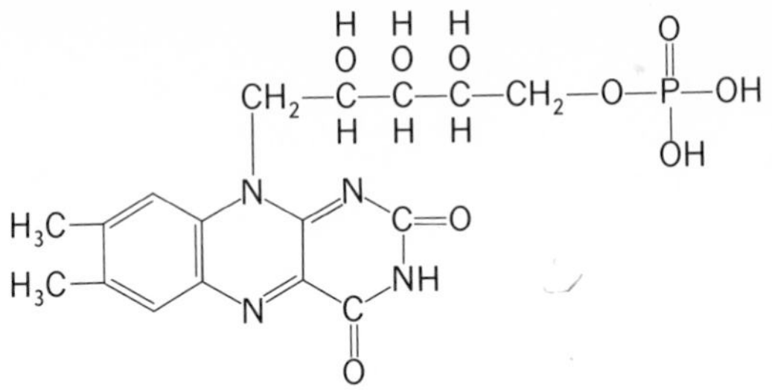

Flavin mononucleotide (FMN)
Riboflavin monophosphate

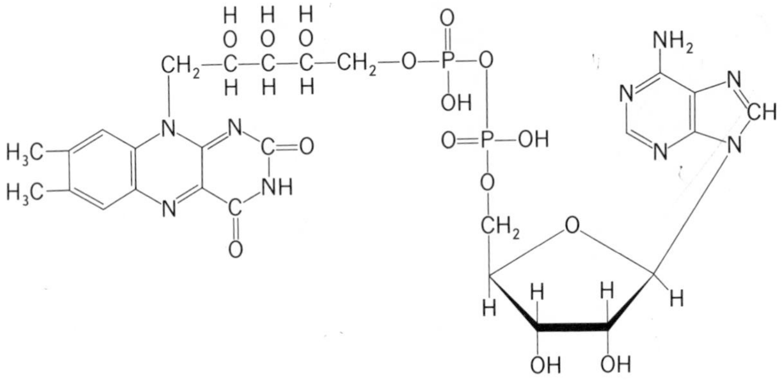

Flavin adenine dinucleotide (FAD)

Structure 9-2

Table 9-2
Some Reactions Catalyzed by Flavoproteins

Enzyme	Electron donor	Product	Coenzyme	Electron acceptor
D-Amino acid oxidase	D-Amino acids	α-Ketoacids + NH_3	FAD	O_2
Glycolic acid oxidase	Glycolate	Glyoxylate	FMN	O_2
NAD^+–cytochrome c reductase	NADH	NAD^+	FAD	Cytochrome c_{ox}
Aldehyde oxidase	Aldehydes	Carboxylic acids	FAD	O_2
Succinic dehydrogenase	Succinate	Fumarate	FAD	Oxidized dyes
Nitrate reductase	NADPH	$NADP^+$	FAD	Nitrate
Nitrite reductase	NADPH	$NADP^+$	FAD	Nitrite
Xanthine oxidase	Xanthine	Uric acid	FAD	O_2
Lipoyl dehydrogenase	Reduced lipoic acid	Oxidized lipoic acid	FAD	NAD^+

the protein nature of the apoenzyme, and the separation is therefore irreversible. The separation by acid in the cold is reversible, and the mixing of flavin coenzyme with the apoenzyme restores activity.

It is difficult to generalize on the types of chemical reaction in which flavoproteins participate. Certainly they function in accepting hydrogen atoms from reduced nicotinamide nucleotides. Another oxidative reaction commonly occurring in intermediary metabolism involves the removal of two hydrogen atoms on adjacent carbon atoms to form a double bond. These reactions frequently involve flavin coenzymes and enzymes. Thus, the enzyme succinic dehydrogenase, which catalyzes the oxidation of succinate to fumarate, contains FAD as a prosthetic group.

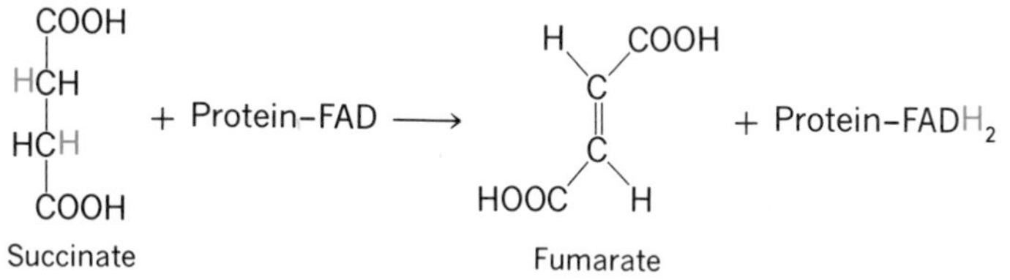

$$\mathrm{HOOC{-}CH_2{-}CH_2{-}COOH} + \text{Protein–FAD} \longrightarrow \mathrm{HOOC{-}CH{=}CH{-}COOH} + \text{Protein–FADH}_2$$

Succinate Fumarate

Some of the flavoproteins listed in Table 9-2 are more complex in that they contain a metal such as molybdenum or iron in addition to the riboflavin derivative. The metals also function by virtue of their ability to be alternately oxidized and reduced, although this may not be their only function.

Various lines of evidence have indicated that the reduction of the flavin coenzyme occurs in two separate steps, each involving the addition of a single electron. If the reaction occurs by the addition of one electron (with its proton) at a time, a semireduced compound known as a semiquinone will be an

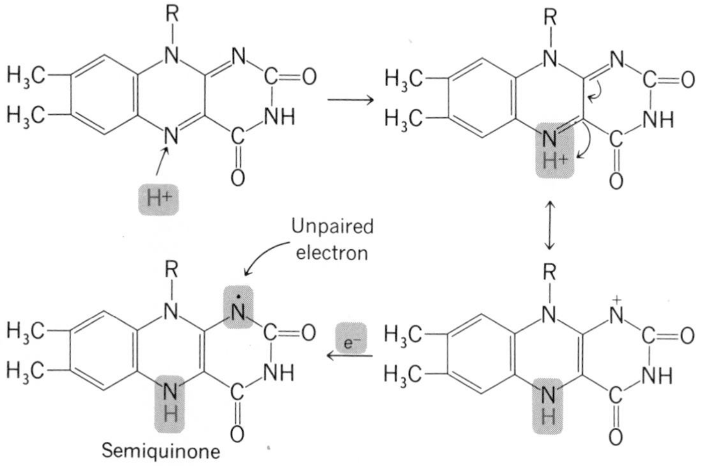

H^+

Unpaired electron

e^-

Semiquinone

intermediate. The reaction may be represented as in the diagram. The semiquinone form of the riboflavin coenzymes may be expected to be reasonably stable because of the possible existence of different resonance forms. In addition, the occurrence of a metal such as molybdenum or iron would be expected to stabilize the semiquinone; such structures possess an unpaired electron which could be shared with the unpaired electrons commonly encountered in metal ions.

Lipoic Acid

Structure.

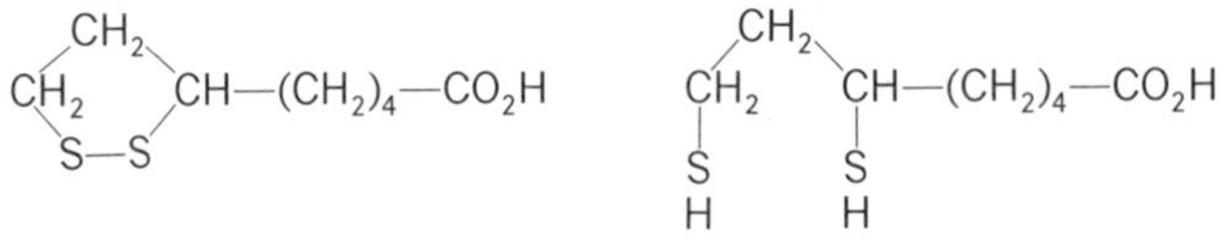

$$CH_2\text{–}CH_2\text{–}CH(\text{–}S\text{–}S\text{–})\text{–}(CH_2)_4\text{–}CO_2H \qquad HS\text{–}CH_2\text{–}CH_2\text{–}CH(SH)\text{–}(CH_2)_4\text{–}CO_2H$$

Oxidized — Reduced

Lipoic acid

Occurrence. Lipoic acid was discovered by virtue of it serving as a growth factor—a vitamin—for certain microorganisms. Liver and yeast are particularly rich sources. The vitamin exists in both oxidized and reduced forms due to the ability of the disulfide linkage to undergo reduction. Bound to protein, lipoic acid is released by acidic, basic, or proteolytic hydrolysis. Careful hydrolysis of lipoyl-protein complexes discloses that lipoate is covalently bound to lysine as ε-*N*-lipoyl-L-lysine. This structure has a striking resemblance to biocytin (ε-*N*-biotinyl-L-lysine), which is isolated as a hydrolysis product of biotin–protein complexes, and indicates that in lipoyl enzymes the lipoic is bonded to lysyl residues of the protein.

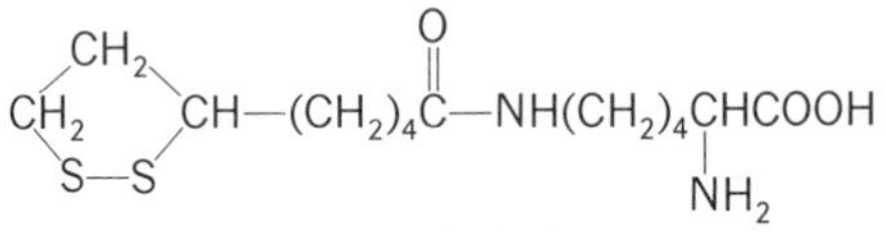

$$\text{(lipoyl ring)}\,CH\text{–}(CH_2)_4\text{–}C(=O)\text{–}NH(CH_2)_4CH(NH_2)COOH$$

ε-*N*-Lipoyl-L-lysine

Biochemical function. Lipoic acid is a cofactor of the multienzyme complexes *pyruvic dehydrogenase* and *α-ketoglutaric dehydrogenase* (Chapter 8). In these complexes, the lipoyl-containing enzymes catalyze the generation and transfer of acyl groups and in the process undergo reduction followed by reoxidation. In the initial step involving the lipoic acid moiety, an acylol–thiamine complex (Scheme 13-1) reacts with the oxidized lipoic residue to form an addition complex which subsequently rearranges to form the free thiamine residue and the acyl–lipoic acid complex. It is in this reaction that the acylol moiety is oxidized to an acyl group and the oxidized lipoic is reduced:

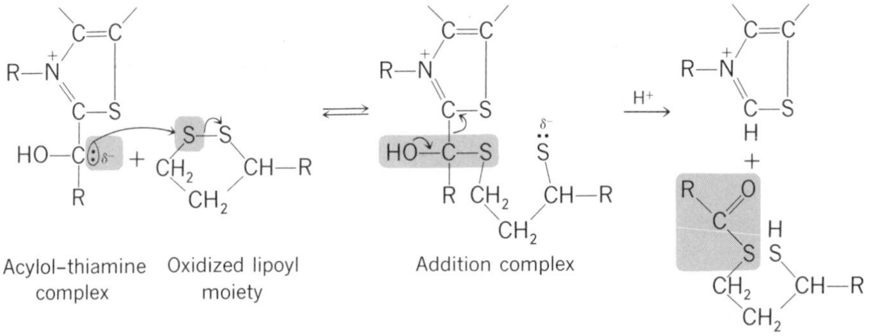

Acylol–thiamine complex + Oxidized lipoyl moiety ⇌ Addition complex $\xrightarrow{H^+}$ (free thiamine, C–H) + Acyl–lipoyl complex ($R\text{–}C(=O)\text{–}S\text{–}CH_2\text{–}CH_2\text{–}CH(SH)\text{–}R$)

Next the acyl group is transferred from the acyl–lipoic acid grouping to coenzyme A (Scheme 13-2) to form acyl–CoA:

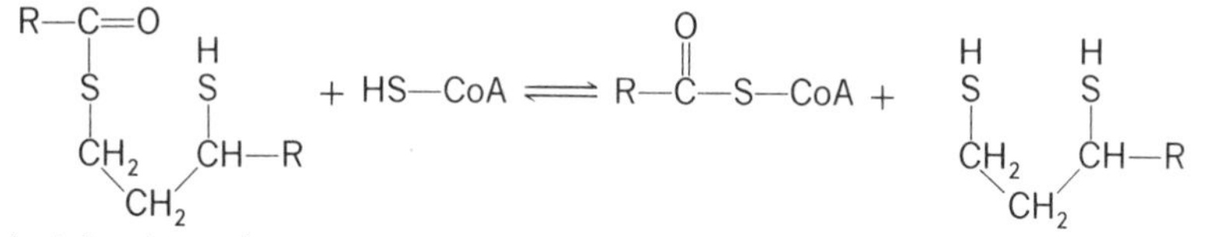

$$\begin{matrix} R{-}C{=}O & & \\ | & H & \\ S & S & \\ | & | & \\ CH_2 & CH{-}R \\ \quad \diagdown & \diagup \\ & CH_2 \end{matrix} + HS{-}CoA \rightleftharpoons R{-}\overset{\overset{O}{\|}}{C}{-}S{-}CoA + \begin{matrix} H & H \\ S & S \\ | & | \\ CH_2 & CH{-}R \\ \diagdown & \diagup \\ & CH_2 \end{matrix}$$

Acyl–lipoyl complex — Acyl–S–CoA — Reduced lipoyl moiety

Finally the reduced lipoic acid moiety is oxidized by a FAD-containing enzyme to regenerate the oxidized lipoyl moiety and allow the process to be repeated:

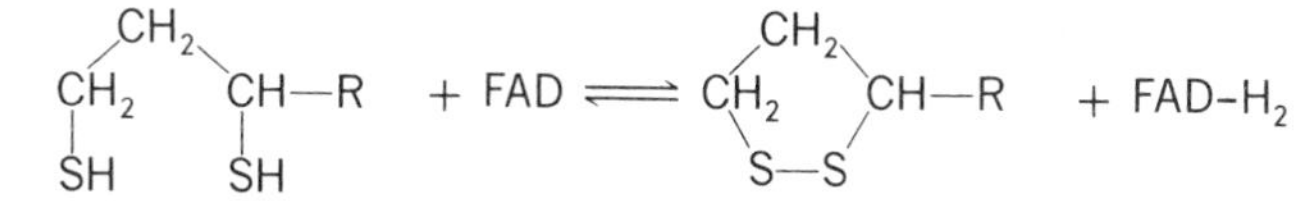

$$\begin{matrix} & CH_2 & \\ \diagup & & \diagdown \\ CH_2 & & CH{-}R \\ | & & | \\ SH & & SH \end{matrix} + FAD \rightleftharpoons \begin{matrix} & CH_2 & \\ \diagup & & \diagdown \\ CH_2 & & CH{-}R \\ \diagdown & & \diagup \\ & S{-}S & \end{matrix} + FAD{-}H_2$$

Reduced lipoyl moiety — Oxidized lipoyl moiety

These enzymes will be considered in more depth in Chapter 13.

Biotin

Structure.

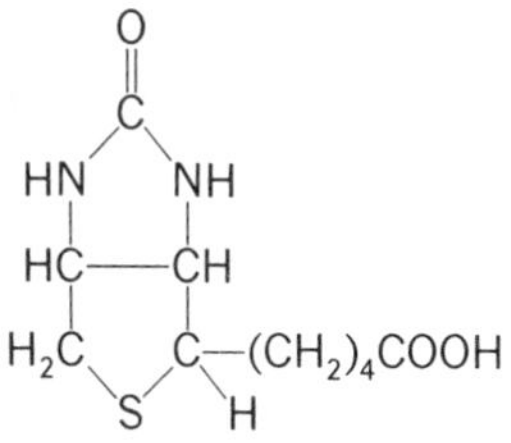

$$\begin{matrix} & O & \\ & \| & \\ & C & \\ \diagup & & \diagdown \\ HN & & NH \\ | & & | \\ HC & \text{---} & CH \\ | & & | \\ H_2C & & C{-}(CH_2)_4COOH \\ \diagdown & & \diagup \quad \diagdown \\ & S & \qquad H \end{matrix}$$

Occurrence. The essential nature of biotin was established by its ability to serve as a growth factor for yeast and certain bacteria, as well as a recognition that it was the "anti-egg white injury factor." The latter term refers to the observation that a nutritional deficiency may be induced in animals by feeding them large amounts of avian egg white. Egg white contains a basic protein known as avidin which has a remarkably high affinity for biotin or its simple derivatives. At 25°C the binding constant is about 10^{15}. Avidin is therefore an extremely effective inhibitor of biotin-requiring systems and is employed by the biochemist to test for possible reactions in which biotin may participate.

Biotin is widely distributed in nature with yeast and liver as excellent sources. The vitamin occurs mainly in combined forms bound to protein through the ε-*N*-lysine moiety. Biocytin, ε-*N*-biotinyl-L-lysine, has been isolated as a hydrolysis product from biotin-containing proteins.

Because of their linkage with proteins through covalent peptide bonds, neither biotin nor lipoic acid is dissociated by dialysis, a technique commonly used to remove readily dissociable groups such as the nicotinamide nucleo-

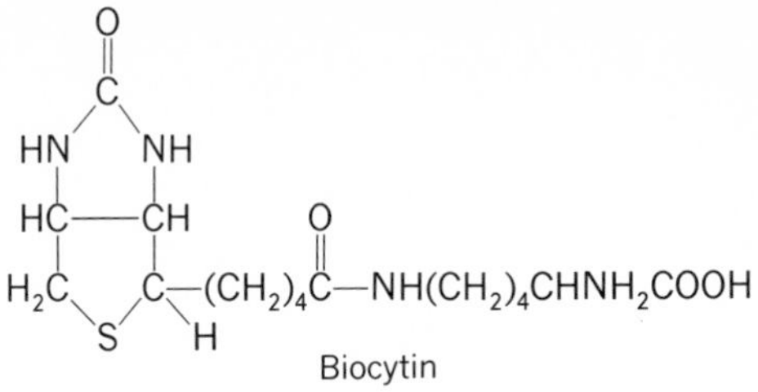

Biocytin

tides. As a result, no enzymes have been described that can be reactivated by the simple expedient of adding biotin to the apoenzyme. A biotin-containing enzyme will be inhibited by the addition of avidin to the reaction, however.

Biochemical function. Biotin bound to its specific enzyme protein is intimately associated with carboxylation reactions. There are three well-described carboxylations catalyzed by different enzymes in which biotin participates. All

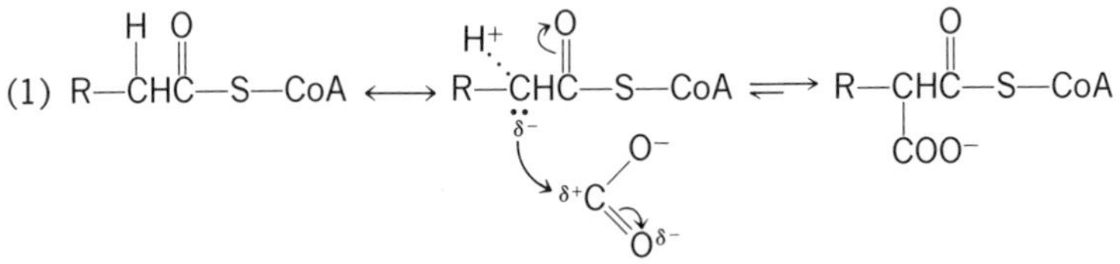

(1) $R{-}CH_2C(=O){-}S{-}CoA \longleftrightarrow R{-}\ddot{C}H^{\delta-}C(=O){-}S{-}CoA + {}^{\delta+}C(O^-)=O^{\delta-} \rightleftharpoons R{-}CH(COO^-)C(=O){-}S{-}CoA$

α-Carboxylation

(2) $H{-}CH_2{-}C(CH_3){=}CHC(=O){-}S{-}CoA \rightleftharpoons {}^-OOCCH_2{-}C(CH_3){=}CHC(=O){-}S{-}CoA$

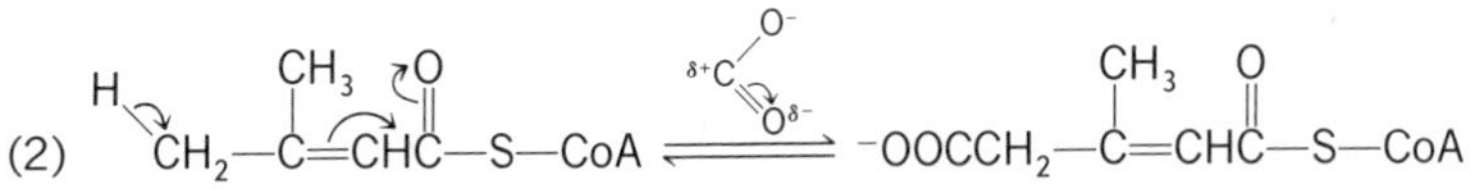

β-Methylcrotonyl-S-CoA → β-Methylglutaconyl-S-CoA

"Conjugated" α-carboxylation

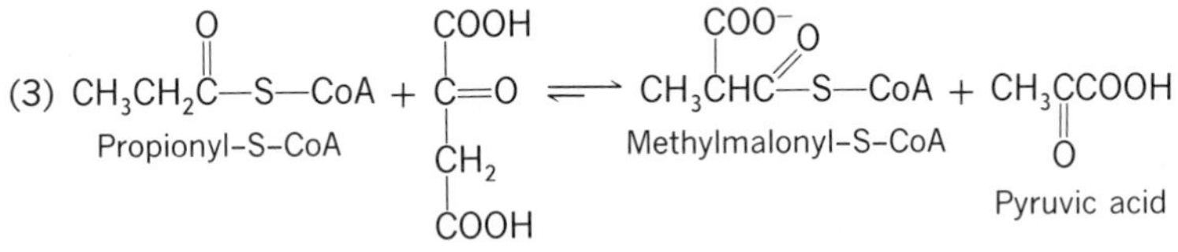

(3) $CH_3CH_2C(=O){-}S{-}CoA + HOOC{-}C(=O){-}CH_2{-}COOH \rightleftharpoons CH_3CH(COO^-)C(=O){-}S{-}CoA + CH_3C(=O)COOH$

Propionyl-S-CoA + Oxaloacetic acid ⇌ Methylmalonyl-S-CoA + Pyruvic acid

Transcarboxylation

three reactions employ biotin as the actual carrier of CO_2. In order that CO_2 can carboxylate the α-carbon atom in reactions 1 and 2, ATP and manganese ion must be present. In reaction 3 no cofactors need be added. The concerted reaction we shall give describes the CO_2-activation step.

This reaction has been further analyzed by P. R. Vagelos, who examined the acetyl-CoA carboxylase of *E. coli* as the model system. The overall

(a)

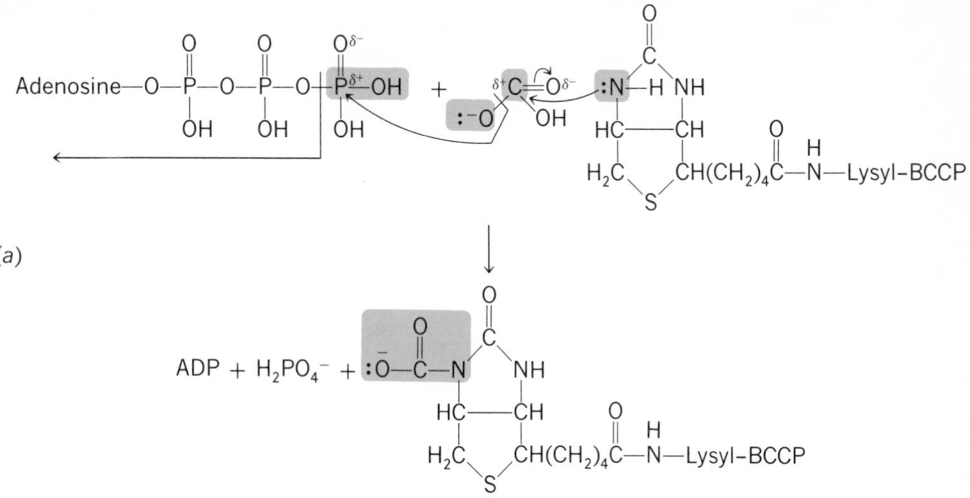

Adenosine—O—P—O—P—O—P—OH + $^{-}$O—C=O + :N (biotin)—CH(CH$_2$)$_4$C—N—Lysyl-BCCP

ADP + $H_2PO_4^-$ + :$\bar{O}$—C—N (biotin)—CH(CH$_2$)$_4$C—N—Lysyl-BCCP

Carboxyl-biotinyl-protein complex

(b)

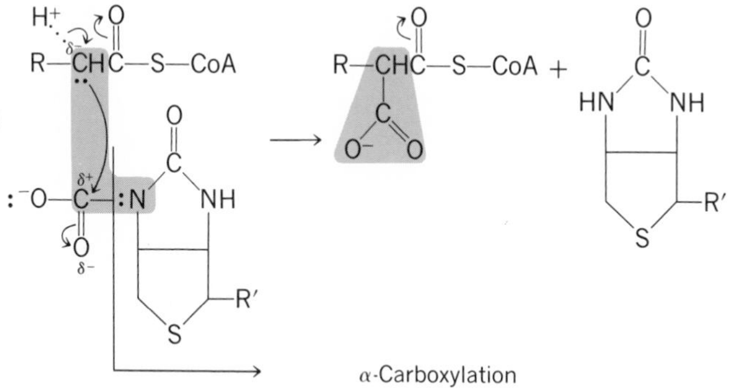

R—CHC—S—CoA → R—CHC—S—CoA + biotin—R′

α-Carboxylation

reaction involves a number of steps and different proteins. Playing a key role is a small protein, the biotin carboxyl carrier protein (BCCP), with a molecular weight of about 20,000, that has 1 mole of biotin linked covalently to the protein through a lysyl bridge. This protein is carboxylated in the presence of *biotin carboxylase* and then transfers the CO_2 group on to the acceptor molecule, acetyl-CoA.

$$\text{(1) ATP} + HCO_3^- + \text{BCCP} \underset{\text{Biotin carboxylase}}{\overset{Mn^{2+}}{\rightleftharpoons}} CO_2^- \text{-BCCP} + \text{ADP} + \text{Pi}$$

$$\text{(2) } CO_2^- \text{-BCCP} + \text{Acetyl-CoA} \overset{\text{Transcarboxylase}}{\rightleftharpoons} \text{BCCP} + \text{Malonyl-CoA}$$

A BCCP-like component is found in all carboxylation systems involving biotin. Indeed, only BCCP contains the functioning biotinyl moiety. The biotin carboxylase and the transcarboxylase are specific proteins which are biotin-free.

Thiamin

Structure. Thiamin, or vitamin B_1, is usually isolated as the free vitamin.

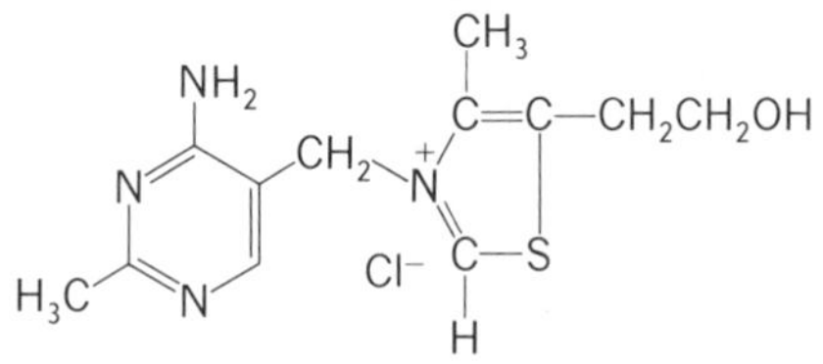

Occurrence. Thiamin occurs free in nature in relatively high concentrations in the cereal grains. In animal tissues and in yeast it occurs chiefly as its coenzyme thiamin pyrophosphate, whose structure was established by Lohman in 1937.

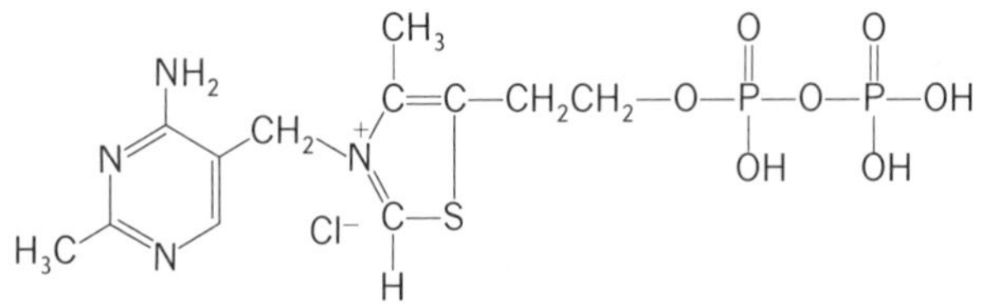

A lack of vitamin B_1 results in the deficiency disease known as beriberi, which is common in the Far East. It is due to the practice of using polished rice from which the outer, thiamin-rich layers of the seed have been removed. Thiamin may also be present in marginal concentrations in the diet of many peoples.

Biochemical function. Thiamin pyrophosphate participates as a coenzyme in the following systems:

(1) α-Keto acid decarboxylases
(2) α-Keto acid oxidases
(3) Transketolase
(4) Phosphoketolase

In all these reactions the common site of action is C-2 of the thiazole ring. The hydrogen atom at this position readily dissociates as a proton with the

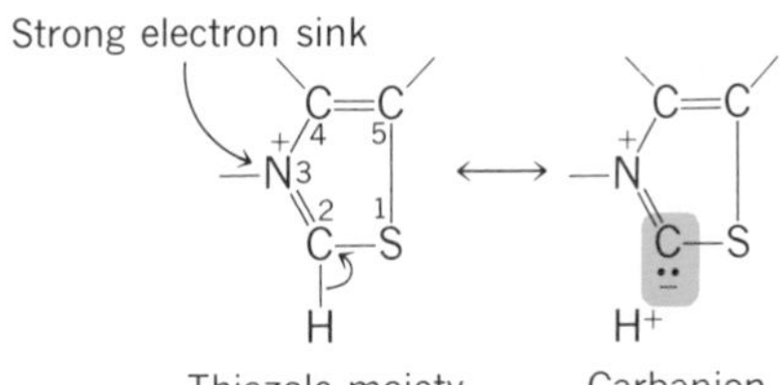

formation of a carbanion. The carbanion participates in the decarboxylation of α-keto acids as shown in the next diagram. The adduct formed, undergoes decarboxylation after the appropriate rearrangement of electrons, and acetaldehyde dissociates with the regeneration of the carbanion.

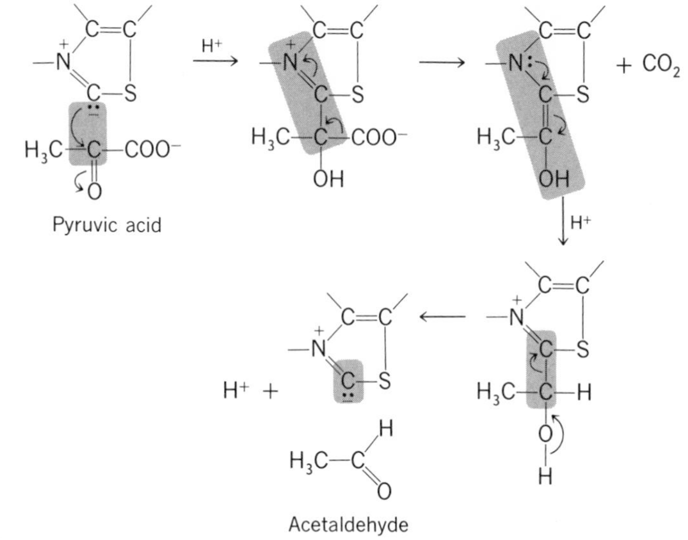

Pyruvic acid

Acetaldehyde

The reaction mechanism for α-keto acid oxidation is discussed in Chapter 13 and in the section on lipoic acid above. A slightly modified mechanism has been proposed for the reaction catalyzed by transketolase, again employing the thiazole carbanion as the attacking species (page 275).

Vitamin B_6

Structure. Three compounds belong to the vitamin group known as B_6. They are *pyridoxal, pyridoxine,* and *pyridoxamine:*

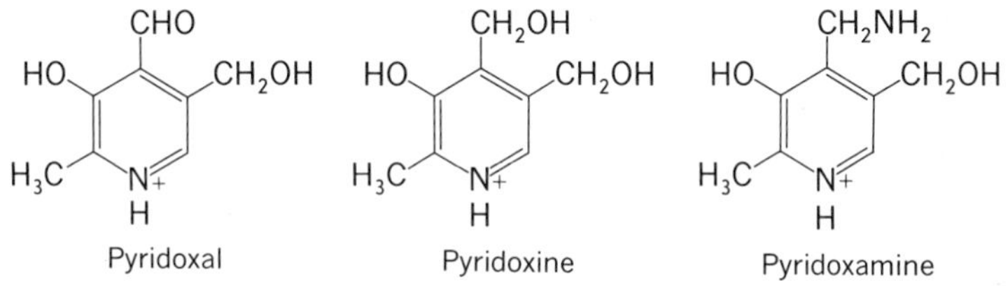

Pyridoxal Pyridoxine Pyridoxamine

Occurrence. The three forms of vitamin B_6 are widely distributed in animal and plant sources; cereal grains are especially rich sources of the vitamin. Pyridoxal and pyridoxamine also occur in nature as their phosphate derivatives which are the coenzyme forms of the vitamin.

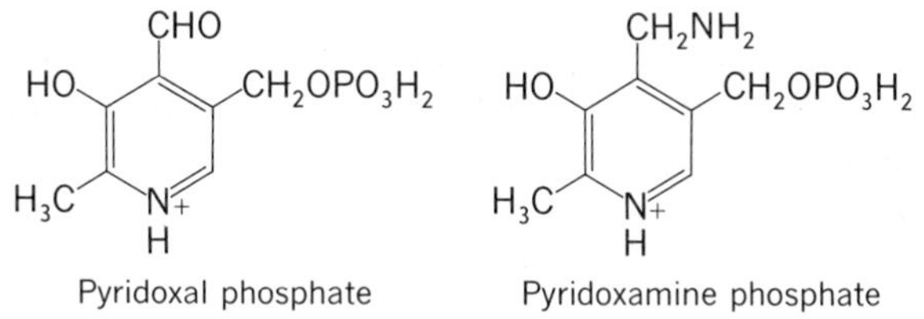

Pyridoxal phosphate Pyridoxamine phosphate

All three forms of the vitamin are effective in preventing vitamin B_6 deficiency symptoms which, in rats, occur initially as a severe dermatitis. Extreme deficiency in animals causes convulsions similar to those of epilepsy and indicates a profound disturbance in the central nervous system. The different forms of vitamin B_6 also serve as growth factors for many bacteria.

Biochemical function. Pyridoxal phosphate is a versatile vitamin derivative which participates in the catalysis of several important reactions of amino acid metabolism, such as transamination, decarboxylation, and racemization.

Transamination:

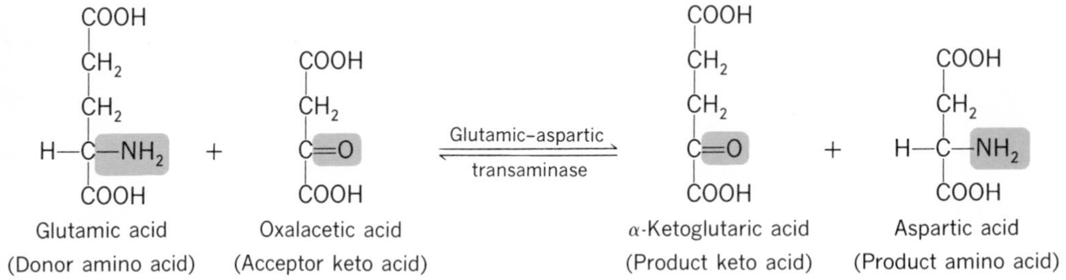

$$HOOC-CH_2-CH_2-CH(NH_2)-COOH + HOOC-CH_2-C(=O)-COOH \xrightleftharpoons[\text{transaminase}]{\text{Glutamic–aspartic}} HOOC-CH_2-CH_2-C(=O)-COOH + HOOC-CH_2-CH(NH_2)-COOH$$

Glutamic acid (Donor amino acid) + Oxalacetic acid (Acceptor keto acid) ⇌ α-Ketoglutaric acid (Product keto acid) + Aspartic acid (Product amino acid)

Decarboxylation:

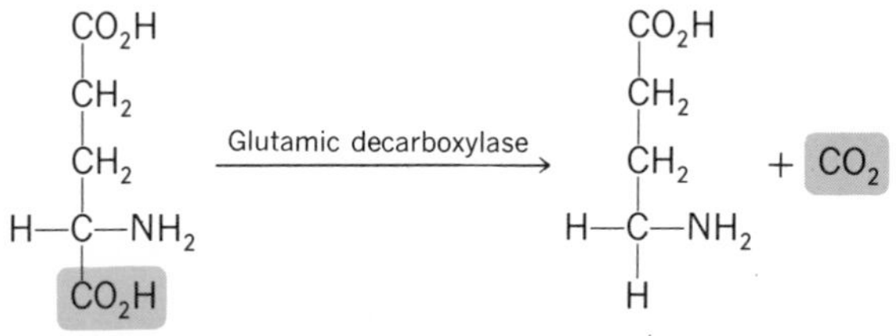

$$CO_2H-CH_2-CH_2-CH(NH_2)-CO_2H \xrightarrow{\text{Glutamic decarboxylase}} CO_2H-CH_2-CH_2-CH_2(NH_2) + CO_2$$

Racemization:

$$\text{L-}CO_2H-CH(NH_2)-CH_2-CH_2-CO_2H \xrightleftharpoons{\text{Glutamic acid racemase}} \text{D-}CO_2H-CH(NH_2)-CH_2-CH_2-CO_2H$$

L-Glutamic acid ⇌ D-Glutamic acid

Each reaction is catalyzed by a different, specific enzyme, but in each case pyridoxal phosphate functions as the coenzyme. There is now good evidence to support the concept developed by Jenkins that pyridoxal phosphate is loosely bound as a Schiff's base to the ε-amino group of a lysyl residue in all enzymes involving pyridoxal phosphate. However, chemical reduction with sodium borohydride reduces the Schiff's base to a secondary amine and binds pyridoxal phosphate irreversibly to the protein.

When a suitable substrate such as an amino acid approaches the Schiff's base, a transaldimation reaction occurs displacing the lysyl amino group and forming a new Schiff's base with the pyridoxal phosphate residue. In the presence of transaminases, sequence (*a*) in Scheme 9-1 will occur; in the presence of specific α-decarboxylases, sequence (*b*) will take place; and with specific racemases, sequence (*c*) will occur.

Approximately twenty other specific reactions of amino acids involving pyridoxal phosphate have been discovered, one of which is the interconversion of serine and glycine. Of unusual interest is the fact that pyridoxal phosphate is found bound to lysine in animal and plant phosphorylases. If the coenzyme is removed from the protein, phosphorylase activity disappears but can be restored by adding pyridoxal phosphate. The precise role of pyridoxal phosphate in this system is unknown.

Folic Acid

Structure.

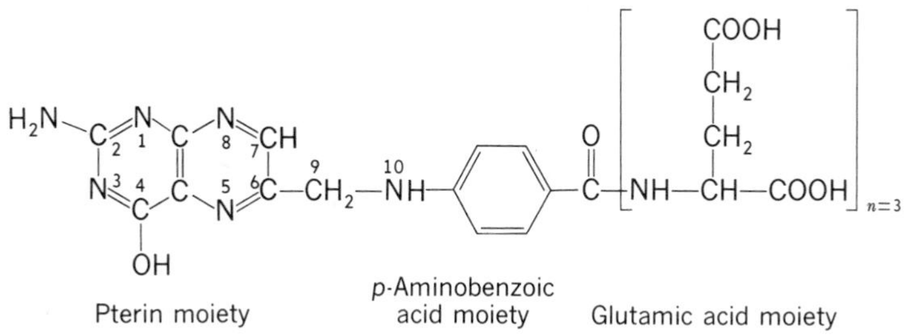

Pterin moiety — *p*-Aminobenzoic acid moiety — Glutamic acid moiety ($n=3$)

Folic acid (Pteroyl-L-glutamic acid, [F])

Occurrence. Folic acid and its derivatives, which are chiefly the tri- and hepta-glutamyl peptides, are widespread in nature. The vitamin cures nutritional anemia in chicks and serves as a specific growth factor in a number of microorganisms. Since extremely small amounts are needed by experimental animals, it is very difficult to produce folic acid deficiencies. Intestinal bacteria provide the small amounts necessary for growth. Derivatives of folic acid play an important but yet unknown role in the formation of normal erythrocytes.

Biochemical function. Although folic acid is the vitamin, its reduction products are the actual coenzyme forms. An enzyme, *folic reductase,* reduces folic acid to dihydrofolic acid (FH_2); this compound is reduced in turn by *dihydro-*

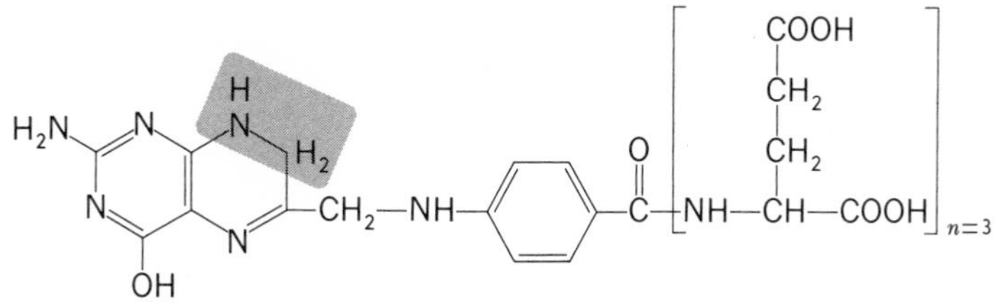

Dihydrofolic acid (FH_2)

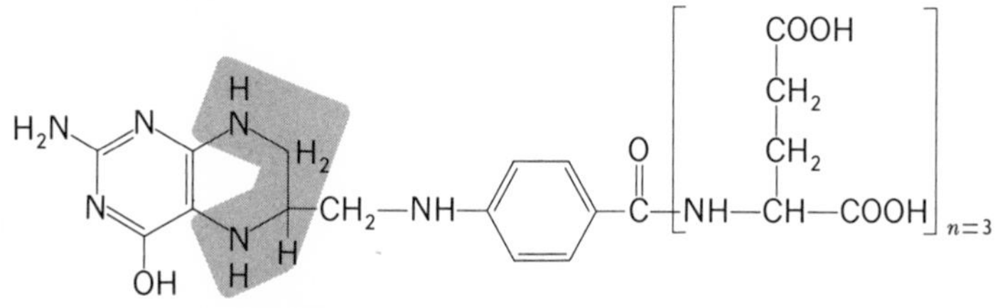

Tetrahydrofolic acid (FH_4)

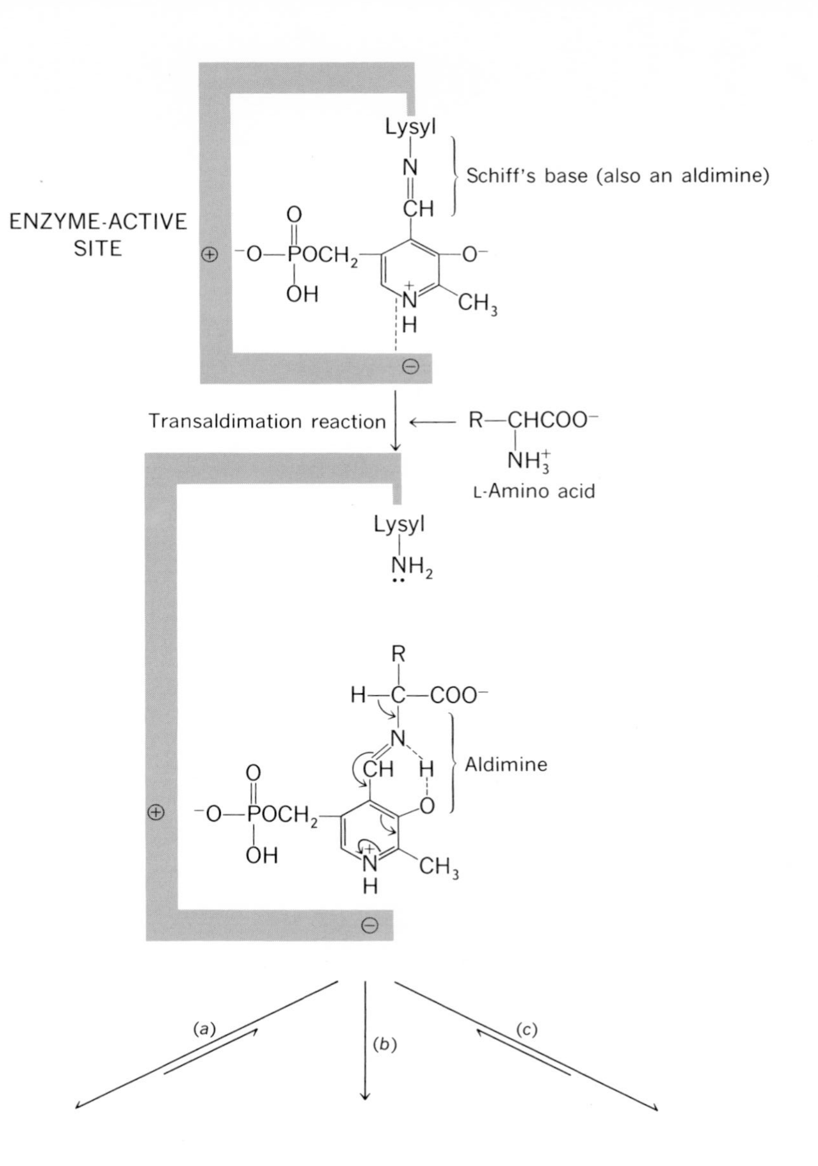
ENZYME-ACTIVE SITE
Lysyl
Schiff's base (also an aldimine)
Transaldimation reaction
R—CHCOO⁻
L-Amino acid
Aldimine
(a)
(b)
(c)

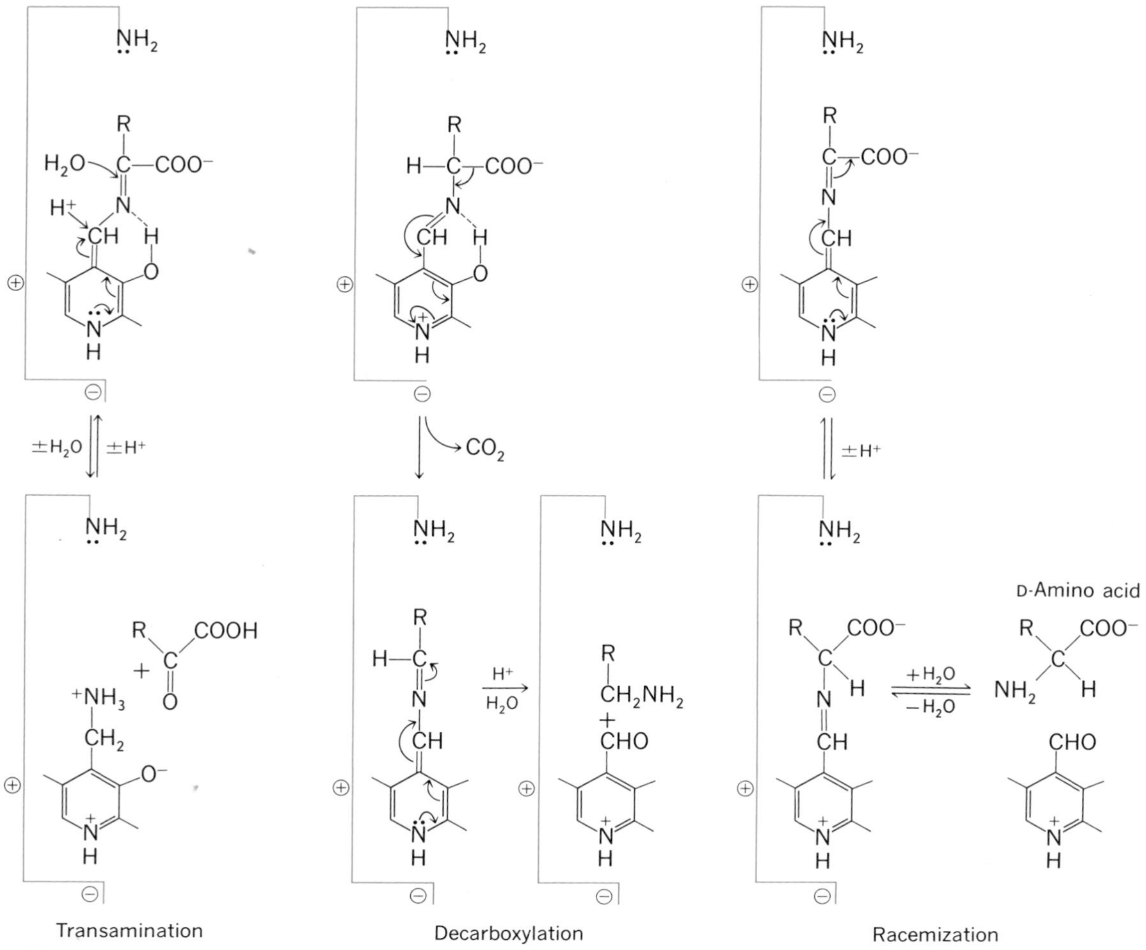

Transamination

Decarboxylation

D-Amino acid

Racemization

Scheme 9-1

folic reductase to tetrahydrofolic acid (FH_4). The reducing agent in both reactions is NADPH:

$$\text{Folic} + \text{NADPH} + H^+ \xrightarrow{\text{Folic reductase}} FH_2 + \text{NADP}^+$$

$$FH_2 + \text{NADPH} + H^+ \xrightarrow{\text{Dihydrofolic reductase}} FH_4 + \text{NADP}^+$$

The central role of FH_4 is that of a carrier for a one-carbon unit at the oxidation level of formate (or formaldehyde). The formate unit is used in the biosynthesis of purines, serine, and glycine. The chemistry of this formate unit is complex but involves initially the activation of formic acid:

$$FH_4 + \text{ATP} + \text{HCOOH} \longrightarrow \text{Formyl } N^{10}FH_4 + \text{ADP} + \text{Pi}$$

Formyl $N^{10}FH_4$ undergoes ring closure to methenyl $N^{5\text{-}10}FH_4$ (as shown in

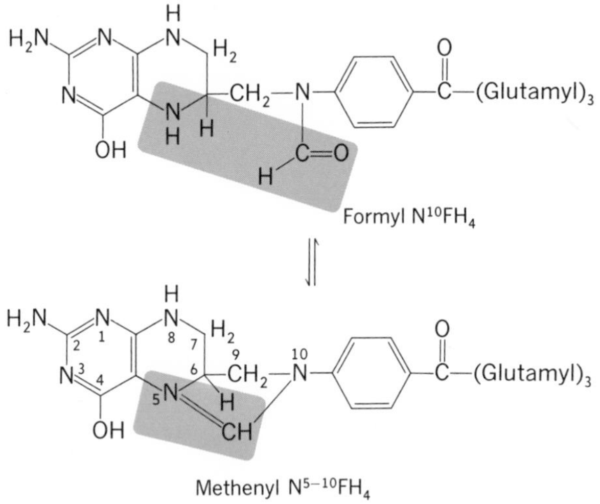

Formyl $N^{10}FH_4$

Methenyl $N^{5-10}FH_4$

the diagram), which then is reduced by NADPH in the presence of a specific dehydrogenase:

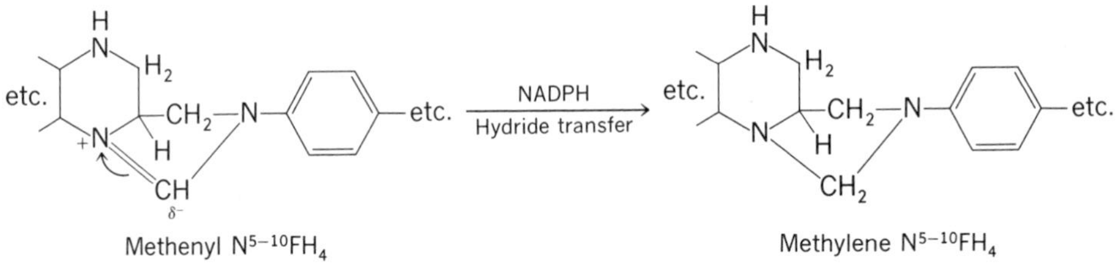

$$\xrightarrow[\text{Hydride transfer}]{\text{NADPH}}$$

Methenyl $N^{5-10}FH_4$ Methylene $N^{5-10}FH_4$

Methylene $N^{5\text{-}10}FH_4$, in the presence of pyridoxal phosphate, serine hydroxyl methylase, and glycine, forms serine. Thus, the reduction of formate to $—CH_2OH$ is complex. Of interest is the observation that not one but two derivatives of important vitamins, folic acid and pyridoxal phosphate, are required cofactors for the utilization of formate to form serine; this is an excellent example of the intermeshing of vitamins in the tissue economy.

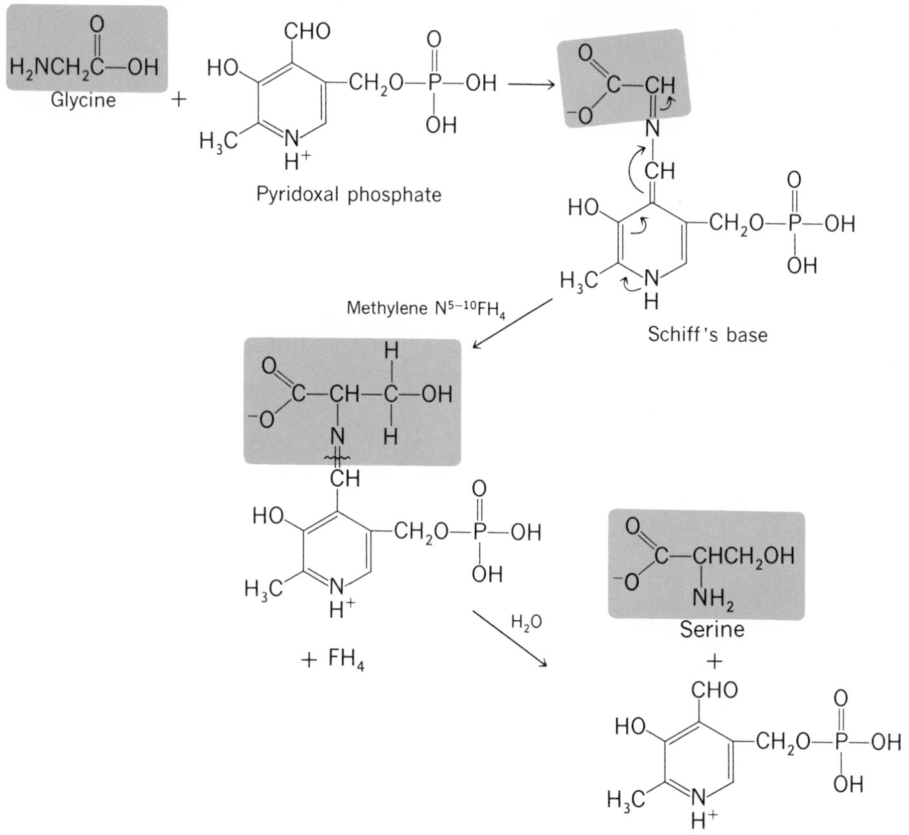

Glycine + Pyridoxal phosphate → Schiff's base

Methylene $N^{5-10}FH_4$

$+ FH_4$ $\xrightarrow{H_2O}$ Serine + Pyridoxal phosphate

Tetrahydrofolic acid is also involved in a most interesting and important reaction by which it functions both as a source of hydrogen atoms and as a source of carbon for CH_3 synthesis. In this reaction dihydrofolic acid (FH_2) is a product.

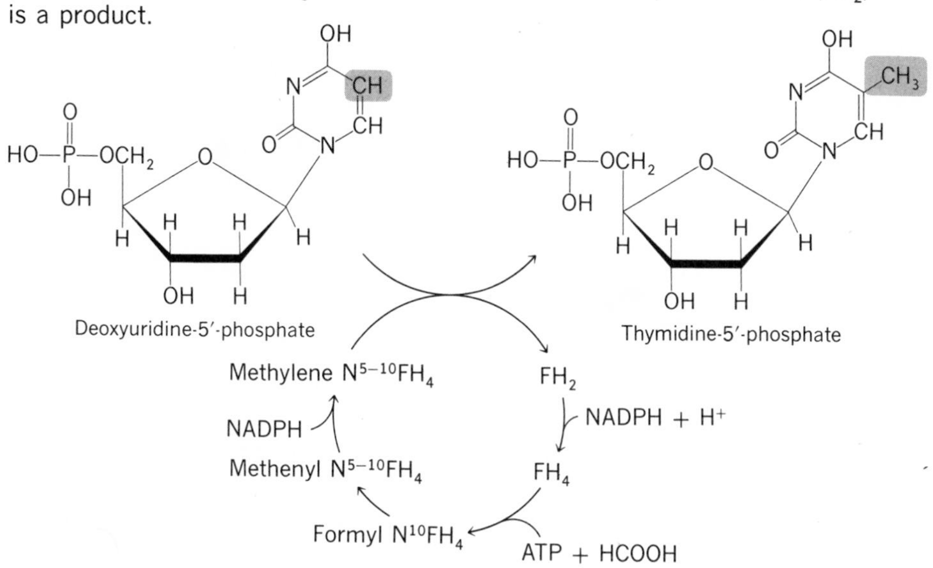

Deoxyuridine-5′-phosphate → Thymidine-5′-phosphate

Methylene $N^{5-10}FH_4$ → FH_2

FH_2 + NADPH + H^+ → FH_4

FH_4 + ATP + HCOOH → Formyl $N^{10}FH_4$

Formyl $N^{10}FH_4$ → Methenyl $N^{5-10}FH_4$

Methenyl $N^{5-10}FH_4$ + NADPH → Methylene $N^{5-10}FH_4$

A summary of the metabolism of C_1 units as related to folic acid reactivities follows:

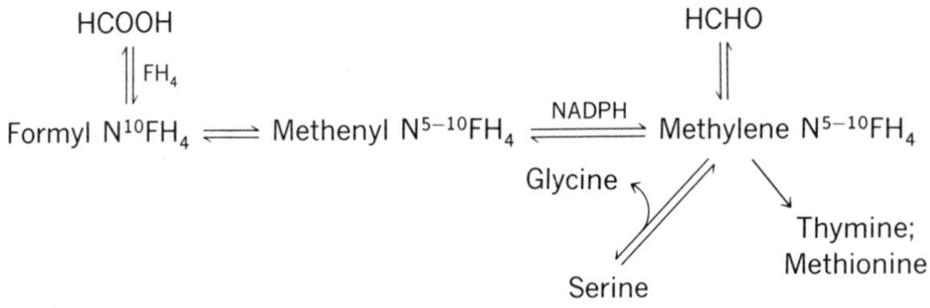

$$HCOOH \underset{}{\overset{FH_4}{\rightleftharpoons}} \text{Formyl } N^{10}FH_4$$

$$\text{Formyl } N^{10}FH_4 \rightleftharpoons \text{Methenyl } N^{5-10}FH_4 \underset{}{\overset{NADPH}{\rightleftharpoons}} \text{Methylene } N^{5-10}FH_4 \rightleftharpoons HCHO$$

$$\text{Serine} \rightleftharpoons \text{Methylene } N^{5-10}FH_4 \text{ (with Glycine)}$$

$$\text{Methylene } N^{5-10}FH_4 \rightarrow \text{Thymine; Methionine}$$

Vitamin B_{12}

Structure. Vitamin B_{12} as it is isolated from liver is a cyanocobalamin whose structure is shown here.

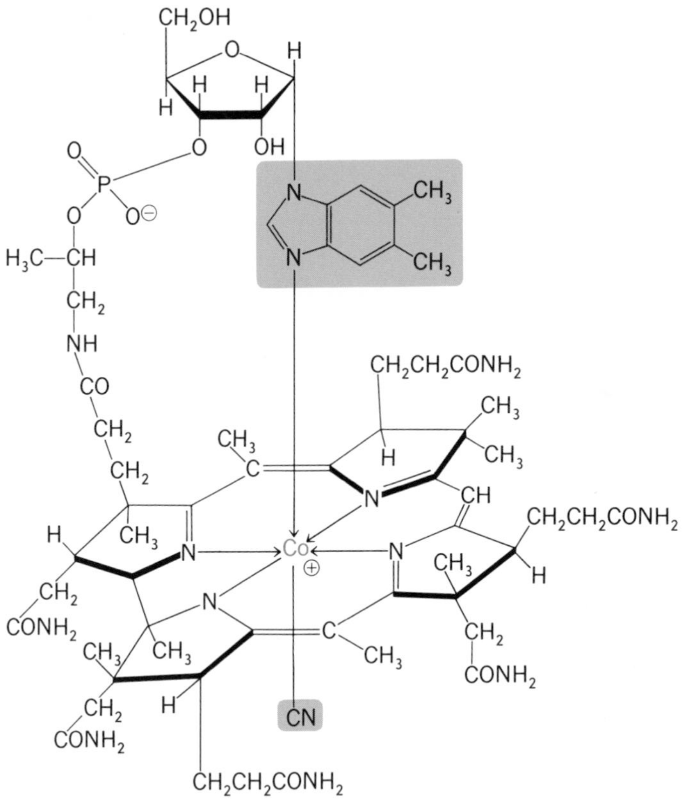

Vitamin B_{12}

Liver extracts also contain a hydroxycobalamin (vitamin B_{12b}) in which the cyanide is replaced by a hydroxyl group, however. Still other vitamin B_{12}-like compounds in which the dimethylbenzimidazole moiety is replaced by other nitrogenous bases have been isolated from bacteria. In pseudo-vitamin B_{12} the nitrogenous base is adenine; in another form of the vitamin, the base is benzimidazole.

Occurrence. Vitamin B_{12}, which has been found only in animals and microorganisms and not in plants, has recently been shown by H. A. Barker to occur

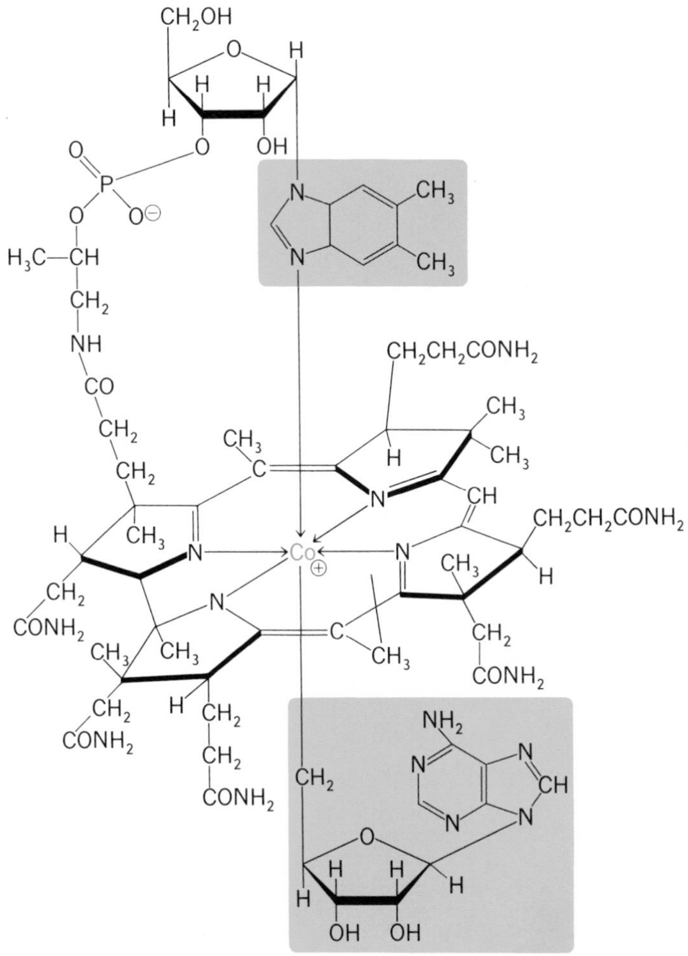

Coenzyme B_{12}

as part of a coenzyme known as coenzyme B_{12}, which has the structure shown. In the coenzyme the position occupied by either a cyanide or a hydroxyl ion in the vitamin is bonded directly to the 5′-carbon atom of the ribose of adenosine.

The coenzyme is relatively unstable, and in the presence of light or cyanide is decomposed, respectively, to the hydroxycobalamin or cyanocobalamin form known as the vitamin. Hence, the very distinct possibility exists that vitamin B_{12} occurs in nature chiefly as coenzyme B_{12}.

Since pseudo-vitamin B_{12} occurs with adenine rather than 5,6-dimethly-benzimidazole as the base attached to ribose, there also exists a coenzyme form of pseudo-vitamin B_{12}. A coenzyme form of the vitamin which contains benzimidazole also occurs.

Vitamin B_{12} was first recognized as an agent useful in the prevention and treatment of pernicious anemia. Previously termed the antipernicious anemia factor, it is also a growth factor for several bacteria, and a protozoan, *Euglena*.

Biochemical function. The coenzyme is synthesized from vitamin B_{12} by a specific B_{12} coenzyme synthetase:

$$B_{12}(Co^{2+}) + \text{Reducing system} + \text{ATP} \longrightarrow B_{12}\text{ coenzyme} + \underset{\text{Triphosphate}}{\text{P—O—P—O—P}}$$

The reducing system is complex in that it involves a NADH–flavoprotein–(S—S) protein system. The reductant, NADH, transfers its electrons via a flavoprotein to the specific (S—S) protein to form a dithiol (SH, SH) protein which converts vitamin $B_{12}(Co^{2+})$ to vitamin $B_{12}(Co^{+})$. This reduced form then becomes the substrate for the alkylation reaction with ATP.

The B_{12} coenzyme participates in approximately eleven distinct biochemical reactions as well as in reactions by which CH_3–vitamin B_{12}–enzyme complex is either reduced to methane or carboxylated by CO_2 to form acetate. Of all these reactions, only that catalyzed by methyl malonyl CoA mutase occurs in animal tissue; all eleven reactions have been discovered and described in bacterial systems. No vitamin B_{12} coenzyme-linked reactions have been observed in higher plants.

Vitamin B_{12} coenzyme reactions can be grouped into four general reactions. Specific examples of these four general reactions are

(1) *General:* carbon–carbon bond cleavage
Specific: methylmalonyl CoA mutase:

① $C(=O)$—S-CoA, ③ CH—CH_3 ②, ④ COOH $\rightleftharpoons$ ① $C(=O)$—S-CoA, ② CH_2, ③ CH_2, ④ COOH

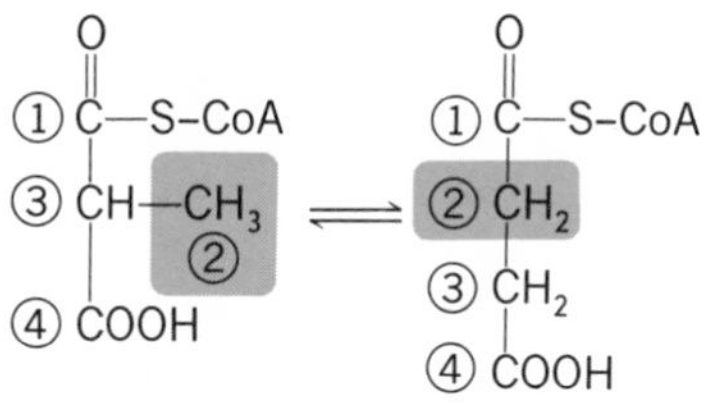

(2) *General:* carbon–oxygen bond cleavage
Specific: (*a*) diol dehydrase:

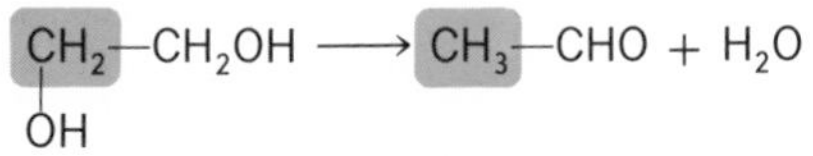

$$\underset{\text{OH}}{CH_2}\text{—}CH_2OH \longrightarrow CH_3\text{—CHO} + H_2O$$

(*b*) ribonucleotide reductase:

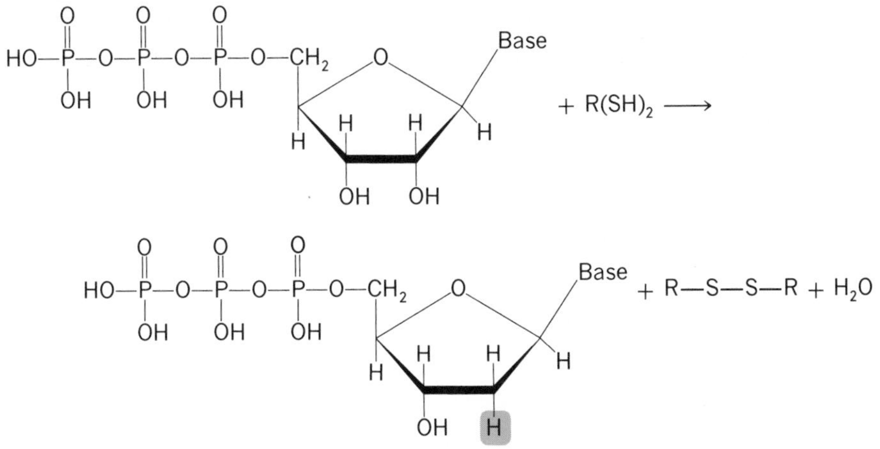

$$\text{HO—}\overset{\overset{\text{O}}{\|}}{\underset{\text{OH}}{\text{P}}}\text{—O—}\overset{\overset{\text{O}}{\|}}{\underset{\text{OH}}{\text{P}}}\text{—O—}\overset{\overset{\text{O}}{\|}}{\underset{\text{OH}}{\text{P}}}\text{—O—CH}_2\text{—(ribose ring: O, Base, H, H, H, H, OH, OH)} + R(SH)_2 \longrightarrow$$

$$\text{HO—}\overset{\overset{\text{O}}{\|}}{\underset{\text{OH}}{\text{P}}}\text{—O—}\overset{\overset{\text{O}}{\|}}{\underset{\text{OH}}{\text{P}}}\text{—O—}\overset{\overset{\text{O}}{\|}}{\underset{\text{OH}}{\text{P}}}\text{—O—CH}_2\text{—(ribose ring: O, Base, H, H, H, H, OH, H)} + R\text{—}S\text{—}S\text{—}R + H_2O$$

(3) *General:* carbon–nitrogen bond cleavage
Specific: D-α-lysine mutase:

$$\underset{NH_2}{CH_2}CH_2CH_2CH_2\underset{NH_2}{CH}COOH \longrightarrow CH_3\underset{NH_2}{CH}CH_2CH_2\underset{NH_2}{CH}COOH$$

(4) Methyl activation:

Methyl donor (CH_3—FH_4)

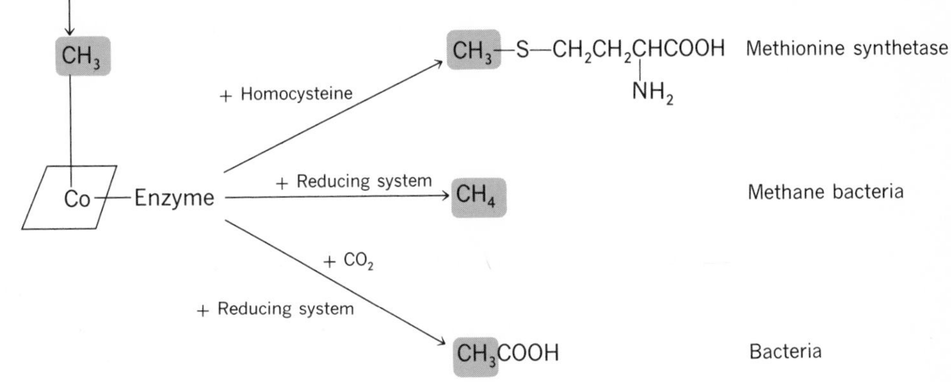

Pantothenic Acid

Structure. Pantothenic acid is required by animals as well as by microorganisms. However, it was first detected because of its ability to stimulate the growth of yeast.

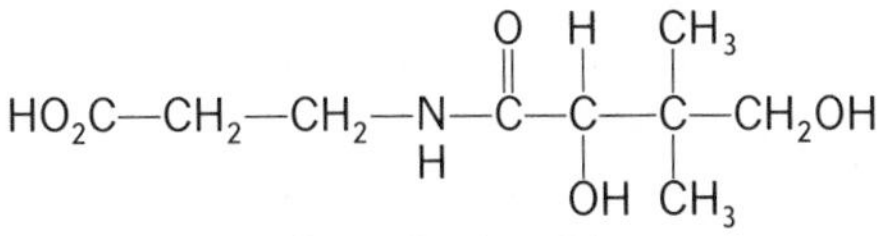

$$HO_2C\text{—}CH_2\text{—}CH_2\text{—}\underset{H}{N}\text{—}\overset{\overset{O}{\|}}{C}\text{—}\overset{H}{\underset{OH}{C}}\text{—}\overset{CH_3}{\underset{CH_3}{C}}\text{—}CH_2OH$$

Pantothenic acid

Occurrence. The vitamin occurs in nature primarily as a component of coenzyme A. This cofactor was discovered and named because of its being required for the enzymic acetylation of aromatic amines, i.e., the coenzyme for acetylation. Coenzyme A was isolated and its structure determined in the

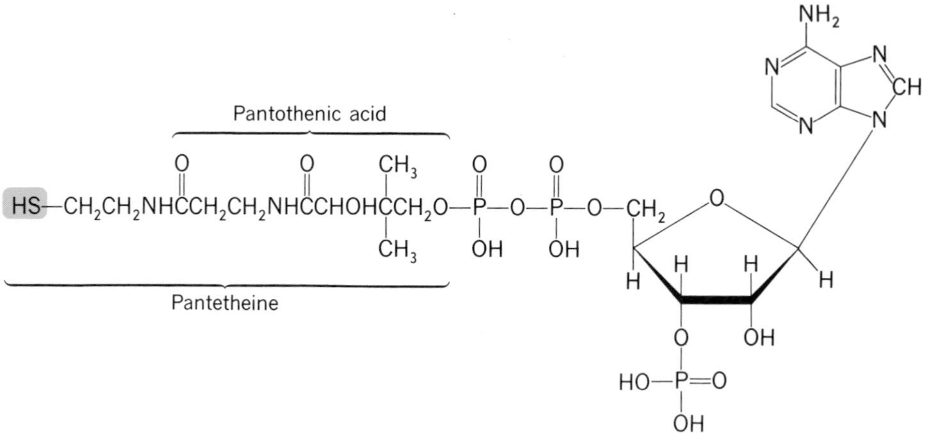

$$\underbrace{HS-\overbrace{CH_2CH_2NH\overset{O}{\overset{\|}{C}}CH_2CH_2NH\overset{O}{\overset{\|}{C}}CHOH\underset{CH_3}{\overset{CH_3}{C}}CH_2O}^{\text{Pantothenic acid}}}_{\text{Pantetheine}}-\underset{OH}{\overset{O}{\overset{\|}{P}}}-O-\underset{OH}{\overset{O}{\overset{\|}{P}}}-O-CH_2-\text{(ribose 3'-phosphate)}-\text{adenine}$$

Coenzyme A (CoA–SH)

late 1940's by F. Lipmann. The complete chemical synthesis of the coenzyme was described by Khorana in 1959.

Biochemical function. Thioesters formed from coenzyme A and carboxylic acids possess unique properties which account for the role the coenzyme plays in biochemistry. These properties are best understood when compared with

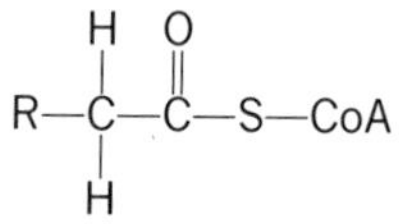

$$R-\underset{H}{\overset{H}{C}}-\overset{O}{\overset{\|}{C}}-S-CoA$$

Thioester

certain properties of oxygen esters. It is possible to write a resonance form of an oxygen ester in which the ester oxygen atom contains a positive change and is double bonded to the carboxylic acid carbon. Sulfur, however, does not readily release its electrons for double bond formation, and thioesters

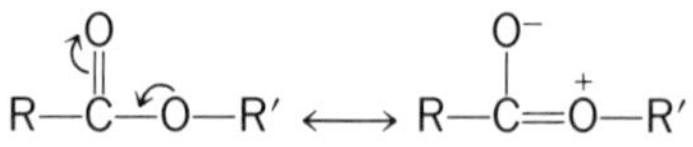

$$R-\overset{O}{\overset{\|}{C}}-O-R' \longleftrightarrow R-\overset{O^-}{\overset{|}{C}}=\overset{+}{O}-R'$$

therefore do not exhibit the resonance forms written for the oxygen ester. Instead, thioesters exhibit considerable carbonyl character in which a fractional positive charge may be represented on the carboxyl carbon and the carboxyl oxygen therefore exhibits a partial negative change. With the fractional positive charge on the carboxyl carbon, the hydrogen atom on the

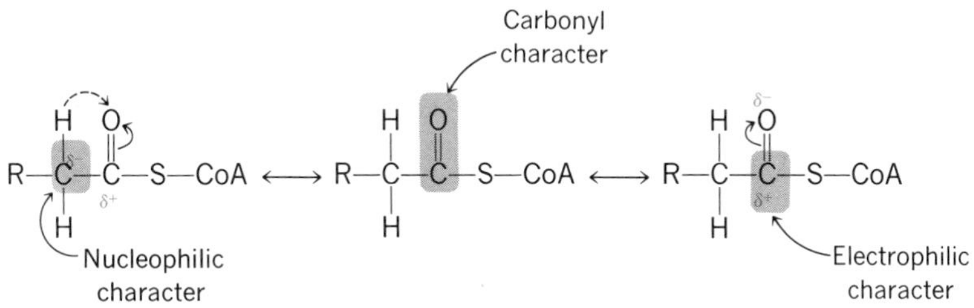

Carbonyl character

$$R-CH_2-\overset{O}{\overset{\|}{C}}-S-CoA \longleftrightarrow R-CH_2-\overset{O}{\overset{\|}{C}}-S-CoA \longleftrightarrow R-CH_2-\overset{O^{\delta-}}{\overset{\|}{C^{\delta+}}}-S-CoA$$

Nucleophilic character

Electrophilic character

adjacent α-carbon will tend to dissociate as a proton leaving a fractional negative charge on that α-carbon. These two possibilities are responsible for the electrophilic character of the carboxyl carbon atom in thioesters as well as the nucleophilic character of the α-carbon atom. Moreover, the inability of thioesters to possess the resonance forms written above for ordinary oxygen atoms explains their significantly greater instability and the higher $\Delta G'$ of hydrolysis exhibited by these compounds.

Nucleophiles such as amines, ammonia, water, thiol compounds, and phosphoric acid can attack the electrophilic site and displace the :S–CoA group. Electrophiles such as CO_2, acyl-CoA, or the CO_2–BCCP complex (page 211) can in turn attack the nucleophilic site. Throughout the text numerous examples are given of the reactivities of thioesters of coenzyme A. Most if not all of these reactions can be explained on the basis of the dual reactivity of these compounds. The student should, in his study of the book, attempt to gather together the many CoA–SH reactions and explain the mechanisms to his own satisfaction. Several examples and further discussion will be found in Chapters 12 and 13.

An interesting heat-stable protein, of low molecular weight and called *acyl carrier protein* (ACP), plays an important role in the biosynthesis of fatty acids. A distinctive feature of this protein is the 4′ phosphoryl pantetheine moiety which is covalently bonded to the hydroxyl group of a serine residue in the protein. Having the pantetheine structure, the molecule can serve as an acyl carrier in a manner analogous to coenzyme A through thioester formation with its sulfhydryl group. The structure and role of this protein is described in detail in Chapter 12. Soluble ACP's occur in plant and bacterial tissues, but in animal tissues part of the ACP molecule is covalently bound to the fatty acid synthetase complex (Chapter 12).

Vitamins Lacking True Coenzyme Function

The vitamin–coenzyme relationships which have been described are those of vitamins which are soluble in water. Those vitamins lacking a known coenzyme function, to be described now, include only one additional water-soluble vitamin, namely ascorbic acid. The remaining compounds in this category are soluble in certain organic solvents and constitute the fat-soluble vitamins. While no coenzyme relationship is established, a significant amount of information regarding the physiological role of these compounds is available in most cases.

Ascorbic Acid (Vitamin C)

Structure.

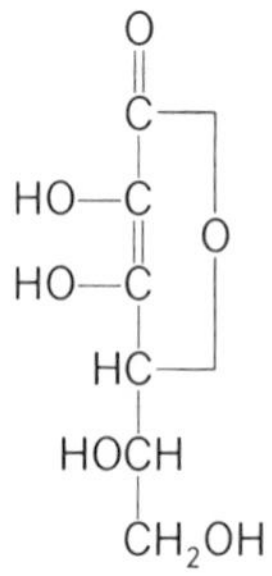

```
        O
        ‖
        C——┐
        |  |
  HO—C     O
        ‖  |
  HO—C     |
        |  |
       HC——┘
        |
     HOCH
        |
     CH₂OH
```

L-Ascorbic acid

Occurrence. Plants and animals, except guinea pigs and primates (including man), can synthesize ascorbic acid from D-glucose. The enzyme which is missing in the species that are unable to produce the vitamin is L-gulonoxidase, which converts L-gulonolactone to 3-keto-L-gulonolactone:

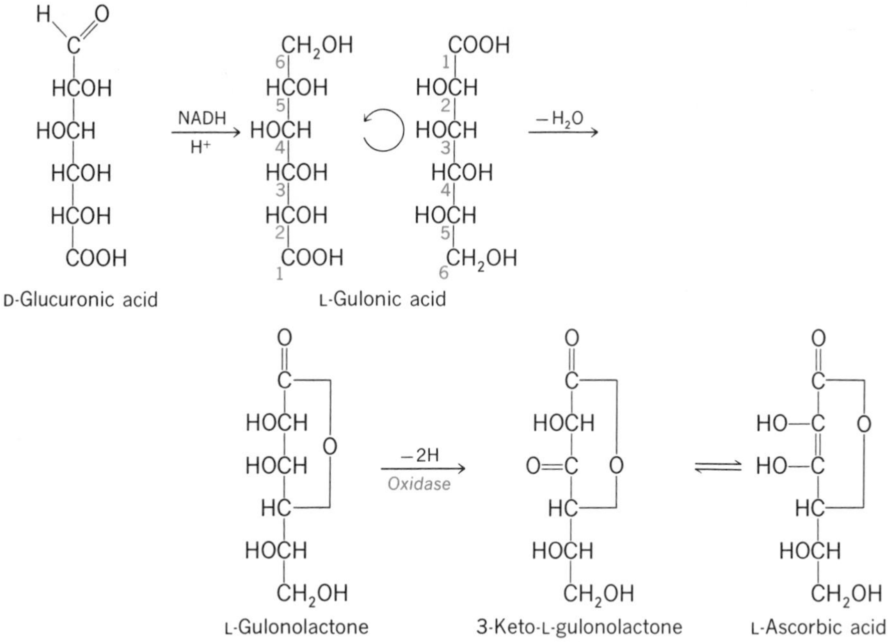

D-Glucuronic acid (H–C=O, HCOH, HOCH, HCOH, HCOH, COOH) $\xrightarrow[H^+]{NADH}$ L-Gulonic acid (6 CH_2OH, 5 HCOH, 4 HOCH, 3 HCOH, 2 HCOH, 1 COOH ⟲ 1 COOH, 2 HOCH, 3 HOCH, 4 HCOH, 5 HOCH, 6 CH_2OH) $\xrightarrow{-H_2O}$

L-Gulonolactone $\xrightarrow[Oxidase]{-2H}$ 3-Keto-L-gulonolactone ⇌ L-Ascorbic acid

Biochemical function. The absence of ascorbic acid in the diet gives rise to scurvy, a disease characterized by pathological changes in the teeth and gums. The disease was known to the ancients, especially among sailors, who often traveled for extended periods of time from sources of fresh fruits and vegetables that were known to prevent scurvy.

A primary characteristic of scurvy is a change in connective tissue. In ascorbic acid deficiency, the mucopolysaccharides of the cell ground substance are abnormal in character, and there are significant changes in the nature of the collagen fibrils that are formed. The presence of ascorbic acid is required for the formation of normal collagen in experimental animals. At the enzyme level, there is an indication that ascorbic acid is involved in the conversion of proline to hydroxyproline, an amino acid found in relatively high concentrations in collagen.

The biochemical role which ascorbic acid plays is undoubtedly related to it being a good reducing agent. Its oxidized form, dehydroascorbic acid, is capable of being reduced again by various reductants including glutathione, and the two forms of ascorbate constitute a reversible oxidation–reduction

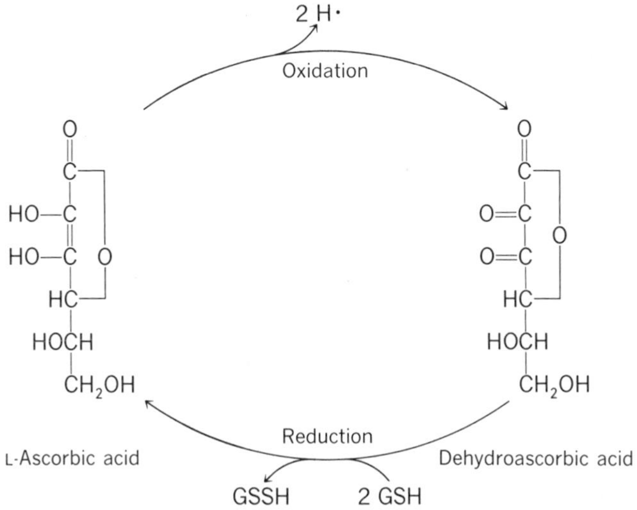

2 H·
Oxidation
L-Ascorbic acid ⇌ Dehydroascorbic acid
Reduction
GSSH 2 GSH

system. In the case of collagen formation, ascorbic acid can function as the external reductant that is required in the conversion of proline to hydroxyproline. Ascorbic acid can function in the hydroxylation of other substances including aromatic compounds and fatty acids, and the rate of hydroxylation of such compounds by ascorbic acid-deficient guinea pigs is markedly reduced. It is quite possible that the biochemical function of ascorbic acid relates to its participation in hydroxylation reactions in the cell.

Vitamin A

Structure. (*a*) Parent substance—α-carotene:

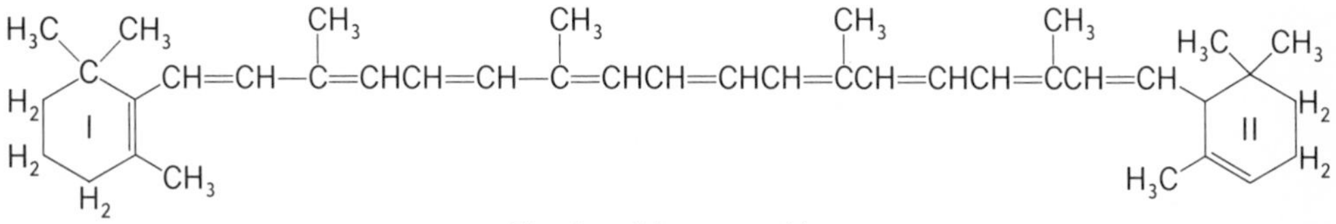

Ring I = β-Ionone residue
Ring II = α-Ionone residue

(*b*) Vitamin A_1 (all *trans*):

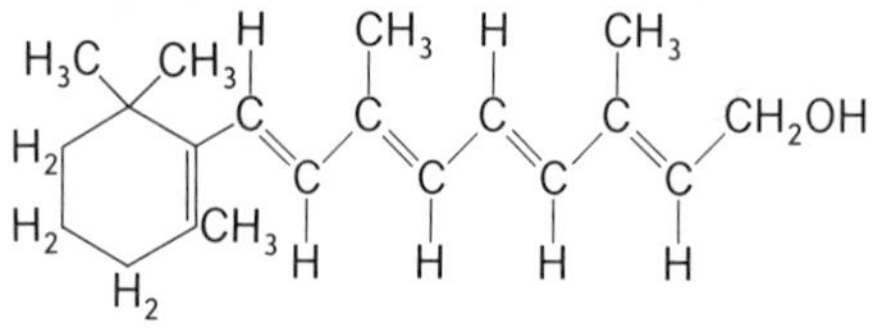

Occurrence. The parent substance is found widespread in the lower and higher forms of plants. Carotenoids, which are more saturated and oxygenated, are also found widely distributed. They occur in tissues of vertebrate and invertebrate animals, particularly in fat deposits, milk, and eye tissue. It is believed that in the small intestine the parent substances are oxidatively broken down to vitamin A.

It is interesting that the major portion of the naturally occurring carotenoid molecules have all their double bonds in the *trans* configuration, although chemical isomerization can occur readily.

Biochemical function. George Wald has made major contributions to our understanding of the role of vitamin A_1 in visual processes. The results can be outlined as in Scheme 9-2.

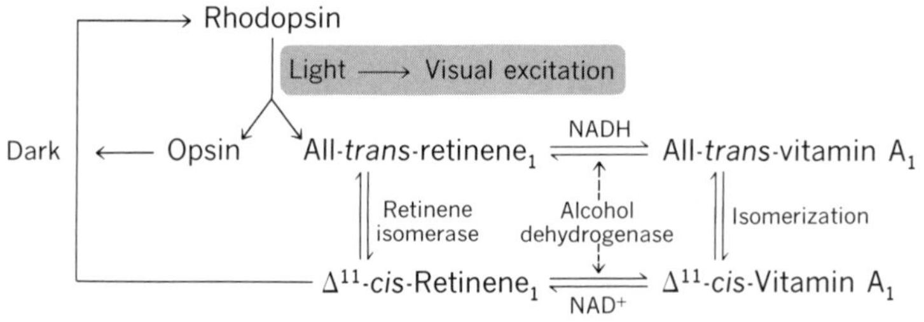

Rhodopsin
Light ⟶ Visual excitation
Dark ⟵ Opsin   All-*trans*-retinene$_1$ ⇌ (NADH) All-*trans*-vitamin A_1
Retinene isomerase   Alcohol dehydrogenase   Isomerization
Δ^{11}-*cis*-Retinene$_1$ ⇌ (NAD$^+$) Δ^{11}-*cis*-Vitamin A_1

Scheme 9-2

Rhodopsin is a visual pigment that consists of a carotenoid–protein complex found in the rods of the retina. The carotenoid is Δ^{11}-*cis*-retinene$_1$, which has five double bonds, all *trans*, except the one between the C-11 and C-12 carbon atoms, which is *cis*. As can be seen in Structure 9-3, Δ^{11}-*cis*-retinene$_1$ is the aldehyde corresponding to Δ^{11}-*cis*-vitamin A_1.

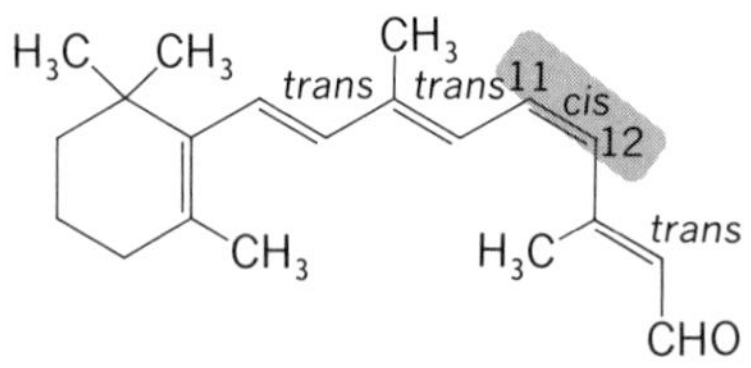

Structure 9-3

When light strikes the visual pigment, isomerization of the Δ^{11}-*cis* double bond to the *trans* configuration occurs with the formation of all *trans* retinene$_1$; the protein–carotenoid complex subsequently dissociates yielding opsin and all-*trans*-retinene$_1$. The latter compound may then be reduced by

NADH and alcohol dehydrogenase in retinal tissue to all *trans* vitamin A_1.

Light has a second role in the rhodopsin cycle; it labilizes the association of Δ^{11}-*cis*-retinene$_1$ with opsin to regenerate rhodopsin. Δ^{11}-*cis*-Retinene$_1$ in turn can be regenerated either by the isomerization of all-*trans*-retinene$_1$ or by the oxidation of Δ^{11}-*cis*-vitamin A_1. This compound may be formed by the enzymic isomerization of all-*trans*-vitamin A. Of primary concern is the problem of how the action of light on rhodopsin results in a nervous excitation leading to vision. It has been suggested that just as trypsinogen is catalytically converted to trypsin, the conversion of rhodopsin to opsin and all-*trans*-retinene may "uncork" an enzymic or other related reaction which is responsible for the actual visual mechanism.

It should be emphasized that avitaminosis A in rats is also characterized by loss of weight, skeletal abnormalities, and disturbances in normal sexual processes. It is therefore obvious that vitamin A has functions other than the visual ones. There is recent evidence that the vitamin may be involved in mucopolysaccharide biosynthesis.

Vitamin D

Structure.

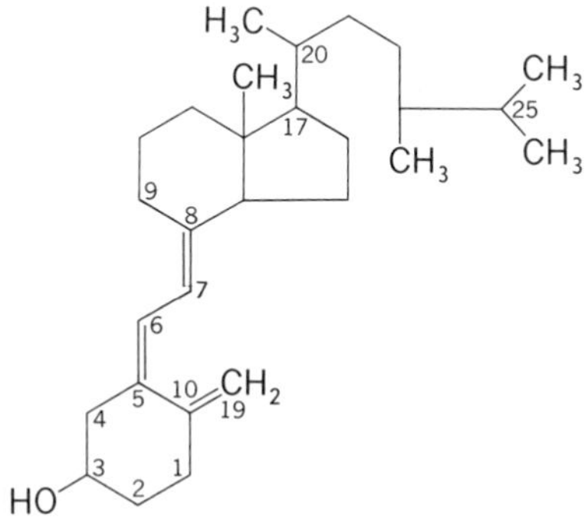

Vitamin D_3
Cholecalciferol

Occurrence. Several compounds are known to be effective in preventing rickets; all are derived by irradiation of different forms of provitamin D; thus, vitamin D_2 (calciferol) is produced commercially by the irradiation of the plant steroid, ergosterol. In animal tissues 7-dehydrocholesterol, which occurs naturally in the epidermal layers, can be converted by ultraviolet irradiation to vitamin D_3. The latter vitamin is also present in fish oil.

Biochemical function. Vitamin D_3 when given to rachitic animals increases the permeability of the intestinal mucosal cells to calcium ion apparently by changing the character of the membrane to calcium permeation. Recently it has been shown that vitamin D_3 induces the appearance of a specific calcium-binding protein in the intestinal mucosa of a number of animals. This protein has been isolated and purified; it has a molecular weight of 24,000 and binds one atom of calcium per molecule of protein. Evidence is rapidly accumulating that vitamin D behaves more like a hormone than as the cofactor of an enzyme. That is, its effect is in controlling the production of a specific calcium-binding protein rather than influencing directly the activity of a specific enzyme.

Recent evidence has also proved conclusively that vitamin D_3 is not the active form of the vitamin. Instead, vitamin D_3 undergoes two chemical modifications, the first in the liver and the second in the kidney, before it is transported in its modified form to the target tissue. These reactions may

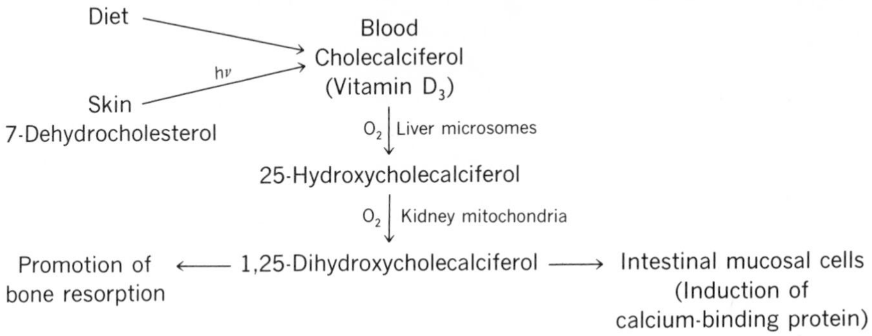

Diet → Blood Cholecalciferol (Vitamin D_3)
Skin 7-Dehydrocholesterol —$h\nu$→ Blood Cholecalciferol (Vitamin D_3)

↓ O_2 Liver microsomes

25-Hydroxycholecalciferol

↓ O_2 Kidney mitochondria

Promotion of bone resorption ← 1,25-Dihydroxycholecalciferol → Intestinal mucosal cells (Induction of calcium-binding protein)

Scheme 9-3

be summarized as in Scheme 9-3. The structure of 1,25-dihydroxycholecalciferol, the active compound eventually formed from vitamin D_3, is also given.

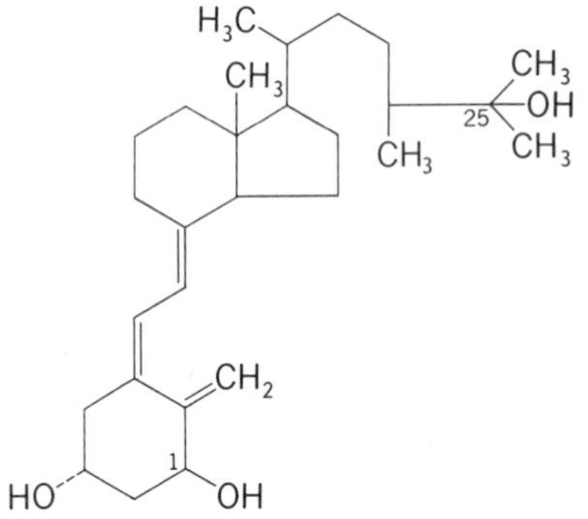

1,25-Dihydroxycholecalciferol

Vitamin E

Structure.

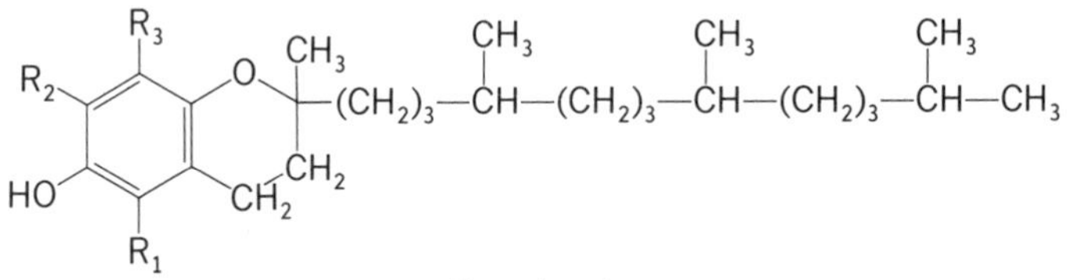

Tocopherol
(R_1, R_2, R_3 may be a combination of methyl groups and H atoms)

Occurrence. The α-tocopherols are widespread in plant oils. Large amounts are found in wheat germ oil and corn oil, for example. Tocopherols are also found in animal body fat. There is some evidence that all α-tocopherol in heart muscle is localized in the mitochondria.

Biochemical function. Characteristic symptoms of avitaminosis E vary with the animal species. In mature female rats, reproductive failure occurs because of fetal reabsorption; with the male rat germinal tissue degenerates. With rabbits and guinea pigs acute muscular dystrophy results; in chickens vascular abnormalities occur. In humans no well-defined syndrome of vitamin E deficiency has been detected.

The most prominent effect that tocopherol has in *in vitro* systems is a strong antioxidant activity. It has been suggested that the biochemical activity of tocopherol is its capacity to protect sensitive mitochondrial systems from irreversible inhibition by lipid peroxides. Thus, in mitochondria prepared from tocopherol-deficient animals, there is a profound deterioration of mitochondrial activity due to hematin-catalyzed peroxidation of highly unsaturated fatty acids normally present in these particles. Addition of tocopherol prevents this deterioration by acting as an antioxidant for peroxidation.

Many workers have observed that tocopherol-deficient muscle shows a high oxygen uptake. Administration of tocopherol to deficient animals lowers the oxygen consumption to normal. The evidence suggests that although the deficiency in some manner interferes with normal oxidative phosphorylation, it may only be a secondary interference, perhaps related to its known antioxidant properties.

Vitamin K

Structure.

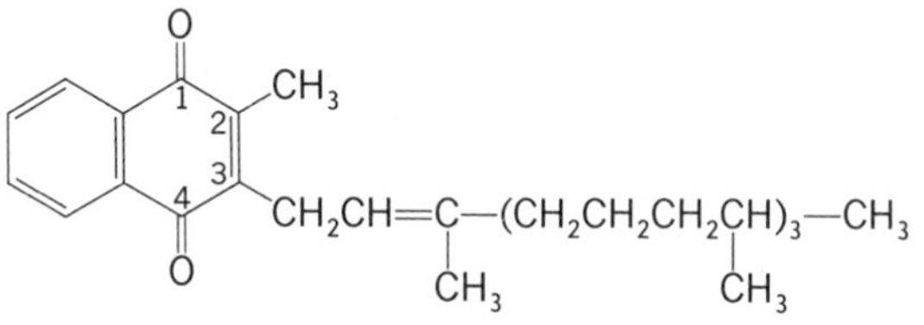

Vitamin K_1

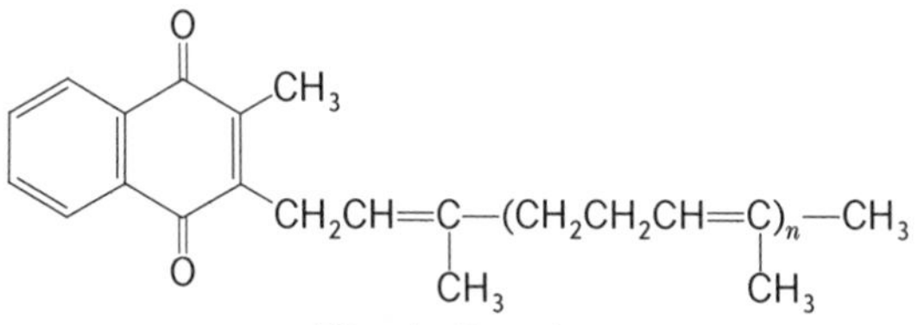

Vitamin K_2 series

Occurrence. Vitamin K_1 (phylloquinone) occurs in green plants, whereas the members of the vitamin K_2 series are found in bacteria; all are fat-soluble. Essential features of the vitamin K napthoquinone derivatives are (*a*) that the methyl group be in the 2 position, (*b*) that there be an unsubstituted benzene ring, and (c) that certain specific requirements be fulfilled regarding the aliphatic substituent in the 3 position. These compounds readily undergo oxidation and reduction but are well-stabilized as quinones. Light rapidly inactivates the K vitamins irreversibly.

Biochemical function. No clear role has been found for vitamin K in any enzyme systems. On the other hand, vitamin K is of fundamental importance in the process of blood clotting. The process is too complicated to describe here and the details are still being clarified. At one point, however, a protein known as prothrombin is converted to thrombin, a proteolytic enzyme involved in subsequent stages of the blood clotting process. If the dietary intake of vitamin K is inadequate, the prothrombin level decreases and the time required for blood to clot is increased. Vitamin K is believed to participate in the synthesis of prothrombin in the liver, and, in this respect, is analogous to vitamin D, acting as a hormone to accelerate some physiological process. One of the preoperative tests carried out routinely in hospitals is the determination of a patient's clotting time. Those individuals with a tendency to slow clotting will be given an injection of vitamin K prior to surgery. The effect of the vitamin K is to increase the level of prothrombin, usually overnight, with a concomitant decrease in the clotting time.

Ubiquinone and Plastoquinone

Structure.

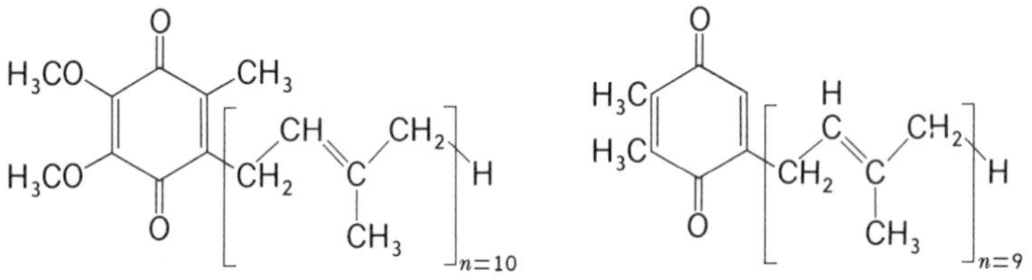

Ubiquinone$_{10}$ ($n=10$)

Plastoquinone ($n=9$)

Occurrence. Two slightly different types of benzoquinones are found in tissues. The ubiquinones (see the structure) are found widespread in animal and bacterial tissues. The mitochondria appear to be the site of localization in animal tissue. The isoprenoid side chain residue number (n) is 10 in mitochondria and 6–9 in lower organisms. In plant tissues, chloroplasts appear to be the site of plastoquinone localization. The isoprenoid residue number is 9. The standard reduction potential (E_0') for both benzoquinones is approximately +0.098 at pH 7.4 and 25°C. The isoprenoid side chain has no effect on the standard redox potential.

Biochemical function. The term CoQ is used in discussing the function of ubiquinone$_{10}$. There is a considerable body of evidence to support the function of CoQ$_{10}$ (ubiquinone) as a necessary component in the electron transport system in mitochondria. This aspect is discussed in more detail in Chapter 14. Similar conclusions have been made concerning the role of plastoquinone in chloroplasts. We shall see that plastoquinone appears to play an important role in the transport of electrons from the photooxidation of water to the cytochrome b_3-f sequence in the chloroplast (see Chapter 15).

Trace Metals

There is good evidence that all metals required nutritionally participate at the molecular level as cofactors for enzymes or as structural components of enzymes. Some of the enzymes that contain metal ions as part of their structure are listed in Table 9-3.

Table 9-3

Typical Metal–Enzymes

Metal	Enzyme
Iron	Cytochromes, peroxidases
Copper	Tyrosinase, ascorbic oxidase
Zinc	Peptidase, carbonic anhydrase
Magnesium	Phosphatases, kinases
Manganese	Kinases, peptidases, arginase
Molybdenum	Xanthine oxidase, nitrate reductase
Cobalt	Vitamin B_{12} coenzyme complexes
Potassium	Pyruvic kinase, β-methyl aspartase

References

1. P. D. Boyer, *The Enzymes*. 3rd ed. New York: Academic Press, 1970.
   This series contains comprehensive reviews on coenzymes.
2. A. F. Wagner and K. Folkers, *Vitamins and Coenzymes*. New York: Interscience, 1964.
   A single volume that discusses the chemistry and biochemistry of vitamins and their derivatives.
3. D. M. Greenberg, ed., *Metabolic Pathways*. 3rd ed., vol. 4. New York: Academic Press, 1970.
   This volume contains chapters on the biosynthesis and metabolism of several water-soluble vitamins.
4. R. S. Harris, ed., *Vitamins and Hormones*. New York: Academic Press. F. F. Nord, ed., *Advances in Enzymology*. New York: Wiley-Interscience.
   These multivolume series contain many articles on the general subject of coenzymes and vitamins.

10

Anaerobic Carbohydrate Metabolism

Introduction

Carbohydrates are a major source of energy for living organisms. In man's food the chief source of carbohydrate is starch, the polysaccharide produced by plants during photosynthesis. Plants may store relatively large amounts of starch within their own cells in time of abundant supply, to be used later when there is a demand for energy production or to be consumed by animals for food. In animals the amount of carbohydrate stored (as glycogen) is much more limited, but it plays a significant role in the rapid production of energy.

Simple sugars such as sucrose, glucose, fructose, mannose, and galactose are also encountered in nature and are utilized by living forms as food. The process by which the chemical energy in these compounds is made available to and utilized by biological organisms is a matter of major concern in this text. The intermediary metabolism of carbohydrates, together with that of lipids, amino acids, and proteins, is indeed the backbone of biochemistry, and we are now prepared to consider the metabolism of the carbohydrates.

It is conceptually useful to separate the intermediary metabolism of carbohydrates into the anaerobic process and aerobic processes which are discussed later (Chapters 11 and 13). The anaerobic process by which glucose is degraded to 2 moles of lactic acid is known as *glycolysis*. This process, which occurs in most microorganisms as well as in most of the cells of higher plants and animals, represents a simple yet elegant mechanism for recovering some of the energy in the glucose molecule as ATP. Polysaccharides, as well as numerous other carbohydrates that serve as fuel molecules, are also degraded by the glycolytic sequence of reactions after they have been converted into a compound in that sequence. *Alcoholic fermentation*, by which yeasts convert glucose into ethanol and CO_2, is identical to glycolysis except for two reactions at the end of the glycolytic sequence.

The sequence of reactions of glycolysis and alcoholic fermentation as it exists today was developed by the pioneers in enzymology. In 1897 the Buchners in Germany obtained a cell-free extract of yeast which fermented sugars to CO_2 and ethanol. Shortly thereafter the work of Harden and Young in England implicated phosphorylated derivatives of the sugars in alcoholic fermentation. Today the glycolytic sequence is recognized as being composed of the reactions diagrammed inside the front cover of this text. A list of the pioneers in the field who were the architects of this scheme includes Embden, Meyerhof, Robison, Neuberg, the Coris, Lipmann, Parnas, and Warburg. Their studies on the enzymatic aspects of glycolysis served as models for later workers examining the metabolism of lipids, amino acids, and proteins, as well as respiration and photosynthesis. Many of the biochemical principles established by workers in the field of glycolysis apply equally well in other areas of intermediary metabolism; efforts will be made to identify these common principles.

Reactions of the Glycolytic Sequence

Ten reactions are involved in the conversion of glucose to lactic acid; these may be conveniently divided into two groups. The first four reactions are concerned with converting glucose into a compound, triose phosphate, whose oxidation subsequently releases energy to the environment. In contrast, the four reactions of the preparative phase require an expenditure of energy as the glucose molecule is phosphorylated prior to the formation of triose phosphate.

Hexokinase. The initial step in the utilization of glucose in glycolysis is its phosphorylation by ATP to yield glucose-6-phosphate:

$$\text{Glucose} + \text{ATP} \xrightarrow{Mg^{2+}} \text{Glucose-6-phosphate} + \text{ADP} \qquad (10\text{-}1)$$

(Glucose: ring with CH_2OH; Glucose-6-phosphate: ring with $CH_2OPO_3H_2$)

$$\Delta G' = -4000 \text{ cal (pH 7.0)}$$

The enzyme hexokinase which catalyzes this reaction was first discovered in yeast by Meyerhof in 1927. The enzyme has been crystallized from yeast and has a molecular weight of 111,000. The yeast enzyme can be dissociated into subunits, two polypeptide chains of 55,000 mol wt, each containing an active site. The yeast enzyme exhibits rather broad specificity in that it will catalyze the transfer of phosphate from ATP not only to glucose, but also to fructose, mannose, glucosamine, and 2-deoxyglucose. The relative rates of reaction depend on the concentration of the sugars in the reaction mixture; fructose is phosphorylated most rapidly at high concentrations. The multi-substrate hexokinase is also found in brain, muscle, and liver. In addition, specific kinases known to catalyze the phosphorylation of glucose (glucokinase), fructose (fructokinase), mannose (mannokinase), and other sugars are known; these enzymes transfer a phosphate to the primary alcoholic group of the sugar. Liver contains a fructokinase which produces fructose-1-phosphate rather than the 6-ester.

It is informative to consider the energy changes that occur in this reaction. When glucose is phosphorylated to form glucose-6-phosphate, a compound having a low-energy phosphate ester grouping has been produced. The free energy of hydrolysis of this compound is about −3300 cal:

$$\text{Glucose-6-phosphate } (CH_2OPO_3H_2) + H_2O \longrightarrow \text{Glucose } (CH_2OH) + H_3PO_4$$

$$\Delta G' = -3300 \text{ cal (pH 7.0)}$$

The phosphate group attached to the sugar was obtained from the terminal phosphate group of ATP. As we have seen, the free energy of hydrolysis of the latter compound is about −7300 cal:

$$\text{ATP} + H_2O \longrightarrow \text{ADP} + H_3PO_4$$

$$\Delta G' = -7300 \text{ cal (pH 7.0)}$$

Inspection reveals that in the hexokinase reaction a high-energy bond of ATP was utilized and a low-energy structure (that of glucose-6-phosphate) was formed. In the terminology of the biochemist, the reaction catalyzed by hexokinase has resulted in the formation of a low-energy phosphate compound by the expenditure of an energy-rich phosphate structure. Normally the loss of a high-energy bond by hydrolysis would result in the liberation of the −7300 cal as heat if changes in entropy are neglected. In the hexokinase reaction, part of that energy (−3300 cal) is conserved in the formation of the low-energy structure, and the remainder (−4000 cal) is liberated as heat, again neglecting entropy changes. Thus, we may estimate the free-energy change for the hexokinase reaction to be −4000 cal; that is, the reaction is strongly exergonic. An equilibrium constant of 2×10^3 at pH 7.0 corresponding to a $\Delta G'$ of −4500 cal, has been obtained experimentally. The equilibrium in this reaction is clearly far to the right.

It is also informative to consider the possibility of reversing the hexokinase reaction (equation 10-1). Given the K_{eq} of 2000, we can calculate with equation 7-4 (Chapter 7) that one needs only a ratio of 200:1 of ADP to ATP to synthesize ATP and glucose if the ratio of glucose-6-phosphate to glucose is 10:1. In addition, the hexokinase reaction is demonstrably reversible in the test tube, and yet in the cell the reaction never goes from right to left. Other factors clearly determine that the phosphorylation of glucose by ATP is a unidirectional process in the intact organism.

One of these is the difference in the maximum velocities (V_{max}) of the forward and reverse reactions; the reverse reaction (synthesis of ATP) has a maximum rate that is only one-fiftieth of that of the forward reaction. Another factor is the affinity of the enzyme for the four compounds. The K_m's, which are a measure of this affinity, are $10^{-4}M$ for both glucose and ATP. The K_m for glucose-6-phosphate is $0.08M$ and for ADP is $3 \times 10^{-3}M$. Since the enzyme exhibits half its maximum velocity at a substrate concentration equal to the K_m, the hexokinase reaction will go more rapidly from left to right when the four components are present at equal concentration because of the greater affinity that the enzyme exhibits for ATP and glucose. Finally, in the case of liver hexokinase, glucose-6-phosphate produced by the enzyme in turn strongly inhibits the enzyme. That is, the enzyme exhibits *product inhibition*. Clearly, the enzyme will cease to function as soon as any significant quantity of glucose-6-phosphate is produced, and it will remain inactive until the level of glucose-6-phosphate decreases as a result of its being used up in other reactions.

Phosphohexoisomerases. The next reaction in glycolysis is the isomerization of glucose-6-phosphate catalyzed by phosphoglucoisomerase:

$$\text{Glucose-6-phosphate} \rightleftharpoons \text{Fructose-6-phosphate} \qquad (10\text{-}2)$$

$$\Delta G' = +400 \text{ cal (pH 7.0)}$$

The enzyme, which has been extensively purified from skeletal muscle and crystallized from yeast, does not require a cofactor; the K_{eq} for the reaction from left to right is approximately 0.5. The human skeletal muscle enzyme has a molecular weight of 130,000; it can be dissociated into subunits of 61,000. An isomerase that catalyzes the conversion of mannose-6-phosphate to fructose-6-phosphate has been isolated from rabbit muscle. Although the three sugars glucose, fructose, and mannose are readily interconverted in dilute alkali (the Lobry de Bruyn-von Ekenstein transformation), it is interesting to note that the two isomerases are highly specific for fructose-6-phosphate and the corresponding hexose-6-phosphate for which they are named. Although reaction 10-2 has been written with the pyranose and furanose structures, the actual isomerization involves the open-chain form of the sugars, and an enediol (page 38) is believed to be an intermediate.

Phosphofructokinase. The kinase that catalyzes the phosphorylation of fructose-6-phosphate by ATP has been purified from both yeast and muscle. The enzyme requires Mg^{2+} and is specific for fructose-6-phosphate (reaction 10-3). As with hexokinase, the high-energy bond of ATP is utilized to synthesize the low-energy ester–phosphate bond of fructose-1,6-diphosphate. Using the arguments presented in the former case, we may expect that this reaction, too, should proceed with a large decrease in free energy and therefore should not be freely reversible. The $\Delta G'$ is -4000 cal/mole.

Phosphofructokinase is an important site for metabolic regulation because the activity of the enzyme may be either increased or decreased by a number of common metabolites. Such effects are of the *allosteric type* (Chapter 8) in that they are the result of an interaction between the metabolite and the

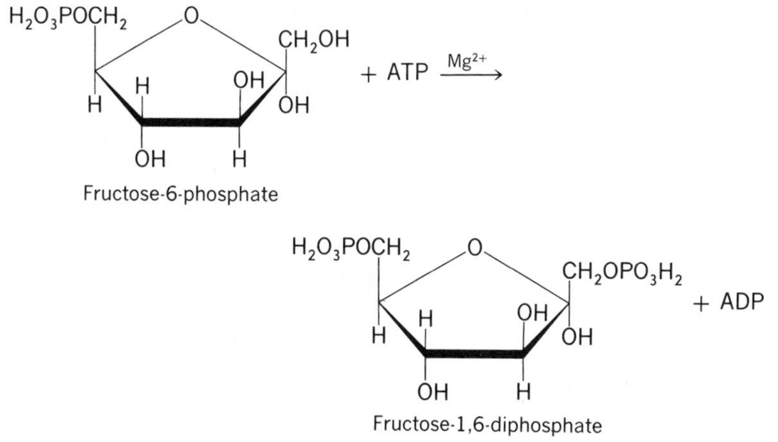

$H_2O_3POCH_2$, O, CH_2OH, H, OH, H, OH, OH, H (furanose ring)

Fructose-6-phosphate

$$+ \text{ATP} \xrightarrow{Mg^{2+}}$$

$H_2O_3POCH_2$, O, $CH_2OPO_3H_2$, H, OH, H, OH, OH, H (furanose ring) + ADP (10-3)

Fructose-1,6-diphosphate

$\Delta G' = -4000$ cal (pH 7.0)

protein catalyst at a site other than the site where catalysis occurs. Thus, excess ATP and citric acid inhibit phosphofructokinase; i.e., they are negative effectors. On the other hand, AMP, ADP, and fructose-6-phosphate stimulate the enzyme and are positive effectors.

As an allosteric enzyme, phosphofructokinase possesses a high molecular weight (~360,000) and is dissociable into four subunits. It also exhibits *sigma* (S-shaped) kinetics typical of many allosteric catalysts (page 475). Moreover, as an allosteric enzyme, the reaction it catalyzes is irreversible. This results not only from the $\Delta G'$ of the reaction and the K_m's of the reactants and products, but also from the nature of the allosteric effectors. The role of this enzyme and its companion, fructose diphosphatase (reaction 10-23) will be discussed later.

Aldolase. The next reaction in the glycolytic sequence involves the cleavage of fructose-1,6-diphosphate to form the two triose phosphate sugars, dihydroxy acetone phosphate and D-glyceraldehyde-3-phosphate. The enzyme aldolase which catalyzes this reaction was first extensively purified from yeast and studied by Warburg. It has now been crystallized from numerous animal, plant, and microbial sources, and is widespread in nature. Indeed, the finding of this enzyme in high concentration in a particular tissue is indicative of a functioning glycolytic pathway.

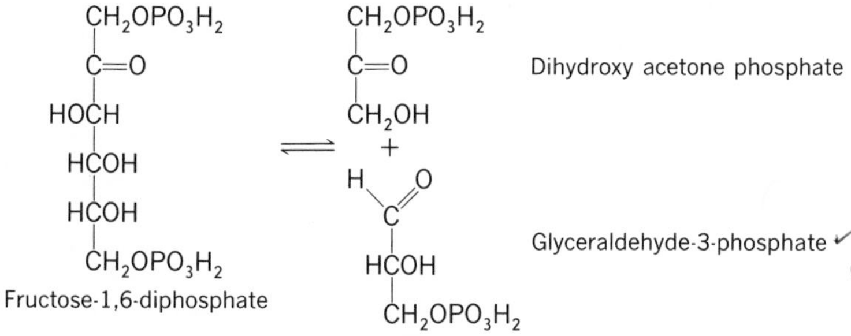

$CH_2OPO_3H_2$–C=O–HOCH–HCOH–HCOH–$CH_2OPO_3H_2$ (Fructose-1,6-diphosphate) $\rightleftharpoons$ $CH_2OPO_3H_2$–C=O–CH_2OH (Dihydroxy acetone phosphate) + H–C(=O)–HCOH–$CH_2OPO_3H_2$ (Glyceraldehyde-3-phosphate) (10-4)

$\Delta G' = +5500$ cal (pH 7.0)

The K_{eq} for reaction 10-4 from left to right is 10^{-4}; this corresponds to a $\Delta G'$ of +5500 cal. Such values for the K_{eq} or $\Delta G'$ would appear to indicate that the reaction does not proceed from left to right. However, a reaction of this sort, in which one reactant gives rise to two products, is strongly influenced by the concentration of the compounds involved. One can readily show by a simple calculation that, as the initial concentration of fructose-1,6-diphosphate is lowered, a progressively larger amount of it will be converted to the triose. Thus, at an initial concentration of 0.1*M* hexose diphosphate, about 97% of it will remain when equilibrium is attained. However, at $10^{-4}M$ hexose diphosphate initially, only 40% will remain at equilibrium.

Aldolase will catalyze the cleavage of a number of ketose di- and monophosphates, e.g., fructose-1,6-diphosphate, sedoheptulose-1,7-diphosphate, fructose-1-phosphate, erythrulose-1-phosphate. In each case, however, dihydroxy acetone phosphate is one of the products. Note that, as reaction 10-4 proceeds from right to left, the aldol condensation results in the formation of two new asymmetric carbon atoms, and theoretically four different isomers of the hexose diphosphate molecule could be formed. Nevertheless, the enzyme specifically catalyzes the formation of only one, namely fructose-1,6-diphosphate.

The mechanism of action of aldolase has been extensively studied as have the properties of the enzyme from many different tissues. From the aspect of comparative biochemistry, therefore, aldolase is one of the best characterized enzymes.

Triose Phosphate Isomerase. The production of D-glyceraldehyde-3-phosphate in the aldolase reaction technically completes the preparative phase of glycolysis. The second or *energy-yielding* phase involves the oxidation of glyceraldehyde-3-phosphate, a reaction which is examined in detail in the next section. Note, however, that only half of the glucose molecule has been converted to D-glyceraldehyde-3-phosphate by reactions 10-1 through 10-4. If cells were unable to convert dihydroxy acetone phosphate to glyceraldehyde-3-phosphate, half of the glucose molecule would accumulate in the cell as the ketose phosphate or be disposed of by other reactions. During evolution, this problem has been solved by the cell acquiring the enzyme *triose phosphate isomerase* which catalyzes the interconversion of these two trioses and permits the subsequent metabolism of all the glucose molecule.

$$\underset{\text{Glyceraldehyde-3-phosphate}}{\begin{array}{c} H\diagdown \;\; \diagup\!\!\!\!/ O \\ C \\ | \\ HCOH \\ | \\ CH_2OPO_3H_2 \end{array}} \rightleftharpoons \underset{\text{Dihydroxy acetone phosphate}}{\begin{array}{c} CH_2OH \\ | \\ C{=}O \\ | \\ CH_2OPO_3H_2 \end{array}} \qquad (10\text{-}4a)$$

$$\Delta G' = -1800 \text{ cal (pH 7.0)}$$

Meyerhof was the first to describe the equilibrium between the triose phosphates. The reaction is analogous to the isomerization of the hexose phosphates (reaction 10-2) in that the interconversion of a ketose and an aldose takes place. Triose phosphate isomerase is an extremely active enzyme; if a molecular weight of 100,000 is assumed, one can demonstrate that 1 mole will catalyze the isomerization of 945,000 moles of substrate per minute. Thus, although the equilibrium constant ($K_{eq} = 22$) favors the ketose derivative, the presence of even a small amount of the isomerase will ensure an immediate conversion of the acetone phosphate into the aldehyde isomer for subsequent degradation.

Glyceraldehyde-3-phosphate Dehydrogenase. This reaction, which is the first in the energy-yielding or second phase of glycolysis, is also the first reaction in the glycolytic sequence to involve oxidation–reduction. As may be seen from equation 10-5, it is also the first reaction in which a high-energy phosphate compound has been formed where none previously existed.

$$\underset{\text{Glyceraldehyde-3-phosphate}}{\begin{matrix} H\diagdown\ \diagup\!\!\!\!\!\!\!=O \\ C \\ | \\ HCOH \\ | \\ CH_2OPO_3H_2 \end{matrix}} + NAD^+ + H_3PO_4 \rightleftharpoons \underset{\text{1,3-Diphosphoglyceric acid}}{\begin{matrix} O \\ \| \\ C{-}OPO_3H_2 \\ | \\ HCOH \\ | \\ CH_2OPO_3H_2 \end{matrix}} + NADH + H^+ \qquad (10\text{-}5)$$

$$\Delta G' = +1500 \text{ cal (pH 7.0)}$$

As a result of the oxidation of an aldehyde group to the level of a carboxylic acid, some of the energy which presumably would have been released in the form of heat has been conserved in the formation of the acyl phosphate group of 1,3-diphosphoglyceric acid. The oxidizing agent involved is NAD^+. The energetics of this reaction together with that of reaction 10-6 (below), which results in the formation of ATP, have been described in more detail in Chapter 7 under coupled reactions.

The $\Delta G'$ for equation 10-5 is about +1500 cal. This corresponds to a K_{eq} of 0.08 and means that the reaction is readily reversible. This is to be expected, since the cell has modified the strongly exergonic oxidation of an aldehyde to a carboxylic acid into a reaction in which much of that energy is conserved as an acyl phosphate.

To consider the mechanism whereby the enzyme brings about this remarkable reaction some of the properties of the protein should be described. The enzyme has been crystallized from rabbit muscle and yeast, and has a molecular weight of 145,000. The enzyme appears to be a tetramer consisting of four identical subunits of approximately 35,000 mol wt each. Each subunit tightly binds one molecule of NAD^+, making four NAD^+ for the intact oligomer. Moreover, these NAD^+ molecules are intimately involved in the enzyme's catalytic action. Thus, glyceraldehyde-3-phosphate dehydrogenase (triose phosphate dehydrogenase) constitutes an important exception to the generalization that nicotinamide nucleotide dehydrogenases are readily isolated free of their coenzyme molecule.

Triose phosphate dehydrogenase possesses sulfhydryl groups (—SH) which must be free (reduced) for catalytic activity. The well-known ability of iodoacetamide to inhibit glycolysis is due to the covalent and irreversible binding

$$R{-}SH + ICH_2CONH_2 \longrightarrow R{-}S{-}CH_2CONH_2 + HI$$

of this reagent with —SH groups of the dehydrogenase, thereby irreversibly blocking its catalytic action. A mechanism which requires the participation of the essential —SH group may be written as follows. In the initial reaction the aldehyde is oxidized to a thioester in the presence of the dehydrogenase–NAD$^+$; the sulfur atom participating in the thioester linkage is represented as a sulfhydryl group of the enzyme:

$$R{-}\overset{H}{C}{=}O + HS{-}Enz{-}NAD^+ \rightleftharpoons R{-}\underset{\overset{\|}{O}}{C}{-}S{-}Enz{-}NADH + H^+$$

The acyl–enzyme compound then exchanges its NADH for NAD$^+$:

$$R{-}\underset{\overset{\|}{O}}{C}{-}S{-}Enz{-}NADH + NAD^+ \rightleftharpoons R{-}\underset{\overset{\|}{O}}{C}{-}S{-}Enz{-}NAD^+ + NADH$$

Finally, the acyl group on the enzyme is transferred to inorganic phosphate:

$$R{-}\underset{\overset{\|}{O}}{C}{-}S{-}Enz{-}NAD^+ + H_3PO_4 \rightleftharpoons R{-}\underset{\overset{\|}{O}}{C}{-}OPO_3H_2 + HS{-}Enz{-}NAD^+$$

The sum of these three reactions is the overall reaction 10-5.

Phosphoglyceryl Kinase. This reaction accomplishes the transfer of the phosphate from the acyl phosphate formed in the preceding reaction to ADP to form ATP. The name of the enzyme is derived from the reverse reaction, in which a high-energy phosphate is transferred from ATP to 3-phosphoglyceric acid:

$$\begin{array}{c} O \\ \| \\ C{-}OPO_3H_2 \\ | \\ HCOH \\ | \\ CH_2OPO_3H_2 \end{array} + ADP \overset{Mg^{2+}}{\rightleftharpoons} \begin{array}{c} CO_2H \\ | \\ HCOH \\ | \\ CH_2OPO_3H_2 \end{array} + ATP \qquad (10\text{-}6)$$

1,3-Diphosphoglyceric acid — 3-Phosphoglyceric acid

$$\Delta G' = -4500 \text{ cal (pH 7.0)}$$

In this reaction, the acyl phosphate group ($\Delta G'$ of hydrolysis of $-11,800$ cal) has been utilized to drive the phosphorylation of ADP and make ATP ($\Delta G'$ of hydrolysis of -7300 cal). From these considerations alone, one would predict that the $\Delta G'$ for reaction 10-6 would be -4500 cal, corresponding to a K_{eq} of 2×10^3. A value of 3.1×10^3 has been reported in the literature. Although, as in reaction 10-1 or 10-3, the thermodynamics favor ATP formation, reaction 10-6 will be reversed if the ratio of ATP:ADP is 10:1 and the ratio of 3-phosphoglycerate to 1,3-diphosphoglycerate exceeds 200:1. In contrast to reactions 10-1 and 10-3, where other factors (product inhibition, allosteric modifiers) determine that those reactions proceed in only one direction, reaction 10-6 catalyzed by phosphoglyceryl kinase *does* proceed from right to left when glycolysis is reversed in the cell. This reversal of reaction 10-6 is undoubtedly aided by the positive $\Delta G'$ for reaction 10-5 described in the preceding section. By combining reactions 10-5 and 10-6 we may write

$$\begin{array}{l}\mathrm{H{-}C{=}O} \\ \mathrm{HCOH} \\ \mathrm{CH_2OPO_3H_2}\end{array} + NAD^+ + ADP + H_3PO_4 \xrightleftharpoons{Mg^{2+}} \begin{array}{l}\mathrm{CO_2H} \\ \mathrm{HCOH} \\ \mathrm{CH_2OPO_3H_2}\end{array} + ATP + NADH + H^+$$

$$\Delta G' = -3000 \text{ cal}$$

and by adding the $\Delta G'$ for reactions 10-5 and 10-6, we obtain an overall $\Delta G'$ of -3000 cal for the combined process. Again the student should note that reactions 10-5 and 10-6 show the formation of 1 mole of ATP from ADP and inorganic phosphate as the result of oxidizing an aldehyde to a carboxylic acid. This is one of the best known examples where the formation of an energy-rich phosphate compound is coupled to a chemical oxidation which normally liberates energy in the form of heat. In the present instance some of that energy has been conserved in the form of ATP and can be subsequently used by the cell to drive other endergonic processes.

Anticipating some interest in determining how much ATP is made available during the conversion of glucose to lactic acid, the reader may note that by the time 1 mole of glucose has been converted to 2 moles of 3-phosphoglycerate by means of reactions 10-1 through 10-6, the 2 moles of ATP expended in reactions 10-1 and 10-3 have been recovered in reaction 10-6. Any additional ATP produced from this point on will represent a net synthesis of that compound. Before additional ATP can be produced, however, 3-phosphoglycerate must be converted to its isomer, 2-phosphoglycerate.

Phosphoglyceryl Mutase. Phosphoglyceryl mutase catalyzes the interconversion of the two phosphoglyceric acids. The equilibrium constant for the reaction from left to right is 0.17:

$$\underset{\text{3-Phosphoglyceric acid}}{\begin{array}{l}\mathrm{CO_2H} \\ \mathrm{HCOH} \\ \mathrm{CH_2OPO_3H_2}\end{array}} \rightleftharpoons \underset{\text{2-Phosphoglyceric acid}}{\begin{array}{l}\mathrm{CO_2H} \\ \mathrm{HCOPO_3H_2} \\ \mathrm{CH_2OH}\end{array}} \qquad (10\text{-}7)$$

$$\Delta G' = +1050 \text{ cal (pH 7.0)}$$

This enzyme falls in the group of catalysts called *phosphomutases* which catalyze the transfer of a phosphoryl group from one carbon atom to a second carbon atom of the same organic compound. The mechanism of action of this group of enzymes is still under active study with the information on phosphoglucomutase (to be discussed later) being voluminous. In the case of the crystalline phosphoglyceryl mutases from rabbit muscle (64,000 mol wt) and yeast (112,000 mol wt), both enzymes require 2,3-diphosphoglyceric acid as a cofactor for activity. This fact and recent studies showing that phosphoglyceryl mutase is a phosphoenzyme, suggest the mechanism

$$\text{3-PGA} + \text{P—ENZ} \rightleftharpoons \text{(2,3-diPGA—ENZ)} \rightleftharpoons \text{P—ENZ} + \text{2-PGA}$$

$$\updownarrows$$

$$\text{ENZ} + \text{2,3-diPGA}$$

The enzyme reacts with 2,3-diphosphoglycerate (2,3-diPGA) to produce the diphosphorylated form of the enzyme which then can dissociate to produce monophosphorylated forms (P–ENZ) and either 3-phosphoglyceric acid (3-PGA) or the 2-isomer (2-PGA). Reading across the diagram from left to right accomplishes the conversion of 3-PGA to 2-PGA.

Enolase. The next reaction in the degradation of glucose involves the dehydration of 2-phosphoglyceric acid to produce phosphoenol pyruvic acid, a compound with a high-energy enolic phosphate group:

$$\underset{\text{2-Phosphoglyceric acid}}{\begin{matrix} CO_2H \\ | \\ HCOPO_3H_2 \\ | \\ CH_2OH \end{matrix}} \xrightleftharpoons{Mg^{2+}} \underset{\text{Phosphoenol pyruvic acid}}{\begin{matrix} CO_2H \\ | \\ C—OPO_3H_2 \\ \| \\ CH_2 \end{matrix}} + H_2O \qquad (10\text{-}8)$$

$$\Delta G' = -650 \text{ cal (pH 7.0)}$$

The equilibrium constant for this reaction is 3; the standard free energy change is small therefore ($\Delta G' = -650$ cal) and the reaction is freely reversible. It is interesting that by this simple process of dehydration an energy-rich enolic phosphate ($\Delta G'$ of hydrolysis is $-14{,}800$ cal) is formed.

Earlier in this chapter (reaction 10-5), the production of an energy-rich acyl phosphate was coupled to an oxidation–reduction reaction, and it was possible to rationalize the synthesis of the energy-rich structure as a consequence of that oxidation. In reaction 10-8, a different chemical reaction, namely dehydration, is involved and it is more difficult to comprehend the synthesis of the enolic phosphate, especially when it is formed in a reaction involving a minimal $\Delta G'$. The difficulty lies primarily in the biochemist's definition of "energy-rich," and another way of looking at 2-phosphoglyceric acid and phosphoenol pyruvic acid is required. Although one compound is "energy-poor" and the other is "energy-rich," when we consider the $\Delta G'$ of hydrolysis of these compounds, about the *same* amount of energy would be produced if they were oxidized to CO_2, H_2O, and H_3PO_4. The key to this puzzle is found in the fact that the dehydration catalyzed by enolase results in a rearrangement of electrons in the two molecules so that a significantly larger amount of the total *potential* energy of the compound is released on hydrolysis. The explanation as to *why* phosphoenol pyruvic acid releases an unusually large amount of energy on hydrolysis was discussed earlier (Chapter 7).

Enolase requires Mg^{2+} for activity. In the presence of Mg^{2+} and phosphate, fluoride ions strongly inhibit the enzyme. This effect is related to the formation of a magnesium–fluorophosphate complex which is only slightly dissociated and thereby effectively removes Mg^{2+} from the reaction mixture.

Pyruvic Kinase. Pyruvic kinase catalyzes the transfer of phosphate from phosphoenol pyruvic acid to ADP to produce ATP and pyruvic acid:

$$\underset{\text{Phosphoenol pyruvic acid}}{\begin{array}{c} CO_2H \\ | \\ C{-}OPO_3H_2 \\ \| \\ CH_2 \end{array}} + ADP \xrightleftharpoons{Mg^{2+},\ K^+} \underset{\text{Pyruvic acid}}{\begin{array}{c} CO_2H \\ | \\ C{=}O \\ | \\ CH_3 \end{array}} + ATP \qquad (10\text{-}9)$$

$$\Delta G' = -6100 \text{ cal}$$

The enzyme has been crystallized from numerous animal sources and from yeast. The latter is a tetramer with a molecular weight of 165,000 that is dissociable into four subunits (42,000 mol wt). It requires both Mg^{2+} and K^+ ions for activity.

Because of the large $\Delta G'$ of hydrolysis of phosphoenol pyruvate (−14,800 cal), one would expect that the $\Delta G'$ for reaction 10-9 would be approximately −7500 cal. The literature reports values for the K_{eq} as high as 3×10^4, corresponding to $\Delta G'$ of −6100 cal. Indeed, the equilibrium is so far to the right that it is difficult to obtain an accurate value for the K_{eq} (and therefore the $\Delta G'$) of the reaction.

Two other factors besides the large K_{eq} contribute their share in determining that reaction 10-9 is irreversible under physiological conditions. One of these is the maximum rate of the forward and reverse reactions; the maximum velocity of the conversion of phosphoenol pyruvate to pyruvate is 200 times that of the reverse reaction. A second factor is the affinity of the enzyme for its four substrates. The K_m's of pyruvic kinase for these compounds are: phosphoenol pyruvate, $7 \times 10^{-5}M$; ADP, $3 \times 10^{-4}M$; ATP, $8 \times 10^{-4}M$; and pyruvate, $1 \times 10^{-2}M$. Thus, while the enzyme could operate at half its maximal catalytic rate or faster when the concentrations of phosphoenol pyruvate, ADP, and ATP reach $0.001M$, the enzyme still requires a 10-fold higher concentration of pyruvate to reach half its maximum velocity. As will be seen, pyruvic acid is an extremely active compound metabolically speaking and has numerous alternative reactions which it can undergo in the cell. The chances, therefore, of it reaching such a high ($0.01M$) concentration in order to permit reaction 10-9 to reverse are not great.

Lactic Dehydrogenase. The last reaction of glycolysis results in the production of L(+)-lactic acid when pyruvic acid is reduced by NADH. Note that because of the production of NADH in reaction 10-5 and its utilization in this reaction, there is no accumulation of NADH in a tissue carrying out glycolysis. The significance of this point will be discussed later. The enzyme has been crystallized from numerous animal sources. It greatly prefers NADH, working

$$\underset{\text{Pyruvic acid}}{\begin{array}{c} CO_2H \\ | \\ C{=}O \\ | \\ CH_3 \end{array}} + NADH + H^+ \rightleftharpoons \underset{\text{L(+)-Lactic acid}}{\begin{array}{c} CO_2H \\ | \\ HOCH \\ | \\ CH_3 \end{array}} + NAD^+ \qquad (10\text{-}10)$$

$$\Delta G' = -6000 \text{ cal (pH 7.0)}$$

170 times more rapidly with that coenzyme than with NADPH. The enzyme is not as specific for pyruvic acid but will catalyze the reduction of a number of other keto acids, including phenylpyruvic acid. The equilibrium is far to the right at pH 7.0 ($K_{eq} = 2.5 \times 10^4$); however, as pointed out in chapter 9, the K_{eq} of reactions involving the nicotinamide coenzymes is greatly dependent on pH, and one can measure pyruvate formation under certain conditions; i.e., one can reverse reaction 10-10, by operating at a higher pH of 8–9.

The isozymic nature of lactic dehydrogenase has been mentioned previously (Chapter 8). The forms that occur in heart (H_4) and skeletal muscle (M_4) have quite different kinetic properties. The heart enzyme (H_4) is active at low levels of pyruvate (and lactic acid) and is inhibited by concentrations of pyruvate exceeding $10^{-3}M$. The muscle enzyme (M_4) does not achieve its maximum velocity until the pyruvate concentration is $3 \times 10^{-3}M$, but maintains its activity in much higher concentrations of pyruvate. Kaplan has pointed out that these properties are consistent with the tasks that the two different tissues have to perform. The heart requires a steady supply of energy that can best be achieved by converting glucose to pyruvate and then oxidizing the pyruvate to CO_2 and H_2O via the Krebs cycle (Chapter 13). This process derives the maximum amount of energy from the glucose molecule and requires an adequate supply of oxygen. In skeletal muscle, there can be sudden demands for energy in the absence of oxygen, relatively speaking. This energy can be supplied by the reactions in glycolysis in which ATP is generated (reactions 10-6 and 10-9) but in which O_2 is not involved. Such demands would, of course, require relatively large amounts of pyruvate to be formed and be reduced to lactic acid.

In support of this thesis, tissues such as the heart which are continually contracting are found to have H_4-lactic dehydrogenase, while the flight muscles of chicken and pheasants which make sporadic short flights have predominantly M_4-lactic dehydrogenase. On the other hand, the storm petrel, which is capable of sustained flight, has predominantly the H_4 type of enzyme in its flight muscles. These examples represent the extremes characteristic of highly aerobic (H_4) or anaerobic (M_4) processes, and many other tissues intermediate between these two extremes are known which contain the isozymes HM_3, MH_3, and M_2H_2.

Pyruvic Decarboxylase. At the same time that the series of reactions constituting the glycolytic sequence was being studied in muscle, alcoholic fermentation was being examined in yeast extracts. Fortunately for students in the biological sciences the sequence of reactions that converts glucose to alcohol and CO_2 (alcoholic fermentation) are identical except for the manner in which pyruvic acid is metabolized.

$$\underset{\text{Pyruvic acid}}{CH_3-C(=O)-CO_2H} \xrightarrow[\text{TPP}]{Mg^{2+}} \underset{\text{Acetaldehyde}}{CH_3-CHO} + CO_2 \qquad (10\text{-}11)$$

Organisms such as yeast, which carry out alcoholic fermentation, contain the enzyme pyruvic decarboxylase—this catalyzes the decarboxylation of pyruvate to acetaldehyde and CO_2. The enzyme requires thiamin pyrophosphate (TPP or cocarboxylase) and Mg^{2+} as cofactors. The mechanism for this reaction was discussed in Chapter 9. The reaction is quite exergonic, which means that, in contrast to glycolysis, the end products of alcoholic fermentation, C_2H_5OH and CO_2, cannot be converted back to glucose.

Alcohol Dehydrogenase. In the final reaction of alcoholic fermentation, acetaldehyde is reduced to ethanol by NADH in the presence of *alcohol dehydrogenase* as follows:

$$\underset{\text{Acetaldehyde}}{CH_3CHO} + NADH + H^+ \rightleftharpoons \underset{\text{Ethanol}}{CH_3CH_2OH} + NAD^+ \qquad (10\text{-}12)$$

The K_{eq} for this reaction (page 200) greatly favors the reduction of acetaldehyde at pH 7.0. Again, being a reaction that involves a nicotinamide nucleotide coenzyme, the process is highly dependent on pH, and one can quantitatively convert alcohol to acetaldehyde, i.e., reverse reaction 10-12, at pH 9.5 in the presence of an excess of NAD^+. Note that the reoxidation of NADH in reaction 10-12 compensates for the production of NADH in reaction 10-5 and means that NADH would not accumulate in a tissue carrying out alcoholic fermentation. Alcohol dehydrogenase is widely distributed, having been found in liver, retina and sera of animals, the seeds and leaves of higher plants, and many microorganisms, including yeast. Clearly, the enzyme is not restricted to tissues which produce large amounts of ethanol.

Important Aspects of Anaerobic Carbohydrate Metabolism

The student will be aided in his mastery of the glycolytic sequence if an overall view of various aspects of this series of reactions can be achieved: its two distinct phases or stages; the balance of the nicotinamide nucleotide coenzymes; the overall energy relationships and the ATP produced; the utilization of carbohydrates other than glucose; the reversal of glycolysis; and finally its regulation.

Two Phases of Glycolysis. In considering the overall features of glycolysis, we have divided the reactions into two groups or phases. In the preparative phase (reactions 10-1 through 10-4; or 10-18 and 10-19) glucose (or the glucosyl unit of a polysaccharide) is converted into a triose phosphate by phosphorylation reactions.

The preliminary phosphorylation is accomplished at the expense of the energy-rich phosphate bonds of ATP (reactions 10-1 and 10-3) or by the action of phosphorylase (reaction 10-18). The second stage starts with the oxidation of the triose phosphate (reaction 10-5) and results in the entrapment of some of the energy of the hexose molecule into a form readily utilized by the organism. The further modification of 3-phosphoglyceric acid results in the formation of another energy-rich compound and eventually leads to the production of pyruvic acid, a key intermediate in glycolysis. The fate of pyruvate in turn depends on the organism under consideration or, more properly, on the enzymes present in that organism.

The enzymes catalyzing the glycolytic sequence, with the exception of enolase and pyruvic decarboxylase, can be classified into four groups: kinases, mutases, isomerases, and dehydrogenases. The kinases catalyze the transfer of a phosphate group from ATP to some acceptor molecule. The mutases catalyze the transfer of phosphate groups at a low-energy level from one position on a carbohydrate molecule to another position on the same molecule. Both classes of enzymes, involving as they do phosphorylated compounds, usually require Mg^{2+} ions. The isomerases, on the other hand, catalyze the isomerization of aldose sugars to ketose sugars; these enzymes, unlike the kinases and mutases, do not require Mg^{2+}. Finally, the dehydrogenases constitute the fourth general class of enzyme encountered in anaerobic carbohydrate metabolism.

Throughout this discussion we have used primarily the *trivial* names for the glycolytic enzyme. The systematic names proposed by the Commission on Enzymes of the International Union of Biochemistry together with the trivial names are listed in Table 10-1.

Balance of Coenzymes. Anaerobic carbohydrate metabolism occurs in the absence of oxygen. How then does the oxidation of D-glyceraldehyde-3-phosphate proceed uninterruptedly in a cell during glycolysis? Inspection of reaction 10-5 shows that NAD^+ is the primary oxidizing agent which accepts the electrons in the oxidation of the triose phosphate. Since the amount of NAD^+ in any cell is limited, the reaction will cease as soon as all the NAD^+ is reduced, *unless* there is a mechanism for reoxidation of the reduced nicotinamide nucleotide. In alcoholic fermentation that reoxidation is accomplished when acetaldehyde is reduced to ethanol in the presence of alcohol dehydrogenase (reaction 10-12). In muscle tissue the reoxidation occurs when pyruvate is reduced to lactate (reaction 10-10). Thus, NAD^+ serves as a carrier of electrons which are transferred from triose phosphate to either acetaldehyde or pyruvate, depending on the tissue involved. This may be represented for the latter compound as in Scheme 10-1.

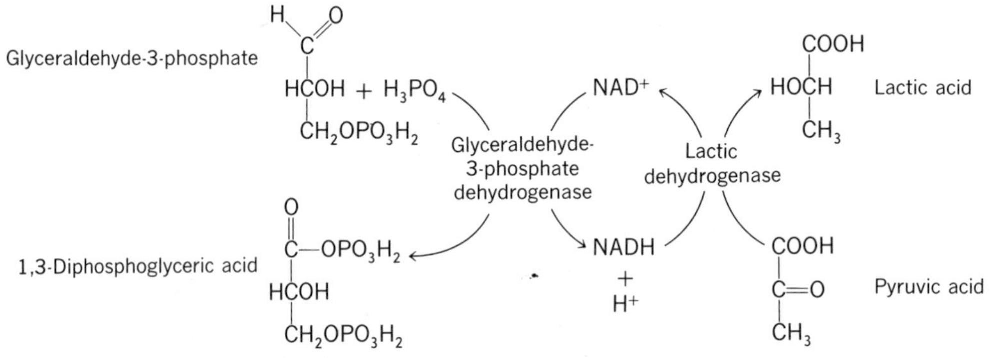

Glyceraldehyde-3-phosphate: $HC(=O)-HCOH-CH_2OPO_3H_2$ + H_3PO_4 → (Glyceraldehyde-3-phosphate dehydrogenase) → 1,3-Diphosphoglyceric acid: $C(=O)(OPO_3H_2)-HCOH-CH_2OPO_3H_2$

NAD^+ → $NADH + H^+$ (Glyceraldehyde-3-phosphate dehydrogenase)

$NADH + H^+$ → NAD^+ (Lactic dehydrogenase)

Pyruvic acid: $COOH-C{=}O-CH_3$ → (Lactic dehydrogenase) → Lactic acid: $COOH-HOCH-CH_3$

Scheme 10-1

Table 10-1

The Trivial and Systematic Names of Enzymes of the Glycolytic Pathway

Reaction	Trivial name	Systematic name
10-1	Hexokinase	ATP: D-Hexose-6-phosphotransferase (EC 2.7.1.1)
	Glucokinase	ATP: D-Glucose-6-phosphotransferase (EC 2.7.1.2)
10-2	Phosphoglucoisomerase	D-Glucose-6-phosphate-ketol isomerase (EC 5.3.1.9)
10-3	Phosphofructokinase	ATP: D-Fructose-6-phosphate-1-phosphotransferase (EC 2.7.1.11)
10-4	Aldolase	Fructose-1,6-diphosphate D-glyceraldehyde-3-phosphate lyase (EC 4.1.2.13)
10-4a	Triose phosphate isomerase	D-Glyceraldehyde-3-phosphate ketol isomerase (EC 5.3.1.1)
10-5	D-Glyceraldehyde 3-phosphate dehydrogenase (Triose phosphate dehydrogenase)	D-Glyceraldehyde-3-phosphate: NAD oxidoreductase (phosphorylating) (EC 1.2.1.12)
10-6	Phosphoglyceryl kinase	ATP: 3-Phospho-D-glycerate-1-phosphotransferase (EC 2.7.2.3)
10-7	Phosphoglyceryl mutase	D-Phosphoglycerate-2,3-phosphomutase (EC 5.4.2.1)
10-8	Enolase	2-phospho-D-glycerate hydro-lyase (EC 4.2.1.11)
10-9	Pyruvic kinase	ATP: Pyruvate phosphotransferase (EC 2.7.1.40)
10-10	Lactic dehydrogenase	L-Lactate: NAD oxidoreductase (EC 1.1.1.27)
10-11	Pyruvic decarboxylase	2-Oxoacid carboxylase (EC 4.1.1.1)
10-12	Alcohol dehydrogenase	Alcohol: NAD oxidoreductase (EC 1.1.1.1)
	Some ancillary enzymes	
10-18	Phosphorylase *a*	α-Glucan phosphorylase (EC 2.4.1.1)
10-19	Phosphoglucomutase	D-Glucose-1,6-diphosphate: D-Glucose-1-phosphate phosphotransferase (EC 2.7.5.1)

Production of High-Energy Phosphate. There is a net formation of high-energy phosphate in the form of ATP both in glycolysis and alcoholic fermentation. In both processes, a high-energy bond is produced (reaction 10-5) when triose phosphate is oxidized and is made available to the cell in the form of ATP in the subsequent reaction (reaction 10-6). Two trioses are produced, moreover, from each molecule of hexose metabolized, and both the trioses are oxidized to 1,3-diphosphoglyceric acid in the presence of the enzymes *triose phosphate isomerase* (reaction 10-4a) and *triose phosphate dehydrogenase* (reaction 10-5). In the absence of the isomerase half of the hexose molecule would remain as dihydroxy acetone phosphate and would not be converted to pyruvate. Since triose phosphate isomerase is widely distributed in nature, however, both trioses can be oxidized, and two high-energy phosphates per mole of hexose will be produced in the oxidative step.

Similarly, in reaction 10-8 a high-energy phosphate bond is produced from 1 mole of triose where none previously existed. This again is transferred to ADP in the subsequent step (reaction 10-9) to make ATP. Thus, for each mole of hexose 2 moles of ATP will be formed at this stage of the pathway. Hence, the total is *four*, but this is not a *net* achievement; glucose must first be phosphorylated to glucose-6-phosphate, and fructose-6-phosphate must in turn be phosphorylated to fructose diphosphate before cleavage into the triose phosphates and subsequent degradation can proceed. Thus, *two* high-energy phosphate bonds are *expended* in the initial step (reactions 10-1 and 10-3) of the preparative phase. The net production of high-energy phosphate is therefore 2 ATP per mole of glucose fermented to either lactic acid or ethanol and CO_2.

Overall Energy Relationships. Now the $\Delta G'$ for the conversion of glucose to 2 moles of lactic acid as it might occur in a test tube can be calculated from various thermodynamic data:

$$\underset{\text{Glucose}}{C_6H_{12}O_6} \xrightarrow{\text{In a test tube}} \underset{\text{Lactic acid}}{2\ CH_3CHOHCOOH}$$
$$\Delta G' = -47{,}000 \text{ cal}$$

Neglecting entropy changes, this amount of energy would be liberated as heat (ΔH). In the biological organism performing glycolysis, however, the reaction must be corrected to show precisely what occurs, namely that, as glucose is converted to 2 moles of lactic acid, 2 moles of ATP are produced from ADP and inorganic phosphate. That is,

$$C_6H_{12}O_6 + 2\ ADP + 2\ H_3PO_4 \xrightarrow{\text{In the cell}} 2\ CH_3CHOHCOOH + 2\ ATP + 2\ H_2O$$
$$\Delta G' = -32{,}400 \text{ cal}$$

Since the two ATP represent a conservation of 14,600 cal (2×7300), the $\Delta G'$ for the second equation is less by that amount ($-47{,}000 - (-14{,}600)$ or $-32{,}400$). Moreover, one can speak of the *efficiency* with which the ATP has been produced in glycolysis. Since $-47{,}000$ cal are available and two ATP's were produced, the efficiency corresponds to $-14{,}600/-47{,}000$ or 31%.

At this point note that ΔG for the hydrolysis of ATP under the conditions that exist in a cell may be more negative, by as much as 4000 cal, than the standard free energy change ($\Delta G' = -7300$ cal). This is due, of course, to the fact that the concentrations of reactants in the formation of ATP are not at standard values (see Chapter 7). If the ΔG for formation of ATP is indeed +12,000 cal, the efficiency of energy conservation will be 24,000/ −47,000 or 51%.

The $\Delta G'$ values for the conversion of glucose to ethanol and CO_2 and the corresponding reaction as it occurs in the yeast cell may be written as

$$\underset{\text{Glucose}}{C_6H_{12}O_6} \xrightarrow{\text{In a test tube}} 2\ \underset{\text{Ethanol}}{CH_3CH_2OH} + 2\ CO_2$$

$$\Delta G' = -40,000 \text{ cal}$$

$$C_6H_{12}O_6 + 2\ ADP + 2\ H_3PO_4 \xrightarrow{\text{In a yeast cell}} 2\ CH_3CH_2OH + 2\ CO_2 + 2\ ATP + 2\ H_2O$$

$$\Delta G' = -25,400 \text{ cal}$$

Utilization of Other Carbohydrates

Sugars other than glucose are metabolized in the glycolytic sequence following their conversion by auxiliary enzymes to intermediates in that sequence. Thus, fructose and mannose can be phosphorylated by ATP in the presence of hexokinase and be converted into fructose-6-phosphate and mannose-6-phosphate. The former is an intermediate in glycolysis; mannose-6-phosphate is converted to fructose-6-phosphate by the enzyme phosphomannose isomerase in a reaction analogous to that catalyzed by phosphoglucoisomerase (reaction 10-2).

Disaccharides such as lactose and sucrose are extremely common sources of carbohydrate in the diet of animals. The initial steps in their utilization involve hydrolysis to the component monosaccharides by specific glycosidases, lactase and sucrase (invertase) found in the animal's digestive tract. The subsequent metabolism of glucose and fructose obtained on hydrolysis of sucrose has been previously discussed. The metabolism of galactose formed (together with glucose) on the hydrolysis of lactose is an interesting story.

Utilization of Galactose. The initial reaction with galactose involves phosphorylation by ATP in the presence of a specific galactokinase that produces galactose-1-phosphate. This enzyme is present in both yeast and animal liver cells.

$$\text{Galactose} + ATP \longrightarrow \text{Galactose-1-phosphate } (\text{C-1}: OPO_3H_2) + ADP \qquad (10\text{-}13)$$

Further metabolism of galactose-1-phosphate involves uridine triphosphate (UTP) and a uracil derivative of that sugar known as uridine diphosphate galactose (UDP-galactose):

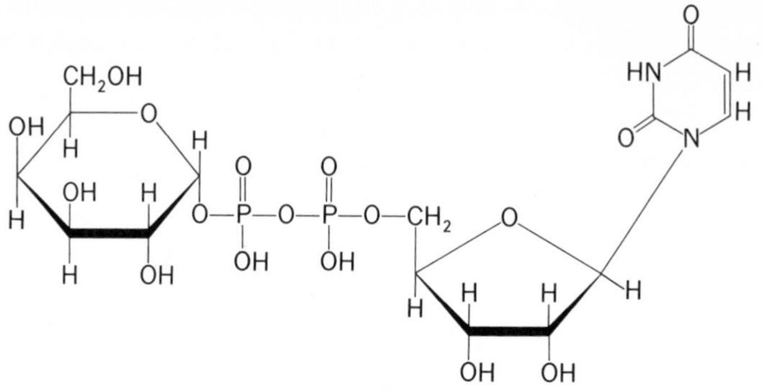

Uridine diphosphate galactose
(UDP-galactose)

The galactose-1-phosphate formed in reaction 10-13 is converted to UDP-galactose by the enzyme *UDP-galactose pyrophosphorylase*, which is present in the liver of adult humans:

$$\underset{\text{Galactose-1-phosphate}}{\text{Gal—P}} + \underset{\text{UTP}}{\text{U—R—P—P—P}} \rightleftharpoons \underset{\text{UDP-galactose}}{\text{U—R—P—P—Gal}} + \underset{\text{Pyrophosphate}}{\text{P—P}} \qquad (10\text{-}14)$$

The various components of the UTP and sugar phosphate molecules have been identified (R = ribose; P = phosphate) to indicate the nature of the reaction. The reaction is readily reversible, as could be anticipated, since the one (interior) pyrophosphate bond is utilized to form the pyrophosphate in the sugar nucleotide; the number of energy-rich structures in the reactants and products is consequently the same. This reaction is a model one for forming these *nucleoside diphosphate sugars* or *sugar nucleotides*. As another example, ADP-glucose would be formed from ATP and glucose-1-phosphate in the presence of the specific pyrophosphorylase.

In the next step, the galactose moiety in UDP-galactose is isomerized to a glucose moiety, thereby forming UDP-glucose. The enzyme that catalyzes this reaction is known as *UDP-glucose epimerase*.

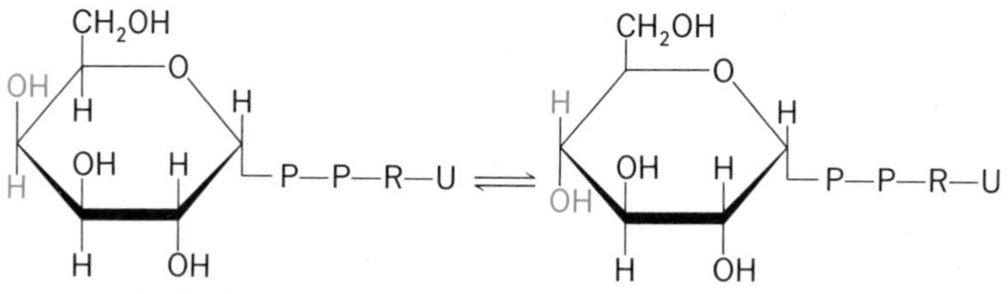

$$\text{UDP-galactose} \rightleftharpoons \text{UDP-glucose} \qquad (10\text{-}15)$$

Finally, the action of a third enzyme *UDP-glucose pyrophosphorylase* liberates the glucose (formerly the galactose) moiety from UDP-glucose as glucose-1-phosphate:

$$\underset{\text{UDP-glucose}}{\text{U—R—P—P—Glu}} + \text{P—P} \rightleftharpoons \underset{\text{UTP}}{\text{U—R—P—P—P}} + \underset{\text{Glucose-1-phosphate}}{\text{Glu—P}} \qquad (10\text{-}16)$$

Note this reaction is the same as reaction 10-14 except that glucose is the sugar involved. The sum of reactions 10-14 through 10-16 is the conversion of galactose-1-phosphate into glucose-1-phosphate. The metabolism of the latter by glycolysis (see below) accounts for the metabolism of galactose in adult humans.

As noted above, the enzyme catalyzing the formation of UDP-galactose from galactose-1-phosphate is found only in the liver of adults. How, then, does an infant metabolize galactose? This is a pertinent question, because one of the major energy sources that an infant has is the sugar lactose in the milk which it consumes.

Studies have shown that fetal and infant liver tissue contains the enzyme *phosphogalactose uridyl transferase:*

$$\text{UDP-glucose} + \text{Galactose-1-phosphate} \rightleftharpoons \text{UDP-galactose} + \text{Glucose-1-phosphate} \qquad (10\text{-}17)$$

The coupling of this reaction with 10-15 accounts for the net conversion of galactose-1-phosphate into glucose-1-phosphate and is the normal route for galactose metabolism in infants. This series of reactions has attracted much attention because of a hereditary disorder known as *galactosemia*. Infants that have this defect can not metabolize galactose and they exhibit a high level of galactose in the blood. The sugar is excreted in the urine and, if the condition is not attended to, the infant can develop cataracts and may become mentally retarded. The simple remedy, once the condition is identified, is to remove the source of galactose, usually the milk in the infant's diet, and supply a galactose-free diet.

Galactosemic individuals lack the uridyl transferase (reaction 10-17), and this accounts for their failure to metabolize galactose. Only after the individual has reached puberty does an adequate amount of UDP-galactose pyrophosphorylase appear in the liver, thereby providing him with the capacity to metabolize lactose.

It should be pointed out that the sugar nucleotides (e.g., UDP-glucose, UDP-galactose) are precursors of important cellular constituents such as glycogen, cell wall components, hyaluronic acids. Since the galactosemic infant needs a source of UDP-galactose to produce these cellular constituents, it will convert glucose-1-phosphate to UDP-galactose by reversing reactions 10-16 and 10-15. The adult human, on the other hand, will have available the pyrophosphorylase (reaction 10-14) for the synthesis of UDP-galactose.

Utilization of Glycerol. One final example of a common metabolite that is metabolized by means of the glycolytic sequence is the compound glycerol. Glycerol is produced during the breakdown of triacylglycerols (Chapter 12) and phosphorylated by ATP in the presence of glycerol kinase:

$$\underset{\text{Glycerol}}{\begin{array}{c} CH_2OH \\ | \\ HOCH \\ | \\ CH_2OH \end{array}} + ATP \xrightarrow{Mg^{2+}} \underset{sn\text{-Glycerol-3-phosphate}}{\begin{array}{c} CH_2OH \\ | \\ HOCH \\ | \\ CH_2OPO_3H_2 \end{array}} + ADP$$

The phosphoglycerol produced in this reaction (also that produced during the breakdown of phosphoglycerides, Chapter 12) can then be oxidized to dihydroxy acetone phosphate by the enzyme *glyceryl phosphate dehydrogenase*. This enzyme of the cytoplasm utilizes NAD^+ as the oxidant. The dihydroxy acetone phosphate can enter directly into the second stage of glycolysis.

$$\underset{sn\text{-Glycerol-3-phosphate}}{\begin{array}{c} CH_2OH \\ | \\ HOCH \\ | \\ CH_2OPO_3H_2 \end{array}} + NAD^+ \longrightarrow \underset{\text{Dihydroxy acetone phosphate}}{\begin{array}{c} CH_2OH \\ | \\ C{=}O \\ | \\ CH_2OPO_3H_2 \end{array}} + NADH + H^+$$

Another glyceryl phosphate dehydrogenase in mitochondria, a flavoprotein, utilizes FAD as the primary oxidant. These two enzymes play an important role in the transport of cytoplasmic NADH into the interior of the mitochondria (see Chapter 14).

Utilization of Polysaccharides. The polysaccharides starch and glycogen encountered in plant and animal, respectively, are important fuel molecules. These polymers of glucose are degraded in two different ways in order that they be accommodated in the glycolytic sequence. In one case, the polysaccharide is hydrolyzed to produce ultimately D-glucose which can be phosphorylated and metabolized in the glycolytic sequence. Two types of enzymes, the α- and β-amylases, accomplish this hydrolysis. α-Amylase is an endoamylase that catalyzes a random hydrolysis of the interior bonds of the branched polysaccharide to produce oligosaccharides and eventually maltose and D-glucose. Limit dextrins still containing the original α-1-6 linkages that are resistant to the enzyme are also formed. The β-amylases are exoamylases which catalyze a stepwise hydrolysis of alternate linkages starting at the nonreducing end. Maltose and a highly branched dextrin are the products of this hydrolysis, since the β-amylase will proceed up to about 2 or 3 residues from the α-1-6 linkage before it will no longer act. The amylases have been found in many sources: animal, plant, and bacterial.

The other way in which the fuel polysaccharide can be degraded is through the action of phosphorylases, enzymes which are widely distributed in nature. Although the phosphorylases from different sources vary in certain respects, they all catalyze the phosphorylytic cleavage of the α-1,4-glucosidic linkage at the nonreducing end of the starch or glycogen chain. The reaction is reversible and is represented as

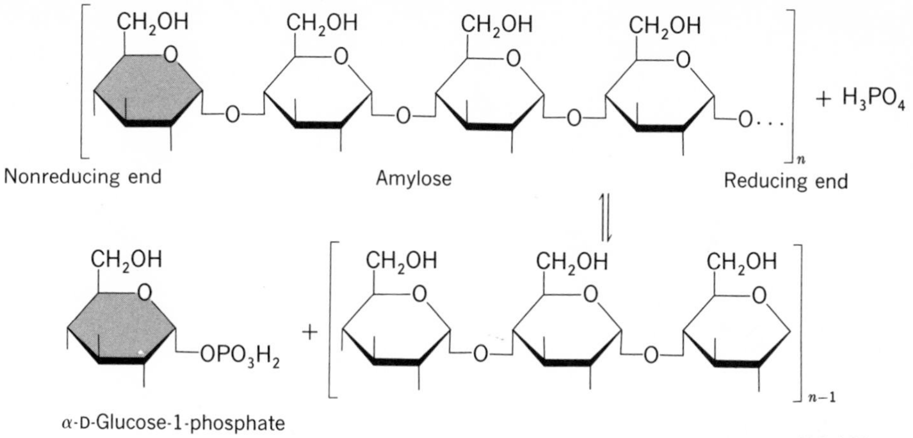

CH_2OH CH_2OH CH_2OH CH_2OH $+ H_3PO_4$

Nonreducing end Amylose Reducing end

CH_2OH $-OPO_3H_2$ + $[\ldots]_{n-1}$

α-D-Glucose-1-phosphate

(10-18)

As written from left to right, the reaction is a phosphorolysis resulting in the formation of α-D-glucose-1-phosphate and the loss of one glucose unit from the nonreducing end of the polysaccharide chain. In the reverse reaction, inorganic phosphate is liberated from α-D-glucose-1-phosphate with a lengthening of the polysaccharide chain at the nonreducing end. Lengthening of the chain cannot occur unless a small amount of starch, glycogen, or dextrin is present as a priming agent. When the primer is absent, the enzyme is either unable to catalyze the direct condensation of glucose-1-phosphate units, or does so very slowly. In the presence of the primer, the enzyme rapidly adds glucose units to the preexisting polysaccharide chain. The nature of the primer required varies with the source of the phosphorylase. The enzyme from muscle requires the highly branched glycogen; the enzyme from potato can utilize the trisaccharide maltotriose as a primer. All phosphorylases are specific in that they react only with α-D-glucopyranosyl-1-phosphate and form only the α-1,4-maltosidic bond.

Phosphorylase will catalyze the stepwise removal of glucose units from a linear portion of a starch or glycogen molecule until it approaches within 4 to 6 units of an α-1-6 branch point. This branch point constitutes an area in which the enzyme is inactive. Highly branched polysaccharides such as amylopectin will therefore be degraded only about 55%, leaving a highly branched residue known as a *limit dextrin* (Figure 10-1). The highly branched dextrin can be further degraded by the action of two additional enzymes that have been found in animals, plants, and yeast. First, a debranching enzyme (an oligo-1,4 $\longrightarrow$ 1,4-glucan transferase) catalyzes the transfer of all but one of the short chains which phosphorylase could not degrade to another part of the molecule (Scheme 10-2). Then, an *α-1,6-glucosidase catalyzes* the removal of the single hexose unit at the α-1-6 branch point, thereby exposing a new linear portion on which phosphorylase can again begin to work.

The equilibrium of the phosphorylase-catalyzed reaction, which is readily reversible, is independent of the polysaccharide concentration, provided a certain minimum concentration is exceeded. Thus, in the following expression for K_{eq} the polysaccharide concentrations represent the number of non-

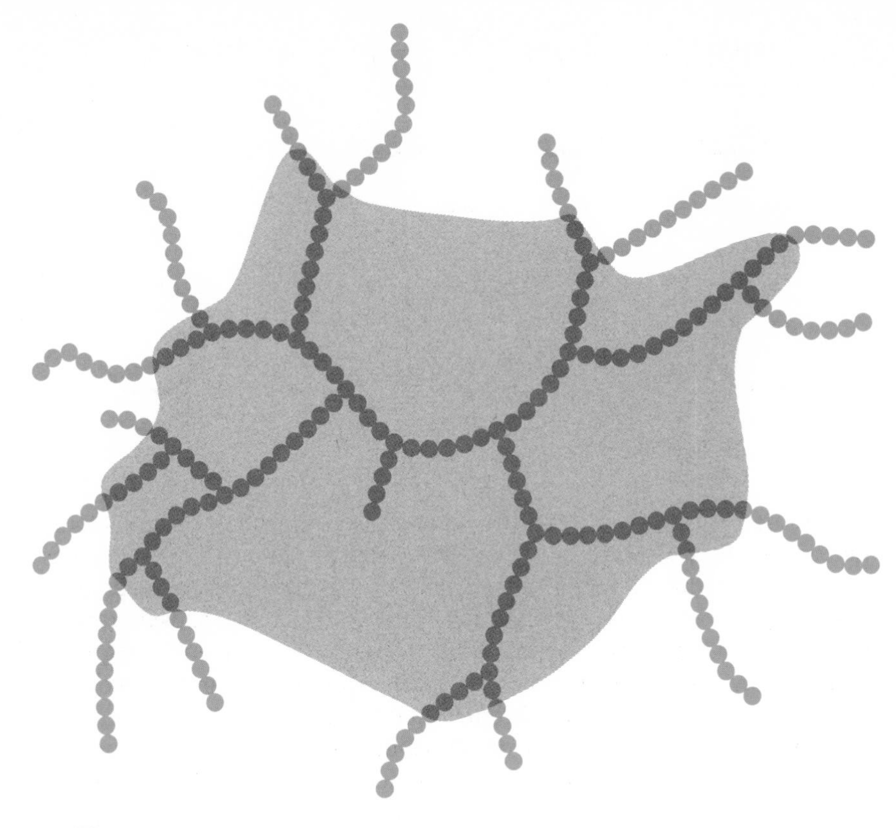

Figure 10-1

Action of phosphorylase on the branched-chain polysaccharide amylopectin. Phosphorylase degrades until the vicinity of a branching point is reached. Within the shaded area is limit dextrin, on which phosphorylase does not act.

reducing chain termini, *a number that does not change*. It then follows that at any pH the K_{eq} is determined by the relative concentrations of glucose-1-phosphate and inorganic phosphate. At pH 7.0,

$$K_{eq} = \frac{[C_6H_{10}O_5]_{n-1}[\text{Glucose-1-phosphate}]}{[C_6H_{10}O_5]_n[H_3PO_4]} = \frac{[\text{Glucose-1-phosphate}]}{[H_3PO_4]} = 0.3$$

Although the reaction catalyzed by phosphorylase is readily reversible, the role of the enzyme is largely degradative in nature. As will be discussed on page 265, a different enzyme and a different reaction sequence is responsible for the synthesis of glycogen.

The muscle phosphorylases exist in two forms, *a* and *b*. The relationships between these two forms and their significance in carbohydrate metabolism

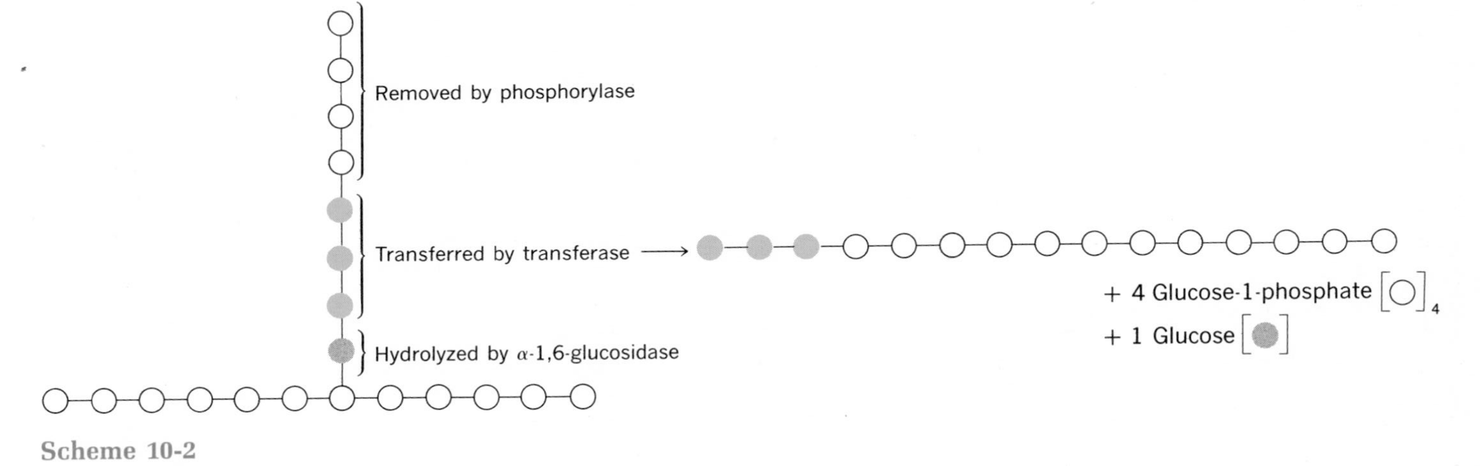
Removed by phosphorylase
Transferred by transferase
Hydrolyzed by α-1,6-glucosidase
+ 4 Glucose-1-phosphate
+ 1 Glucose

Scheme 10-2

have been extensively investigated by E. H. Fischer and E. G. Krebs. Phosphorylase *a* possesses 60–70% of its maximal activity in the absence of adenosine-5′-phosphate (AMP), while phosphorylase *b* has an absolute requirement for this nucleotide. Rabbit muscle phosphorylase *a* is a tetramer with a molecular weight of 400,000, consisting of four identical polypeptide chains. Each chain contains a serine residue whose hydroxyl group is esterified with phosphate and a lysine residue whose free amino group (ε) is bound in a Schiff's base with pyridoxal phosphate. While the roles of these groups is not understood, their removal results in inactivation of phosphorylase *a*.

The phosphate groups may be removed through hydrolysis in the presence of the enzyme *phosphorylase phosphatase* found in muscle:

$$\text{Phosphorylase } a \xrightarrow[\text{Phosphatase}]{\text{Phosphorylase}} 2 \text{ Phosphorylase } b + 4\ H_3PO_4$$

As shown, this involves the release of 4 moles of inorganic phosphate and the formation of 2 moles of phosphorylase *b* with molecular weight of 200,000. Phosphorylase *b*, which does catalyze reaction 10-18 in the presence of AMP, is converted back to phosphorylase *a* by ATP in the presence of the enzyme *phosphorylase b kinase:*

$$2 \text{ Phosphorylase } b + 4 \text{ ATP} \xrightarrow{\text{Kinase, } Mg^{2+}} \text{Phosphorylase } a + 4 \text{ ADP}$$

The interconversions of the *a* and *b* forms is of prime importance in regulating glycogen breakdown in intact tissues.

That the physiological control of glycogenolysis is more complex may be seen by citing some of the additional factors influencing the enzymes just described. Phosphorylase *b* kinase itself exists in active and inactive forms. The latter is converted into the active form by still another enzyme, *phosphorylase kinase kinase,* that requires Mg^{2+} ions and adenosine-3′,5′-phosphate (cyclic adenylic acid or cyclic AMP). Cyclic AMP, in turn, is produced from ATP by an enzyme which is stimulated by the hormone adrenaline.

The effect of some of these factors in the regulation of carbohydrate

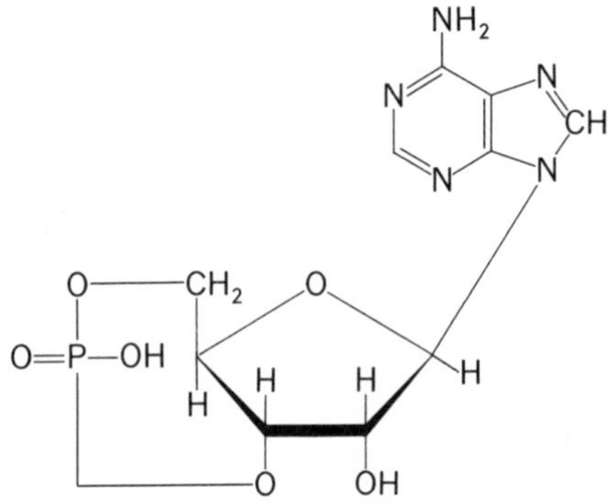

Adenosine-3′,5′-phosphate
(Cyclic AMP)

metabolism is discussed briefly later in this chapter. The role of cyclic AMP will be discussed further in Chapter 20.

The entry of the glucose unit produced by the action of phosphorylase into the glycolytic sequence requires the action of one additional enzyme, phosphoglucomutase. This enzyme catalyzes the interconversion of glucose-1-phosphate and glucose-6-phosphate:

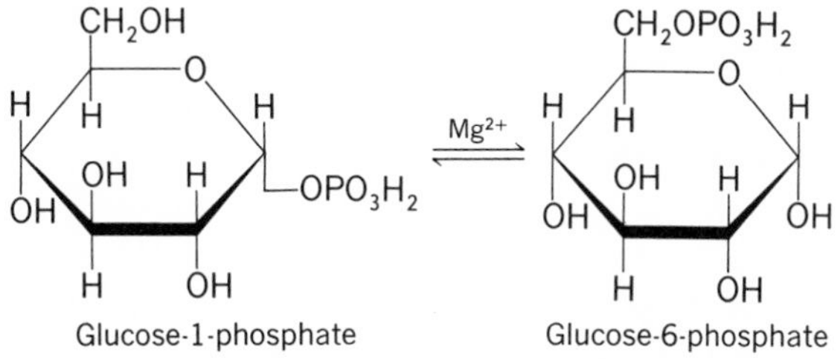

$$\text{Glucose-1-phosphate} \underset{}{\overset{Mg^{2+}}{\rightleftharpoons}} \text{Glucose-6-phosphate} \qquad (10\text{-}19)$$

The K_{eq} of the reaction from left to right is 19 at pH 7 and favors the formation of glucose-6-phosphate. This is in agreement with the observation that the $\Delta G'$ of hydrolysis of glucose-1-phosphate is intermediate between that of the energy-rich pyrophosphates and simple phosphate esters.

Phosphoglucomutase has been crystallized (74,000 mol wt) from rabbit skeletal muscle, of which it constitutes almost 2% of the water-soluble protein. Studies by the Argentinean biochemist Leloir and his colleages on the mechanism of the reaction led to the discovery of glucose-1,6-diphosphate. These workers originally proposed that the role of the glucose diphosphate was to donate phosphate reversibly to glucose-1-phosphate or glucose-6-phosphate. Later studies by Najjar and by Milstein have required revision of the mechanism and the role of the diphosphate. A mechanism similar to that proposed for phosphoglyceryl mutase (reaction 10-7) may be assumed at this time.

Reversal of Glycolysis

It is an experimental fact that lactic acid in a resting muscle can be converted to glycogen provided an energy source is available. This resynthesis of glycogen (or glucose) from smaller, nonglucose molecules (lactic acid, pyruvate) is termed *glyconeogenesis* (*gluconeogenesis*). While one might assume that gluconeogenesis could occur by reversal of each step of the glycolytic pathway, it has been pointed out, and amply demonstrated experimentally, that three reactions (reactions 10-1, 10-3, and 10-9) are physiologically irreversible in the cell. As a consequence, these reactions must be bypassed in gluconeogenesis.

The bypassing of reaction 10-9 involves the participation of two new enzymes that catalyze reactions known as CO_2-*fixation* reactions, a group of reactions which have an important role in the functioning of the Krebs cycle, soon to be discussed (Chapter 13). The first of these enzymes, *pyruvic carboxylase*, is located in the mitochondria; thus, pyruvic acid produced in the cytosol either from lactate or phosphoenol pyruvate must enter the mitochondria as a first step. The reaction catalyzed is

$$\underset{\text{Pyruvic acid}}{\begin{matrix}CO_2H\\ |\\ C{=}O\\ |\\ CH_3\end{matrix}} + CO_2 + ATP \xrightleftharpoons[Mg^{2+}]{\text{Acetyl–CoA}} \underset{\text{Oxalacetic acid}}{\begin{matrix}CO_2H\\ |\\ C{=}O\\ |\\ CH_2\\ |\\ CO_2H\end{matrix}} + ADP + H_3PO_4 \qquad (10\text{-}20)$$

$$\Delta G' = -500 \text{ cal (pH 7.0)}$$

The $\Delta G'$ is quite small, and therefore the reaction is readily reversible. Pyruvic carboxylase of chicken liver is a large (660,000 mol wt) allosteric protein that has tetrameric structure. It has an absolute requirement for acetyl–CoA as an activator; the significance of this will be mentioned later in this chapter and in Chapter 13. The enzyme is also a biotin-containing protein, each monomeric unit (150,000 mol wt) containing 1 mole of biotin.

The second enzyme involved in reversing this part of the glycolytic sequence is known as phosphoenol pyruvic (PEP) carboxykinase:

$$\underset{\text{Oxalacetic acid}}{\begin{matrix}CO_2H\\ |\\ C{=}O\\ |\\ CH_2\\ |\\ CO_2H\end{matrix}} + GTP \xrightleftharpoons{Mg^{2+}} \underset{\text{Phosphoenol pyruvic acid}}{\begin{matrix}CO_2H\\ |\\ C{-}OPO_3H_2\\ \|\\ CH_2\end{matrix}} + CO_2 + GDP \qquad (10\text{-}21)$$

$$\Delta G' = +700 \text{ cal (pH 7.0)}$$

In this reaction oxalacetate produced in reaction 10-20 is converted to PEP by a reaction involving little change in free energy, but one in which CO_2 is produced, a reverse "CO_2 fixation." The cellular distribution of this enzyme varies greatly in different species. In those tissues (the liver of pig, guinea pig, and rabbit) where it is mitochondrial, the PEP produced subsequently diffuses out and then can be taken up to fructose-1,6-diphosphate, provided the ATP and NADH needed to reverse reactions 10-6 and 10-5, respectively, are available.

The overall reaction for converting pyruvate to PEP can be obtained by adding reactions 10-20 and 10-21:

$$\begin{matrix}CO_2H\\ |\\ C{=}O\\ |\\ CH_3\end{matrix} + ATP + GTP \rightleftharpoons \begin{matrix}CO_2H\\ |\\ C{-}OPO_3H_2\\ \|\\ CH_2\end{matrix} + ADP + GDP + H_3PO_4 \qquad (10\text{-}22)$$

$$\Delta G' = +200 \text{ cal}$$

Note that the large $\Delta G'$ of reaction 10-9 has now been overcome but that 2 moles of nucleoside triphosphate have been expended in order for the overall reaction (reaction 10-22) to have a negligible $\Delta G'$ of +200 cal.

In many species, PEP carboxykinase is located in the cytoplasm, providing further complications since the oxalacetate produced in reaction 10-20 is not able to pass through the mitochondrial membrane. It is generally agreed now that, in order for oxalacetate to be made available to the cytoplasmic PEP carboxykinase, it is first reduced to malic acid by the malic dehydrogenase in the mitochondria (Chapter 13). The mitochondrial inner membrane is permeable to malate which then diffuses out into the cytoplasm and is reoxidized to oxalacetate by a cytoplasmic malic dehydrogenase and converted to PEP by the PEP carboxykinase:

$$\text{Oxalacetate} + \text{NADH} + \text{H}^+ \xrightarrow{\text{Mitochondrial malic dehydrogenase}} \text{Malate} + \text{NAD}^+$$

$$\text{Malate (mitochondrial)} \xrightarrow{\text{Mitochondrial inner membrane}} \text{Malate (cytoplasmic)}$$

$$\text{Malate} + \text{NAD}^+ \xrightarrow{\text{Cytoplasmic malic dehydrogenase}} \text{Oxalacetate} + \text{NADH} + \text{H}^+$$

In gluconeogenesis, reactions 10-3 and 10-1 involving ATP and ADP must also be bypassed. In the former instance this is accomplished by a *phosphatase* which catalyzes the *hydrolysis* of fructose-1,6-diphosphate to form fructose-6-phosphate.

$$\text{Fructose-1,6-diphosphate} + H_2O \longrightarrow \text{Fructose-6-phosphate} + H_3PO_4 \qquad (10\text{-}23)$$

$\Delta G' = -4000$ cal

The presence of the phosphatase in a wide variety of tissues accounts for the formation of fructose-6-phosphate from hexose diphosphate and thus provides a means by which glycogen or glucose can subsequently be formed from hexose diphosphate.

Fructose diphosphate phosphatase is a regulatory enzyme which, together with phosphofructokinase (reaction 10-3), plays a key role in regulating the flow of carbon up and down the glycolytic sequence. In this oligomeric protein, the number of monomers depends on the source. Regardless of the source, however, the phosphatase is strongly inhibited by AMP.

The production of glucose from glucose-6-phosphate requires a second phosphatase that catalyzes the following exergonic reaction:

$$\text{Glucose-6-phosphate} + H_2O \longrightarrow \text{Glucose} + H_3PO_4 \qquad (10\text{-}24)$$

Glucose-6-phosphatase is characteristically associated with the endoplasmic reticulum and is contained in particles (microsomes) obtained from that structure. It is present in tissues (e.g., mammalian liver) that can produce free glucose.

Starting with 2 moles of lactate, then, and proceeding through reactions 10-20 to 10-22, reactions 10-8 through 10-4, and reactions 10-23, 10-2, and

10-24, we can write the overall equation accounting for the reversal of glycolysis:

$$2\ \text{Lactate} + 4\ \text{ATP} + 2\ \text{GTP} + 6\ H_2O \longrightarrow \text{Glucose} + 4\ \text{ADP} + 2\ \text{GDP} + 6\ H_3PO_4$$

From this it is apparent that a total of six energy-rich phosphates are required to make glucose. This equation is clearly not the reverse of that on page 250 in which glucose was converted to 2 moles of lactate and serves again to emphasize the energy relationships of the glycolytic sequence.

Biosynthesis of Some Carbohydrates

A limited number of homopolysaccharides such as inulin, a fructosan found in artichokes, are made by specific transglycosidases which transfer fructosyl units directly from a donor such as sucrose to an acceptor such as the growing chain of inulin. However, the majority of the important disaccharides and polysaccharides found in nature, such as sucrose, glycogen, starch, and cellulose, are synthesized by the transfer of glycosyl units from nucleoside diphosphate sugars to suitable acceptors.

Two important general equations constitute the basic mechanism for this synthetic process. The first of these involves the formation of the nucleoside diphosphate sugar (or sugar nucleotide):

$$\text{NTP} + \text{Sugar-1-phosphate} \rightleftharpoons \underset{\substack{\text{Nucleoside}\\ \text{diphosphate sugar}\\ \text{(Sugar nucleotide)}}}{\text{NDP—Sugar}} + \text{Pyrophosphate} \qquad (10\text{-}25)$$

The second involves the transfer of the sugar or glycosyl moiety to an acceptor:

$$\text{NDP—Sugar} + \underset{\text{(Primer)}}{\text{Acceptor}} \rightleftharpoons \text{NDP} + \text{Sugar—Acceptor} \qquad (10\text{-}26)$$

The flow of glycosyl-1-phosphates and nucleoside triphosphates into reaction 10-25 are important requirements for the smooth synthesis of polysaccharides. We have already described the synthesis of glucose-1-phosphate by glucokinase coupled to phosphoglucomutase:

$$\text{Glucose} + \text{ATP} \xrightarrow{\text{Glucokinase}} \text{Glucose-6-phosphate} \xrightarrow{\text{Phosphoglucomutase}} \text{Glucose-1-phosphate}$$

and by starch phosphorylase:

$$(\text{Glucose})_n + H_3PO_4 \rightleftharpoons (\text{Glucose})_{n-1} + \text{Glucose-1-phosphate}$$

The nucleoside triphosphates are generated by a widespread enzyme called *nucleoside diphosphate kinase:*

$$\text{NDP} + \text{ATP} \rightleftharpoons \text{NTP} + \text{ADP}$$

The enzymes responsible for the formation of the sugar nucleotide donor (reaction 10-25) are known as *nucleoside diphosphate sugar pyrophosphorylases*. The synthesis of UDP-glucose by a specific UDP-glucose pyrophosphorylase has been described:

$$\underset{\text{Glucose-1-phosphate}}{\text{Glu—P}} + \underset{\text{UTP}}{\text{U—R—P—P—P}} \rightleftharpoons \underset{\text{UDP-glucose}}{\text{U—R—P—P—Glu}} + \underset{\text{Pyrophosphate}}{\text{P—P}} \quad (10\text{-}16')$$

Similarly, GDP-mannose is synthesized by a specific GDP-mannose pyrophosphorylase:

$$\underset{\text{Mannose-1-phosphate}}{\text{Man—P}} + \underset{\text{GTP}}{\text{G—R—P—P—P}} \rightleftharpoons \underset{\text{GDP-mannose}}{\text{G—R—P—P—Man}} + \underset{\text{Pyrophosphate}}{\text{P—P}}$$

Once the correct nucleoside diphosphate sugar has been synthesized, the sugar moiety is transferred in the presence of the appropriate nucleoside diphosphate sugar transferase to the suitable acceptor by the general reaction 10-26.

These steps can be summarized as in Scheme 10-3.

A few examples will suffice to illustrate the general scheme of synthesis.

Two enzymes have been discovered by Leloir which are concerned with the synthesis of sucrose in plants; *UDP-glucose fructose transglycosylase* catalyzes the reaction:

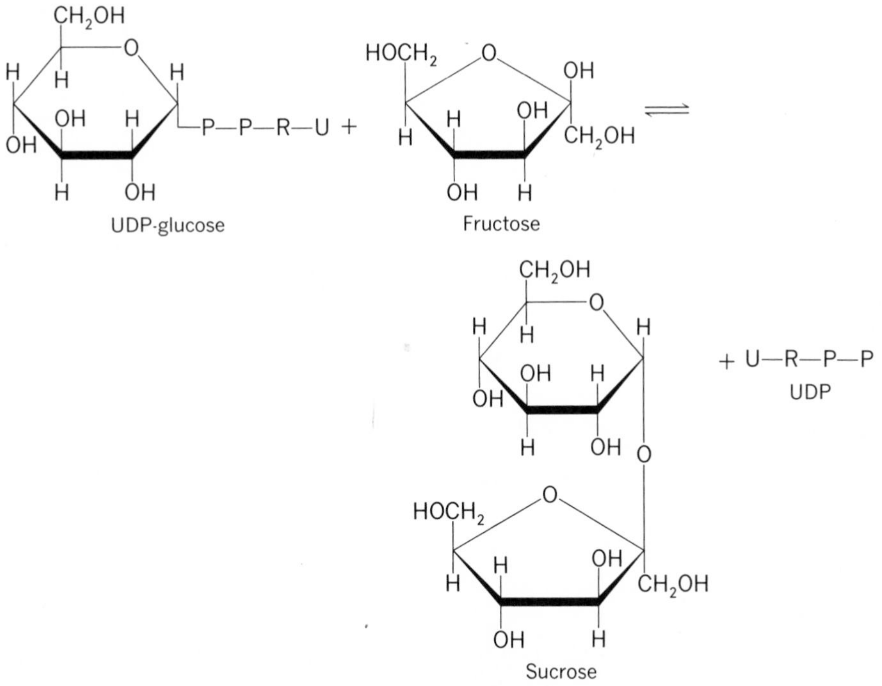

UDP-glucose + Fructose ⇌ Sucrose + U—R—P—P (UDP)

There is good evidence that this enzyme, although it synthesizes sucrose because of its favorable K_{eq}, is actually involved in the degradation of sucrose with the formation of UDP-glucose (reaction going to the left) with the preservation of glycosidic bond energy. This provides a mechanism whereby sucrose can be broken down to UDP-glucose which can then enter other synthetic pathways.

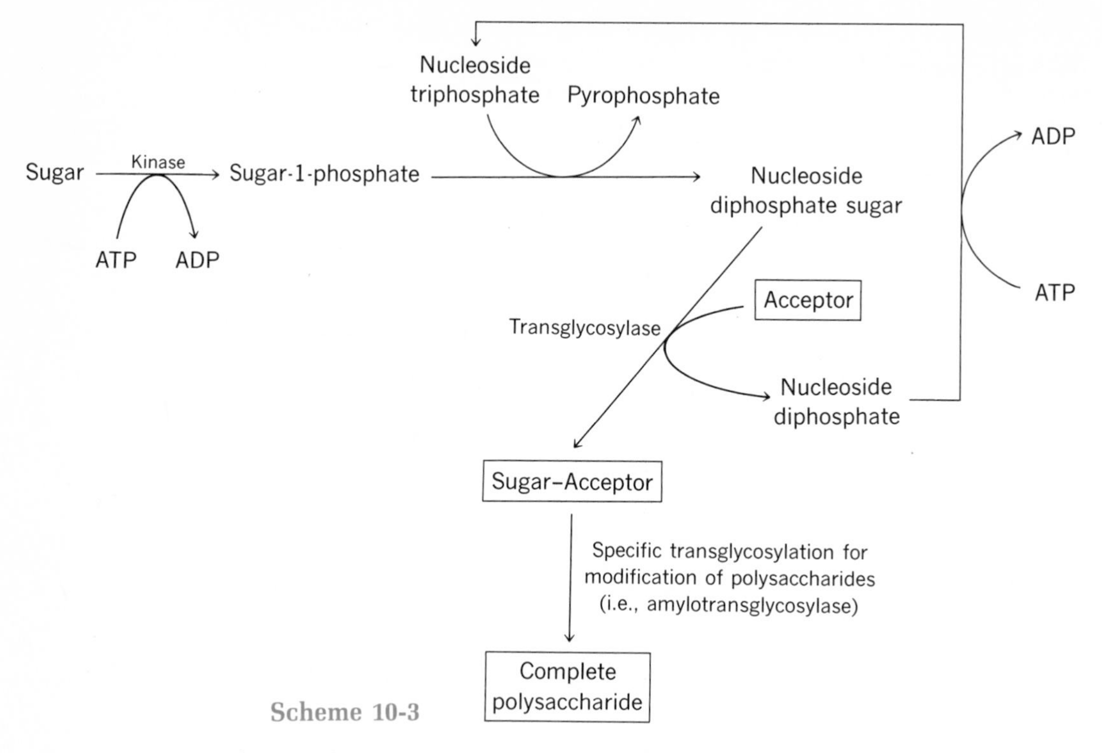

Scheme 10-3

A second enzyme concerned with sucrose synthesis, *UDP-glucose fructose 6-phosphate transglycosylase*, catalyzes the reaction:

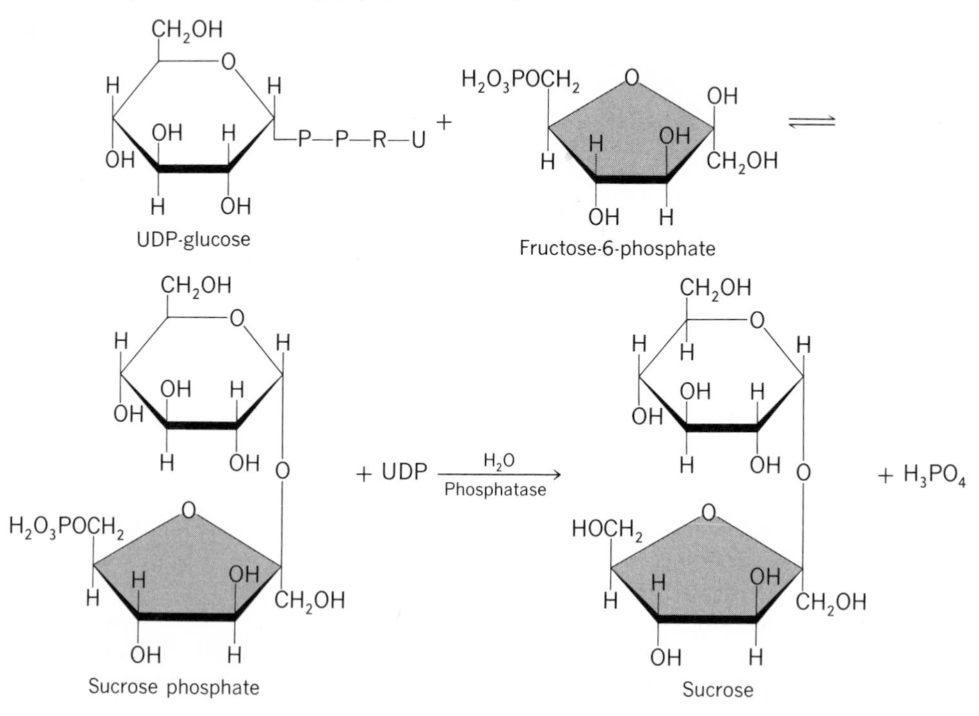

In the presence of a specific phosphatase, sucrose-6-phosphate is dephosphorylated to form sucrose. This enzyme is responsible for the synthesis of sucrose in plants, since it involves the phosphatase step which is irreversible.

The enzyme *glycogen synthase* catalyzes the transfer of glucose moieties from UDP-glucose in animal tissues to the nonreducing end of an α-1,4-glucan to form a new α-1,4-glucosyl–glucan:

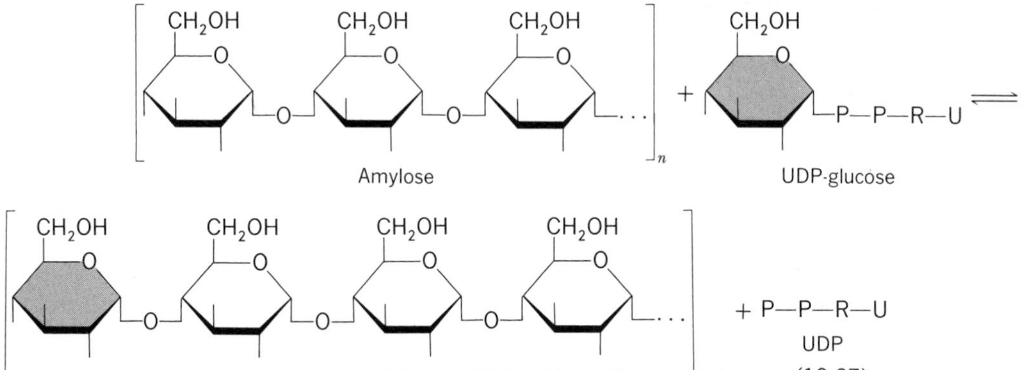

$$[\text{Amylose}]_n + \text{UDP-glucose (P—P—R—U)} \rightleftharpoons [\ldots]_{n+1} + \text{P—P—R—U (UDP)} \qquad \Delta G' = -3000 \text{ cal (pH 7.0)} \qquad (10\text{-}27)$$

The $\Delta G'$ for this reaction is comparatively large since the linkage between the C-1 carbon atom of glucose and phosphate in the UDP-glucose molecule is an energy-rich structure ($\Delta G'$ of hydrolysis is -8000). Glycogen synthase requires a primer molecule to accept the glucose units; another enzyme is required to form the α-1-6 linkage found in glycogen. This enzyme, known as an *amylotransglycosylase,* catalyzes the transfer of an oligosaccharide unit of six or seven residues in length to another point in the amylose chain to make the α-1-6 branch point:

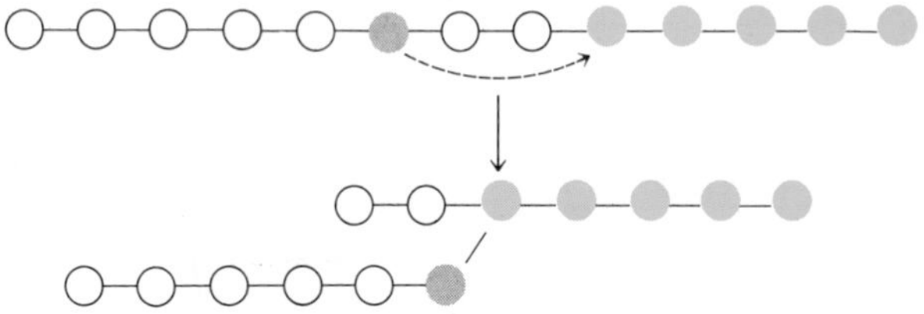

The regulation of glycogen synthesis in animal and yeast cells is at the glycogen synthase level since this enzyme has allosteric properties with glucose-6-phosphate as the positive allosteric effector. In contrast, ADP-glucose is the donor in all plant and bacterial α-1,4-glucan synthesizing systems, and ADP-glucose pyrophosphorylase is under allosteric regulation.

The complex heteropolysaccharides found in bacterial cell walls (Chapters 2 and 6) are synthesized by variations of the schemes presented here. However, because of the complex nature of the reactions, they are not discussed in this text.

The sugar nucleotides also function as important substrates for a number of enzymes which modify the sugars to important sugar derivatives found as components of polysaccharides. Three types of reactions can be listed.

(1) Epimerization of a glycosyl moiety (e.g., reaction 10-15):

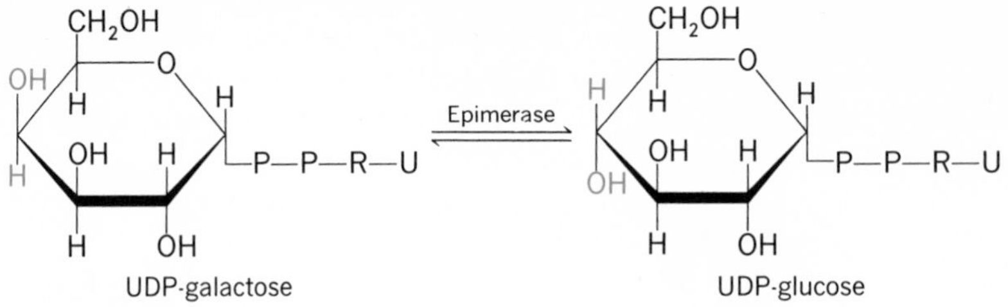

CH_2OH O OH H H OH H H H OH —P—P—R—U

UDP-galactose

Epimerase ⇌

CH_2OH O H H H OH OH H H H OH —P—P—R—U

UDP-glucose

(2) Dehydrogenation:

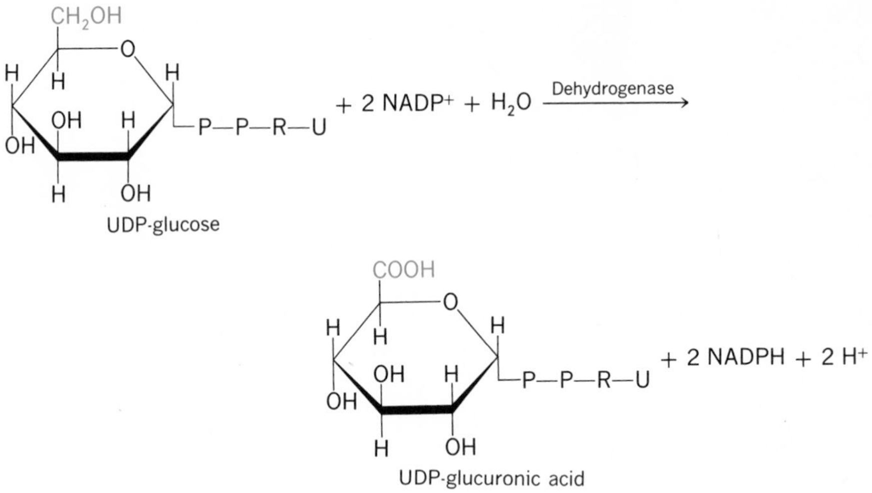

CH_2OH O H H H OH OH H H H OH —P—P—R—U

UDP-glucose

+ 2 NADP$^+$ + H_2O —Dehydrogenase→

COOH O H H H OH OH H H H OH —P—P—R—U

UDP-glucuronic acid

+ 2 NADPH + 2 H$^+$

(3) Decarboxylation:

CO_2H O H H H OH OH H H H OH —P—P—R—U

UDP-glucuronic acid

Decarboxylase →

H O H H H OH OH H H H OH —P—P—R—U

UDP-xylose

+ CO_2

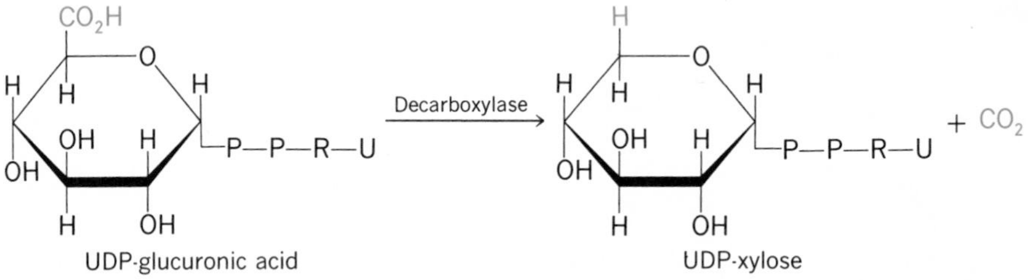

The Regulation of Glycolysis

The glycolytic sequence is probably carefully regulated in all cells and tissues. That this is so is indicated by the effect of O_2, first noted by Pasteur, on tissues that possess not only the capacity to convert glucose to lactate by glycolysis, but also can oxidize pyruvic acid completely to CO_2 and H_2O via the Krebs cycle (Chapter 13). Such tissues utilize glucose much more rapidly in the absence of O_2 than they do when O_2 is present. The functional significance of this inhibition of glucose consumption by oxygen—known as the Pasteur effect—is appreciated when we recognize that much more energy is made available as ATP when glucose is oxidized aerobically to CO_2 and

H_2O than when it is anaerobically converted only to lactic acid or alcohol and CO_2. Since more ATP is formed under aerobic conditions, less glucose needs to be consumed to do the same amount of work in the cell.

One of the regulatory processes resulting in the Pasteur effect is the control exerted by the *energy charge* of the cell. In glycolysis we have encountered the interconversion of ATP and ADP by reactions that produce and consume these compounds. There are, however, numerous reactions in lipid and nucleic acid metabolism as well as in protein synthesis in which ATP is converted to AMP instead of ADP when it provides the driving force for a particular reaction. That is, ATP undergoes a pyrophosphate cleavage (reaction 7-11) in these reactions. These three derivatives of adenosine are further interconvertible due to the presence of the enzyme *adenylic kinase* which is widely distributed in nature:

$$2\ \text{ADP} \xrightleftharpoons{\text{Adenylic kinase}} \text{ATP} + \text{AMP} \qquad (10\text{-}28)$$

This enzyme, operating from left to right, provides a mechanism for converting *half* of the ADP in a cell back to ATP which can then be used for further endergonic reactions. Obviously, in the cell the relative amounts of ATP, ADP, and AMP will depend on the metabolic activities which predominate at any one time.

Atkinson has proposed the term *energy charge* to define the energy state of the ATP–ADP–AMP system and has pointed out the analogy with the charge of an electromotive cell. He has derived an equation which permits the calculation of the energy charge provided the relative amounts of ATP, ADP, and AMP are known:

$$\text{Energy charge} = \frac{1}{2}\left(\frac{[\text{ADP}] + 2[\text{ATP}]}{[\text{AMP}] + [\text{ADP}] + [\text{ATP}]}\right)$$

From this equation we can see that the energy charge will be 1.0 when all of the AMP and ADP in the cell have been converted to ATP. This condition would be approached, for example, when a cell is carrying out oxidative phosphorylation at a rapid rate and few biosynthetic reactions were occuring. Under these conditions the maximum number of energy-rich phosphate bonds would be available in the adenylic system. When all of the adenosine compounds are present as ADP, the energy charge will be 0.5 and half as many energy-rich bonds will be contained in the adenylic system. When all the ATP and ADP have been converted to AMP, the energy charge will be 0 and the adenylic system will be devoid of energy-rich structures.

The energy charge of the cell exerts its control of metabolism through allosteric regulation of specific enzymes by ATP, ADP, and AMP. The major locus for control of glycolysis by means of the ATP–AMP system is at the interconversion of fructose-6-phosphate and fructose-1,6-diphosphate. The allosteric enzyme phosphofructokinase (reaction 10-3) is strongly inhibited by ATP but is stimulated by AMP and ADP. On the other hand, fructose diphosphate phosphatase (reaction 10-23) is stimulated by ATP and inhibited by AMP.

Consider the processes that can occur when a tissue that is capable of burning glucose completely to CO_2 and H_2O via glycolysis and the Krebs cycle is exposed to oxygen after having been functioning anaerobically. Glucose will now be oxidized to CO_2 and H_2O by the tricarboxylic acid cycle (as described in Chapter 13), and ATP will be produced from ADP and inorganic phosphate by oxidative phosphorylation (Chapter 14). The uptake of ADP together with the formation of ATP will drive reaction 10-28 from right to left, thereby using up AMP that is present. The ADP produced will in turn be oxidatively phosphorylated, and the energy charge of the cell will approach 1.0. The major effect will be that of the high concentration of ATP inhibiting phosphofructokinase and shutting down the flow of carbon from glucose to triose phosphate. (Indeed, carbon could flow from pyruvate back up the glycolytic sequence since the higher concentration of ATP would also stimulate the activity of fructose diphosphate phosphatase.) A secondary control can be expected due to the effect of the low levels of ADP and inorganic phosphate on the enzymes glyceraldehyde-3-phosphate dehydrogenase, phosphoglyceric kinase, and pyruvic kinase. Requiring as they do either inorganic phosphate or ADP, these enzymes must compete for the available and limited amounts of ADP and inorganic phosphate, and presumably will not be operating at maximum rates. Finally, the low AMP concentration resulting from the strongly aerobic state will tend to retard the action of glycogen phosphorylase (reaction 10-18) because AMP is a positive effector of this enzyme.

Should conditions change so that ATP utilization is increased, its concentration is lowered, and that of ADP and AMP are increased, the activity of phosphofructokinase and glycogen phosphorylase will increase. Similarly, a rise in AMP concentration, signaling the need for more ATP, will inhibit fructose diphosphate phosphatase, thereby impeding the flow of triose back toward polysaccharide.

It should be obvious that conditions other than high oxygen tension can increase the energy charge of a cell. For example, an abundant supply of carbohydrate or other energy-yielding substrate coupled with a deficiency of nitrogen could result in a tissue being unable to synthesize proteins. Under such conditions ATP could be expected to accumulate.

Note also that AMP does not act as a regulator only in reactions of the glycotic sequence; it can also serve as a regulator of isocitric dehydrogenase activity in the tricarboxylic acid cycle (Chapter 13). Moreover, compounds other than adenylic acid derivatives can function as regulators. Citric acid, an intermediate of the tricarboxylic acid cycle, is a negative effector of phosphofructokinase (reaction 10-3) and a positive effector of acetyl–CoA carboxylase (Chapter 12), while acetyl–CoA is a positive effector of pyruvic carboxylase (reaction 10-20). The action of these compounds can be integrated with those of the adenylic acid system in a meaningful manner, but a discussion of these effects will be deferred until Chapter 14.

In recent years it has been possible to relate the action of several hormones which influence carbohydrate metabolism with the actions of the adenylic acid effectors. An example is *glucagon*, a polypeptide of small molecular weight produced by the α cells of the islets of Langerhans in the pancreas.

Glucagon stimulates the production of cyclic AMP, a positive modulator of phosphorylase kinase kinase (page 258). Epinephrine (adrenalin), produced by the adrenal medulla, also stimulates the production of cyclic AMP. The ultimate effect is the production of phosphorylase *a* and the rapid conversion of liver and muscle glycogen to glucose and subsequently lactic acid. These hormones thus play a major role in making glucose rapidly available to tissues under stress.

The hormone insulin, secreted by the β cells of the islets of Langerhans, has a profound effect on the metabolism of carbohydrates. Although the clinical condition of insulin deficiency (diabetes) and its symptoms have been known since the last century, there is no adequate explanation of the action of this hormone. A diabetic animal is unable to metabolize carbohydrate normally and there are many associated symptoms (hyperglycemia, glucosuria, ketosis, dehydration). The administration of insulin can correct all these symptoms. One theory of the action of insulin is based on the observation that insulin increases the rate of transport of glucose across cellular membranes.

In addition to insulin, hormones secreted by the adrenal cortex, thyroxin produced by the thyroid, and hormones from the anterior pituitary can also influence carbohydrate metabolism. In the last case, the pituitary secretions exert their action indirectly by influencing the production of the cortical and thyroid hormones. The mechanism of action of these agents is highly complex and not yet well-understood.

References

1. B. Axelrod, "Glycolysis," in *Metabolic Pathways,* D. M. Greenberg, ed. 3rd ed., vol. I. New York: Academic Press, 1967.
2. M. Florkin and E. H. Stotz, eds., *Comprehensive Biochemistry*. vol. 17. New York: American Elsevier, 1967.

   These chapters in two multivolume works review the subject of glycolysis thoroughly.
3. D. E. Atkinson, "Enzymes as Control Elements in Metabolic Regulation," in *The Enzymes*. P. D. Boyer, ed. 3rd ed., vol. I. New York: Academic Press, 1970.

   A thorough discussion of the energy-charge concept together with other aspects of metabolic regulation.
4. H. A. Krebs and H. L. Kornberg, *Energy Transformations in Living Matter*. Berlin: Springer, 1957.

   A masterful survey of the energy transformations encountered in glycolysis and other metabolic routes. Required reading for the advanced student in biochemistry.

Alternate Routes of Glucose Catabolism

Introduction

The Embden–Meyerhof scheme is a major route for the anaerobic degradation of hexoses to pyruvate. In view of the great emphasis that is placed on this sequence of reactions, we must take pains to stress that this is not the only pathway for the degradation of hexoses. For example, in all the forms of plant and animal life which have now been studied it is evident that there are several routes for the metabolism of glucose. It is impossible for reasons of space, and indeed it is unnecessary, to describe all of them, but the new student of biochemistry needs to be familiar with the details of one of the more important of these routes. This is the pentose phosphate pathway, also known as the hexose monophosphate shunt.

Pentose Phosphate Pathway

It was early recognized that an alternate route existed for the metabolism of glucose. Its existence was indicated by the fact that in some tissues the classical inhibitors of glycolysis, iodoacetate and fluoride, had no effect on the utilization of glucose. In addition, the experiments of Warburg, resulting in the discovery of $NADP^+$ and the oxidation of glucose-6-phosphate to 6-phosphogluconic acid, led the glucose molecule into an unfamiliar area of metabolism. Moreover, with the advent of carbon-14, it could be shown in some instances that glucose labeled in the C-1 carbon atom was more readily oxidized to $^{14}CO_2$ than was glucose labeled in the C-6 position. If the glycolytic sequence were the only means whereby glucose could be converted to pyruvate-3-^{14}C and subsequently broken down to CO_2, then $^{14}CO_2$ should have been produced at an equal rate from glucose-1-^{14}C and glucose-6-^{14}C. These observations stimulated work, and the work has resulted in the delineation of the pentose phosphate pathway. The pathway in its entirety is shown inside the back cover of this book. The chief architects of the pathway are B. Horecker and E. Racker; among the earlier workers three who should be mentioned are Warburg, Lipmann, and Dickens.

Enzymes of the Pentose Phosphate Pathway

Glucose-6-phosphate Dehydrogenase. Warburg's discovery of this enzyme and its coenzyme $NADP^+$ is one of the classic episodes of biochemistry. From this work resulted our present knowledge that the function of a vitamin is established by its role as a constituent of a coenzyme.

Glucose-6-phosphate dehydrogenase, originally called *Zwischenferment*, catalyzes the following reaction:

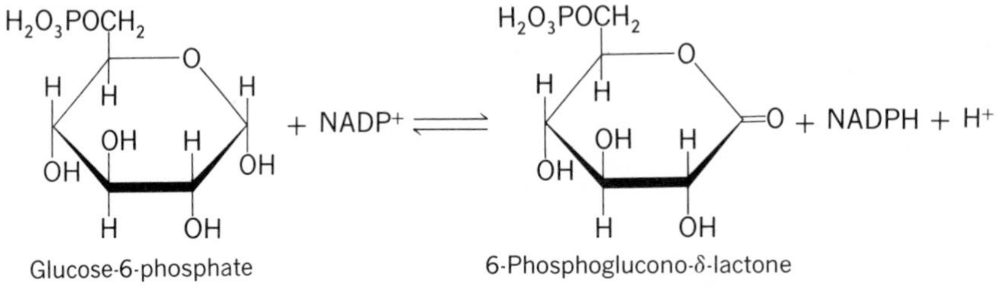

$$\text{Glucose-6-phosphate} + NADP^+ \rightleftharpoons \text{6-Phosphoglucono-}\delta\text{-lactone} + NADPH + H^+ \qquad (11\text{-}1)$$

Glucose-6-phosphate

6-Phosphoglucono-δ-lactone

Although the product was initially believed to be phosphogluconic acid, there is good evidence that the δ-lactone of this acid is the first product. The reaction is reversible because the oxidation of NADPH occurs in the presence of the enzyme and the lactone. It is easy to visualize that the oxidation of the pyranosyl form of the substrate involves the removal of two hydrogen atoms and results in the formation of the lactone. *Zwischenferment* is widely distributed in nature and in some sources appears to require a divalent cation for activity.

6-Phosphogluconolactonase. An enzyme that catalyzes the hydrolysis of the lactone produced in reaction 11-1 has been described, although the nonenzymatic hydrolysis of the lactone is extremely rapid:

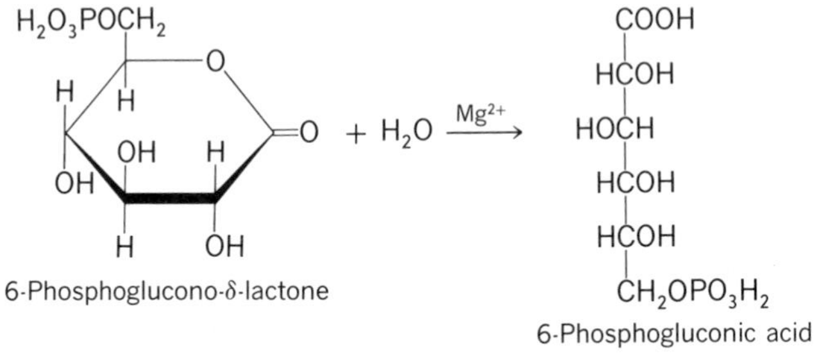

$$\text{6-Phosphoglucono-}\delta\text{-lactone} + H_2O \xrightarrow{Mg^{2+}} COOH\text{-}HCOH\text{-}HOCH\text{-}HCOH\text{-}HCOH\text{-}CH_2OPO_3H_2 \qquad (11\text{-}2)$$

6-Phosphoglucono-δ-lactone

6-Phosphogluconic acid

The $\Delta G'$ for the hydrolysis of the lactone is large; therefore, the overall oxidation of glucose-6-phosphate to phosphogluconic acid is irreversible. Moreover, the next reaction also is irreversible, and together with reactions 11-1 and 11-2 constitute the *irreversible phase* of the pentose phosphate pathway.

6-Phosphogluconic Acid Dehydrogenase. This dehydrogenase was also included in the early work of Warburg, who showed that CO_2 was a product of a crude yeast extract which contained *Zwischenferment*.

Because the reaction involves both an oxidation and decarboxylation, it was first suggested that a 3-keto-6-phosphohexonic acid might be an intermediate product prior to decarboxylation. No direct evidence in support of

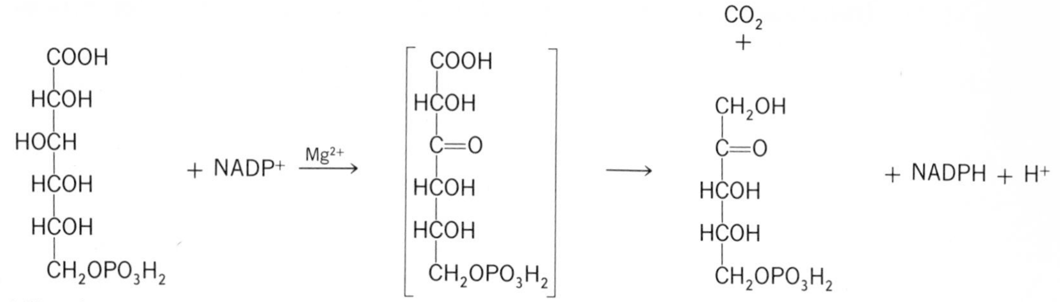

$$\begin{array}{c} COOH \\ | \\ HCOH \\ | \\ HOCH \\ | \\ HCOH \\ | \\ HCOH \\ | \\ CH_2OPO_3H_2 \end{array} + NADP^+ \xrightarrow{Mg^{2+}} \left[\begin{array}{c} COOH \\ | \\ HCOH \\ | \\ C{=}O \\ | \\ HCOH \\ | \\ HCOH \\ | \\ CH_2OPO_3H_2 \end{array} \right] \longrightarrow \begin{array}{c} CO_2 \\ + \\ CH_2OH \\ | \\ C{=}O \\ | \\ HCOH \\ | \\ HCOH \\ | \\ CH_2OPO_3H_2 \end{array} + NADPH + H^+ \qquad (11\text{-}3)$$

6-Phosphogluconic acid — 3-Ketohexonic acid (Postulated intermediate) — D-Ribulose-5-phosphate

such a compound has been offered, and the reaction is hence believed to be a single-step oxidative decarboxylation resulting in the formation of ribulose-5-phosphate. The dehydrogenase, which is widely distributed, requires Mg^{2+} or other divalent cations for activity. The reaction is not reversible.

Phosphoriboisomerase. At the level of ribulose-5-phosphate, the carbon atoms of glucose enter the second or *reversible* part of the pentose phosphate pathway; all subsequent reactions of this part are readily reversible.

Initially, ribulose-5-phosphate undergoes two isomerization reactions to form products subsequently utilized in the pathway. Phosphoriboisomerase catalyzes the interconversion of the keto sugar and the aldopentose phosphate, ribose-5-phosphate. This reaction is analogous in its action to the phosphohexose isomerase encountered in glycolysis. The K_{eq} for the reaction from left to right is approximately 3:

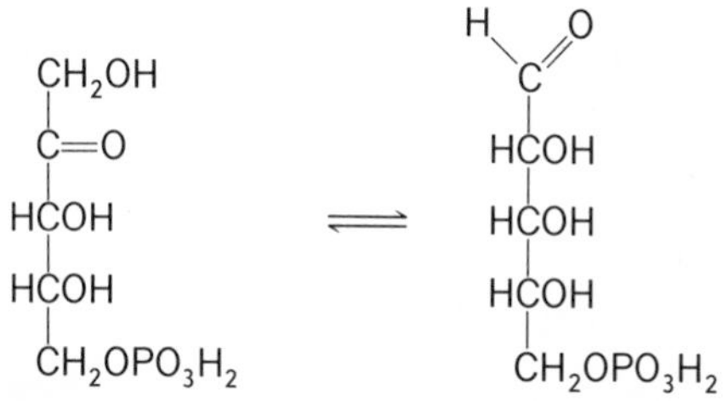

$$\begin{array}{c} CH_2OH \\ | \\ C{=}O \\ | \\ HCOH \\ | \\ HCOH \\ | \\ CH_2OPO_3H_2 \end{array} \rightleftharpoons \begin{array}{c} H \diagdown \;\; \diagup\!\!\!\!= O \\ C \\ | \\ HCOH \\ | \\ HCOH \\ | \\ HCOH \\ | \\ CH_2OPO_3H_2 \end{array} \qquad (11\text{-}4)$$

D-Ribulose-5-phosphate — D-Ribose-5-phosphate

Phosphoketopentoepimerase. The second isomerization involving ribulose-5-phosphate is catalyzed by the enzyme phosphoketopentoepimerase. The K_{eq} is 0.8:

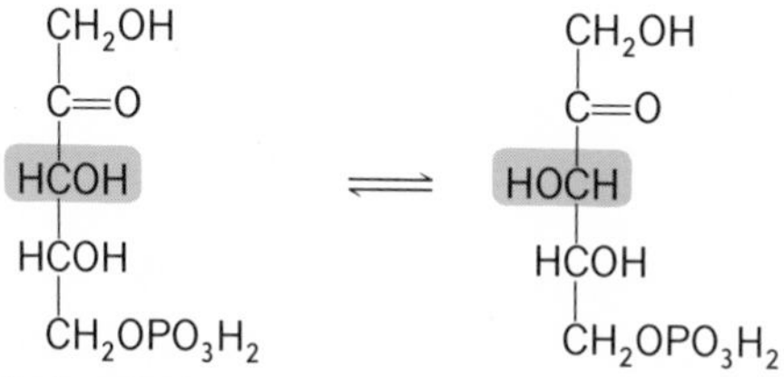

$$\begin{array}{c} CH_2OH \\ | \\ C{=}O \\ | \\ \boxed{HCOH} \\ | \\ HCOH \\ | \\ CH_2OPO_3H_2 \end{array} \rightleftharpoons \begin{array}{c} CH_2OH \\ | \\ C{=}O \\ | \\ \boxed{HOCH} \\ | \\ HCOH \\ | \\ CH_2OPO_3H_2 \end{array} \qquad (11\text{-}5)$$

D-Ribulose-5-phosphate — D-Xylulose-5-phosphate

The mechanism for this reaction is not known although it probably involves the enediol as an intermediate.

Transketolase. Up to this point the enzymes have dealt with the oxidative degradation of the hexose chain of glucose-6-phosphate and the subsequent interrelations of the pentose phosphates produced. During the period in which these reactions were being studied it was apparent that other sugars, including heptoses, tetroses, and trioses, were also formed. Some clarification of the relations between the pentoses and these other sugars resulted when the enzyme *transketolase* was discovered and described. This enzyme catalyzes the transfer of the ketol group from a donor molecule to an acceptor aldehyde. The generalized reaction may be written as

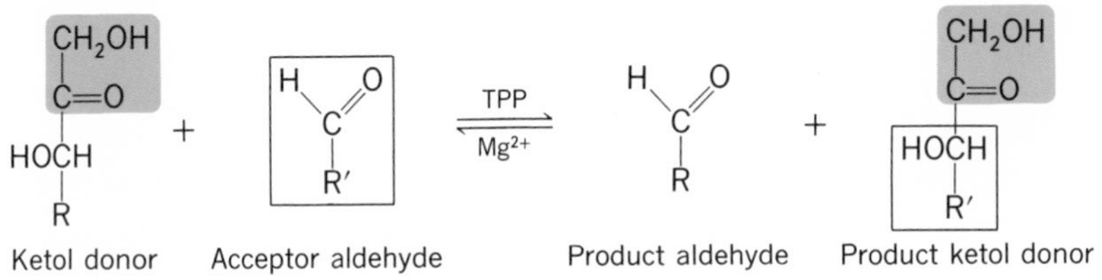

CH_2OH | C=O | HOCH | R (Ketol donor) + H–C(=O)–R′ (Acceptor aldehyde) $\underset{Mg^{2+}}{\overset{TPP}{\rightleftharpoons}}$ H–C(=O)–R (Product aldehyde) + CH_2OH | C=O | HOCH | R′ (Product ketol donor) (11-6)

Ketol donor · Acceptor aldehyde · Product aldehyde · Product ketol donor

In the specific instance, transketolase catalyzes the transfer of a ketol group from xylulose-5-phosphate to ribose-5-phosphate to form sedoheptulose-7-phosphate and glyceraldehyde-3-phosphate.

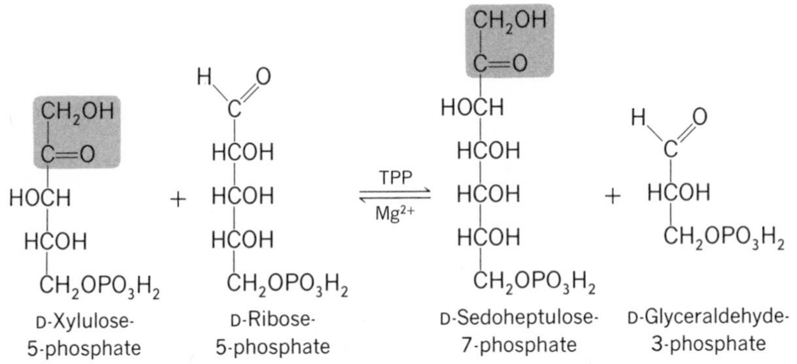

CH_2OH | C=O | HOCH | HCOH | $CH_2OPO_3H_2$ + H–C(=O) | HCOH | HCOH | HCOH | $CH_2OPO_3H_2$ $\underset{Mg^{2+}}{\overset{TPP}{\rightleftharpoons}}$ CH_2OH | C=O | HOCH | HCOH | HCOH | HCOH | $CH_2OPO_3H_2$ + H–C(=O) | HCOH | $CH_2OPO_3H_2$ (11-7)

D-Xylulose-5-phosphate · D-Ribose-5-phosphate · D-Sedoheptulose-7-phosphate · D-Glyceraldehyde-3-phosphate

Transketolase requires thiamin pyrophosphate (TPP) and Mg^{2+} as cofactors. The TPP functions because it is able to form a carbanion by dissociation of a proton at the C-2 carbon atom of the thiazole ring:

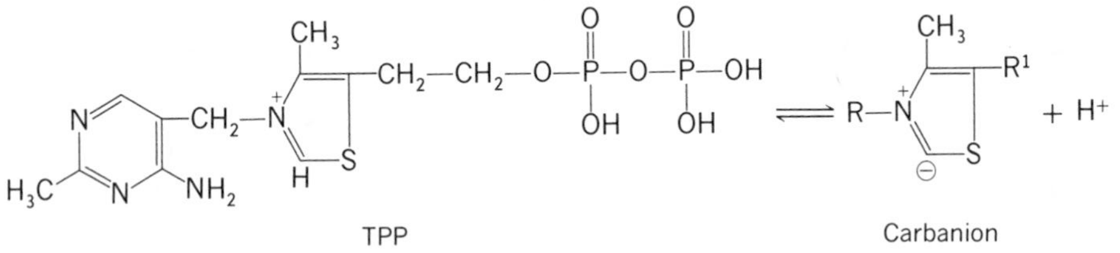

TPP (2-methyl-4-amino-pyrimidine–CH_2–thiazolium ring with CH_3 and $-CH_2-CH_2-O-P(=O)(OH)-O-P(=O)(OH)-OH$, H at C-2) $\rightleftharpoons$ Carbanion (R–$\overset{+}{N}$ thiazole ring with CH_3, R^1, $\ominus$ at C-2) + H^+

TPP · Carbanion

The resultant carbanion can, in turn, react with the ketol donor to form an addition product (I), which by appropriate rearrangement of electrons, can dissociate in another manner to form the product aldehyde and leave the ketol group on the TPP forming α,β-dihydroxyethyl thiamine pyrophosphate (II):

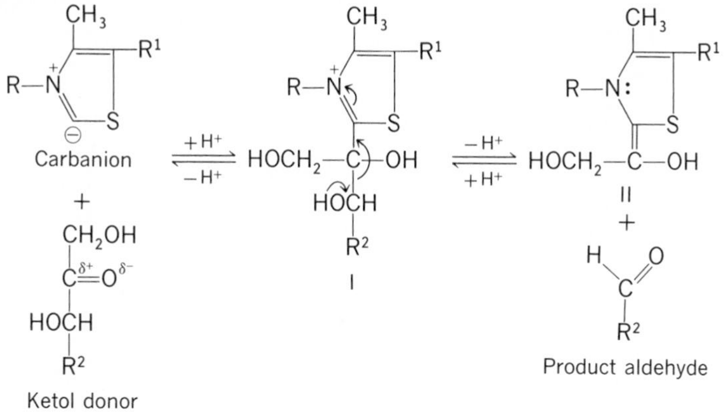

Carbanion + Ketol donor $\underset{-H^+}{\overset{+H^+}{\rightleftharpoons}}$ I $\underset{+H^+}{\overset{-H^+}{\rightleftharpoons}}$ II + Product aldehyde

The ketol–TPP addition product (II) can then react with an acceptor aldehyde to form the product ketol donor and regenerate the carbanion:

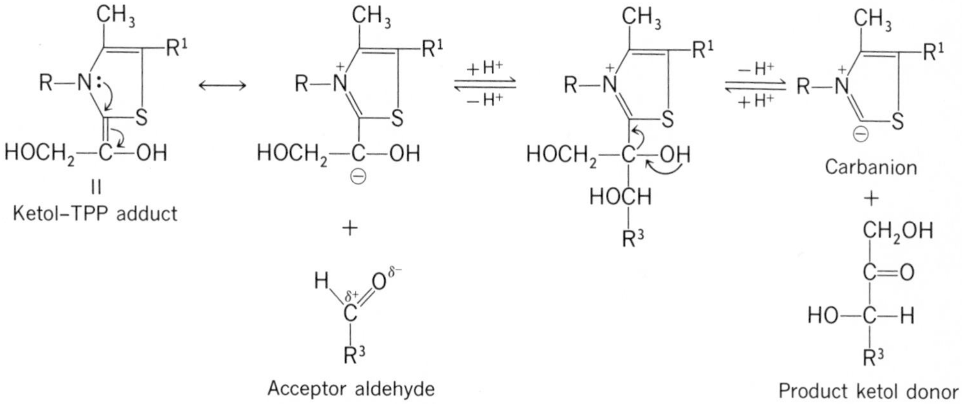

Ketol–TPP adduct $\longleftrightarrow$ + Acceptor aldehyde $\underset{-H^+}{\overset{+H^+}{\rightleftharpoons}}$ $\underset{+H^+}{\overset{-H^+}{\rightleftharpoons}}$ Carbanion + Product ketol donor

Transketolase may also catalyze the transfer of a ketol group from xylulose-5-phosphate to erythrose-4-phosphate to form fructose-6-phosphate and glyceraldehyde-3-phosphate (see inside of cover). Since this reaction as well as reaction 11-7 is readily reversible, we can list the following compounds which will serve as donor molecules and acceptor aldehydes for the enzyme:

Ketol donors (ketoses)	Acceptor aldehydes (aldoses)
D-Xylulose-5-phosphate	D-Ribose-5-phosphate
D-Fructose-6-phosphate	D-Glyceraldehyde-3-phosphate
D-Sedoheptulose-7-phosphate	D-Erythrose-4-phosphate

It is worthwhile to note that all the donor ketoses have the L configuration at the C-3 position:

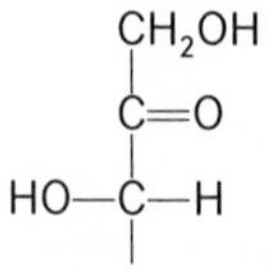

CH_2OH
$C{=}O$
$HO{-}C{-}H$

Transaldolase. This enzyme, like transketolase, functions as a transferring enzyme by catalyzing the transfer of the dihydroxy acetone moiety of fructose-6-phosphate or sedoheptulose-7-phosphate to a suitable aldose. As represented in the scheme for pentose phosphate metabolism, the acceptor aldose may be glyceraldehyde-3-phosphate or, in the reverse direction, erythrose-4-phosphate:

$$\begin{array}{c} CH_2OH \\ | \\ C{=}O \\ | \\ HOCH \\ | \\ HCOH \\ | \\ HCOH \\ | \\ HCOH \\ | \\ CH_2OPO_3H_2 \\ \text{D-Sedoheptulose-} \\ \text{7-phosphate} \end{array} + \begin{array}{c} H \diagdown \ \diagup O \\ C \\ | \\ HCOH \\ | \\ CH_2OPO_3H_2 \\ \text{D-Glyceraldehyde-} \\ \text{3-phosphate} \end{array} \rightleftharpoons \begin{array}{c} CH_2OH \\ | \\ C{=}O \\ | \\ HOCH \\ | \\ HCOH \\ | \\ HCOH \\ | \\ CH_2OPO_3H_2 \\ \text{D-Fructose-} \\ \text{6-phosphate} \end{array} + \begin{array}{c} H \diagdown \ \diagup O \\ C \\ | \\ HCOH \\ | \\ HCOH \\ | \\ CH_2OPO_3H_2 \\ \text{D-Erythrose-} \\ \text{4-phosphate} \end{array} \qquad (11\text{-}8)$$

Ribose-5-phosphate may also be an acceptor, in which case an octose, octulose-8-phosphate, is formed.

Finally, to complete the pentose phosphate pathway, the erythrose-4-phosphate produced in reaction 11-8 can accept a C_2-unit from xylulose-5-phosphate in a reaction also catalyzed by transketolase to form fructose-6-phosphate and glyceraldehyde-3-phosphate:

$$\begin{array}{c} H \diagdown \ \diagup O \\ C \\ | \\ HCOH \\ | \\ HCOH \\ | \\ CH_2OPO_3H_2 \\ \text{D-Erythrose-} \\ \text{4-phosphate} \end{array} + \begin{array}{c} CH_2OH \\ | \\ C{=}O \\ | \\ HOCH \\ | \\ HCOH \\ | \\ CH_2OPO_3H_2 \\ \text{D-Xylulose-} \\ \text{5-phosphate} \end{array} \rightleftharpoons \begin{array}{c} CH_2OH \\ | \\ C{=}O \\ | \\ HOCH \\ | \\ HCOH \\ | \\ HCOH \\ | \\ CH_2OPO_3H_2 \\ \text{D-Fructose-} \\ \text{6-phosphate} \end{array} + \begin{array}{c} H \diagdown \ \diagup O \\ C \\ | \\ HCOH \\ | \\ CH_2OPO_3H_2 \\ \text{D-Glyceraldehyde-} \\ \text{3-phosphate} \end{array} \qquad (11\text{-}9)$$

Summary and Significance of the Pentose Phosphate Pathway

Although many of the reactions of the pentose phosphate pathway were initially worked out in yeast and bacteria, the pathway exists in animals as well, where it performs, as in other organisms, two important roles. The first of these is the production of NADPH; the oxidation of glucose-6-phosphate to ribulose-5-phosphate and CO_2 produces 2 moles of NADPH for each mole of glucose ester oxidized.

In contrast to NADH, NADPH plays a special role in biosynthetic processes within the cell. Whenever a biosynthetic step involves a reduction with a nicotinamide nucleotide, the coenzyme is, with few exceptions, NADPH. For example, NADPH is specifically utilized in the biosynthesis of long-chain fatty acids. It is the reducing agent employed in the reduction of glucose to sorbitol, the reduction of dihydrofolic acid to tetrahydrofolic acid, and the reduction of glucuronic acid to L-gulonic acid. In addition, NADPH is used in the reductive carboxylation of pyruvic acid to malic acid by the malic enzyme. Finally, NADPH plays a unique role in hydroxylation reactions involved in the formation of unsaturated fatty acids, the conversion of phenylalanine to tyrosine, and the formation of certain steroids. Further evidence of the role of the pentose phosphate pathway in producing NADPH for biosynthetic purposes is found in the fact that the pathway enzymes are especially prominent in tissues such as adipose tissues, mammary gland, or adrenal cortex that carry out these biosyntheses.

The second important role of the pentose phosphate pathway is the biosynthesis of ribose-5-phosphate, an essential component of nucleotides and RNA. In this case, the ribose-5-phosphate can be formed by the action of phosphoriboisomerase on ribulose-5-phosphate produced by the irreversible portion of the pathway. Equally possible is the formation of ribose-5-phosphate from fructose-6-phosphate and glyceraldehyde-3-phosphate by reversal of reactions 11-9, 11-8, and 11-7, in that order. Thus, the cell possessing the pentose phosphate pathway has the option of producing ribose-5-phosphate by an oxidative route or by a nonoxidative route.

The interconversion of triose, tetrose, pentose, hexose, and heptose phosphate esters encountered in reactions 11-7, 11-8, and 11-9 can be confusing. Another way of representing the reactions involving these compounds is

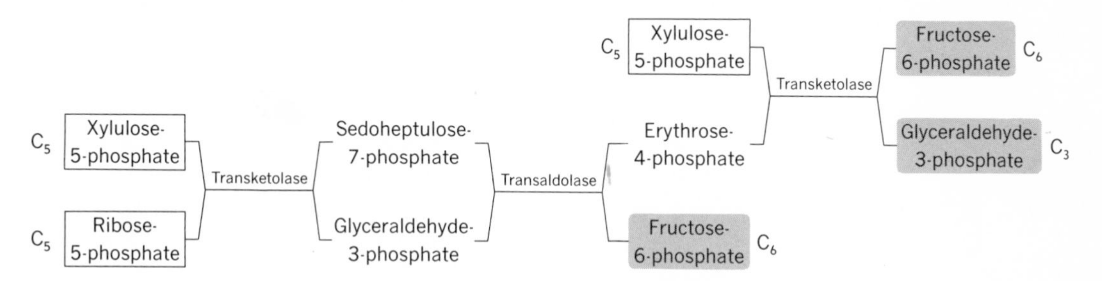

By drawing rectangles around three pentose molecules which can be considered as reactants in this scheme and by shading three molecules produced thereby, we see that this scheme constitutes a readily reversible mechanism for making hexose and triose intermediates from pentoses derived from the oxidation of glucose-6-phosphate. That is, from left to right, 15 carbon atoms in 3 pentose molecules give rise to 15 carbon atoms in 2 molecules of hexose and 1 of triose. In the reverse direction the scheme can account for the formation of pentose derivatives from intermediates of glycolysis. As we shall see, these reactions together with reaction 11-5 are intimately involved in the assimilation of CO_2 during photosynthesis.

Entner-Doudoroff Pathway

Some bacteria (e.g., *Pseudomonads, Azotobacter* sp.) lack phosphofructokinase and therefore cannot degrade glucose by the glycolytic sequence. These organisms instead initiate glucose catabolism by producing 6-phosphogluconic acid by reactions 11-1 and 11-2. The acid then undergoes a dehydration and rearrangement to form an α-ketodeoxy sugar phosphate which in turn is cleaved by an aldolase–type enzyme into pyruvate and glyceraldehyde-3- phosphate:

$$\begin{array}{c} COOH \\ | \\ HCOH \\ | \\ HOCH \\ | \\ HCOH \\ | \\ HCOH \\ | \\ CH_2OPO_3H_2 \end{array} \xrightarrow[\text{Dehydrase}]{-H_2O} \begin{array}{c} COOH \\ | \\ C{=}O \\ | \\ CH_2 \\ | \\ HCOH \\ | \\ HCOH \\ | \\ CH_2OPO_3H_2 \end{array} \xrightarrow{\substack{\text{An} \\ \text{Aldolase}}} \begin{array}{cl} COOH & \\ | & \\ C{=}O & \text{Pyruvate} \\ | & \\ CH_3 & \\ + & \\ CHO & \\ | & \\ HCOH & \text{Glyceraldehyde-3-phosphate} \\ | & \\ CH_2OPO_3H_2 & \end{array}$$

Modification of this scheme permits other sugars (galactose) and sugar acids (D-glucuronic acid, D-galacturonic acid) to be metabolized, but an essential feature is the production of a 2-keto-3-deoxy intermediate which can be cleaved after phosphorylation.

Other routes are known for the metabolism of glucose and other sugars, but they are beyond the purpose of this text.

References

1. B. Axelrod, "Other Pathways of Carbohydrate Metabolism," in *Metabolic Pathways*. D. M. Greenberg, ed. 3rd ed. New York: Academic Press, 1967.
   A well-written review of the pentose phosphate pathway as well as other routes for the metabolism of glucose and related sugars.
2. R. Y. Stanier, M. Doudoroff, and E. A. Adelberg, *The Microbial World*. 3rd ed. Englewood Cliffs, N.J.: Prentice-Hall, 1970.
   This standard text of microbiology has several chapters devoted to metabolism, including the diverse metabolic schemes utilized by microorganisms.
3. B. L. Horecker, *Pentose Metabolism in Bacteria*. New York: Wiley, 1962.
   A summary of the pentose cycle by a major contributor to the field.

12

Lipid Metabolism

Introduction

In both plants and animals lipids are stored in large amounts as neutral, highly insoluble triacyl glycerols; they can be rapidly mobilized and degraded to meet the cell's demands for energy. In the complete combustion of a typical fatty acid, palmitic acid, there is a large negative free energy change:

$$C_{16}H_{32}O_2 + 23\ O_2 \longrightarrow 16\ CO_2 + 16\ H_2O$$
$$\Delta G' = -2338 \text{ kcal/mole}$$

This negative change is due to the oxidation of the highly reduced hydrocarbon radical attached to the carboxyl group of the fatty acid. Of all the common foodstuffs, only the long-chain fatty acids possess this important chemical feature. Thus, lipids have quantitatively the best caloric value of all foods.

Lipids also function as important insulators of delicate internal organs. Nerve tissue, cell membranes, and membranes of subcellular particles such as mitochondria, microsomes, and nuclei have complex lipids as essential components. In addition, the vital electron transport system in mitochondria and the intricate structures found in chloroplasts, the site of photosynthesis, contain lipid derivatives in their basic architecture.

As we have indicated, the chief storage form of available energy in the animal cell is the lipid molecule. When the caloric intake exceeds utilization, excess food is invariably stored as fat; the body cannot store any other form of food in such large amounts. Carbohydrates are converted to glycogen, for example, but the capacity of the body to store this polysaccharide as a potential source of energy is strictly limited. In a normal liver the average amount of glycogen is 5–6% of the total weight, and in skeletal muscle the glycogen content averages only 0.4–0.6%. Blood glucose, a source of glycogen units, is present at a level of 60–100 mg per 100 ml of whole blood. Only under pathological conditions are these values drastically altered. The normal animal therefore very carefully regulates, by hormonal and metabolic controls, the carbohydrate concentration in its various tissues, and this class of compound can serve only to a limited extent as a storage form of energy.

Proteins, the third major class of foodstuffs, differ considerably from carbohydrates and fats in their biological function; they serve as a source of twenty-odd amino acids required for *de novo* protein synthesis and as a source of the carbon skeletons essential for the synthesis of purines, pyrimidines, and other nitrogenous compounds. Moreover, in an adult organism in which active growth has ceased, nitrogen output is more or less geared to nitrogen intake, and the organism shows no tendency to store surplus proteins from the diet.

History

As early as 1882, Shotten fed the sodium salts of fatty acids to animals and examined the urine for fatty acid derivatives. The higher members of the fatty acids were completely oxidized, but the animals excreted large amounts of acetic acid when they were fed acetic acid. In 1904, Knoop conceived the now classical idea of introducing the phenyl structure, which was not easily changed in the body, into the terminal methyl groups of the fatty acids. In this manner the degradation products of the fatty acid could be easily identified in the urine.

Phenyl derivatives of fatty acids containing from one to five carbon atoms were administered to dogs, and the excretion products in the urine were examined. Knoop discovered that the phenyl derivatives of the even-numbered fatty acids always led to the excretion of phenylacetic acid (actually the glycine conjugate, phenylaceturic acid), whereas the phenyl derivatives of the odd-numbered fatty acids were degraded to benzoic acid (that is, to the glycine conjugate, hippuric acid). From these results Knoop postulated that a successive removal of C_2 units could readily explain the experimental observations. Termed the *β-oxidation* theory, this postulate has played a dominant role in lipid metabolism. The success of his experimental approach was also a strong impetus to apply stable isotopes and radioisotopes to similar biochemical problems.

Odd:

C_6H_5—C⫶C—C⫶C—COOH →→ C_6H_5—COOH + 2 "C_2"

Benzoic acid

Even:

C_6H_5—C—C⫶C—C⫶C—COOH →→ C_6H_5—C—COOH + 2 "C_2"

Phenylacetic acid

Knoop's classic experiments stimulated much work in the field, but the science of modern biochemistry had not developed sufficiently to handle the complex problem of fatty acid oxidation. Then, in 1943, Munoz and Leloir of Argentina showed that homogenized guinea pig liver could oxidize butyric acid, provided that adenosine-5′-phosphate, inorganic phosphate, magnesium ion, cytochrome c, and succinic acid were present. When radioactive carbon (^{14}C) was discovered by Kamen soon after World War II and new fractionation procedures for cell homogenates were developed by Albert Claude, the solution to Knoop's 1904 observations was at hand. The following observations, listed chronologically, proved of great importance to the ultimate solution:

(1) In 1944, Weinhouse incubated octanoic acid-1-^{14}C with rat liver slices and isolated acetoacetic acid with a ^{14}C label in both the carboxyl and the carbonyl carbons of the acid. This observation strongly suggested that the even-chain fatty acids were degraded to a C_2 unit that could then combine with another C_2 unit to form acetoacetic acid.
(2) In 1950, Lehninger and Kennedy demonstrated that the exclusive site of fatty acid oxidation was the subcellular unit, the mitochondrion. All other subcellular fractions were inert.
(3) In 1950, Stadtman and Barker demonstrated a completely water-soluble enzyme system from *Clostridium kluvyeri* which could catalyze either the degradation or synthesis of fatty acids.
(4) The discovery of coenzyme A by Lipmann and of acetyl–coenzyme A by Lynen at Munich opened wide the door to the rapid solution of the problem of fatty acid degradation.

The "C_2" Unit

Lynen isolated acetyl–coenzyme A (acetyl–CoA) from a yeast suspension which was oxidizing ethanol aerobically. With the elucidation of the structure of acetyl–CoA, the unique chemistry of the "C_2" unit became evident. The structure is depicted as

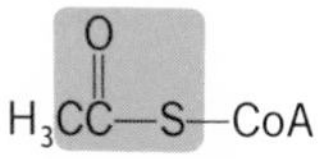

$$H_3C\overset{\overset{\displaystyle O}{\|}}{C}\text{—}S\text{—}CoA$$

and represents a class of compounds known as thioesters. As discussed in Chapters 7 and 9, thioesters are known as energy-rich compounds because they have a relatively large negative $\Delta G'$ of hydrolysis of approximately -8000 cal/mole. Some of their unique properties are now described.

Thioesters differ from oxygen esters in several important ways. Oxygen esters can exist in two forms that are stabilized by resonance:

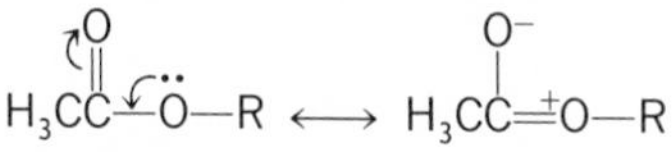

$$H_3C\overset{\overset{\displaystyle O}{\|}}{C}\text{—}\ddot{O}\text{—}R \longleftrightarrow H_3C\overset{\overset{\displaystyle O^-}{|}}{C}=\overset{+}{O}\text{—}R$$

Oxygen ester: resonance-stabilized

The sulfur of a thioester does not readily release its electrons for double bond formation, and cannot exist in the resonance-stabilized form found with the oxygen esters. Instead, because both the carboxyl oxygen and the sulfur atom are electronegative, thioesters possess considerable carbonyl character and may be depicted as

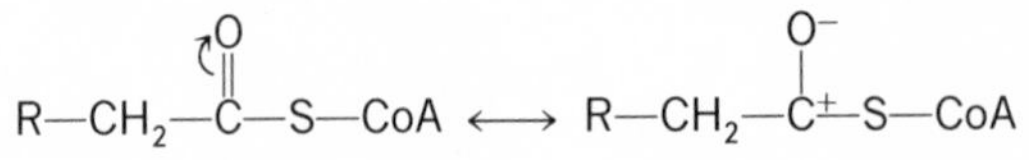

$$R{-}CH_2{-}\overset{\overset{O}{\|}}{C}{-}S{-}CoA \longleftrightarrow R{-}CH_2{-}\overset{\overset{O^-}{|}}{C^{+}}{-}S{-}CoA$$

Since the carbonyl carbon possesses a fractional positive charge (δ^+), one of the hydrogens in the α-methylene carbon will tend to dissociate as a proton leading to a potential carbanion in which the α-carbon has a fractional negative charge (δ^-). We see in the diagram that acetyl-CoA has a unique

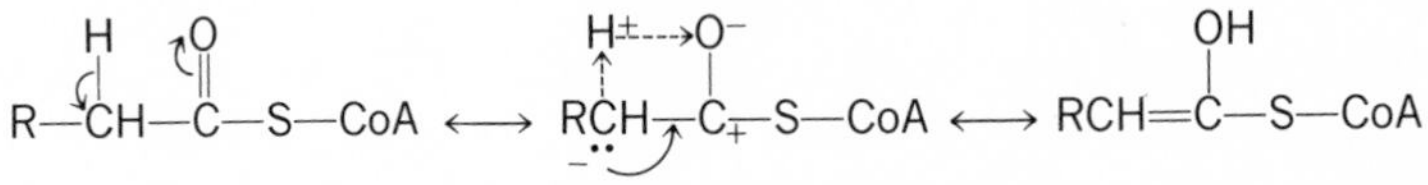

$$R{-}\overset{\overset{H}{|}}{C}H{-}\overset{\overset{O}{\|}}{C}{-}S{-}CoA \longleftrightarrow R\overset{H^{\pm}\cdots\to O^-}{\ddot{C}H^{-}{-}C^{+}}{-}S{-}CoA \longleftrightarrow RCH{=}\overset{\overset{OH}{|}}{C}{-}S{-}CoA$$

structure in that a *nucleophile* such as H_2O, R—S:$^-$, or the α-carbon of acetyl-CoA can attack the site of the fractional positive charge. In addition, an *electrophile* such as CO_2 or the carbonyl carbon of acetyl-CoA can also approach the site of the negative charge with its pair of unshared electrons. With these basic facts in mind we are now prepared to examine the modern concept of β-oxidation.

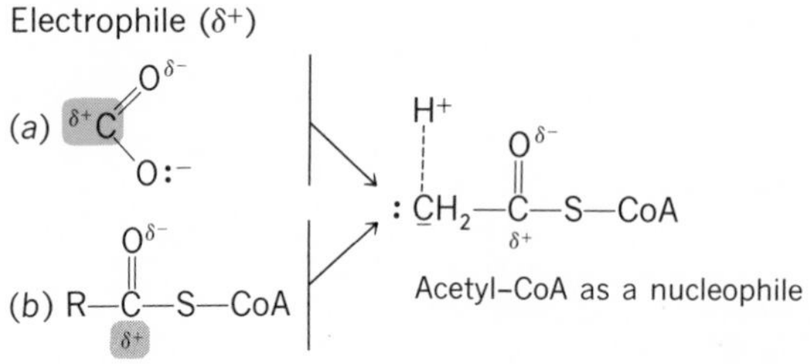

Electrophile (δ^+)

(*a*) $^{\delta+}C(=O^{\delta-})O:^-$

(*b*) $R{-}\overset{\overset{O^{\delta-}}{\|}}{\underset{\delta+}{C}}{-}S{-}CoA$

→ $H^+ \; : \underline{C}H_2{-}\overset{\overset{O^{\delta-}}{\|}}{\underset{\delta+}{C}}{-}S{-}CoA$

Acetyl-CoA as a nucleophile

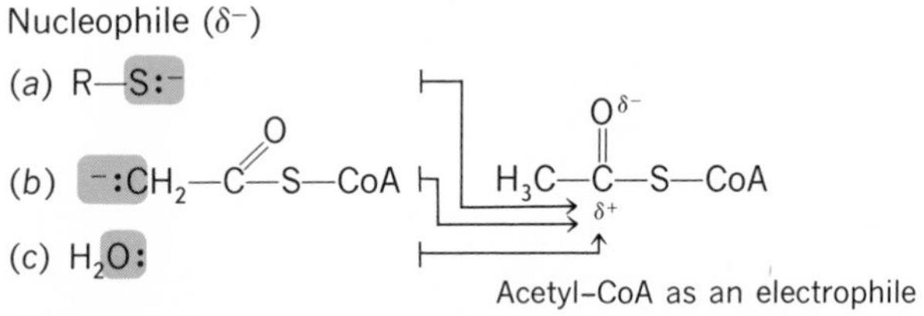

Nucleophile (δ^-)

(*a*) R—S:$^-$

(*b*) $^-$:$CH_2{-}\overset{\overset{O}{\|}}{C}{-}S{-}CoA$

(*c*) H_2O:

→ $H_3C{-}\overset{\overset{O^{\delta-}}{\|}}{\underset{\delta+}{C}}{-}S{-}CoA$

Acetyl-CoA as an electrophile

β-Oxidation Scheme

In 1952, Green in Wisconsin and Lynen in Munich announced the separation, isolation, and purification of the five enzymes responsible for the β-oxidation of fatty acids. They are:

Acyl-CoA synthetase type of reaction:

$$R{-}COOH + ATP + CoA{-}SH \xrightleftharpoons{Mg^{2+}} R{-}\overset{\overset{O}{\|}}{C}{-}S{-}CoA + AMP + P{-}P$$

Acyl–CoA dehydrogenase type of reaction:

$$RCH_2CH_2\overset{O}{\overset{\|}{C}}—S—CoA + FAD \rightleftharpoons \underset{\beta}{RCH}\overset{trans}{=}\underset{\alpha}{CH}—\overset{O}{\overset{\|}{C}}—S—CoA + FADH_2$$

Enoyl–CoA hydrase type of reaction:

$$RCH{=}CH\overset{O}{\overset{\|}{C}}—S—CoA + H_2O \rightleftharpoons \underset{\text{L}(+)\text{-}\beta\text{-Hydroxyacyl–CoA}}{RCHOHCH_2—\overset{O}{\overset{\|}{C}}—S—CoA}$$

β-Hydroxyacyl–CoA dehydrogenase type of reaction:

$$RCHOHCH_2\overset{O}{\overset{\|}{C}}—S—CoA + NAD^+ \rightleftharpoons \underset{\beta\text{-Ketoacyl–CoA}}{RCOCH_2\overset{O}{\overset{\|}{C}}—S—CoA} + NADH + H^+$$

β-Ketoacyl–CoA thiolase type of reaction:

$$R—\underset{O}{\underset{\|}{C}}—CH_2—\overset{O}{\overset{\|}{C}}—S—CoA + CoA—SH \rightleftharpoons R—\underset{O}{\underset{\|}{C}}—S—CoA + H_3C—\overset{O}{\overset{\|}{C}}—S—CoA$$

Here R is the saturated aliphatic chain of the fatty acids. These five reactions are integrated into the helical scheme (Figure 12-1). Each turn of the cycle removes a two-carbon unit.

There are several important features of Figure 12-1 that should be emphasized:

(1) Only the CoA derivatives of the fatty acids serve as substrates for the enzymes.
(2) The free energy of hydrolysis of the thioester is of the order of −8000 cal, which places this type of ester in the group of high-energy or energy-rich compounds. The driving potential built into the thioester bond confers on these compounds some of their unique properties.
(3) Only one molecule of ATP is required to activate a fatty acid for its complete degradation to acetyl–CoA regardless of the number of carbon atoms in its hydrocarbon chain. In other words, whether we wish to oxidize either a C_4 acid or a C_{16} acid, only one equivalent of ATP is needed for activation. This makes for great economy and efficiency in the oxidation of fatty acids.
(4) Several derivatives of vitamins such as riboflavin, pantothenic acid, nicotinamide, adenine nucleotide, and trace metals such as Mn^{2+} or Mg^{2+} play essential roles in fatty acid oxidation. If there were a deficiency in any one of these substances, serious blocks would occur in the degradation of fatty acids.

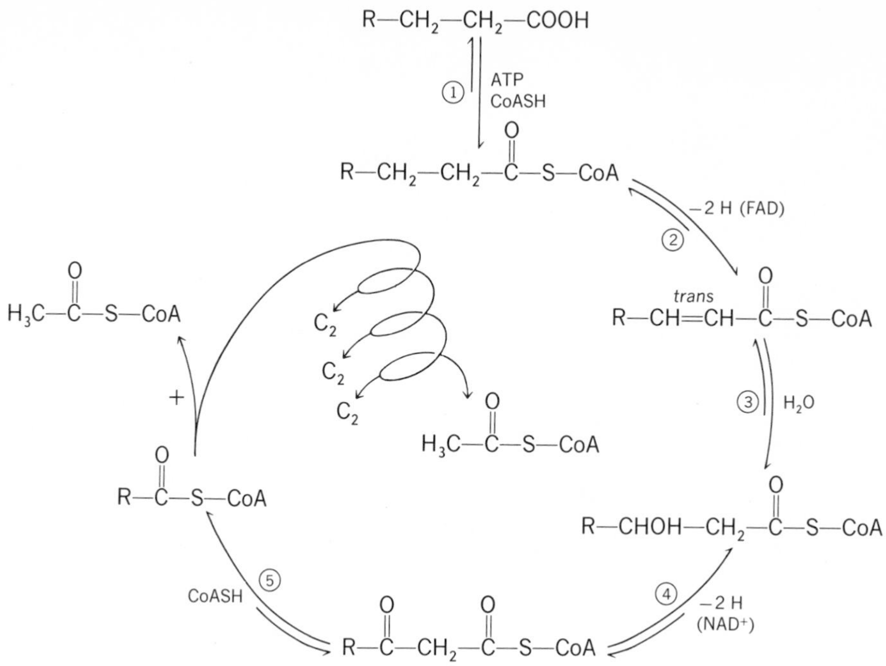

R—CH_2—CH_2—COOH

(1) ATP, CoASH

R—CH_2—CH_2—C(=O)—S—CoA

(2) −2 H (FAD)

R—CH=CH—C(=O)—S—CoA (*trans*)

(3) H_2O

R—CHOH—CH_2—C(=O)—S—CoA

(4) −2 H (NAD^+)

R—C(=O)—CH_2—C(=O)—S—CoA

(5) CoASH

R—C(=O)—S—CoA + H_3C—C(=O)—S—CoA

Figure 12-1

The β-oxidation helical scheme: (1) fatty acid–CoA synthetase; (2) fatty acyl–CoA dehydrogenases; (3) enoyl–CoA hydrase; (4) β-hydroxyacyl–CoA dehydrogenase; (5) β-ketoacyl–CoA thiolase.

(5) All enzymes associated with the β-oxidation system are localized in the inner membranes and the matrix of animal mitochondria. Since the inner membrane is also the site of the electron transport and oxidative phosphorylation systems, this arrangement is of fundamental importance to the efficient release and conservation of the potential energy stored in the long-chain fatty acid. When acetyl–CoA is produced in the breakdown of fatty acids, it may be subsequently oxidized to CO_2 and H_2O by means of the tricarboxylic acid cycle enzymes which are localized as soluble enzymes in the matrix. An unusual property of animal mitochondria is their inability to oxidize fatty acids or fatty acyl–CoA's unless (−)-carnitine (3-hydroxy-4-trimethyl ammonium butyrate) is added in catalytic amounts. Evidently, free fatty acids or fatty acyl–CoA's cannot penetrate the inner membranes of animal mitochondria, whereas palmityl carnitine readily passes through the membrane and is then converted to palmityl–CoA in the matrix. The following steps catalyze the translocation of palmityl–CoA from outside the mitochondrion to the internal site of the β-oxidation system.

Outer membrane:

$$\text{Fatty acid} + \text{ATP} + \text{CoA} \xrightarrow[\text{synthetase}]{\text{Fatty acid}} \text{Acyl–CoA} + \text{AMP} + \text{PP}$$

Inner membrane:

$$\text{Acyl–CoA} + \text{Carnitine} \underset{\text{Acyltransferase}}{\overset{\text{Acyl–CoA:Carnitine}}{\rightleftharpoons}} \text{Acyl carnitine} + \text{CoA}$$

Inner face of inner membrane:

$$\text{Acyl carnitine} + \text{CoA} \underset{\text{Acyltransferase}}{\overset{\text{Acyl–CoA:Carnitine}}{\rightleftharpoons}} \text{Acyl–CoA} + \text{Carnitine}$$

$$\downarrow \beta\text{-Oxidation (Matrix enzymes)}$$

$$\text{Acetyl–CoA}$$

(6) The β-oxidative system is found in all organisms. However, in bacteria grown in the absence of fatty acids, the β-oxidative system is practically absent but is readily induced by the presence of fatty acids in the growth medium. The bacterial β-oxidation system is completely soluble and hence is not membrane-bound. Curiously, in germinating seeds possessing a high lipid content, the β-oxidation system is exclusively located in microbodies called glyoxysomes (see Chapter 6), but in seeds with a low lipid content, the enzymes are associated with mitochondria. The important function of the glyoxysomes is considered in more detail in Chapter 6.

The universality of the system implies the prime importance of this sequence as a means of degrading fatty acids.

Specialized Oxidative Systems

While the β-oxidative system undoubtedly is the primary mechanism for degrading fatty acids, the student should be aware of a number of other systems which attack the hydrocarbon chain oxidatively. A brief survey of these mechanisms and possible functions will now be given.

α-Oxidation. This system was observed first in a number of germinating seeds, later in leaf tissue and mammalian liver and brain tissue. Fatty acids can undergo a stepwise α-oxidation sequence as illustrated in Scheme 12-1.

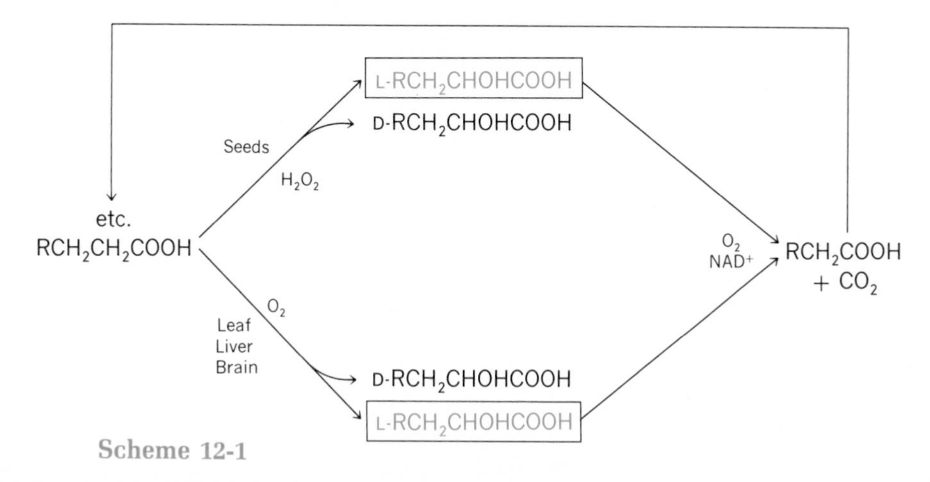

Scheme 12-1

Whereas the seed system appears to be a fatty acid peroxidase with H_2O_2 as the hydroxylating reagent, the leaf and mammalian systems employ molecular oxygen to hydroxylate the α-carbon of fatty acids to an L-α-hydroxy fatty acid. Interestingly, the plant systems form both the D and the L isomers of an α-hydroxy fatty acid. The L isomer is the substrate for further degradation, but the D isomer accumulates. Notice that NAD^+ is required in the reaction sequence for the further degradation of the L-hydroxy fatty acid. The α-oxidation system has recently been shown to play a key role in the capacity

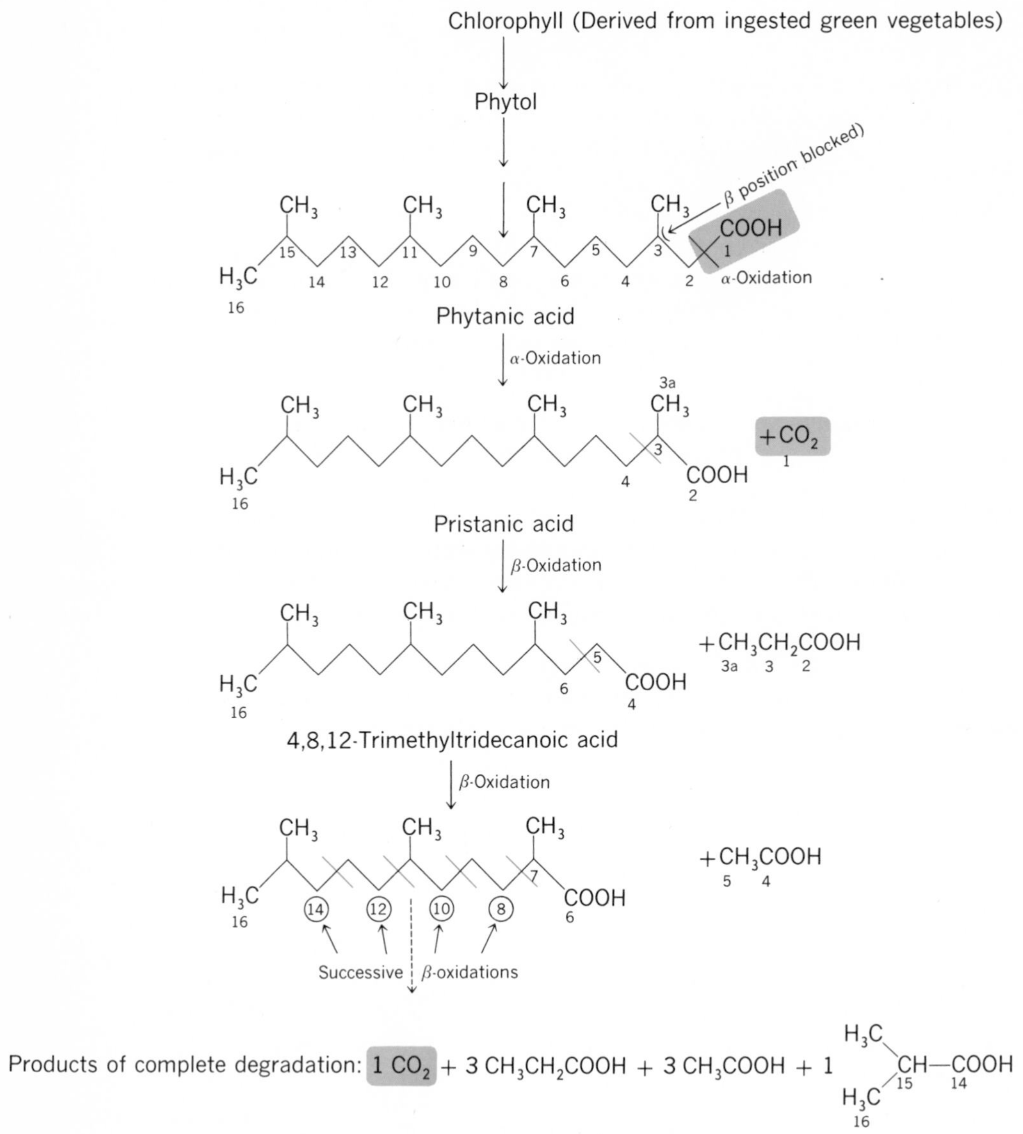

Chlorophyll (Derived from ingested green vegetables)

↓

Phytol

↓

Phytanic acid (β position blocked); α-Oxidation

↓ α-Oxidation

Pristanic acid $+ CO_2$

↓ β-Oxidation

4,8,12-Trimethyltridecanoic acid $+ CH_3CH_2COOH$

↓ β-Oxidation

$+ CH_3COOH$

Successive β-oxidations

Products of complete degradation: $1\ CO_2 + 3\ CH_3CH_2COOH + 3\ CH_3COOH + 1\ (H_3C)_2CH{-}COOH$

Scheme 12-2

of mammalian tissues to oxidize phytanic acid, the oxidation product of phytol, to CO_2 and water. Normally, phytanic acid is rarely found in serum lipids because of the ability of normal tissue to degrade the acid very rapidly. It has now been observed that patients with Refsum's disease, a rare inheritable disease, have lost their α-oxidation system and hence their normally functioning β-oxidation system cannot cope with the degradation of the phytanic acid. It is believed that the sequence shown in Scheme 12-2 explains the disease on a molecular level. α-Oxidation therefore makes possible the bypassing of blocking groups in a hydrocarbon chain which otherwise could prevent the participation of the β-oxidation system. α-Oxidation also synthesizes α-hydroxy fatty acids which are components of complex lipids.

ω-Oxidation. A number of aerobic bacteria have been isolated from oil-soaked soil which rapidly degrade hydrocarbons or fatty acids to water-soluble products. The reactions involve an initial hydroxylation of a terminal methyl group to a primary alcohol and subsequent oxidation to a carboxylic acid (Scheme 12-3). Thus, straight-chain hydrocarbons are oxidized to fatty acids and fatty acids in turn are β-oxidized to acetyl–CoA. Hepatic tissues also rapidly catalyze the oxidation of hexanoic, octanoic, decanoic, and lauric acids to corresponding dicarboxylic acids via a similar mechanism. These series of reactions, which at first glance were of mild interest, now have assumed an extremely important scavenging role in the bacterial biodegradation of both detergents derived from fatty acids and even more important the large amounts of oil spilled over the ocean surface. It has been estimated that the rate of bacterial oxidation of floating oil under aerobic conditions may be as high as 0.5 g/day per square meter of oil surface. The mechanism of the oxidation of oils is primarily by the ω-oxidation mechanism.

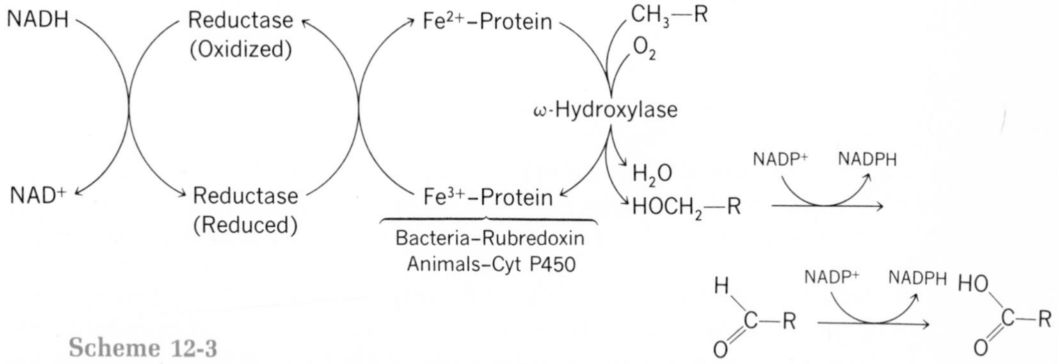

NADH → NAD^+; Reductase (Oxidized) ⇄ Reductase (Reduced); Fe^{2+}–Protein ⇄ Fe^{3+}–Protein (Bacteria–Rubredoxin; Animals–Cyt P450); CH_3—R + O_2 → ω-Hydroxylase → H_2O + $HOCH_2$—R

$HOCH_2$—R → ($NADP^+$ → NADPH) → HC(=O)—R → ($NADP^+$ → NADPH) → HOC(=O)—R

Scheme 12-3

Oxidation of Unsaturated Fatty Acids. Although the β-oxidation system readily explains the degradation of saturated fatty acids, it offers no explanation for the oxidation of mono- or polyunsaturated fatty acids. W. Stoffel of the University of Cologne has recently resolved this problem by describing two important enzymes which make possible the β-oxidation of these acids (Scheme 12-4).

With these enzymes incorporated in an extended β-oxidation scheme, the

I

Inactive substrate in normal β-oxidation system

$\Delta^{3\,cis}$-$\Delta^{2\,trans}$-Enoyl-CoA isomerase

Normal β-oxidation substrate

II

(a)

$+H_2O$ Enoyl-CoA hydrase

D(−)-3-Hydroxy acyl-CoA

Inactive β-oxidation substrate

(b) D(−)-3-Hydroxy acyl-CoA

D(−)-3-Hydroxy acyl-CoA epimerase

L(+)-3-Hydroxy acyl-CoA

Normal β-oxidation substrate

Scheme 12-4

student can readily construct series of reactions for the β-oxidations of oleic, linoleic, and α-linolenic acids. For example, with linoleic acid as substrate, we would employ three normal β-oxidation enzymes for three cycles (3 C_2), then use reaction I; again two more β-oxidation cycles (2 C_2), then reactions II*a* and *b*; and conclude with three cycles of the β-oxidation scheme (4 C_2).

Oxidation of Odd-Number Carbon Chain Fatty Acids, viz., Propionic Acid. The oxidation of propionic acid presents an interesting problem, since at first glance the acid would appear to be a substrate unsuitable for β-oxidation. However, the substrate is handled by two strikingly dissimilar mechanisms, the first pathway is found only in animal tissues and some bacteria and involves biotin and vitamin B_{12}, while the second pathway, found widespread in plants, is a modified β-oxidation pathway (Scheme 12-5).

The plant pathway, which is ubiquitous, nicely resolves the problem of how plants can cope with propionic acid, the product of oxidation degradation of valine and isoleucine, by a system not involving vitamin B_{12} (a cobalamine

I *The animal system:*

$$\text{Propionyl–CoA} + CO_2 + \text{ATP} \xrightarrow[①]{Mg^{2+}} \text{L-Methylmalonyl–CoA}$$

$$\updownarrow ②$$

$$\text{D-Methylmalonyl–CoA}$$

$$\updownarrow ③$$

$$\text{Succinyl–CoA} \longrightarrow \text{etc.}$$

① Propionyl–CoA carboxylase (A biotinyl enzyme)
② Methylmalonyl–CoA racemase
③ Methylmalonyl–CoA mutase (A cobalamine (B_{12}) enzyme)

II *The plant system:*

$$\text{Propionyl–CoA} \underset{①}{\rightleftharpoons} \text{Acrylyl–CoA}$$

$$\updownarrow ②$$

$$\beta\text{-Hydroxypropionyl–CoA}$$

$$\downarrow ③$$

$$\text{Malonyl semialdehyde} \overset{④}{\rightleftharpoons} \beta\text{-Hydroxypropionic acid}$$

$$\text{Malonyl semialdehyde} \xrightarrow[\text{CoA}]{\text{NAD}^+}⑤ \text{Malonyl–CoA}$$

$$\text{Malonyl–CoA} \xrightarrow{⑥} CO_2 + \text{Acetyl–CoA}$$

① Acyl–CoA dehydrogenase
② Enoyl–CoA hydrase
③ Acyl–CoA thioesterase
④ β-Hydroxypropionic dehydrogenase
⑤ Malonyl semialdehyde dehydrogenase
⑥ Malonyl–CoA decarboxylase

Scheme 12-5

coenzyme). Since plants have no B_{12} functional enzymes, the animal system is absent; thus, the modified β-oxidation system of plant tissues bypasses the B_{12} barrier in an effective manner.

Energetics of β-Oxidation

In the total combustion of palmitic acid, considerable energy is released:

$$C_{16}H_{32}O_2 + 23\ O_2 \longrightarrow 16\ CO_2 + 16\ H_2O$$
$$\Delta G' = -2338\ \text{kcal/mole}$$

How much of this potential energy is actually made available to the cell? When palmitic acid is degraded enzymically, one energy-rich bond of ATP is required for the primary activation, and eight energy-rich thioester bonds are formed. Each time the helical cycle (Figure 12-1) is traversed, 1 mole of FAD–H_2 and 1 mole of NADH are formed; they may be reoxidized by the electron-transport chain. Since, in the final turn of the helix, 2 moles of acetyl–CoA are produced, the helical scheme must be traversed only *seven* times to degrade palmitic acid completely. In this process 7 moles each of reduced flavin and pyridine nucleotide are formed. The sequence can be divided into two steps:

Step 1:

Palmitic acid $\longrightarrow$	8 Acetyl-S-CoA + 14 Electron pairs
7 Electron pairs $\longrightarrow$	Flavin system $\times$ 2 = 14 Energy-rich bonds
7 Electron pairs $\longrightarrow$	NAD^+ system $\times$ 3 = 21 Energy-rich bonds
	Total = 35
	Net = 35 − 1
	= 34 Energy-rich bonds

Step 2:

$$8 \text{ Acetyl–CoA} + 16\ O_2 \xrightarrow{\text{TCA cycle}} 16\ CO_2 + 8\ H_2O + 8 \text{ CoA–SH}$$

If we assume that for each oxygen atom consumed three energy-rich bonds are formed during oxidative phosphorylation, then

$$32 \times 3 = 96 \text{ Energy-rich bonds}$$

Thus, step 1 (34 bonds) and step 2 (96 bonds) = 130 energy-rich bonds; and

$$\frac{130 \times 8000 \times 100}{2{,}338{,}000} = 48\%$$

In the complete oxidation of palmitic acid to CO_2 and H_2O, 48% of the available energy can theoretically be conserved in a form (ATP) that is utilized by the cell for work. The remaining energy is lost, probably as heat. It hence becomes clear why, as a food, fat is an effective source of available energy. In this calculation we neglect the combustion of glycerol, the other component of a triglyceride.

Synthesis of Fatty Acids

Ever since Knoop postulated his β-oxidation sequence in 1904, biochemists have been intrigued with the idea of *β-multiple condensation* of the same C_2 units obtained from the oxidation of a fatty acid for the synthesis of even-chain fatty acids. The answer to this problem had to wait for the complete definition of the degradative sequence.

In 1954, Stansly and Beinert, employing the five purified enzymes of β-oxidation and acetyl–CoA, performed the crucial experiment to test the β-multiple condensation theory. Their results indicated that neither palmitic nor stearic acid were synthesized from acetyl–CoA by enzymes of the β-oxidation sequence. Earlier, however, Brady and Gurin had described homogenates that had the capacity to synthesize fatty acids from acetate. Therefore, acetate could be converted to long-chain fatty acids. The dilemma of how acetate units were joined together to form fatty acids was solved in 1959, when S. Wakil showed that rat liver extracts could only synthesize fatty acids from acetyl–CoA when a high concentration of bicarbonate was present. He showed that the condensing unit for fatty acid synthesis was not acetyl–CoA but malonyl–CoA, resulting from the carboxylation of acetyl–CoA by an avidin-sensitive acetyl–CoA carboxylase with biotin as the functional cofactor:

$$CO_2 + ATP + \text{Biotin–Enzyme} \rightleftharpoons CO_2\text{–Biotin–Enzyme} + ADP + H_3PO_4$$
$$CO_2\text{–Biotin–Enzyme} + \text{Acetyl–CoA} \rightleftharpoons \text{Malonyl–CoA} + \text{Biotin–Enzyme}$$

All synthesizing extracts obtained from plant, bacteria, and animal tissues require CO_2 if acetyl-CoA is the initial substrate.

The details of this carboxylation reaction are discussed in Chapter 9. The inclusion of malonyl-CoA as an essential substrate for fatty acid synthesis immediately points out a major difference between the β-oxidation sequence and the synthetic pathway. In addition, a heat-stable protein is another key component in bacterial and plant synthetase systems. The heat-stable protein, acyl carrier protein or ACP, has a molecular weight of 8700, and has one thiol group per mole of protein. The thiol group is associated with 4′-phosphopantetheine, which in turn is bonded to the hydroxyl group of serine, one of the amino acids of the protein. The complete structure of the protein is shown in Scheme 12-6.

Acyl carrier proteins have been isolated and purified from a number of plant and bacterial sources. Although there are minor differences in amino acid compositions, all are heat-stable, small-molecular-weight (7500–10,000) proteins with the same core amino acids directly associated with the 4′-phosphopantetheine component, namely

```
                    4′-Phosphopantetheine
                    |
etc.–Gly–Ala–Asp–Ser–Leu–Asp–etc.
```

Recently, *E. coli* ACP has been chemically synthesized by the Merrifield procedure (Chapter 19) and found to be completely equivalent to the biological ACP. In animal and yeast synthetases, 4′-phosphopantetheine is bound via serine to a protein which in turn is firmly associated with the fatty acid synthetase protein complex. Bacterial ACP is essentially inactive in the animal synthetase systems but can substitute equally well for plant ACP's in a number of plant fatty acid synthetases. The bacterial synthetase enzymes, when carefully purified, will only employ as their substrates the acyl-S-ACP derivatives. The acyl-CoA thioesters are ineffective as substrates. The se-

```
    1                                      10
NH2–Ser–Thr–Ile–Glu–Glu–Arg–Val–Lys–Lys–Ile–Ile–Gly–Glu–

                         20
Gln–Leu–Gly–Val–Lys–Gln–Glu–Glu–Val–Thr–Asp–Asn–Ala–Ser–

        30                          *                 40
Phe–Val–Glu–Asp–Leu–Gly–Ala–Asp–Ser–Leu–Asp–Thr–Val–Glu–

                                 50
Leu–Val–Met–Ala–Leu–Glu–Glu–Glu–Phe–Asp–Thr–Glu–Ile–Pro–

                60
Asp–Glu–Glu–Ala–Glu–Lys–Ile–Thr–Thr–Val–Gln–Ala–Ala–Ile–

 70                         77
Asp–Tyr–Ile–Asn–Gly–His–Gln–Ala–COOH
```

Scheme 12-6

The complete structure of *E. coli* ACP according to Vanaman, Hill, and Wakil. Ser* is the site for 4′-phosphopantetheine.

quence of reactions which now have been revealed for the synthesis of long-chain fatty acids is as follows:

$$\text{Acetyl–CoA} + \text{ACP–SH} \xrightarrow{①} \text{Acetyl–S–ACP} + \text{CoA}$$

$$\text{Acetyl–S–ACP} + \text{Enz③} \longrightarrow \text{Acetyl–S–Enz③} + \text{ACP}$$

$$\text{Malonyl–CoA} + \text{ACP–SH} \xrightarrow{②} \text{Malonyl–S–ACP} + \text{CoA}$$

$$\text{Acetyl–S–Enz③} + \text{Malonyl–S–ACP} \longrightarrow \text{Acetoacetyl–S–ACP} + \text{Enz③} + CO_2$$

$$\text{Acetoacetyl–S–ACP} + \text{NADPH} + H^+ \xrightarrow{④} \text{D}(-)\text{-}\beta\text{-Hydroxybutyryl–S–ACP} + \text{NADP}^+$$

$$\text{D}(-)\text{-}\beta\text{-Hydroxybutyryl–S–ACP} \xrightarrow{⑤} \Delta^2\text{-}trans\text{-Crotonyl–S–ACP} + H_2O$$

$$\Delta^2\text{-}trans\text{-Crotonyl–S–ACP} + \text{NADPH} + H^+ \xrightarrow{⑥} \text{Butyryl–S–ACP} + \text{NADP}^+$$

$$\text{Butyryl–S–ACP} + \text{Enz③} \longrightarrow \text{Butyryl–S–Enz③} + \text{ACP}$$

$$\text{Butyryl–S–Enz③} + \text{Malonyl–S–ACP} \longrightarrow \beta\text{-Ketohexanoyl–S–ACP} + \text{Enz③} + CO_2 \text{ etc.}$$

① Acetyl transacylase
② Malonyl transacylase
③ β-Ketoacyl ACP synthetase
④ β-Ketoacyl ACP reductase
⑤ Enoyl ACP hydrase
⑥ Enoyl ACP reductase

The newly formed butyryl–S–ACP reacts now with another molecule of malonyl–S–ACP and β-ketoacyl–ACP synthetase and the sequence outlined above is repeated until the desired length of fatty acid is attained. The important aspect to note is that acyl–CoA thioesters are not the true substrates for fatty acid synthesis, and this fact explains some of the earlier observations in which intermediate-chain CoA thioesters were ineffective as substrates for fatty acid synthesis.

Although the chemical events in the synthesis of long-chain fatty acids from acetyl–CoA and malonyl–CoA are identical in all organisms with the reactions described above, two general forms of the synthetases are now recognized.

Table 12-1

Source	Molecular weight	Products	ACP requirement
	Prototype I—Multienzyme complexes		
Yeast	2.3×10^6	C_{14}, C_{16}, C_{18} CoA	None
Pigeon liver	4.5×10^5	C_{16} free	None
Euglena gracilis (heterotrophic)	1×10^6	C_{16} CoA	None
	Prototype II—Freely dissociated enzymes		
E. coli	—	C_{16}, C_{18}, $C_{16:1}$ free	+
Spinach chloroplast	—	C_{16}, C_{18} CoA	+
Potato tuber	—	C_{12}, C_{14}, C_{16} free	+
Euglena gracilis (photoautrophic)	—	C_{12}, C_{14}, C_{16}, C_{18} ACP	+

The first consists of closely associated multienzyme complexes which behave as single functional units (prototype I) and the second consists of the individual enzymes which are separable and devoid of any tendency to associate in *in vitro* conditions. Table 12-1 summarizes the physical characteristics of some of these synthetases.

In summary, there are major differences between the β-oxidative sequence and the synthetase systems. The correct thioester substrates for β-oxidation are the CoA esters, while with the prototype II systems of plants and bacteria, the ACP thioesters are the true substrates. With the prototype I systems, a modified ACP type of protein is built into the highly aggregated complex and the elongating acyl component is associated via thioester linkage to this protein. Figure 12-2 depicts what is considered to be a prototype I fatty acid

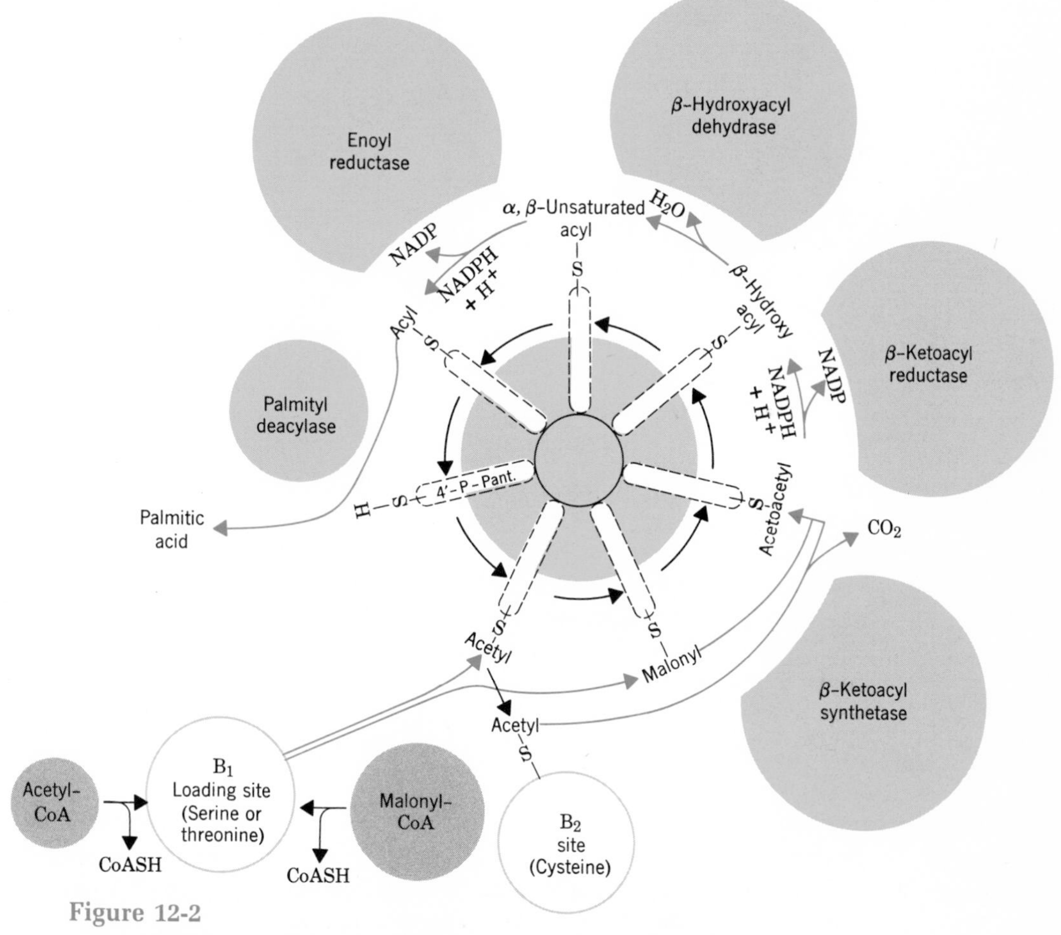

Figure 12-2

A mechanism of fatty acid synthesis by a mammalian prototype I system according to J. Porter. Note the flow of acetyl–CoA and malonyl–CoA to the loading site and the subsequent movement of the C_2 and C_3 units attached to a central ACP-like protein which serves as the substrate component for the peripherally oriented enzyme systems.

synthetase. Furthermore, the L(+)-β-hydroxy acyl derivatives of CoA are the substrates for the β-hydroxy acyl–CoA dehydrogenase and the enoyl–CoA hydrase of the β-oxidation enzymes, whereas the D(−)-β-hydroxy acyl–S–ACP derivatives are the proper substrates for the β-ketoacyl–ACP reductase and the enoyl–ACP hydrase of the prototype II synthetase system. By these differences the cell can readily separate the anabolic and catabolic reactions of fatty acids and their derivatives.

The overall reaction for fatty acid synthesis may then be written as

$$H_3C\overset{\displaystyle O}{\overset{\|}{C}}\text{–S–CoA} + n\ \text{Malonyl–S–CoA} + 2n\ \text{NADPH} + 2n\ H^+ \xrightarrow{\text{ACP}}$$

$$n\ \text{CoA–SH} + CH_3CH_2(CH_2CH_2)_{n-1}CH_2\overset{\displaystyle O}{\overset{\|}{C}}\text{–S–CoA} + 2n\ \text{NADP}^+ + n\ CO_2 \quad (12\text{-}1)$$

And thus the origin of carbon atoms in palmitic acid is as follows:

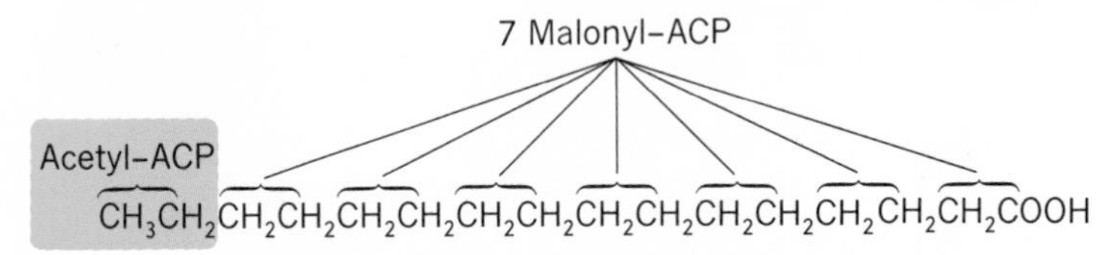

Why should the initial reaction catalyzed by the synthetase unit differ so greatly from the fifth reaction in the β-oxidation sequence? A consideration of the K_{eq} for the thiolase reaction in the β-oxidation sequence provides important information. The equilibrium for the thiolase reaction is

$$\frac{(\text{Acetyl–CoA})^2}{(\text{Acetoacetyl–CoA})(\text{CoASH})} = 10^5$$

The equilibrium is therefore greatly in favor of the breakdown of acetoacetyl–CoA. In other words, if we have 1000 molecules of acetyl–CoA, we can calculate the number of molecules of acetoacetyl–CoA that would be in equilibrium. Starting with 1003 molecules of acetyl–CoA, only 3 molecules of acetoacetyl–CoA (0.3%) would be synthesized. This is a highly unfavorable equilibrium for the initial step in synthesis. With malonyl–ACP as the condensing unit, we have a very favorable condition, in that a decarboxylation occurs as the condensation takes place. The loss of CO_2 as a gas therefore "drives" the reaction in the direction of synthesis.

Biosynthesis of Unsaturated Fatty Acids

It is now recognized that unsaturated fatty acids play important roles in the structure of membrane systems in all living cells. All membrane systems contain in their complex lipids fatty acids with different degrees of unsaturation. In procaryotic cells, for example, although polyunsaturated fatty acids are absent, monoenoic acids are important components of the membrane lipids, while in eucaryotic cells, polyunsaturated fatty acids are the key acyl moieties. In photosynthetic tissues in which molecular oxygen is released by the photooxidation of water, highly unsaturated fatty acids are found with very few exceptions to be associated with the lamellar membrane lipids. Certain unsaturated fatty acids have been classified as "essential fatty acids," since their omission in normal diets leads to pathological changes which can be reversed by the reintroduction of the acids to the diet. Recently, the precursors of a class of important hormones, the prostaglandins, have been shown to be arachidonic acid and homo-γ-linolenic acid. With the advent of new and highly sensitive techniques to detect and determine the structures of fatty acids, a very large number of unsaturated fatty acids have been described, including both nonconjugated, and conjugated fatty acids; also, all-*trans*, *cis*-*trans*, and all-*cis* double bond systems as well as acetylenic fatty acids have been reported.

Introduction of a Single Double Bond. Two principal biosynthetic mechanisms are involved in the formation of a single double bond, namely the anaerobic and the aerobic mechanisms. These will now be described.

Anaerobic pathway. Since a large number of eubacteria can synthesize monoenoic acids under anaerobic conditions, the direct desaturation mechanism by molecular oxygen is inoperative in these organisms. The work of Bloch and his group clearly showed that the single *cis* double bond is introduced at the C_{10} level and positioned in the hydrocarbon chain by chain elongation. Specifically, in the synthesis of C_{18} fatty acids, the branching point is at the D(−)-β-hydroxydecanoyl–S–ACP level. Although this thioester can serve as a substrate for several dehydrases, it serves as a highly specific substrate for β-hydroxydecanoyl–S–ACP dehydrase, which introduces a single *cis*-β,γ-double bond to form a *cis*-3,4-decanoyl-S–ACP, which is then extended as indicated below to form a monoenoic acid. Other dehydrases will form a *trans*-α,β double bond system which is, however, readily reduced to form saturated fatty acids.

Scheme 12-7 summarizes these observations. Each arrow ($\longrightarrow$) represents the sequential events of reduction, dehydration, reduction, and further condensation, as summarized on page 292. As already indicated, three other dehydrases operate in bacteria in addition to the D(−)-β-hydroxydecanoyl–ACP dehydrases. They are:

(1) β-Hydroxybutyryl–ACP dehydrase, whose substrates are the C_4, C_6, and C_8 acyl–ACP derivatives, with the C_4 substrate the most active and the C_8 the least.
(2) β-Hydroxyoctanoyl–ACP dehydrase, whose substrates include (in decreasing activity) $C_8 > C_{10} > C_{12} > C_6 > C_4$.
(3) β-Hydroxypalmityl–ACP dehydrase, whose substrates include (in decreasing activity) $C_{16} > C_{12} > C_{14} > C_{10} >$.

These dehydrases remove the element of water to form exclusively the α,β-*trans*-monoenoyl–ACP derivative.

Thus, in bacterial organisms there is a balance of four dehydrases which must be maintained by the organism to allow the synthesis of the required fatty acids. The key enzyme, the D(−)-β-hydroxydecanoyl–ACP dehydrase,

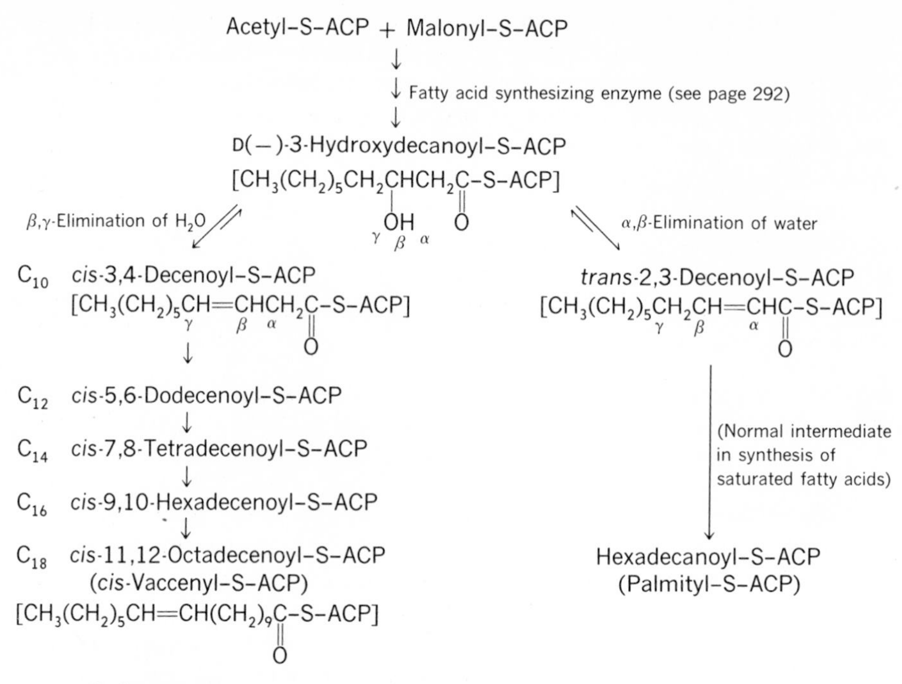

Scheme 12-7

redirects the flow of the alkyl chain lengthening process from the saturated to the monounsaturated fatty acid.

Aerobic pathway. While it has been known for many years that animals are able to synthesize monoenoic acid, it was only in recent years that Bloch defined the aerobic pathway by which a double bond is inserted in the 9,10 position of stearic acid in the presence of molecular oxygen. In higher plants, oxygen is also an absolute requirement for the unsaturation reaction. In recent years, evidence clearly indicates that even in the orders *Pseudomonodales* and *Eubacteriales*, representatives of these bacterial families have the aerobic pathway. These include *Corynebacterium diptheriae, Mycobacterium phlei,* and *Micrococcus lysodeikticus,* all of which synthesize the Δ^9-monoenoic fatty acids. A number of *Bacilli,* including *B. subtilis*, and *B. megatherium,* synthesize both the Δ^5-monoenoic fatty acid and the Δ^{10}-monoenoic fatty acid by an oxygen-dependent system. In all bacteria so far studied either one system or the other is the exclusive system; that is, in bacterial systems, either the anaerobic or the aerobic pathway occurs as the sole system. The mutual coexistence of the two pathways in a single bacterial specie has not been observed.

In the synthesis of oleic acid by the aerobic pathway, the enzyme stearyl-CoA desaturase, requires NADH, and molecular oxygen.

$$\text{NADH} \longrightarrow \text{FAD} \longrightarrow \text{Cyt } b_5 \longrightarrow \text{CN}^-\text{-sensitive factor} \longrightarrow \left\{ \begin{array}{l} -CH_2-CH_2- \\ O_2 + \text{Desaturase} \\ -CH{=}CH- + 2\,H_2O \end{array} \right.$$

Cytochrome b_5 is part of the electron-transport system between NADH and oxygen. The desaturase is not carbon monoxide-sensitive. Most desaturase enzymes are closely associated with microsomal particles.

The two hydrogens are removed in the desaturation process in a stereospecific manner. While the details of the unsaturation process are not known, the mechanisms probably involve a simultaneous concerted removal of the hydrogen atom pair.

Introduction of Additional Double Bonds. One of the remarkable metabolic blocks in animal tissue is their inability to desaturate a monoenoic acid, oleic acid, toward the methyl end of the fatty acid, whereas the plant kingdom carries out this reaction readily. Thus, linoleic acid, which is required by animals, must be obtained by animals from plant dietary sources. The major differences between the plants and mammals in terms of polyunsaturation is illustrated in Chapter 3. Once the animal has obtained linoleic acid from plant dietary sources, the acid is desaturated toward the COOH end to form γ-linoleic acid. An elongation step occurs to form homo-γ-linoleic, which is finally desaturated to form the typical polyunsaturated fatty acid, arachidonic acid. In all desaturation steps in the mammalian system, the enzyme desaturases are microsomal, the substrates acyl–CoA's, and NADPH or NADH and O_2 are essential components. Recent work by Morris, in England, has shown that, as with oleic synthesis, in polyunsaturation mechanisms, the two hydrogen atoms with the D absolute configuration are stereospecifically removed.

The role of polyunsaturated fatty acids is probably related to their occurrence in membrane lipids. Another very important function in the animal kingdom is the role of a number of polyunsaturated fatty acids as precursors for the synthesis of prostaglandins. Thus:

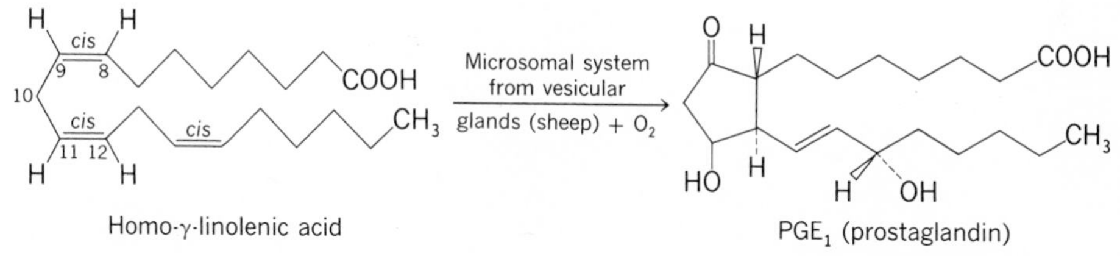

Homo-γ-linolenic acid $\xrightarrow{\text{Microsomal system from vesicular glands (sheep) } + O_2}$ PGE_1 (prostaglandin)

Finally, procaryotic organisms are incapable of synthesizing polyunsaturated fatty acids. It is tempting to correlate this observation with the absence of organelles such as mitochondria, chloroplasts, nuclei, etc., in the procaryotic organisms. Further investigations are necessary to bring these relationships into sharper focus (see Chapter 6 for further discussion).

Triacylglycerol and Phospholipid Biosynthesis

These complex lipids are readily synthesized in the animal cell by enzymes attached to the endoplasmic reticulum. In the *de novo* synthesis of triacylglycerol, three pathways may be listed, although the relative importance of one to the other is as yet not too well-defined.

(1) Dietary triacylglycerols are converted in the intestinal lumen by lipases to free fatty acids and 2-monoacylglycerol. On absorption, the monoacylglycerols are acylated by acyl–CoA to triacylglycerols.

(2) Glycerol, derived from glycolysis, is phosphorylated and then acylated by acyl–CoA's to form phosphatidic acid. A specific phosphatase then hydrolyzes the phosphatidic acid to a diacylglycerol and inorganic phosphate. Finally, the diacylglycerol is fully acylated to yield a triacylglycerol. 3-*sn*-Phosphatidic acid is a central intermediate for both phospholipid biosynthesis and triacylglycerol synthesis. The final assembling of the units is accomplished as in Scheme 12-8.

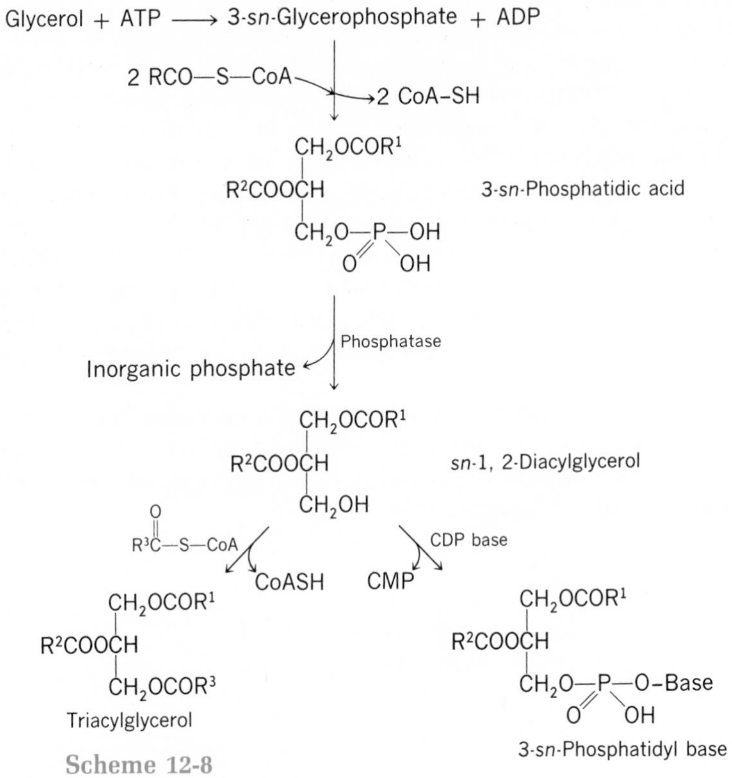

Glycerol + ATP ⟶ 3-*sn*-Glycerophosphate + ADP

2 RCO—S—CoA ⟶ 2 CoA-SH

CH_2OCOR^1 / R^2COOCH / $CH_2O—P(=O)(OH)—OH$ 3-*sn*-Phosphatidic acid

Phosphatase → Inorganic phosphate

CH_2OCOR^1 / R^2COOCH / CH_2OH *sn*-1, 2-Diacylglycerol

$R^3C(=O)—S—CoA$ → CoASH

CH_2OCOR^1 / R^2COOCH / CH_2OCOR^3

Triacylglycerol

CDP base → CMP

CH_2OCOR^1 / R^2COOCH / $CH_2O—P(=O)(OH)—O\text{-Base}$

3-*sn*-Phosphatidyl base

Scheme 12-8

(3) Recently a dihydroxyacetone phosphate pathway for phosphatidic acid biosynthesis has been described:

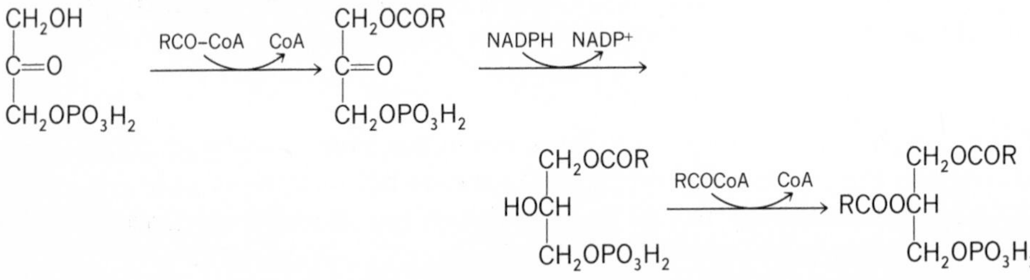

CH_2OH / $C{=}O$ / $CH_2OPO_3H_2$ —(RCO–CoA → CoA)→ CH_2OCOR / $C{=}O$ / $CH_2OPO_3H_2$ —(NADPH → $NADP^+$)→

CH_2OCOR / $HOCH$ / $CH_2OPO_3H_2$ —(RCOCoA → CoA)→ CH_2OCOR / $RCOOCH$ / $CH_2OPO_3H_2$

Phosphatidic acid can then undergo the same reaction as cited above in (2) to yield triacylglycerol.

If we examine the structure of phosphatidyl choline (lecithin), we note that its basic units are long-chain fatty acids (—OCOR[1] and —OCOR[2]), glycerol, phosphate, and choline. How are they assembled?

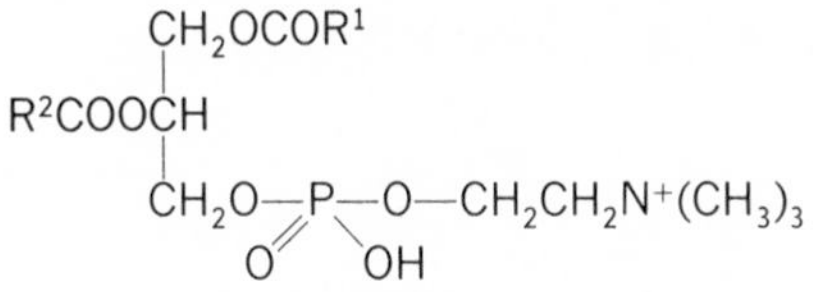

$$\begin{array}{l} \quad\quad\quad\; CH_2OCOR^1 \\ \quad\quad\quad\;\; | \\ R^2COOCH \\ \quad\quad\quad\;\; | \\ \quad\quad\quad\; CH_2O{-}\overset{}{P}(=O)(OH){-}O{-}CH_2CH_2N^+(CH_3)_3 \end{array}$$

Lecithin (3-*sn*-Phosphatidyl choline)

In the cell an orderly sequence of events, all catalyzed by specific enzymes, brings the separate units together. The first reaction is the phosphorylation of choline:

$$\text{Choline} + \text{ATP} \longrightarrow \text{Phosphorylcholine} + \text{ADP}$$

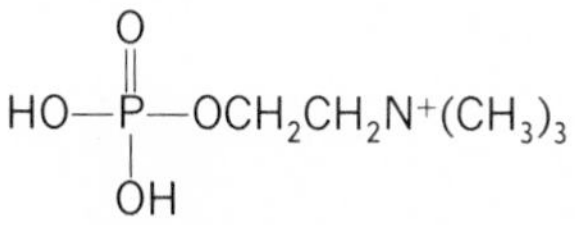

$$HO{-}\overset{O}{\overset{\|}{P}}(OH){-}OCH_2CH_2N^+(CH_3)_3$$

Phosphorylcholine then reacts with CTP to form cytidine diphosphate choline (CDP-choline).

$$\text{CTP} + \text{Phosphorylcholine} \longrightarrow \text{CDP-choline} + \text{Pyrophosphate}$$

The complete structure of CDP-choline is

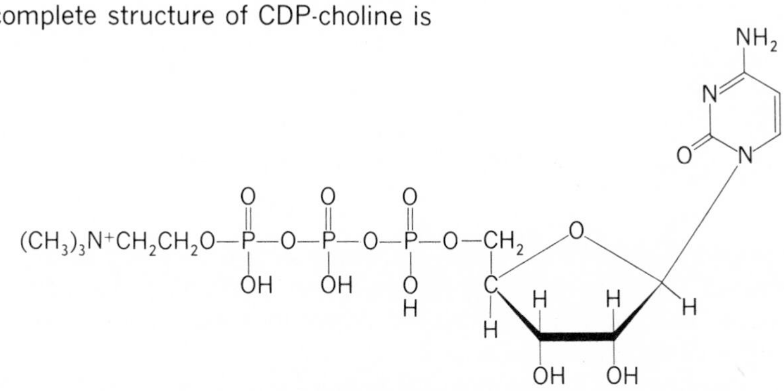

$$(CH_3)_3N^+CH_2CH_2O{-}\overset{O}{\overset{\|}{P}}(OH){-}O{-}\overset{O}{\overset{\|}{P}}(OH){-}O{-}\overset{O}{\overset{\|}{P}}(OH){-}O{-}CH_2{-}\text{ribose(OH, OH)}{-}\text{cytosine}(NH_2)$$

Cytidine diphosphate choline
(CDP-choline)

The final assemblage is depicted in Scheme 12-8 with the transfer of the phosphorylcholine moiety to diacylglycerol to form phosphatidyl choline. Phosphatidyl ethanolamine and phosphatidyl serine are also formed by similar mechanisms.

It should be noted however that while these mechanisms are important, alternative mechanisms for the synthesis of the phospholipids have been described and can be examined by referring to the references listed at the end of this chapter.

The details for the biosynthesis of sphingomyelin have been recently described by E. E. Snell and by W. Stoffel of the University of Cologne. A brief description of this synthesis will show how the assemblage of another complex lipid can take place by rather different series of events from those for the triacylglycerols and phospholipids.

Step I:

Conversion of serine to 3-ketosphinganine:

$$CH_2OHCHNH_2COOH + CH_3(CH_2)_{14}\overset{O}{\overset{\|}{C}}{-}S{-}CoA \xrightarrow[\text{3-Ketosphinganine synthetase}]{\text{Pyridoxal phosphate}} CH_3(CH_2)_{14}\overset{O}{\overset{\|}{C}}{-}CHNH_2CH_2OH + CO_2$$

Step II:

Conversion to 4*t*-sphingenine:

$$\xrightarrow[\text{NADPH}]{\text{3-Ketosphinganine reductase}} CH_3(CH_2)_{12}CH_2CH_2{-}\underset{H}{\overset{OH}{C}}{-}CHNH_2CH_2OH$$

$$\xrightarrow[-2\,H]{\text{3-Hydroxysphinganine dehydrogenase}} CH_3(CH_2)_{12}CH\overset{t}{=}CHCHOHCHNH_2CH_2OH$$

Step III:

Final conversion to sphingomyelin:

$$\xrightarrow{\text{RCO-CoA}} CH_3(CH_2)_{12}CH\overset{t}{=}CHCHOH\underset{NHCOR}{\underset{|}{C}}HCH_2OH$$

$$\xrightarrow{\text{CDP-choline}} CH_3(CH_2)_{12}CH{=}CHCHOH\underset{NHCOR}{\underset{|}{C}}HCH_2O\underset{\underset{O}{\|}}{\overset{\overset{O^-}{|}}{P}}O{-}Choline$$

Biosynthesis of Cholesterol

Ever since the early 1930s, when the chemical structure of cholesterol was determined, the biogenesis of this complex ring system has been an intriguing puzzle. The solution was achieved almost forty years later by the efforts of many investigators.

The ring structure of the molecule is planar, and can be written as

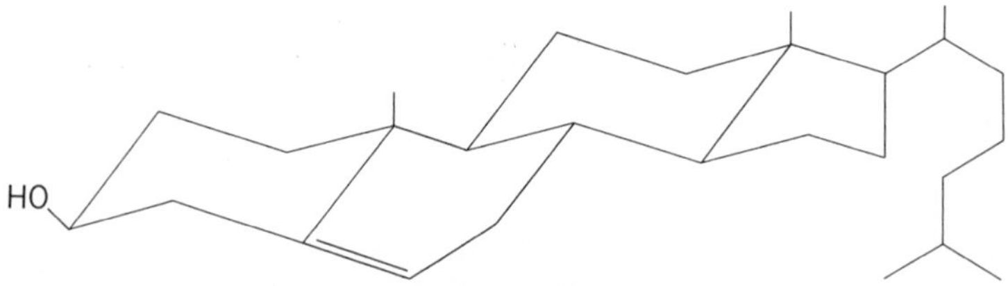

HO

Cholesterol

All the carbon atoms of cholesterol derive directly from acetate. Reactions very different from those involved in straight-chain fatty acid synthesis were discovered in its biosynthesis. Its biosynthesis can be divided into three groups of reactions:

(1) Formation of mevalonic acid
(2) Conversion of mevalonic acid to squalene
(3) Conversion of squalene into lanosterol and then to cholesterol

(1) Acetate $\longrightarrow$ mevalonate (microsomal enzymes):

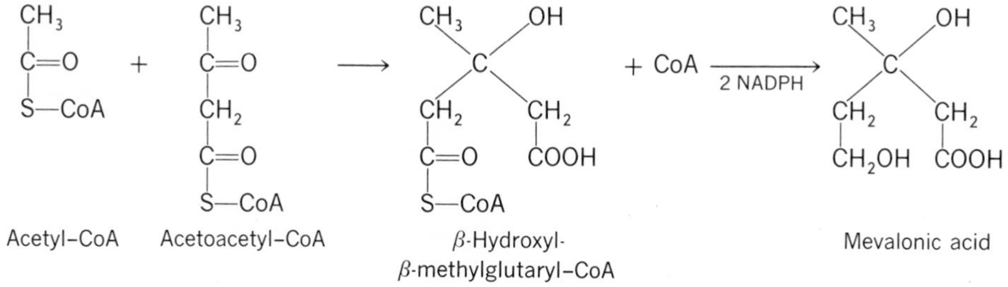

CH_3–C(=O)–S–CoA + CH_3–C(=O)–CH_2–C(=O)–S–CoA $\longrightarrow$ $(CH_3)(OH)C(CH_2COOH)(CH_2$–C(=O)–S–CoA$)$ + CoA $\xrightarrow{\text{2 NADPH}}$ $(CH_3)(OH)C(CH_2COOH)(CH_2CH_2OH)$

Acetyl–CoA; Acetoacetyl–CoA; β-Hydroxyl-β-methylglutaryl–CoA; Mevalonic acid

(2) Mevalonate to squalene (soluble enzymes):

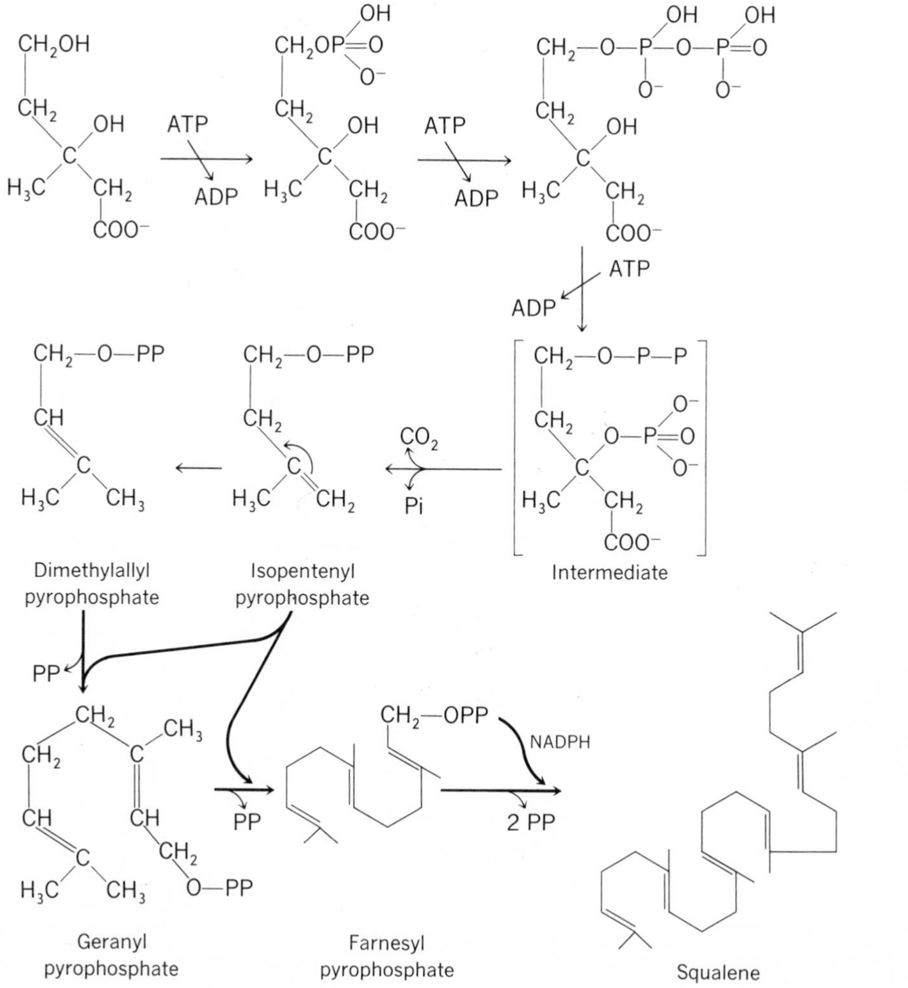

CH_2OH–CH_2–C(OH)(H_3C)–CH_2–COO^- $\xrightarrow{\text{ATP} \rightarrow \text{ADP}}$ $CH_2OP(=O)(OH)O^-$–CH_2–C(OH)(H_3C)–CH_2–COO^- $\xrightarrow{\text{ATP} \rightarrow \text{ADP}}$ CH_2–O–P(OH)(O^-)–O–P(=O)(OH)(O^-)–CH_2–C(OH)(H_3C)–CH_2–COO^-

$\xrightarrow{\text{ATP} \rightarrow \text{ADP}}$ [CH_2–O–P–P–CH_2–C(O–P(=O)(O^-)$_2$)(H_3C)–CH_2–COO^-] Intermediate $\xrightarrow{\rightarrow CO_2,\ Pi}$ CH_2–O–PP–CH_2–C(H_3C)=CH_2 Isopentenyl pyrophosphate $\longrightarrow$ CH_2–O–PP–CH=C(H_3C)–CH_3 Dimethylallyl pyrophosphate

Dimethylallyl pyrophosphate + Isopentenyl pyrophosphate $\xrightarrow{\rightarrow PP}$ Geranyl pyrophosphate (H_3C)(CH_3)C=CH–CH_2–CH_2–C(CH_3)=CH–CH_2–O–PP

Geranyl pyrophosphate + Isopentenyl pyrophosphate $\xrightarrow{\rightarrow PP}$ Farnesyl pyrophosphate (CH_2–OPP) $\xrightarrow[\rightarrow \text{2 PP}]{\text{NADPH}}$ Squalene

Geranyl pyrophosphate; Farnesyl pyrophosphate; Squalene

(3) Squalene ⟶ lanosterol ⟶ cholesterol (aerobic and microsomal):

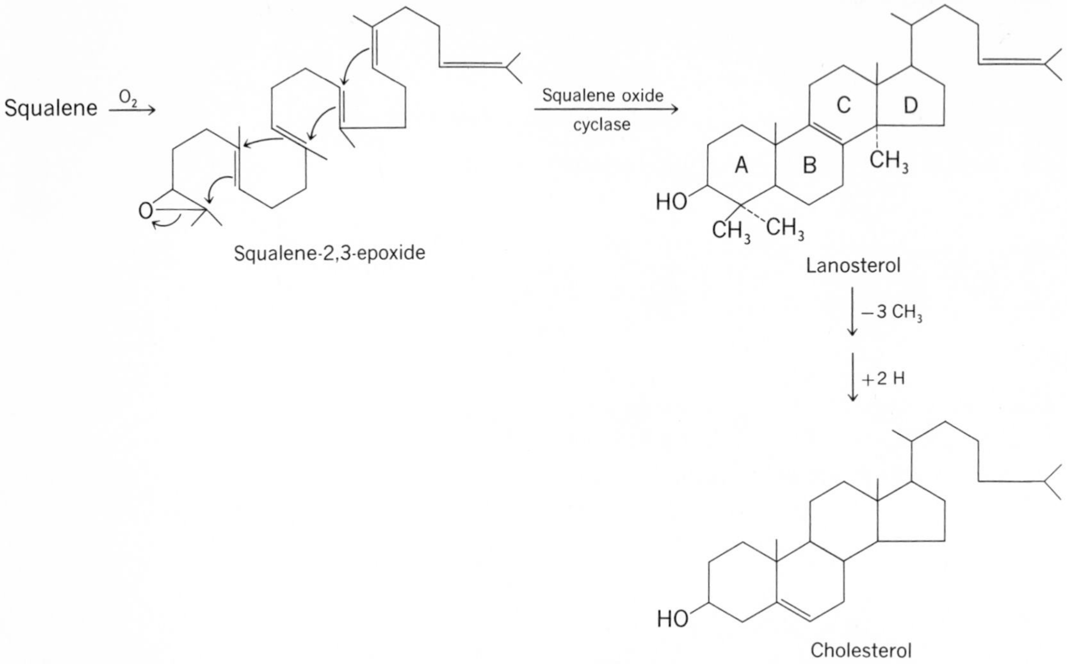

Squalene $\xrightarrow{O_2}$

Squalene-2,3-epoxide

Squalene oxide cyclase

A B C D

HO

CH_3 CH_3 CH_3

Lanosterol

$-3\ CH_3$

$+2\ H$

HO

Cholesterol

Regulation of Cholesterol Synthesis

Cholesterol is synthesized in all animal tissues. In addition, a variable quantity of dietary cholesterol contributes to the total concentration in man.

When the diet is rich in cholesterol, *de novo* synthesis is markedly inhibited; when deficient, *de novo* synthesis occurs. There is good evidence that cholesterol itself inhibits the enzyme responsible for the conversion of β-hydroxy-β-methylglutaryl–CoA to mevalonic acid.

$$\beta\text{-Hydroxy-}\beta\text{-methylglutaryl–CoA} \xrightarrow{NADPH} \text{Mevalonate} + \text{CoA}$$

Fasting will also inhibit cholesterol synthesis, although the details of this inhibition are not known. Although the steroid ring is readily synthesized in plants and animals, procaryotic organisms cannot synthesize the ring system although they readily form polyisoprenoid pigments. Insects have lost the ability to synthesize sterols and hence employ exogenous sources for further conversion to important insect hormones such as ecdysome, the oxygenated derivative of cholesterol. In vertebrates, cholesterol is the substrate for a complex of modifications of the side chains and the ring system to form progesterone, androgens, estrogens, and corticosteroid, all extremely important mammalian hormones.

References

1. M. Florkin and E. H. Stotz, eds., *Lipid Metabolism*. Vol. 18 of *Comprehensive Biochemistry*. Amsterdam: Elsevier, 1970.

   A useful compilation of specialized chapters on lipid metabolism.

2. S. Wakil, ed., *Lipid Metabolism*. New York: Academic Press, 1971.

   The most recent thorough discussion on various topics on lipid metabolism for the advanced student.

13

The Tricarboxylic Acid Cycle

Introduction

The failure of lactic acid to accumulate in a stimulated muscle which is exposed to air indicates a further metabolic breakdown of lactic acid in this tissue. Other organic acids were also known to be metabolized in muscle; by 1920 Thunberg had shown that some forty compounds underwent oxidation by air in the presence of tissue homogenates. Some of the most rapidly oxidized were succinic, fumaric, malic, and citric acids. A more complex relation was also implied by the studies of Szent-Gyorgi, who reported that some of these acids appeared to catalyze the oxidation of unknown substrates in the homogenates. Thus, an amount of fumarate that should have caused an uptake of 20 μliters of O_2, if it were completely oxidized by the homogenate of pigeon breast muscle to which it was added, instead caused seven times that amount of oxygen consumption.

With the elaboration of the glycolytic sequence in yeast and muscle it appeared that the compounds which were being oxidized further to CO_2 and H_2O by animal tissues were pyruvate and lactate. Szent-Gyorgi subsequently showed that minced pigeon breast muscle did oxidize pyruvic acid to completion. The biochemists Keilin, Martius, Knoop, Baumann, Ochoa, and Lipmann have also contributed to our understanding of the metabolic pathway which accomplishes the aerobic oxidation of pyruvate and lactate.

The most important single contributor was the distinguished English biochemist Sir Hans Krebs. His extensive studies allowed him to postulate in 1937 the cycle of reactions which accounted for the oxidation of pyruvic acid to CO_2 and water. Although some slight modification has occurred since then, the scheme as shown on the inside of the front cover of this book is essentially that proposed by Krebs in 1937. His contributions to the problem were of such magnitude that the cycle is frequently referred to as the Krebs cycle. Krebs himself prefers to call it the *tricarboxylic acid cycle*. In 1953 he was awarded the Nobel Prize in medicine for his important discoveries.

A landmark discovery was made in 1948 by E. P. Kennedy and A. L. Lehninger when they found that rat liver mitochondria could catalyze the oxidation of pyruvate and all the intermediates of the tricarboxylic acid cycle by molecular oxygen. Since only Mg^{2+} and an adenylic acid (ATP, ADP, or AMP) had to be added, this finding meant that mitochondria contain not only all the enzymes of the tricarboxylic acid cycle, but also those required to transport the electrons from the substrate to molecular oxygen. Subsequent work has shown that some enzymes (e.g., malic dehydrogenase, fumarase, aconitase) required in the cycle are also found in the cytoplasm, but the reactions they catalyze are independent of the mitochondrial oxidation process (Chapter 6).

Before considering the cycle in detail it is necessary to point out that the further oxidation of pyruvate or lactate by a living organism is of considerable significance from the standpoint of energy production. The free-energy change for the complete oxidation of glucose to CO_2 and H_2O has been given as -686 kcal:

$$C_6H_{12}O_6 + 6\ O_2 \longrightarrow 6\ CO_2 + 6\ H_2O$$
$$\Delta G' = -686{,}000 \text{ cal (pH 7.0)}$$

In Chapter 10 we indicated that the $\Delta G'$ for the formation of lactic acid from glucose was about -47 kcal:

$$C_6H_{12}O_6 \longrightarrow 2\ CH_3CHOHCOOH$$
$$\Delta G' = -47{,}000 \text{ cal (pH 7.0)}$$

This means that only about 7% of the available energy of the glucose molecule has been released when lactic acid is formed in glycolysis and about $-639{,}000$ cal remain to be released when the 2 moles of lactate from the original glucose molecule are oxidized to completion. Thus, the $\Delta G'$ per mole of lactate oxidized can be estimated as $-319{,}500$ cal.

$$CH_3CHOHCOOH + 3\ O_2 \longrightarrow 3\ CO_2 + 3\ H_2O$$
$$\Delta G' = -319{,}500 \text{ cal (pH 7.0)}$$

The student should also appreciate that the aerobic oxidation of glucose to CO_2 and H_2O by the living organism does not necessarily involve the formation of lactic acid as an intermediate step. Instead, the key compound produced in glycolysis, which can be either reduced to lactic acid or instead be oxidized completely to CO_2 and H_2O, is pyruvic acid. By 1935, Szent-Gyorgi had shown that pyruvate could be oxidized to completion by a muscle mince provided catalytic quantities of dicarboxylic acids such as succinate, malate, and oxalacetate were added. The individual reactions of the sequence that accomplishes this oxidation will now be considered.

Oxidation of Pyruvate to Acetyl-CoA

Strictly speaking, pyruvic acid is not an intermediate in the tricarboxylic acid cycle. The α-keto acid is first converted to acetyl–CoA by the multienzyme complex known as the *pyruvic dehydrogenase complex*.

This reaction, which is an oxidative decarboxylation, involves six cofactors: coenzyme A, NAD^+, lipoic acid, FAD, Mg^{2+}, and thiamin pyrophosphate

$$H_3C{-}\underset{\underset{O}{\|}}{C}{-}CO_2H + CoA\text{-}SH + NAD^+ \xrightarrow[\text{TPP, FAD}]{\text{Lipoic acid, } Mg^{2+}} H_3C{-}\underset{\underset{O}{\|}}{C}{-}S{-}CoA + NADH + H^+ + CO_2 \quad (13\text{-}1)$$

$$\Delta G' = -8000 \text{ cal (pH 7.0)}$$

(TPP). The overall reaction can be broken down into partial reactions catalyzed by three separate enzymes that constitute the multienzyme complex. In the initial reaction, pyruvic acid is decarboxylated to form CO_2 and a acetol complex of TPP that is tightly bound to one of these enzymes, *pyruvic dehydrogenase:*

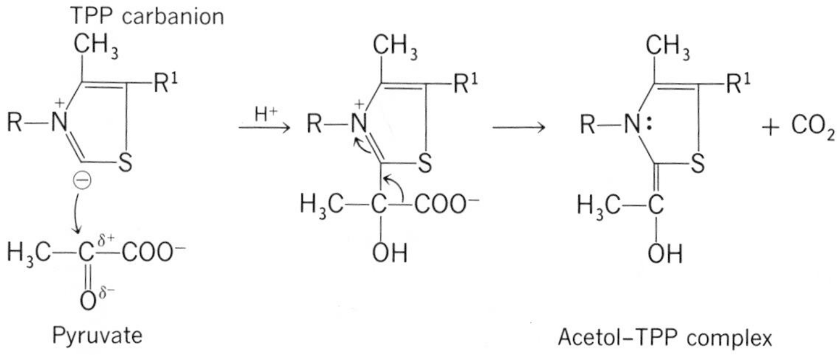

TPP carbanion + Pyruvate $\xrightarrow{H^+}$ → Acetol–TPP complex + CO_2

The two-carbon acetol group is next transferred to an oxidized lipoic acid moiety that is covalently bound to the second enzyme of the complex, *dihydrolipoyl transacetylase* (Scheme 13-1):

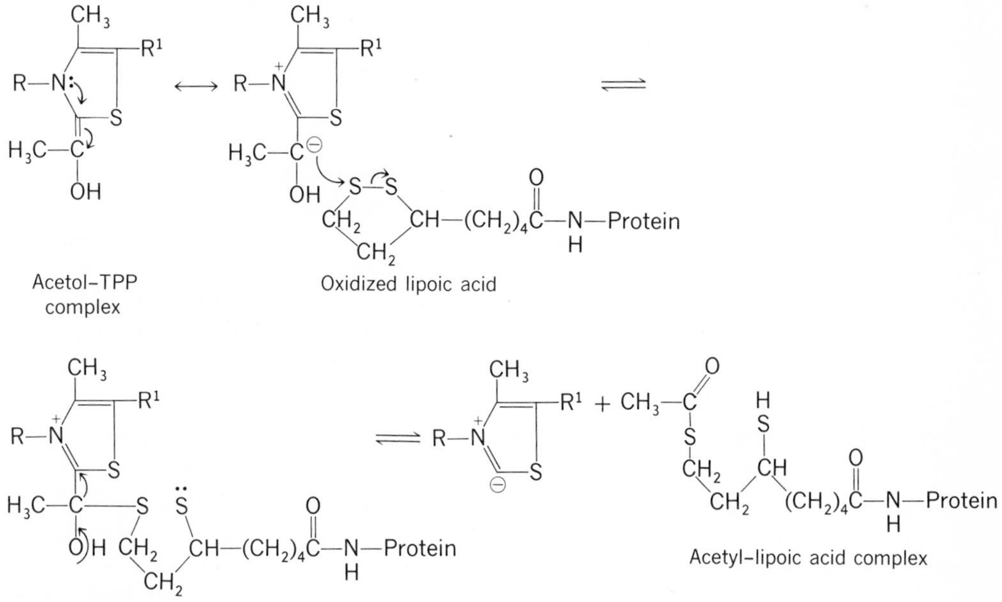

Acetol–TPP complex ⟷ ⇌ (with Oxidized lipoic acid, $CH-(CH_2)_4C(=O)-NH-Protein$) ⇌ TPP carbanion + Acetyl–lipoic acid complex ($CH_3-C(=O)-S-CH_2-CH_2-CH(SH)-(CH_2)_4C(=O)-NH-Protein$)

Scheme 13-1

Note that, as a result of this reaction, a high-energy thioester (of reduced lipoic acid) has been formed, and that the two-carbon unit is now at the oxidation level of acetic acid rather than acetaldehyde.

In a third reaction (Scheme 13-2), the acetyl group is transferred to coenzyme A to form acetyl-CoA, which dissociates from the enzyme in a free form, being one of the products of the overall reaction (reaction 13-1):

$$\text{Acetyl lipoic acid} + HS{-}CoA \rightleftharpoons H_3C{-}\overset{\overset{\displaystyle O}{\|}}{C}{-}S{-}CoA\ (\text{Acetyl-CoA}) + \text{Reduced lipoic acid}$$

Acetyl lipoic acid

Acetyl-CoA

Reduced lipoic acid

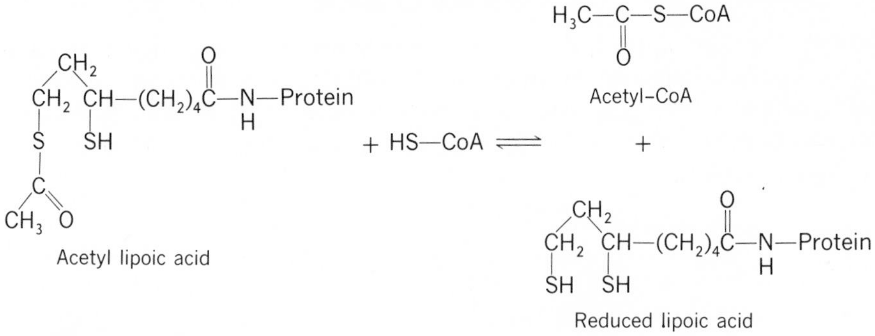

Scheme 13-2

The reduced lipoic acid moiety of the dihydrolipoyl dehydrogenase is then reoxidized to the cyclic lipoyl form by the third enzyme of the complex, *dihydrolipoyl dehydrogenase,* a flavoprotein that contains FAD:

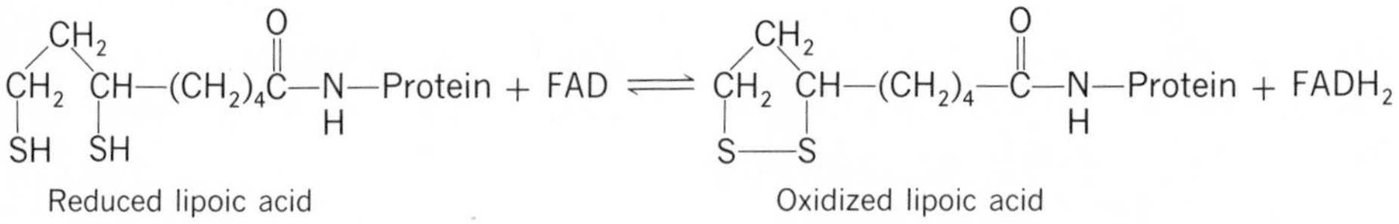

$$\text{Reduced lipoic acid-}C(O){-}NH{-}\text{Protein} + FAD \rightleftharpoons \text{Oxidized lipoic acid-}C(O){-}NH{-}\text{Protein} + FADH_2$$

Reduced lipoic acid

Oxidized lipoic acid

Finally, the reduced flavin coenzyme is reoxidized by NAD^+, one of the reactants in the overall process (reaction 13-1), and NADH is produced:

$$FADH_2 + NAD^+ \rightleftharpoons FAD + NADH + H^+$$

Note that all of the partial reactions catalyzed by the pyruvic dehydrogenase complex are reversible except the initial decarboxylation. The irreversible nature of this process makes the overall reaction (reaction 13-1) irreversible and the $\Delta G'$ has been estimated as approximately -8000 cal.

The pyruvic dehydrogenase complex has been isolated from pig heart and *E. coli* where it has a molecular weight of 4.8×10^6. Some of the properties of this enzyme complex are discussed on page 192.

Reactions of the Tricarboxylic Acid Cycle

Citrate Synthase. The enzyme that catalyzes the entry of acetyl-CoA into the tricarboxylic acid cycle is known as citrate synthase (formerly condensing enzyme). The two carbon atoms which originate from the acetyl-CoA are shaded in the reaction shown here and in subsequent reactions. The equilibrium constant for the reaction is 3×10^5. (Note in reaction 13-2 that there is a formation of a carbon–carbon bond and free coenzyme A at the expense of the thioester.) Indirect evidence indicates that citryl-CoA is formed as an

$$\underset{\text{Acetyl-CoA}}{H_3C{-}\overset{\overset{}{}}{\underset{\|}{C}}{-}S{-}CoA} \rightleftharpoons \ominus{:}CH_2{-}C({=}O){-}S{-}CoA \;\; (H^+) \;+\; \underset{\text{Oxalacetate}}{\overset{\delta^-\;\delta^+}{O{=}C}{-}CO_2H \,/\, CH_2{-}CO_2H} \rightleftharpoons \left[\underset{\text{Citryl-CoA}}{CH_2{-}C({=}O){-}S{-}CoA \,/\, HO{-}C{-}CO_2H \,/\, CH_2{-}CO_2H}\right] \xrightleftharpoons{H_2O}$$

$$\underset{\text{Citric acid}}{CH_2{-}CO_2H \,/\, HO{-}C{-}CO_2H \,/\, CH_2{-}CO_2H} + CoA{-}SH \qquad (13\text{-}2)$$

intermediate on the enzyme but does not dissociate as such until it is cleaved to free citrate and coenzyme A. Citrate synthase is also regulated by the energy charge of the cell; high concentrations of ATP inhibit the synthase; NADH is also inhibitory and its inhibition can be reversed by AMP.

Aconitase. The reaction of interest that is catalyzed by aconitase is the interconversion of citric and isocitric acid:

$$\underset{\text{Citric acid}}{CH_2CO_2H \,/\, HOC{-}CO_2H \,/\, CH_2{-}CO_2H} \rightleftharpoons \underset{\text{Isocitric acid}}{CH_2CO_2H \,/\, HC{-}CO_2H \,/\, HC{-}CO_2H \,/\, OH} \qquad (13\text{-}3)$$

At equilibrium, the ratio of citric acid to isocitric acid is about 15.

Aconitase, which requires Fe^{2+}, also catalyzes an isomerization between citric acid, isocitric acid, and a third acid, *cis*-aconitic acid. Indeed, *cis*-aconitic acid is frequently indicated as an intermediate in the conversion of citric to isocitric acid. Speyer and Dickman have proposed, however, that the carbonium ion of a tricarboxylic acid is the true intermediate and that this ion is in ready equilibrium with all three tricarboxylic acids interconverted by aconitase. The requirement for Fe^{2+} ion by the enzyme suggests that its role is in the formation of the carbonium ion by promoting the dissociation of the hydroxyl group.

Note that when isocitric acid is formed from citric acid (reaction 13-3) the symmetric molecule citric acid is acted upon in an asymmetric manner by the enzyme aconitase. That is, the hydroxyl group in isocitric acid is located on a carbon atom derived initially from oxalacetate rather than the methyl group of acetyl–CoA. Ogston, in Australia, explained this asymmetry of action by his three-point attachment theory; this theory is discussed in detail in Chapter 8.

Isocitric Dehydrogenase. Isocitric dehydrogenase catalyzes the oxidative decarboxylation of isocitric acid to α-ketoglutaric acid and CO_2 in the presence of a divalent cation (Mg^{2+} or Mn^{2+}); a nicotinamide nucleotide is the oxidant. It would be logical to consider this reaction as the result of an initial oxidation which produces oxalosuccinic acid and then a decarboxylation reaction on this β-keto acid to form CO_2 and α-ketoglutarate.

$$\underset{\text{Isocitric acid}}{\begin{array}{l} CH_2CO_2H \\ HC{-}CO_2H \\ HC{-}CO_2H \\ \;\;OH \end{array}} + \underset{(NADP^+)}{NAD^+} \rightleftharpoons \underset{(NADPH)}{NADH} + H^+ + \underset{\text{Oxalosuccinic acid}}{\left[\begin{array}{l} CH_2CO_2H \\ HC{-}CO_2H \\ O{=}C{-}CO_2H \end{array}\right]} \xrightleftharpoons{Mg^{2+}} \underset{\alpha\text{-Ketoglutaric acid}}{\begin{array}{l} CH_2CO_2H \\ HCH \\ O{=}C{-}CO_2H \end{array}} + CO_2 \quad (13\text{-}4)$$

The evidence, however, indicates that oxalosuccinate, if formed, is firmly bound to the surface of the enzyme and is not released as a free intermediate in either the oxidative decarboxylation of isocitrate or the reverse reaction, the reductive carboxylation of α-ketoglutarate. For this reason, the name *isocitric enzyme* has been proposed in analogy with the malic enzyme (page 315).

Most tissues contain two kinds of isocitric dehydrogenases. One of these requires NAD^+ and is found only in the mitochondria. The other enzyme requires $NADP^+$ and occurs both in mitochondria and in the cytoplasm. The NAD^+-specific enzyme is involved in the functioning of the tricarboxylic acid cycle; the mitochondrial $NADP^+$ requiring enzyme is associated with other, anabolic activities of the cycle to be described later.

The regulation of the tricarboxylic acid cycle centers on the NAD^+-specific isocitric dehydrogenase since this enzyme catalyzes the rate-limiting step of the cycle. Moreover, the dehydrogenase is affected in an allosteric manner by AMP as a positive effector. (In some tissues ADP is a positive effector.) Thus, under conditions where the ATP concentration is high in cells or mitochondria, the corresponding low level of AMP will decrease the activity of isocitric dehydrogenase and the Krebs cycle. But, where ATP consumption is increased and its level is lower, the reciprocal increase in AMP concentration will stimulate the rate of isocitric dehydrogenase and, in turn, the flow of carbon through the cycle.

This control mechanism may be further enhanced by the fact that the concentration of citric acid, as well as isocitric acid, increases when the activity of isocitric dehydrogenase is diminished. This is related to the equilibrium constant for the aconitase reaction (reaction 13-3) which greatly favors citrate accumulation. Citric acid, in turn, has been shown to act in an allosteric manner to stimulate the activity of fructose-1,6-diphosphate phosphatase and acetyl–CoA carboxylase, both enzymes thereby decreasing the flow of substrate into the tricarboxylic acid cycle.

α-Ketoglutaric Acid Dehydrogenase. The next step of the tricarboxylic acid cycle involves the formation of succinyl–CoA by the oxidative decarboxylation of α-ketoglutaric acid. This reaction is catalyzed by the *α-ketoglutaric dehydrogenase* complex which requires TPP, Mg^{2+}, NAD^+, FAD, lipoic acid, and coenzyme A as cofactors. The mechanism is analogous to that of the pyruvic acid dehydrogenase complex. The overall process can be written as the sum of individual reactions in a manner entirely analogous to the partial reactions written for reaction 13-1:

$$\underset{\alpha\text{-Ketoglutaric acid}}{\begin{array}{l} CH_2CO_2H \\ | \\ CH_2 \\ | \\ C(=O)-CO_2H \end{array}} + NAD^+ + CoA{-}SH \xrightarrow[\text{Lipoic acid, FAD}]{TPP,\ Mg^{2+}} \underset{\text{Succinyl-CoA}}{\begin{array}{l} CH_2CO_2H \\ | \\ CH_2 \\ | \\ C(=O)-S-CoA \end{array}} + NADH + H^+ + CO_2 \qquad (13\text{-}5)$$

$$\Delta G' = -8000 \text{ cal}$$

The reaction as a whole is not readily reversible because of the decarboxylation step. The molecular weight of the *E. coli* enzyme is 2×10^6.

Succinic Thiokinase. In the preceding reaction the high-energy bond of a thioester has been formed as the result of an oxidative decarboxylation. The enzyme *succinic thiokinase* catalyzes the formation of a high-energy phosphate structure at the expense of the thioester (reaction 13-6).

Since reaction 13-6 involves the formation of a new high-energy phosphate structure and the utilization of a thioester, the total number of high-energy

$$\underset{\text{Succinyl-CoA}}{\begin{array}{l} CH_2CO_2H \\ | \\ CH_2 \\ | \\ C(=O)-S-CoA \end{array}} + GDP + H_3PO_4 \rightleftharpoons \underset{\text{Succinic acid}}{\begin{array}{l} CH_2CO_2H \\ | \\ CH_2CO_2H \end{array}} + GTP + CoA{-}SH \qquad (13\text{-}6)$$

(Randomization of carbon atoms)

structures on each side of the reaction is equal. Therefore the reaction is readily reversible; the K_{eq} is 3.7. The GTP formed in reaction 13-6 can in turn react with ADP to form ATP and GDP. Since the pyrophosphate linkages in GTP and ATP have approximately the same $\Delta G'$ of hydrolysis, the reaction is readily reversible, with an K_{eq} of about 1:

$$GTP + ADP \rightleftharpoons GDP + ATP$$

Succinic Dehydrogenase. This enzyme catalyzes the removal of two hydrogen atoms from succinic acid to form fumaric acid:

$$\underset{\text{Succinic acid}}{\begin{array}{c} CO_2H \\ | \\ HCH \\ | \\ HCH \\ | \\ CO_2H \end{array}} + \text{FAD-Enz} \rightleftharpoons \underset{\text{Fumaric acid}}{\begin{array}{c} H \quad CO_2H \\ C \\ \| \\ C \\ HO_2C \quad H \end{array}} + \text{FADH}_2\text{-Enz} \qquad (13\text{-}7)$$

The immediate acceptor (oxidizing agent) of the electrons is a flavin coenzyme (FAD) which, in contrast to other flavin enzymes, is bound to succinic dehydrogenase through a covalent bond. Succinic dehydrogenase is firmly associated with the inner mitochondrial membrane and is rendered soluble only with difficulty. The "solubilized" preparations from beef heart and yeast contain 1 mole of flavin per mole of enzyme (200,000 mol wt) and four atoms of iron described as nonheme iron. Since the iron is not associated with a heme, as in the cytochromes, it is spoken of as "nonheme" iron. Several such proteins are known now, and it is clear that they function in oxidation–reduction reactions, the iron atom being alternately oxidized and reduced. Succinic dehydrogenase is competitively inhibited by malonic acid, a fact which was most useful to those who were concerned initially with working out the details of the tricarboxylic acid cycle.

Fumarase. The next reaction is the addition of H_2O to fumaric acid to form L-malic acid:

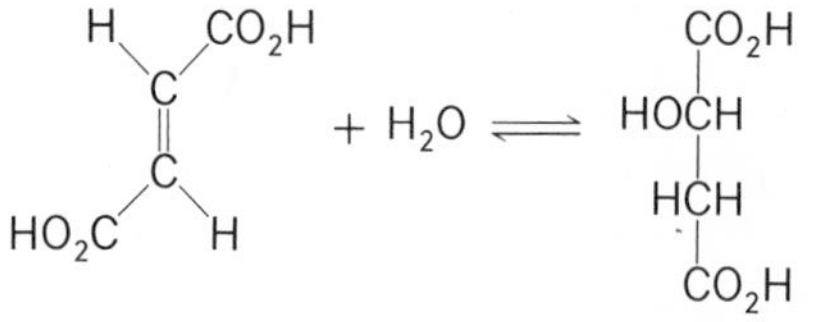

$$\underset{\text{Fumaric acid}}{\ce{HO2C-CH=CH-CO2H}} \text{ (trans: H and CO}_2\text{H on each C)} + H_2O \rightleftharpoons \underset{\text{L-Malic acid}}{\ce{HO2C-CH(OH)-CH2-CO2H}} \quad (13\text{-}8)$$

The equilibrium for this reaction is about 4.5. The enzyme that catalyzes the reaction, *fumarase*, has been crystallized (200,000 mol wt) from pig heart. Its kinetics and mechanism of action have been extensively studied.

Malic Dehydrogenase. The tricarboxylic cycle is completed when the oxidation of L-malic acid to oxalacetic acid is accomplished by the enzyme *malic dehydrogenase*. The reaction is the fourth oxidation–reduction reaction to be encountered in the cycle; the oxidizing agent for the enzyme from pig heart is NAD^+.

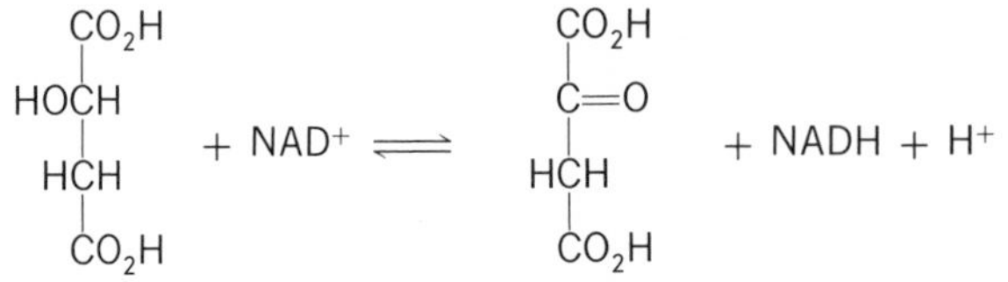

$$\underset{\text{L-Malic acid}}{\ce{HO2C-CH(OH)-CH2-CO2H}} + NAD^+ \rightleftharpoons \underset{\text{Oxalacetic acid}}{\ce{HO2C-C(=O)-CH2-CO2H}} + NADH + H^+ \quad (13\text{-}9)$$

At pH 7.0 the equilibrium constant is 1.3×10^{-5}; thus, the equilibrium is very much to the left. The further reaction of acetyl–CoA with oxalacetate in the condensation reaction (reaction 13-2) is strongly exergonic, however, in the direction of citrate synthesis. This tends to drive the conversion of malate to oxalacetate by displacing the equilibrium through the continuous removal of oxalacetate.

Features of the Tricarboxylic Acid Cycle

Stoichiometry. The balanced equation for the complete oxidation of pyruvate to CO_2 and H_2O may be written:

$$CH_3\overset{\displaystyle O}{\overset{\|}{C}}CO_2H + 2\tfrac{1}{2}\,O_2 \longrightarrow 3\,CO_2 + 2\,H_2O \quad (13\text{-}10)$$

Since this is accomplished in a stepwise manner by the reactions of the tricarboxylic acid cycle (reactions 13-1 through 13-9), it is useful to examine the stoichiometry in detail.

(1) There are five oxidation steps: reactions 13-1, 13-4, 13-5, 13-7, and 13-9. In each of these a pair of hydrogen atoms is removed from the substrate and transferred to either a nicotinamide coenzyme or a flavin coenzyme. As we shall see in Chapter 14, the reoxidation of these reduced coenzymes, five in all, by means of the cytochrome electron-transport system results in the reduction of five atoms or $2\frac{1}{2}$ moles of oxygen.
(2) When the five electron pairs are used to reduce O_2, 5 moles of H_2O are formed:

$$\frac{1}{2}\,O_2 + 2\,H^+ + 2\,e^- \longrightarrow H_2O$$

By inspection, one can see that 2 moles of H_2O have been consumed directly in reactions 13-2 and 13-8. To account for the net production of only 2 moles of H_2O in pyruvate oxidation (reaction 13-10) a third mole of H_2O must be accounted for. This is done by noting that GTP is produced from GDP and H_3PO_4 in reaction 13-6 of the cycle and that, in order to write the overall reaction of 13-10 as corresponding to the sum of reactions 13-1 through 13-9, the GTP produced in 13-6 must be balanced out—it does not appear in reaction 13-10—by consuming a third mole of H_2O to convert the GTP back to GDP and H_3PO_4.
(3) Finally, 3 moles of CO_2 are produced in the tricarboxylic acid cycle. These are equivalent to the three carbon atoms in the pyruvic acid, but note that only the CO_2 produced in reaction 13-1 arises directly from the pyruvic acid. The other two CO_2 (reactions 13-4 and 13-5) have as their origin the two carboxylic groups of oxalacetate (note shading).

All the reactions of the tricarboxylic acid cycle are reversible except the oxidative decarboxylation of α-ketoglutarate (reaction 13-5). As pointed out earlier, this reaction is entirely analogous to the irreversible oxidative decarboxylation of pyruvic acid. This then means that the cycle cannot be made to proceed in a reverse direction, although individual sections are reversible (from oxalacetate to succinate or from α-ketoglutarate to citrate, for example). Similarly, acetyl–CoA and CO_2 cannot be converted to pyruvate by a reversal of reaction 13-1.

The Effect of Inhibitors. Several compounds are known to serve as inhibitors of specific reactions of the tricarboxylic acid cycle. One of these, malonic acid, was instrumental in establishing the cyclical nature of the sequence of reactions. In the presence of 0.01*M* malonate the oxidation of succinate by succinic dehydrogenase is strongly inhibited. Therefore, in a muscle mince which can oxidize acids of the cycle but to which has been added 0.01*M* malonate the reactions will proceed only until succinate is formed, and this acid will accumulate.

The effect of malonate on the oxidation of pyruvate is important to understand. As described earlier, the addition of certain dicarboxylic and tricarboxylic acids to muscle homogenates stimulated the respiration of these tissues. Subsequently, it was shown that the oxidation of pyruvic acid by a muscle homogenate was catalyzed by the addition of the di- or tricarboxylic acid intermediates of the cycle to the system. On oxidation these acids give rise to oxalacetate which in turn condenses with the acetyl–CoA formed from pyruvate. Only a catalytic amount of the cycle intermediate is required, however, since once present it can traverse the cycle many times, with each passage disposing of a molecule of acetyl–CoA. This is the way the cycle normally functions in intact tissue.

It is clear that succinic dehydrogenase in the presence of malonate cannot oxidize succinate to fumarate; succinate will hence accumulate. Under these conditions the utilization of acetyl–CoA can proceed only if there is a supply of oxalacetate with which it can condense. Clearly then, fumarate, malate, and oxalacetate are the only compounds which can provide for the utilization of acetyl–CoA, since they are the intermediates in the cycle after succinic dehydrogenase which can be converted to oxalacetate. Moreover, they must be added in *stoichiometric amounts* equivalent to the acetyl–CoA which is to be utilized. The addition of the tricarboxylic acids or α-ketoglutaric acid in a malonate-inhibited system is of no help, since they can only be converted to oxalacetate through succinic dehydrogenase.

Another inhibitor of one of the enzymes of the tricarboxylic acid cycle is fluorocitrate, which inhibits aconitase. Fluorocitrate is an interesting inhibitor because it can be *synthesized* within the living cell and at that point accomplish its inhibitory action. Certain plants in South Africa are known to be toxic because they contain monofluoroacetic acid (FCH_2COOH). This compound, which has been used as a rodentocide, is acted on by the condensing enzyme to form fluorocitrate, apparently because the condensing enzyme is able to utilize fluoroacetyl–CoA as a substrate instead of acetyl–CoA. Once the fluorocitrate is formed, however, it strongly inhibits aconitase, and large quantities of citric acid accumulate in the tissues of poisoned animals.

Demonstration of the Tricarboxylic Acid Cycle. If the question of the occurrence of the cycle in an unstudied tissue is raised, the following characteristics should be evident: (*a*) the intermediates of the cycle should be oxidized by the particles; (*b*) the oxidation of pyruvic acid should be strongly stimulated by the addition of catalytic quantities of di- and tricarboxylic acids; (*c*) the oxidation of succinate should be inhibited by malonate; and (*d*) the oxidation of pyruvate in such an inhibited system should require stoichiometric quantities of dicarboxylic acids. It should be possible to detect the enzymes in the mitochondria and, if the particles are not isolated with considerable care, it may be necessary to add back certain of the cofactors required (for instance, NAD^+ or Mg^{2+}). Although all of these observations can be made with mitochondria, it is important to stress that the initial observations on animal tissues were made with tissue homogenates.

Oxidation of Krebs Cycle Intermediates. Up to this point we have stressed the catabolic nature of the Krebs cycle, i.e., its ability to accomplish the complete oxidation of pyruvic acid, or more precisely acetyl–CoA derived from pyruvate (reaction 13-1), to CO_2 and H_2O. It should be noted that the acetyl–CoA can be derived from other sources, for example from the breakdown of fatty acids (Chapter 12) or certain amino acids (Chapter 16).

The Krebs cycle obviously can serve as a mechanism for oxidizing the seven tri- and dicarboxylic acid intermediates of the cycle itself. As an example, consider the sequence of reactions by which succinate, produced in the breakdown of isoleucine, would be oxidized completely to CO_2 and H_2O. Initially, the succinate can be converted to oxalacetate by reactions 13-7 through 13-9. At this point the oxalacetate could then condense with a mole of acetyl–CoA, but this, in effect, would simply represent accelerated oxidation of acetate by the cycle, because of the increased levels of oxalacetate introduced from succinate. To accomplish the complete oxidation of oxalacetate, this compound would undergo decarboxylation in the presence of the mitochondrial PEP–carboxy kinase and GTP (reactions 10-21 and 13-12) to form phosphoenol pyruvic acid (PEP). This compound, after diffusing out of the mitochondria, could then be converted to pyruvate (reaction 10-9) in the cytoplasm. The latter compound would then reenter the mitochondria and be oxidized to CO_2 and H_2O by the cycle of reactions we have been discussing. Note that this process accomplishes the production of CO_2 from one of the four carbon atoms of oxalacetate in reaction 10-19, the other three being converted to CO_2 during the oxidation of pyruvate itself. The oxidation of oxalacetate by this combination of CO_2-fixation (reaction 10-21), glycolytic (reaction 10-9), and Krebs cycle enzymes obviously calls for coordination between the cytoplasm and the mitochondria since reaction 10-9 is catalyzed by a cytoplasmic enzyme. This is discussed on page 259.

The Amphibolic Nature of the Krebs Cycle

As will be seen in Chapter 16 the Krebs cycle is the primary source of certain key biosynthetic intermediates of the cell. A prominent example is α-ketoglutarate formed in reaction 13-4. α-Ketoglutarate provides the carbon skeleton for the biosynthesis of glutamic acid, glutamine, ornithine (and therefore $\frac{5}{6}$ of the carbon of citrulline and arginine), proline, and hydroxyproline. Another essential intermediate is succinyl–CoA, which is utilized in the synthesis of the porphyrins found in hemoglobin, myoglobin, and the cytochromes. Other more specialized examples can be cited: the citric acid that accumulates in the vacuoles of citrus species, or the isocitric and malic acids that are found in high concentration in certain Sedums and apple fruit would have their origin in the cycle.

In order for α-ketoglutarate, or any of the other Krebs cycle intermediates just mentioned, to function in an anabolic role, one essential condition must obviously be met. Both the C_2 unit (acetyl–CoA) and the C_4 unit (oxalacetate) that combine and give rise in the Krebs cycle to α-ketoglutarate (or other intermediate) must be provided in a stoichiometric amount equivalent to the α-ketoglutarate (or other intermediate) being removed for anabolic purposes. Thus, if a cell over a period of time needs to make 5.76 μmoles of glutamic acid from α-ketoglutarate, it must provide 5.76 μmoles of acetyl–CoA *and* 5.76 μmoles of oxalacetate to "balance the books," so to speak.

This consideration immediately introduces the question of the "normal" sources of the C_2 and C_4 units, questions which we have already considered. As noted in this chapter, the acetyl–CoA can be derived from pyruvate (reaction 13-1) and therefore have its origin in carbohydrates that give rise to pyruvate in glycolysis. The C_2 unit can also be derived from fatty acids during β-oxidation (Chapter 12). The C_4 unit oxalacetate can be derived from a number of sources; we have considered its production from pyruvate through the action of pyruvic carboxylase (reactions 10-20 and 13-11). It could also be produced by the action of PEP-carboxykinase (reactions 10-21 and 13-12) although physiologically this reaction appears to take carbon atoms out of the Krebs cycle rather than into it. A very important source of OAA is from the intermediates of the Krebs cycle itself. Thus, succinate, produced in the glyoxylic acid cycle soon to be described, could provide the OAA by its conversion to the C_4 unit via reactions 13-7, 13-8, and 13-9. Finally, oxalacetate can be produced by the transamination of aspartic acid (Chapter 16).

Anaplerotic Reactions. The preceding section on the anabolic nature of the Krebs cycle has raised the question of how the level of intermediates can be *replenished* when, for example, certain of those intermediates are removed for anabolic purposes. H. L. Kornberg has proposed the term *anaplerotic* for these replenishing or "filling-up" reactions. A relisting of the ones we have already considered and additional ones not yet described is appropriate at this point.

(1) Pyruvic carboxylase:

$$CO_2 + \underset{\text{Pyruvic acid}}{\begin{matrix} CO_2H \\ | \\ C{=}O \\ | \\ CH_3 \end{matrix}} + ATP + H_2O \xrightleftharpoons[\text{}]{\substack{Mg^{2+} \\ \text{Acetyl-CoA}}} \underset{\text{Oxalacetic acid}}{\begin{matrix} CO_2H \\ | \\ C{=}O \\ | \\ CH_2 \\ | \\ CO_2H \end{matrix}} + ADP + H_3PO_4 \quad (13\text{-}11)$$

The properties of this enzyme were described in detail in Chapter 10. Together with PEP–carboxy kinase (reaction 13-12) and PEP–carboxylase (reaction 13-13), it catalyzes reactions linking intermediates of the glycolytic sequence and the tricarboxylic acid cycle.

(2) Phosphoenol pyruvic acid carboxykinase is described on page 260:

$$CO_2 + \underset{\text{Phosphoenol pyruvic acid}}{\begin{matrix} COOH \\ | \\ C{-}OPO_3H_2 \\ \| \\ CH_2 \end{matrix}} + GDP \rightleftharpoons \underset{\text{Oxalacetic acid}}{\begin{matrix} COOH \\ | \\ C{=}O \\ | \\ CH_2 \\ | \\ CO_2H \end{matrix}} + GTP \quad (13\text{-}12)$$

(3) The enzyme phosphoenol pyruvic acid carboxylase (PEP–carboxylase) catalyzes reaction 13-13:

$$CO_2 + \underset{\text{Phosphoenol pyruvic acid}}{\begin{matrix} CO_2H \\ | \\ C{-}OPO_3H_2 \\ \| \\ CH_2 \end{matrix}} + H_2O \longrightarrow \underset{\text{Oxalacetic acid}}{\begin{matrix} CO_2H \\ | \\ C{=}O \\ | \\ CH_2 \\ | \\ CO_2H \end{matrix}} + H_3PO_4 \quad (13\text{-}13)$$

The enzyme requires Mg^{2+} for activity; the reaction is irreversible. Phosphoenol pyruvic acid carboxylase occurs in higher plants, yeast, and bacteria (except pseudomonads), but not in animals. It presumably has the same function as pyruvic carboxylase, namely, to ensure that the Krebs cycle has an adequate supply of oxalacetate. The enzyme in some species is activated by fructose-1,6-diphosphate; this is consistent with the function of seeing that the Krebs cycle can adequately oxidize the pyruvate being formed from glucose. Phosphoenol pyruvic acid carboxylase is inhibited by aspartic acid; this effect is understandable when it is recognized that oxalacetate is the direct precursor of aspartic acid (by transamination). Thus, the biosynthetic sequence

$$\text{Phosphoenol pyruvate} \longrightarrow \text{Oxalacetate} \longrightarrow \text{Aspartic acid}$$

is a simple means for synthesizing aspartate from PEP, and aspartate can control its own production by inhibiting the first step in the sequence.

(4) Malic enzyme catalyzes the reversible formation of L-malate from pyruvate and CO_2; the K_{eq} for the reaction at pH 7 is 1.6. Malic enzyme is

$$CO_2 + \underset{\text{Pyruvic acid}}{\begin{array}{c} CO_2H \\ | \\ C{=}O \\ | \\ CH_3 \end{array}} + NADPH + H^+ \rightleftharpoons \underset{\text{L-Malic acid}}{\begin{array}{c} CO_2H \\ | \\ HOCH \\ | \\ CH_2 \\ | \\ CO_2H \end{array}} + NADP^+ \qquad (13\text{-}14)$$

found in plants, in some tissues of animals (liver) and in some bacteria grown on malic acid (i.e., it is an adaptive enzyme in such organisms). The enzyme, which is found in the cytoplasm, is believed to be significant because of its ability to produce NADPH required for biosynthetic purposes. It, together with the two dehydrogenases of the pentose phosphate pathway (page 272) and the $NADP^+$-specific isocitric dehydrogenase, provides a means for producing the reduced coenzyme from either glycolytic or Krebs cycle intermediates. In bacteria the malic enzyme can account for the production of acetate (via pyruvate) from an intermediate of the Krebs cycle.

Reactions 13-11 through 13-14 are examples of "CO_2-fixation reactions." The initial observations that stimulated work on these reactions were made by Wood and Werkman in 1936. They observed that when propionic acid bacteria fermented glycerol to propionic and succinic acids, more carbon was found in the products than had been added as glycerol. Carbon dioxide, moreover, proved to be the source of the extra carbon atoms or the carbon that was "fixed." Today, the physiological significance of CO_2-fixation extends beyond the metabolism of propionic acid bacteria and includes not only the anaplerotic reactions listed above but also such enzymes as *acetyl–CoA carboxylase* (page 291), *propionyl carboxylase* (page 289), and *ribulose-1,5-diphosphate carboxylase* (page 354).

The Glyoxylic Acid Cycle

Two major roles of the tricarboxylic acid cycle have now been described: the complete oxidation of acetyl–CoA (and compounds convertible to acetyl–CoA) and multiple anabolic activities—e.g., the synthesis of glutamic acid, succinyl–CoA, aspartic acid. Since the reactions of the cycle (reaction 13-2 through 13-9) can only degrade acetate, there remains the basic question of how some organisms (many bacteria, algae, and certain higher plants at a certain stage in their life cycle) can utilize acetate as the only carbon source for all the carbon compounds of the cell. Or put another way, since acetate can only be oxidized to CO_2 and H_2O by the Krebs cycle, how can it, in some organisms, give rise to carbohydrates as well as amino acids derivable from the tricarboxylic acid cycle?

This challenging problem was most successfully pursued by H. L. Kornberg, University of Leicester, who together with others showed that acetate undergoes, in those organisms that convert acetate to carbohydrate, an *anabolic* sequence called the *glyoxylate cycle*. In effect, the glyoxylate cycle bypasses reactions 13-4 through 13-8 of the Krebs cycle, thereby omitting the two reactions in which CO_2 is produced (reactions 13-4 and 13-5). The bypass consists of two reactions that convert isocitrate to malic acid; in the process 1 mole of acetyl–CoA is assimilated and 1 mole of succinic acid is formed.

Consider first the two reactions. Instead of isocitrate being oxidized, it is cleaved by the enzyme isocitritase (isocitrate lyase) to form succinic and glyoxylic acids:

$$\underset{\text{Isocitrate}}{\begin{matrix} CH_2\text{—}COOH \\ | \\ HC\text{—}COOH \\ | \\ HOC\text{—}COOH \\ H \end{matrix}} \rightleftharpoons \underset{\text{Succinate}}{\begin{matrix} CH_2\text{—}COOH \\ | \\ CH_2\text{—}COOH \end{matrix}} + \underset{\text{Glyoxylate}}{\begin{matrix} O{=}C\text{—}COOH \\ H \end{matrix}} \qquad (13\text{-}15)$$

The glyoxylic acid formed is then condensed with 1 mole of acetyl–CoA to produce L-malic acid in a reaction analogous to that of citrate synthase (reaction 13-2) discussed earlier. The enzyme involved is called malate synthase:

$$\underset{\text{Acetyl–CoA}}{\begin{matrix} H_3C\text{—}C\text{—}S\text{—}CoA \\ \| \\ O \end{matrix}} + \underset{\text{Glyoxylate}}{\begin{matrix} O{=}C\text{—}COOH \\ H \end{matrix}} + H_2O \longrightarrow \underset{\text{L-Malate}}{\begin{matrix} CH_2\text{—}COOH \\ | \\ HOC\text{—}COOH \\ H \end{matrix}} + CoA\text{—}SH \qquad (13\text{-}16)$$

While these two reactions bypass the decarboxylation steps of the Krebs cycle, this in itself does not constitute a cycle until they are written together with the reactions catalyzed by malic dehydrogenase (reaction 13-9), citrate synthase (13-2), and aconitase (13-3). Together the five enzymes constitute the glyoxylate cycle and accomplish the conversion of 2 moles of acetate (as acetyl–CoA) to succinic acid (Scheme 13-3).

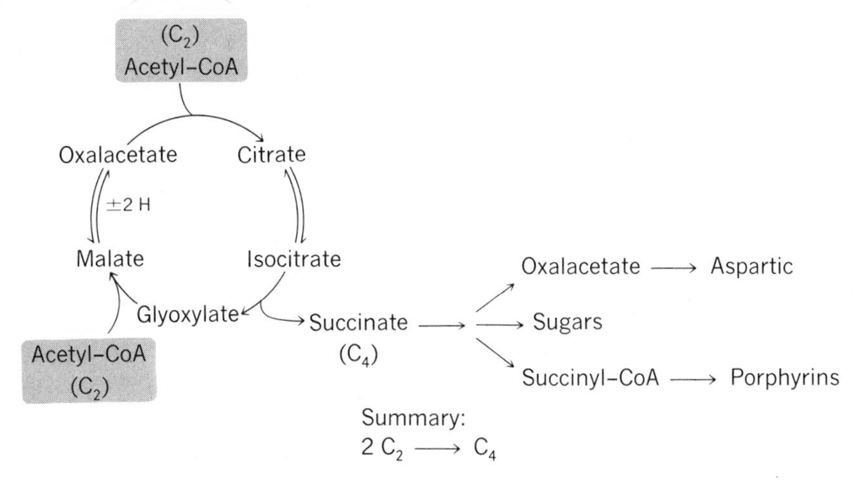

Scheme 13-3

The full significance of the glyoxylate cycle can now be appreciated by realizing that succinate, as a product of the cycle, can undergo reactions which have been previously described. For example, the succinate can be converted to succinyl–CoA (reaction 13-6) and serve as a precursor of porphyrins. The succinate can be oxidized to oxalacetate via reactions 13-7 through 13-9 and can be utilized for the synthesis of aspartic acid (page 365) and other compounds (e.g., pyrimidines) derived from aspartic acid. The oxalacetate can be converted to PEP and can undergo the reactions of the glycolytic sequence. Finally, the oxalacetate could condense with acetyl–CoA (reaction 13-2) and meet the requirements earlier specified (page 313) for the Krebs cycle to function in an anabolic manner.

Those tissues of higher plants that have a functioning glyoxylate cycle contain organelles (glyoxysomes) having the five enzymes required for the cycle to operate. It is interesting to note that the glyoxysome appears in the cotyledons of high-lipid seeds shortly after germination begins and at a time when lipids are being utilized as the major source of carbon for carbohydrate synthesis. Thus, the high-lipid seed (e.g., peanut, castor bean) can convert lipid to carbohydrate, a synthesis which animals are incapable of performing since they lack the glyoxylate cycle.

Mitochondrial Compartmentation

Reference was made in Chapter 6 to the localization of the enzymes of the tricarboxylic acid cycle either in the inner membrane or in the inner matrix. The numerous coenzymes (NAD^+, FAD, TPP, lipoic acid, CoA–SH) and cofactors (Mg^{2+}, ADP, GDP) associated with the cycle are also located in the membrane or matrix separated from pools of these compounds in the cytoplasm. The oxidation of pyruvic acid proceeds only after that α-keto acid has penetrated, which it freely does, through the inner membrane into the matrix where the necessary enzymes are located.

Entry of the Krebs cycle acids into the mitochondrial matrix is accomplished by an exchange mechanism. Thus, for malate and succinate to enter, an equivalent amount of inorganic phosphate must leave the mitochondrion. The same exchange mechanism applies for citrate and isocitrate. In contrast, fumarate and oxaloacetate cannot penetrate the mitochondrial membranes from either direction.

α-Ketoglutarate can move in or out of the organelle, but an equivalent amount of malate must pass in the opposite direction (see page 334 for further discussion).

An important source of acetate units for fatty acid synthesis is the acetyl–CoA produced by the oxidative decarboxylation of pyruvate inside the mitochondria. However, in order to participate in fatty acid synthesis, the acetyl–CoA must be transferred to the cytoplasm where the process occurs. The inability of acetyl–CoA to pass through the inner membrane was referred to in Chapter 12 as was the role of carnitine in transporting fatty acids through that barrier. While acetyl units can be transported out of the mitochondria in the form of acetyl–carnitine, citrate can also serve a similar role.

Citrate produced by reaction 13-2 can leave the mitochondria and, in the cytoplasm, be cleaved by the ATP-dependent citrate cleavage enzyme to produce acetyl–CoA:

$$\underset{\text{Citrate}}{\begin{array}{c} CH_2{-}CO_2H \\ | \\ HO{-}C{-}CO_2H \\ | \\ CH_2{-}CO_2H \end{array}} + ATP + CoASH \longrightarrow \underset{\text{Acetyl–CoA}}{C_3H{-}\overset{\displaystyle O}{\overset{\|}{C}}{-}S{-}CoA} \;\; ADP + H_3PO_4 + \underset{\text{Oxalacetate}}{\begin{array}{c} COOH \\ | \\ C{=}O \\ | \\ CH_2 \\ | \\ COOH \end{array}} \qquad (13\text{-}17)$$

The carbon atoms in the oxalacetate produced in this reaction can make their way back into the mitochondria after being reduced to malate by the NAD^+ malic dehydroplase that occurs in the cytoplasm. In the mitochondrial matrix the malate will be oxidized to oxalacetate, which can then pick up another mole of acetyl–CoA to form citrate and repeat the process.

References

1. T. W. Goodwin, ed., *The Metabolic Roles of Citrate*. London: Academic Press, 1968.
   The book contains eight articles covering numerous aspects of the tricarboxylic acid cycle. The articles were presented at a symposium of the Biochemical Society honoring Sir Hans Krebs, the chief contributor to the metabolic cycle that often bears his name.
2. J. M. Lowenstein, "The Tricarboxylic Acid Cycle," in *Metabolic Pathways,* D. M. Greenberg, ed. 3rd ed., vol. 1. New York: Academic Press, 1967.
   A recent review of the tricarboxylic acid cycle, including stereochemical aspects.
3. H. L. Kornberg, "Anaplerotic Sequences and Their Role in Metabolism," in *Essays in Biochemistry*. vol. 2. London: Academic Press, 1966.
   The operation of the glyoxylate cycle and its anaplerotic role is reviewed by the authority in the subject.
4. H. Beevers, *Ann. N.Y. Acad. Sci.* **168,** 313 (1969).
   The glyoxylate cycle as it occurs in certain plant tissues is described.
5. H. G. Wood and M. F. Utter, "The Role of CO_2 Fixation in Metabolism," in *Essays in Biochemistry*. vol. 1. London: Academic Press, 1965.
   A lucid review of the subject of CO_2-fixation and the role it plays in replenishing the tricarboxylic acid cycle.

Electron Transport and Oxidative Phosphorylation

Introduction

In Chapter 13 the oxidation of pyruvate and acetyl–CoA as it occurs in mitochondria was described. Although the oxidation is frequently called the *aerobic* phase of carbohydrate metabolism, the title is misleading, since both pyruvate and acetyl–CoA can also be obtained from noncarbohydrate sources. In addition, the term aerobic is not strictly precise when the reactions are described as on the inside of the front cover, since the immediate oxidizing agents are nicotinamide and flavin nucleotides rather than oxygen. As in the glycolytic sequence, the amount of nicotinamide and flavin nucleotides in the cell is limited, and the reactions cease when the supply of oxidized nucleotides is exhausted. Hence, in order for the oxidation of organic substrates to continue, the reduced nicotinamide and flavin nucleotides must be reoxidized.

In procaryotic cells, the reoxidation is accomplished by enzymes located on the cell membrane; in eucaryotes the necessary catalysts are in the inner membrane of the mitochondrion adjacent to the matrix where the nucleotides are reduced (see page 145). In all aerobic organisms the ultimate oxidizing agent is molecular oxygen and, for the case of NADH we may write the overall reaction

$$\text{NADH} + \text{H}^+ + \tfrac{1}{2}\,\text{O}_2 \longrightarrow \text{NAD}^+ + \text{H}_2\text{O} \qquad (14\text{-}1)$$

$$\Delta G' = -52{,}500 \text{ cal (pH 7.0)}$$

The enzymes accomplishing this oxidation constitute an *electron-transport chain* in which a series of electron carriers are alternately reduced and oxidized.

This reoxidation of NADH by O_2 is accompanied by a large decrease in free energy (page 169). The amount is sufficient to produce several moles of ATP per mole of NADH oxidized. The enzymes which catalyze the production of ATP as the NADH is oxidized are also localized in the mitochondrial inner membrane. Although the process is known as *oxidative phosphoryla-*

tion, it is perhaps better described as *respiratory-chain phosphorylation.* This process will be described after the electron-transport chain is discussed.

Composition of the Electron-Transport Chain

Pyridine Nucleotide Dehydrogenases. Four of the five dehydrogenases involved in the oxidation of pyruvic acid by means of the tricarboxylic acid cycle utilize nicotinamide nucleotides as oxidants. In two, pyruvic dehydrogenase and α-ketoglutaric dehydrogenase, lipoic acid is first reduced and it then reduces NAD^+ in the presence of a specific flavoprotein. The other two, malic and isocitric dehydrogenases, catalyze the direct removal of the equivalent of two hydrogen atoms from their respective substrates; the oxidized nucleotide in turn accepts the equivalent of two electrons and one proton and is thereby reduced. For malic dehydrogenase we may write

$$\underset{\text{L-Malate}}{\text{HO}_2\text{C-CHOH-CH}_2\text{-CO}_2\text{H}} + \underset{\text{NAD}^+}{\text{NAD}^+} \rightleftharpoons \underset{\text{Oxalacetate}}{\text{HO}_2\text{C-C(=O)-CH}_2\text{-CO}_2\text{H}} + \underset{\text{NADH}}{\text{NADH}} + \text{H}^+ \qquad (14\text{-}2)$$

The properties of the nicotinamide nucleotides and their dehydrogenases were described in Chapter 9. Although the reactions catalyzed by these enzymes are, in general, reversible, the ability of the electron-transport chain to link the reoxidation of the NADH to the reduction of O_2 means that the oxidation of malate (and not the reduction of oxalacetate) is favored.

Flavin Dehydrogenases. Several flavoproteins are involved in the electron-transport chain; all of them utilize FAD as an oxidant. The *dihydrolipoyl dehydrogenase* components of pyruvic and α-ketoglutaric dehydrogenase (page 306) transfer two electrons from dihydrolipoic acid to NAD^+ via FAD.

The flavoprotein *NADH dehydrogenase* catalyzes the transfer of two electrons from NADH to the next carrier in the chain, probably a quinone called coenzyme Q. The prosthetic group of this dehydrogenase, FAD, is reduced in an intermediate step.

$$\text{NADH} + \text{H}^+ + \text{FAD} \rightleftharpoons \text{NAD}^+ + \text{FADH}_2 \qquad (14\text{-}3)$$

Succinic dehydrogenase catalyzes the removal of two hydrogen atoms from the succinate molecule. These are transferred to oxidized FAD, thereby reducing it.

In recent years, a number of flavoproteins have been shown to contain metals such as molybdenum and iron. The latter is firmly associated with a specific protein called a *nonheme iron protein,* because the iron is present in a form other than heme (see page 372). These proteins are relatively small, having molecular weights of 6000–12,000. Whether they function as carriers by the iron being alternately reduced and oxidized is not clear. The nonheme iron proteins in the mitochondria are similar in several respects to *ferredoxin* (page 350), which is involved in photosynthesis.

Quinones. There is good evidence that a quinone serves as a carrier between flavoproteins and the cytochromes. Mitochondria contain a quinone called ubiquinone which has the general structure

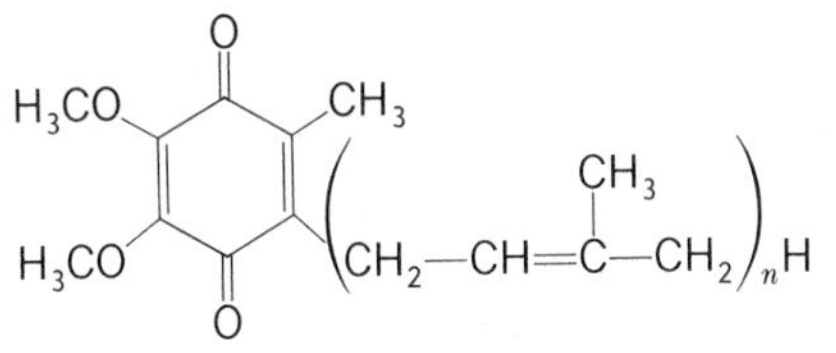

Ubiquinone

The length of the side chain varies with the source of the mitochondria; in animal tissues the quinone possesses ten isoprenoid units in its side chain and is called coenzyme Q_{10} (CoQ_{10}). If the quinone is extracted from the mitochondria, the transport of electrons from substrates to oxygen is inhibited; the activity is restored when the quinone is added back. Because it is easily reduced and oxidized, it may serve as an additional carrier between the flavin coenzymes and the cytochromes.

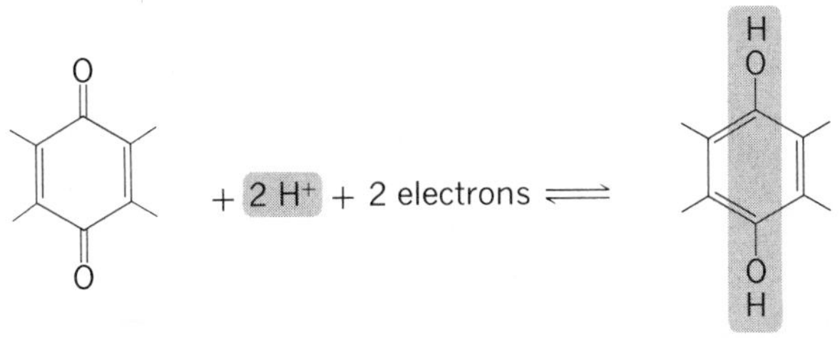

$$+ \; 2\,H^+ + 2 \text{ electrons} \rightleftharpoons$$

Oxidized quinone

Hydroquinone
or reduced quinone

Coenzyme Q_{10} probably serves as an acceptor of electrons not only from NADH dehydrogenase, but also from the flavin components of succinic dehydrogenase, glycerol phosphate dehydrogenase, and fatty acyl-CoA dehydrogenase, as illustrated in Figure 14-2.

The Cytochromes. These respiratory carriers were among the earliest components of the electron-transport chain to be studied. The classical experiments of D. Keilin in England in 1926–1927 demonstrated that these cell pigments (cytochromes) were found in almost all living tissues which implied an essential role for these substances in cellular respiration. Indeed, Keilin's studies

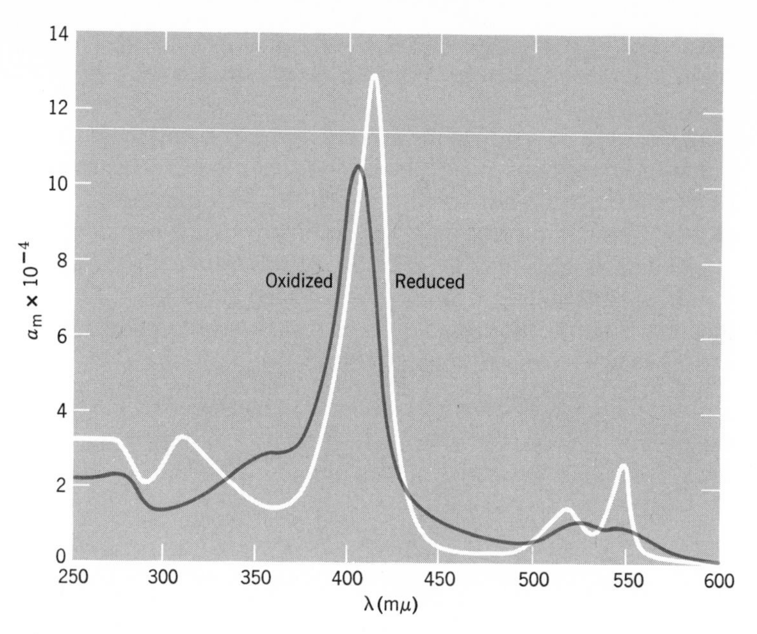

Figure 14-1

Absorption spectra of oxidized and reduced cytochrome c. [Data of E. Margoliash, reproduced from D. Keilin and E. C. Slater, *British Medical Bulletin* 9, 89 (1953), The British Council, London.]

showed that there were at least three cytochromes, to which he assigned the letters a, b, and c. The research on the cytochromes was facilitated by the fact that they absorb light of different wavelengths in a characteristic manner. The absorption spectra of oxidized and reduced cytochrome c are shown in Figure 14-1. This cytochrome is the only one which is readily

HOOC—CH$_2$—CH$_2$
HOOC—CH$_2$—CH$_2$
H$_3$C
H$_3$C
H$_3$C
CH$_3$
CH$_3$
CH$_3$
Fe
N
HC
CH
NH
S—CH$_2$—CH
C=O
Protein

Cytochrome c

solubilized from the mitochondrial membrane. Its structure has been studied extensively (page 94).

The absorption spectra together with other properties of the cytochromes indicate that these compounds are conjugated proteins having an iron porphyrin as a prosthetic group. The structure of the prosthetic group for cytochrome c is shown; it is a derivative of iron–protoporphyrin IX (page 373), and it is bound through thioether linkages with cysteine residues in the protein component. The iron porphyrins associated with cytochromes a and b are known to be different because of differences in their absorption spectra; this has subsequently been confirmed by chemical studies on the structure of the porphyrins. The prosthetic group of cytochrome b is known to be iron-protoporphyrin IX. The porphyrin of cytochrome a is porphyrin A, characterized chiefly by a long hydrophobic isoprenoid chain. In this regard porphyrin A resembles the porphyrin of chlorophyll (page 343).

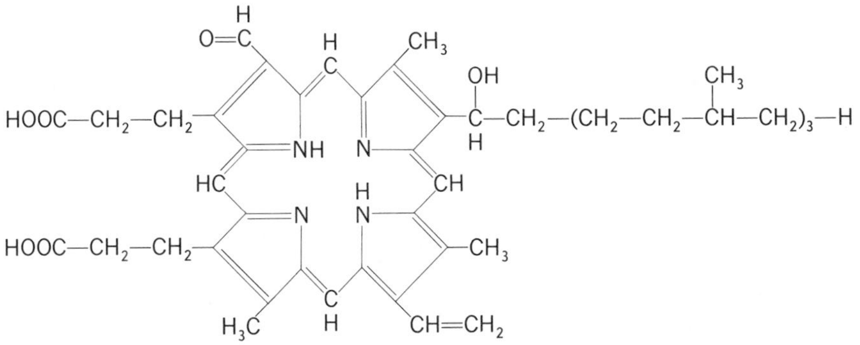

Porphyrin A

The cytochromes form complexes with substances such as HCN, CO, and H_2S; these complexes were first identified by their characteristic absorption spectra. It is the combination of carbon monoxide with the ferrous iron of cytochrome a-a_3 which accounts for the extreme toxicity of this compound to biological organisms.

The studies on soluble cytochrome c confirmed what Keilin had originally observed in intact tissues, that the cytochromes are capable of being alternately reduced and oxidized. The iron of the oxidized cytochromes is ferric iron; it is reduced to ferrous iron by the incorporation of one electron into the valence shell of the iron atom. Indeed, it is this property that allows the cytochromes to function as carriers in the electron-transport process. As indicated above, cytochromes are reduced when CoQ_{10}–H_2 is reoxidized. Since each reduced quinone can furnish two electrons for the reduction of the iron pigment, two molecules of cytochrome are required to react with one molecule of reduced quinone:

$$CoQ_{10}—H_2 + 2\ \text{Cytochrome–}(Fe^{3+}) \longrightarrow CoQ_{10} + 2\ \text{Cytochrome–}(Fe^{2+}) + 2\ H^+ \quad (14\text{-}4)$$

The reaction is balanced by the release of two protons into the medium.

There are five cytochromes of the a, b, and c types in the mitochondria of aerobic cells. These react in a sequence, with cytochrome b the first to be reduced, then cytochrome c_1, cytochrome c, cytochrome a, and finally cytochrome a_3 (Scheme 14-1).

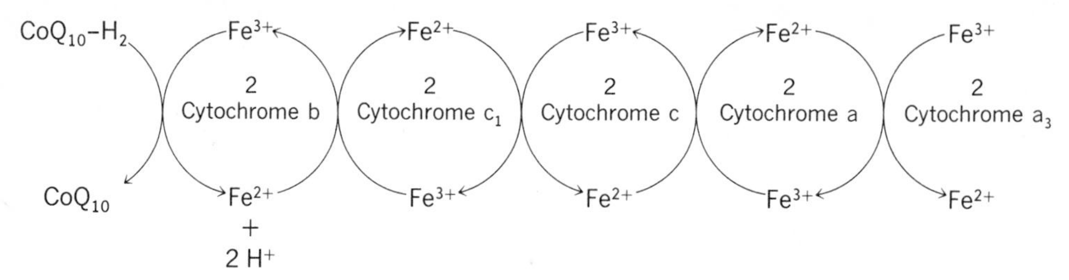

Scheme 14-1

In discussing the cytochromes some extra attention should be paid to cytochromes a and a_3. Together they constitute *cytochrome oxidase*, a term used for many years to describe the last carrier or *terminal oxidase* in the chain of electron transport in aerobes. The reduced form of cytochrome oxidase was known to be capable of reducing molecular O_2 to H_2O, a process requiring a total of four electrons for each mole of O_2 reduced.

The firm association of cytochrome oxidase with the inner mitochondrial membrane has made study of this iron–porphyrin system difficult. Evidence now indicates that the enzyme is a complex (240,000 mol wt) consisting of six subunits, each containing a heme A group and one atom of copper. The hexamer has two units which differ in their absorption spectrum from the rest; these, called cytochrome a, are not able to react directly with O_2. The other four units of the hexamer are called a_3 and their reduced forms can react with O_2. Without implying anything about the mechanism of the reaction, we may write it as

$$4 \text{ Cytochrome } a_3\text{—}Fe^{2+} + O_2 + 4\,H^+ \longrightarrow 4 \text{ Cytochrome } a_3\text{—}Fe^{3+} + 2\,H_2O$$

Cytochrome a_3 combines with carbon monoxide to form a complex that is dissociable by light. In one of the classic experiments in biochemistry, Warburg determined the action spectrum of this complex to establish the iron–porphyrin nature of this enzyme.

The Respiratory Chain

The components of the respiratory chain may be arranged as shown in Figure 14-2. Note that electrons may enter at NAD^+ direct from substrates such as malate, isocitrate, glutamate, and coenzyme A esters of β-hydroxy fatty acids. Electrons may also enter at NAD^+ from pyruvate and α-ketoglutarate dehydrogenase via the flavoproteins associated with these multienzyme complexes. Finally, electrons may bypass NAD^+ and enter the chain at the level of CoQ, in every case having first been passed through a flavin carrier. In anticipation of the next section, the sites of ATP formation are indicated as are the sites of action of some of the more common inhibitors of electron transport. What, then, is the evidence for the arrangement shown in Figure

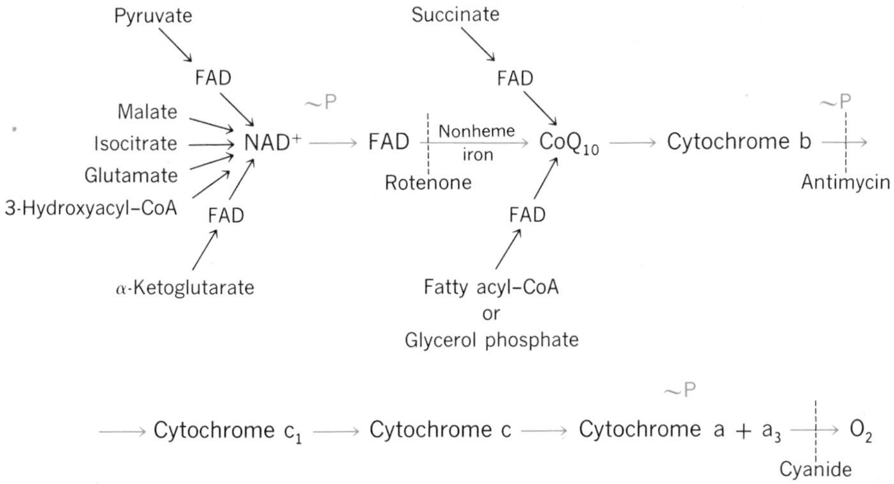

Pyruvate → FAD → NAD+
Malate, Isocitrate, Glutamate, 3-Hydroxyacyl-CoA → NAD+
α-Ketoglutarate → FAD → NAD+

NAD^+ —(~P)→ FAD —(Nonheme iron; Rotenone)→ CoQ_{10} → Cytochrome b —(~P; Antimycin)→

Succinate → FAD → CoQ_{10}
Fatty acyl–CoA or Glycerol phosphate → FAD → CoQ_{10}

→ Cytochrome c_1 → Cytochrome c → Cytochrome a + a_3 —(~P; Cyanide)→ O_2

Figure 14-2

The respiratory chain showing the coenzymes, heme proteins, and oxidation–reduction carriers (including nonheme iron) that are involved.

14-2? Among the many pieces of evidence we can cite the following:

(1) The sequence presented is consistent with the reduction potentials of the carriers in that as one moves along the chain, from NAD^+ to oxygen, the E_0' becomes more positive (Table 7-3).

(2) The *respiratory assembly* that contains the electron-transport chain has been broken into smaller units containing portions of the chain; we have already mentioned the close association of cytochromes a and a_3. Complexes containing cytochromes b and c_1 as well as NADH dehydrogenase, nonheme iron, CoQ, and cytochrome b have been prepared.

(3) With these fragments, the electron-transport chain can be partially reconstructed. Thus, reduced cytochrome c cannot be reoxidized directly by molecular O_2 and requires a preparation of cytochrome oxidase (a + a_3). Similarly, NADH will not reduce cytochrome c in the absence of the NADH dehydrogenase complex.

(4) Examination of the absorption spectra of intact tissue by sensitive spectrophotometric equipment supports the arrangement shown in Figure 14-2. With this technique, B. Chance has measured the relative amounts of the oxidized and reduced forms of the different carriers under two different conditions. He has shown that when electrons are being fed into the chain, NAD^+ is the most reduced component, whereas cytochrome a + a_3 are the most fully oxidized. Nearly all the other carriers can be observed *in vivo* and were present in successively more reduced states as one goes from O_2 to the source of the electrons. Moreover, when electron transport is blocked by one of the inhibitors indicated in Figure 14-2, all the carriers to the left or reducing side of the site of inhibition become reduced, while those on the oxidizing side become more oxidized.

The clarification of the sequence of carriers in the electron-transport chain aided the study of the phosphorylation reactions that accompany the transport of electron from substrate to O_2. This process of respiratory-chain phosphorylation can now be discussed.

Oxidative Phosphorylation

A major objective of the degradation of carbon substrates by a living organism is the production of energy for the development and growth of that organism. In the anaerobic degradation of sugars to lactic acid, some of the energy available in the sugar molecule was conserved in the formation of energy-rich phosphate compounds, which are made available to the organism. As pointed out in Chapter 13, however, over 90% of the energy available in glucose is released when pyruvate is oxidized to CO_2 and H_2O through the reactions of the tricarboxylic acid cycle. In that process there was only one energy-rich compound, namely succinyl–CoA, synthesized by reactions involving the substrates of the cycle itself; in the presence of succinic thiokinase, this thioester was utilized to convert GDP to GTP.

When the production of energy-rich compounds in biological organisms was investigated in more detail, two different types of phosphorylation process were recognized. In one of the processes, phosphorylated or thioester derivatives of the substrate were produced initially and were subsequently utilized to produce ATP. Examples of these are the reactions of glycolysis in which phosphoenol pyruvic acid and 1,3-diphosphoglyceric acid are formed and react with ADP to form ATP, as well as the reaction catalyzed by succinic thiokinase in the Krebs cycle. These phosphorylation processes have been referred to as *substrate-level phosphorylations* and are to be distinguished from the phosphorylations associated with electron transport which are usually referred to as *oxidative phosphorylation*.

In 1937, Belitzer in Russia and Kalckar in the United States observed that phosphorylation occurred during the oxidation of pyruvic acid by muscle homogenates. Although the subsequent fate of the pyruvate molecule was not clear at that time, oxygen was consumed by the homogenate, and inorganic phosphate was esterified as hexose phosphates. If the reaction were inhibited by cyanide or by the removal of O_2, both the phosphorylation and the oxidation ceased. Thus, the synthesis of a sugar phosphate bond was dependent on a biological oxidation in which molecular oxygen was consumed.

Several important advances occurred which simplified the study of this important process. First, in 1948, Kennedy and Lehninger showed that isolated rat liver mitochondria catalyzed oxidative phosphorylation coupled to the oxidation of Krebs cycle intermediates. Today, it is recognized that the inner membrane of mitochondria is the locus of the phosphorylation enzymes. In bacteria, smaller units in the cell membrane contain the phosphorylation assemblies.

Second, it was found that the only phosphorylation reaction that could be identified was the incorporation of inorganic phosphate into ADP to form ATP:

$$ADP + H_3PO_4 \longrightarrow ATP + H_2O$$

This is clearly a reaction which requires energy; if all reactants are in the standard state, the ΔG ($\Delta G'$ by definition) would be +7300 cal/mole. Since the reactants are undoubtedly not present at concentrations of 1*M*, the ΔG will be considerably larger, perhaps as much as +12,000 cal/mole.

Third, the composition of the electron-transport chain of mitochondria was investigated in some detail; and fourth, the oxidation of NADH itself by O_2 in the presence of mitochondria was shown to lead to the formation of ATP by the esterification of inorganic phosphate. The importance of this extremely significant observation by Friedkin and Lehninger should be emphasized.

If NADH is added to a reaction mixture containing ADP, inorganic phosphate, Mg^{2+}, and animal or plant mitochondria which have been properly prepared (see discussion of mitochondrial permeability below), the NADH will be oxidized to NAD^+, and one atom of O_2 will be reduced. This occurs because, as described earlier, mitochondria contain the intact electron-transport chain. Simultaneously with this oxidation, inorganic phosphate will react with ADP to form ATP. Under ideal conditions, between 2 and 3 moles of ATP will be formed per atom of O_2 consumed. Since the mitochondria contain ATP-ase and also can catalyze side reactions which utilize ATP, it is believed that 3 moles of ATP are formed per mole of NADH oxidized or atom of oxygen consumed. This may be represented schematically as

$$\text{NADH} + \text{H}^+ + \tfrac{1}{2}\,\text{O}_2 + 3\,\text{ADP} + 3\,\text{H}_3\text{PO}_4 \longrightarrow \text{NAD}^+ + 3\,\text{ATP} + 4\,\text{H}_2\text{O}$$

This reaction is also said to have a *P:O ratio* of 3.0, a term used to describe the ratio of the atoms of phosphate esterified to the atoms of oxygen consumed in the oxidation. Since the oxidation of malate and isocitrate exhibited P:O ratios of 3.0, the phosphorylations associated with these substrates were assumed to occur after the NAD^+ that serves as oxidant was reduced. A P:O ratio of 2.0 for succinate similarly indicated that one fewer phosphorylation step was involved when this compound is oxidized.

Much experimental evidence supports the conclusion that the phosphorylations occur as a pair of electrons makes its way along the electron-transport chain pictured in Figure 14-2. Since only one phosphorylation occurs when reduced cytochrome c is oxidized by molecular oxygen, only one phosphorylation site is shown to the right of cytochrome c in Figure 14-2. When NADH is oxidized by cytochrome c, two phosphorylations occur (two ATP's are formed per mole of NADH oxidized) and their postulated sites of formation in the chain are shown. Since one of these occurs before CoQ in the chain, and since electrons from succinate oxidation enter at the level of CoQ, the P:O ratio of 2.0 for succinate oxidation is to be expected.

The sites of phosphorylation in the electron-transport chain are not yet settled. Experimental evidence based on the use of inhibitors and *uncoupling agents* support the positions indicated in Figure 14-2. Uncoupling agents are compounds which uncouple the synthesis of ATP from the transport of electrons through the cytochrome system. In the intact mitochondria these two processes are closely associated. When they are uncoupled, the transport of electrons may actually speed up, thereby indicating that the phosphorylation of ADP has been a rate-limiting process. 2,4-Dinitrophenol is one of the most effective agents for uncoupling respiratory-chain phosphorylation.

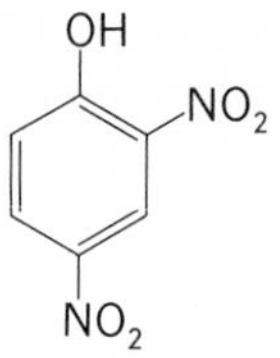

OH
NO2
NO2
2,4-Dinitrophenol

It does not have any effect on the substrate-level phosphorylations that occur in glycolysis. Uncoupling agents have also been utilized in studying the mechanism of oxidative phosphorylation. The three theories which are currently held will be discussed after the energy relationships of this process are summarized.

Energetics of Oxidative Phosphorylation

In Chapter 7 the $\Delta G'$ for the oxidation of 1 mole of NADH by molecular O_2 was calculated as approximately $-52{,}000$ cal from the reduction potentials of $NAD^+/NADH$ and O_2/H_2O. Since the oxidation of NADH by O_2 through the cytochrome electron-transport system leads to the formation of three high-energy phosphate bonds, the efficiency of the process of energy conservation may be calculated as $-21{,}900$ or (3×-7300) divided by $-52{,}000$, or 42%.

It is now possible to summarize the esterification of inorganic phosphate that accompanies the oxidation of pyruvic acid to CO_2 and H_2O by means of the tricarboxylic acid cycle. The oxidation steps in the process lead to the production of reduced nicotinamide and flavin coenzymes; when these are reoxidized by means of the electron-transport system of the mitochondria, the process of oxidative phosphorylation leads to the production of ATP from ADP and inorganic phosphate.

Table 14-1 lists the different reactions which result in the formation of energy-rich phosphate compounds; the total number of high-energy phosphate bonds synthesized per mole of pyruvate oxidized is 15. Since the oxidation of pyruvate to CO_2 and H_2O results in a free-energy change of $-273{,}000$ cal (Chapter 13), the efficiency of energy conservation in this process is at least $-109{,}000$ or (-7300×15) divided by $-273{,}000$, or 40%.

In line with this calculation, it is possible to estimate the total number of high-energy phosphate bonds which may be synthesized when glucose is oxidized to CO_2 and H_2O aerobically. The conversion of 1 mole of glucose to 2 moles of pyruvic acid forms two high-energy phosphates as a result of substrate-level phosphorylation in the glycolytic sequence. The further oxidation of the 2 moles of pyruvic acid in the tricarboxylic acid cycle forms thirty high-energy phosphates. In addition, there are four to six more high-energy phosphates to be added to the thirty-two just listed. When glucose is converted to two molecules of pyruvate in glycolysis and the latter is not reduced to lactic acid, two molecules of NADH remain in the cytoplasm to be accounted for. While the NADH might be reoxidized by other cytoplasmic dehydrogenases, in a tissue that is actively oxidizing glucose completely to CO_2 and H_2O, these two molecules of NADH would be oxidized by the

Table 14-1

Formation of Energy-Rich Phosphate During the Oxidation of Pyruvate by the Tricarboxylic Acid Cycle

Enzyme or process	Reaction	Energy-rich phosphate produced
Pyruvic dehydrogenase	Pyruvate + NAD^+ + CoASH ⟶ Acetyl–CoA + NADH + H^+ + CO_2	0
Electron transport	NADH + H^+ + $\frac{1}{2}O_2$ ⟶ NAD^+ + H_2O	3
Isocitric dehydrogenase	Isocitrate + NAD^+ ⟶ α-Ketoglutarate + CO_2 + NADH + H^+	0
Electron transport	NADH + H^+ + $\frac{1}{2}O_2$ ⟶ NAD^+ + H_2O	3
α-Ketoglutaric dehydrogenase	α-Ketoglutarate + NAD^+ + CoASH ⟶ Succinyl–CoA + NADH + H^+ + CO_2	0
Electron transport	NADH + H^+ + $\frac{1}{2}O_2$ ⟶ NAD^+ + H_2O	3
Succinic thiokinase	Succinyl–CoA + GDP + H_3PO_4 ⟶ Succinate + GTP + CoASH	1
Succinic dehydrogenase	Succinate + FAD ⟶ Fumarate + $FADH_2$	0
Electron transport	$FADH_2$ + $\frac{1}{2}O_2$ ⟶ FAD + H_2O	2
Malic dehydrogenase	Malate + NAD^+ ⟶ Oxalacetate + NADH + H^+	0
Electron transport	NADH + H^+ + $\frac{1}{2}O_2$ ⟶ NAD^+ + H_2O	3
	Sum	15

electron-transport chain of the organism just as is the NADH produced by oxidation of Krebs cycle intermediates.

In procaryotic organisms, there would be no particular problem, as the NADH would presumably have ready access to the respiratory assemblies containing the electron-transport chain and phosphorylation enzymes. A total of 38 ATP would therefore be formed in the complete oxidation of glucose to CO_2 and H_2O. However, the inner membrane of the mitochondria of eucaryotic organisms is not permeable to NADH, and a *shuttle* process involving *sn*-glycerol-3-phosphate is employed. In this process, NADH produced in glycolysis (or in any other cytoplasmic oxidation–reduction reaction) is first reoxidized by dihydroxy acetone phosphate in the presence of the cytoplasmic *sn*-glycerol-3-phosphate dehydrogenase.

$$\underset{\text{Dihydroxy acetone phosphate}}{\begin{array}{c} CH_2OH \\ | \\ C{=}O \\ | \\ CH_2OPO_3H_2 \end{array}} + NADH + H^+ \longrightarrow \underset{sn\text{-Glycerol-3-phosphate}}{\begin{array}{c} CH_2OH \\ | \\ HOCH \\ | \\ CH_2OPO_3H_2 \end{array}} + NAD^+ \quad (14\text{-}5)$$

The *sn*-glycerol-3-phosphate formed is readily permeable to the mitochondrial membranes and enters through the inner membrane to the matrix where it is oxidized, this time by a dehydrogenase that utilizes FAD instead of NAD^+:

$$\underset{sn\text{-Glycerol-3-phosphate}}{\begin{array}{c} CH_2OH \\ | \\ HOCH \\ | \\ CH_2OPO_3H_2 \end{array}} + FAD \longrightarrow \underset{\text{Dihydroxy acetone phosphate}}{\begin{array}{c} CH_2OH \\ | \\ C{=}O \\ | \\ CH_2OPO_3H_2 \end{array}} + FADH_2 \qquad (14\text{-}6)$$

The $FADH_2$ produced by this flavoprotein then contributes electrons to the electron-transport chain at the level of CoQ (see Figure 14-2) and, as with succinate, 2 moles of ATP are formed when the CoQ–H_2 is oxidized. To keep the shuttle operating, the dihydroxy acetone phosphate produced in reaction 14-6 then passes out of the mitochondria into the cytoplasm, where it can repeat the process. This shuttle operates only in the manner described, namely to transport reducing equivalents *into* the mitochondria, probably because of the largely unidirectional movement of electrons along the electron-transport chain. The transfer of reducing equivalents from the two cytoplasmic NADH's produced as 1 mole of glucose is converted to pyruvate in animal mitochondria will therefore result in the formation of four high-energy phosphates or a total of 36 ATP for the complete oxidation of glucose to $CO_2 + H_2O$ in a eucaryote.

As discussed previously, the $\Delta G'$ for the oxidation of glucose by O_2 to CO_2 and H_2O has been estimated from calorimetric data:

$$C_6H_{12}O_6 + 6\ O_2 \longrightarrow 6\ CO_2 + 6\ H_2O \qquad (14\text{-}7)$$
$$\Delta G' = -686{,}000 \text{ cal (pH 7.0)}$$

If there were no mechanism for trapping any of this energy, it would be released to the environment as heat, for the entropy term (see Chapter 7) is negligible. The cell can conserve a large portion of this energy, however, by coupling the energy released to the synthesis of the energy-rich ATP from ADP and H_3PO_4. If 38 moles of ATP are formed during the oxidation of glucose, this represents a total of 38×-7300 or $-277{,}000$ cal. The amount of energy that would be liberated as heat in reaction 14-7 is hence reduced by this amount, and the overall oxidation and phosphorylation may now be written as

$$C_6H_{12}O_6 + 6\ O_2 + 38\ ADP + 38\ H_3PO_4 \longrightarrow 6\ CO_2 + 38\ ATP + 44\ H_2O \qquad (14\text{-}8)$$
$$\Delta G' = -409{,}000 \text{ cal (pH 7.0)}$$

The conservation of 277,000 cal as energy-rich phosphate represents an efficiency of conservation of 277,000 divided by $-686{,}000$, or 40%. The trapping of this amount of energy is a noteworthy achievement for the living cell.

The Mechanism of Oxidative Phosphorylation

The phosphorylation mechanism associated with electron transport is fundamentally different from the phosphorylation step in glycolysis in that energy-rich phosphorylated forms of substrates (e.g., 1,3-diphosphoglyceric acid or phosphoenol pyruvate) have not been identified. Despite the efforts made in several productive laboratories (those of Chance, Green, Lardy, Lehninger, Racker, Slater, to mention a few alphabetically!) the problem has not been solved. To recount all the approaches is beyond the scope of this text, but two can be described.

One of these involves the study of certain partial reactions catalyzed by mitochondria. One partial reaction is the hydrolysis of ATP:

$$\text{ATP} + H_2O \xrightarrow[\text{ATP-ase activity}]{} \text{ADP} + H_3PO_4 \tag{14-9}$$

This "ATP-ase" activity of intact mitochondria is normally low, but it can be greatly stimulated by 2,4-dinitrophenol, the uncoupling reagent previously mentioned. A second reaction is the exchange of inorganic phosphate with the terminal phosphate group of ATP:

$$\underset{\text{ATP}}{\text{A—R—P—P—P}} + \underset{\substack{\text{Labeled}\\\text{inorganic phosphate}}}{{}^{32}\text{Pi}} \rightleftharpoons \text{A—R—P—P—}{}^{32}\text{P} + \text{Pi} \tag{14-10}$$

This reaction, which can be followed by using inorganic phosphate labeled with radioactive ^{32}P, is strongly inhibited by dinitrophenol and by an antibiotic called oligomycin. Other partial reactions together with these are known to be dependent on the mitochondrial structure being relatively intact.

Another approach has been to fractionate active mitochondria into protein components which by themselves are unable to carry out oxidative phosphorylation, but which, when recombined, are again active. Racker has obtained two proteins, called *coupling factors,* which when added back to inactive mitochondria, are able to restore partially their ability to phosphorylate. One of these factors, F_1, is an ATP-ase that has a molecular weight of 280,000 (see page 144). It has been extensively purified and contains none of the components of the electron-transport chain.

Chemical Coupling Hypothesis. These studies have led over the years to three hypotheses, the oldest of which employs the concept already seen in glycolysis, in which the synthesis of an energy-rich intermediate, produced as the result of an oxidation–reduction reaction, is coupled to the generation of ATP. Specifically, as an oxidation–reduction reaction occurs between A_{red} and B_{ox}, the factor I is incorporated into the formation of an energy-rich structure $A_{ox} \sim I$, where the $\sim$ indicates a linkage having an energy-rich nature:

$$A_{red} + I + B_{ox} \rightleftharpoons A_{ox}{\sim}I + B_{red} \tag{14-11}$$

In subsequent reactions, A_{ox} is replaced by E (an enzyme) to form an energy-rich $E \sim I$ complex; inorganic phosphate next reacts to form a phosphoenzyme complex $E \sim P$ containing the energy-rich enzyme–phosphate bond:

$$A_{ox}{\sim}I + E \rightleftharpoons A_{ox} + E{\sim}I \tag{14-12}$$

$$E{\sim}I + Pi \rightleftharpoons E{\sim}P + I \tag{14-13}$$

This enzyme–phosphate component finally reacts with ADP to form ATP:

$$E{\sim}P + ADP \rightleftharpoons E + ATP \qquad (14\text{-}14)$$

The sum of reactions 14-12 through 14-14 would account for both the ATP-ase activity of mitochondria referred to earlier as well as the exchange of ^{32}P-labeled inorganic phosphate in the terminal phosphate group of ATP.

Chemiosmotic Coupling Hypothesis. The failure of workers in oxidative phosphorylation to identify any of the energy-rich intermediates in reactions 14-11 through 14-14 stimulated other theories to explain this process. P. Mitchell of England is the chief exponent of the chemiosmotic hypothesis which has, as its major experimental observation, the fact that H^+ ions are transported out of the mitochondria as electron transport occurs in these particles. Mitchell has proposed that the inner membrane of the mitochondria is impermeable to protons (H^+). Moreover, as those steps in the electron-transport chain that produce or consume H^+ occur, the spatial arrangement of those enzymes in the inner membrane is such that the proton (H^+) consumed comes from the matrix, and the proton (H^+) produced is always exterior to the matrix (Figure 14-3). As a consequence, there would be a gradient of H^+ across the membrane, with the lower H^+ concentration

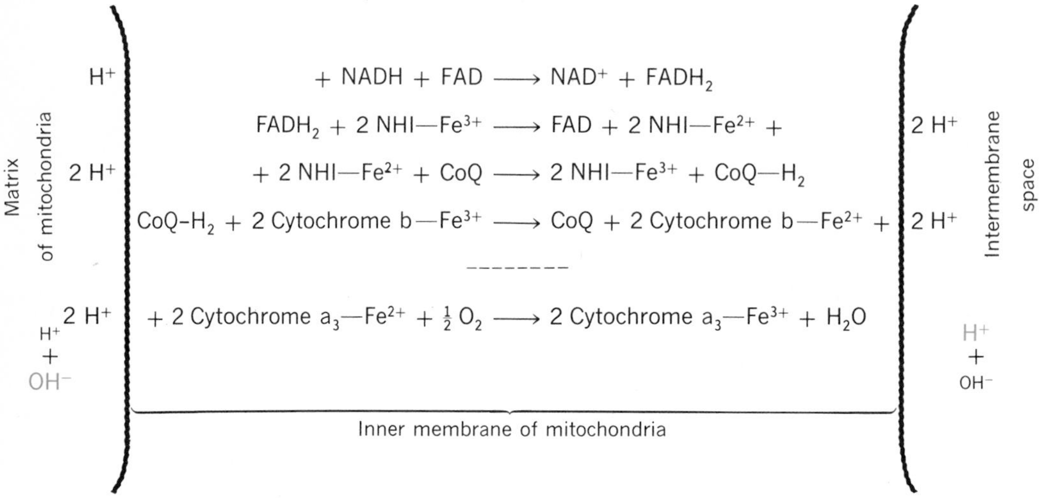

Matrix of mitochondria | Intermembrane space

H^+ + NADH + FAD ⟶ $NAD^+ + FADH_2$

$FADH_2$ + 2 NHI—Fe^{3+} ⟶ FAD + 2 NHI—Fe^{2+} + 2 H^+

2 H^+ + 2 NHI—Fe^{2+} + CoQ ⟶ 2 NHI—Fe^{3+} + CoQ—H_2

CoQ-H_2 + 2 Cytochrome b—Fe^{3+} ⟶ CoQ + 2 Cytochrome b—Fe^{2+} + 2 H^+

----

2 H^+ + 2 Cytochrome a_3—Fe^{2+} + $\frac{1}{2}O_2$ ⟶ 2 Cytochrome a_3—Fe^{3+} + H_2O

Matrix side: H^+ + OH^- | Intermembrane side: H^+ + OH^-

Inner membrane of mitochondria

Figure 14-3

Representation of spatial arrangement of some of the respiratory chain carriers in the inner mitochondrial membrane to account for the migration of (H^+) from the matrix through the membrane to the intermembrane space: NHI = nonheme iron protein; CoQ = Coenzyme Q_{10}. The relative amounts of H^+ and OH^- on the two sides of the membrane are indicated by the relative sizes and colors of the symbols used (H^+, H^+, OH^-, OH^-).

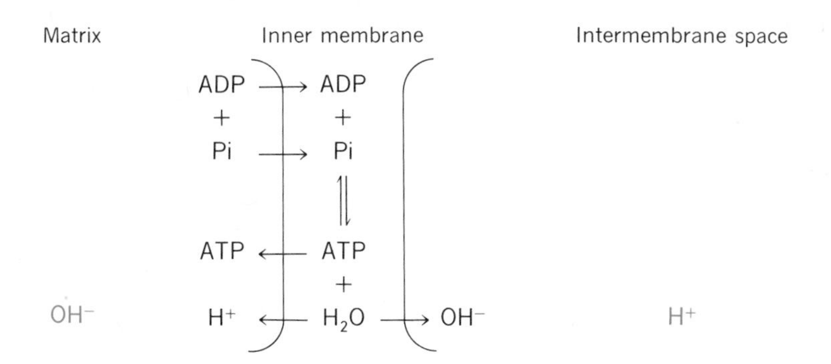

Figure 14-4

Scheme showing formation of ATP from ADP and Pi (reverse ATP-ase reaction) in the inner membrane. The H_2O produced in the reaction dissociates in such a way that the H^+ are released into the matrix where a localized, higher concentration of OH^- has been established as a result of electron transport. Simultaneously, the OH^- produced on ionization of the H_2O passes into intermembrane space, where a higher concentration of H^+ exists.

(equivalent to a higher OH^- concentration) being inside and the higher concentration being outside.

In support of this hypothesis is the established semipermeable nature of the mitochondrial membrane and the fact that H^+ ions are transferred into the medium as mitochondria carry out electron transport.

Assuming these localized differences in pH, Mitchell proposes that the hydrolysis of ATP which requires H_2O can be driven in the direction of ATP synthesis provided the H_2O formed is removed. It is a key feature of this hypothesis that the H_2O can be removed by dissociating into H^+ and OH^- and reacting, respectively, with the localized higher concentrations of OH^- and H^+ on the two sides of the inner membrane (Figure 14-4).

Conformational Coupling. The third hypothesis for phosphorylation rests in the observation that the inner membrane undergoes certain structural changes during electron transport and that these changes can be inhibited by 2,4-dinitrophenol and oligomycin. In mitochondria that are actively phosphorylating in the presence of an excess of ADP, the inner membrane will pull away from the outer membrane and assume a "condensed" state. In the absence of ADP, the mitochondria have the usual structural features or "swollen state" in which the cristae project into the large matrix (Figure 6-6). The proponents of this hypothesis suggest that the energy released in electron transport is translated into the conformational changes just described, and that this energy-rich condensed structure in turn can be utilized for ATP synthesis as it converts to the swollen, energy-poor conformation. Precisely how these changes in the shape of a membrane can be coupled to the interconversion of ADP and ATP is not yet clear.

The three hypotheses just described provide numerous suggestions for experimentation. Whether any one of these will prove to be an accurate description of oxidative phosphorylation remains to be established.

The Permeability of Mitochondria

As pointed out in Chapter 6, the permeability of the inner and outer mitochondrial membranes is quite different. While the outer membrane is fully permeable to molecules up to 10,000 in molecular weight, the inner membrane exhibits great selectivity. It has already been pointed out that pyruvic acid can freely enter the inner membrane for oxidation, but that oxalacetate cannot. The consequence of this selectivity on the process whereby pyruvate (or lactate) is reversibly converted back to glucose in gluconeogenesis has already been discussed (page 261). Thus, while OAA, produced from pyruvic carboxylase in mitochondria, cannot readily leave the particle, malate can, and both mitochondrial and cytoplasmic malic dehydrogenase are utilized in reversing glycolysis (see Figure 14-5).

The inner membrane is also impermeable to fumarate, but citrate and isocitrate can enter into the matrix. As pointed out in Chapter 6, carrier systems are responsible for the transport of the tricarboxylic and dicarboxylic acids that can enter the matrix. The role of citrate in the transport of acetyl–CoA out of the mitochondria, where it is formed in β-oxidation, and into the cytoplasm, where it is utilized in fatty acid synthesis, has been discussed (page 318).

The inability of NAD^+ and NADH to pass from the matrix to the cytoplasm was discussed earlier in this chapter. The use of the *sn*-glycerol-3-phosphate shuttle in transferring reducing equivalents from cytoplasmic NADH to the inner membrane where they can participate in electron transport has been described. Malate, oxalacetate, and aspartate, together with the mitochondrial and cytoplasmic malic dehydrogenases and aspartic–oxalacetate transaminases, constitute another shuttle system for reducing equivalents. This one, in contrast to the *sn*-glycerol phosphate shuttle, can operate in both directions and can transfer reducing equivalents either into or out of the mitochondria (Figure 14-5).

The question of transport of ADP and ATP across the inner membrane is a key one in view of the localization of the Krebs cycle enzymes, electron-transport carriers, and phosphorylation enzymes either in the matrix or on the inner membrane. Studies have shown that ADP and ATP can pass through the membrane provided there is an exchange of the two molecules. Thus, one molecule of ADP can enter the matrix provided one molecule of ATP simultaneously leaves the mitochondria. This requirement then establishes a metabolic pool of adenine nucleotide in the matrix which is distinct from that in the cytoplasm. The two pools are interconnected, however, through the exchange process that occurs in the inner membrane.

When the metabolism of amino acids is described, we shall see that cytoplasmic transaminases appear to account for the transfer of amino nitrogen to α-ketoglutarate to form glutamic acid. This amino acid can then enter the mitochondrial matrix, where it can transaminate with oxalacetate (formed by pyruvic carboxylase), regenerate α-ketoglutarate, and be oxidized.

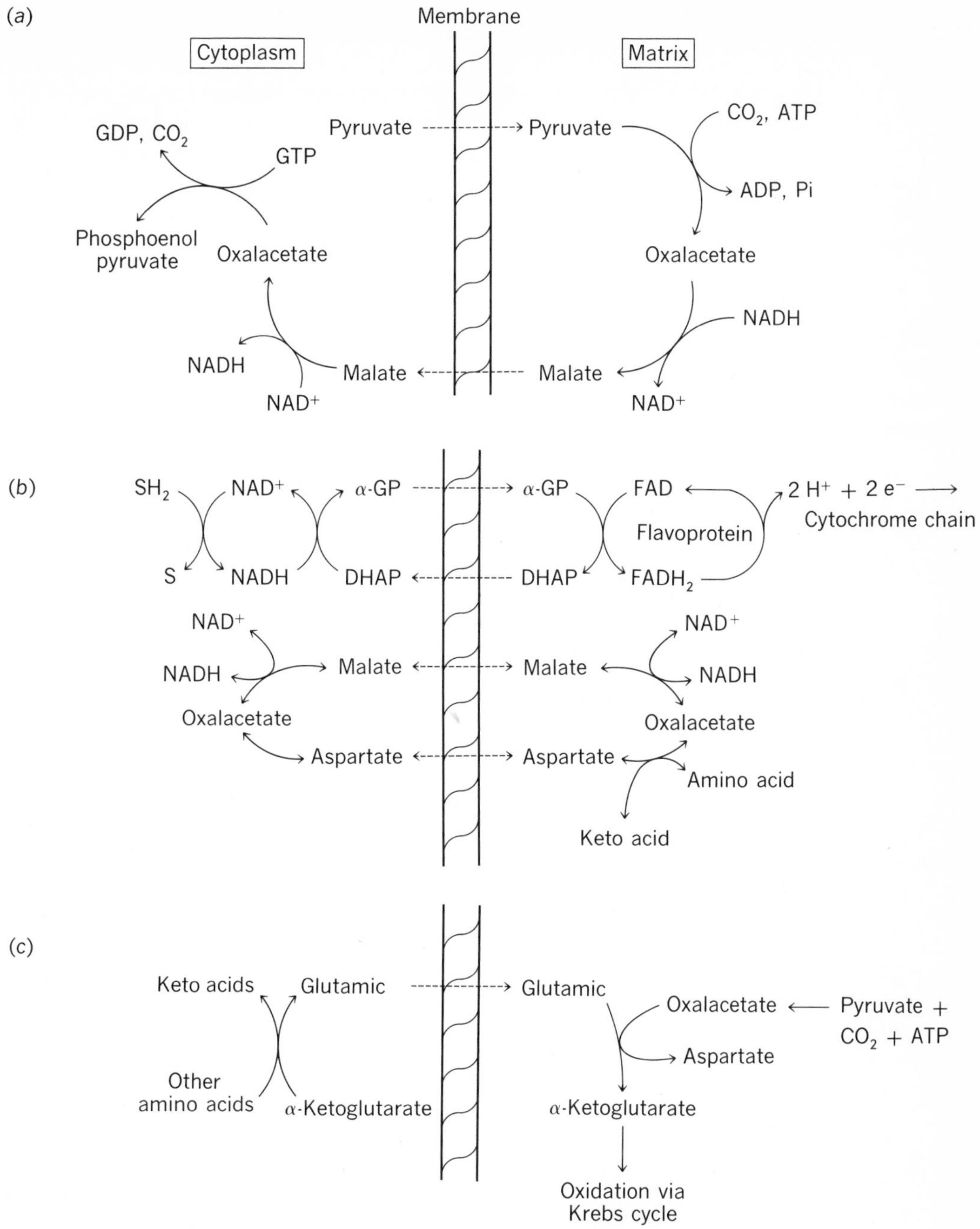

Figure 14-5

Four shuttle systems operating across the mitochondrial inner membrane. (*a*) System required to convert pyruvate (or lactate) to phosphoenol pyruvate in gluconeogenesis. (*b*) The unidirectional α-glycerol phosphate (α-GP) and dihydroxyacetophosphate (DHAP) shuttle that transports reducing equivalents only *into* the mitochondrial matrix. The reversible malate–oxalate–aspartate shuttle for reducing equivalents is also given. (*c*) The glutamate shuttle for transport of amino nitrogen.

Thus, the five carbon atoms of α-ketoglutarate cannot directly enter the mitochondria as such but must first be disguised as glutamate which can freely enter. These and the other metabolism restrictions created because of the selectivity of the mitochondrial membranes are still being determined and evaluated, but it appears that such devices are of fundamental significance in regulating metabolism.

Integration of Carbohydrate, Lipid, and Amino Acid Metabolism

At this point it will be useful to integrate some of the information on energy production from carbohydrates and lipids that has been discussed. In addition, we will anticipate some of the general features of amino acid metabolism, although this subject is not treated until Chapter 16.

Krebs and Kornberg have pointed out that many different compounds which may be classified roughly as carbohydrates, lipids, or proteins can serve as sources of energy for living organisms. These authors have also emphasized that the number of reactions involved in obtaining energy from these compounds is astonishingly small, whether the organism involved is animal, higher plant, or microorganism. Thus, nature has practiced great economy in the processes developed for handling these compounds. These authors divide substrate degradation into three phases, as indicated in Figure 14-6.

In phase 1, polysaccharides, which serve as an energy source for many organisms, are hydrolyzed to monosaccharides, usually hexoses. Similarly, proteins can be hydrolyzed to their component amino acids, and triacylglycerols, which make up the major fraction of the lipid food sources, are hydrolyzed to glycerol and fatty acids. These processes are, for the most part, hydrolytic, and the energy released as the reactions occur is made available to the organism as heat.

In phase 2, the monosaccharides, glycerol, fatty acids, and amino acids are further degraded to three compounds by processes which may result in the formation of some energy-rich phosphate compounds. Thus, in glycolysis, the hexoses are converted to pyruvate and then to acetyl–CoA by reactions involving the formation of a limited number of high-energy phosphate bonds, as described in Chapter 10. Similarly, in phase 2, the long-chain fatty acids are oxidized to acetyl–CoA (Chapter 12), while glycerol, obtained from hydrolysis of triacylglycerols, is converted to pyruvate and acetyl–CoA by means of the glycolytic sequence.

For the amino acids the situation is somewhat different. In phase 2 some amino acids (alanine, serine, cysteine) are converted to pyruvate on degradation, and thus, acetyl–CoA formation is predicted if these amino acids are utilized by an organism for energy production. Other amino acids (the prolines, histidine, arginine) are converted to glutamic acid on degradation; this amino acid in turn undergoes transamination to yield α-ketoglutarate, a member of the tricarboxylic acid cycle. Aspartic acid is readily transaminated to form oxalacetate, another intermediate of the cycle. Some amino acids (the leucines) yield acetyl–CoA on degradation, and phenylalanine and tyrosine, on oxidative degradation, produce both acetyl–CoA and oxalacetic acid through fumaric acid.

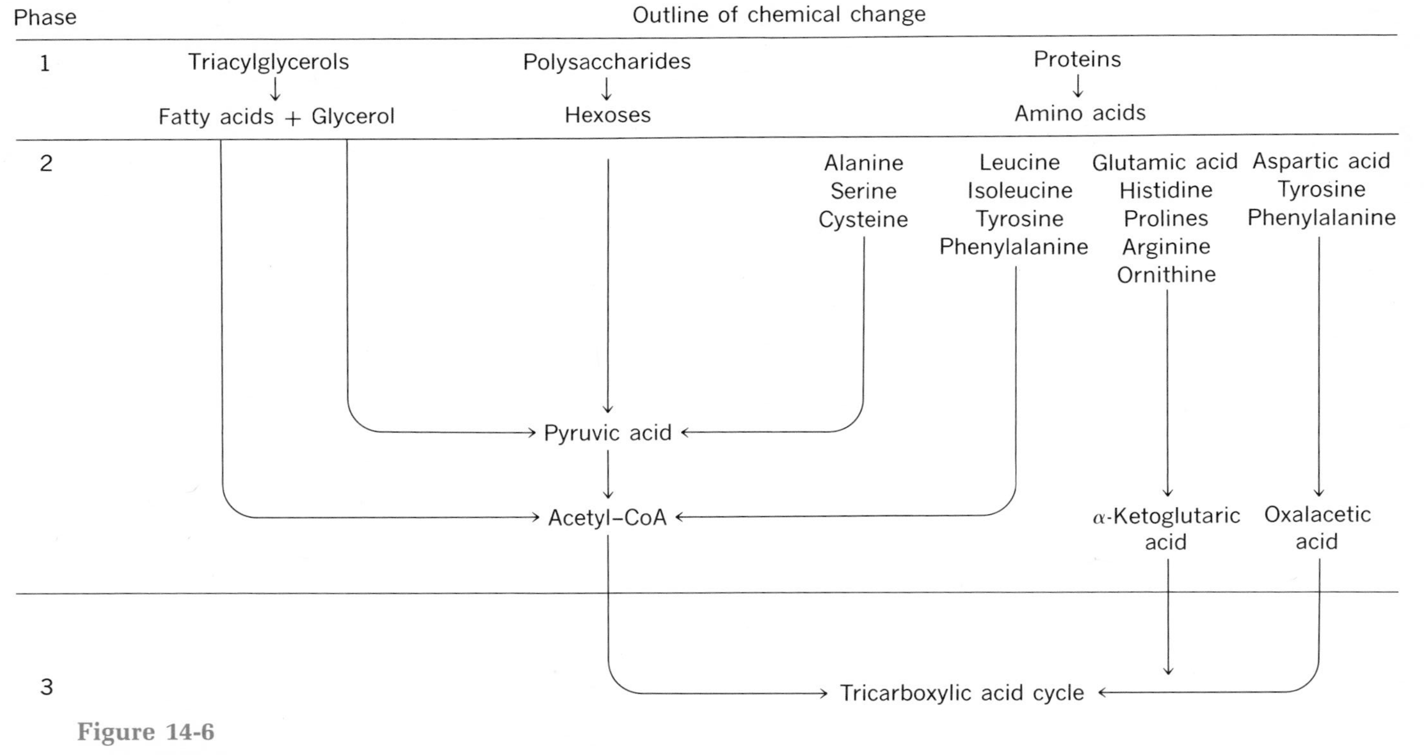

Figure 14-6

The main phases of energy production from foodstuffs.

Thus, the carbon skeletons of essentially all the amino acids yield either an intermediate of the tricarboxylic acid cycle (oxalacetate, or α-ketoglutarate) or acetyl–CoA, which is in turn oxidized by means of the cycle. During the oxidation of these compounds in phase 3, energy-rich ATP is produced by oxidative phosphorylation. Specifically, twelve energy-rich bonds are produced for each mole of acetyl–CoA oxidized. Hence, hundreds of organic compounds that can conceivably serve as food for biological organisms are utilized by their conversion to acetyl–CoA or an intermediate of the tricarboxylic acid cycle and their subsequent oxidation by the cycle.

In considering the actual steps involved in making energy available to the organisms, the reactions of oxidative phosphorylation that occur during electron transport through the cytochrome system are quantitatively the most significant. Even here an economy in the number of reactions is involved. As discussed in Chapter 13, the oxidation of substrates in the tricarboxylic acid cycle is accompanied by the reduction of either a nicotinamide or a flavin nucleotide. It is the oxidation of the reduced nucleotide by molecular oxygen in the presence of mitochondria that results in the formation of the energy-rich ATP. As pointed out, three phosphorylations occur during the transfer of a pair of electrons from NADH to O_2. We have discussed only three other reactions leading to the production of energy-rich compounds where none previously existed before. These are (*a*) the formation of acylphosphate in the oxidation of triose phosphate (Chapter 10), (*b*) the formation of phosphoenol pyruvate (Chapter 10), and (*c*) the formation of thioesters (Chapter 12). It is indeed a beautiful design which permits the energy in the myriad foodstuffs to be trapped in only six different processes. Even here a single compound, ATP, is the energy-rich substance formed.

Interconversion of Carbohydrate, Lipid, and Protein. The interconversions among the three major foodstuffs may be summarized with the help of Figure 14-7 as follows: In this figure two reactions that are effectively irreversible are indicated by heavy unidirectional arrows. (*a*) Carbohydrates are convertible to fats through the formation of acetyl–CoA. (*b*) Carbohydrates may also be converted to certain amino acids (alanine, aspartic, and glutamic acids), provided a supply of dicarboxylic acid is available for formation of the keto acid analogs of those amino acids. Specifically, a supply of both oxalacetate (or other C_4-dicarboxylic acid) *and* acetyl–CoA are required in an amount stoichiometrically equivalent to the amino acid being synthesized. Several reactions exist for forming the C_4-decarboxylic acids; the principal one is the formation of oxalacetic acid from pyruvic acid, the reaction catalyzed by pyruvic carboxylase. Another is the formation of malic acid from pyruvic acid, the reaction catalyzed by malic enzyme. These reactions have been described in detail in Chapters 10 and 13. (*c*) Fatty acids may be similarly converted to certain amino acids provided a source of dicarboxylic acid is available. (*d*) Fatty acids *cannot* be converted to carbohydrate by the reactions shown in Figure 14-7. This inability is due to the fact that the equivalent of the two carbon atoms acquired in acetyl–CoA has been lost as CO_2 prior to the production of the dicarboxylic acids. Note, however, that the glyoxylate cycle

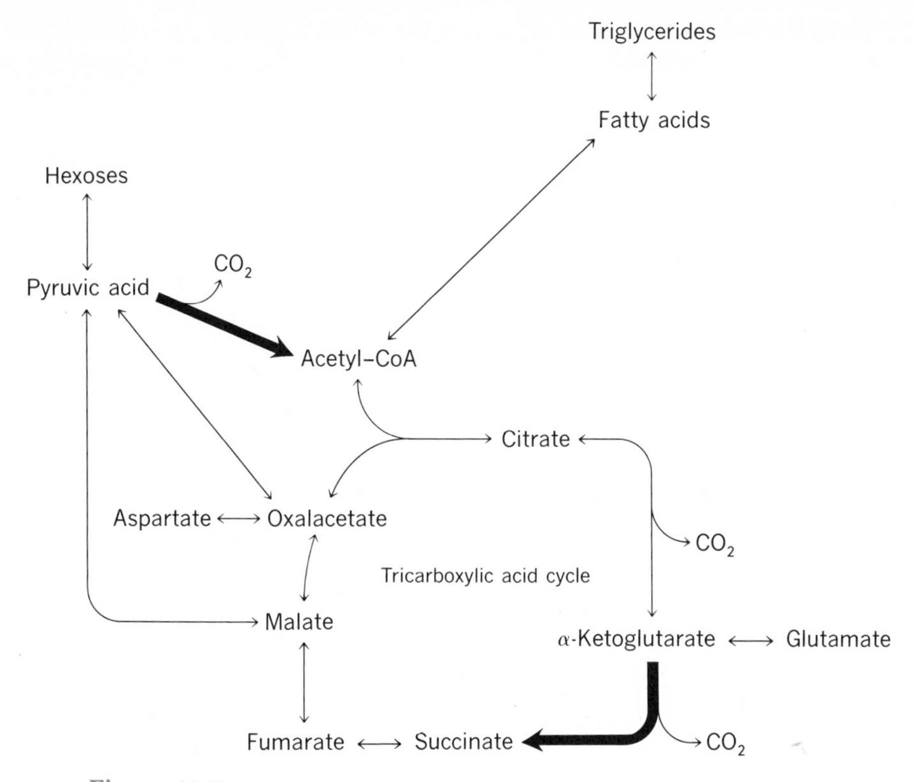

Figure 14-7

The possible interconversions between carbohydrates, lipids, and certain amino acids.

(discussed in Chapter 13) can enable an organism to form carbohydrate from fat, as it does, for instance, in plants, some bacteria, and some molds. (e) The naturally occurring amino acids are convertible to carbohydrates and lipids. Each of the twenty amino acids may be classified as *glucogenic, ketogenic,* or *both glucogenic and ketogenic,* depending on the specific metabolism of the amino acid. As an example, aspartic acid is glucogenic through formation of oxalacetic acid and its subsequent conversion to phosphoenol pyruvic acid. Similarly, glutamic acid is glucogenic by virtue of its conversion to oxalacetic acid in the tricarboxylic acid cycle and the conversion of oxalacetate to phosphoenol pyruvic acid.

Interrelationships in Metabolic Control. The interconversions of lipids, carbohydrates, and amino acids just described appear reasonable when discussed in terms of known enzymatic reaction. It is now apparent that these interrelationships exist in the area of metabolic regulation as well. While some of the following control processes have already been discussed elsewhere, they will be repeated here to emphasize the interrelation of regulation.

Consider a cell or tissue in which the energy charge value is approaching 1.0. The resulting low level of AMP will decrease the activity of the tricarboxylic acid cycle by lowering the activity of isocitric dehydrogenase. As pointed out on page 308, an immediate decrease in ATP production by oxidative phosphorylation will occur. At the same time, citric acid can be expected to accumulate. Since this acid is known to increase the activity of acetyl–CoA carboxylase which catalyzes the first step in the conversion of acetyl–CoA to fatty acids (see Chapter 12), the cell can shunt the acetyl–CoA being produced from glucose from energy production into fat storage. When ATP utilization is resumed, as it would be in fatty acid synthesis, the corresponding increase in AMP production would lower the citric acid concentration and the ability of fatty acid synthesis to compete for the acetyl–CoA.

The interrelationships possible in control can also be extended back into reactions of glycolysis. Thus, in the "energy-saturated" cell under discussion, the low level of AMP (and high level of ATP) will decrease the glycolytic degradation of glucose because of the action of these nucleotides on phosphofructokinase and fructose-1,6-diphosphate phosphatase (see page 268). Instead, fructose-6-phosphate and its precursor, glucose-6-phosphate, would accumulate, and the action of the latter ester, as a positive effector of UDPG–glycogen glucosyl transferase, would be to stimulate polysaccharide formation. Again, when the level of ATP is lowered (and AMP concentration is increased), the glycolytic degradation of glucose would increase and oxidation of pyruvate through the tricarboxylic acid cycle would provide a renewed supply of ATP.

It should be stressed that not all of the control mechanisms cited above and elsewhere in this text have been demonstrated in a single type of tissue. Thus, unequivocal proof that all of these controls actually function in a single tissue is lacking. Nevertheless, there is no shortage of evidence that the intact living organism possesses an amazing ability to regulate its metabolism. Knowledge of the manner in which it does so will only result from further experimentation.

References

1. D. E. Green and D. H. MacLennan, "The Mitochondrial System of Enzymes," in *Metabolic Pathways*. D. M. Greenberg, ed. vol. I. New York: Academic Press, 1967.

   A succinct review of electron transfer and oxidative phosphorylation and the organelle which carries out these processes.

2. A. L. Lehninger, *The Mitochondrion: Molecular Basis of Structure and Function*. New York: Benjamin, 1964.

   E. Racker, *Mechanisms in Bioenergetics*. New York: Academic Press, 1965.

   Two well-written monographs dealing with mitochondria and oxidative phosphorylation.

3. H. A. Lardy and S. M. Ferguson, "Oxidative Phosphorylation in Mitochondria," *Ann. Rev. Biochem*. **38,** 991 (1969).

   A comprehensive review of recent progress in the field.

4. H. A. Krebs and H. L. Kornberg, *Energy Transformation in Living Matter, A Survey*. Berlin: Springer, 1957.

   An excellent summary of the interrelations among carbohydrates, lipids, and proteins with emphasis on the energy transformations involved.

Photosynthesis

Introduction

All life on the planet earth is dependent on *photosynthesis*, the process by which CO_2 and H_2O are converted into the organic compounds associated with living cells. The conversion of CO_2 and H_2O to glucose, for example, may be presented as the reverse of reaction 14-7 and will require the input, as a minimum, of the same amount of energy which was released when glucose was oxidized to CO_2 and H_2O:

$$6\,CO_2 + 6\,H_2O \longrightarrow C_6H_{12}O_6 + 6\,O_2 \qquad (15\text{-}1)$$
$$\Delta G' = +686{,}000 \text{ cal}$$

As the term photosynthesis implies, the energy for this process is provided by light.

Properties of Light

The study of radiant energy has disclosed that light may be treated as a wave of particles known as *photons*. The energy of these photons may be calculated from the equation

$$E = \frac{N\hbar c}{\lambda}$$

where E is the energy (in kilocalories) of 1 mole or Einstein of photons; N is Avogadro's number (6.023×10^{23}); h is Planck's constant (1.58×10^{-34} cal-sec); c is the velocity of light (3×10^{10} cm/sec); and λ is the wavelength (in nanometers). As the equation indicates, the energy of photons is inversely proportional to the wavelength of the wave of particles. Thus, the energy of blue light of short wavelength is greater than that of a corresponding amount of red light of longer wavelength. Table 15-1 lists the energy contents of Einsteins (6.023×10^{23}, of photons) of different types of light.

One of the important properties of matter is its ability to absorb light. Briefly, the ability of a substance to absorb light is dependent upon its atomic

Table 15-1
Energy Content of Light of Different Wavelengths

Color of light	Wavelength (Å)	Energy (cal/Einstein)
Far red	7500	37,800
Red	6500	43,480
Yellow	5900	48,060
Blue	4900	57,880
Ultraviolet	3950	71,800

structure. In a stable atom the number of electrons surrounding the nucleus is equal to the positive charges (the atomic number) in the nucleus. These electrons are arranged in different orbitals around the nucleus and those in the outer orbitals are less strongly attracted to the nucleus. Still other orbitals further out from the nucleus can be occupied by these electrons, but energy is required to place an electron into these outer, unoccupied orbitals, because the placement involves moving a negative charge further away from the positively charged nucleus.

One way in which the electron can acquire this energy and be moved into an outer or higher orbital is to absorb a photon of light. When this occurs the atom is said to be in an excited state. The amount of energy required to excite the atom depends on the energy difference between the two orbitals. Thus, only certain wavelengths of light—those having sufficient energy—will be effective in exciting the atom.

An atom in an excited state is not stable; the tendency is for the electron to return from the outer orbital to the lower energy level. Its return is done in stages and is accompanied by the release of some of the energy acquired in excitation. The initial act is to return to a slightly lower energy level (transitional level) from the excited state, a process accompanied by the production of heat. When the electron returns to its original or ground state, the remainder of the excitation energy is released in a form of light known as fluorescence. Since some of the energy of the excited state is released as heat, the amount represented by fluorescence is less than the energy acquired initially on excitation. Therefore the fluorescing light will be of longer wavelength (and hence lower energy) than the absorbed light.

The Photosynthetic Apparatus

Photosynthesis is carried out by both procaryotic and eucaryotic cells. The procaryotes include the blue-green algae, and the purple and green bacteria; in these organisms the light-trapping process takes place in small structures called *chromatophores*. In eucaryotic organisms that photosynthesize (higher green plants, the multicellular red, green and brown algae, dinoflagellates, and diatoms), the chloroplast is the site of the photosynthetic process.

The chloroplasts, whose structure and composition was described on page 147, contain the photosynthetic pigments; these are chlorophylls *a* and *b* in the higher green plants together with certain carotenoids, one of which is β-carotene.

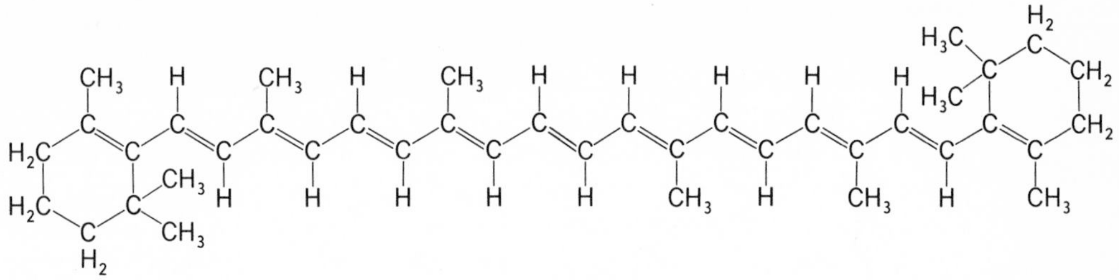

β-Carotene

Chlorophyll is a magnesium-containing porphyrin which has an aliphatic alcohol phytol esterified to a propionic acid residue on ring IV of a tetrapyrrole. The structures for both chlorophylls *a* and *b* are given here. The red and

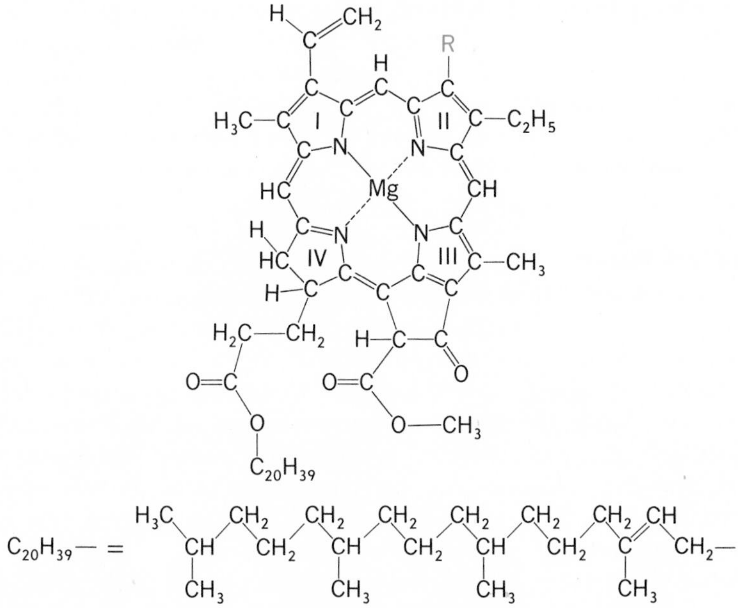

Chlorophyll *a*, R = CH_3
Chlorophyll *b*, R = CHO

blue green algae contain, in addition to chlorophyll *a*, blue or red pigments known as phycobilins, tetrapyrroles related to the chlorophylls but lacking Mg^{2+} and the *cyclic* structure of these compounds.

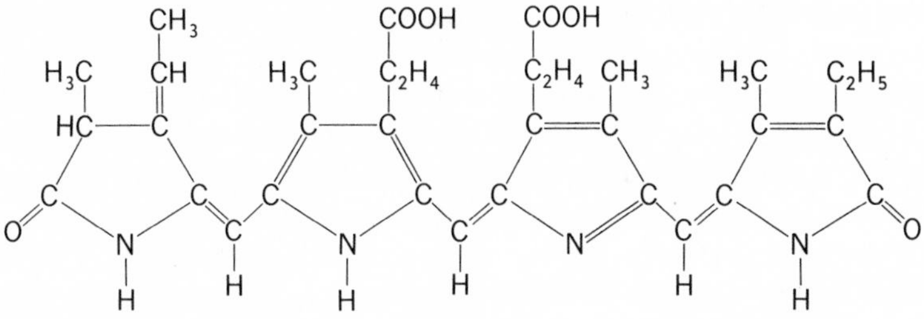

Phycobilin

While photosynthetic organisms contain a variety of photosynthetic pigments, a distinction can be made between those which play a *primary* role in the photosynthetic act and those which perform a *secondary* function. Only chlorophyll *a* (or the bacterial chlorophyll of the *a* type that is found in the photosynthetic bacteria) undergoes the excitation and subsequent fluorescence characteristic of the process in which light energy is converted into energy-rich chemical compounds. The other pigments do not participate directly in this energy-conversion process but instead collect light (of shorter wavelength and higher energy content) and pass it along to chlorophyll *a* by processes not yet well-understood.

The photosynthetic pigments together with the necessary enzymes and structural components are organized into photosynthetic units within the lamellae of the chloroplast. These units contain the different components of the energy-conversion system in definite proportions. A single unit contains 400 molecules of chlorophyll *a*; one molecule each of a special form of chlorophyll *a* known as P-700 and chlorophyll *a*-682; one molecule each of cytochrome f and plastocyanin; and two molecules each of cytochrome b_6 and cytochrome b_3. The roles played by these different components will become apparent when the details of the energy-conversion process are described.

Early Studies on Photosynthesis

Light and Dark Reactions. In studies initiated in 1905, Blackman showed that photosynthesis consists of two processes, a *light-dependent* phase that is limited in its rate by light-independent or *dark reactions*. The light-dependent processes exhibit the usual independence of temperature characteristic of photochemical reactions, while the dark reactions are sensitive to different temperatures. Today, the light-dependent processes are recognized as those in which light energy is converted into chemical energy, actually ATP and NADPH. The dark reactions, on the other hand, refer to the enzymatic reactions in which CO_2 is incorporated into reduced carbon compounds previously encountered in carbohydrate metabolism.

Evidence was first provided by Robert Emerson in the 1930s to show that the light-dependent phase of photosynthesis consisted of at least two light reactions. When Emerson measured the amount of photosynthesis carried out by the green algae, *Scenedesmus,* as a function of the wavelength of light, he observed that photosynthesis did not proceed at wavelengths greater than 700 nm. This was surprising, since light of this far red wavelength was still being absorbed by the algae cells. Emerson subsequently showed that this drop in the far red region could be reversed to varying amounts if he supplemented the light at 700 nm with a second source of light having a wavelength of 650 nm. This enhancement of the amount of photosynthesis caused Emerson to postulate that, in the case of *Scenedesmus*, the assimilation of CO_2 in photosynthesis required light of two different wavelengths. Today, these requirements are met by postulating that each photosynthetic unit has two photosystems (PS I and PS II) and that these are activated by light of far red wavelength (680–700 nm) and shorter wavelength (650 nm), respectively.

Bacterial Photosynthesis. Studies on photosynthetic bacteria provided much useful information and were the basis for one of the classic hypotheses that stimulated research in photosynthesis for many years. The two classes of purple bacteria, sulfur and nonsulfur, have been extensively used. To compare the process of photosynthesis in these organisms consider writing the overall reaction of photosynthesis as carried out in green plants on the basis of 1 mole of CO_2. This may be done by dividing reaction 15-1 by six to give

$$CO_2 + H_2O \xrightarrow{h\nu} C(H_2O) + O_2 \qquad (15\text{-}2)$$

(Oxidized: $H_2O \rightarrow O_2$; Reduced: $CO_2 \rightarrow C(H_2O)$)

$$\Delta G' = +118{,}000 \text{ cal}$$

Further note that this is an oxidation–reduction reaction in which the oxidizing agent CO_2 is reduced to the level of carbohydrate represented by $C(H_2O)$. The electrons for this reduction come from the oxygen of H_2O which in turn is oxidized to O_2. Since the reaction is highly endergonic, it will only proceed when the necessary energy is supplied by light ($h\nu$).

The purple sulfur bacteria, e.g., *Chromatium,* utilize H_2S instead of H_2O as a reducing agent in photosynthesis. Elemental sulfur, S, is produced, but no oxygen is formed:

$$CO_2 + 2\,H_2S \xrightarrow{h\nu} C(H_2O) + 2\,S + H_2O \qquad (15\text{-}3)$$

Note that 2 moles of H_2S are required to balance the equation, the S^{2-} ions in the H_2S furnishing the total of four electrons required to reduce CO_2 to $C(H_2O)$. Thiosulfate can also serve as the reductant for photosynthesis by purple sulfur bacteria:

$$2\,CO_2 + Na_2S_2O_3 + 5\,H_2O \xrightarrow{h\nu} 2\,C(H_2O) + 2\,H_2O + 2\,NaHSO_4$$

This reaction demonstrates that the reducing agent need not contain hydrogen itself but simply be capable of furnishing electrons.

The nonsulfur purple bacteria (e.g., *Rhodospirillum rubrum*) require organic compounds such as ethanol, isopropanol, or succinate as electron donors. The balanced equation with ethanol, for example, may be written as

$$CO_2 + 2\,CH_3CH_2OH \xrightarrow{h\nu} C(H_2O) + 2\,CH_3CHO + H_2O \qquad (15\text{-}4)$$

C. B. van Niel has pointed out the similarity of this reaction to the one which occurs in green plants and he has suggested that a general reaction for photosynthesis may be represented as

$$CO_2 + 2\,H_2A \xrightarrow{h\nu} C(H_2O) + 2\,A + H_2O \qquad (15\text{-}5)$$

where H_2A is a general expression for a reducing agent which, as we have seen, may be a variety of compounds.

Since H_2S is a much stronger reducing agent than $Na_2S_2O_3$ or H_2O, we might expect that less light energy would be required for photosynthesis with H_2S as the reducing agent than with $Na_2S_2O_3$ or H_2O. Experimentally, however, the same amount of light energy is required, regardless of the nature of the external reducing agent. This caused van Niel to postulate that the primary reaction is the same in all organisms and that it consists of the splitting of a molecule of H_2O to yield both a reducing agent [H] and an oxidizing agent [OH].

$$H_2O \xrightarrow{h\nu} [H] + [OH] \qquad (15\text{-}6)$$

This hypothesis stimulated much experimental work that led to greater understanding of the process of photosynthesis. The realization that four electrons are required to reduce CO_2 to $C(H_2O)$ meant that equation 15-2 must be rewritten to involve 2 moles of H_2O as reductant, each atom of oxygen providing two electrons:

$$CO_2 + 2\,H_2{}^{18}O \xrightarrow{h\nu} C(H_2O) + H_2O + {}^{18}O_2 \qquad (15\text{-}7)$$

Further, this revised equation would indicate that the two oxygen atoms produced in green plant photosynthesis should come only from H_2O. This was confirmed experimentally by Ruben and Kamen in a classical experiment, in which H_2O labeled with the isotope ^{18}O was utilized in photosynthesis by algae. The oxygen produced under these conditions contained the same concentration of ^{18}O as the H_2O. Recent developments concerning the role of H_2O in photosynthesis have made it necessary to abandon van Niel's hypothesis of the photolytic cleavage of H_2O. However, his proposal that the initial photosynthetic act involves the production of an oxidant and a reductant is retained in the current descriptions of the energy-conversion process.

The Hill Reaction. In 1937, Robin Hill of Cambridge University attempted to study the reactions of photosynthesis by working with isolated chloroplasts rather than intact plants. He reasoned that more information might be obtained if grana or chloroplasts, which contain the chlorophylls, were studied separately from the cell. It would have been ideal if the chloroplasts could have carried out both the oxidation of H_2O and the reduction of CO_2 to organic carbon compounds. This was not accomplished at that time. Nevertheless, chloroplasts were able to produce O_2 photochemically in the presence of a suitable oxidizing agent, potassium ferric oxalate. In this reaction the ferric ion substitutes for CO_2 as an oxidizing agent during the photooxidation of H_2O:

$$4\,Fe^{3+} + 2\,H_2O \xrightarrow[\text{Chloroplasts}]{h\nu} 4\,Fe^{2+} + 4\,H^+ + O_2$$

Molecular oxygen is evolved in an amount stoichiometrically equivalent to the oxidizing agent added. This observation was of fundamental importance, for it permitted the study of the role of H_2O as a reducing agent in photosynthesis. The reaction is known as the *Hill reaction*, and potassium ferric oxalate is known as a *Hill reagent*. Other compounds were subsequently shown to serve as Hill reagents in studies on isolated chloroplasts; Warburg showed that benzoquinone could function as such:

$$2\ \text{Benzoquinone} + 2\,H_2O \xrightarrow[\text{Chloroplasts}]{\hbar\nu} 2\ \text{Hydroquinone} + O_2$$

Benzoquinone Hydroquinone

Oxidized dyes were later shown to function as Hill reagents by being reduced. Although this approach was criticized because the substances that could serve as Hill reagents were not physiologically important compounds, the properties of these reactions were extensively studied.

In 1952, three American laboratories reported that $NADP^+$ (and NAD^+) could serve as Hill reagents in the presence of spinach grana and light. With intact chloroplasts $NADP^+$ was preferentially reduced. Thus, for the first time a physiologically important compound could function as a Hill reagent. This observation was of prime importance; it constituted a mechanism whereby reduced nicotinamide nucleotides were produced as the result of a light-dependent reaction.

$$2\,NADP^+ + 2\,H_2O \xrightarrow[\text{Chloroplasts}]{\hbar\nu} 2\,NADPH + 2\,H^+ + O_2 \qquad (15\text{-}8)$$

Numerous examples have been given earlier in this text of the ability of NADPH and NADH to reduce various substrates in the presence of the proper enzyme. As will be described, both NADPH and ATP are needed in the conversion of CO_2 to carbohydrates in photosynthesis. Having obtained the NADPH via a Hill reaction, it was assumed that reoxidation of the reduced nicotinamide nucleotide by oxygen through the cytochrome electron-transport system of plant mitochondria would produce ATP. In the intact plant cell containing chloroplasts and mitochondria, both of these organelles were believed to be involved in the production of the two coenzymes NADPH and ATP needed to drive the photosynthetic carbon reduction cycle (Figure 15-2). In 1954, Arnon and his associates questioned whether ATP is so produced when they discovered that chloroplasts alone, when isolated by special techniques, could convert CO_2 to carbohydrates in the light. Further studies in Arnon's laboratory showed that chloroplasts, in the absence of mitochondria, could synthesize ATP in two types of light-dependent phosphorylation reactions. The first type, *cyclic photophosphorylation,* yields ATP only and produces no net change in any external electron donor or acceptor:

$$ADP + H_3PO_4 \xrightarrow{\hbar\nu} ATP + H_2O \qquad (15\text{-}9)$$

The second type, *noncyclic photophosphorylation,* involves a process in which ATP formation is coupled with a light-driven transfer of electrons from water to a terminal electron acceptor, such as $NADP^+$, with the resultant evolution of oxygen:

$$2\,NADP^+ + 2\,H_2O + 2\,ADP + 2\,H_3PO_4 \xrightarrow{\hbar\nu} 2\,NADPH + 2\,H^+ + O_2 + 2\,ATP + 2\,H_2O \qquad (15\text{-}10)$$

This reaction deserves further comment for two reasons: First note that the movement of electrons would appear to be the opposite of that encountered in the electron-transport system of mitochondria. In the latter, electrons flow from NADH ($E_0' = -0.320$) to O_2 ($E_0' = 0.820$) along a potential gradient that releases energy, some of which is trapped in the form of ATP. According to reaction 15-10, electrons arising in the oxygen atom of H_2O make their way to $NADP^+$ and reduce it to NADPH. This movement of electrons *against* the potential gradient clearly requires energy; this is the function of light in photosynthesis. Second, reaction 15-10 is even more remarkable in that, as electrons are apparently made to flow from H_2O to $NADP^+$, energy is also made available as ATP. These observations may be explained by considering the energy-conversion scheme (Figure 15-1) first proposed by Hill and Bendall of Cambridge University and subsequently modified by other investigators in the field.

The Energy-Conversion Process

In higher green plants and any other organisms that utilize H_2O as the reducing agent, the photosynthetic unit contains the two photosystems, PS I and PS II, that are activated by far-red (680–700 nm) and red light (650 nm), respectively. The 400 molecules of chlorophyll a in the basic unit are divided approximately equally between these two systems. Light energy absorbed by these chlorophylls, or by accesory pigments and transferred to

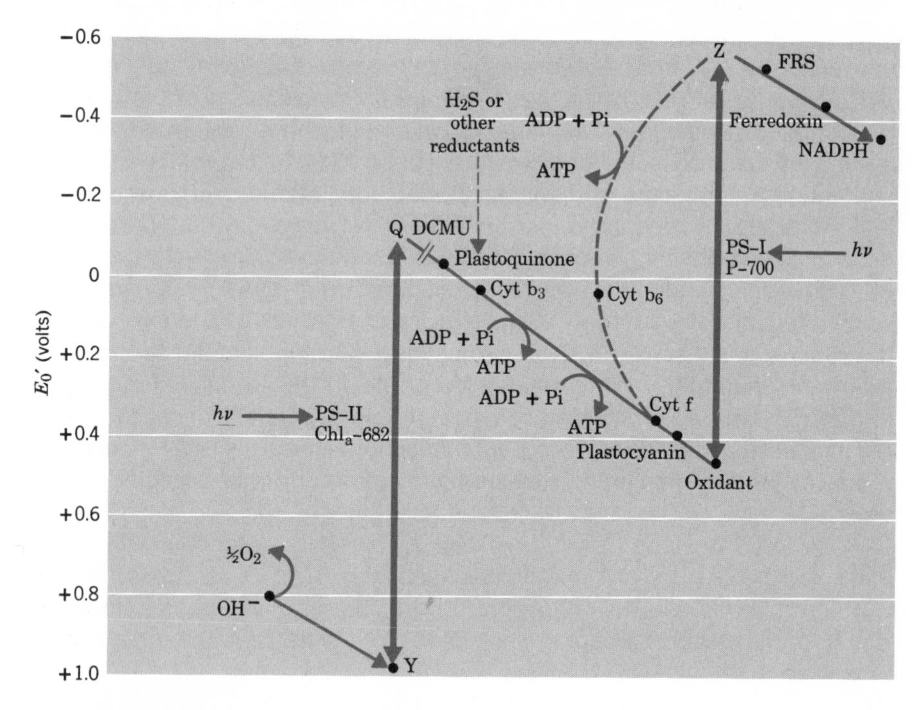

Figure 15-1

The energy-conversion process of photosynthesis, based on a scheme originally proposed by R. Hill and F. Bendall.

these chlorophylls, drives a photochemical reaction in which both an oxidant and a reductant are formed. In the case of PS I, absorbance and fluorescence studies indicate that a special form of chlorophyll a, known as P-700, is a key pigment in the excitation process in which a strong reductant, Z, capable of reducing $NADP^+$ and ferredoxin (see below), is formed together with a weak oxidant ($E_0' \cong 0.400$ V). The chlorophyll a (chl_a-682) involved in the trapping process in PS II has an absorption maximum at 682 nm; excitation of this pigment gives rise to a reductant, Q ($E_0' \cong -0.200$ V), and a strong oxidant, Y, that is capable of oxidizing H_2O to O_2.

An essential feature of the Hill–Bendall scheme is the interaction of the relatively weak reductant Q produced by PS II with the weak oxidant produced by PS I. Thus, while light reactions were postulated to provide the energy to produce these reactants, the flow of electrons (from Q to PS I) which was proposed is actually *with* the potential gradient; i.e., from a reductant with an E_0' of approximately -0.200 V to an oxidant with a potential E_0' of 0.400 V. Note that the Hill–Bendall scheme also contains elements of the van Niel hypothesis in that the action of light on the two photosystems is to produce *both* a reductant and an oxidant. Hill and his associates had discovered cytochrome f ($E_0' = 0.365$ V) in leaves many years earlier. The subsequent detection of other compounds such as plastoquinone, cytochrome b_3, and plastocyanin made it possible to propose that an electron-transport chain formally analogous to that existing in mitochondria also occurs in chloroplasts. The various components mentioned have been positioned in this chain on the basis of their reduction potentials, if known, as well as spectral changes undergone in red and far-red light. Moreover, as electrons flow along this chain from PS II to PS I, Hill and Bendall postulated that energy-rich phosphate in the form of ATP is generated.

Information is available regarding some of the carriers in this chain. Cytochrome f (from the Latin, *frons*, leaf) is a c type of cytochrome; its E_0' is $+0.365$ V and it has an absorption maximum at 555 nm. Cytochrome b_3, with an absorption maximum at 559 nm, has a lower E_0' characteristic of b-type cytochromes. Plastocyanin is a blue, copper-containing protein that undergoes one-electron reductions; its E_0' is about 0.400. Plastoquinone, similar in structure to coenzyme Q (page 321), has an E_0' of approximately 0.000 V. The nature of Q, the weak reductant produced when PS II is activated, remains to be established. It is known primarily from its ability to quench the fluorescence produced by illuminating PS II. Some workers have proposed that it may be identical to plastoquinone. The nature of the weak oxidant produced by irradiation of PS I is not completely clear, although it may be the unique form of chlorophyll a (P-700) that absorbs at 700 nm.

There is much evidence in support of this electron-transport chain linking the two photosystems. Boardman, in Australia, has disrupted chloroplasts and obtained fractions enriched in their capacity to carry out either the oxidation of H_2O (PS II) or the reduction of $NADP^+$ (PS I) when properly supplemented with suitable electron acceptors or donors. Then, too, when chloroplasts are illuminated by light of longer wavelength (the kind which activates PS I), the oxidant produced accepts electrons from cytochrome f and plastocyanin, and the oxidized forms of those pigments predominate in the plastid. When light of shorter wavelength activates PS II, the reductant Q feeds electrons into the chain leading to PS I, and all the carriers including cytochrome f become reduced.

The herbicide dichlorophenyldimethyl urea (DCMU) exerts its action as a weed killer by blocking the flow of electrons along the electron chain at the point indicated in Figure 15-1. In the presence of DCMU, chloroplasts will oxidize the carriers in that chain in the presence of far-red light (PS I), but when shorter-wavelength light is used (PS II), the carriers are not reduced.

The nature of the strong oxidant Y produced by PS II is not known, and the details of the mechanism whereby that oxidant oxidizes H_2O (more likely the hydroxyl ion OH^-) remain unclear. More is known about the process in which the reductant Z accomplishes the reduction of $NADP^+$. When PS I is activated, a substance—*ferredoxin-reducing substance* (FRS)—that has a reduction potential of -0.600 V is reduced. This compound transfers its electron to the nonheme iron protein known as ferredoxin ($E_0' = -0.420$ V). When reduced, this protein can in turn, in the presence of ferredoxin–$NADP^+$ reductase, reduce $NADP^+$. Again, the flow of electrons is from compounds of lower potential (FRS) to those of higher potential ($NADP^+$) or along the potential gradient; FRS and Z may be identical.

The protein ferredoxin has been isolated from a large number of photosynthetic organisms—bacteria, algae, higher plants. It was discovered, however, by Carnahan and Mortenson, who were studying nitrogen-fixation in *Clostridium pasteurianum*. Ferredoxin from this organism contains seven atoms of iron per molecule of protein and seven sulfide groups per mole of protein. It has a molecular weight of 6000 and the remarkably low redox potential at pH 7.55 of -0.417 V. When isolated from spinach chloroplasts, it contains two iron atoms linked to two specific sulfur atoms that are released as H_2S on acidification. In spinach its molecular weight is 11,600 and its redox potential is -0.432 V. Oxidized ferredoxin has characteristic absorption bands at 420 and 463 nm. When acidified, these bands disappear, and the protein loses its biochemical activity.

Noncyclic Phosphorylation

The flow of electrons in noncyclic phosphorylation can now be outlined in terms of the energy-conversion process first described. According to reaction 15-10, both ATP and NADPH are produced as H_2O is oxidized to O_2. Starting then with PS II, an electron from OH^- can reduce the oxidant produced there as another electron reduces Q and is passed along the chain to the oxidant at PS I. To complete the process, electrons flow from Z or FRS to $NADP^+$ and accomplish the reduction of that compound.

As electrons flow between PS II and PS I, phosphorylation of ADP takes place, presumably by energy-coupling processes analogous to those occurring in mitochondria during respiratory-chain phosphorylation. Whether the mechanisms are identical obviously remains unsettled. There are ion movements that occur during the illumination of chloroplasts that support the chemiosmotic hypothesis. There are conformational changes in chloroplast membranes that support the conformation hypothesis. Observations favoring the chemical-coupling hypothesis include the isolation from chloroplasts of lamellar fragments enriched in ATP-ase, and coupling factors. More than one phosphorylation step are believed to occur as a pair of electrons flow from PS II to PS I; two are indicated in Figure 15-1.

Cyclic Phosphorylation

The essential feature of cyclic phosphorylation is that ATP is produced in this process without any net transfer of electrons to $NADP^+$. This process, studied extensively by Arnon and his associates, is accomplished by light of longer wavelengths that activates PS I. Since neither NADPH nor any other reduced compound accumulates in this process, electrons made available by the action of light on PS I are postulated to make their way back to the oxidant (P-700) produced when PS I is activated. Thus, a *cyclic* flow of electrons occurs involving still another type of cytochrome—b_6. During the flow of electrons over this cyclic path, at least one ATP is produced from ADP and H_3PO_4, presumably by mechanisms also analogous to those occurring in oxidative phosphorylation.

There is evidence to suggest that cyclic photophosphorylation and cyclic electron flow do not share a common photosystem (PS I) with noncyclic electron flow and phosphorylation. Recent proposals by Arnon and his associates have greatly modified the scheme proposed in Figure 15-1, but these are yet too tentative for inclusion in this text.

Electron Flow in Bacterial Photosynthesis

Those photosynthetic bacteria which utilize inorganic (H_2S, $Na_2S_2O_3$) and organic (succinate, acetate) reducing agents instead of H_2O do not require PS II that is necessary in organisms using H_2O. The electron path utilized by these organisms is represented in Figure 15-1, where electrons feed into the electron-transport chain at the level of plastoquinone. Thus, these organisms require light (for PS I) in order to reduce ferredoxin. The oxidant (P-700) produced by light will accept electrons originally contributed by the primary reductant.

These processes, then, describe how NADPH and ATP are generated by light during photosynthesis. The section below discusses the dark reactions responsible for the assimilation of CO_2 into the organic compounds of photosynthetic organisms.

Path of Carbon

The series of reactions whereby CO_2 is eventually converted to carbohydrates and other organic compounds has been largely worked out in the laboratories of Calvin, Horecker, and Racker. The problem was not extensively pursued, however, until the first product into which CO_2 is incorporated in photosynthesis was identified by Calvin and his associates. This research is an outstanding example of the application of new techniques to the solution of an extremely complicated problem.

The basic experimental approach was as follows: In a plant which is carrying out photosynthesis at a steady rate, CO_2 is being converted to glucose through a series of intermediates:

$$CO_2 \longrightarrow \text{Compound A} \longrightarrow \text{Compound B} \longrightarrow \text{Compound C} \longrightarrow \text{Glucose}$$

If, at time zero, radioactive CO_2 ($^{14}CO_2$) is introduced into the system, some of the labeled carbon atoms will be converted to glucose, and during the time it takes for this to occur all the intermediates will be labeled. If, after a relatively short period of time, the photosynthesizing plant is plunged into hot alcohol to inactivate its enzymes and stop all reactions, the labeled carbon atom will have had time to make its way through only the first few intermediates. If the time interval is short enough, the labeled carbon atoms will have made their way only into the first stable intermediate, compound A, and only the first product of CO_2 fixation will be labeled.

In 1946, carbon-14 was made available in appreciable amounts from the Atomic Energy Commission. Moreover, the technique of paper chromatography (see Appendix 2 for description) was in full development and provided a means for separating the large number of cell constituents which occur in a plant. With these tools Calvin's group was able to identify the early stable intermediates in the path of carbon from CO_2 to glucose. They used suspensions of algae, *Scenedesmus* or *Chlorella,* which were grown at a constant rate in the presence of light and CO_2. Radioactive CO_2 was introduced into the reaction mixture at zero time, and a period of time was allowed to elapse. The cells were then extracted with boiling alcohol, and the soluble constituents of the alcohol solution were analyzed by paper chromatography. When the algae were exposed to $^{14}CO_2$ for 30 seconds, hexose phosphates, triose phosphates, and phosphoglyceric acid were labeled. With longer periods, these compounds as well as amino acids and organic acids were labeled. With 5 sec. exposure most of the radioactive carbon was located in 3-phosphoglyceric acid and, within this compound, the carboxyl group contained the majority of the radioactivity.

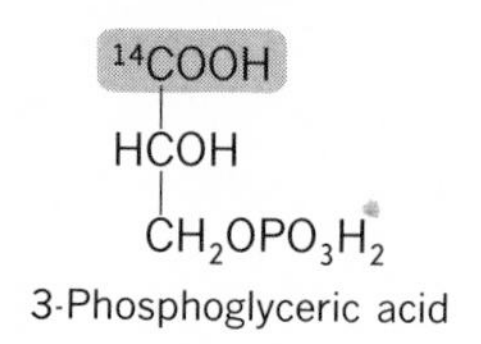

3-Phosphoglyceric acid

This result suggested that 3-phosphoglyceric acid was formed by the carboxylation of some unknown compound containing two carbon atoms. Attempts to demonstrate any such acceptor molecule failed, however. More careful examination of the early products of photosynthesis disclosed that sedoheptulose-7-phosphate and ribulose-1,5-diphosphate were also present as labeled compounds, and this in turn suggested that the sugars might be involved in forming the acceptor molecule for CO_2.

During this period the reactions of the pentose phosphate pathway (Chapter 11) were being clarified in other laboratories, and the relationships between trioses, tetroses, pentoses, hexoses, and heptoses were being established. More careful study of the ^{14}C labeling of the sugars produced during short periods of photosynthesis permitted Calvin's laboratory to postulate the operation of a carbon reduction cycle (Figure 15-2) during photosynthesis. This cycle involves only one unfamiliar reaction, the carboxylation of ribulose-1,5-diphosphate, discussed below; the remainder of the reactions

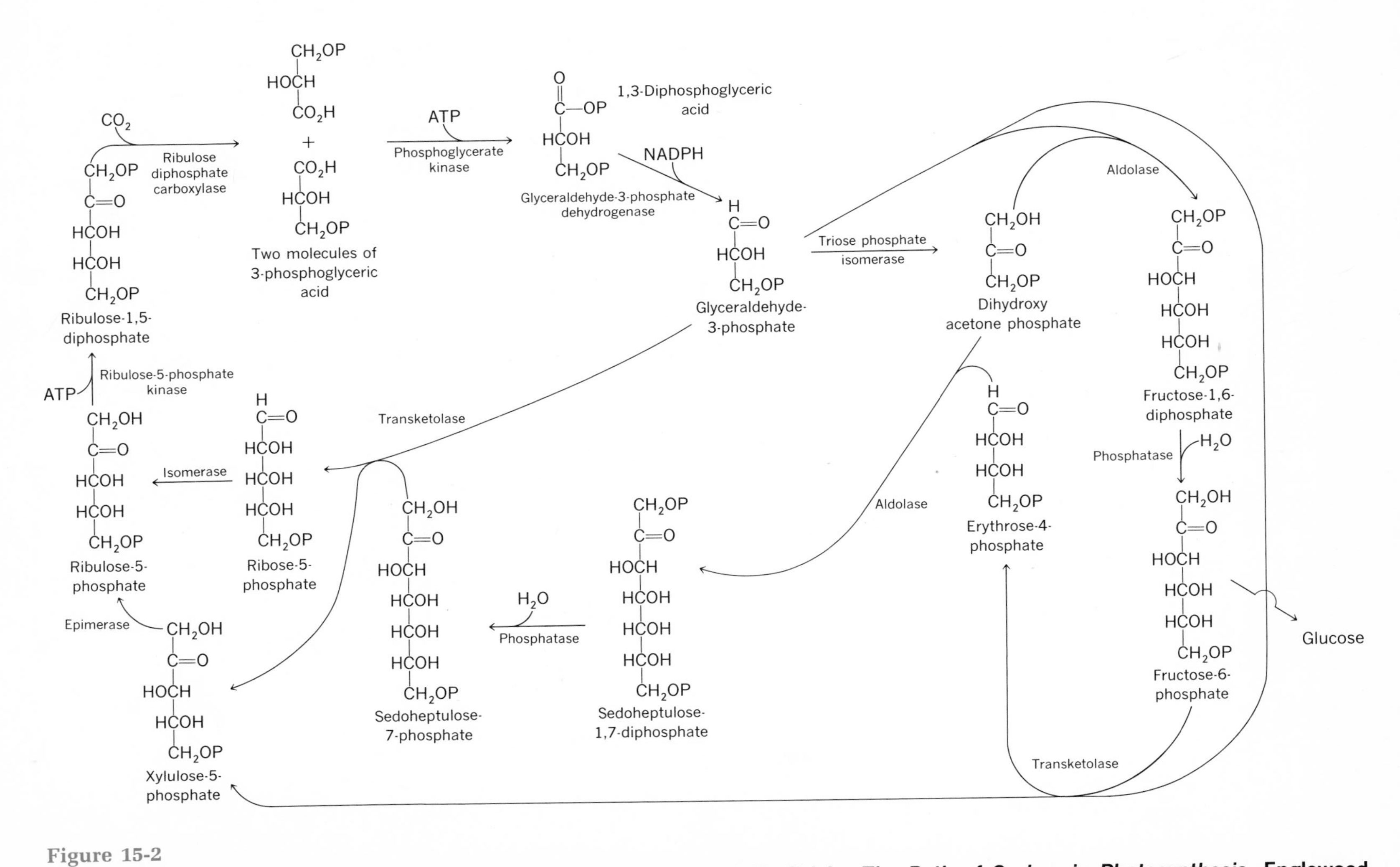

Figure 15-2

The photosynthetic carbon reduction cycle. From J. A. Bassham and M. Calvin, *The Path of Carbon in Photosynthesis*, Englewood Cliffs, N.J.: Prentice-Hall, 1957. Reprinted by permission.

are identical or similar to reactions encountered previously in glycolysis and pentose phosphate metabolism. All of the enzymes required to catalyze the reactions postulated in Calvin's carbon reduction scheme are known to occur in photosynthetic tissues.

This rather bewildering scheme can be better understood when the reactions are grouped into three phases. The first of these, the *carboxylation phase,* involves the single reaction catalyzed by ribulose-1,5-diphosphate carboxylase (also called carboxydismutase). This key reaction involves the carboxylation *not of a two-carbon compound, but of a five-carbon compound,* ribulose-1,5-diphosphate, to yield 2 moles of 3-phosphoglyceric acid.

$$\begin{array}{c} CH_2OPO_3H_2 \\ | \\ C{=}O \\ | \\ H{-}C{-}OH \\ | \\ H{-}C{-}OH \\ | \\ CH_2OPO_3H_2 \end{array} \rightleftharpoons \left[\begin{array}{c} CH_2OPO_3H_2 \\ | \\ C{-}OH \\ \| \\ C{-}OH \\ | \\ H{-}C{-}OH \\ | \\ CH_2OPO_3H_2 \end{array}\right] \xrightarrow{CO_2} \left[\begin{array}{c} CH_2OPO_3H_2 \\ | \\ HCO_2{-}C{-}OH \\ | \\ C{=}O \\ | \\ H{-}C{-}OH \\ | \\ CH_2OPO_3H_2 \end{array}\right] \xrightarrow{H_2O} \begin{array}{c} CH_2OPO_3H_2 \\ | \\ HCO_2{-}C{-}OH \\ | \\ H \\ + \\ OH \\ | \\ C{=}O \\ | \\ H{-}C{-}OH \\ | \\ CH_2OPO_3H_2 \end{array}$$

D-Ribulose-1,5-diphosphate; Enediol; β-Keto acid intermediate; Two molecules of 3-phosphoglyceric acid

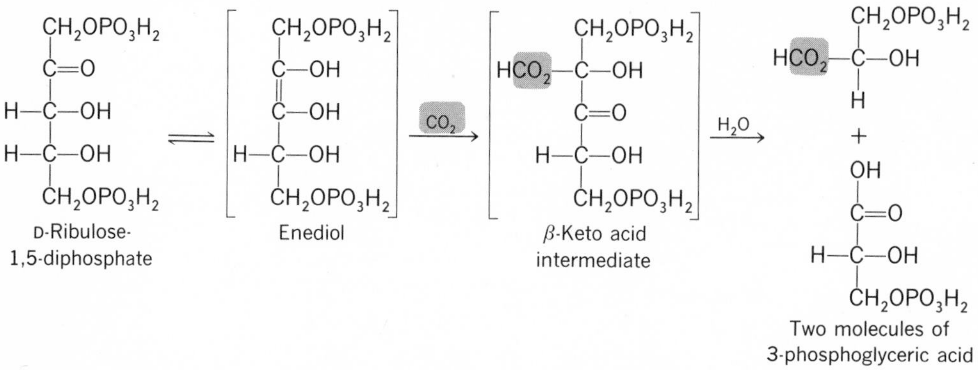

In the presence of the enzyme ribulose-1,5-diphosphate carboxylase, CO_2 adds to the enediol form of ribulose diphosphate to form an unstable β-keto acid which undergoes hydrolytic cleavage to form two molecules of phosphoglyceric acid. The equilibrium of the reaction is far to the right. The carboxylase was first purified as a homogenous protein by Horecker from spinach leaves, where it constitutes 5–10% of the soluble protein. It has a molecular weight of 550,000 and is an oligomer composed of two different kinds of subunits.

A second phase of the carbon reduction cycle, termed the *reduction phase,* consists of two reactions previously encountered in glycolysis. In these reactions ATP and a reduced nicotinamide nucleotide are consumed. The first involves the phosphorylation of 3-phosphoglycerate by ATP to form 1,3-diphosphoglycerate:

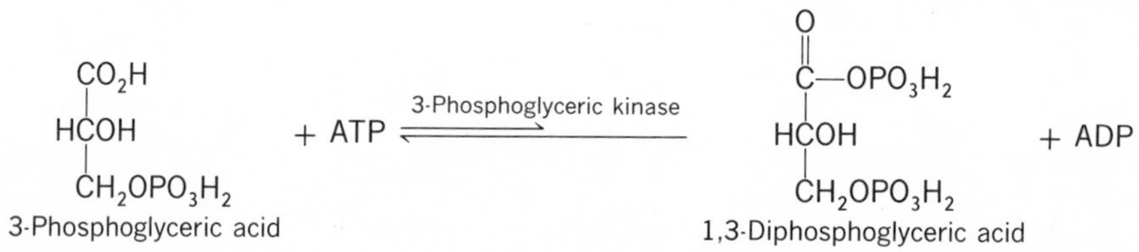

$$\begin{array}{c} CO_2H \\ | \\ HCOH \\ | \\ CH_2OPO_3H_2 \end{array} + ATP \xrightleftharpoons{\text{3-Phosphoglyceric kinase}} \begin{array}{c} O \\ \| \\ C{-}OPO_3H_2 \\ | \\ HCOH \\ | \\ CH_2OPO_3H_2 \end{array} + ADP$$

3-Phosphoglyceric acid; 1,3-Diphosphoglyceric acid

The second reaction involves reduction of the 1,3-diphosphoglyceric acid by NADPH in the presence of a NADP-specific glyceraldehyde-3-phosphate dehydrogenase:

$$\begin{array}{c} O \\ \| \\ C{-}OPO_3H_2 \\ | \\ HCOH \\ | \\ CH_2OPO_3H_2 \end{array} + NADPH + H^+ \rightleftharpoons \begin{array}{c} CHO \\ | \\ HCOH \\ | \\ CH_2OPO_3H_2 \end{array} + NADP^+ + H_3PO_4$$

1,3-Diphosphoglyceric acid — Glyceraldehyde-3-phosphate

The chloroplasts of higher green plants contain the $NADP^+$-specific enzyme. It is in these two reactions that the NADPH and half of the ATP required to drive the carbon reduction cycle are utilized.

The remainder of the reactions in the cycle compose a third, or *regeneration phase,* which accomplishes the regeneration of ribulose-1,5-diphosphate necessary to keep the cycle operating. In Table 15-2 are listed the reactions of this phase with the stoichiometry required. Note that a total of thirty-six carbon atoms present in twelve molecules of glyceraldehyde-3-phosphate at the end of the reduction phase are converted, by the reactions of the regeneration phase, into one molecule of fructose-6-phosphate (six carbon atoms) and six molecules of ribulose-1,5-diphosphate (thirty carbon atoms) at the end of the regeneration phase. The last reaction of the regeneration phase also requires ATP. The six molecules of ribulose diphosphate produced

Table 15-2

Stoichiometry of the Carbon Reduction Cycle

Carboxylation phase

6 Ribulose-1,5-diphosphate + **6** CO_2 + **6** H_2O ⟶ **12** 3-Phosphoglycerate

Reduction phase

12 3-Phosphoglycerate + **12** ATP ⟶ **12** 1,3-Diphosphoglycerate + **12** ADP

12 1,3-Diphosphoglycerate + **12** NADPH + **12** H^+ ⟶ **12** Glyceraldehyde-3-phosphate + **12** $NADP^+$ + **12** H_3PO_4

Regeneration phase

5 Glyceraldehyde-3-phosphate ⟶ **5** Dihydroxy acetone phosphate

3 Glyceraldehyde-3-phosphate + **3** Dihydroxy acetone phosphate ⟶ **3** Fructose-1,6-diphosphate

3 Fructose-1,6-diphosphate + **3** H_2O ⟶ **3** Fructose-6-phosphate + **3** H_3PO_4

2 Fructose-6-phosphate + **2** Glyceraldehyde-3-phosphate ⟶ **2** Xylulose-5-phosphate + **2** Erythrose-4-phosphate

2 Erythrose-4-phosphate +**2** Dihydroxy acetone phosphate ⟶ **2** Sedoheptulose-1,7-diphosphate

2 Sedoheptulose-1,7-diphosphate + **2** H_2O ⟶ **2** Sedoheptulose-7-phosphate + **2** H_3PO_4

2 Sedoheptulose-7-phosphate + **2** Glyceraldehyde-3-phosphate ⟶ **2** Ribose-5-phosphate + **2** Xylulose-5-phosphate

2 Ribose-5-phosphate ⟶ **2** Ribulose-5-phosphate

4 Xylulose-5-phosphate ⟶ **4** Ribulose-5-phosphate

6 Ribulose-5-phosphate + **6** ATP ⟶ **6** Ribulose-1,5-diphosphate + **6** ADP

SUM

6 CO_2 + **18** ATP + **12** NADPH + **12** H^+ + **11** H_2O ⟶ Fructose-6-phosphate + **18** ADP + **12** $NADP^+$ + **17** H_3PO_4

by this phase are then available for the carboxylation process, and they can keep the cycle functioning.

The overall stoichiometry of the carbon reduction cycle is given in the sum in Table 15-2. The fructose-6-phosphate produced can in turn be converted to glucose by reversal of the reactions encountered in the early stages of glycolysis (Chapter 10):

$$\text{Fructose-6-phosphate} \longrightarrow \text{Glucose-6-phosphate} \xrightarrow{H_2O} \text{Glucose} + H_3PO_4$$

When this is done, the overall carbon reduction cycle becomes

$$6\,CO_2 + 18\,\text{ATP} + 12\,\text{NADPH} + 12\,H^+ + 12\,H_2O \longrightarrow \text{Glucose} + 18\,\text{ADP} + 18\,H_3PO_4 + 12\,\text{NADP}^+ \quad (15\text{-}11)$$

Dividing this equation by six illustrates an important fact regarding the energetics of photosynthesis:

$$CO_2 + 3\,\text{ATP} + 2\,\text{NADPH} + 2\,H^+ + 2\,H_2O \longrightarrow C(H_2O) + 3\,\text{ADP} + 3\,H_3PO_4 + 2\,\text{NADP}^+ \quad (15\text{-}12)$$

Reaction 15-12 shows that photosynthesis requires 3 moles of ATP and 2 moles of NADPH to convert 1 mole of CO_2 to the level of carbohydrate.

Returning to the energy-conversion process in Figure 15-1, we can now place a lower limit on the number of light quanta necessary to make the two molecules of NADPH required for equation 15-12. In green plants that utilize H_2O, both PS I and PS II will have to be activated four times each to produce the four electrons required to reduce 2 $NADP^+$. Therefore, a total of eight quanta of light would appear to be required as a minimum, since it is generally assumed that at least one quantum is required for each photoactivation process that makes an electron available at Q and FRS in the energy-conversion scheme. It should also be apparent that noncyclic photophosphorylation can only produce two-thirds of the ATP required if only one phosphorylation occurs when a pair of electrons travels the path connecting PS II and PS I. Under those conditions cyclic phosphorylation presumably could furnish the additional ATP, provided further light is supplied. If, however, two phosphorylation sites exist in the chain linking PS I and PS II, sufficient ATP would then be available.

One final aspect of the energy requirements of photosynthesis deserves comment. In Table 15-1, light of 6500 Å was shown to have an energy content of 43,480 cal/Einstein. It is generally assumed that about 75% of this energy (approximately 30,000 cal/Einstein) is available for photosynthesis, the remaining being dissipated as heat as electrons pass from the triplet state to transitional levels. If, however, eight quanta per two molecules of NADPH (8 Einsteins for 2 moles of NADPH) are the minimum required to drive the energy-conversion process, we see that $8 \times 30{,}000$ or 240,000 cal are available to convert 1 mole of CO_2 into carbohydrate. From equation 15-2 we have seen that 118,000 cal are required as a minimum to accomplish this conversion. The overall efficiency of photosynthesis with these calculations would therefore be 118,000/240,000 or 49%.

CO_2-Fixation into the C_4-Dicarboxylic Acids

Considerable attention has centered on the path of carbon in several important crop plants of tropical origin—sugar cane, corn, and sorghum—that are characterized by high rates of photosynthesis and growth and by a lack of photorespiration (see below). The Australian workers M. D. Hatch and C. R. Slack showed in 1966 that the initial product of CO_2-fixation during short-term photosynthesis by leaves of sugar cane is oxalacetate rather than 3-phosphoglyceric acid. The enzyme that catalyzes this reaction is phosphoenol pyruvic acid carboxylase (PEP–carboxylase), one of the CO_2-fixation enzymes discussed earlier (page 260):

$$\underset{\text{Phosphoenol pyruvate}}{\begin{array}{l} CO_2H \\ | \\ C{-}OPO_3H_2 \\ \| \\ CH_2 \end{array}} + CO_2 + H_2O \xrightarrow{\text{PEP–carboxylase}} \underset{\text{Oxalacetate}}{\begin{array}{l} CO_2H \\ | \\ C{=}O \\ | \\ CH_2 \\ | \\ CO_2H \end{array}} + H_3PO_4 \qquad (15\text{-}13)$$

The enzyme, which has a much higher affinity for CO_2 than does ribulose-1,5-diphosphate carboxylase, is highly active in these plants and was believed to be involved in a pathway of carbon assimilation that was fundamentally different from that proposed by Calvin and his associates.

This important observation stimulated work in many laboratories which disclosed that those plants exhibiting the C_4-fixation phenomenon are characterized by a common feature of leaf anatomy in which bundle sheath cells are surrounded by layers of mesophyll cells (Figure 15-3). The chloroplasts of mesophyll cells contain relatively large amounts of PEP–carboxylase

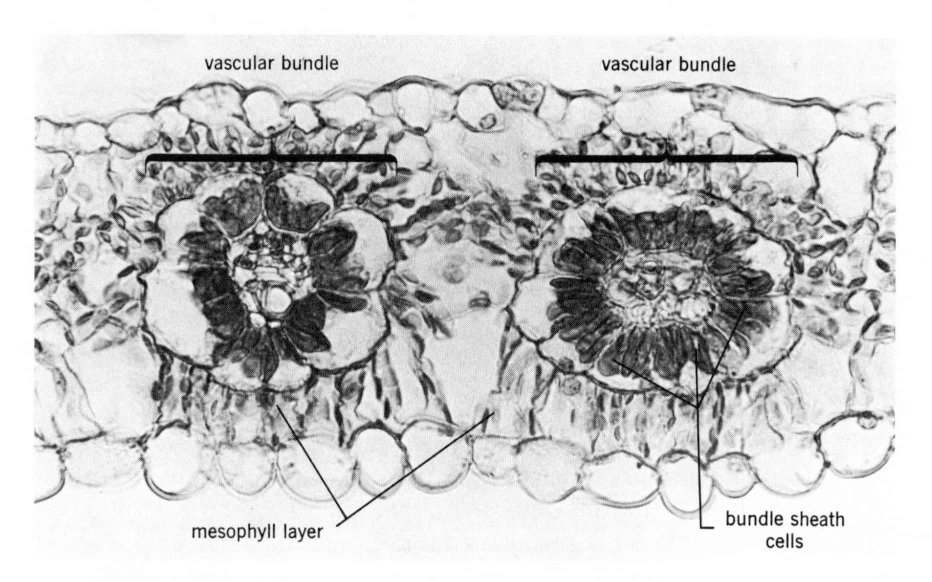

Figure 15-3

Microphotograph of the vascular bundle, bundle sheath cells, and mesophyll layer of *Amaranthus edulis*. Courtesy of W. M. Laetsch.

together with malic dehydrogenase that produces malate by reduction of the oxalacetate formed in reaction 15-13. An aspartic transaminase present in these chloroplasts can also convert oxalacetate to aspartic acid which together with the two dicarboxylic acids contain the majority of the CO_2 fixed in short-term photosynthesis.

Rather than constituting an entirely new scheme for the assimilation of carbon, the C_4-fixation phenomenon is an extremely efficient mechanism for the initial fixation of CO_2. In one group of C_4 species, the malate produced in the mesophyll cells is transferred to the chloroplasts of the adjacent bundle sheath cells which contain a high concentration of malic enzyme:

$$\underset{\text{L-Malate}}{HO_2C{-}CH_2{-}CH(OH){-}CO_2H} + NADP^+ \xrightarrow{\text{Malic enzyme}} \underset{\text{Pyruvate}}{CH_3{-}C({=}O){-}CO_2H} + CO_2 + NADPH + H^+ \quad (15\text{-}14)$$

This enzyme releases the carbon atom fixed as CO_2 into oxalacetate; the CO_2 is then taken into the Calvin cycle by ribulose-1,5-diphosphate carboxylase that is present in bundle sheath chloroplasts. In another group of C_4 species, the aspartate produced in the mesophyll cells is transported into the bundle sheath cells, where aspartate transaminase, present in high concentrations, gives rise to oxalacetate which is presumably decarboxylated:

$$\underset{\text{L-Aspartate}}{HO_2C{-}CH_2{-}CH(NH_2){-}CO_2H} \xrightarrow{\text{Transamination}} \underset{\text{Oxalacetate}}{HO_2C{-}CH_2{-}C({=}O){-}CO_2H} \xrightarrow{\text{Decarboxylation}} \underset{\text{Pyruvate}}{CH_3{-}C({=}O){-}CO_2H} + CO_2 \quad (15\text{-}15)$$

The pyruvic acid produced in reactions 15-14 and 15-15 is then transferred back to the mesophyll cells where a final enzyme, pyruvate phosphate dikinase, present only in the mesophyll chloroplasts, converts the pyruvate back to phosphoenol pyruvate:

$$CH_3{-}C({=}O){-}CO_2H + ATP + H_3PO_4 \xrightarrow[\text{dikinase}]{\text{Pyruvate phosphate}} CH_2{=}C(OPO_3H_2){-}CO_2H + AMP + HO{-}P(O)(OH){-}O{-}P(O)(OH){-}OH$$

The fixation of CO_2 into the C_4-dicarboxylic acids may be viewed, therefore, as a very efficient mechanism for trapping and converting CO_2 into a familiar metabolite. There is now much evidence that indicates that most if not all of the CO_2 fixed into 3-phosphoglycerate in C_4-pathway species is derived from the decarboxylation of C_4 acids; that is, ribulose-1,5-diphosphate carboxylase does not contribute significantly to the initial fixation of CO_2 derived directly from the atmosphere.

The C_4 pathway has apparently evolved in response to ecological situations characterized by the combination of high radiation, higher temperatures, and the limited supply of water. The capacity of C_4-pathway species to survive and grow under these conditions is primarily due to the high activity of PEP–carboxylase and its capacity to operate with low concentrations of CO_2. This allows plants to continue photosynthesis with very restricted stomatal opening and explains why C_4-pathway species are much more efficient in terms of water loss compared with Calvin cycle species.

Reductive Carboxylation in Bacteria

Evidence is available that indicates that certain photosynthetic bacteria (*Chorobium thiosulfatophillum* and *Chromatium*) do not utilize the scheme proposed by Calvin and his associates for the assimilation of CO_2. Instead, these organisms utilize the tricarboxylic acid cycle by operating it in the *reverse* direction. From what has been said in Chapter 13 regarding the operation of this cycle, the reader will realize that the one irreversible reaction catalyzed by α-ketoglutaric dehydrogenase must be modified in order for the cycle to go backwards.

Arnon and Buchanan have observed that these organisms contain an enzyme called *α-ketoglutaric synthase* that in effect reverses the oxidative decarboxylation of α-ketoglutarate by utilizing reduced ferredoxin as the reductant instead of NADH:

$$\begin{matrix} CO_2H \\ | \\ C{=}O \\ | \\ CH_2 \\ | \\ CH_2 \\ | \\ CO_2H \end{matrix} + 2\ \text{Ferredoxin-}Fe^{3+} + \text{CoASH} \rightleftharpoons \begin{matrix} O \\ \| \\ C{-}S{-}CoA \\ | \\ CH_2 \\ | \\ CH_2 \\ | \\ CO_2H \end{matrix} + 2\ \text{Ferredoxin-}Fe^{2+} + CO_2 + 2\ H^+$$

$$\Delta G' = -3400 \text{ cal (pH 7.0)}$$

On page 309, the $\Delta G'$ for the reaction catalyzed by α-ketoglutaric dehydrogenase was calculated to be -8000 kcal. The E_0' of ferredoxin at pH 7.0 is given as -0.420 V, while that for $NAD^+/NADH$ is -0.320. The difference in E_0' of the two oxidants (-0.100 V) can be calculated (from equation 7-28) to be equivalent to -4600 cal. Therefore, the $\Delta G'$ for the reaction catalyzed by α-ketoglutaric synthase in the photosynthetic bacteria can be estimated as $-8000 - (-4600)$ or -3400 cal. This value is in the region of other reactions which we have seen are reversible, and thus the reaction can occur from right to left. Since all the other reactions of the Krebs cycle are reversible, and since these organisms contain a *pyruvic synthase* that requires ferredoxin rather than NAD^+, the net synthesis of pyruvate from 3 moles of CO_2 can occur as shown in Figure 15-4. These organisms presumably utilize citrate cleavage enzyme (page 318) rather than citrate synthase to cleave the citric acid and regenerate oxalacetate required for the next turn of the cycle. Once pyruvic acid is formed, it may be converted to carbohydrate and lipid by reactions previously considered. Alternately, it can be converted to

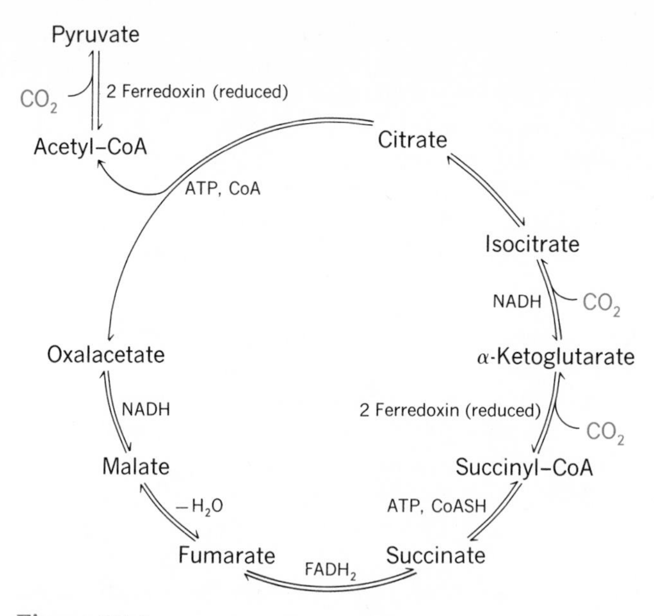

Figure 15-4

The reductive carboxylation cycle of photosynthetic bacteria.

oxalacetate by pyruvic carboxylase (page 314) as an anaplerotic process, and the Krebs cycle can then be used, as discussed on page 313, for anabolic purposes.

The energy requirements of this reductive carboxylation cycle are rather formidable and considerably more than the Calvin cycle as it operates in green plants. A total 2 moles of ATP and five reducing equivalents of two electrons each (as $FADH_2$, 2 NADH, and 4 ferredoxins–Fe^{2+}) are required to produce 1 mole of pyruvic acid from 3 moles of CO_2.

Photorespiration

Plants of course carry out the same general respiratory processes as animals and microorganisms in that they degrade carbohydrates by means of glycolysis and the Krebs cycle. They also exhibit β-oxidation and catalyze the general reactions of protein and amino acid catabolism. Moreover, these reactions occur in the same parts of the plant cell (mitochondria, cytoplasm, microsomes, etc.) as in animal cells. However, many plants also exhibit an additional metabolic activity termed *photorespiration* which occurs only when those plants are illuminated. Since this respiration gives rise to CO_2 evolution in light, it has the net effect of decreasing photosynthesis, which of course consumes CO_2 in light.

The reactions constituting photorespiration and the cellular organelles that carry it out are only now being clarified. The initial substrate for photorespiration is glycolic acid, a major product formed in chloroplasts during photosynthesis under conditions (high O_2 and low CO_2 concentrations) that favor photorespiration. Glycolic acid is postulated to leave the chloroplasts and undergo reactions catalyzed by enzymes in both peroxisomes, microbodies that occur in green leaves, and mitochondria. The overall reaction is one in which two molecules of glycolic acid are converted to 3-phosphoglyceric acid and CO_2.

$$2\begin{array}{l}CO_2H\\|\\CH_2OH\end{array} \longrightarrow \longrightarrow \longrightarrow \begin{array}{l}CO_2H\\|\\HCOH\\|\\CH_2OPO_3H_2\end{array} + CO_2$$

Glycolic acid — 3-Phosphoglyceric acid

The overall process results in the conversion of 25% of the carbon of a major photosynthetic product into CO_2 by reactions that are dependent on light. This is in direct contrast to the carbon reduction cycle in which light provides the energy for assimilation of CO_2 into organic compounds. The 3-phosphoglyceric acid formed in the peroxisomes can then return to the chloroplasts, where it can reenter the carbon reduction cycle.

The origin of the glycolic acid produced in the chloroplasts is not clear. It may be formed (as phosphoglycolate) from the C_2 addition product formed when sedoheptulose-1,7-diphosphate reacts with transketolase. It has also been proposed as a product of the oxidation of ribulose-1,5-diphosphate. Either of these reactions would lower the level of intermediates in the carbon reduction cycle and thereby inhibit CO_2 assimilation by that process.

The list of plants that exhibit photorespiration includes wheat, rice, and other cereals, many legumes, and sugar beets, crops that are important from the standpoint of the world's food supply. In some of these it has been estimated that the net assimilation of CO_2 by photosynthesis may be reduced by as much as 50% by photorespiration. Some workers have proposed that it might be practical to increase the crop yield of such plants by finding a means to inhibit the photorespiration which they exhibit. Other equally important food crops—corn, sorghum, sugar cane—do not exhibit the phenomenon of photorespiration. Since, however, the leaves of such plants do contain peroxisomes and presumably can still perform the individual reactions of photorespiration, some explanation is required. Some evidence suggests that these plants may rely on the C_4-fixation process described above, which, because of its greater efficiency of CO_2-fixation, can reutilize the CO_2 produced in photorespiration and retain effectively all of the CO_2.

This brief discussion of the primary life process of photosynthesis can only serve to disclose how limited is the knowledge of the subject. Hopefully, further research will clarify the details of the energy-conversion process and provide basic information that will allow man to regulate photosynthesis for optimum production of his food.

References

1. R. Hill, "The Biochemists' Green Mansions: the Photosynthetic Electron Transport Chain in Plants," *Essays in Biochemistry*. vol. I. London: Academic Press, 1965.
   An excellent review of the energy-conversion processes in photosynthesis by a major figure in the field.

2. N. K. Boardman, "The Photochemical Systems of Photosynthesis," *Adv. Enzymol.* **30,** 1 (1968).
   A critical review of recent work on the energy-conversion processes in photosynthesis.
3. J. A. Bassham and M. Calvin, *The Path of Carbon in Photosynthesis*. Englewood Cliffs, N.J.: Prentice-Hall, 1957.
   An account of the experimental approach and results obtained in the authors' studies on the path of carbon in photosynthesis.
4. M. D. Hatch and C. R. Slack, in *Progress in Phytochemistry*, L. Reinhold and Y. Liwschitz, eds. vol. 2. London: Interscience, 1970.
   The role of the dicarboxylic acids in the assimilation of CO_2 by sugar cane, maize, and other tropical grasses.
5. I. Zelitch, *Photosynthesis, Photorespiration and Plant Productivity*. New York: Academic Press, 1971.
   This excellent new book is a rich source of information on the three topics found in its title.

16

The Metabolism of Ammonia and Nitrogen-Containing Monomers

Introduction

The amino acids share with the purine and pyrimidine nucleotides the fact that they are nitrogen-containing building blocks of large, informational molecules—the proteins and nucleic acids. Higher plants and many microorganisms regularly obtain the nitrogen required for biosynthesis of these compounds in the form of nitrate ion. Plants and most microorganisms will also utilize NH_3 when it is available as a source of nitrogen for synthesis of amino acids, proteins, and nucleic acids. While the higher animal can also utilize NH_3 for synthesis of its nitrogen-containing compounds, the animal's principal source of nitrogen is the protein it consumes in its diet. The protein is hydrolyzed to amino acids by enzymes in the gastrointestinal tract, and these are absorbed into the blood and transported to the liver. This organ will remove a portion of the amino acids for specific biosynthetic tasks, while the remainder pass on to extra-hepatic tissues where they can be synthesized into proteins. The liver is the site of synthesis of several blood proteins (plasma albumin, the globulins, fibrinogen, and prothrombin). It also metabolizes any amino acids in excess of hepatic needs for protein synthesis, converting the nitrogen atoms into urea and the carbon skeleton into intermediates previously encountered in the metabolism of carbohydrates and lipids. While much is known about the detailed metabolism of the twenty amino acids found in most proteins, we have space only to treat those reactions which, in general, apply to all amino acids and to discuss the role of NH_3 in the formation of urea and the purines and pyrimidines.

Nitrogen Balance Studies

Our knowledge of the intermediary metabolism of amino acids and proteins has its foundations in early nutritional investigations. Osborne and Mendel demonstrated in 1914 that the growing rat required tryptophan and lysine in its diet. Subsequently, W. C. Rose showed that eight other amino acids were required by the rat for growth and development. World War II provided the stimulus and the research funds for identifying the amino acids required by man in experiments which involved the feeding of gram quantities of highly purified amino acids to male volunteers. These experiments, which were performed by keeping the subjects in *nitrogen equilibrium*, demonstrated that lysine, tryptophan, phenylalanine, threonine, valine, methionine, leucine, and isoleucine were *indispensable*.

An individual (man or other animal) is said to be in nitrogen equilibrium when the nitrogen consumed per day in the diet is equal to the amount of nitrogen excreted. The former is easily measured, especially if the diet is a synthetic one consisting of a mixture of amino acids; the nitrogen excreted is that found in the urine and feces. An *adult* animal can be maintained in nitrogen equilibrium provided an amount of nitrogen is supplied which is adequate to meet its minimum metabolic needs. This nitrogen, however, cannot be furnished simply as NH_3 but must be provided in the form of the indispensable amino acids. If one of these amino acids is omitted from the diet, the animal will degrade tissue proteins to meet its requirements and will go into negative nitrogen balance. That is, the nitrogen excreted in the urine and feces exceeds that in the diet. When the omitted amino acid is restored to the diet, the individual attains equilibrium again.

Fevers and wasting diseases place an individual in negative nitrogen balance as does inadequate dietary nitrogen. On the other hand, a growing animal which is continually increasing the amount of its body protein will be in positive nitrogen balance; that is, it takes in more nitrogen than it excretes.

There are two important consequences of the nutritional work on the indispensable amino acids we have described; first, it is clear that the animal cannot make these amino acids—at least in the amounts it requires. We may ask, then, whether the animal lacks the ability to make the carbon skeleton of the indispensible amino acid. The answer apparently is yes, for if an animal which is being furnished a diet which is deficient in phenylalanine, for example, is supplied with phenylpyruvic acid, the keto analog of phenylalanine, and extra nitrogen in the form of the other indispensable amino acids, it goes into equilibrium. These results are interpreted as meaning that the problem is not one of supplying nitrogen but rather one of synthesis of the carbon skeleton. In the case of phenylalanine, the difficulty is in the synthesis of the aromatic ring the amino acid possesses. Thus, it may be concluded that certain types of carbon skeleton are not readily synthesized by higher animals.

Since only about half of the naturally occurring amino acids are indispensable to animals, it is clear that animals can synthesize the remaining amino acids. These amino acids which can be synthesized are known as *dispensable* amino acids. Their synthesis involves not only the manufacture of the carbon skeleton, but also includes the transfer of nitrogen atoms from some nitrogen donor, usually glutamic or aspartic acid, to complete the dispensable amino acid. This transfer of the nitrogen atom is accomplished by *transamination,* a general reaction of amino acids that is involved both in the breakdown and the synthesis of many amino acids.

Transamination

The reaction of transamination involves the transfer of the amino group from one amino acid to a keto acid (the carbon skeleton) to form the analogous amino acid and to produce the keto acid (the carbon skeleton) of the original amino donor.

$$\underset{\substack{\text{Amino acid}_1\\ \text{(donor)}}}{R^1\text{—}CH(NH_2)\text{—}CO_2H} + \underset{\substack{\text{Keto acid}_2\\ \text{(acceptor)}}}{R^2\text{—}C(=O)\text{—}CO_2H} \rightleftharpoons \underset{\text{Keto acid}_1}{R^1\text{—}C(=O)\text{—}CO_2H} + \underset{\text{Amino acid}_2}{R^2\text{—}CH(NH_2)\text{—}CO_2H}$$

Transaminases capable of reacting with nearly all of the amino acids have been reported, but especially important are *glutamic transaminase* and *alanine transaminase*. Glutamate transaminase is specific for glutamic and α-ketoglutaric acid as one of its two substrate pairs but will react, at different rates, with nearly all of the other proteineous amino acids:

$$\underset{\text{Donor amino acid}}{R^1\text{—}CH(NH_2)\text{—}CO_2H} + \underset{\alpha\text{-Ketoglutaric acid}}{HO_2C\text{—}C(=O)\text{—}CH_2\text{—}CH_2\text{—}CO_2H} \xrightleftharpoons{\text{Glutamic transaminase}} \underset{\text{Keto acid}}{R^1\text{—}C(=O)\text{—}CO_2H} + \underset{\text{Glutamic acid}}{HO_2C\text{—}CH(NH_2)\text{—}CH_2\text{—}CH_2\text{—}CO_2H} \qquad (16\text{-}1)$$

Similarly, *alanine transaminase* is specific for alanine and pyruvic acid as one of its substrate pairs but reacts with almost any other amino acid. Finally, a highly specific glutamic–alanine transaminase, found in many organisms, catalyzes the transamination between these two amino acids (reaction 16-2):

$$\underset{\text{Alanine}}{HO_2C\text{—}CH(NH_2)\text{—}CH_3} + \underset{\alpha\text{-Ketoglutaric acid}}{HO_2C\text{—}C(=O)\text{—}CH_2\text{—}CH_2\text{—}CO_2H} \xrightleftharpoons{\text{Glutamic–alanine transaminase}} \underset{\text{Pyruvic acid}}{HO_2C\text{—}C(=O)\text{—}CH_3} + \underset{\text{Glutamic acid}}{HO_2C\text{—}CH(NH_2)\text{—}CH_2\text{—}CH_2\text{—}CO_2H} \qquad (16\text{-}2)$$

The reactions catalyzed by the transaminases have, as would be expected, an equilibrium constant of approximately 1.0; therefore, the reactions are readily reversible. The transaminases require pyridoxal phosphate as a cofactor and, in the presence of the enzyme, the coenzyme forms a Schiff's base with the amino acid. By subsequent electron rearrangements (see page 217) the amino group is transferred to the coenzyme to form pyridoxamine phosphate. The latter compound can then react with the acceptor keto acid to regenerate the pyridoxal phosphate and the product amino acid.

The significance of the transamination process is better appreciated if one understands that reaction 16-1 (or alanine transaminase together with reaction 16-2) serves to collect amino groups of many other amino acids as glutamic acid. These reactions occur primarily in the cytoplasm, and it is the glutamic acid which, being specifically permeable to the inner mitochondrial membrane, enters the matrix of the mitochondria. There it can transaminate again with a mitochondrial aspartate transaminase or alternatively be oxidatively deaminated by the mitochondrial glutamic dehydrogenase. (In the next section the deamination of glutamic acid is described and the significance of the combined processes of transamination and deamination are discussed.) Transaminases therefore are found both in the cytoplasm and the mitochondria of eucaryotic cells, the enzymes in each region of the cell having characteristic properties.

Deamination

L-Glutamic acid plays a key role in the metabolism of amino acids because of the widespread occurrence of the enzyme *glutamic dehydrogenase*. This enzyme catalyzes the reversible *oxidative deamination* by NAD^+ of L-glutamate to form α-ketoglutaric acid, NH_3, and NADH:

$$\begin{array}{c} COOH \\ | \\ H_2NCH \\ | \\ HCH \\ | \\ HCH \\ | \\ COOH \end{array} + NAD^+ + H_2O \rightleftharpoons \begin{array}{c} COOH \\ | \\ C{=}O \\ | \\ CH_2 \\ | \\ CH_2 \\ | \\ COOH \end{array} + NADH + H^+ + NH_3 \qquad (16\text{-}3)$$

L-Glutamic acid $\qquad$ α-Ketoglutaric acid

The significance of this enzyme, which can also use $NADP^+$ as a coenzyme, in the deamination of other amino acids is readily seen when it is coupled with glutamic transaminase (reaction 16-1 and Scheme 16-1). The two

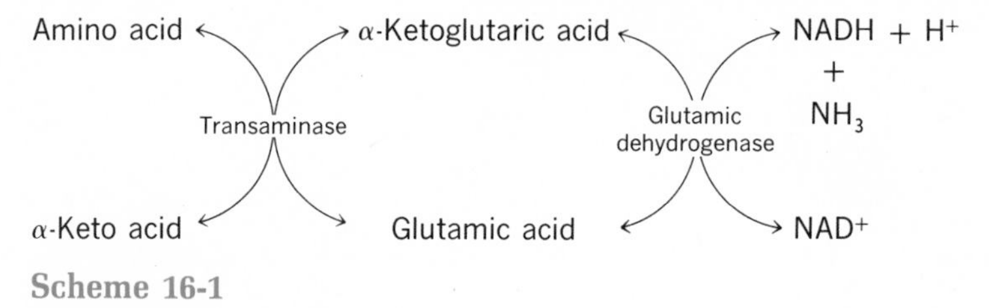

Scheme 16-1

enzymes acting in sequence accomplish the release as NH_3 of the amino nitrogen atom of any amino acid that can transaminate with glutamic transaminase. Moreover, since both the transamination and oxidative deamination reactions are readily reversible, the two enzymes acting in reverse can accomplish the synthesis of those amino acids whose α-keto acid analog can be synthesized by the organism under consideration. In the case of higher plants and many microorganisms, this means the keto acid of every proteinaceous amino acid. In the case of higher animals, this includes only the keto acid analogs of the dispensible amino acids.

Because of its importance in the metabolism of other amino acids, as well as the fact that glutamic acid serves as a precursor of proline and ornithine (and indirectly therefore of hydroxyproline, citrulline, and arginine (see page 386)), it is not surprising to learn that glutamic dehydrogenase is an allosteric enzyme. The beef liver enzyme, as an example, is inhibited by ATP and NADH and is stimulated by ADP and AMP.

Although the enzyme was originally believed to be specific for L-glutamic acid, other L-amino acids (L-leucine, L-valine) can also be oxidized by the crystalline enzyme. Since the maximum rates of oxidation with these amino acids are much lower, however, the reaction with L-glutamic acid appears to be the only significant reaction.

Oxidative deamination reactions are also catalyzed by a group of flavin enzymes known as *amino acid oxidases*. In 1935, Krebs showed that kidney and liver slices catalyzed the formation of NH_3 from different amino acids and that oxygen was consumed. Subsequently, both enantiomers of racemic mixtures of amino acids were shown to be acted on by the slices, and the enzyme which catalyzed the oxidative deamination of the D isomer was shown to be soluble. The details of the reaction were elucidated with a partially purified D-amino oxidase prepared from sheep kidney. The overall reaction is

$$R{-}\overset{\overset{NH_3^+}{|}}{\underset{\underset{H}{|}}{C}}{-}COO^- + O_2 + H_2O \longrightarrow R{-}\overset{\overset{O}{\|}}{C}{-}COO^- + NH_4^+ + H_2O_2 \qquad (16\text{-}4)$$

Most of the proteinaceous amino acids serve as substrates, the enzyme requires FAD as a prosthetic group, and the reaction is not readily reversible. The overall reaction may be broken down into individual steps for which there is experimental evidence. In the first step, oxidation of the amino acid leads to the corresponding imino acid:

$$R{-}\overset{\overset{NH_3^+}{|}}{\underset{\underset{H}{|}}{C}}{-}COO^- + FAD \rightleftharpoons R{-}\overset{\overset{\overset{H}{N}}{\|}}{C}{-}COO^- + FADH_2 + H^+ \qquad (16\text{-}5)$$

The imino acid in turn is spontaneously hydrolyzed in the presence of H_2O:

$$R{-}\overset{\overset{\overset{H}{N}}{\|}}{C}{-}COO^- + H_2O + H^+ \rightleftharpoons R{-}\overset{\overset{O}{\|}}{C}{-}COO^- + NH_4^+ \qquad (16\text{-}6)$$

The reduced flavin formed will in turn be reoxidized by molecular oxygen to form H_2O_2:

$$FADH_2 + O_2 \longrightarrow FAD + H_2O_2 \qquad (16\text{-}7)$$

This last reaction, the oxidation of $FADH_2$ by O_2, is not reversible. Therefore, the overall reaction 16-4, which is the sum of reactions 16-5 through 16-7, is not. In the presence of a highly purified enzyme that contains no impurities to destroy H_2O_2 the reaction proceeds further. In this case the H_2O_2 formed in reaction 16-7 reacts nonenzymically with the keto acid:

$$R{-}\underset{\underset{O}{\|}}{C}{-}COO^- + H_2O_2 \longrightarrow R{-}COO^- + CO_2 + H_2O$$

Recent studies have shown that the D-amino acid oxidase of liver is located in the microbody known as the peroxisome together with some other oxidative enzymes that utilize O_2 as an oxidant. The function of the D-amino acid oxidase remains obscure, although D-amino acids do have a limited occurrence in nature as components of peptidoglycans and cyclic peptides.

Animal tissues also contain an L-amino acid oxidase which can oxidize a variety of L-amino acids. In kidney and liver, this enzyme is firmly associated with the endoplasmic reticulum of the cell. The activity is so low that its physiological significance is doubtful, however. D-Amino acid oxidases have been found in *Neurospora crassa,* and L-amino acid oxidase has been purified from snake venom, *Proteus vulgaris,* and *Neurospora*. This latter enzyme, unlike those of animal tissues, clearly provides a means for oxidative deamination of a large number of amino acids and perhaps serves as a first step in the degradation of amino acids to NH_3, CO_2, and H_2O in bacteria and fungi. Since the reaction is not readily reversible, the enzyme is of no significance in the biosynthesis of amino acids.

In contrast to the reactions of oxidative deamination is the process of *nonoxidative deamination*. One type of nonoxidative deamination is the reaction catalyzed by the α-deaminases (systematic name, amino acid ammonia lyases). Aspartase, which belongs to this group of enzymes, catalyzes the following reaction:

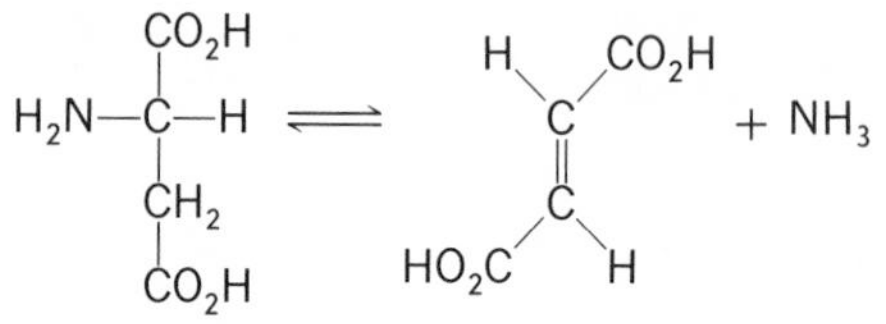

$$H_2N{-}\overset{\overset{CO_2H}{|}}{\underset{\underset{\underset{CO_2H}{|}}{CH_2}}{\underset{|}{C}}}{-}H \rightleftharpoons \begin{matrix} H & & CO_2H \\ & C & \\ & \| & \\ & C & \\ HO_2C & & H \end{matrix} + NH_3 \qquad (16\text{-}8)$$

L-Aspartic acid — Fumaric acid

The enzyme, which is specific for L-aspartic acid and fumaric acid, has been found in *E. coli* and a few other microorganisms. Since the reaction catalyzed is readily reversible, this reaction, as well as the one catalyzed by glutamic dehydrogenase, constitutes a mechanism for incorporating inorganic nitrogen in the form of NH_3 into the α-amino position of an amino acid in the organisms cited. Other ammonia lyases catalyze the deamination of histidine, phenylalanine, and tyrosine in animal and plant tissues. However, these reactions, in contrast to reaction 16-8, are not reversible and therefore are of no significance in the biosynthesis of the amino acids whose deamination they catalyze.

A somewhat different type of deamination is catalyzed by an enzyme in liver termed serine dehydratase (systematic name, L-serine hydro-lyase (deaminating)). The reaction, which is specific for L-serine, involves the loss of NH_3 and rearrangement of the remaining atoms to yield pyruvate:

$$H_2N{-}\overset{\displaystyle CO_2H}{\underset{\displaystyle CH_2OH}{C}}{-}H \longrightarrow \overset{\displaystyle CO_2H}{\underset{\displaystyle CH_3}{C{=}O}} + NH_3$$

L-Serine → Pyruvic acid

Previously it was assumed that amino acrylic acid and its isomer, an imino acid (CH_3—C(=NH)COOH), were intermediates in this process. It has been subsequently shown that this enzyme requires pyridoxal phosphate as a coenzyme, and that a Schiff's base of the coenzyme with the amino acid (see page 217) is the true intermediate. A similar deamination of threonine is catalyzed by the enzyme *threonine dehydratase,* and α-ketobutyric acid is the product formed.

The deamination of cysteine is catalyzed by an enzyme found in animals, plants, and microorganisms:

$$H_2N{-}\overset{\displaystyle CO_2H}{\underset{\displaystyle CH_2SH}{C}}{-}H + H_2O \longrightarrow \overset{\displaystyle CO_2H}{\underset{\displaystyle CH_3}{C{=}O}} + NH_3 + H_2S$$

L-Cysteine → Pyruvic acid

This enzyme, *cysteine desulfhydrase,* also requires pyridoxal phosphate as a coenzyme and presumably operates by a mechanism similar to serine dehydratase.

In addition to these reactions, in which the α-amino groups of amino acids are released as NH_3, mention should be made of the reactions in which the amide nitrogen of glutamine and asparagine are liberated as ammonia. Specific hydrolytic enzymes catalyze the hydrolysis of these two amides and produce NH_3. What role these hydrolytic enzymes play in the metabolism

$$H_2N{-}\overset{\displaystyle CO_2H}{\underset{\displaystyle CH_2CH_2CONH_2}{C}}{-}H + H_2O \xrightarrow{\text{Glutaminase}} H_2N{-}\overset{\displaystyle CO_2H}{\underset{\displaystyle CH_2CH_2COOH}{C}}{-}H + NH_3$$

L-Glutamine → L-Glutamic acid

$$H_2N{-}\overset{\displaystyle CO_2H}{\underset{\displaystyle CH_2CONH_2}{C}}{-}H + H_2O \xrightarrow{\text{Asparaginase}} H_2N{-}\overset{\displaystyle CO_2H}{\underset{\displaystyle CH_2COOH}{C}}{-}H + NH_3$$

L-Asparagine → L-Aspartic acid

of the amides is not entirely clear. The amide group of glutamine is an important source of nitrogen atoms in the synthesis of purines (see page 390), pyrimidines, and other compounds, and enzymes exist that catalyze the direct transfer of nitrogen from the amide to the acceptor molecule. Asparagine, on the other hand, does not appear to serve as a source of nitrogen in biosynthetic reactions, and in plants it is metabolically inert except for its incorporation into proteins.

Decarboxylation A third type of enzymatic reaction which many amino acids undergo is decarboxylation:

$$R\text{—}\overset{\displaystyle H}{\underset{\displaystyle NH_3^+}{C}}\text{—}COO^- \longrightarrow R\text{—}\overset{\displaystyle H}{\underset{\displaystyle NH_2}{C}}\text{—}H + CO_2$$

In contrast to the deamination and transamination reactions which are involved in the catabolism of amino acids, the anabolic aspect of decarboxylation reactions should be noted. Some of the amines formed as a result of decarboxylation have important physiological effects. Thus, a histidine decarboxylase, found in animal tissues, can produce histamine, a substance which among other effects stimulates gastric secretion. The reaction catalyzed is

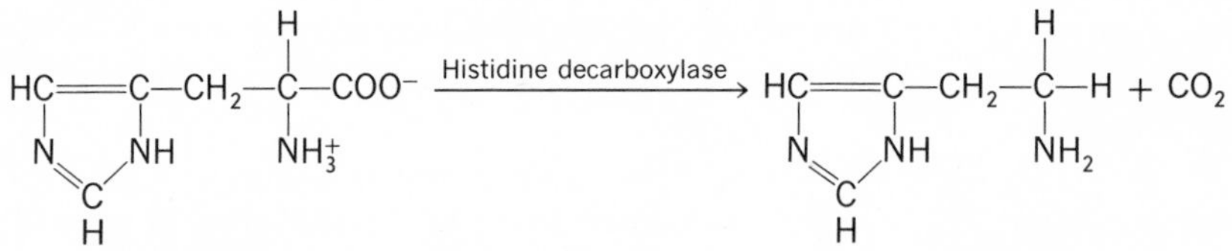

$$\text{L-Histidine} \xrightarrow{\text{Histidine decarboxylase}} \text{Histamine} + CO_2$$

L-Histidine Histamine

Another decarboxylase will decarboxylate 3,4-dihydroxyphenylalanine to form dopamine. This substance in turn is an intermediate in the formation of adrenaline, a vasoconstrictor which is released into the blood stream when an individual is frightened or startled. While the release of adrenaline is under other control mechanisms, the decarboxylase must function in the formation of the amine precursor.

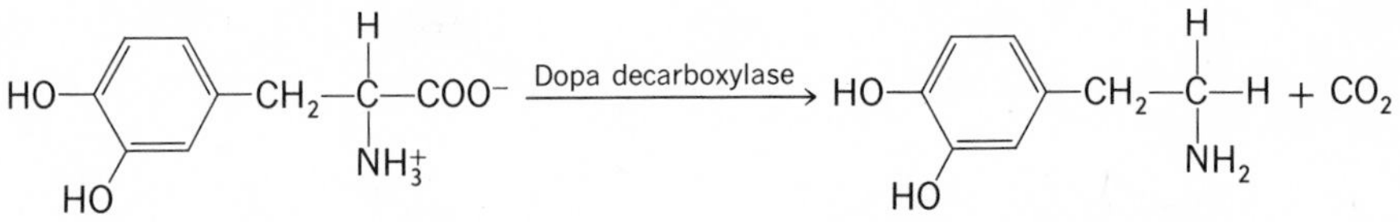

$$\text{Dopa} \xrightarrow{\text{Dopa decarboxylase}} \text{Dopamine} + CO_2$$

3,4-Dihydroxyphenylalanine (Dopa) 3,4-Dihydroxyphenylethylamine (Dopamine)

Serotonin (5-hydroxytryptamine) is formed by the action of a specific decarboxylase on 5-hydroxytryptophan.

Other examples of amines which can be formed from amino acids by decarboxylase activity include γ-amino butyric acid. This substance, which is abundant in potato tubers and found in other plants, can be formed by the enzymatic decarboxylation of the α-COOH group of glutamic acid:

$$
\underset{\text{L-Glutamate}}{{}^{+}H_3N-\overset{\displaystyle COO^-}{\underset{\displaystyle \underset{\displaystyle \underset{\displaystyle COO^-}{|}}{\underset{\displaystyle CH_2}{|}}}{\underset{\displaystyle CH_2}{|}}}{\overset{|}{\underset{|}{C}}}-H}
\xrightarrow{\text{Glutamic decarboxylase}}
\underset{\gamma\text{-Amino butyrate}}{H_2N-\overset{\displaystyle H}{\underset{\displaystyle \underset{\displaystyle \underset{\displaystyle COO^-}{|}}{\underset{\displaystyle CH_2}{|}}}{\underset{\displaystyle CH_2}{|}}}{\overset{|}{\underset{|}{C}}}-H}
+ CO_2
$$

This amine appears to be a highly important compound in the central nervous system of animals.

The amino acid decarboxylases require pyridoxal phosphate as a cofactor. A Schiff's base is again an intermediate, and it is possible to write a detailed mechanism resulting in decarboxylation. (Again, see page 217 for a detailed mechanism.) A common source of amino acid decarboxylases is bacteria, although the enzymes are widely distributed in nature. In bacteria the enzymes, which are inducible, are formed when the bacteria are grown with amino acids in the culture medium. For example, it is possible to obtain a specific decarboxylase for L-glutamic acid in *Clostridium perfringens* by growing the bacterium with this amino acid as its chief source of carbon. Because of specificity of the induction of the enzyme, dried cells of *Clostridium perfringens* grown in this way can be used as a means for specifically decarboxylating L-glutamic acid. Since CO_2 can be accurately measured manometrically, this constitutes a means for quantitative analysis of L-glutamic acid.

Fate of the Carbon Skeletons

The transamination and deamination processes described above are primarily concerned with the nitrogen atom found in the amino acids. Investigations carried out by many workers have examined the fate of the carbon skeleton during metabolism. Thus, while it is possible to write detailed catabolic sequences for each proteinaceous amino acid, the description of such sequences falls outside the purpose of this text. Instead, Table 16-1 lists the catabolic end product of the twenty protein amino acids. It may be seen that nearly all of the amino acids yield on breakdown either an intermediate of the tricarboxylic acid cycle or pyruvate and acetyl–CoA. The exceptions are the five amino acids which give rise to acetoacetic acid. Since, however, this compound also forms acetyl–CoA, all of the amino acids are ultimately oxidized via the tricarboxylic acid cycle. Those amino acids which give rise to an intermediate of the cycle (or to pyruvic acid) can in turn be converted to glucose (see page 317). For this reason, those amino acids have been described as *glucogenic* amino acids. On the other hand, those which on degradation produce acetyl–CoA or acetoacetic acid will, under some conditions, give rise to ketone bodies in the animal and they have therefore been described as *ketogenic* amino acids. Some, such as phenylalanine and tyrosine, are both glucogenic and ketogenic by virtue of the fact that part of their carbon atoms are converted to fumarate while the remainder are converted to acetoacetate.

Table 16-1

Metabolic Fate of Carbon Skeletons of the Amino Acids

Amino acids[a]	End product
Alanine, serine, cysteine (cystine), glycine, and threonine (2)	Pyruvic acid
Leucine (2)	Acetyl–CoA
Phenylalanine (4), tyrosine (4), leucine (4), lysine (4), and tryptophan (4)	Acetoacetic acid (or CoA–ester)
Arginine (5), proline, histidine (5), glutamine, and glutamic acid	α-Ketoglutaric acid
Methionine, isoleucine (4), and valine (4)	Succinyl–CoA
Phenylalanine (4) and tyrosine (4)	Fumarate
Asparagine and aspartic acid	Oxaloacetic acid

[a] The figures in parentheses specify the number of carbon atoms in the amino acid that are actually converted to the end product listed.

Amino Acids as Precursors of Other Compounds

In addition to being oxidized to NH_3, CO_2, and H_2O, certain amino acids play important roles as precursors of cellular components that contain nitrogen. We have already referred to the synthesis of physiologically active amines by the decarboxylation of amino acids (page 370). The plant alkaloids are derived from lysine, tryptophan, tyrosine, and lysine. Another important example is the role of glycine in the biosynthesis of the porphyrin molecule.

Chemistry. The biochemically important compounds chlorophyll, hemoglobin, and the cytochromes have in common a cyclic tetrapyrrole structure called a *porphyrin*. The parent compound, *porphin*, contains four pyrrole rings linked by methine bridges (—CH=). Before considering their chemistry we shall outline a useful method for writing a porphyrin ring. Figure 16-1 illustrates the sequence. In the figure, rings I, II, III, and IV are connected by methine bridges, α, β, γ, and δ. Note that the double bond system is highly conjugated. In actual fact, the double bonds are not definitely assigned, since the structure is a resonating system with several possible structures. Protoporphyrin IX is one of fifteen possible isomers and is the most common in nature. The porphyrin ring is sterically a flat structure with a specific metal very firmly chelated by the electron pairs of the nitrogen atoms of the four pyrrole residues. The only metals found in the biologically functional tetrapyrroles are magnesium (in chlorophyll), iron (in heme), and cobalt (in the cobalamines, modified tetrapyrroles).

Biosynthesis. David Shemin and S. Granick have made major contributions to the problem of biosynthesis of these important cyclic pyrrole structures. Isotopic data showed that all the carbon and nitrogen atoms of the porphyrin ring are derived from glycine and succinic acid. The biosynthetic sequence can be divided into four steps.

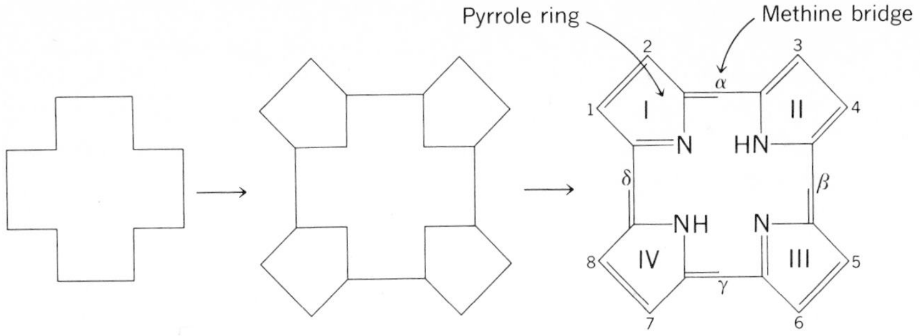

Pyrrole ring

Methine bridge

Porphin

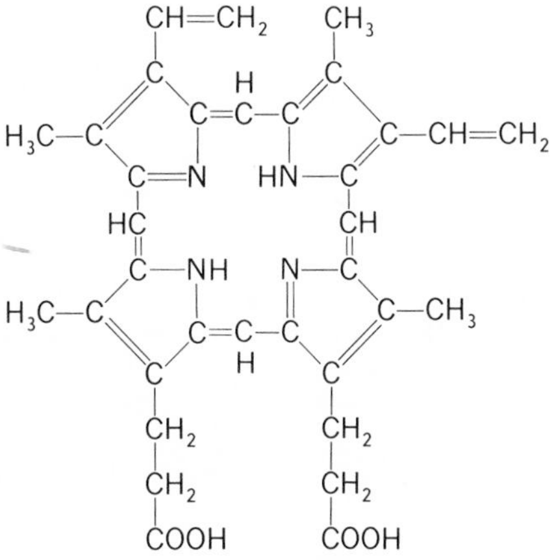

Protoporphyrin IX

Figure 16-1

At the top is shown a simple procedure for drawing a porphin ring. First draw a symmetrical cross. Add the rings. Then complete the structure to give the compound porphin. Below is shown protoporphyrin IX.

Step 1. Glycine and succinyl–CoA (the activated form of succinic acid) condense in the presence of a widespread enzyme ALA synthase to form δ-aminolevulinic acid (ALA). The enzyme requires pyridoxal phosphate. The enzyme is very unstable and, hence, is difficult to purify.

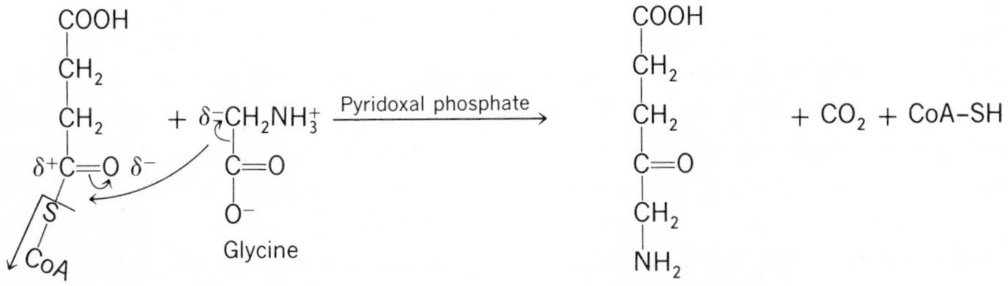

$$HOOC{-}CH_2{-}CH_2{-}C(=O){-}S{-}CoA + {}^{+}H_3N{-}CH_2{-}COO^- \xrightarrow{\text{Pyridoxal phosphate}} HOOC{-}CH_2{-}CH_2{-}C(=O){-}CH_2{-}NH_2 + CO_2 + CoA{-}SH$$

Succinyl–CoA + Glycine → δ-Aminolevulinic acid

Step 2. The second step involves the condensation of two molecules of δ-aminolevulinic acid by ALA dehydrase to yield the pyrrole derivative *porphobilinogen*. Note the distribution of the glycine (plain) and succinate (shaded) residues in the ring. The enzyme is inhibited by low concentrations of heme, an end product, and thus is an important regulatory enzyme in the biosynthesis of heme.

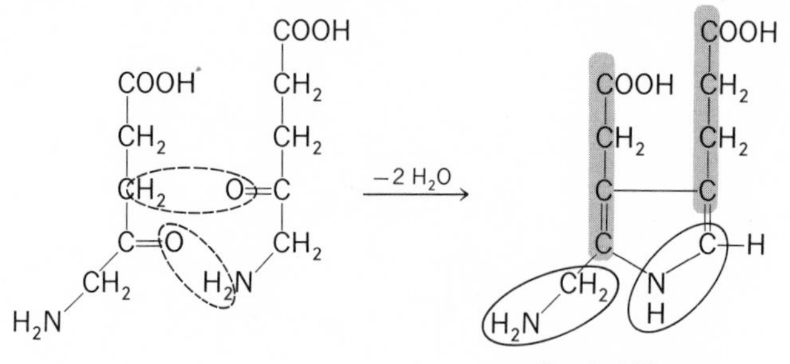

$\xrightarrow{-2\,H_2O}$

Porphobilinogen

Step 3. Although this reaction is not very well-understood, it is important, since in this sequence only the correct isomer, uroporphrinogen III, of the four possible isomers of uroporphyrinogen, is synthesized.

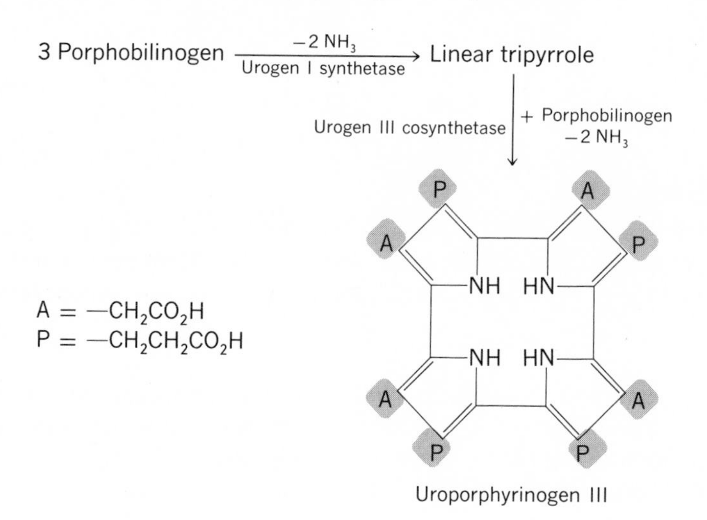

Uroporphyrinogen III

Step 4. In these series of reactions, the acetyl side chains in rings I, II, III, and IV are decarboxylated to the methyl groups by a widespread decarboxylase to coproporphyrinogen III. Next, the propionyl residues in rings I and II and the methane bridges in the α, β, γ, and δ positions are oxidized by a particulate system to protoporphyrin IX. Finally, a specific ferrochelatase in mitochondria inserts the ferrous ion into the tetrapyrrole ring to form heme. Other modifications catalyzed by specific enzymes are believed to convert protoporphyrin IX to chlorophyll in green plants.

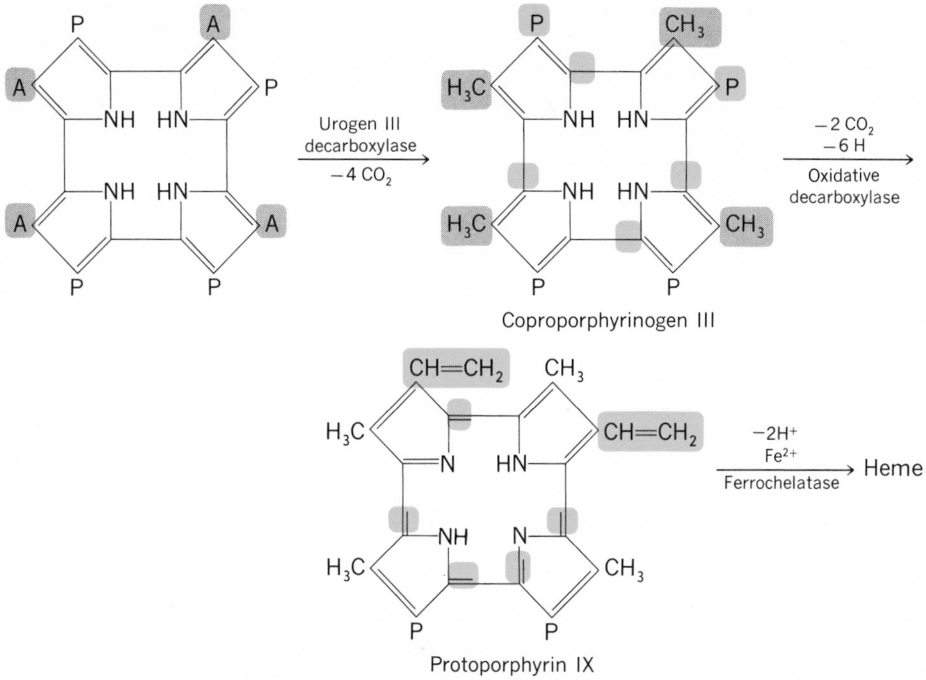

Urogen III decarboxylase, $-4\ CO_2$

Coproporphyrinogen III

$-2\ CO_2$, $-6\ H$, Oxidative decarboxylase

Protoporphyrin IX

$-2H^+$, Fe^{2+}, Ferrochelatase → Heme

The heme moiety in cytochrome c is held to its specific protein by thioether bonds to cysteine residue, and by, methionyl and histidyl residues as described on page 95.

The role of the porphyrins is extremely important in the economy of the cell. Whenever the student reads about photosynthesis, the transport of oxygen (by hemoglobin and myoglobin), the transport of electrons to oxygen (by the cytochrome systems), or the catalytic activity of catalase and peroxidase, he should be aware that the porphyrin system is the key structure in all these functions.

Regulation of Tetrapyrrole Synthesis

Several factors are involved in the regulation of porphyrin synthesis. Compartmentation of the several enzymes involved in porphyrin biosynthesis must be considered in the complete picture. Thus, while ALA synthase and coproporphyrinogen oxidase are in the mitochondria, the other enzymes are located in the cytoplasm.

Two enzymes that occur early in the biosynthetic sequence appear to be the control points for heme synthesis, namely, ALA synthase and ALA dehydrase. The dehydrase is inhibited 50% by 40 μM heme while the synthase is inhibited 50% by 1 μM heme by a feedback mechanism. Since ALA synthase is present in low concentrations, it probably is the rate-limiting enzyme in porphyrin synthesis. However, ALA dehydrase appears to be a second control point.

In addition to the control just described, the formation of ALA synthase is repressed by low concentrations of heme in growing cultures of a number of bacteria and embryonic tissues. Another example of control is the remarkable effect of oxygen on hemoprotein synthesis in yeast. When yeast cells are grown anaerobically, these cells are devoid of mitochondria and do not contain significant amounts of cytochrome. When the cells are exposed to oxygen, there is a rapid appearance of mitochondria, and the complete cytochrome complex is formed. The cytochrome c content increases 50-fold during this adaptation from anaerobic to aerobic conditions.

$$\left.\begin{array}{l}\text{Glycine}\\ \text{Succinyl–CoA}\end{array}\right\rangle \longrightarrow \text{ALA} \longrightarrow \text{PBG} \longrightarrow \text{Protoporphyrin IX} \longrightarrow \text{Heme}$$

Scheme 16-2

In summary, heme exerts both a feedback control and a repression control on ALA synthase. In addition, ALA dehydrase is inhibited by heme. Finally, oxygen and a host of chemicals can markedly affect the levels of heme and hemoproteins in both procaryotic and eucaryotic cells.

Metabolism of the Sulfur-Containing Amino Acids

Another interesting group of reactions involving amino acids are those interrelating the amino acids cysteine (cystine), homocysteine, and methionine. The unique aspect of the metabolism of these compounds involves the sulfur atoms which they contain.

The form of sulfur normally available to biological organisms is the highly oxidized form, the sulfate ion, SO_4^{2-}. Before it can be utilized for synthesis of the sulfur-containing amino acids, the SO_4^{2-} must be reduced to H_2S. This process, somewhat analogous to nitrate reduction, is discussed on page 411.

Microorganisms and higher plants, presented with a source of H_2S, will utilize this compound to make cysteine. The enzyme responsible is known as *cysteine synthase*. The enzyme requires an "activated" form of serine

$$\begin{array}{l}CH_2-O-\overset{\overset{\displaystyle O}{\|}}{C}-CH_3\\ |\\ CHNH_2\\ |\\ CO_2H\end{array} + H_2S \xrightarrow[\text{PALP}]{} \begin{array}{l}CH_2-SH\\ |\\ CHNH_2\\ |\\ CO_2H\end{array} + CH_3COOH$$

O-Acetyl serine — Cysteine — Acetic acid

that is produced by the acetylation of serine with acetyl–CoA. The acetyl group is a good leaving group which can be readily displaced by H_2S to form cysteine.

Bacteria and higher plants can now utilize the cysteine as a sulfur source to form methionine; the process known as *transsulfurylation* has its analogy in the manner which the nitrogen atom of aspartic acid makes its way into the guanidinium group in arginine. In the presence of the enzyme *cystathionine synthase* I a sulfur-containing addition product known as cystathionine is formed:

$$\underset{\text{Cysteine}}{\begin{array}{l}CH_2\text{—}SH\\ CHNH_2\\ CO_2H\end{array}} + \underset{\text{Homoserine}}{\begin{array}{l}HO\text{—}CH_2\\ \quad\; CH_2\\ \quad\; CHNH_2\\ \quad\; CO_2H\end{array}} \xrightarrow[\text{synthase I (Bacteria; plants)}]{\text{Cystathionine}} \underset{\text{Cystathionine}}{\begin{array}{ll}CH_2\text{—}S\text{—} & CH_2\\ CHNH_2 & CH_2\\ CO_2H & CHNH_2\\ & CO_2H\end{array}} + H_2O$$

Note that the three-carbon unit is contributed by cysteine and the four-carbon unit by homoserine (derived from aspartic acid). In the presence of cystathionase, the cystathionine is hydrolytically cleaved on the opposite side of the sulfur atom to produce *homocysteine*, pyruvic acid, and NH_3:

$$\underset{\text{Cystathionine}}{\begin{array}{ll}CH_2\text{—}S\text{—} & CH_2\\ CHNH_2 & CH_2\\ CO_2H & CHNH_2\\ & CO_2H\end{array}} + H_2O \xrightarrow[\text{(bacterial)}]{\text{Cystathionase}} \underset{\text{Pyruvate}}{\begin{array}{l}CH_3\\ C{=}O\\ CO_2H\end{array}} + NH_3 + \underset{\text{Homocysteine}}{\begin{array}{l}HS\text{—}CH_2\\ \quad\; CH_2\\ \quad\; CHNH_2\\ \quad\; CO_2H\end{array}}$$

The homocysteine is subsequently methylated (by tetrahydrofolic acid or vitamin B_{12}) to form methionine.

The reactions and relationships just described for bacteria and plants are almost reversed in animals, which cannot make cysteine (or homocysteine) from H_2S and SO_4^{2-}. Instead, animals synthesize their cysteine from methionine, which is an indispensible amino acid. Indeed, the "essentiality" of methionine is due to this inability. Since the sulfur of methionine can be utilized to make cysteine, however, the latter is not an essential amino acid. Cystathionine encountered above is again an intermediate in this process. In the presence of mammalian cystathionine synthase II homocysteine (derived from the demethylation of methionine) reacts with serine to produce cystathionine:

$$\underset{\text{Homocysteine}}{\begin{array}{l}CH_2\text{—}SH\\ CH_2\\ CHNH_2\\ CO_2H\end{array}} + \underset{\text{Serine}}{\begin{array}{l}HO\text{—}CH_2\\ \quad\; CHNH_2\\ \quad\; CO_2H\end{array}} \xrightarrow[\text{synthase II (mammalian)}]{\text{Cystathionine}} \underset{\text{Cystathionine}}{\begin{array}{ll}CH_2\text{—}S\text{—} & CH_2\\ CH_2 & CHNH_2\\ CHNH_2 & CO_2H\\ CO_2H & \end{array}} + H_2O$$

This compound is then hydrolyzed to yield cysteine, α-ketobutyric acid, and NH_3:

$$\underset{\text{Cystathionine}}{\begin{array}{ll}CH_2\text{—}S\text{—} & CH_2\\ CH_2 & CHNH_2\\ CHNH_2 & CO_2H\\ CO_2H & \end{array}} + H_2O \xrightarrow[\text{(mammalian)}]{\text{Cystathionase}} \underset{\alpha\text{-Ketobutyrate}}{\begin{array}{l}CH_3\\ CH_2\\ C{=}O\\ CO_2H\end{array}} + NH_3 + \underset{\text{Cysteine}}{\begin{array}{l}HS\text{—}CH_2\\ \quad\; CHNH_2\\ \quad\; CO_2H\end{array}}$$

Note that this time the three carbon atoms of cysteine originate in serine, while the sulfur atom comes from homocysteine and indirectly from methionine. The four enzymes just described that are involved in the formation and hydrolysis of cystathionine are enzymes that contain pyridoxal phosphate as a cofactor.

Two additional reactions will round out the relationships between the sulfur amino acids just described and, in addition, illustrate a different way in which ATP can serve to activate a substrate. There is much evidence to indicate that the methyl groups of methionine are transferred to acceptor molecules to form methylated derivatives. In the presence of an activating enzyme, ATP reacts with methionine to form the active methionine, *S*-adenosyl methionine:

$$CH_3-S-CH_2-CH_2-CHNH_2-CO_2H \;+\; HO-\overset{O}{\overset{\|}{P}}(OH)-O-\overset{O}{\overset{\|}{P}}(OH)-O-\overset{O}{\overset{\|}{P}}(OH)-O-CH_2\text{-ribose(H, H, H, H; OH, OH)-Adenine} \xrightarrow{\text{Activating enzyme}}$$

Methionine — ATP

$$CH_3-S^{+}(-CH_2\text{-ribose(H, H, H, H; OH, OH)-Adenine})-CH_2-CH_2-CHNH_2-CO_2H \;+\; HO-\overset{O}{\overset{\|}{P}}(OH)-O-\overset{O}{\overset{\|}{P}}(OH)-OH \;+\; H_3PO_4$$

S-Adenosyl methionine

In this reaction, the three phosphate groups of ATP are removed, as inorganic phosphate and pyrophosphate, and the adenosine residue is attached to the sulfur atom to form a sulfonium derivative. This compound is a high-energy compound that readily transfers its methyl group to acceptor molecules (e.g., guanidoacetic acid); in the process, *S*-adenosyl homocysteine is formed:

$$CH_3-S^{+}(-CH_2\text{-ribose(H, H, H, H; OH, OH)-Adenine})-CH_2-CH_2-CHNH_2-CO_2H \;+\; H-N(-C(=NH)NH_2)-CH_2-CO_2H \longrightarrow S(-CH_2\text{-ribose(H, H, H, H; OH, OH)-Adenine})-CH_2-CH_2-CHNH_2-CO_2H \;+\; H_3C-N(-C(=NH)NH_2)-CH_2-CO_2H$$

S-Adenosyl methionine — Guanido-acetic acid — *S*-Adenosyl homocysteine — Creatine

The *S*-adenosyl homocysteine can be hydrolyzed to form adenosine and homocysteine, which can serve in turn in the synthesis of cysteine:

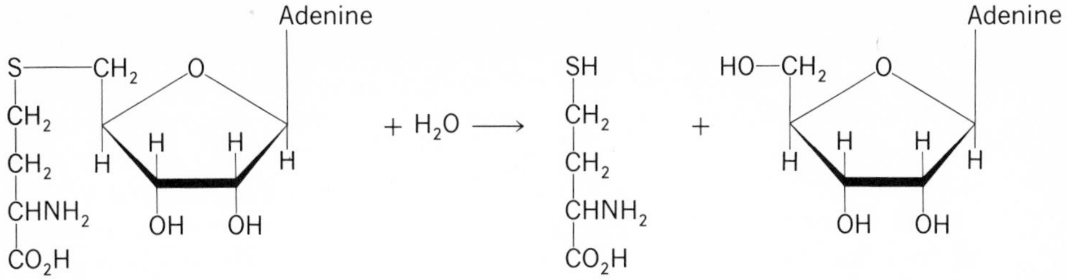

S-Adenosyl homocysteine $+ H_2O \longrightarrow$ Homocysteine $+$ Adenosine

The Dynamic Metabolism of Nitrogen Compounds

Prior to 1937, the body proteins, in contrast to carbohydrates and lipids, were considered to be relatively inert metabolically. Once synthesized in the cell, they were believed to remain intact until the death of the animal or plant occurred and the process of decay was initiated. It is easy to understand why this concept prevailed in the case of proteins, in contrast to body lipids (and carbohydrates) whose amount in the animal obviously varies directly with the nutritional status of the animal. Fat deposition occurs during excessive caloric intake and fat stores are depleted in a calory-deficient diet. In contrast, body proteins are not utilized by the animal for energy production until all other reserves are depleted and starvation is extreme.

R. Schoenheimer and his associates performed a series of experiments in the 1930s which drastically modified this concept. They fed amino acids labeled with isotopic nitrogen (^{15}N-nitrogen) to adult rats and mice and expected the animals which were in nitrogen equilibrium to oxidize the dietary amino acids and excrete the labeled nitrogen. Instead, the isotope was incorporated into the proteins of the liver (an organ extremely active in protein synthesis) and other tissues as well. Moreover, the label was found not only in the amino acid originally administered but in many other amino acids. The transfer of label from one amino acid to many others might be expected in view of the widespread occurrence of transaminases that could catalyze this exchange. The finding of isotope in the proteins of animals that were not increasing in size was unexpected and caused Schoenheimer to conclude that the proteins of animals, as well as lipids and carbohydrates, are in a *dynamic* rather than a static state. He proposed that, in an adult nongrowing animal, a relatively high rate of protein synthesis is counterbalanced by an equal rate of breakdown, and that the two provide for an active metabolic turnover of those molecules in the animal. Schoenheimer's work introduced the concept of a *metabolic pool* of amino acids and NH_3 molecules in the case of proteins, the origin of which could not be specified. The components of these pools could be utilized for resynthesis of the body proteins.

It should be stressed that animal protein molecules differ in their rate of turnover. Proteins of the blood, liver, kidney, and other vital organs have half-lives (i.e., the time required for half of that protein to enter the metabolic pool) ranging from 2 to 10 days. Hemoglobin of the red blood cell has a half-life of about 30 days; muscle protein, 180 days, and collagen 1000 days. For the adult human, the rate has been estimated at 1.2 g of protein per kilogram of body weight per day. As much as one-fourth of the amino acids from this protein undergoes oxidative degradation and must be replaced by protein in the diet. Since about half of the amino acids that are degraded are indispensable, we can understand the need for a diet containing the indispensible amino acids in adequate amount (see Chapter 19 for comparison of Schoenheimer's concept between procaryotic and eucaryotic cells).

One of the major nutritional problems facing the world is the provision of an adequate (in the sense of containing the essential amino acids) protein diet for its human population. Carbohydrate, in the form of plant starch, is readily available from the major food crops—rice, corn, wheat, and cassava—but an adequate protein supply is much more difficult to obtain. In countries where meat proteins are consumed, the problem is insignificant because those proteins contain the essential amino acids. However, most of the world's population subsists on plant food, and if the amount of protein is low or inadequate in quality (as in corn which is low in lysine), the diet is inferior. It is a tragic fact that growing children are particularly susceptible to protein-deficiency, perhaps because as they are growing, they require a condition of positive nitrogen balance. As long as a baby is nursed by its mother its protein intake will be adequate in amount and quality. When, however, the older child is displaced by a new baby, the effects of an inadequate plant diet are noticed and the disease *Kwashiorkor* develops, characterized by bloated stomachs, discolored hair and skin, and a generalized malaise. Such children are weakened in their resistance to the normal complement of diseases and infections in their environment and usually die in their first few years of life.

Anabolic Aspects of Amino Acid Metabolism. The synthesis of amino acids from simpler building blocks can also be considered from the processes whereby (*a*) NH_3 is introduced into the carbon skeleton, and (*b*) the carbon skeleton is produced from cellular metabolites. Again, the description of the detailed biosynthetic sequence for each of the proteinaceous amino acids is beyond the purpose of this text, but certain obvious sources of those carbon skeletons can be mentioned.

The keto acids, pyruvic, oxalacetic, and α-ketoglutaric, have been previously encountered; by transamination, these compounds are converted to alanine, aspartic, and glutamic acids, respectively. Since their keto acids can be produced from carbohydrate precursors (see pages 313 and 260 for specific conditions for α-ketoglutarate and oxalacetate formation), it is not surprising that alanine, aspartic, and glutamic acids are dispensible amino acids. Since aspartic and glutamic acids can be converted to their amides (see page 382) in the presence of ATP, NH_3, and the appropriate enzymes, the amides are dispensible. In addition, glutamic acid can be converted to proline and ornithine (and indirectly therefore to hydroxyproline, citrulline, and arginine (see page 386)), and these amino acids are classified as dispensible. The biosynthesis of indispensible amino acids has been examined in microorganisms and higher plants which can make these compounds from simple precursors (e.g., glucose, acetate, or CO_2), and the student can find this information elsewhere.

Incorporation of Ammonia into Organic Compounds

Some of the ammonia produced by deamination processes discussed earlier (page 366) is eliminated as a nitrogenous waste product, as NH_3 urea, or uric acid. However, the dynamic state of the nitrogen compounds requires that much of the NH_3 be reutilized by the cell in the resynthesis of nitrogenous compounds. A major reaction in this reutilization is that catalyzed by the enzyme, glutamic dehydrogenase. As pointed out earlier, the reaction catalyzed by this enzyme is readily reversible, and in the presence of NADH, NH_3, and α-ketoglutarate, glutamate will be formed. By subsequent transamination

$$\underset{\alpha\text{-Ketoglutarate}}{HO_2C{-}C(=O){-}CH_2{-}CH_2{-}CO_2H} + NH_3 + NADH + H^+ \rightleftharpoons \underset{\text{L-Glutamate}}{H_2N{-}CH(CO_2H){-}CH_2{-}CH_2{-}CO_2H} + NAD^+ + H_2O$$

reactions, the nitrogen atom can be transferred on to form the other amino acids.

The amide group of glutamine serves as the precursor of the nitrogen atom in numerous compounds (purines, pyrimidines, glucosamine, NAD^+); glutamine therefore is an important intermediate in the incorporation of ammonia into other organic compounds.

Glutamine is synthesized by *glutamine synthetase,* an enzyme found in plants, animals, and bacteria. The first step is believed to be the formation of a γ-glutamyl–phosphate–enzyme complex. In the second step, ammonia, a good nucleophile, attacks the complex and displaces the phosphate group to form glutamine and inorganic phosphate. Note that only the γ-amide is formed. Isoglutamine, the compound with the α-carboxyl group amidated, is never formed. The enzyme is also highly specific, since aspartic acid cannot replace glutamic acid as a substrate (Scheme 16-3).

Because glutamine is a multifunctional precursor, it is not surprising that the enzyme catalyzing its synthesis is an allosteric enzyme subject to control by a variety of different compounds (pages 474). No fewer than eight products of glutamine metabolism—tryptophan, histidine, glycine, alanine, glucoseamine-6-phosphate, carbamyl phosphate, AMP, CTP—have been shown to serve as independent negative feedback inhibitors of the enzyme in *E. coli.*

The third major route for the uptake of NH_3 into organic compounds is catalyzed by the enzyme, *carbamyl kinase;* the reaction catalyzed by the bacterial enzyme is

$$NH_3 + CO_2 + ATP \xrightarrow{Mg^{2+}} \underset{\text{Carbamyl phosphate}}{NH_2{-}C(=O){-}OPO_3H_2} + ADP$$

A somewhat more complex reaction is catalyzed by the mammalian enzyme (page 383). Carbamyl phosphate made by either of these processes serves as a donor of the carbamyl group in citrulline biosynthesis and in pyrimidine biosynthesis.

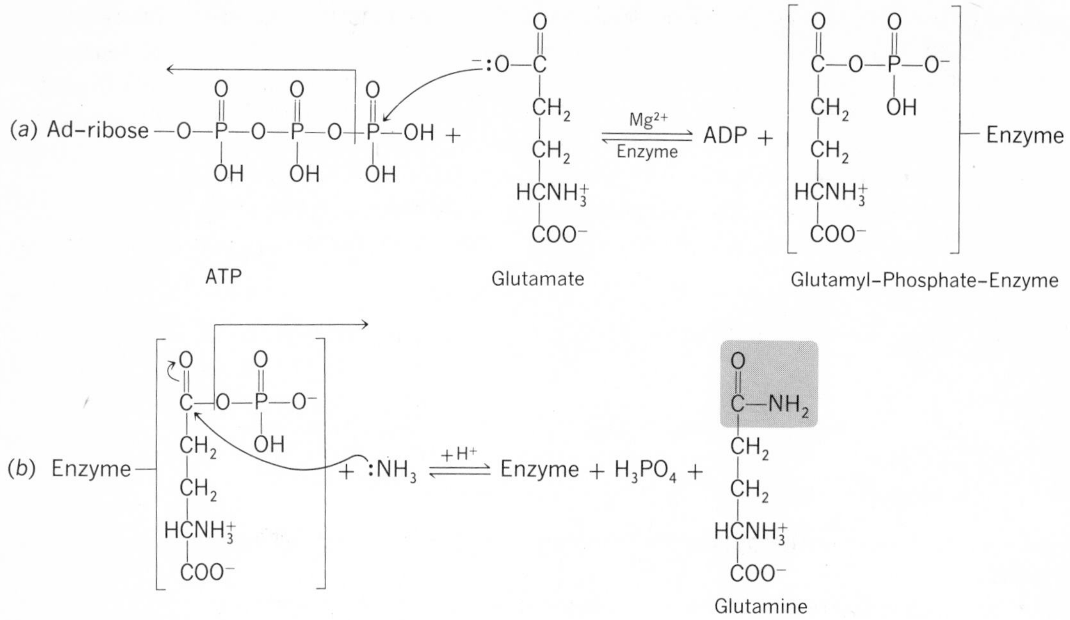

(a) Ad-ribose—O—P(=O)(OH)—O—P(=O)(OH)—O—P(=O)(OH)—OH + $^{-}$:O—C(=O)—CH_2—CH_2—$HCNH_3^+$—COO^- $\xrightleftharpoons[\text{Enzyme}]{Mg^{2+}}$ ADP + [C(=O)—O—P(=O)(OH)—O^- / CH_2 / CH_2 / $HCNH_3^+$ / COO^-]—Enzyme

ATP Glutamate Glutamyl-Phosphate-Enzyme

(b) Enzyme—[C(=O)—O—P(=O)(OH)—O^- / CH_2 / CH_2 / $HCNH_3^+$ / COO^-] + :NH_3 $\xrightleftharpoons{+H^+}$ Enzyme + H_3PO_4 + C(=O)—NH_2 / CH_2 / CH_2 / $HCNH_3^+$ / COO^-

Glutamine

Overall reaction:

$$\text{Glutamate} + \text{ATP} + NH_3 \xrightarrow[\text{synthetase}]{\text{Glutamine}} \text{Glutamine} + \text{ADP} + H_3PO_4$$

Scheme 16-3

Ammonia in excess of that required for synthesis of organic nitrogen compounds is disposed of by animals in different ways (see page 386). Mammals convert the nitrogen atoms of NH_3 into urea which is secreted in the urine. The cycle of reactions that accomplishes urea synthesis is also, except for one reaction, the biosynthetic pathway for arginine formation.

The Urea Cycle

Sir Hans Krebs, then of Germany, and K. Henseleit were among the first to study the formation of urea in animal tissues. They observed that rat liver slices could convert CO_2 and NH_3 (2 moles/mole of CO_2) to urea, provided some energy source was available. The requirement for some oxidizable substance such as lactic acid or glucose was understandable, since the formation of urea from NH_3 and CO_2 required energy.

The amino acid arginine was also implicated in this process, since the enzyme arginase, which catalyzes reaction 16-14 (below), was known to form urea and ornithine on the hydrolysis of arginine. The exact relationship was indicated, however, when Krebs showed that catalytic quantities of arginine, and ornithine, or citrulline as well, stimulated the formation of appreciable amounts of urea from ammonia. In 1932, Krebs proposed a cycle of reactions which accounted for the production of urea from NH_3 and CO_2 and explained the catalytic action of arginine, ornithine, and citrulline. That cycle, known as the urea or ornithine cycle, is shown in Scheme 16-4. Although the essen-

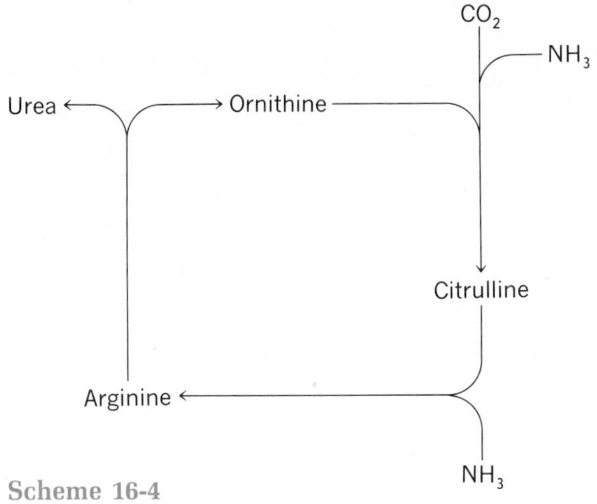

Scheme 16-4

The urea cycle.

tial features of this cycle remain unchanged, it is possible to write out some of the reactions in greater detail.

In bacteria an enzyme *carbamyl kinase* has been shown to catalyze the phosphorylation of the ammonium salt of carbamic acid (reaction 16-10). The chemistry of carbamic acid is complex, but carbamyl phosphate has been identified as the product of the action of carbamyl kinase. The reaction is relatively endergonic; its $\Delta G'$ is +2000 cal/mole.

$$2\,NH_3 + CO_2 \longrightarrow \underset{\text{Ammonium carbamate}}{[NH_4]^+ \left[O{-}\overset{\overset{\displaystyle O}{\|}}{C}{-}NH_2\right]^-} \tag{16-9}$$

$$\underset{\text{Ammonium carbamate}}{[NH_4]^+ \left[O{-}\overset{\overset{\displaystyle O}{\|}}{C}{-}NH_2\right]^-} + ATP \underset{\substack{\text{Carbamyl}\\ \text{kinase}}}{\overset{Mg^{2+}}{\rightleftharpoons}} \underset{\text{Carbamyl phosphate}}{H_2O_3PO{-}\overset{\overset{\displaystyle O}{\|}}{C}{-}NH_2} + ADP + NH_3 \tag{16-10}$$

In frog and mammalian livers, another enzyme, *carbamyl phosphate synthetase,* catalyzes the formation of carbamyl phosphate from NH_3 and CO_2; in this reaction, 2 moles of ATP and a cofactor, *N*-acetyl glutamic acid, are required. The details of the reaction are not clear, but the stoichiometry has been established:

$$NH_3 + CO_2 + 2\,ATP \xrightarrow[\substack{\text{Carbamyl}\\ \text{phosphate}\\ \text{synthetase}}]{\text{Cofactor}} H_2O_3PO{-}\overset{\overset{\displaystyle O}{\|}}{C}{-}NH_2 + 2\,ADP + H_3PO_4$$

This reaction is not readily reversible due to the fact that there is a decrease of one energy-rich bond as the reaction proceeds from left to right.

Carbamyl phosphate, which has been synthesized chemically, will react with ornithine to form citrulline in the presence of the enzyme *ornithine transcarbamylase*. The enzyme has been purified from beef liver; it has no cofactors and exhibits extreme substrate specificity. The equilibrium is in the direction of citrulline synthesis.

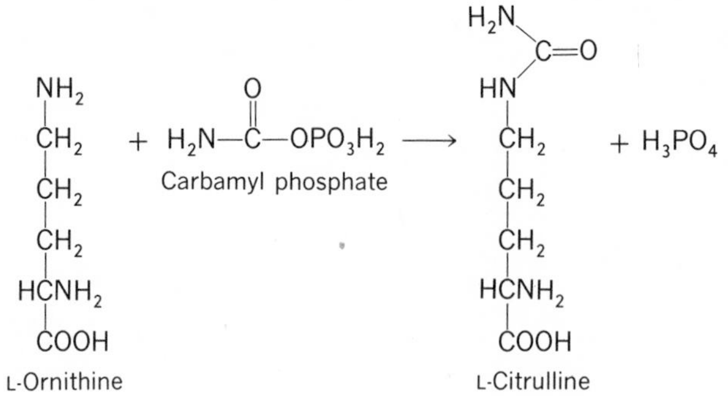

$$NH_2CH_2CH_2CH_2CH(NH_2)COOH \text{ (L-Ornithine)} + H_2N{-}C({=}O){-}OPO_3H_2 \text{ (Carbamyl phosphate)} \longrightarrow H_2N{-}C({=}O){-}NH{-}CH_2CH_2CH_2CH(NH_2)COOH \text{ (L-Citrulline)} + H_3PO_4 \qquad (16\text{-}11)$$

The next step in the cycle, the formation of arginine from citrulline, was largely worked out by Sarah Ratner, who first showed that two enzymes were involved. The first of these, *argininosuccinic synthetase,* catalyzes the formation of argininosuccinic acid from citrulline and aspartic acid. This may be conveniently represented by picturing the enolic form of citrulline as reacting with the aspartic acid to form a new compound, argininosuccinic acid. Reac-

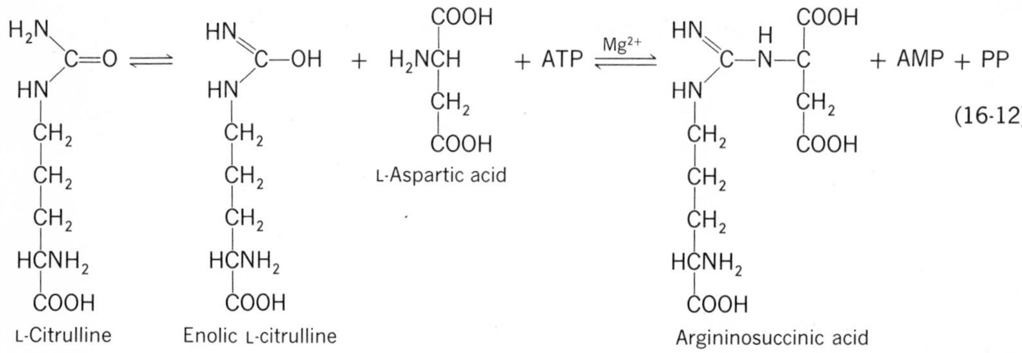

$$H_2N{-}C({=}O){-}NH{-}CH_2CH_2CH_2CH(NH_2)COOH \text{ (L-Citrulline)} \rightleftharpoons HN{=}C(OH){-}NH{-}CH_2CH_2CH_2CH(NH_2)COOH \text{ (Enolic L-citrulline)} + H_2NCH(COOH)CH_2COOH \text{ (L-Aspartic acid)} + ATP \xrightleftharpoons{Mg^{2+}} HN{=}C(NH{-}CH(COOH)CH_2COOH){-}NH{-}CH_2CH_2CH_2CH(NH_2)COOH \text{ (Argininosuccinic acid)} + AMP + PP \qquad (16\text{-}12)$$

tion 16-12 requires ATP and Mg^{2+}. The K_{eq} for this reaction is approximately 9 at pH 7.5; therefore, the reaction is readily reversible. Note that the nitrogen atom which eventually becomes one of the two such atoms in urea is contributed by aspartic acid in this reaction and not by NH_3. Other examples of reactions in which aspartic acid contributes its nitrogen atom in the biosynthesis of a new nitrogenous compound will be seen later in this chapter.

The subsequent cleavage of argininosuccinic acid is catalyzed by the *argininosuccinic cleavage enzyme,* which has been purified from ox liver; it has also been observed in plant tissues and microorganisms. Reaction 16-13 is a reaction formally analogous to the aspartic ammonia lyase reaction (page 368), in which NH_3 or a substituted amine is eliminated to form fumaric acid. The K_{eq} for the reaction is 11.4×10^{-3} at pH 7.5. Since the reaction as

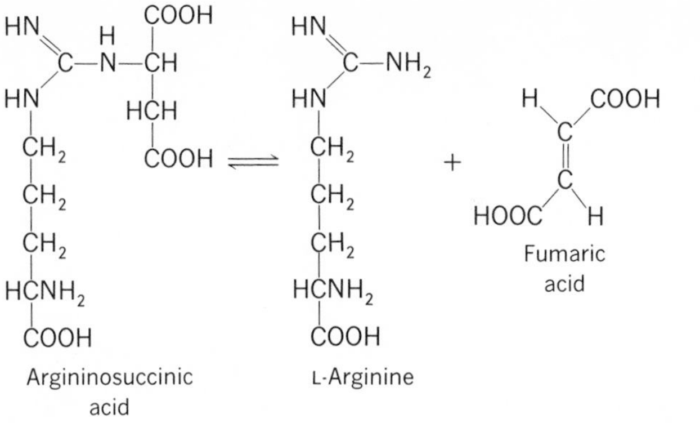

$$\mathrm{HN{=}C(NH{-}CH(COOH)CH_2COOH)NH{-}CH_2{-}CH_2{-}CH_2{-}HCNH_2{-}COOH} \rightleftharpoons \mathrm{HN{=}C(NH_2)NH{-}CH_2{-}CH_2{-}CH_2{-}HCNH_2{-}COOH} + \mathrm{HOOC{-}CH{=}CH{-}COOH} \qquad (16\text{-}13)$$

Argininosuccinic acid; L-Arginine; Fumaric acid

written from left to right results in the formation of two products from a single reactant, this value of K_{eq} determines that argininosuccinic acid will predominate in concentrated solutions, whereas arginine and fumaric acid will predominate in dilute solution.

Arginase, which catalyzes the irreversible hydrolysis of L-arginine to ornithine and urea, is the enzyme which converts the unidirectional sequence for biosynthesis of arginine into a cyclic process for making urea:

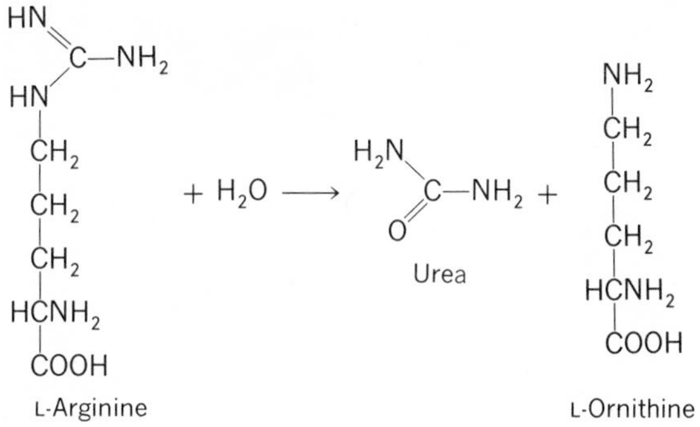

$$\mathrm{HN{=}C(NH_2)NH{-}CH_2{-}CH_2{-}CH_2{-}HCNH_2{-}COOH} + H_2O \longrightarrow \mathrm{H_2N{-}C({=}O){-}NH_2} + \mathrm{NH_2{-}CH_2{-}CH_2{-}CH_2{-}HCNH_2{-}COOH} \qquad (16\text{-}14)$$

L-Arginine; Urea; L-Ornithine

Thus, reactions through 16-13 accomplish the formation of arginine, a widely occurring amino acid, from ornithine, NH_3, and CO_2. The enzymes catalyzing these reactions presumably occur in a wide number of tissues in animals, plants, and microorganisms. Arginase, whose activity makes urea formation possible, is found in the liver of animals known to excrete urea together with the other enzymes for arginine biosynthesis. Liver is the major site of urea formation in mammals, although some urea synthesis can occur in brain and kidney.

The sequence of reactions just discussed is shown in Figure 16-2. The cycle accounts for the formation of urea from NH_3, CO_2, and the amino group of aspartic acid. The requirement for the oxidizable substrates reported by Krebs is explained by the participation of ATP in the formation of carbamyl phosphate and argininosuccinic acid. By the eventual conversion of fumaric acid back to aspartic acid, another mole of amino nitrogen can be brought to the point of reaction in the cycle.

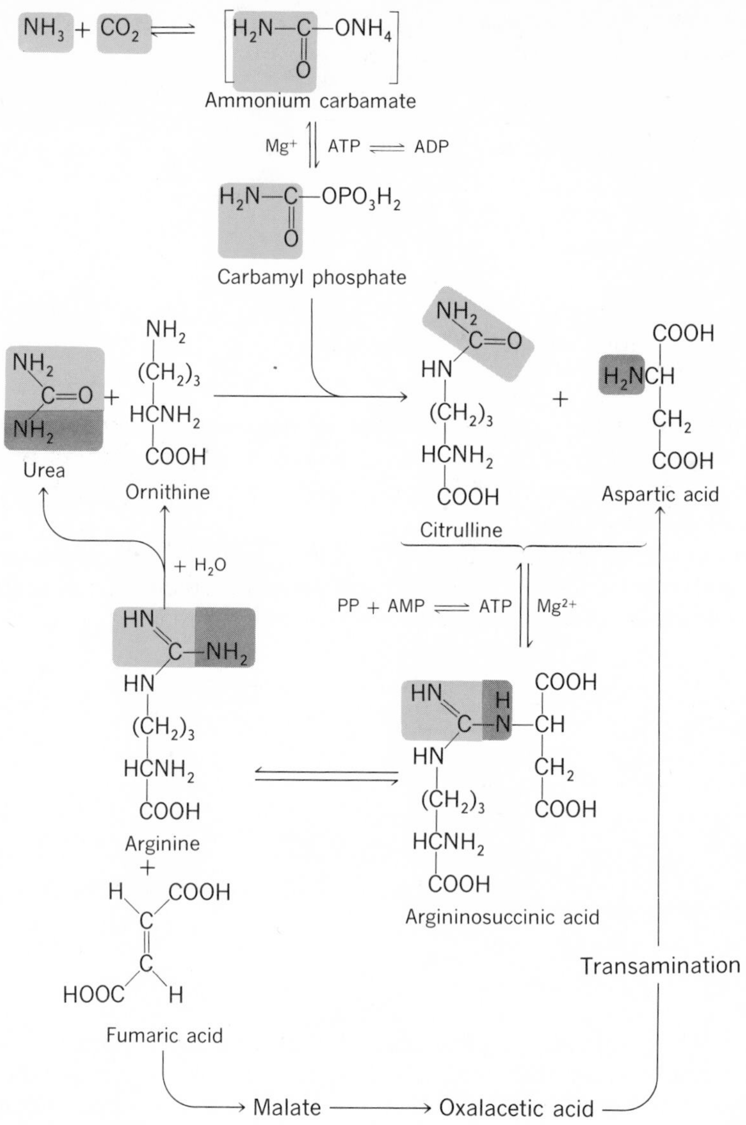

$NH_3 + CO_2 \rightleftharpoons [H_2N-C(=O)-ONH_4]$

Ammonium carbamate

Mg^+ ⇅ $ATP \rightleftharpoons ADP$

$H_2N-C(=O)-OPO_3H_2$

Carbamyl phosphate

Ornithine: $NH_2-(CH_2)_3-HCNH_2-COOH$

Citrulline: $NH_2-C(=O)-HN-(CH_2)_3-HCNH_2-COOH$

\+ Aspartic acid: $COOH-H_2NCH-CH_2-COOH$

$PP + AMP \rightleftharpoons ATP$ ⇅ Mg^{2+}

Argininosuccinic acid

⇌ Arginine + Fumaric acid

Arginine $+ H_2O$ → Urea + Ornithine

Urea: $NH_2-C(=O)-NH_2$

Fumaric acid → Malate → Oxalacetic acid → Transamination → Aspartic acid

Figure 16-2

The urea cycle.

Comparative Biochemistry of Nitrogen Excretion

If we survey the animal kingdom, we find that three nitrogen excretory products are common: NH_3, urea, and uric acid. An organism's choice of one of these forms depends in part on certain properties of the compounds: NH_3 is very toxic but it is also extremely soluble in H_2O; urea is far less toxic and is appreciably soluble in H_2O; uric acid is quite insoluble and, as such, is fairly nontoxic. There is abundant evidence to suggest that the form in which nitrogen is excreted by an organism is determined largely by the accessibility of H_2O to that organism.

Marine animals, living in H_2O, have large amounts of H_2O into which their waste products can be excreted. Although NH_3 is fairly toxic, it can be excreted and will be diluted out instantly in the H_2O of the environment. As a result, many marine forms excrete NH_3 as the major nitrogenous end product, although there are important exceptions to this among the bony fishes.

Land-dwelling animals no longer have an unlimited supply of H_2O in intimate contact with their tissues. Since NH_3 is toxic, it cannot be conveniently accumulated. As a result, most terrestrial animals have developed procedures for converting NH_3 into either urea or uric acid.

According to Needham, the English biochemist, the choice between urea and uric acid is determined by the conditions under which the embryo develops. The mammalian embryo develops in close contact with the circulatory system of the mother. Thus, urea, which is quite soluble, can be removed from the embryo and excreted. On the other hand, the embryos of birds and reptiles develop in a hard-shelled egg in an external environment. The eggs are laid with enough water to see them through the hatching period. Production of NH_3 or even urea in such a closed system would be fatal because they are so toxic. Instead, uric acid is produced by these embryos and precipitates out as a solid in a small sac on the interior surface of the shell. These characteristics, which are so necessary for development of the embryo, are then carried over to the adult organism.

There are interesting examples in support of the principles we have cited. The tadpole, which is aquatic, excretes chiefly NH_3. When it undergoes metamorphosis into the frog, however, it becomes a true amphibian and spends much of its time away from water. During the metamorphosis the animal begins excreting urea instead of NH_3, and by the time the change is complete, urea is the predominant nitrogen-excretory product.

Lungfish are another interesting example. While in water they excrete chiefly NH_3, but as the river or lake runs dry, the lungfish settles down in the mud, begins to aestivate, and accumulates urea as the nitrogen end product. When the rains return, the lungfish excretes a massive amount of urea and sets about excreting NH_3 again.

Within one group of animals, the chelonia (tortoises and turtles), there are totally aquatic species, semiterrestrial species, and a third group (the tortoises) which is wholly terrestrial. The aquatic forms secrete a mixture of urea and ammonia; the semiterrestrial species, on the other hand, excrete urea; and the tortoises excrete almost all their nitrogen as uric acid.

The topic of nitrogen excretion is one of the best examples of comparative biochemistry that has been developed.

Formation of Uric Acid

Uric acid, referred to in the preceding section, is the form in which birds and terrestrial reptiles excrete the NH_3 produced in protein metabolism. It is also the chief end product of metabolism of purines in man and other primates, the Dalmatian coach hound, birds, and some reptiles. Thus, the birds and reptiles which have uric acid as their chief nitrogen waste product first must convert NH_3 into purines by reactions to be considered shortly.

The free purine bases are converted to uric acid as shown in Figure 16-3. Xanthine oxidase, which catalyzes the formation of uric acid, is found in the peroxisomes of the kidney along with other oxidases (see page 368). Mammals other than the primates and most reptiles produce allantoin as their end product of purine metabolism. Such organisms contain the enzyme uricase which converts uric acid to allantoin. The teleost fish convert allantoin on to allantoic acid, while most fish and the amphybia degrade the allantoic acid further to urea and glyoxylic acid. The pyrimidine bases are broken down, by reactions not to be discussed here, into NH_3, CO_2, and propionic and succinic acids.

While the free nitrogen bases can be degraded by the reactions and processes just described, it should be noted that living organisms have mechanisms for salvaging these free bases rather than losing them to catabolic sequences. This salvage operation is sensible in view of the involved, endergonic sequence of reactions (see section below) which must be carried

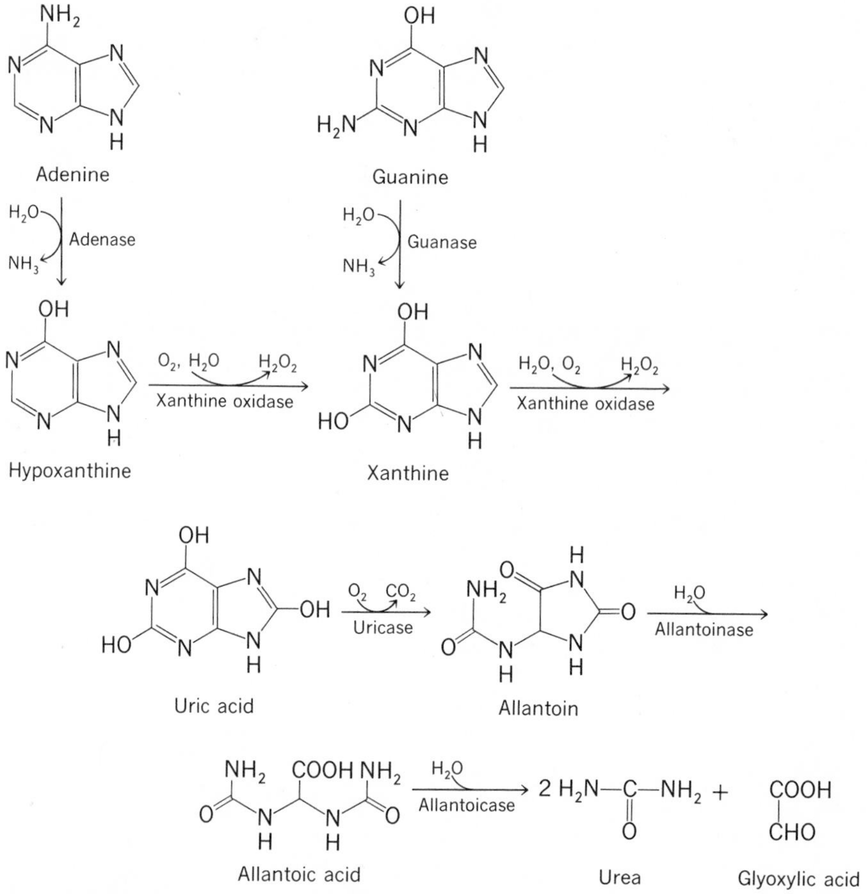

Figure 16-3

The metabolic degradation of adenine and guanine.

out in order to form these compounds from simple precursors. Enzymes known as phosphoribosyl transferases carry out this operation:

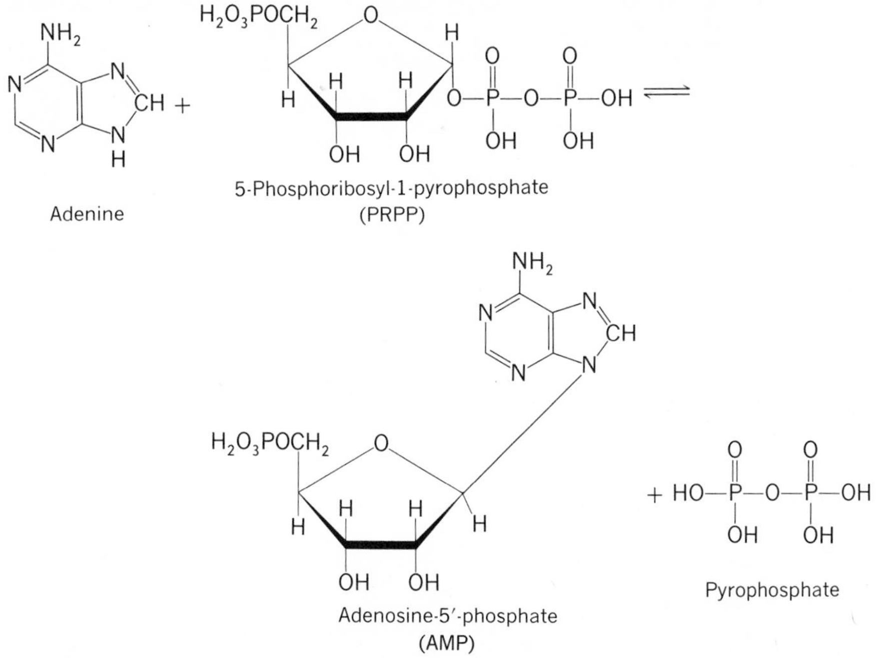

Adenine + 5-Phosphoribosyl-1-pyrophosphate (PRPP) $\rightleftharpoons$ Adenosine-5′-phosphate (AMP) + Pyrophosphate

In this reaction the ribose-5-phosphate moiety is transferred to the free base to form the corresponding mononucleotide. While this is a useful reaction for utilizing free bases obtained on breakdown of preexisting mononucleotides, it will be seen that the synthesis *de novo* of the purine and pyrimidine nucleotides involves the attachment of the sugar phosphate to the corresponding nitrogen at a much earlier step; indeed, in the case of purine biosynthesis, it is at the very beginning of the formation of the purine molecule that the N—C linkage is synthesized.

Purine Biosynthesis

In contrast to the essential or indispensible amino acids, the purines (and pyrimidines) can be formed from simple precursors by both plants and animals. Experiments with radioisotopes have shown that the nine atoms of the purine nucleus are derived from five different precursors, each precursor contributing the atoms indicated in Scheme 16-5. Work by a number of different researchers with mammals, birds, and bacteria has shown that essentially the same biosynthetic pathway is followed in these diverse living forms. As will be seen, the pathway consists of a stepwise addition of individual atoms to the carbon-1 of ribose-5-phosphate to produce the key intermediate, *inosinic acid*. It is possible to describe most of the reactions in detail, and these will be given, not to confuse the student but to demonstrate that certain principles of biochemical reactions, already encountered in the metabolism of carbohydrates, lipids, and the amino acids, can be applied to the synthesis of nucleic acids and their derivatives.

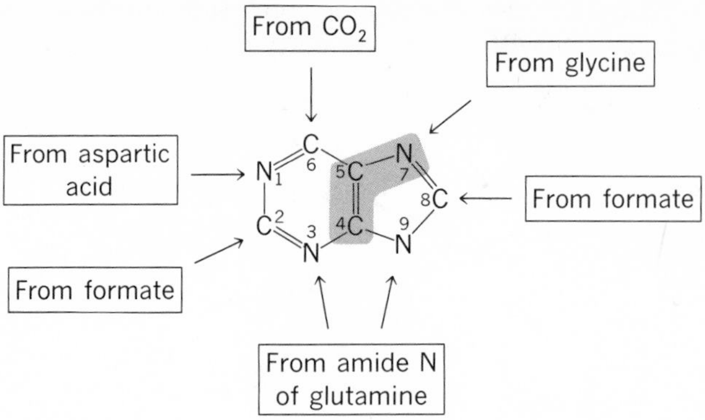

From CO_2

From glycine

From aspartic acid

From formate

From formate

From amide N of glutamine

Scheme 16-5

The starting point for purine biosynthesis is the compound α-5-phosphoribosyl-1-pyrophosphate (PRPP), which is obtained from ATP and ribose-5-phosphate:

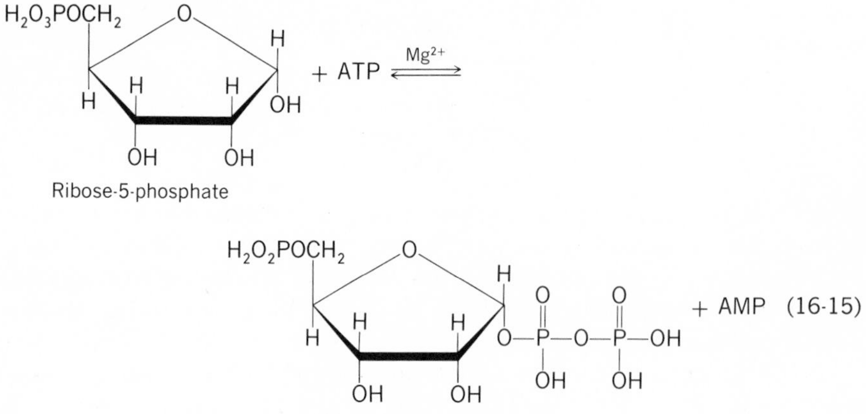

Ribose-5-phosphate + ATP $\xrightleftharpoons{Mg^{2+}}$

α-5-Phosphoribosyl-1-pyrophosphate (PRPP) + AMP (16-15)

The enzyme which is involved in this reaction is interesting in that it catalyzes the transfer of the pyrophosphate moiety of ATP rather than the terminal phosphate group to the acceptor molecule, ribose-5-phosphate.

α-5-Phosphoribosyl-1-pyrophosphate then participates in the initial step in purine biosynthesis by reacting with glutamine to form 5-phosphoribosyl-1-amine, glutamic acid, and pyrophosphate. This is a reaction in which glutamine donates its amide nitrogen atom into organic combination. The enzyme which catalyzes reaction 16-16 is inhibited by purine nucleotides produced in later steps of the biosynthetic pathway. For this reason, reaction 16-16 is then the site of *feedback inhibition* by later products of the pathway. Note that the configuration at the hemiacetal carbon of ribose is inverted during the reaction; the linkage of the pyrophosphate group in PRPP is α, while the amine group has the β configuration. This, of course, is the configuration of the *N*-ribosyl bond in the purine nucleotides eventually formed by the

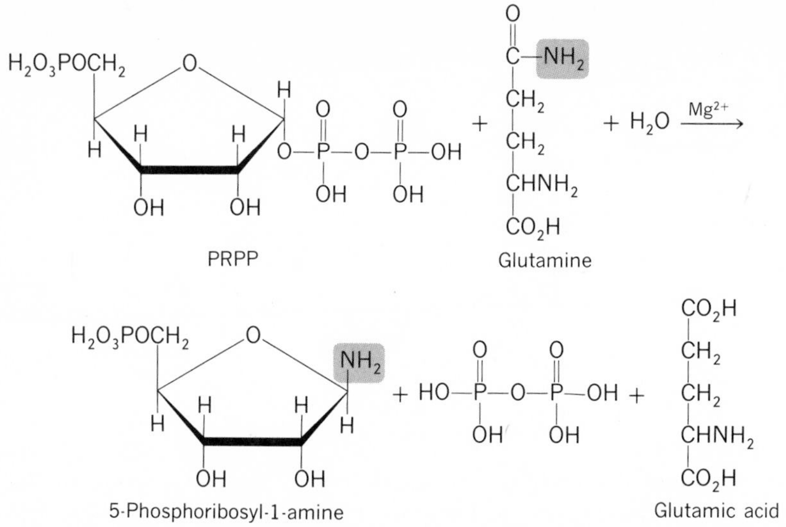

$H_2O_3POCH_2$ (ribose ring) –O–P(=O)(OH)–O–P(=O)(OH)–OH + $O{=}C(NH_2)–CH_2–CH_2–CHNH_2–CO_2H$ + H_2O $\xrightarrow{Mg^{2+}}$

PRPP Glutamine

$H_2O_3POCH_2$ (ribose ring) –NH_2 + $HO–P(=O)(OH)–O–P(=O)(OH)–OH$ + $CO_2H–CH_2–CH_2–CHNH_2–CO_2H$ (16-16)

5-Phosphoribosyl-1-amine Glutamic acid

pathway. Reaction 16-16 is inhibited by the antibiotic azaserine, a structural analog of glutamine required by this reaction.

In the next step, the amino acid glycine is linked to the ribosylamine in an amide linkage. It is not surprising, therefore, that the reaction should require a source of energy which is supplied by ATP.

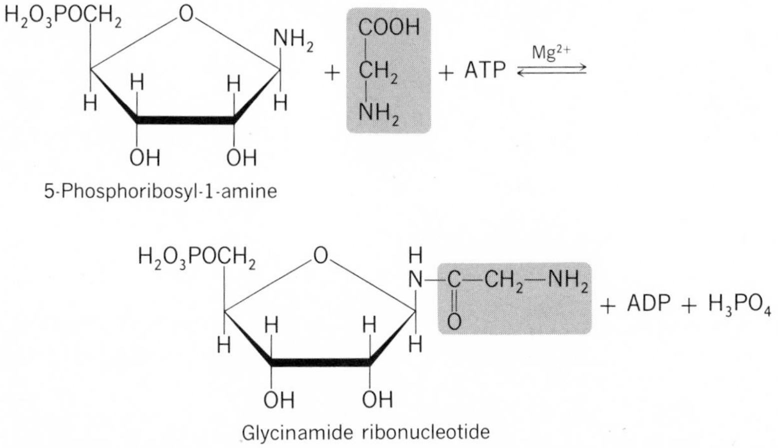

$H_2O_3POCH_2$ (ribose ring) –NH_2 + $COOH–CH_2–NH_2$ + ATP $\overset{Mg^{2+}}{\rightleftharpoons}$

5-Phosphoribosyl-1-amine

$H_2O_3POCH_2$ (ribose ring) –N(H)–C(=O)–CH_2–NH_2 + ADP + H_3PO_4 (16-17)

Glycinamide ribonucleotide

The glycinamide ribonucleotide formed in reaction 16-17 then reacts in the presence of a *transformylase* which catalyzes the transfer of a *formyl* group from the formyl transfer coenzyme, methenyl N^{5-10} tetrahydrofolic acid (see page 218) to produce formylglycinamide ribonucleotide. At this point all the atoms of the imidazole ring of the purine nucleus have been attached to the phosphoribose moiety, and the latter will be represented in subsequent reactions by Ribose—PO_3H_2.

While it would be reasonable to have ring closure at this point, the next reaction involves the addition of the nitrogen atom located at position 3 of

$$H_2O_3POCH_2\text{-ribose-}NH\text{-}C(=O)\text{-}CH_2\text{-}NH_2 + \text{Methenyl-}N^{5\text{-}10}\text{-}FH_4 + H_2O \longrightarrow$$

Glycinamide ribonucleotide

$$\alpha\text{-}N\text{-Formylglycinamide ribonucleotide} + FH_4 + H^+ \quad (16\text{-}18)$$

α-N-Formylglycinamide ribonucleotide

the purine structure. As might be predicted, the nitrogen is provided by the amide group of glutamine in the presence of an energy source, ATP. The mechanism of reaction 16-19, and several others to 16-19 follow in which a nitrogen atom is transferred to the purine pyrimidive skeleton, is not well-understood. It may be similar to the synthesis of glutamine (page 382) in that a phosphorylated intermediate may be involved.

$$\alpha\text{-}N\text{-Formylglycinamide ribonucleotide} + \text{Glutamine} + ATP + H_2O \xrightarrow{Mg^{2+}}$$

α-N-Formylglycinamide ribonucleotide   Glutamine

$$\alpha\text{-}N\text{-Formylglycinamidine ribonucleotide} + \text{Glutamic acid} + ADP + H_3PO_4 \quad (16\text{-}19)$$

α-N-Formylglycinamidine ribonucleotide   Glutamic acid

The α-N-formylglycinamide ribonucleotide then undergoes ring closure by a poorly understood dehydration reaction which requires ATP. In this step, the imidazole ring of the purine nucleus is realized, and ATP is hydrolyzed to ADP and H_3PO_4. As the three remaining atoms of the purine skeleton have yet to be acquired, the carbon atom at position 6 is formed next by carbox-

α-*N*-Formylglycinamidine ribonucleotide + ATP $\xrightarrow{Mg^{2+}}$ 5-Aminoimidazole ribonucleotide + ADP + H_3PO_4 (16-20)

ylation of the imidazole nucleus with CO_2. This reaction would be expected to require a biotin coenzyme system (page 209), and there is evidence for a biotin requirement by the bacterial enzyme system which catalyzes reaction 16-21.

5-Aminoimidazole ribonucleotide + CO_2 ⇌ 5-Aminoimidazole-4-carboxyribonucleotide (16-21)

The next step in the pathway is one of several reactions in intermediary metabolism where a nitrogen atom is contributed by aspartic acid. The process is quite analogous to the synthesis of argininosuccinic acid in the urea cycle (page 384), where ATP is required as an energy source. In the

5-Aminoimidazole-4-carboxyribonucleotide + Aspartic acid + ATP ⇌ 5-Aminoimidazole-4-*N*-succinocarboxamide ribonucleotide + ADP + H_3PO_4 (16-22)

present instance, however, ATP is cleaved to ADP and H_3PO_4. Subsequently, the succinocarboxamide derivative is cleaved in an *aspartase* type of reaction (page 368) to form fumaric acid in a manner quite analogous to the cleavage of argininosuccinic acid.

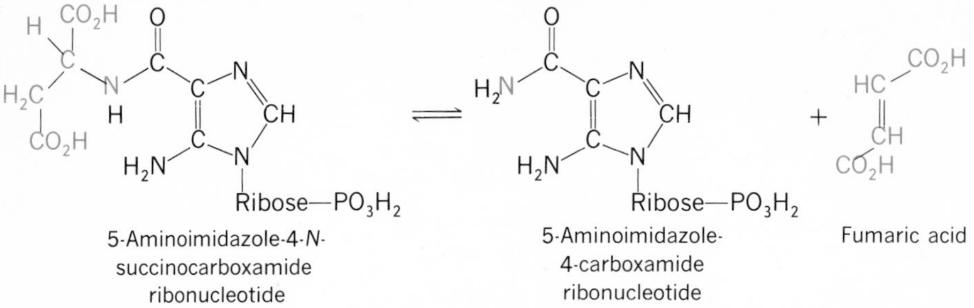

5-Aminoimidazole-4-*N*-succinocarboxamide ribonucleotide ⇌ 5-Aminoimidazole-4-carboxamide ribonucleotide + Fumaric acid (16-23)

One final carbon atom must now be acquired before the six-membered ring of the purine can be formed by ring closure. This atom is provided through the one-carbon metabolism of the folic acid system (see page 218) in the form of a formyl group:

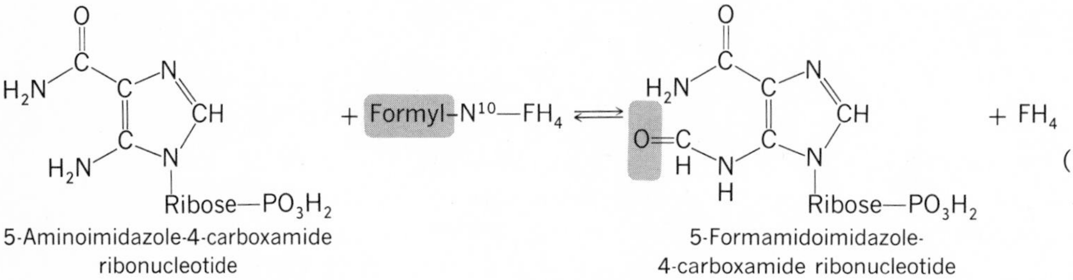

5-Aminoimidazole-4-carboxamide ribonucleotide + Formyl-N^{10}—FH_4 ⇌ 5-Formamidoimidazole-4-carboxamide ribonucleotide + FH_4 (16-24)

The reaction is also inhibited by sulfonamide antibiotics.

Ring closure then occurs in the presence of an enzyme which catalyzes the removal of H_2O in a reversible reaction. The product is *inosinic acid* which does occur free in biclogical materials, but of course is not a component of RNA or DNA.

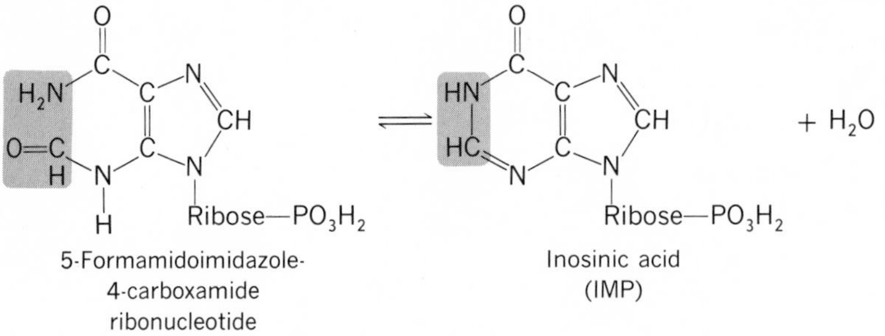

5-Formamidoimidazole-4-carboxamide ribonucleotide ⇌ Inosinic acid (IMP) + H_2O (16-25)

If we express reactions 16-16 through 16-25 in terms of the overall reaction, we can write

$$2\ NH_3 + 2\ HCOOH + CO_2 + \text{Glycine} + \text{Aspartic acid} + \text{Ribose-5-phosphate} \longrightarrow \text{Inosinic acid} + \text{Fumaric acid} + 9\ H_2O$$

The energy required to accomplish this process is, of course, provided by ATP molecules, all but one of which are cleaved to produce ADP and H_3PO_4. Thus, we have another example of the means by which the energy-richness

$$9\,\text{ATP} + 9\,\text{H}_2\text{O} \longrightarrow 8\,\text{ADP} + 8\,\text{H}_3\text{PO}_4 + \text{AMP} + \text{HO—P(=O)(OH)—O—P(=O)(OH)—OH}$$

of ATP can be utilized in discrete reactions to accomplish a biosynthetic sequence requiring energy.

Purine Nucleotide Interconversions

The two purine nucleotides AMP and GMP are subsequently formed from inosinic acid, the initial product of the purine pathway. In the case of AMP, the nitrogen atom at position 6 is contributed by aspartic acid in a reaction

$$\text{IMP} + \text{Aspartic acid} + \text{GTP} \xrightarrow{Mg^{2+}} \text{Adenylosuccinic acid} + \text{GDP} + \text{H}_3\text{PO}_4 \quad (16\text{-}26)$$

IMP: hypoxanthine ring (C=O at position 6) attached to Ribose—PO_3H_2

Aspartic acid: CO_2H—$CH(NH_2)$—CH_2—CO_2H (H_2NCH)

Adenylosuccinic acid: $HO_2C—CH_2—CH(CO_2H)—NH$— attached to C-6 of the purine ring, Ribose—PO_3H_2

involving the formation of a substituted succinic acid. While this reaction is strictly analogous to reaction 16-22, note that a different nucleoside triphosphate (GTP) is required. The substituted succinic acid is then cleaved by an aspartase type of reaction to yield fumaric acid and AMP:

$$\text{Adenylosuccinic acid} \rightleftharpoons \text{AMP} + \text{Fumaric acid} \quad (16\text{-}27)$$

Adenylosuccinic acid: $HO_2C—CH_2—CH—CO_2H$ / NH attached to purine, Ribose—PO_3H_2

AMP: NH_2 attached to purine, Ribose—PO_3H_2

Fumaric acid: $HO_2C—CH{=}CH—CO_2H$

This reaction is similar to reaction 16-23, and the enzymes catalyzing the two reactions are probably identical.

In the case of GMP synthesis, a nitrogen atom must be introduced in position 2 of the purine nucleus. To do this, the carbon atom at that position in inosinic acid must first be oxidized to the higher oxidation state in order to acquire the amino group which is already at that oxidation level. The oxidation is accomplished by a dehydrogenase which requires the nicotinamide nucleotide NAD^+:

$$\text{IMP} + NAD^+ + H_2O \rightleftharpoons \text{Xanthylic acid} + NADH + H^+ \quad (16\text{-}28)$$

IMP (Ribose—PO_3H_2) Xanthylic acid (Ribose—PO_3H_2)

Once xanthylic acid is obtained, the nitrogen atom is acquired from glutamine in a reaction utilizing ATP:

$$\text{Xanthylic acid} + \text{Glutamine} + ATP \xrightarrow[Mg^{2+}]{+H_2O} \text{GMP} + \text{Glutamic acid} + AMP + HO{-}PO(OH){-}O{-}PO(OH){-}OH \quad (16\text{-}29)$$

Xanthylic acid (Ribose—PO_3H_2) Glutamine ($H_2NOC{-}CH_2{-}CH_2{-}CHNH_2{-}CO_2H$)

GMP (Ribose—PO_3H_2) Glutamic acid ($CO_2H{-}CH_2{-}CH_2{-}CHNH_2{-}CO_2H$)

Although the mechanism of this reaction which yields GMP has not been completely clarified, AMP and pyrophosphate are the products fromed from ATP.

Regulation of Purine Nucleotide Biosynthesis

The regulation of purine mononucleotide biosynthesis occurs at two different levels in the biosynthetic sequence. The first is at the synthesis of 5-phosphoribosylamine (reaction 16-16), which can be considered the initial step in the biosynthetic pathway. The enzyme catalyzing this reaction is inhibited by AMP, ADP, ATP, GMP, GDP, and GTP, each compound acting independently of the other.

The other control is at the branching compound, inosinic acid. It may be seen that reaction 16-26, leading ulimately to AMP, requires GTP, while reactions 16-28 and 16-29, leading to GMP, require ATP in the latter reaction. Thus, when ATP is in excess, the higher concentration available simply leads to more GMP (and eventually GTP) being produced. Conversely, an excess of GTP leads to a higher production of AMP and therefore ATP.

Pyrimidine Biosynthesis

The atoms of the pyrimidine nucleus are derived from three simple precursors, CO_2, NH_3, and aspartic acid (Scheme 16-6).

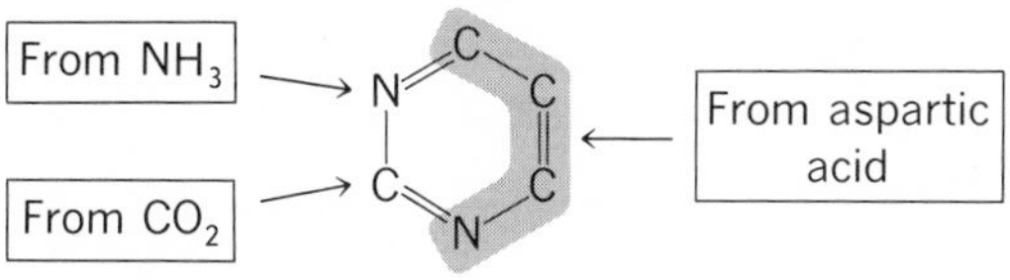

From NH_3 → N; From CO_2 → C; From aspartic acid → C–C=C–N (pyrimidine ring: N, C, N, C, C, C)

Scheme 16-6

When the individual biosynthetic steps were studied, the first was shown to involve the transfer of a carbamyl group from carbamyl phosphate to aspartic acid to form *N*-carbamyl aspartic acid (ureidosuccinic acid):

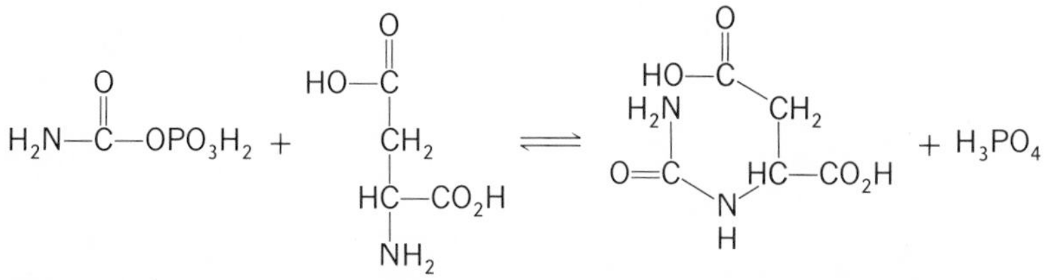

$$\underset{\text{Carbamyl phosphate}}{H_2N{-}\overset{O}{\overset{\|}{C}}{-}OPO_3H_2} + \underset{\text{Aspartic acid}}{HO{-}\overset{O}{\overset{\|}{C}}{-}CH_2{-}CH(NH_2){-}CO_2H} \rightleftharpoons \underset{\substack{N\text{-Carbamyl aspartic acid}\\ \text{(Ureidosuccinic acid)}}}{HO{-}\overset{O}{\overset{\|}{C}}{-}CH_2{-}CH(CO_2H){-}NH{-}\overset{O}{\overset{\|}{C}}{-}NH_2} + H_3PO_4 \qquad (16\text{-}30)$$

Carbamyl phosphate is one of the three compounds through which the nitrogen atoms enter organic combination (page 381). Its biosynthesis and its role in the urea cycle are discussed on page 383. The enzyme which catalyzes reaction 16-30 is known as *aspartic transcarbamylase;* it is the site of feedback inhibition by CTP which will be shown to be a product of the pathway under consideration. A detailed discussion of this enzyme as it relates to feedback inhibition is presented elsewhere (page 470).

Ring closure of *N*-carbamyl aspartic acid catalyzed by the enzyme *dihydroorotase* leads to the formation of dihydroorotic acid:

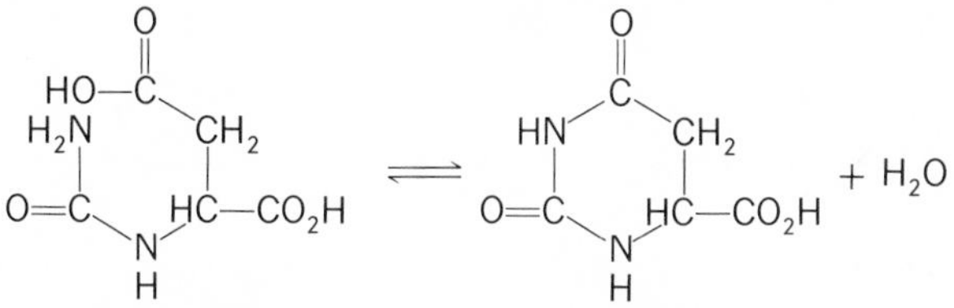

$$\underset{N\text{-Carbamyl aspartic acid}}{HO{-}\overset{O}{\overset{\|}{C}}{-}CH_2{-}CH(CO_2H){-}NH{-}\overset{O}{\overset{\|}{C}}{-}NH_2} \rightleftharpoons \underset{\text{Dihydroorotic acid}}{\text{cyclo}[HN{-}\overset{O}{\overset{\|}{C}}{-}CH_2{-}CH(CO_2H){-}NH{-}\overset{O}{\overset{\|}{C}}]} + H_2O \qquad (16\text{-}31)$$

This dehydration is freely reversible, with the open-chain compound predominating in the ratio of 2:1 at equilibrium.

In the next step, a flavin enzyme, *dihydroorotic acid dehydrogenase,* catalyzes the formation of orotic acid by removing two hydrogen atoms from adjacent carbon atoms to form a carbon–carbon double bond:

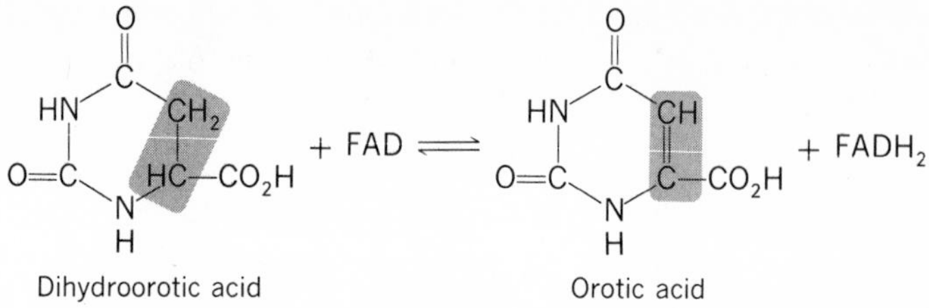

Dihydroorotic acid + FAD ⇌ Orotic acid + $FADH_2$ (16-32)

When this enzyme was first discovered, a nicotinamide nucleotide was observed to be the oxidizing agent. However, since the formation of such double bonds were known to require flavin cofactors (for example, in the formation of unsaturated acylthioesters in lipid metabolism, page 283), this question was examined carefully by Birgit Vennesland. She and her associates demonstrated that the dihydroorotic acid enzyme was a flavoprotein requiring FAD as a prosthetic group which, when reduced, contributed its electrons to NAD^+ as the physiological acceptor. The overall reaction may therefore be represented as in Scheme 16-7.

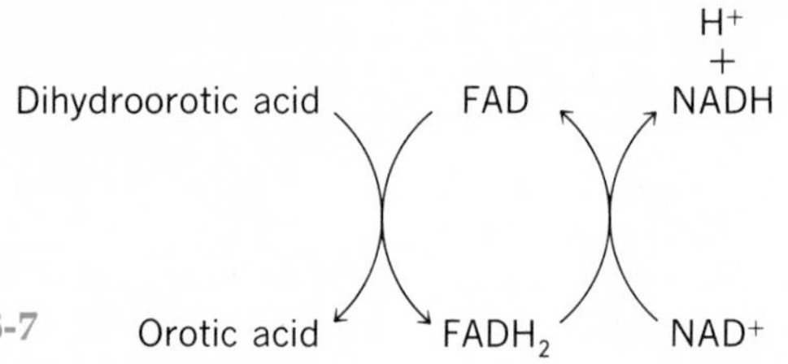

Scheme 16-7

Dihydroorotic acid → Orotic acid; FAD → $FADH_2$; $FADH_2$ → FAD; NAD^+ → H^+ + NADH

Orotic acid reacts with PRPP to acquire the 5-phosphoribosyl moiety (—Ribose—PO_3H_2) and become the nucleotide orotidine-5′-phosphate. In this process the ring nitrogen of orotic acid reacts as a nucleophile to displace the pyrophosphate group of PRPP and form the β-*N*-glycosyl bond.

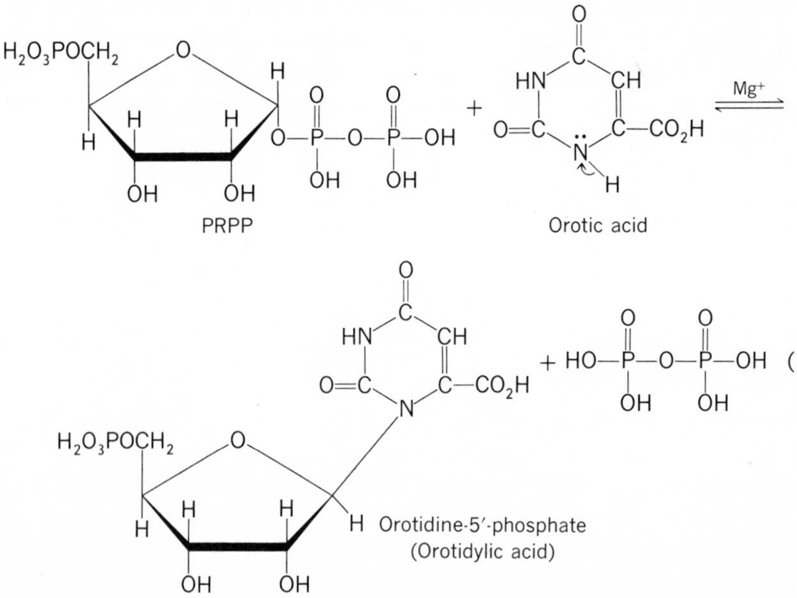

PRPP + Orotic acid $\xrightleftharpoons{Mg^+}$ Orotidine-5′-phosphate (Orotidylic acid) + HO—P(O)(OH)—O—P(O)(OH)—OH (16-33)

Finally, orotidylic acid is decarboxylated in the presence of a specific decarboxylase to yield uridine-5′-phosphate (UMP), the starting point for synthesis of the cytidine and thymidine nucleotides.

$$\text{Orotidylic acid (Ribose—PO}_3\text{H}_2\text{; C—CO}_2\text{H)} \longrightarrow \text{Uridine-5'-phosphate (UMP) (Ribose—PO}_3\text{H}_2\text{)} + CO_2 \qquad (16\text{-}34)$$

We might logically expect that some mechanism would exist for the amination of UMP to yield CMP, these monophosphates subsequently being phosphorylated to produce the di- and triphosphates. Instead, the formation of the cytidine derivative requires the triphosphate form of uridine. An enzyme from bacteria can then catalyze the direct amination by ammonia of UTP to form CTP, energy for the process being furnished by ATP. While we might also expect intermediates in such a reaction, none have been detected. In animal tissues the nitrogen atom is obtained from glutamine. This difference between bacteria and animals illustrates the frequently observed fact that bacteria can utilize ammonia directly where, in the same reaction, animal systems will require glutamine. This may be related to the observation that animals readily dispose of the fairly toxic ammonia molecule by the elaborate biosynthetic sequence of the urea cycle (Figure 16-2), while bacteria and other lower forms can tolerate and utilize the NH_3.

$$\text{Uridine-5'-triphosphate (UTP)} + NH_3 + ATP \xrightarrow{Mg^{2+}} \text{Cytidine-5'-triphosphate (CTP)} + ADP + H_3PO_4 \qquad (16\text{-}35)$$

Uridine-5′-triphosphate (UTP): OH—C; Ribose—P(=O)(OH)—O—P(=O)(OH)—O—P(=O)(OH)—OH

Cytidine-5′-triphosphate (CTP): NH_2—C; Ribose—O—P(=O)(OH)—O—P(=O)(OH)—O—P(=O)(OH)—OH

The biosynthesis of pyrimidine nucleotides is primarily regulated through the action of CTP on the enzyme aspartate transcarbamylase that catalyzes the first reaction in the biosynthetic sequence (reaction 16-30).

Formation of Deoxyribotides

The biosynthesis of deoxyribose derivatives of the purines and pyrimidines involves enzymatic reactions whose details are still being examined. There is good evidence that two separate systems are able to form the deoxyribosyl derivatives. One of these, found in extracts of *Lactobacillus leichmannii,* is called ribonucleoside triphosphate reductase. The system requires a vitamin B_{12} coenzyme and a thiol such as dihydrolipoic acid for the reduction of ribonucleoside triphosphate to the deoxyribonucleoside derivatives.

A second system for deoxyribotide synthesis has been discovered by P. Reichard and his colleagues at the University of Upsala. This system utilizes the ribonucleoside diphosphates rather than the triphosphates as substrates and does not require a cobamide coenzyme. The reductant of the ribose moiety is the reduced form of a flavoprotein, called *thioredoxin,* a small protein (68,000 mol wt) that contains two free –SH, and two FAD molecules per 108 amino acid residues. The reduction of the nucleoside diphosphate (XDP) requires the participation of two separate proteins and can be represented as

$$\text{XDP} + \text{Thioredoxin–(SH)}_2 \xrightarrow{\text{2 Enzymes}} d\text{XDP} + \text{Thioredoxin–S}_2 \qquad (16\text{-}36)$$

The original thioredoxin can, in turn, be regenerated by NADPH in the presence of *thioredoxin reductase:*

$$\text{NADPH} + \text{H}^+ + \text{Thioredoxin–S}_2 \xrightarrow{\text{Thioredoxin reductase}} \text{NADP}^+ + \text{Thioredoxin–(SH)}_2 \qquad (16\text{-}37)$$

and can, in turn, reduce another mole of XDP.

Synthesis of the Diphosphates and Triphosphates

Once the synthesis of the monophosphates of purine and pyrimidine nucleosides is achieved, formation of the di- and triphosphates is readily accomplished. There are transphosphorylases which will catalyze the transfer of phosphate from ATP to the individual riboside and deoxyriboside monophosphates:

$$\text{ZMP} + \text{ATP} \xrightleftharpoons{\text{Mg}^{2+}} \text{ZDP} + \text{ADP} \qquad (16\text{-}38)$$

Here, Z stands for any purine or pyrimidine, and the enzyme involved, a monophosphokinase, may show little or no specificity for the individual base. The deoxyriboside monophosphates (dZMP) can also participate in reaction 16-38.

The triphosphates can, in turn, be formed by the phosphorylation of the diphosphate in the presence of nucleoside diphosphokinases:

$$\text{ZDP} + \text{ATP} \rightleftharpoons \text{ZTP} + \text{ADP}$$
$$d\text{ZDP} + \text{ATP} \rightleftharpoons d\text{ZTP} + \text{ADP}$$

As a result of these reactions, the ribosyl and deoxyribosyl triphosphates needed for the synthesis of RNA and DNA have been produced.

Figure 16-4 summarizes the many participating reactions for the synthesis of the nucleotide triphosphates required for the synthesis of RNA and DNA.

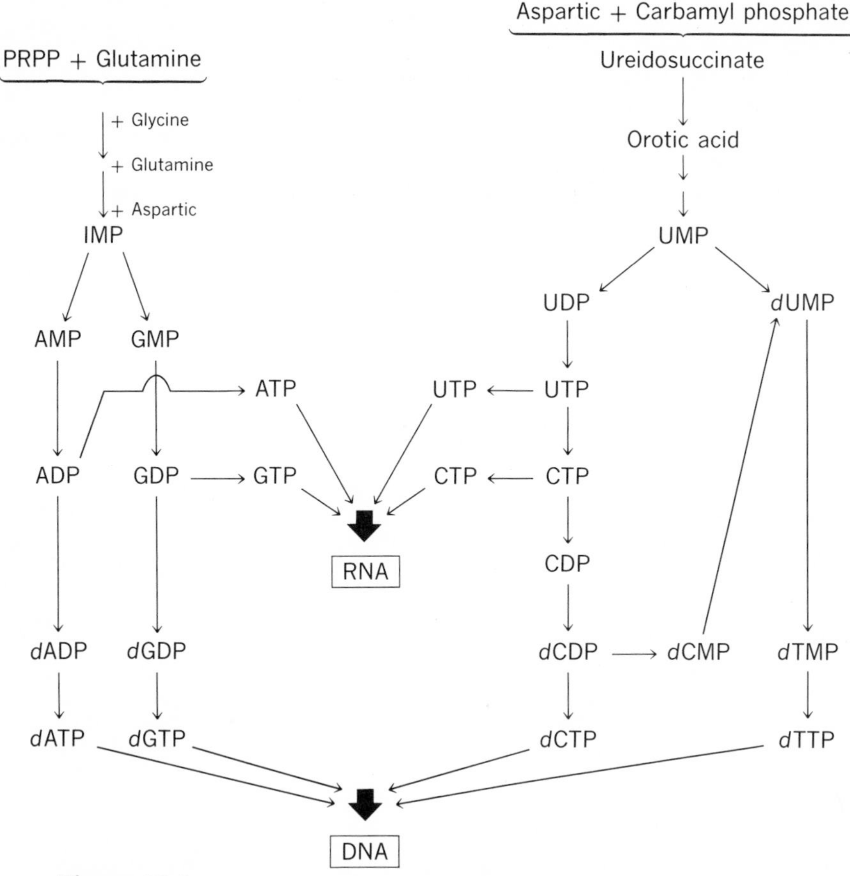

Figure 16-4

Relationship of synthesis of nucleotides and nucleic acids. The prefix *d* signifies deoxy-; thus, AMP contains the ribosyl moiety, but *d*AMP contains a deoxyribosyl moiety. Although in some organisms the ribonucleoside triphosphates are directly reduced to their deoxy derivatives, for simplicity these reactions are not included here.

References

1. A. Meister, *Biochemistry of the Amino Acids*. 2nd ed. New York: Academic Press, 1965.
   This two-volume work is the standard reference in this area of biochemical research.
2. D. M. Greenberg, *Metabolic Pathways*. vol. 3. New York: Academic Press, 1967.
   The several chapters in this volume of Greenberg's series on metabolism are

concerned with the metabolism of specific amino acids. The volume covers both catabolism and biosynthesis.

3. J. O. Stanbury, J. B. Wyngaarden, and D. S. Fredrickson (eds.), *The Metabolic Basis of Inherited Diseases*. 2nd ed. New York: McGraw-Hill, 1966.

   This volume provides thorough coverage of the metabolic diseases so often associated with amino acid metabolism.

4. P. P. Cohen and G. W. Brown, Jr., "Ammonia Metabolism and Urea Biosynthesis," in *Comparative Biochemistry*, M. Florkin and H. S. Mason, eds. vol. 2. New York: Academic Press, 1960.

   E. Baldwin, *An Introduction to Comparative Biochemistry*. 4th ed. Cambridge: Cambridge University Press, 1964.

   The comparative and developmental aspects of nitrogen excretion are discussed.

17

The Nitrogen Cycle

Introduction

A third fundamental process in addition to photosynthesis and respiration is that of *nitrogen fixation*. This process in turn is part of the cycle of reactions known as the *nitrogen cycle*. Many constituents of the living cell contain nitrogen; they include proteins, amino acids, nucleic acids, purines, pyrimidines, porphyrins, alkaloids, and vitamins. The nitrogen atoms of these compounds eventually travel the nitrogen cycle, in which the nitrogen of the atmosphere serves as a reservoir. Nitrogen is removed from the reservoir by the process of fixation; it is returned by the process of denitrification.

To give some measure of the magnitude of the chemical processes that occur, it has recently been estimated that 25×10^6 tons of nitrogen are removed yearly from the soils of the United States, chiefly by the harvesting of crops and to a small extent by the leaching of soils. To restore the fertility of the soil it is estimated that 3×10^6 tons of nitrogen are returned in the form of fertilizers (manure, urine, and commercial fertilizers). The restoration of an equal amount is accomplished by rainfall with the hydration of nitrogen oxides produced in the atmosphere by lightning storms. The most significant amount (10×10^6 tons of nitrogen) is returned through nitrogen fixation by biological organisms. Even so, it is apparent that a nitrogen deficit is developing and must be remedied if the fertility of the soil is to be maintained.

Several inorganic nitrogen compounds, as well as a myriad number of organic nitrogen compounds, can be considered as components of the nitrogen cycle. The former include N_2 gas, NH_3, nitrate ion (NO_3^-), nitrite ion (NO_2^-), and hydroxylamine (NH_2OH). At a glance it is apparent that the nitrogen atom can possess a variety of oxidation numbers. Some of these are the following:

	Nitrate ion	Nitrite ion	Hyponitrite ion	Nitrogen gas	Hydroxylamine	Ammonia
	NO_3^-	NO_2^-	$N_2O_2^{2-}$	N_2	NH_2OH	NH_3
Oxidation number	+5	+3	+1	0	−1	−3

Thus, in nature, nitrogen may exist in either a highly oxidized form (NO_3^-) or a highly reduced state (NH_3).

Nonbiological Nitrogen Fixation

The term *fixation* is defined as the conversion of molecular N_2 into one of the inorganic forms listed above. The distinguishing feature of this process is the separation of the two atoms of N_2 which are triply bonded ($N{\equiv}N$); N_2 is an extremely stable molecule. An indication of the difficult nature of this reaction is seen in the conditions for the fixation of nitrogen in the Haber process, developed in Germany during World War I. The English naval blockade of Germany prevented German access to the Chilean nitrate fields, and it was necessary to develop another source of nitrate for their explosives. The Haber process involves the reaction of N_2 and H_2 at extreme temperatures and pressures to form NH_3. The latter then can be oxidized to HNO_3. The Haber process is used today for the fixation of N_2 by the chemical industry in the production of commercial fertilizer.

$$N_2 + 3\,H_2 \xrightarrow[200\text{ atm}]{450^\circ\text{C}} 2\,NH_3$$

$$\Delta H = -24 \text{ kcal}$$

A second manner in which nitrogen may be fixed is through the electrical discharges that occur during lightning storms. During the discharge, oxides of nitrogen are formed which are subsequently hydrated by water vapor and carried to earth as nitrites and nitrates:

$$N_2 + O_2 \longrightarrow 2\,NO \xrightarrow{O_2} 2\,NO_2$$

Although these processes are significant in the nitrogen economy, the major amount of N_2 fixed is fixed by living organisms.

Biological Nitrogen Fixation

In sharp contrast to the chemical fixation of nitrogen is the biological fixation:

$$N_2 + 3\,H_2 \xrightarrow[1\text{ atm}]{25^\circ\text{C}} 2\,NH_3$$

Biological fixation of nitrogen is accomplished either by nonsymbiotic microorganisms which can live independently or by certain bacteria living in *symbiosis* with higher plants. The former group includes aerobic organisms of the soil (e.g., *Azotobacter*), soil anaerobes (e.g., *Clostridium* sp.), photosynthetic bacteria (e.g., *Rhodospirillum rubrum*), and algae (e.g., *Myxophyceae*). The symbiotic system consists of bacteria (*Rhizobia*) living in symbiosis with members of the *Leguminoseae* such as clover, alfalfa, and soy beans. Legumes are not the only higher plants that can fix nitrogen symbiotically; 190 species of shrubs and trees, including the Sierra Sweet Bay, ceanothus, and alder, are nitrogen fixers. Indeed, the fertility of high-altitude mountain lakes may be determined by the number of alder trees growing near their inlets.

An essential feature of symbiotic fixation are the nodules, which form on the roots of the plant. The nodules are formed by the joint action of the host plant, a legume and the bacterium, always a specific strain of the *Rhizobia*. Neither the plant nor the bacteria can fix nitrogen when grown separately. When the plants are grown in soil inoculated with the bacteria, root nodulation occurs and nitrogen fixation is possible. These nodules contain a pigment (leghemoglobin) similar to hemoglobin. Its function has not been defined. There has been much speculation about the symbiosis of a specific legume with a specific Rhizobium species. One theory suggests that during the evolution of the symbiotic relationship, the synthesis of nitrogenase passed to the host plant in whose genomes the information was encoded. Following infection, this information is passed to the bacterium by a specific host mRNA which then allows the bacteroid to synthesize the complete nitrogenase.

Until 1960, many workers were totally unsuccessful in obtaining cell-free preparations that could fix nitrogen. In that year, J. E. Carnahan and his group made the historical announcement of the first successful *in vitro* reduction of nitrogen gas to ammonia by a water-soluble extract of *Clostridium pasteurianum*. They discovered that in order for fixation to occur, large amounts of pyruvic acid had to be added to the extracts, whereupon the keto acid underwent phosphoroclastic dissimilation to acetyl phosphate, CO_2, and H_2. It was soon discovered that the extract could be fractionated into two systems; one, the HD or *hydrogen-donating* component, was responsible for the flow of electrons from the dissimilation of pyruvic acid via ferredoxin to the second component, called the *nitrogenase* system, which participated in the conversion of nitrogen to ammonia. The HD component also catalyzed the generation of ATP from acetyl phosphate by the following reactions:

$$\text{Acetyl–CoA} + \text{Pi} \xrightleftharpoons{\text{Phosphotransacetylase}} \text{Acetyl phosphate} + \text{CoA}$$

$$\text{Acetyl phosphate} + \text{ADP} \xrightleftharpoons[\text{Acetic kinase}]{Mg^{2+}} \text{ATP} + \text{Acetate}$$

Thus, pyruvate did not participate directly in nitrogen fixation but served as a source of electrons and ATP. The other important observation made was that the nitrogenase system was absent in extracts of *Clostridia* grown in the presence of NH_3 as the sole source of nitrogen although the HD system was still present in normal amounts.

The research by Carnahan and his colleagues stimulated a new thrust by a number of investigators into a detailed analysis of this important series of reactions. As a result, the present knowledge of nitrogen fixation has revealed that regardless whether extracts are prepared from anaerobes, aerobes, facultative anaerobes, blue-green algae, or legume nodules, the essential reaction components are (*a*) an electron donor, (*b*) an electron acceptor (i.e., nitrogen gas), (*c*) ATP together with divalent cation such as Mg^{2+}, and (*d*) two protein components, the first being a molybdenum, nonheme iron protein of about 270,000 molecular weight (molybdoferredoxin) and a nonheme iron protein component of 57,000 molecular weight (azoferredoxin). Each component alone is ineffective in catalyzing nitrogen fixation, but combined they form the nitrogenase complex. Another curious result is the specific requirement for ATP. In the absence of nitrogen as an electron acceptor, ATP can readily undergo hydrolysis to ADP and inorganic phosphate (ATP-ase activity), with the evolution of hydrogen gas, by the reaction

$$5\ ATP + 2\ e^- + 2\ H^+ \longrightarrow 5\ ADP + 5\ Pi + H_2$$

This reaction is somehow involved in the nitrogenase reaction, since the ATP-ase activity is absent in NH_3-grown cells and only present in N_2-grown cells. Both molybdoferredoxin and azoferredoxin are required. Since six electrons are required for the reduction of N_2 to ammonia, we can write the reaction

$$15\ ATP + 6\ e^- + 6\ H^+ + N_2 \longrightarrow 2\ NH_3 + 15\ ADP + 15\ Pi$$

and this result would explain the high concentration of pyruvate required by Carnahan in his early studies on nitrogen fixation.

We can now write a generalized scheme for nitrogen fixation that is common to all nitrogen-fixing systems, including the free-living as well as the symbiotic (root nodule) systems (Scheme 17-1).

Thus, whatever the source of electrons, one of the metals in the nitrogenase complex is reduced and activated ($X_{ox} \longrightarrow X^*_{red}$) by a cascade of electrons and ATP. Then, X^*_{red} may transfer its electrons to a suitable acceptor, normally nitrogen gas. If nitrogen gas is absent, hydrogen gas is released; that is, protons are reduced. The astonishing observation was made a few years ago that a whole host of compounds can accept electrons. Thus,

Reaction	Relative rate
$N_2 \xrightarrow{6\ e^-} 2\ NH_3$	1.0
$C_2H_2 \xrightarrow{2\ e^-} C_2H_4$ (Acetylene → Ethylene)	3.4
$HCN \xrightarrow{6\ e^-} CH_4 + NH_3$	0.6
$N_2O \xrightarrow{2\ e^-} N_2 + H_2O$	3.0

R. W. F. Hardy and co-workers have applied these observations to the development of an ingenious microassay for nitrogen fixation. By measuring the rate of reduction of acetylene to ethylene by a soil or water sample under standard conditions, a field analysis can be rapidly conducted to reveal the capacity of that sample to fix nitrogen. Thus, this information can provide a basis for evaluating the effect of different environment (bacterial and plant) factors on N_2 fixation. This information, in turn, can be of great value in agricultural practices.

Although NH_3 is an early, if not the first, product of biological nitrogen fixation, it is not accumulated by organisms performing this process. Rather, these organisms utilize the NH_3 to produce the nitrogenous components (proteins, nucleic acids, pigments) of their tissues. The excess nitrogen fixed

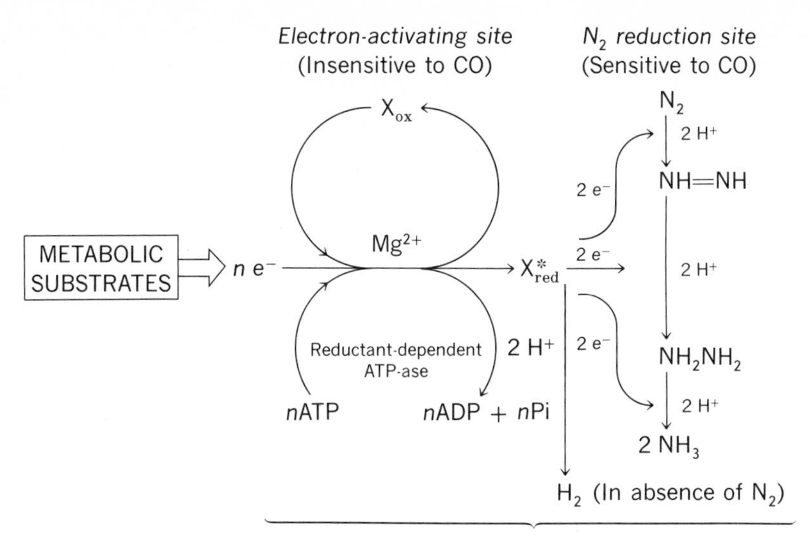

Scheme 17-1

may be excreted into the soil or other medium in which the nitrogen fixer is growing. For example, there is evidence that legumes and alders growing in sand culture excrete NH_3 and some amino acids into the sand surrounding their roots. The blue-green algae also excrete NH_3 as well as amino acids and peptides. If NH_3 is excreted into the soil, it can be subjected to the process of nitrification described below, or it may be utilized by other living forms (soil bacteria or higher plants) incapable of nitrogen fixation. If the fixing organism is a higher plant, the excess fixed nitrogen may be synthesized into asparagine or glutamine and stored in this form.

When the nitrogen-fixing organism perishes, the proteins of its cells will be hydrolyzed to amino acids and these subsequently deaminated by decay bacteria through the action of amino acid oxidases, or transaminases and glutamic dehydrogenase. The reactions of interest, which result in the formation of NH_3, have been described in Chapter 16. Obviously, the nitrogenous components of the nonfixing organisms encounter the same fate on death of the organism. Thus, the fertility of the soil is built up by the acquisition of NH_3 directly from nitrogen-fixing systems and indirectly after the nitrogen atom has made a cycle into the amino acids and proteins of the nitrogen fixers.

Nitrification

Despite the fact that NH_3 is the form in which nitrogen is normally added to the soil, little NH_3 is found there. Studies have shown that it is rapidly oxidized to nitrate ion; the latter represents the chief source of nitrogen for nonfixing organisms. The oxidation of NH_3 is carried out by two groups of bacteria called the nitrifying bacteria. One group, *Nitrosomonas,* converts NH_3

to nitrite ion with O_2 as the oxidizing agent:

$$NH_3 + \frac{3}{2} O_2 \longrightarrow NO_2^- + H_2O + H^+$$
$$\Delta G' = -66{,}500 \text{ cal}$$

The other group, *Nitrobacter*, oxidizes nitrite to nitrate:

$$NO_2^- + \frac{1}{2} O_2 \longrightarrow NO_3^-$$
$$\Delta G' = -17{,}500 \text{ cal}$$

Both reactions are exergonic; the first involves the oxidation of nitrogen from -3 to $+3$; the second is a two-electron oxidation from $+3$ to $+5$. Both groups of organisms are *autotrophs*, that is, they make all their cellular carbon compounds (protein, lipids, carbohydrates) from CO_2. As was indicated in Chapter 15, the conversion of CO_2 to carbohydrate requires energy. In photosynthesis, that energy is supplied by light; in the cases of *Nitrosomonas* and *Nitrobacter*, the energy for the reduction of CO_2 to carbohydrate and other carbon compounds is furnished by the oxidation of NH_3 and NO_2^- ion, respectively. Since the organisms obtain their energy for growth by the oxidation of simple inorganic compounds, they are termed *chemoautotrophs*.

Little is known about the intermediates in the oxidation of NH_3 to NO_2^- by *Nitrosomonas*, nor is there much information on the intermediary metabolism of the carbon compounds found in these bacteria. The lack of knowledge is due chiefly to the difficulty encountered in growing adequate amounts of the bacteria for experimentation. From the standpoint of comparative biochemistry, it may be predicted that the carbon compounds will undergo reactions resembling those described for animals, plants, and other microorganisms. The unique reactions, if any, may be expected to involve NH_3 and NO_2^-, the compounds that supply the energy for the growth of these bacteria.

Utilization of Nitrate Ion

With NO_3^- as the most abundant form of nitrogen in the soil, plants and soil organisms have developed an ability to utilize the anion as the nitrogen source required for their growth and development. In Chapter 16, however, it was pointed out that a major route for the incorporation of inorganic nitrogen into organic nitrogen is the reaction catalyzed by glutamic dehydrogenase. It is hence not surprising to find that higher plants and microorganisms which use NO_3^- must first reduce it to the valence level of NH_3. There is considerable information on the intermediates in this process: for example, the first step is the reduction of NO_3^- to NO_2^-, which is catalyzed by the enzyme *nitrate reductase*. The balanced reaction is

$$NO_3^- + NADPH + H^+ \longrightarrow NO_2^- + NADP^+ + H_2O$$

Nitrate reductases have been purified from bacteria, higher plants (soya beans), and the bread mold, *Neurospora*. In each case, one of the reduced nicotinamide nucleotides (NADPH or NADH) serves as a source of electrons for the reduction. The enzymes are flavoproteins which require FAD and the metal molybdenum as cofactors which undergo oxidation–reduction during the reaction.

The process is apparently repeated in the further reduction of nitrite through the intermediates, hyponitrite and hydroxylamine, to NH_3. The enzymes involved are indicated in the complete sequence

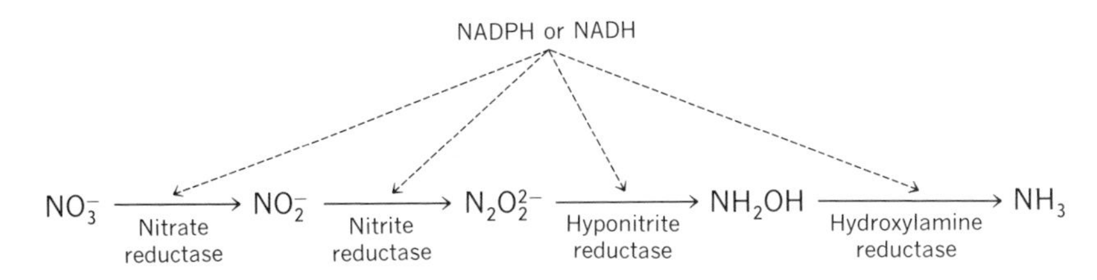

Each reaction involves the addition to the nitrogen atom of two electrons, which are furnished by reduced nicotinamide nucleotide. Thus, the reactions proceed as do most biological reductions. There is evidence that each enzyme requires flavin nucleotides and a metal as cofactors.

This utilization of nitrogen, in which aerobic microorganisms and higher plants reduce nitrate ion to NH_3 in order to incorporate it into cell protein, is referred to as *nitrate assimilation*. It is perhaps difficult to understand why in nature NH_3 is readily oxidized to NO_3^- which, in turn, must be again reduced to NH_3 before incorporation into amino acids. One advantage, of course, is that NO_3^- represents a more stable storage form than the somewhat volatile NH_3, although the existence of the latter as NH_4^+ is more likely in neutral and acid soils. A second advantage is that the ammonia molecule is rather toxic and therefore cannot be stored as such in tissue, whereas nitrate is relatively less toxic and can accumulate in large amounts in plant sap.

Some microorganisms, including *E. coli* and *B. subtilis*, reduce NO_3^- to NH_3 for another purpose; they utilize NO_3^- as a terminal electron acceptor instead of O_2. The NO_3^-, with its high oxidation reduction potential of 0.96 V at pH 7.0, can accept electrons released during the oxidation of organic substrates. The intermediates are NO_2^-, $N_2O_2^{2-}$, and NH_2OH, as in nitrate assimilation, but this process is known as *nitrate respiration*. Moreover, the enzymes involved are firmly associated with the insoluble matter of the cell (cell wall, endoplasmic reticulum). In the case of *Achromobacter fischeri*, the reduction of NO_3^- has been coupled with the oxidation of reduced cytochrome c; the presence of a cytochrome electron-transport chain which can react with NO_3^- rather than O_2 is therefore indicated.

Many bacteria (*Pseudomonas denitrificans, Denitrobacillus*) that carry out nitrate respiration produce N_2 instead of NH_3. In this case, the return of the nitrogen atom to the nitrogen of the atmosphere is accomplished. This sequence is referred to as *denitrification*. There is little detailed information on the enzyme systems involved.

The different processes that constitute the nitrogen cycle are diagrammed in Figure 17-1.

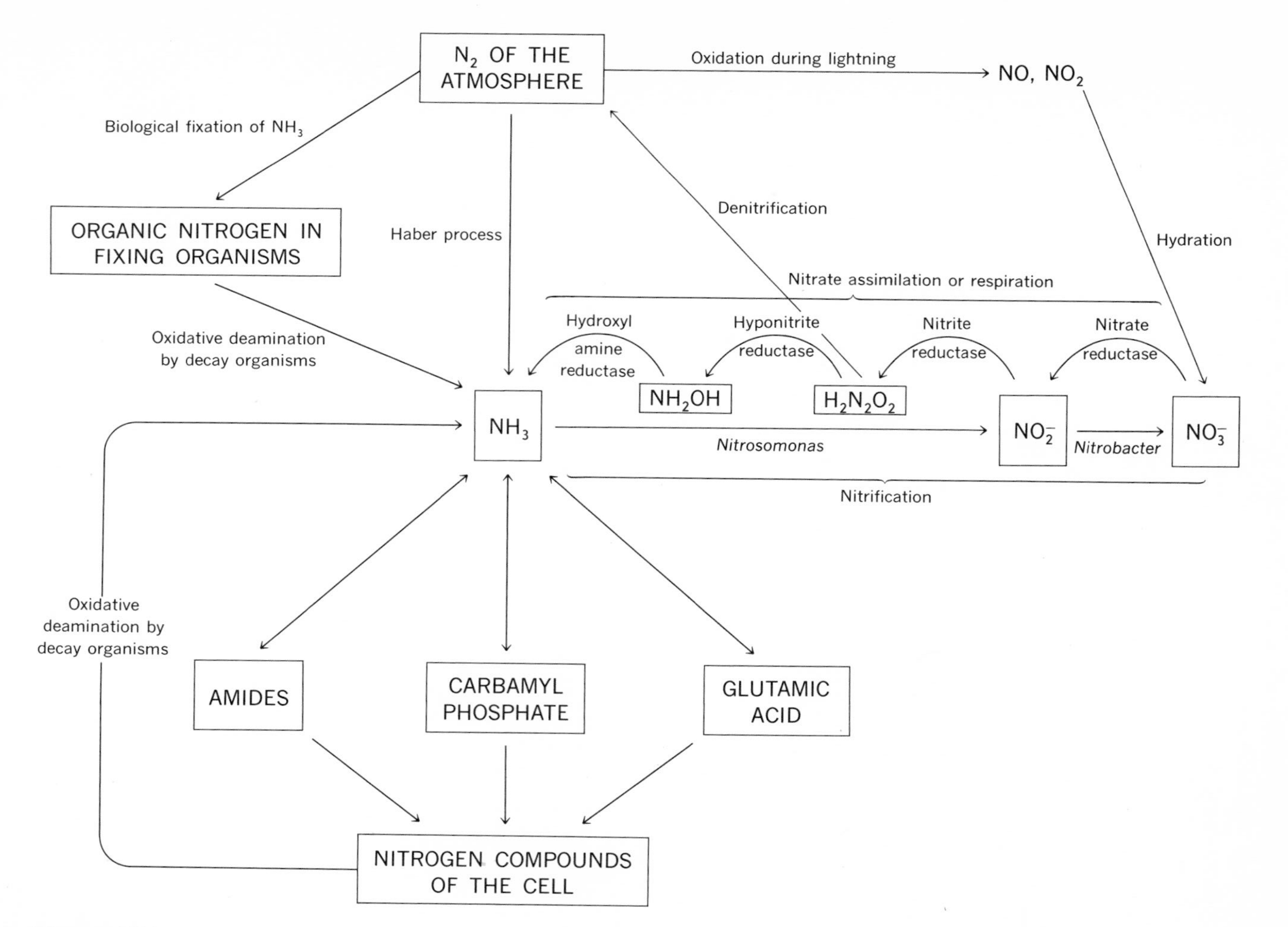

Figure 17-1
The nitrogen cycle.

Incorporation of Ammonia into Organic Compounds

It is appropriate at this point to reemphasize that there are three major enzyme reactions responsible for the incorporation of ammonia into organic nitrogen compounds. While these reactions are discussed in other chapters, the enzymes involved will be relisted here.

Glutamic dehydrogenase (page 381):

$$\underset{\alpha\text{-Ketoglutaric acid}}{HOOC{-}C(=O){-}CH_2{-}CH_2{-}COOH} + NH_3 + NADH + H^+ \rightleftharpoons \underset{\text{L-Glutamic acid}}{HOOC{-}CH(NH_2){-}CH_2{-}CH_2{-}COOH} + NAD^+ + H_2O$$

Glutamine synthase (page 382):

$$\underset{\text{Glutamic acid}}{HOOC{-}CH(NH_2){-}CH_2{-}CH_2{-}COOH} + NH_3 + ATP \xrightleftharpoons{Mg^{2+}} \underset{\text{Glutamine}}{HOOC{-}CH(NH_2){-}CH_2{-}CH_2{-}CONH_2} + ADP + H_3PO_4$$

Carbamyl kinase (page 383):

$$NH_3 + CO_2 + ATP \xrightleftharpoons{Mg^{2+}} H_2N{-}C(=O){-}OPO_3H_2 + ADP$$

The Sulfur Cycle

There are many similarities between the biochemistry of the sulfur atom and that of nitrogen. Because of the occurrence of sulfur in many essential biochemical compounds, the metabolism of this element will be discussed briefly.

The sulfur atom exists in several inorganic forms, as sulfate (SO_4^{2-}), sulfite (SO_3^{2-}), thiosulfate ($S_2O_3^{2-}$), elemental sulfur (S), and sulfide (S^{2-}). The oxidation state for these compounds ranges from +6 for sulfate to −2 for sulfide. Before the sulfur atom can enter into organic combination, it must be reduced to the level of sulfide (H_2S). When the sulfur atom is released from organic combination, it can be oxidized by soil organisms, to its highest oxidation state (SO_4^{2-}). Indeed, one may speak of the *sulfur cycle* in nature and identify certain aspects of that cycle.

Sulfate Reduction. Higher plants and many microorganisms utilize sulfate as their source of sulfur. *Sulfate assimilation,* anologous to nitrate assimilation, refers to the series of reactions whereby sulfate is reduced to sulfide and utilized in the formation of cysteine. The initial step involves the activation of sulfate by ATP to form adenosine-5′-phosphosulfate (APS) and then 3′-phosphoadenosine-5′-phosphosulfate (PAPS).

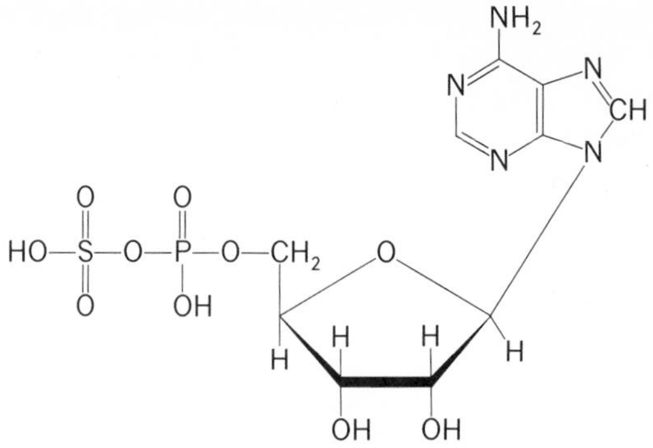

Adenosine-5′-phosphosulfate
(APS)

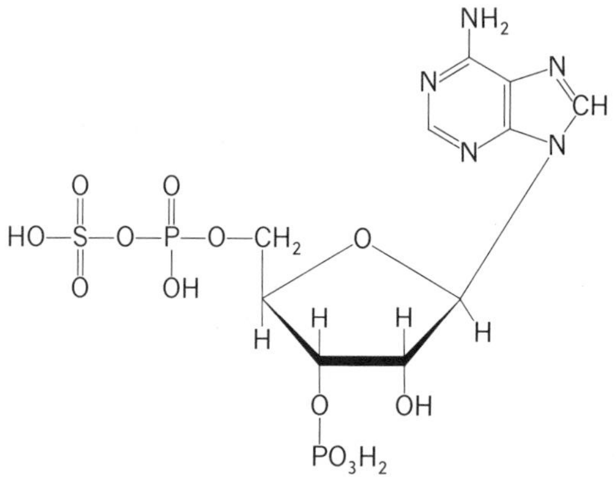

3′-Phosphoadenosine-5′-phosphosulfate
(PAPS)

The initial reaction catalyzed by *ATP sulfurylase* is analogous to the activation of the carboxyl group of fatty acids or amino acids in that the adenylic acid moiety of ATP is transferred to the sulfate and inorganic pyrophosphate is formed:

$$SO_4^{2-} + \underset{\text{ATP}}{A{-}R{-}P{-}P{-}P} \rightleftharpoons \underset{\text{APS}}{A{-}R{-}P{-}S} + P{-}P \tag{17-1}$$

Although the K_{eq} greatly favors the synthesis of ATP, the reaction is pulled to the right by the subsequent hydrolysis of pyrophosphate by the ubiquitous pyrophosphatases.

The second step in the activation of the sulfate is catalyzed by the *APS kinase,* which has a high affinity for APS and tends to pull reaction 17-1 to the right:

$$\text{APS} + \text{ATP} \rightleftharpoons \text{PAPS} + \text{ADP} \tag{17-2}$$

3′-Phosphoadenosine-5′-phosphosulfate is an energy-rich compound. The $\Delta G'$ of hydrolysis of the sulfate–phosphate anhydride is $-11{,}000$ kcal/mole. An important role for this compound is found in its ability to serve as a "sulfate donor" in the sulfurylation of compounds such as carbohydrates, phenols, and sterols. The sulfate esters of these compounds have increased solubility in H_2O, and numerous examples of their natural occurrence may be cited.

3′-Phosphoadenosine-5′-phosphosulfate is also the substrate for the reduction of +6 sulfur to +4 sulfur as sulfite. The enzyme complex that catalyzes this 2 electron reduction is "PAPS reductase," and the process may be separated into three partial reactions. Initially the sulfate of PAPS is transferred to a heat-stable dithiol protein having a low molecular weight:

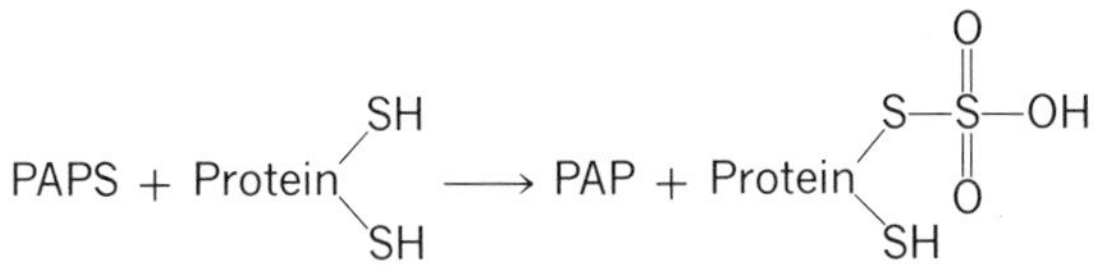

$$\text{PAPS} + \text{Protein}\begin{matrix}\diagup SH\\ \diagdown SH\end{matrix} \longrightarrow \text{PAP} + \text{Protein}\begin{matrix}\diagup S{-}S(=O)_2{-}OH\\ \diagdown SH\end{matrix}$$

The protein–thiosulfate that is formed then undergoes an internal oxidation–reduction reaction that liberates bisulfite in which the oxidation state of sulfur is +4:

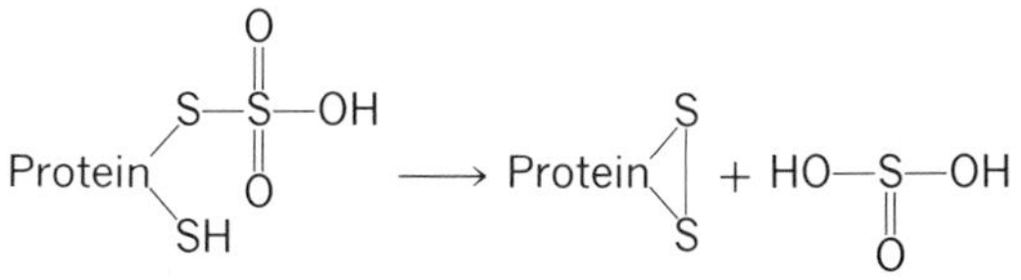

$$\text{Protein}\begin{matrix}\diagup S{-}S(=O)_2{-}OH\\ \diagdown SH\end{matrix} \longrightarrow \text{Protein}\begin{matrix}\diagup S\\ \diagdown S\end{matrix}\!\Big| + HO{-}S(=O){-}OH$$

The disulfide form of the protein is then reduced to the dithiol form by NADPH, and now another molecule of PAPS can react and be reduced:

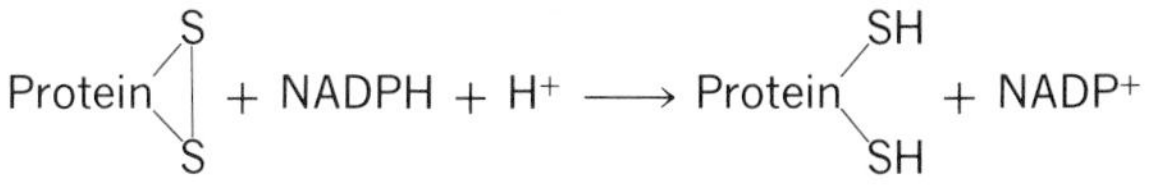

$$\text{Protein}\begin{matrix}\diagup S\\ \diagdown S\end{matrix}\!\Big| + \text{NADPH} + H^+ \longrightarrow \text{Protein}\begin{matrix}\diagup SH\\ \diagdown SH\end{matrix} + \text{NADP}^+$$

The sulfite (or bisulfite) produced above can then be further reduced to H_2S by an enzyme complex known as *sulfite reductase*. A total of 6 electrons required for this reduction are furnished by 3 moles of NADPH. The details of this overall process remain to be clarified, however.

$$H_2SO_3 + 3\ \text{NADPH} + 3\ H^+ \longrightarrow H_2S + 3\ \text{NADP}^+ + 3\ H_2O$$

The final step in the assimilation of sulfur involves the reaction of H_2S with O-acetyl serine (activated serine) to form cysteine. This enzyme, *cysteine synthase*, is described on page 376.

An analogous reaction involving homoserine (as O-succinyl homoserine) and H_2S to form homocysteine is catalyzed by a separate enzyme found in higher plants and some bacteria.

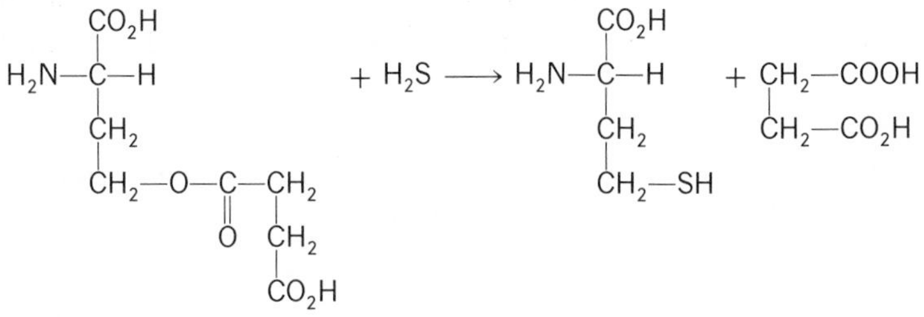

$$H_2N{-}CH(CO_2H){-}CH_2{-}CH_2{-}O{-}C(=O){-}CH_2{-}CH_2{-}CO_2H + H_2S \longrightarrow H_2N{-}CH(CO_2H){-}CH_2{-}CH_2{-}SH + CH_2{-}COOH\ |\ CH_2{-}CO_2H$$

O-Succinyl homoserine — Homocysteine — Succinic acid

Cysteine as a Primary Sulfur Source. Cysteine serves as the primary source of sulfur for the formation of methionine in plants and microorganisms. Intermediates in this process described in Chapter 16 are cystathionine and homocysteine, and the two enzymes, *cystathionine synthase I* and *cystathionase* are required. The synthesis of cysteine from methionine in animals has also been discussed (Chapter 16).

Sulfate Respiration. In strict analogy with nitrate respiration, sulfate can serve as a terminal oxidant for certain anaerobic bacteria (e.g., *Desulfovibrio desulfuricans*). When it does, the sulfur atom is reduced stepwise from SO_4^{2-} to S^{2-} by electrons that have their origin in the organic substrates oxidized by the bacteria. Presumably, an electron-transport chain similar to that in aerobic cells is involved and ATP is formed as electrons pass along the chain.

The Release of Sulfur from Organic Compounds. Animals, plants and many microorganisms contain the enzyme *cysteine desulfurylase* that catalyzes the following reaction:

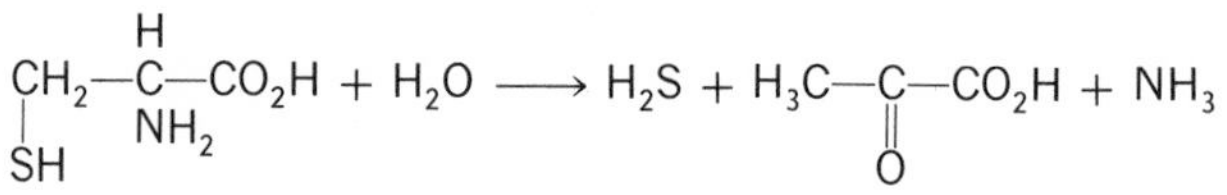

$$\underset{\text{Cysteine}}{HS{-}CH_2{-}CH(NH_2){-}CO_2H} + H_2O \longrightarrow H_2S + \underset{\text{Pyruvic acid}}{H_3C{-}C({=}O){-}CO_2H} + NH_3$$

The H_2S so produced can then be oxidized by a *sulfide oxidase* complex found in soil organisms:

$$2\,H_2S + 2\,O_2 \longrightarrow \underset{\text{Thiosulfate}}{S_2O_3^{2-}} + H_2O + 2\,H^+$$

The sulfur atom can also be oxidized while remaining in organic combination; the enzyme responsible is *cysteine oxidase* and cysteine sulfinic acid is the product:

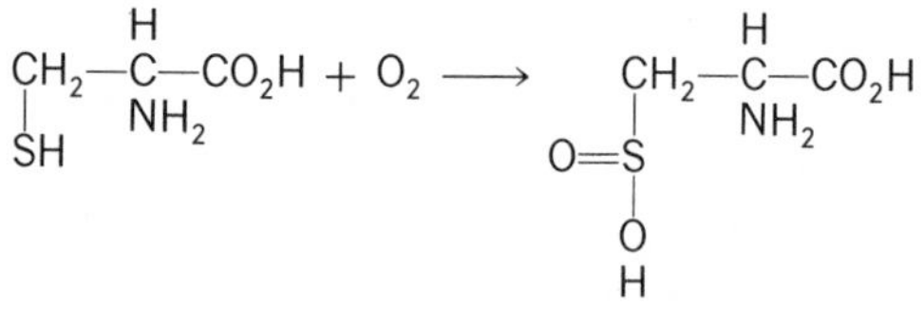

$$\underset{\text{Cysteine}}{HS{-}CH_2{-}CH(NH_2){-}CO_2H} + O_2 \longrightarrow \underset{\text{Cysteine sulfinic acid}}{HO{-}S({=}O){-}CH_2{-}CH(NH_2){-}CO_2H}$$

This reaction, which has no common analogy in the metabolism of amino acids, can be followed by transamination to yield pyruvic acid, SO_2, and the product amino acid:

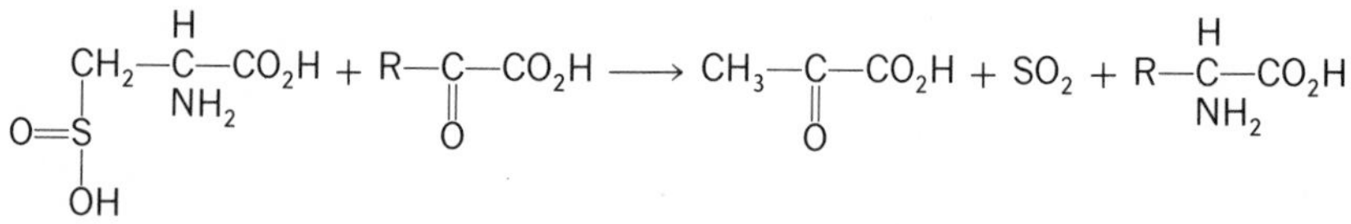

$$\underset{\text{Cysteine sulfinic acid}}{HO{-}S({=}O){-}CH_2{-}CH(NH_2){-}CO_2H} + \underset{\text{Keto acid}}{R{-}C({=}O){-}CO_2H} \longrightarrow \underset{\text{Pyruvic acid}}{CH_3{-}C({=}O){-}CO_2H} + SO_2 + \underset{\text{Amino acid}}{R{-}CH(NH_2){-}CO_2H}$$

The SO_2 produced is readily oxidized by air to sulfate thereby completing the sulfur cycle in nature.

Since sulfur is included among the six most abundant elements found in living forms, it is not surprising that those forms have evolved mechanisms for conserving this element within the biosphere.

References

1. John Postgate, ed. *Chemistry and Biochemistry of Nitrogen Fixation*. New York: Plenum Press, 1971.
   An excellent compilation of articles on all aspects of the subject by specialists.
2. A. B. Roy and P. A. Trudinger. *The Biochemistry of Inorganic Compounds of Sulphur*. Cambridge: Cambridge University Press, 1970.
   A recent treatment of the subject.

PART III

Metabolism of Informational Molecules

Biosynthesis of Nucleic Acids

18

Introduction

In Chapters 18–20 we shall consider the various steps which are involved in the synthesis of informational biopolymers, the mechanism the cell employs for the flow of information from DNA, the primary carrier, to the proteins, the ultimate products of that information, and the means by which the cell can regulate its metabolism.

These ideas in part are depicted as

$$\text{Replication}\circlearrowleft \text{DNA} \xrightarrow{\text{Transcription}} \text{RNA}\begin{bmatrix}\text{mRNA}\\ \text{tRNA}\\ \text{rRNA}\end{bmatrix} \xrightarrow{\text{Translation}} \text{Proteins} \longrightarrow \text{Metabolic reactions and their regulation}$$

$$\text{DNA} \xleftarrow{\text{Reverse transcription}} \text{RNA (Viruses)}$$

Biosynthesis of DNA

In 1956, Arthur Kornberg described the *de novo* synthesis of DNA by a DNA polymerase from *E. coli*. The four deoxyribonucleoside triphosphates, dATP, dGTP, dCTP, and dTTP, Mg^{2+} and a DNA template are the components of the reaction system. Thus:

$$\left.\begin{matrix} n\,d\text{ATP}\\ n\,d\text{GTP}\\ n\,d\text{CTP}\\ n\,d\text{TTP}\end{matrix}\right\} + \text{DNA} \longrightarrow \text{DNA}-\left\{\begin{matrix} d\text{AMP}\\ d\text{GMP}\\ d\text{CMP}\\ d\text{TMP}\end{matrix}\right. + 4n\,pp$$

The newly synthesized DNA has a base composition of (A + T)/(G + C), which corresponds very closely to that of its template. The polymerase takes direction from the template and faithfully reproduces the base pattern of the primer. In addition, when a synthetic DNA polymer, consisting of only A and T bases (poly-dAT), is employed as the primer, the product contains only deoxyadenylic and deoxythymidylic acids, even when the four deoxyribonucleoside triphosphates are present. Thus, the polymerase employing poly-

*d*AT as a template rejects deoxyguanylic and deoxycytidylic triphosphates and uses only the deoxyadenylic and deoxythymidylic triphosphates as substrates. The DNA obtained from a wide variety of sources, including animal, plant, bacterial, and viral, will serve as an effective primer, particularly if it is denatured before use.

The newly formed polymer has the double helix structure conforming to the Watson–Crick model, and its deoxyribonucleotide chains are joined by 3′,5′-diester bonds. The daughter chain has opposite polarity to the parent strand (see Chapter 5 for structure of DNA).

The *E. coli* polymerase has been extensively investigated. It has a molecular weight of 109,000 and is monomeric. The protein contains one binding site for all four deoxyribonucleoside triphosphates and one binding site for the template DNA. Single-stranded circular DNA binds a number of polymerase molecules, but closed, circular double-stranded DNA does not bind to the enzyme unless it is first denatured, i.e., a region of the double-stranded template opens by a separation of base pairing. Although double-stranded DNA is a poor primer, the polymerase activity is greatly increased by introducing breaks into the DNA strand by limited nuclease action. Curiously, the pure enzyme has nuclease activity, namely exonuclease activity, which removes mononucleotides sequentially either in the 3′ ⟶ 5′ or in the 5′ ⟶ 3′ direction. The activity in the 3′ ⟶ 5′ direction requires a 3′-hydroxyl terminus and will not act on a phosphodiester bridge adjacent to a 3′-phosphorus terminus. The product is exclusively 5′-mononucleotides. The second nuclease activity involves a 5′ ⟶ 3′ hydrolysis and will function whether the 5′ terminus is 5′-phosphorus or 5′-hydroxyl. The product may be 5′-mononucleotides, di- or oligonucleotides, as well as dimers of the thymine type.

(*a*) 3′ ⟶ 5′ nuclease activity

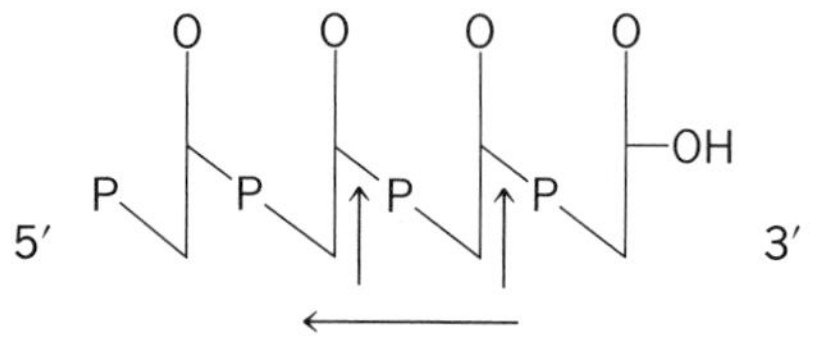

(*b*) 5′ ⟶ 3′ nuclease activity

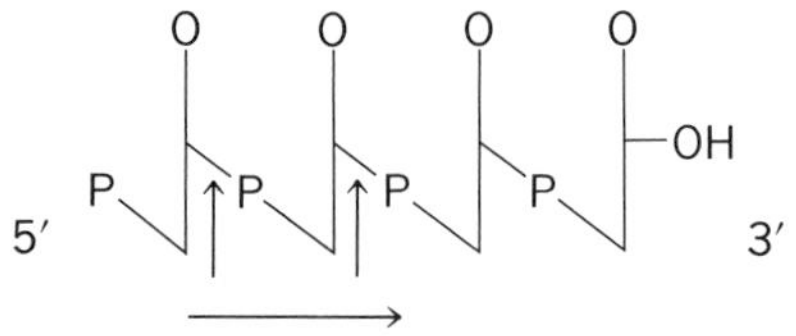

A reasonable summary of DNA polymerase functions can now be written, keeping in mind the three activities, namely (*a*) the polymerase activity involving the joining of nucleoside triphosphate to the exposed 3′-hydroxyl terminus in a 5′ ⟶ 3′ direction with the primer strand in duplex structure with a template strand, (*b*) the 3′ ⟶ 5′ nuclease activity, and (*c*) the 5′ ⟶ 3′ nuclease activity.

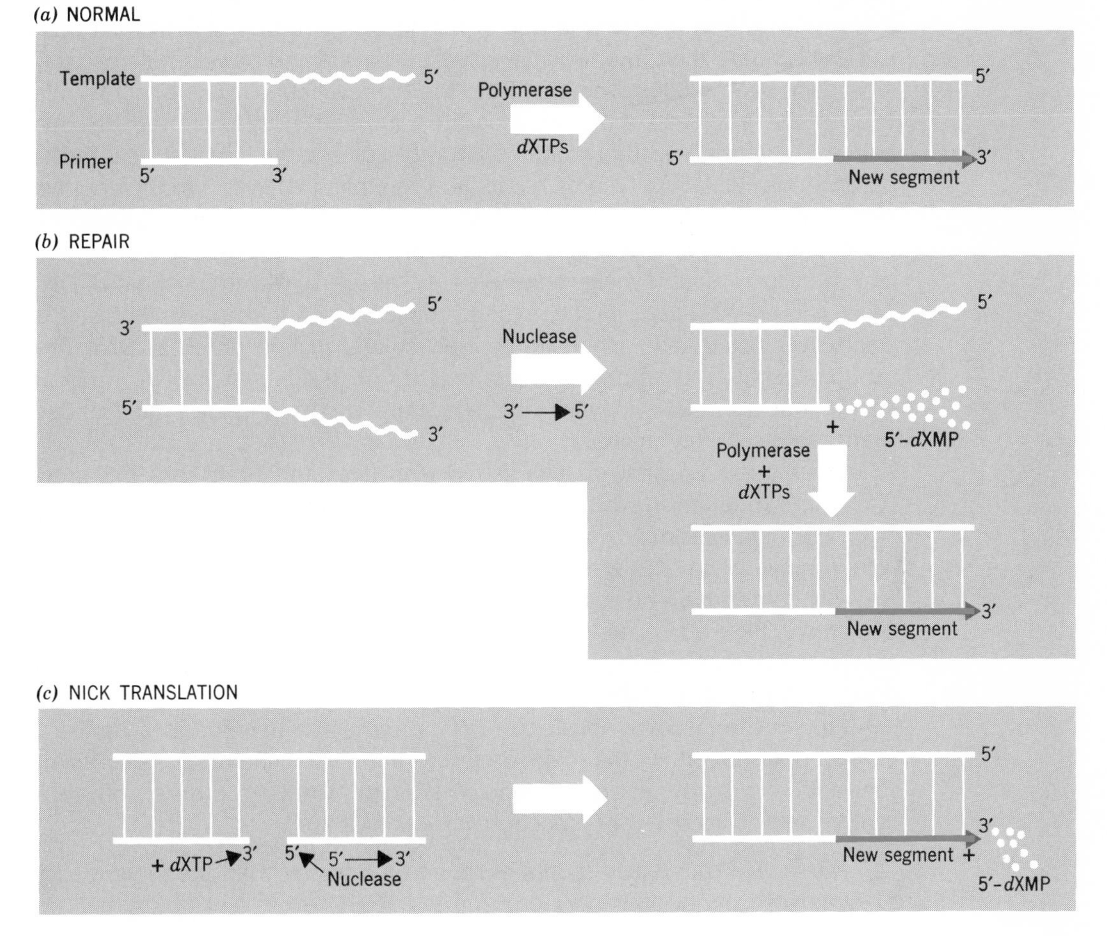

Figure 18-1

The trifunctional activity of *E. coli* DNA polymerase.

The first activity is depicted in Figure 18.1*a*. The second activity, involving the 3′ ⟶ 5′ nuclease activity as well as the polymerase activity, occurs in a repair mechanism in which an unpaired 3′ segment is digested back to the first paired base by the 3′ ⟶ 5′ nuclease activity of the polymerase, with a subsequent resynthesis by the polymerase as in Figures 18.1*a* and *b*. The third activity, called *nick translation,* involves a very interesting synthesis and degradation at equivalent rates with a nicked DNA (Figure 18.1*c*) involving both the polymerase and the 5′ ⟶ 3′ nuclease and resulting in no net synthesis but a repair of the defective region.

Replication

We have already noted that DNA directs the sequence of bases when RNA as well as DNA polymerases catalyze the condensation of nucleotides to form RNA and DNA. Because the bonding rules of the Watson–Crick hypothesis require the base adenine to hydrogen bond with thymine, and cytosine with guanine, the sequence of bases in the first strand of DNA automatically determines the sequence of bases in the second strand in a DNA molecule.

In replication, then, the two strands will separate and new complementary strands of DNA will be assembled from the four available deoxyribonucleotide triphosphates on each of the two separate parent strands. Assuming the base pairing to be precise, the two new DNA molecules should be identical to the parent molecule. This type of replication has been called *semiconservative* and is illustrated in Figures 18-2 and 18-3. Another possibility is that the final duplication product consists of a double helix of the original two strands and a second double helix consisting of newly synthesized chains. This process is called a *conservative* type of replication. A third possibility, called *dispersive*, could take place if the nucleotides of the parental DNA are randomly scattered among the components of the daughter DNA material so that the new DNA consists of a mixture of old and new nucleotides scattered along the chains.

To test these possible mechanisms, Meselson and Stahl in 1958 grew *E. coli* in a medium in which the sole source of nitrogen was $^{15}NH_4Cl$. After several generations of growth, $^{14}NH_4Cl$ was added, and at short intervals cells were removed; the DNA was carefully extracted and analyzed for relative ^{14}N and ^{15}N content by equilibrium density gradient centrifugation. The results depicted in Figure 18-3 definitely eliminated the dispersive mechanism. With other evidence, these results gave strong support to the semiconservative mechanism of replication.

Although the discovery of *E. coli* DNA polymerase ushered in a new era of experimentation in the field of informational biochemistry, DNA synthesis *in vitro* by the *E. coli* DNA polymerase of Kornberg differs from replication *in vivo* in several respects, two of which are:

(1) The *in vivo* rate is two orders of magnitude higher than the *in vitro* rate catalyzed by the Kornberg enzyme.

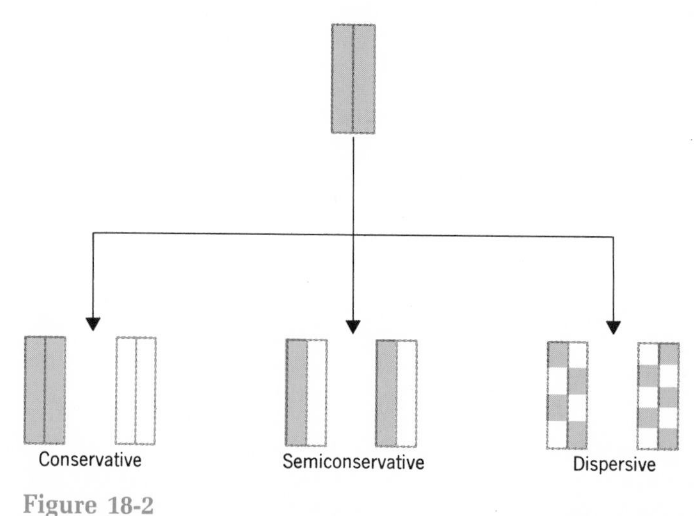

Figure 18-2

Types of replication.

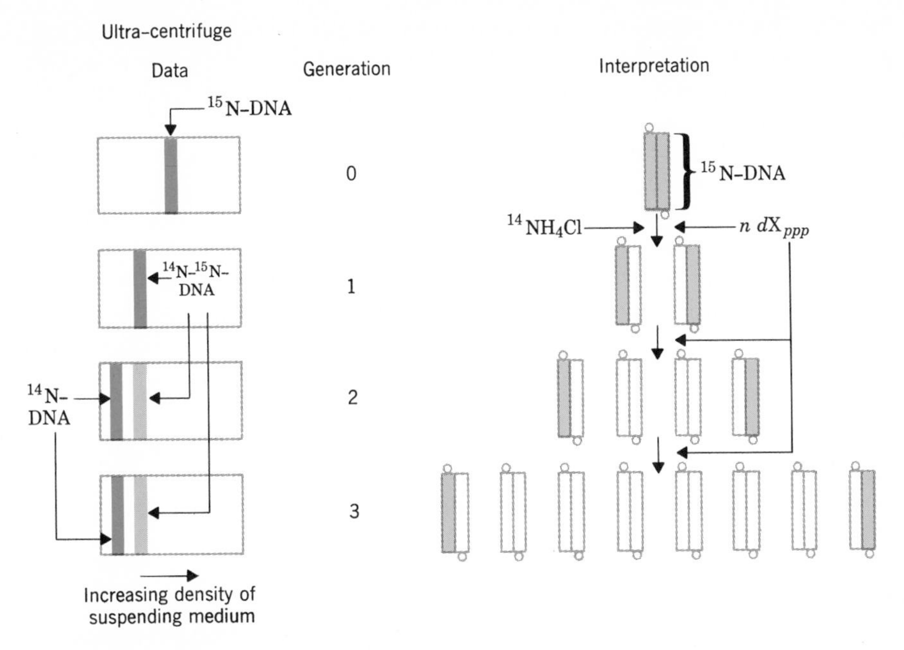

Figure 18-3

Meselson–Stahl experiment to demonstrate semiconservative replication of DNA. Small circles at each end of diagrammatic DNA indicate antiparallel nature of DNA (see page 120 for antiparallel rule).

(2) No net semiconservative synthesis is achieved *in vitro* with double-stranded DNA as template by the Kornberg polymerase.

Recently, however, a membrane-associated DNA polymerase has been described in an *E. coli* bacterial mutant lacking Kornberg's DNA polymerase but growing like a wild type of strain and thus containing a normal replication system. In contrast to the membrane-bound system, Kornberg's polymerase is soluble and thus present in the cytoplasm of the cell. The membrane-associated polymerase has been solubilized by detergent treatment and partially purified to a molecular weight of 60,000–90,000. Anti-sera against Kornberg's polymerase does not inhibit the new polymerase, but does greatly reduce the activity of the Kornberg enzyme. The new polymerase, like all DNA polymerases, adds to the 3′-hydroxyl end of a polynucleotide strand by a semiconservative mechanism and requires a complementary template strand for its function. Although its function is not known, ATP is required. Moreover, newly synthesized DNA can be shown to be biologically active.

A number of DNA-dependent DNA polymerases have been isolated from a variety of eucaryotic organisms, and their properties have been described. Since their properties show a mixed pattern of specificities and requirements, we will not describe these further. The DNA polymerase has also been isolated from rat liver and yeast mitochondria, and in both species the catalytic and physical parameters are distinctly different from their counterparts in the nucleus. Thus, in eucaryotic cells a complex pattern of polymerases emerges with most of the polymerase activity in the nucleoplasm of the nucleus, but superimposed on this activity are the polymerases in the mitochondria and chloroplasts.

Very recently, Temin has reported on the discovery of a new, unique RNA-directed DNA polymerase (also called reverse transcriptase) in oncogenic RNA viruses. The DNA synthesized is complementary to the viral RNA and DNA–RNA hybrids are early intermediates of the reaction. The startling conclusion is that viral RNA can serve as a template in the synthesis of DNA. All oncogenic viruses so far examined contain DNA polymerase activities directed effectively by a single-stranded RNA. Recently, a few nononcogenic RNA viruses have also been found to have reverse transcriptase activity. The reaction then proceeds with the synthesis of double-stranded DNA. The important aspect of the problem is that the newly synthesized DNA contains base sequences complementary to the viral RNA, and this observation suggests a possible relation to cancer and its subsequent biochemical understanding.

Mutation

A process of great importance, *mutation,* is defined as an abrupt, stable change of a gene which is expressed in some unusual phenotypic character, frequently as a biochemical modification. In a mutation there may be a loss of the capacity to carry out some specific biochemical function.

In general there are three classes of mutation (shown below) which re-

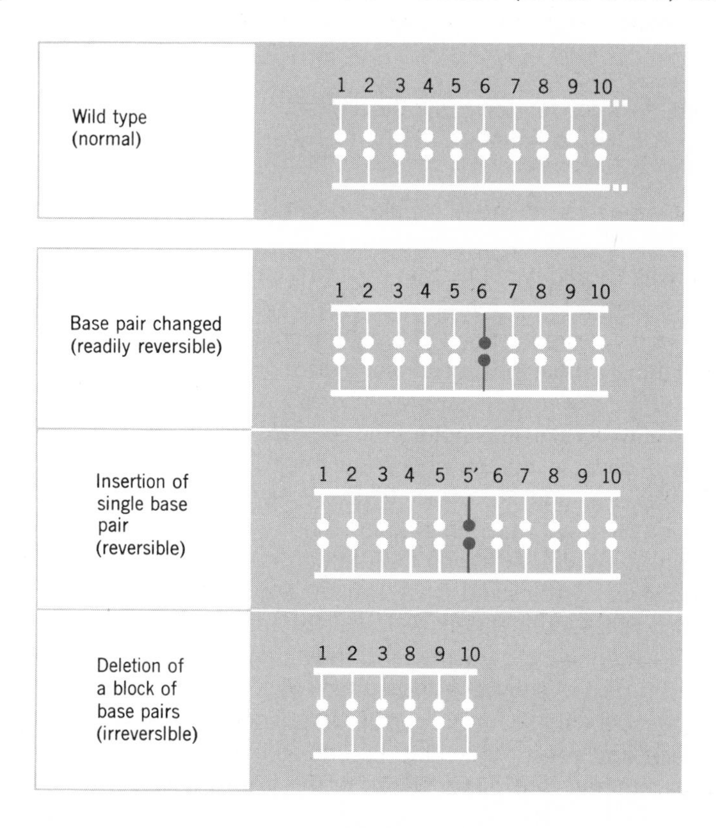

sult by an introduction of defects or changes in the sequence of the bases, A, T, G, C in the DNA molecule.

Physical and Chemical Mutagenesis. In nature, mutations may occur by accident either by physical or chemical changes of bases in DNA or by shifting the codon reading frame by the deletion, addition, or modification of DNA base. Examples of chemical mutagenesis include the following:

(1) HNO_2 as a deaminating reagent:

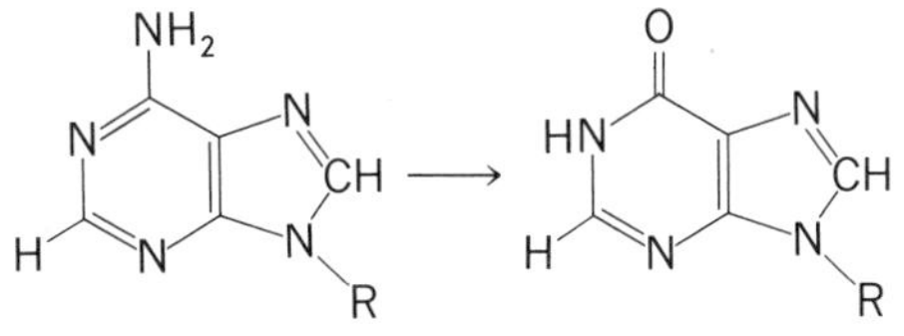

Adenine ⟶ Hypoxanthine

Guanine ⟶ Xanthine
Cytosine ⟶ Uracil

Conversion of adenine to hypoxanthine will result in incorrect pairing with cytosine; change of cytosine to uracil leads to adenine pairing, while guanine to xanthine results in cytosine pairing which is normal.

(2) Hydroxylamine is a very powerful mutagen but only with isolated systems, since the normal components of a cell would readily scavenger the reagent. The reagent reacts specifically with cytosine:

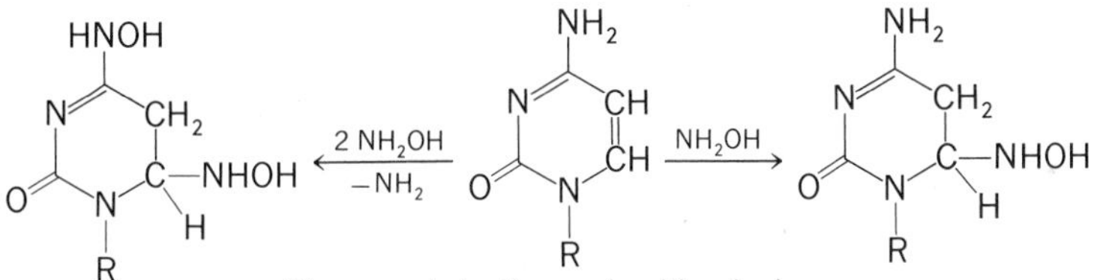

$\xleftarrow[-NH_2]{2\ NH_2OH}$ $\xrightarrow{NH_2OH}$

The new derivatives pair with adenine

(3) Alkylating reagents—dimethyl sulfate and ethyl methane sulfonate—are specific for guanine:

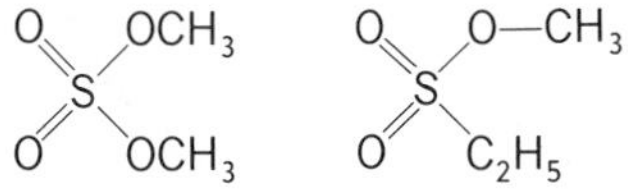

DMS EMS

The reaction which follows methylation leads to the formation of a quaternary nitrogen which destabilizes the deoxyriboside link and releases the deoxyriboside. The loss of the base can lead to replacement by any of four bases or even rupture of the DNA chain:

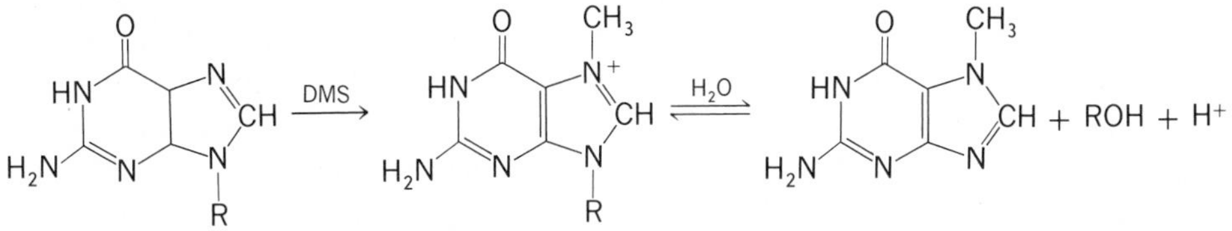

Guanine $\xrightarrow{DMS}$ $\underset{}{\overset{H_2O}{\rightleftharpoons}}$ + ROH + H^+

These compounds as well as another alkylating mutagen, β-chloroethylamine, a nitrogen mustard, are very toxic.

(4) Methylating reagents—extremely mutagenic compounds and thus very dangerous to use unless very careful precautions are taken—include *N*-methyl-*N′*-nitro-*N*-nitrosoguanidine:

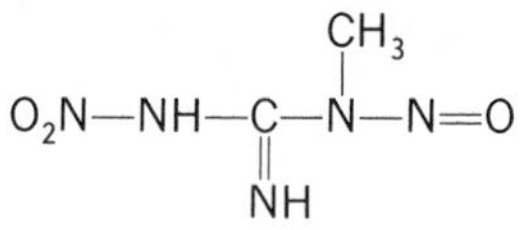

This reagent probably converts to diazomethane,

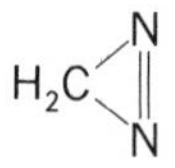

an extremely effective methylating reagent. This compound, used commonly for the methylation of carboxylic acid and amino groups, must be treated with great respect! The reagent methylates nucleic acids.

(5) Mutagenic action by x-ray irradiation and ultraviolet-irradiation is very effective in inducing mutagenesis. x Rays probably react with DNA by a free radical mechanism to cause single-stranded chain breaks in the DNA chain. Ultraviolet irradiation with a strong absorption at 260 nm leads frequently to a photochemical dimerization of two adjacent thymines, thymine–cytosine, or two cytosines to dimers. Thymine residues are particularly susceptible to the following reaction:

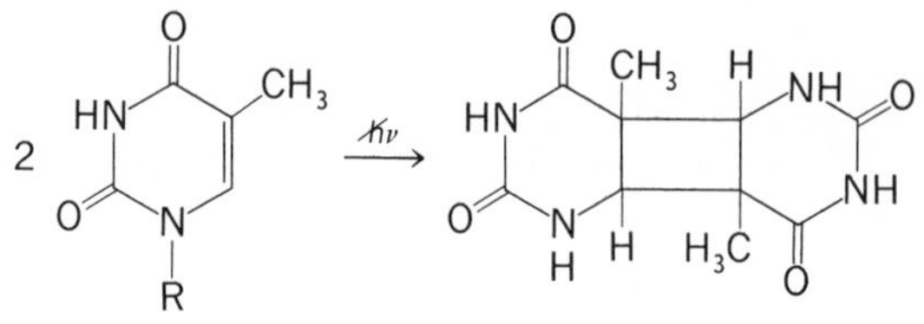

Repair Mechanisms. The important observation that *E. coli* mutants which are defective in polymerase activity can be isolated has pointed to a search not only for the correct DNA replicase but also for the true role of Kornberg's DNA polymerase. Since these defective mutants have enhanced sensitivity to ultraviolet irradiation and are very sensitive to x irradiation, the conclusion is now being drawn that the soluble polymerase is indeed an intimate component of the DNA repair mechanism of a cell.

When a cell is exposed to x irradiation, single-chain breaks occur in the DNA strands; in the intact cell these breaks are very rapidly repaired in the interval of a few minutes. The remainder of the breaks are repaired slowly (hours), suggesting two general types of repair, one rapid and the second slow. There is now good evidence that Kornberg's DNA polymerase functions as a repair enzyme in about 85% of the chain breaks, whereas the second mechanism involves a recombinational repair system which we shall not discuss here.

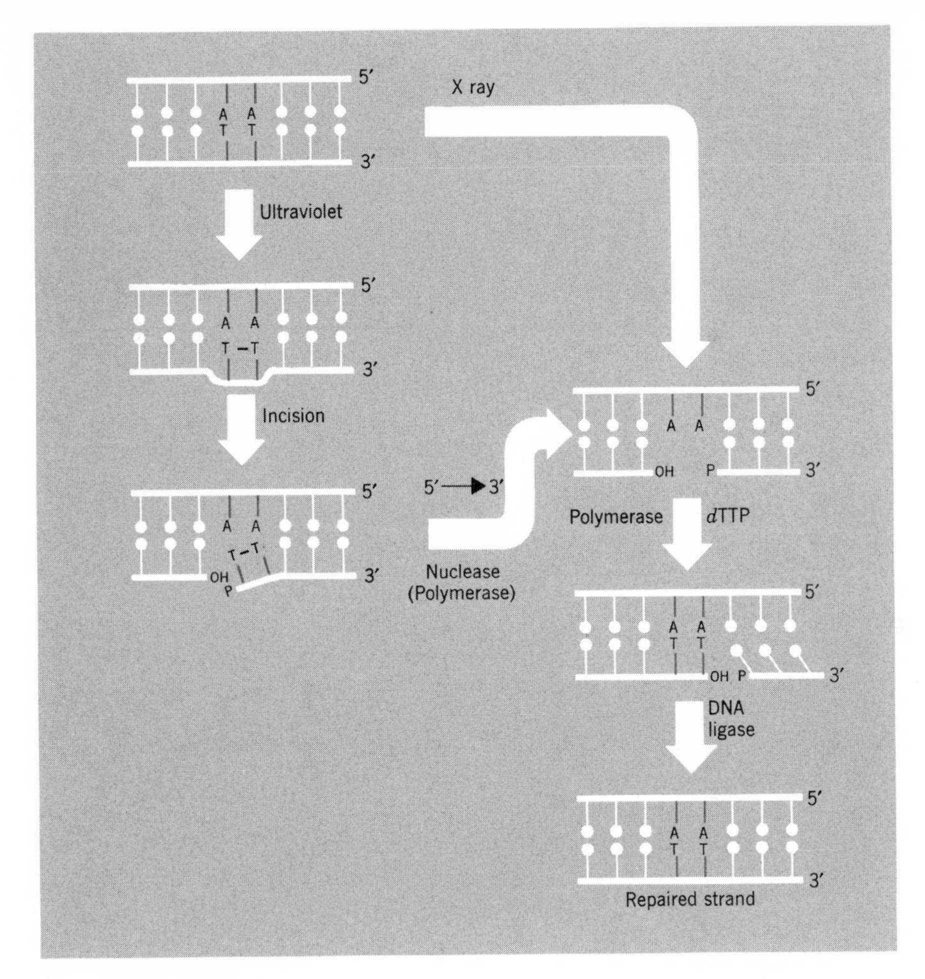

Figure 18-4
Repair mechanisms.

The DNA ligase. If the chain break occurs, the damaged section must be removed by excission and the gap must be restored by the addition of the suitable bases by action of the Kornberg enzyme using the opposite strand of DNA as the template (see Figure 18-4). Finally, the single-strand interruption is closed enzymatically by a polynucleotide ligase. A ligase-deficient mutant has been isolated and has been shown to be sensitive to ultraviolet and x irradiation. The mechanism of joining a broken strand together is of interest. In *E. coli* the enzyme requires NAD^+, the end products being 5′-adenylate and nicotinamide mononucleotide, and in the animal cell ATP is required with the formation of 5′-adenylate and pyrophosphate. The 5′-adenylate is at first covalently linked to the enzyme and the reaction sequence is believed to be that shown in Scheme 18-1.

The tremendous amount of research activity in the field of information biochemistry is of course directly related to the enormous importance of a

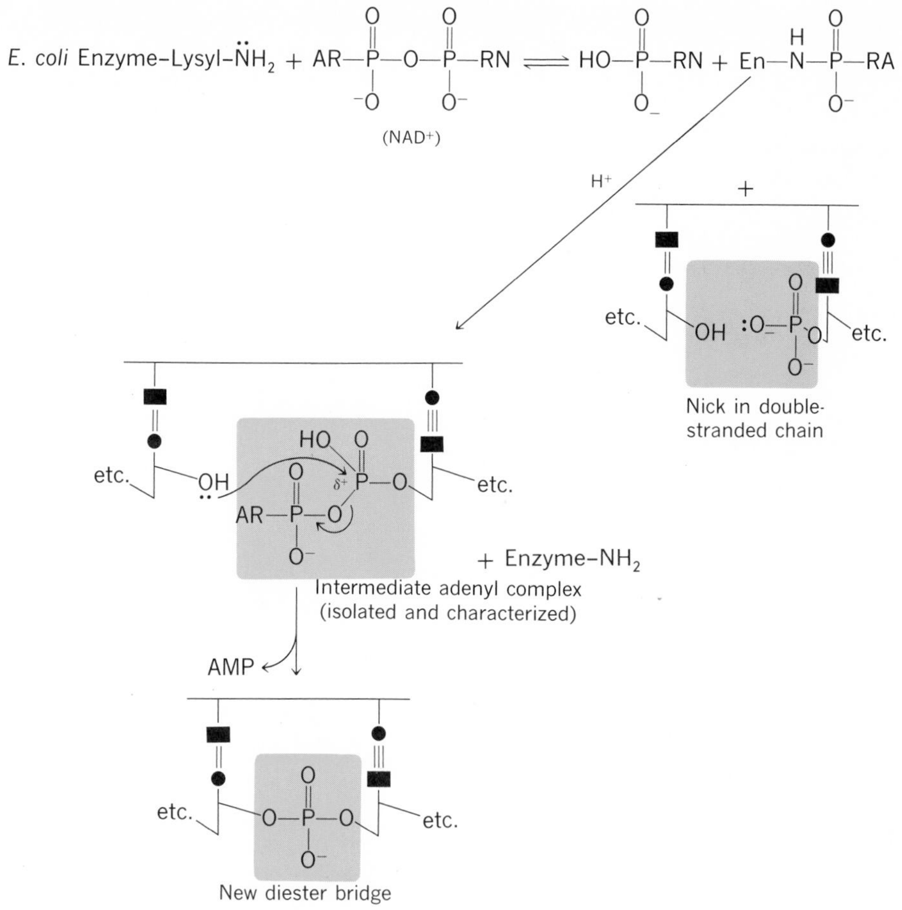

$$E.\ coli\ \text{Enzyme–Lysyl–}\ddot{N}H_2 + AR\text{—}\overset{O}{\overset{\|}{P}}(O^-)\text{—O—}\overset{O}{\overset{\|}{P}}(O^-)\text{—RN} \rightleftharpoons HO\text{—}\overset{O}{\overset{\|}{P}}(O^-)\text{—RN} + En\text{—}\overset{H}{N}\text{—}\overset{O}{\overset{\|}{P}}(O^-)\text{—RA}$$

(NAD+)

H+

+

etc. OH :O—P(=O)(O⁻)—O etc.

Nick in double-stranded chain

etc. OH HO O P—O etc. AR—P(=O)(O⁻)—O

+ Enzyme–NH_2

Intermediate adenyl complex (isolated and characterized)

AMP

etc. O—P(=O)(O⁻)—O etc.

New diester bridge

Scheme 18-1

full understanding of the process of DNA replication in normal and abnormal cells. Once such an understanding is achieved, the possibilities of genetic repair in the mammalian cell will become apparent and the implications will be profound.

Biosynthesis of RNA—Transcription

With the exception of the biosynthesis of RNA's in such organelles as mitochondria and chloroplasts, in eucaryotic cells the site of DNA-dependent RNA biosynthesis (transcription) is the nucleus. While the nucleolus appears to contain the enzymes and genes for ribosomal RNA biosynthesis, the enzymes responsible for the synthesis of messenger and transfer RNA's are localized in the nucleoplasm (see Chapter 6 for further details). In procaryotic organisms RNA polymerase occurs in the cytoplasm.

In a RNA polymerase-catalyzed reaction several sequential processes are involved: (*a*) association of the DNA template with the polymerase, (*b*) initiation of the chain, (*c*) chain elongation, and (*d*) termination by liberation of the newly synthesized RNA and enzyme from its DNA template.

Since the *E. coli* RNA polymerase has been most extensively examined, we shall describe its structure and properties in some detail. The RNA polymerase from *E. coli* as well as other organisms catalyzes the net synthesis of RNA with base sequences complementary to the strand of DNA that serves as template, as indicated in Scheme 18-2.

In vivo only one strand of DNA is copied. This must be so since if each of the two DNA strands served as a RNA template, two RNA products of complementary sequences would be transcribed which would code for two different proteins! Since genetic evidence demonstrates that one gene codes for only one protein, only one DNA strand functions as a template, while the other strand does not function as a template. The term *asymmetric*

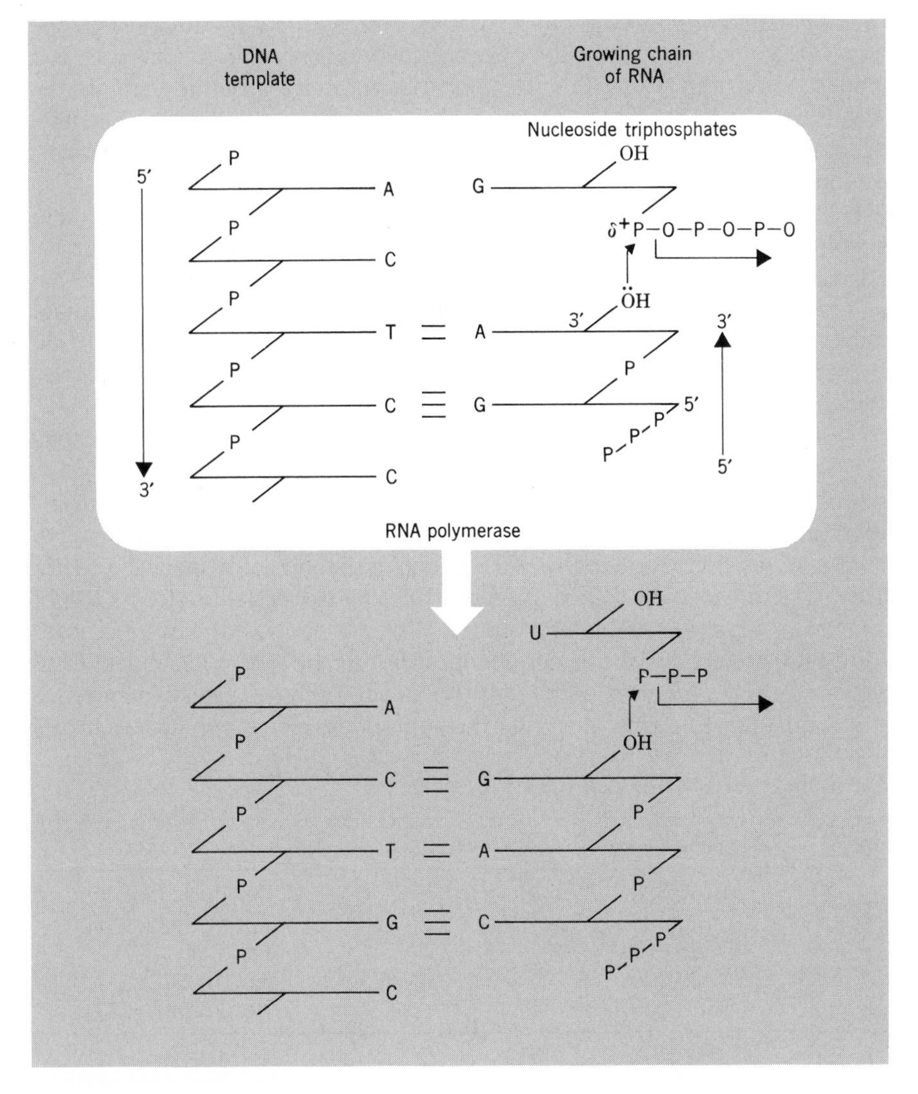

Scheme 18-2

transcription is employed for the copying of one strand of DNA and *symmetrical transcription* for the copying of both template strands.

In the synthesis of RNA the polymerase joins incoming nucleoside triphosphate to the 3′-hydroxyl of the terminal nucleoside of the growing chain. The chains are said to grow in the 5′ to the 3′ direction as indicated above.

The synthesis of all types of RNA in procaryotic organisms is probably mediated by a single type of polymerase of considerable complexity. Extensive purification of the *E. coli* polymerase has revealed that the enzyme consists of five separable subunits in the proportions shown in the table. Although the functions of all the subunits are not completely understood, it is clear that the β' subunit is required for binding of the RNA polymerase to the DNA template, while the β subunit is the site of interaction of rifamyicin, an antibiotic that inhibits initiation of RNA synthesis *in vivo* and *in vitro*. The sigma factor (σ) is required for the correct initiation of RNA synthesis at a specific site of the DNA template. The RNA polymerase binding site on the DNA is known as the promotor site. There may be a number of specific σ proteins which alter the affinity of the polymerase for different promotor sites. Without the σ protein, the enzyme ($\alpha_2\beta'\beta\omega$) is called the *core* polymerase; with the σ factor it is called the holoenzyme ($\alpha_2\beta'\beta\omega\sigma$). The function of ω is not known, since enzymes lacking this factor appear to have normal activity. The core enzyme will transcribe a DNA template symmetrically; that is, both strands of DNA can serve as templates. However, the transcription reaction is slow and nonspecific. With the σ factor, the holoenzyme transcribes DNA asymmetrically, initiating RNA chains at specific promotor sites. This process is believed to duplicate *in vivo* reactions.

Figure 18-5 outlines the mechanism of transcription involving the holoenzyme with its initiation σ factor and a termination factor, ρ, which is required only in the terminal stage of RNA synthesis and is not considered a formal component of the complete RNA polymerase. In brief, the mechanism requires that the holoenzyme with the assistance of a specific σ factor selects the correct binding site (promotor site) for transcription, forms a tight complex in which the double strands of DNA are opened or separated over a short region, and synthesis commences with an input of nucleotides. After synthesis has begun, the σ factor dissociates and becomes available for reuse with another core polymerase. After the appropriate elongation has occurred,

Components of *E. coli* RNA Polymerase

Subunit	Molecular weight	Number	Function
β'	165,000	1	Polynucleotide binding
β	155,000	1	Polynucleotide binding
σ	95,000	1	Initiation
α	39,000	2	Not known
ω	9,000	1	Not known
ρ	200,000		Termination factor

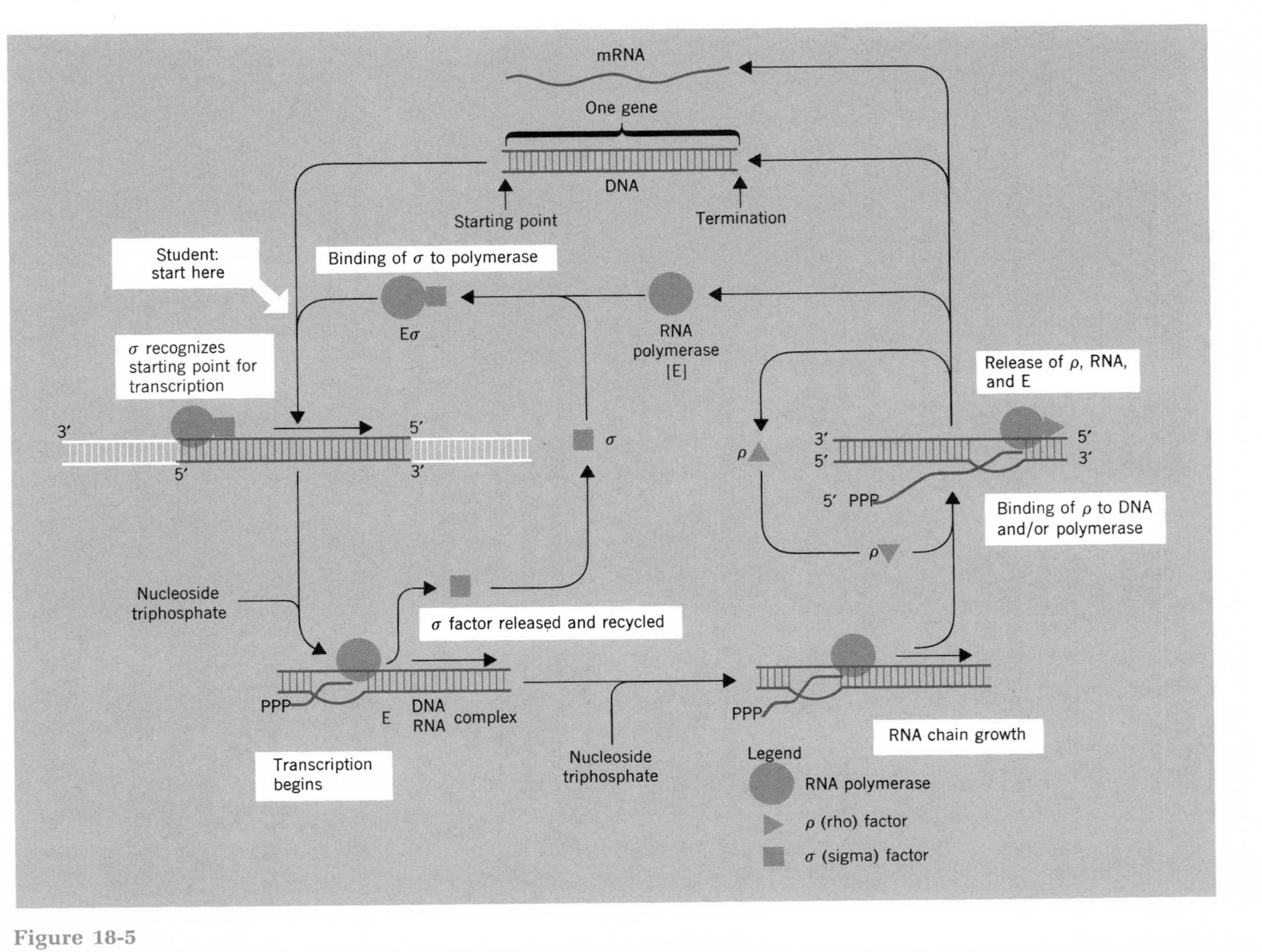

Figure 18-5

Synthesis of RNA by *E. coli* RNA polymerase. (Reproduced with permission of James D. Watson, Harvard University.)

an additional protein, the ρ factor, associates with the polymerase or the termination sequence on the DNA to terminate the transcription with the release of the newly synthesized RNA.

Although up to 40% of the RNA synthesized by rapidly growing *E. coli* cells is ribosomal RNA, under *in vitro* conditions with *E. coli* DNA as template, less than 0.2% of the RNA synthesized by the holoenzyme is rRNA. Recently, a new transcription factor, psi (ψ), a protein obtained from the cytosol of *E. coli* when added to the holoenzyme, results in a several hundredfold stimulation of rRNA synthesis, i.e., 30 or 40% of the total RNA is now rRNA.

The newly synthesized single-stranded RNA may now undergo a number of modifications. In the procaryotic cell, the modifier enzymes presumably occur in the cytoplasm and reduce the length of RNA's for final conversion to ribosomal, messenger, and transfer RNA. Methylation of bases of tRNA and mRNA's by specific enzymes involving *S*-adenosyl methionine further modifies nucleic acid strands in preparation for their final structures. In addition, sulfur-containing bases, while minor components of bacterial and mammalian tRNA's, are formed by special thiolase enzymes. The sulfur donor is usually cysteine.

In eucaryotic cells, nuclear transcription is a process of great complexity. The mass of nuclear DNA is extremely large and contains a great complexity of nucleotide sequences. Unlike the bacterial systems with a single species of DNA-dependent RNA polymerase, eucaryotic cells contain at least three different nuclear RNA polymerases. The RNA polymerase I is associated with the nucleolus, requires either Mn^{2+} or Mg^{2+}, has a molecular weight of 500,000–700,000, and appears to be oligomeric. It is presumably responsible for the synthesis of ribosomal RNA's. The RNA polymerases II and III are in the nucleoplasm. Enzyme II, which has been extensively purified, requires Mn^{2+}, has a molecular weight of about 700,000, and unlike enzyme I is highly sensitive to the bicyclic peptide toxin from a toadstool, α-amanitin. It appears to be also oligomeric. Enzyme III has not been studied in detail. Polymerase IV is localized in the inner mitochondrial membrane and is concerned with the asymmetric transcription of mitochondrial DNA. The product appears to become associated with mitochondrial ribosomes. Finally, in chloroplast-containing plants, a chloroplast DNA-dependent RNA polymerase has also been characterized. All these polymerases may have σ-like initiation factors associated with them, although highly purified *E. coli* σ factor is completely inactive in cross-reactions with eucaryotic RNA polymerases.

Polynucleotide Phosphorylase. Although RNA polymerases are the correct enzymes for transcription of RNA from the DNA template, another enzyme, called polynucleotide phosphorylase, catalyzes the reversible reaction.

$$n\ ppX \underset{Mg^{2+}}{\overset{\text{RNA primer}}{\rightleftharpoons}} (pX)_n + n\ \text{Pi}$$

Homopolymers are formed from ADP, IDP, GDP, CDP, UDP to yield, respectively, poly-A, poly-I, poly-G, poly-C, and poly-U. With mixtures of nucleoside diphosphate, random copolymers are formed. No template nucleic acid is required, although the presence of a primer greatly stimulates the reaction.

The enzyme is of historical importance since it was the first enzyme observed in bacterial cells to allow a net synthesis of RNA from nucleoside diphosphates. The enzyme appears to function primarily as a degradation enzyme which rapidly cleaves RNA by a phosphorylytic reaction to form nucleoside diphosphates.

References

1. J. D. Watson. *Molecular Biology of the Gene*. 2nd ed. New York: Benjamin, 1970.
   An excellent summary of the synthesis of RNA and DNA is found in this book.
2. E. E. Snell, ed., *Annual Review of Biochemistry*. Vol. 40 (1971); Vol. 41 (1972). Palo Alto, Ca.: Annual Reviews.
   The advanced student should consult this reference book for the most recent account of research in this area.
3. J. N. Davidson. *Biochemistry of Nucleic Acids*. 6th ed. London: Methuen, Wiley, 1969.
   A very good account of the general aspects of nucleic acid chemistry and metabolism.

Biosynthesis of Proteins

Introduction

The elucidation of the mechanism of protein synthesis in both procaryotic and eucaryotic cells, has occurred in the past 15 years and ranks as one of the major triumphs in modern biochemistry. The synthesis of the peptide bond is the principal chemical event in protein synthesis and, as we shall see, is coupled to a sophisticated machinery for the precise translation of the specific sequences programmed in the informational nucleic acids. However, a number of peptides are synthesized in the cell which do not require such an elegant machinery and we shall outline these first.

Glutamine

Glutamine is synthesized by *glutamine synthetase,* an enzyme found in plants, animals, and bacteria. The first step is believed to be the formation of a γ-glutamyl phosphate–enzyme complex. In the second step, ammonia, a good nucleophile, attacks the complex and displaces the phosphate group

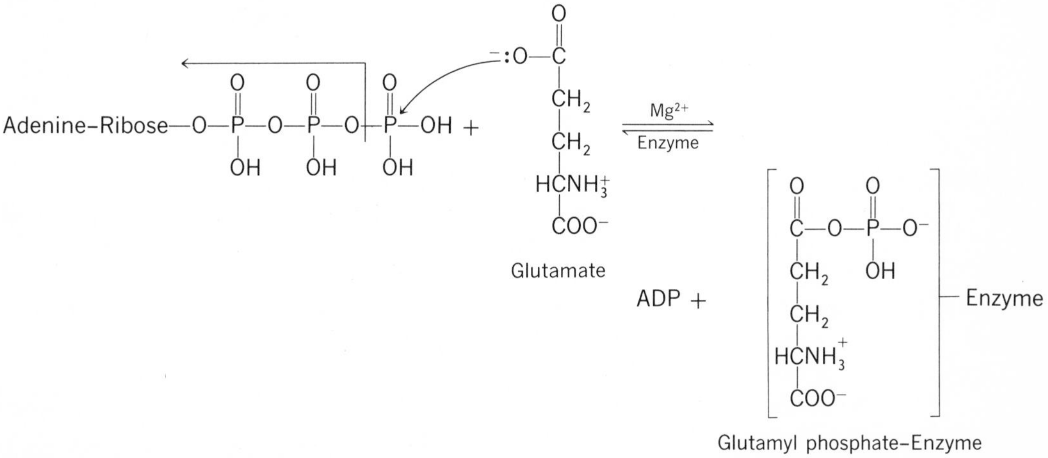

$$\text{Adenine–Ribose—O—}\underset{\text{OH}}{\overset{\text{O}}{\overset{\|}{\text{P}}}}\text{—O—}\underset{\text{OH}}{\overset{\text{O}}{\overset{\|}{\text{P}}}}\text{—O—}\underset{\text{OH}}{\overset{\text{O}}{\overset{\|}{\text{P}}}}\text{—OH} + {}^{-}\!:\!\text{O—C(=O)—CH}_2\text{—CH}_2\text{—HCNH}_3^+\text{—COO}^- \underset{\text{Enzyme}}{\overset{\text{Mg}^{2+}}{\rightleftharpoons}} \text{ADP} + \left[\text{COO}^-\text{—HCNH}_3^+\text{—CH}_2\text{—CH}_2\text{—C(=O)—O—}\underset{\text{OH}}{\overset{\text{O}}{\overset{\|}{\text{P}}}}\text{—O}^-\right]\text{—Enzyme}$$

Glutamate

Glutamyl phosphate–Enzyme

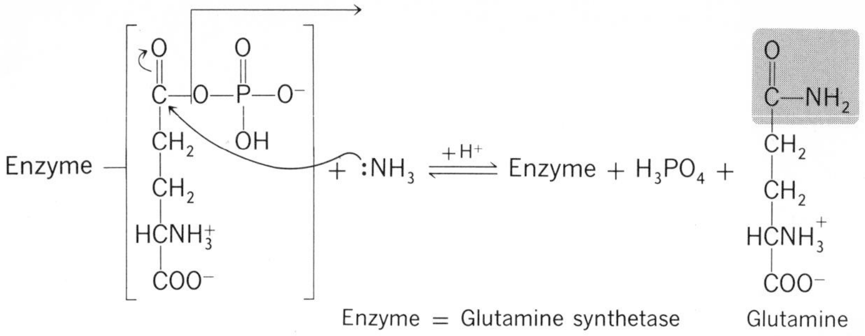

Enzyme—[$C(=O)$—O—$P(=O)(OH)$—O^- / CH_2 / CH_2 / $HCNH_3^+$ / COO^-] + :NH_3 $\xrightleftharpoons{+H^+}$ Enzyme + H_3PO_4 + $C(=O)$—NH_2 / CH_2 / CH_2 / $HCNH_3^+$ / COO^-

Enzyme = Glutamine synthetase

Glutamine

to form glutamine and inorganic phosphate. Note that only the γ-amide is formed. Isoglutamine, the compound with the α-carboxyl group amidated, is never formed. The enzyme is not only highly specific, since aspartic acid cannot replace glutamic acid as a substrate, but also an extremely complex protein which is regulated by an interesting number of mechanisms which we shall discuss in detail in Chapter 20.

Hippuric Acid

The enzymic synthesis of hippuric acid, a common urinary product in mammalian animals, involves the sequence shown in Scheme 19-1.

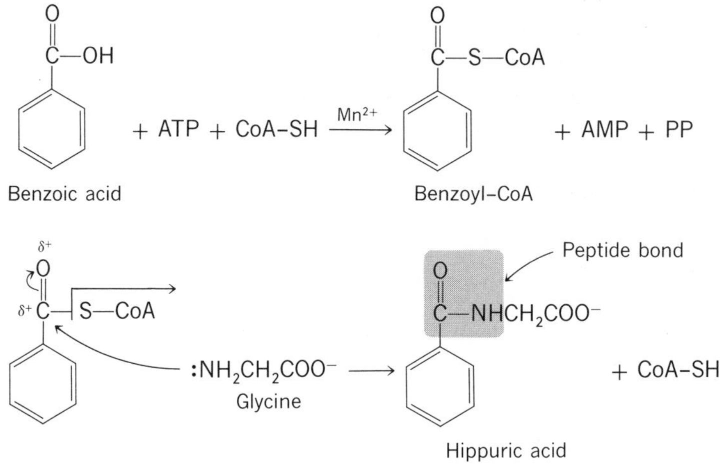

Benzoic acid (C_6H_5—$C(=O)$—OH) + ATP + CoA–SH $\xrightarrow{Mn^{2+}}$ Benzoyl-CoA (C_6H_5—$C(=O)$—S—CoA) + AMP + PP

C_6H_5—$\delta^+C(=O^{\delta^+})$—S—CoA + :$NH_2CH_2COO^-$ (Glycine) $\longrightarrow$ Hippuric acid (C_6H_5—$C(=O)$—$NHCH_2COO^-$) + CoA–SH

Peptide bond

Scheme 19-1

Glutathione

Glutathione a tripeptide which occurs in yeast, plants, and animal tissues, requires two discrete enzyme systems to form its two peptide bonds. The first enzyme, *γ-glutamyl cysteine synthetase* (*a*), catalyzes the condensation of glutamic acid and cysteine with the formation of the first peptide bond. Then a second enzyme, *glutathione synthetase* (*b*), adds glycine to the previously synthesized dipeptide to form the second peptide bond. In each step, the carboxyl group is presumably activated by ATP as already outlined for glutamine synthesis. Cysteine does not directly attack the γ-carboxyl group of glutamic acid since the —O^- of the carboxyl group is a poor leaving group;

if a phosphate group is placed on the carboxyl carbon at the expense of ATP, then, as with glutamine synthesis, we have an excellent leaving group.

(*a*)

First activated by ATP

$$\gamma\text{-}OOC\text{-}CH_2\text{-}CH_2\text{-}HCNH_3^+\text{-}COO^- \text{ (Glutamic)} + :NH_2CH(CH_2SH)COO^- \text{ (Cysteine)} + ATP \xrightarrow[(a)]{Mn^{2+}} H_3PO_4 + ADP + \gamma\text{-}C(=O)\text{-}NHCH(CH_2SH)\text{-}COO^-\text{ on }CH_2\text{-}CH_2\text{-}HCNH_3^+\text{-}COO^-$$

γ-Glutamyl cysteine

(*b*)

First activated by ATP

$$O{=}C\text{-}NHCH(CH_2SH)\text{-}COO^-\text{ on }CH_2\text{-}CH_2\text{-}HCNH_3^+\text{-}COO^- + :NH_2CH_2COO^- \text{ (Glycine)} + ATP \xrightarrow[(b)]{Mn^{2+}} H_3PO_4 + ADP + O{=}C\text{-}NHCH(CH_2SH)CONHCH_2COO^-\text{ on }CH_2\text{-}CH_2\text{-}HCNH_3^+\text{-}COO^-$$

γ-Glutamyl-cysteinyl-glycine or Glutathione

Note that CoASH is required in the synthesis of hippuric acid and that the reaction products include AMP and pyrophosphate; in the synthesis of glutamine and glutathione, ADP and inorganic phosphate are the reaction products, and CoASH is not required. This indicates strongly that the peptide bond in hippuric acid is synthesized by a mechanism unlike that found in the latter two examples. Note, moreover, that the sequence of these simple steps is controlled by the specificity of the enzymes involved. That is, in glutathione synthesis, the reverse peptide glycyl–glutamyl–cysteine is not produced because the specificities of the two enzyme systems control the order of addition. Thus, enzyme *a* catalyzes *only* reaction *a* and enzyme *b* catalyzes only reaction *b*. Hence, the order of reaction is glutamic with cysteine and not glutamic with glycine.

Cyclic Polypeptides

The biosynthesis of antibiotic cyclic polypeptides, like the biosynthesis of glutathione, occurs in the complete absence of polynucleotides, which suggests sequence determination by enzyme specificity.

As an example, let us examine the biosynthesis of gramicidin-S, a cyclic decapeptide with the structure

D-Phe ⟶ Pro ⟶ Val ⟶ Orn ⟶ Leu
↑ ↓
Leu ⟵ Orn ⟵ Val ⟵ Pro ⟵ D-Phe

Extracts of *Bacillus brevis* strains which synthesize gramicidin-S contain two protein fractions which on addition of ATP, Mg^{2+}, and the appropriate amino acids readily catalyze the synthesis of the cyclic polypeptide. Protein I has a molecular weight of 280,000 and protein II, 100,000. Protein II activates and racemizes D- or L-phenylalanine. It also becomes charged with D-phenylalanine via a thioester linkage. Protein I activates the other four amino acids, namely proline, ornithine, valine, and leucine, via the sequence

$$\text{Amino acid} + \text{ATP} \rightleftharpoons \text{Amino acyladenylate} + \text{PP}$$
$$\text{Amino acyladenylate} + \text{HS–Protein} \rightleftharpoons \text{Amino acyl–}S\text{–Protein} + \text{AMP}$$

The following steps then occur: The D-phenylalanyl–*S*–protein II complex initiates polypeptide synthesis by transfer to L-prolyl–*S*–protein I, releasing free protein II and forming a dipeptidyl–*S*–protein I complex. This intermediate then reacts rapidly with the amino group of the next amino acid, valine, which is also linked by a thioester to protein I, to form NH_2·Phe–Pro–Val–thioenzyme complex. Ornithine and leucine are next inserted until the pentapeptidyl stage is reached. As soon as this stage is achieved, two pentapeptidyl thioenzyme complexes interact to form the completed cyclic polypeptide (Scheme 19-2).

Although more primitive than the nucleic acid-directed ribosomal system for protein synthesis, this system is severely restrictive in that only the correct stereoisomeric amino acids, namely D-phenylalanine, L-proline, L-valine, L-ornithine and L-leucine are utilized; all other amino acids or the incorrect optical isomers are rejected. In addition, instead of forming a highly reactive amino acyl oxygen ester–tRNA complex, an equally reactive amino acyl thioester–protein complex is employed for activation of the carboxyl groups necessary for peptide bond formation. There is now good evidence that the sulfhydryl group essential for these reactions in protein I is associated with a phosphopantethine residue which is covalently linked with protein I. The student should note the parallelism of amino acid polymerization and fatty acid synthesis in which covalent linkages are formed via a thioester activation.

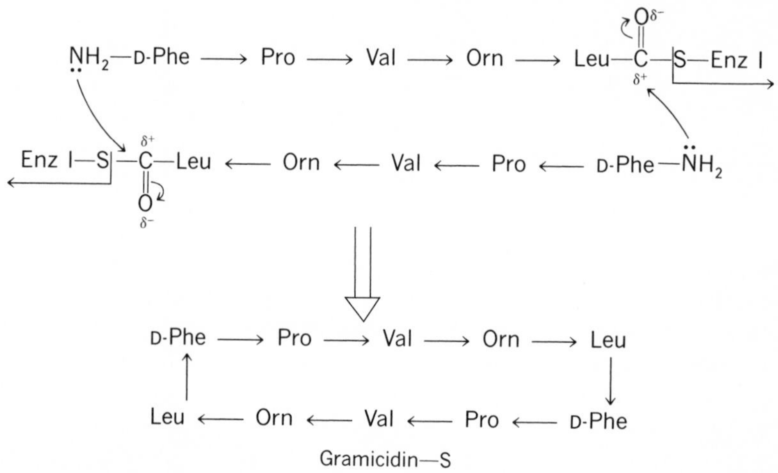

NH_2—D-Phe ⟶ Pro ⟶ Val ⟶ Orn ⟶ Leu—C(=$O^{\delta-}$)$^{\delta+}$—S—Enz I

Enz I—S—C(=$O^{\delta-}$)$^{\delta+}$—Leu ⟵ Orn ⟵ Val ⟵ Pro ⟵ D-Phe—NH_2

⇓

D-Phe ⟶ Pro ⟶ Val ⟶ Orn ⟶ Leu
↑ ↓
Leu ⟵ Orn ⟵ Val ⟵ Pro ⟵ D-Phe

Gramicidin—S

Scheme 19-2

Components of Protein Synthesis

In examining the problem of protein synthesis we are immediately faced with the gigantic task of synthesizing a very complex molecule containing hundreds of L-amino acid residues in exactly the same sequence each time the molecule is produced. In other words, the mechanism of synthesis must have a precise coding system which automatically programs the insertion of only one specific amino acid residue in a specific position in the protein chain. The coding system, in determining the primary structure precisely, in turn establishes the secondary and tertiary structures of a given protein. This problem obviously does not exist in the area of simple peptides or amide synthesis.

Of primary interest is the mechanism by which information from DNA, the genetic carrier of the coding system, is used in a precise manner for the biosynthesis of proteins. We have already discussed DNA and RNA synthesis and will now attempt to summarize the orderly processes by which proteins are formed.

Genetic information from DNA is programmed in RNA for the orderly synthesis of new proteins as indicated:

$$\begin{array}{l} \text{DNA} \xrightarrow{\text{Transcription}} \text{RNA} \xrightarrow{\text{Translation}} \text{Protein} \\ \big\downarrow \text{Replication} \\ \text{DNA} \end{array}$$

Transcription is defined as the process whereby the genetic information in DNA is employed to order a complementary sequence of bases in a new RNA chain. A vital role is played by DNA-dependent RNA polymerase in this transcription process (see Chapter 18 for discussion of this reaction). If the nucleotide sequence in a specific RNA is to be translated into an amino acid sequence, the RNA is referred to as messenger RNA (mRNA). *Translation* is thus the process whereby the genetic information in a mRNA molecule directs the order of insertion of the specific amino acid during protein synthesis. We shall now examine in detail the intricacies of the translation process.

The machinery of protein synthesis consists of a large number of components, the important ones being mRNA, the ribosomes, which are the actual sites of protein synthesis, amino acyl tRNA's, and a number of enzymes and cofactors. The whole machinery operates in the cytoplasm of procaryotic and eucaryotic organisms. Exceptions are mitochondria and chloroplasts, which have a somewhat limited but rather complete machinery for protein synthesis.

Activation of Amino Acid. The twenty amino acids commonly found in protein structures must undergo an initial activation step, which also involves a selection and preliminary screening of amino acids. Thus, D isomers and certain amino acids such as ornithine, citrulline, β-alanine, and diamino pimelic acid, which are used for other purposes in the cell, are rejected at this stage. Each amino acid of the twenty normally found in proteins has its own specific activation enzyme system called amino acyl-tRNA synthetase. The step involves the following:

$$\underset{(aa)}{\text{ATP} + \text{Amino acid}} \xrightleftharpoons[\text{Mn}^{2+}]{\text{Enzyme}} \underset{\substack{\text{Amino acid–Adenylate–}\\ \text{Enzyme complex}}}{\text{aa-AMP–Enz}} + \underset{\text{Pyrophosphate}}{\text{P—P}} \qquad (19\text{-}1)$$

Mechanism:

$$\text{Adenine–Ribose—O—}\overset{O}{\overset{\|}{\underset{OH}{\underset{|}{P}}}}\text{—O—}\overset{O}{\overset{\|}{\underset{OH}{\underset{|}{P}}}}\text{—O—}\overset{O}{\overset{\|}{\underset{OH}{\underset{|}{P}}}}\text{—OH} + {}^{-}\text{:O—}\overset{O}{\overset{\|}{C}}\text{—}\overset{\overset{+}{NH_3}}{\overset{|}{C}}\text{HR} \xrightleftharpoons{\text{Enzyme}}$$

$$\left[\text{Adenine–Ribose—O—}\overset{O^{\delta-}}{\overset{\|}{\underset{OH}{\underset{|}{P^{\delta+}}}}}\sim\text{O—}\overset{O^{\delta-}}{\overset{\|}{C^{\delta+}}}\text{—}\overset{\overset{+}{NH_3}}{\overset{|}{C}}\text{HR}\right]\text{Enzyme} + \text{P—P}$$

Very labile bond
because of electrostatic repulsion
by positive charges

Amino acyl adenylates are extremely reactive but are stabilized by remaining associated with the parent enzyme. The great lability is associated with the large positive charge on the amino group adjacent to the positive phosphorus atom resulting in a strong electrostatic repulsion and a subsequent labilization of the P–O–C bond. While the activation step, namely the formation of an amino acid–adenylate–enzyme complex, has considerable specificity built into it, some activating enzymes have a limited capacity for activating a number of amino acids. Thus, the L-isoleucine–tRNA synthetase will also activate L-valine, and the valine-tRNA synthetase will also react with threonine as measured by pyrophosphate-^{32}P exchange as illustrated in reaction 19-1. However, while these enzymes do not have strict substrate specificity, they do recognize *only* $\text{tRNA}^{\text{ileu}}$ and tRNA^{val}, respectively. Thus, specificity is exercised at two levels, (*a*) the activation step, and (*b*) the transfer step to tRNA, suggesting that the synthetase protein must have had at least two recognition sites, one for its specific amino acid and the other for its specific tRNA.

A large number of amino acid tRNA synthetases have been purified from a large number of tissues. Their average molecular weights are approximately 100,000, and they appear to be oligomeric proteins.

The mechanism of transfer of the amino acid–adenylate–enzyme complex to its specific tRNA is believed to be as shown in Scheme 19-3.

Transfer RNA. As we have already indicated, the special acceptors of activated amino acids are the specific transfer RNA's (tRNA's). Each cell contains about sixty different tRNA species which can be separated by a variety of techniques. Of these, over twenty tRNA's have had their base sequences determined. A number have been crystallized. Chain lengths of known tRNA species vary between 75 and 87 nucleoside residues, with about 10–20% of the bases being modified. The average molecular weight ranges between

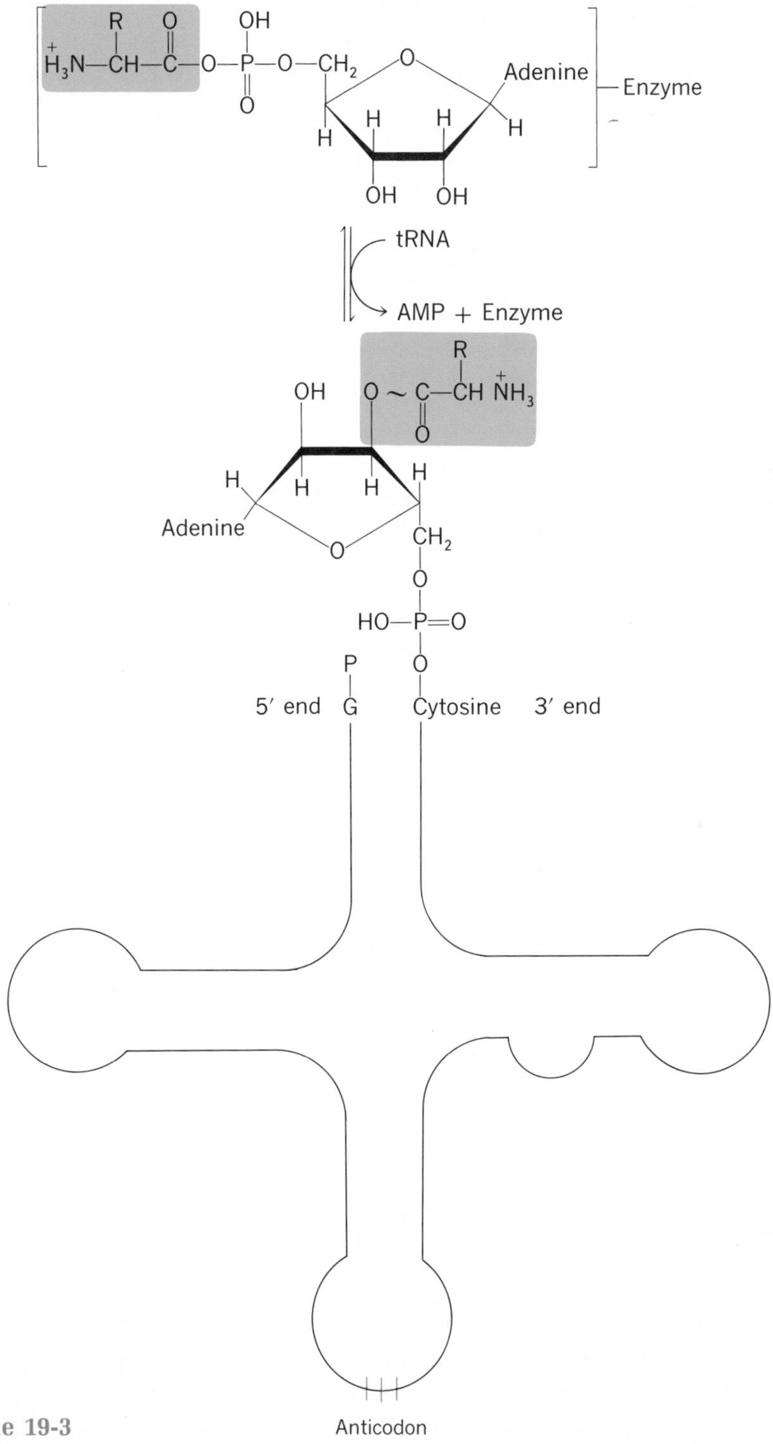

$[H_3\overset{+}{N}-CH(R)-C(=O)-O-P(=O)(OH)-O-CH_2$-ribose(2'-OH, 3'-OH)-Adenine$]$-Enzyme

tRNA

AMP + Enzyme

Adenine-ribose(2'-OH, 3'-$O \sim C(=O)-CH(R)\overset{+}{N}H_3$)-$CH_2-O-P(=O)(OH)-O$-

P

5′ end G Cytosine 3′ end

Anticodon

Scheme 19-3

25,000 and 30,000. The nomenclature $tRNA^{ala}_{yeast}$ signifies a transfer RNA specific for alanine and obtained from yeast.

Over 85% of the tRNA's have as their 5′ terminus the base guanine, while the remainder have cytosine as the terminal base. All tRNA's which are capable of being charged with amino acids have as their 3′ terminal sequence cytosine–cytosine–adenine.

A generalized secondary structure of tRNA is shown in Scheme 19-4. It is of considerable interest that when the primary sequence of 16–20 tRNA's was determined, and attempts were made to write logical secondary structures based on the base-pairing rules of Watson and Crick, namely G—C and

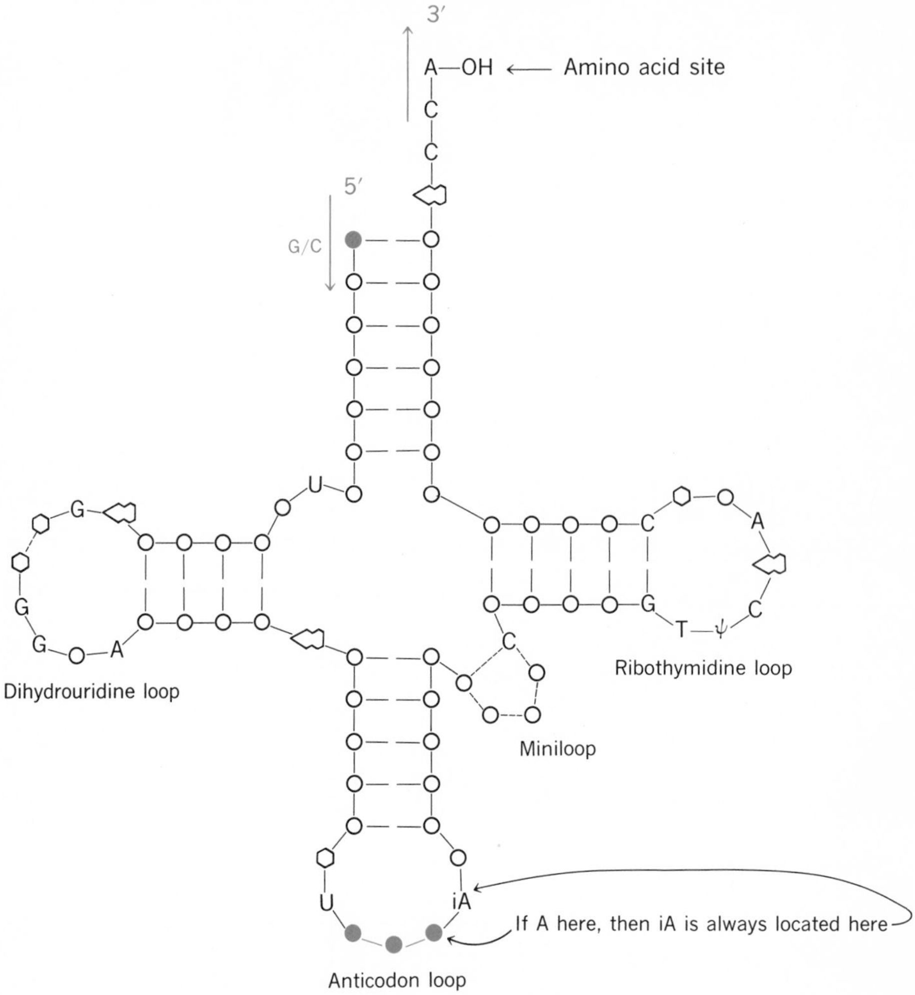

3′
A—OH ⟵ Amino acid site
5′
G/C
Dihydrouridine loop
Ribothymidine loop
Miniloop
If A here, then iA is always located here
Anticodon loop

Scheme 19-4

Generalized cloverleaf model for tRNA. The sugar phosphate backbone is represented as a solid line; a dotted line replaces it in sections of variable chain length. Bases: ○, variable; ⬡, pyrimidine; purine; ●, anticodon; iA, isopentenyl adenine.

A—U, the most satisfying structure common to all tRNA's with known sequence was the cloverleaf structure. Although this structure fulfills many of the requirements for a functional tRNA, namely the exposure of the anticodon in the anticodon loop and the availability of the 3′-hydroxyl of the ribosyl component of adenylic acid residue on the 3′ terminus, no experimental evidence is at present available concerning the precise secondary and tertiary structures of tRNA's.

At present the sequence of eucaryotic biosynthesis of tRNA species can be depicted as

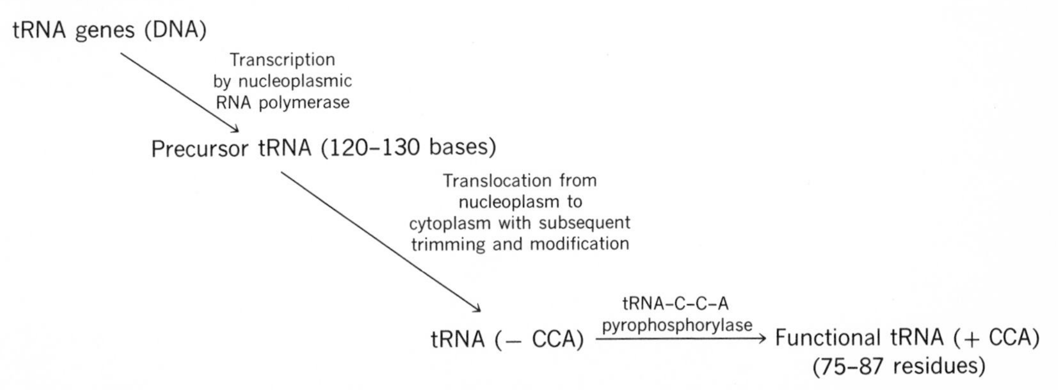

In eucaryotic organisms an unmodified precursor tRNA is presumably formed by RNA polymerase located in the nucleoplasm transcribing from a specific tRNA cistron of the nuclear DNA. Once translocated into the cytoplasm, the precursor tRNA is degraded to a tRNA minus the -CCA 3′ terminal bases. The precursor tRNA presumably has few, if any, modified bases, but in the cytoplasm, specific tRNA methylases transfer the methyl group from *S*-adenosyl methionine to suitable bases which are modified by alkylation of the carbon, nitrogen, and oxygen atoms to form the modified bases so characteristic of tRNA species. Some of these structures are depicted on page 106. An interesting correlation seems to exist between the extent of modified nucleosides contained in a tRNA of a particular organism and the evolutionary development of that organism. Thus, tRNA of *Mycoplasma*, the smallest free living organism known, contains only a low amount of modified nucleosides. However, *E. coli*, yeast, wheat germ, rat liver, and tumor cells contain an increasing number of modified bases in their tRNA species.

In general there is a cluster of modified nucleosides residing in the loop regions of the cloverleaf structure where only single strands exist. Perhaps their presence stabilizes the single-stranded loops to ensure a particular three-dimensional structure. An interesting modified nucleoside contains isopentenyl adenosine (iA):

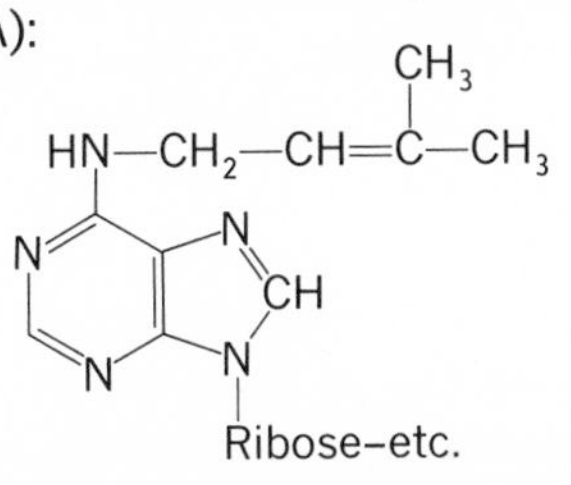

CH_3
$HN—CH_2—CH=C—CH_3$
N N CH N N
Ribose–etc.

Isopentenyl adenosine itself belongs to a class of cytokinins which are potent plant growth factors that promote cell division, growth, and organ formation in plants. Although the relationship of iA to these activities is not clear, iA always occurs adjacent to the base adenine in the anticodon region when adenine is the third base of the anticodon triplet sequence. Isopentenyl adenosine is readily synthesized by a cytoplasmic enzyme, isopentenyl pyrophosphate:tRNA isopentenyl transferase:

$$\text{Isopentenyl pyrophosphate} + \text{Adenine–tRNA} \longrightarrow \text{P—P} + \text{Isopentenyl–adenine–tRNA}$$

Mention was made earlier of about sixty different tRNA species that can be separated and purified. The existence of several tRNA's for the same amino acids is called multiplicity. For example, one type is caused by a degeneracy of the genetic code, that is, three tRNA^{ser} species are required for the six serine codons (see the genetic code, Table 19-3). A second type involves the mitochondrial tRNA species in animal cells and the chloroplast tRNA's in chloroplasts of plants.

A third type is a specific tRNA species employed by the cell for a highly specialized function. For example, gram-positive procaryotic organisms synthesize large amounts of a cell wall component called a peptidoglycan (see Chapter 2 for structure and Chapter 6 for function). The interpeptide bridge which joins the separate strands of peptidoglycans in some organisms is a pentaglycyl peptide. *Staphylococcus epidermidis* is a typical organism which possesses a unique tRNA^{gly} at relatively high concentrations, has about 85 residues, but contains only one modified nucleoside, namely 4-thiouridine. Although three additional isoaccepting tRNA^{gly} species were isolated from this organism, which together with the specific tRNA^{gly} were charged with glycine and could participate in the formation of the interpeptide bridge, the specific tRNA$^{gly}_{I}$ did not participate in protein synthesis and did not possess a glycine anticodon, whereas the three other tRNA^{gly} species readily participated in protein synthesis. One can thus speculate that the organism, by synthesizing a highly specific tRNA, could depend on this tRNA species for the all-important task of forming the interpeptide bridges so essential for the synthesis of a complete peptidoglycan without competition from the protein-synthesizing machinery of the cell.

In summary, transfer RNA's serve as adaptors in directing the proper placement of amino acids according to the nucleotide sequence of mRNA. Transfer RNA must have several recognition sites in its structure, namely (*a*) the anticodon site, that is, the three-base site responsible for the recognition of the complementary triplet encoded in mRNA; (*b*) the synthetase site by which the specific amino acyl tRNA synthetase recognizes and charges the specific amino acid with the tRNA; (*c*) amino acid attachment site, which in all tRNA's is the 3′ terminal–CCA nucleoside sequence; and (*d*) the ribosome recognition site. Future investigation hopefully will define these sites with precision.

Messenger RNA. A key component of the translation process is mRNA (see Chapter 5), which comprises only a few percent of the total RNA of a cell.

It carries the genetic message from DNA to the site of protein synthesis, the ribosomes, and hence is called messenger RNA. In 1957 Volkin and Astrachan noticed that a small fraction of the total RNA of *E. coli* infected with T-2 phage had a high turnover rate. In addition, the base ratio of this RNA complemented that of the DNA from T-2 phage rather than that of the host DNA. It furthermore did not resemble in base composition the ribosomal RNA (rRNA) of the *E. coli* host. This evidence was later interpreted to mean that when T-2 DNA entered the host cell, it served as a template for new RNA which copied the genetic message of T-2 DNA faithfully and in turn served as a template for new viral protein synthesis. Since no new rRNA was formed during infection and since no new ribosomes were made, cell rRNA could not possibly have served as a template for the new synthesis of protein. The transient fraction of the new RNA must have been serving this function. It was shown later that, in general, rRNA cannot serve as a template but that mRNA does.

Messenger RNA varies greatly in chain length and thus in molecular weight. This great variation might be related to the heterogenous nature of protein chain lengths. Since few proteins contain less than 100 amino acids, the mRNA coding for these proteins must have at least 100×3 or 300 nucleotide residues. In *E. coli* the average size of mRNA is 900–1500 nucleotide units. The instability of mRNA is characteristic of bacterial systems with a half-life of about 2 minutes. In mammalian systems, however, mRNA molecules are considerably more stable. This has been interpreted to mean that bacteria, which must have greater flexibility in adjusting to the ever-changing environment, must be able to synthesize different enzymes to cope with their surroundings and hence require mRNA's of a short lifetime.

In eucaryotic cells, mRNA is synthesized in the nucleoplasm by DNA-dependent RNA polymerase and is then translocated to the cytoplasm where it becomes associated to the ribosomal system. We shall have more to say about the functions of mRNA later in this chapter and when we discuss metabolic regulation in Chapter 20.

Ribosomes. In the early 1950s, evidence began to accumulate suggesting that ribosomes were the site for protein synthesis. For example, Zamecnik injected radioactive amino acids into a rat and then, within a short interval of time, homogenized the liver and fractionated the homogenate by differential centrifugation into nuclei, mitochondria, microsomes, and supernatant protein. The microsomal fraction had the highest specific activity. When these microsomes were treated with detergent to free the ribosomes from the vesicular matrix and again assayed for radioactivity, the ribosomes contained up to seven times more radioactivity per milligram of protein than the remainder of the microsomes. Clearly, then, the ribosomes served in some capacity as the locus for protein synthesis. At first, biochemists concluded that because of the high RNA content of ribosomes they could serve admirably as carriers of the template RNA. In 1956, this view was somewhat modified with the discovery of tRNA. With the discovery of mRNA in 1957–58, further modifications of the earlier views were necessary. What role do the ribosomes play? Let us first examine in some detail the chemistry of these particles.

Ribosomes are large ribonucleoprotein particles on which the actual process of translation occurs. In procaryotic cells they occur in the free form as monosomes or are associated with mRNA as polysomes. An average bacterial cell contains about 10^4 ribosomes. In eucaryotic cells they occur in forms similar to those found in procaryotic cells and also are associated with membranes of the rough endoplasmic reticulum (Chapter 6). About 10^6–10^7 ribosomes occur in these cells. Mitochondria and chloroplasts also possess these particles. Table 19-1 summarizes the physical properties of procaryotic ribosomes, plant cytoplasmic, and animal cytoplasmic ribosomes.

Readily isolated by prolonged centrifugation of tissue extracts at high centrifugal speeds (100,000 $\times$ g for several hours), all ribosomes consist of a larger subunit and a smaller subunit which associate at 10m*M* $MgCl_2$ concentration to the 70S unit and completely dissociate at 0.1m*M* $MgCl_2$. The 30S subunit of procaryotic cells contains 21 distinct proteins, most of which are basic in nature. A similar number of proteins are found in the 40S subunits of eucaryotic cells. The larger subunits, namely the 50S of procaryotic cells, contain about 35 distinct proteins and the 60S subunit of eucaryotic cells contain about 35–40 proteins. Evidence suggests that all these proteins are involved functionally or structurally with the translation process. No proteins are common to both the large and small subunits.

The two ribosomal subunits have different binding properties. Thus, the *E. coli* 30S subunit binds mRNA in the absence of the 50S subunit and the

Table 19-1

Sedimentation Values for Ribosomes

Ribosomes	Subunits	rRNA (mol wt)
Procaryotic		
Bacteria, actinomycetes, blue-green algae, mitochondria from eucaryotes		
70S →	30S	16S (550,000)
→	50S	5S (40,000) 23S (1,100,000)
Eucaryotic		
Plant kingdom[a]		
~80S →	40S	16–18S (~700,000)
→	60S	5S (40,000) 25S (~1,300,000)
Animal kingdom[a]		
~80S →	40S	18S (~700,000)
→	60S	5S (40,000) 28–29S (1,400,000–1,800,000)

[a] In general, organelle ribosomes (mitochondrial or chloroplastic) are in the 70S category.

30S–mRNA complex binds specific tRNA's. The 50S subunit does not associate with mRNA in the absence of the 30S subunit, but will nonspecifically bind tRNA. Each 70S ribosome contains two different binding sites for tRNA molecules. Site A (amino acyl site) is involved in the positioning of the specific incoming amino acyl tRNA with its matching codon on the mRNA. Site P (peptidyl site) binds the growing polypeptidyl tRNA. The peptidyl transferase which is responsible for the formation of peptide bond is located on the 50S ribosomal particle, presumably near the P site.

A few comments should be made concerning the biosynthesis of ribosomes. In eucaryotes the site of synthesis of the ribosomal RNA is the nucleolus (Chapter 6). In the nucleolus an RNA polymerase transcribes a large precursor rRNA from the RNA cistron region of the nuclear DNA. Precursor rRNA has a sedimentation value of 45S (4.1×10^6 mol wt) and is rapidly cleaved to two smaller RNA's, namely a 32S and a 20S component. The 32S component is further modified to yield a 28S rRNA, while the 20S component is cleaved to the 18S rRNA. In the meantime, specific ribosomal proteins are translocated to the nucleolus and become associated with each rRNA to form the completed 40S and the 60S ribosomal subunits. At this point, the 40S unit is transferred to the cytoplasm, where it becomes associated with mRNA; shortly thereafter, the 60S subunit is translocated to the cytoplasm to become associated with the 40S–mRNA complex to form the complete 80S–mRNA complex. The function of the 5S RNA is presently not known. The residual fragments from the precursor rRNA are presumably converted back to nucleotides and recycled into the nucleolar machinery. The precursors of the 16 and 23S subunit rRNA's in procaryotic cells are only slightly larger than the final product and require apparently a minimum of trimming to achieve the correct structures.

Protein Synthesis

The translation of coded information from DNA via mRNA into the amino acid sequences of proteins involves the orderly interactions of over 100 different macromolecules. We shall now describe the current status of this highly complex sequence of reactions in which tRNA, mRNA, ribosomes, and many ancilliary enzymes and proteins are involved. Since the mechanism of protein synthesis has been investigated in great depth in extracts of *E. coli*, we shall use this organism as the model for our description but will introduce comments concerning eucaryotic systems at the appropriate points.

There are four broad steps in the synthesis of a protein: (*a*) activation of amino acids, (*b*) initiation of the synthesis of the polypeptide chain, (*c*) its elongation, and (*d*) termination. Table 19-2 summarizes the important components of these reactions. We have already described step *a* in some detail. Steps *b*, *c*, and *d* are components of the ribosome cycle which is illustrated in Figure 19-1 and which summarizes these events. The student should refer to Table 19-2 and Figure 19-1 as we describe the process of protein synthesis.

Initiation. The initiation of protein synthesis requires the formation of an initiation complex consisting of the 30S ribosomal subunit, mRNA, and three proteins F1, F2, and F3 which are obtained by washing crude *E. coli* ribosomes

Table 19-2

Important Components of the *E. coli* Protein-Synthesizing System

Step	Components
Activation	tRNA's ATP Mg^{2+} L-Amino acids Amino acyl–tRNA synthetases
Initiation	30S Ribosomal unit 50S Ribosomal unit mRNA with initiator codon AUG F1, 9400 mol wt; F2, 80,000 mol wt; F3, 21,000 mol wt proteins Initiator tRNA = fMET–$tRNA_f$ Amino acyl tRNA's Formyl tetrahydrofolic acid MET–$tRNA_f$ formylase GTP, Mg^{2+}
Elongation	Ts factor (19,000) Tu factor (40,000) G factor (80,000) Amino acyl–tRNA's
Termination	Terminator codons UAA, UAG, UGA R1 (44,000) factor R2 (47,000) factor S factor TR factor
Deformylmethionylation	Deformylase Amino peptidase

with 1*M* NH_4Cl solutions. This mRNA–30S subunit complex in the presence of F1 and F2, GTP, and Mg^{2+} will specifically bind a special initiation tRNA, *N*-formyl methionyl $tRNA_f$ (fMET–$tRNA_f$ *E. coli*), at the initiation codon, AUG, on mRNA.

A few comments should now be made about the role of fMET–$tRNA_f$. Some years ago it was observed that the major NH_2-terminal amino acid of total *E. coli* proteins was methionine. Later it was noted that the addition of methionine and in particular *N*-formyl methionine greatly stimulated protein synthesis in crude *E. coli* extracts. It was finally discovered that the starting amino acid in the synthesis of all proteins in procaryotic organisms is *N*-formyl methionine, which in turn is associated with a specific $tRNA_f^{met}$. Thus, the reaction

$$\text{Formyl tetrahydrofolate} + NH_2\text{–MET–}tRNA_f \xrightarrow{\text{Formylase}} \text{N-Formyl MET–}tRNA_f$$

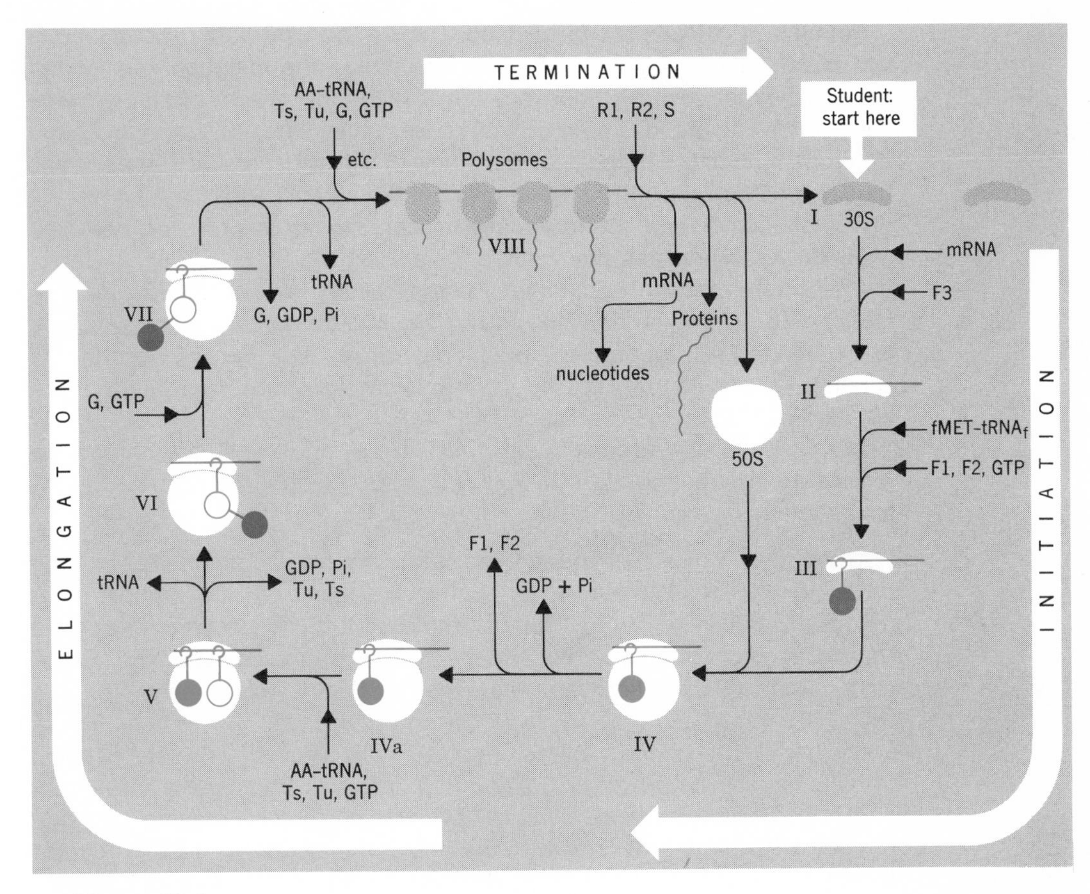

Figure 19-1

The ribosomal cycle showing the process of chain initiation, elongation, and termination. Steps I–II indicate the binding of mRNA to the 30S ribosome subunits requiring F3. Step III indicates the binding of fMET–tRNA$_f$ to mRNA–30S ribosome complex which requires F1 and F2. Steps IV–IVa show the joining of 50S to the mRNA–30S ribosome subunit to complete the 70S ribosome particle complex. Steps V–VII indicate chain elongation. Step VIII indicates chain termination and dissociation of 70S ribosome with deformylation (deformylase) and demethioninylation (aminopeptidase). (Reproduced with permission of S. Pestka and the *Annual Reviews, Inc.*, 1971.)

is an extremely important reaction in which the α-NH_2 group of methionine is formylated:

$$\underbrace{H-C(=O)}_{\text{Formyl}}-\underbrace{NH-CH(CH_2-CH_2-S-CH_3)-C(=O)}_{\text{Methionine residue}}-O-tRNA$$

Not all MET–tRNA's are formylated. The second type with a slightly modified base sequence, MET–$tRNA^{met}$, is inactive in the initiation step and is the specific methionine carrier for internal methionyl residues in the growing polypeptide chain. Since both tRNA's possess the same anticodon, AUG, the differentiation between the two amino acyl tRNA's must reside in the specific binding of F1 to the fMET–$tRNA_f$ and not to the MET–tRNA.

It is of considerable interest that in eucaryotic organisms, an unblocked MET–$tRNA^{met}$ serves as the specific initiator tRNA.

After the formation of the initiator complex consisting of the 30S subunit, mRNA, fMET–$tRNA_f$, and F1, F2, and F3, the 50S ribosomal subunit becomes associated to form the full 70S initiation complex. With the entry of the 50S subunit, F1 and F2 are released from the complex. In the formation of the 70S complex GTP is converted to GDP + Pi.

The P site is now occupied with fMET–$tRNA_f$ and the A site is ready to receive the first amino acyl–tRNA which is specified by the codon adjacent to the initiator codon AUG.

Elongation Step. This step involves three stages: (*a*) the codon-directed binding of amino acyl–tRNA to site A, (*b*) peptidyl transfer from the peptidyl tRNA (or the P site) to the newly bound amino acyl–tRNA (on site A), and (*c*) translocation of both the mRNA and the newly synthesized $\text{peptidyl}_{(n+1)}$ tRNA from site A to P, thus liberating the A site for the incoming new amino acyl tRNA.

The translocation step involves two protein factors, Tu, Ts, and the peptidyl transferase activity on the 50S subunit; T refers to transferase activity; Tu and Ts occur in large amounts in the cytoplasm of *E. coli*, making up 2–3% of the total supernatant protein of the cell, and they interact rapidly to form a very stable complex. The Tu factor also binds both GDP and GTP very strongly. The following steps describe the association of the incoming amino acyl–tRNA with the A site:

$$\text{Tu–GDP} \xrightleftharpoons{\text{Ts}} \text{Tu–Ts} + \text{GDP}$$
$$\text{Tu–Ts} + \text{GTP} + \text{Amino acyl–tRNA} \rightleftharpoons (\text{Amino acyl–tRNA–Tu–GTP}) + \text{Ts}$$
$$\text{AA–tRNA–Tu–GTP} + (\text{mRNA–Ribosome–fMET–tRNA}_f) \rightleftharpoons$$
$$(\text{AA–tRNA–mRNA–Ribosome–fMET–tRNA}_f) + \text{Tu–GDP} + \text{Pi}$$

After the amino acyl–tRNA is bound to the ribosome at the A site, the α-amino group of the amino acyl–tRNA rapidly reacts by a nucleophilic attack with the activated carboxyl terminal residue of the neighboring initiator or peptidyl–tRNA component to form the new peptide bond. The growing peptide is now attached to the tRNA which had carried the new amino acid on the A site, and the tRNA which had carried the initiator (fMET) or the peptidyl chain on the P site must now be discharged to provide space for the incoming tRNA (on the A site).

An additional protein, the G factor or translocase, enters the picture by catalyzing the shift of the bound peptidyl–tRNA from site A to site P, thereby liberating the A site for the entry of a new amino acyl–tRNA. In the translocation GTP is converted to GDP + Pi. A generalized picture is depicted in Figure 19-2.

In the eucaryotic cells, similar translocation factors have been found. Of considerable interest, the T2 factor of eucaryotic cells, which is identical in function with the G factor of procaryotic cells, is rapidly inactivated by diphtheria toxin. Apparently, in the presence of NAD^+, the following reaction occurs with free T2:

$$\underset{\text{(Active)}}{\text{T2}} + \underset{\text{NAD}^+}{\text{ARPPR–Nicotinamide}} \xrightarrow{\text{Diphtheria toxin}} \underset{\text{(Inactive)}}{\text{T2–RPPRA}} + \text{Nicotinamide}$$

In the intact eucaryotic cell, the toxin is bound to the cell membrane. However, T2, bound to the ribosomal system, when released into the cytoplasm rapidly diffuses to the periphery of the cell and becomes inactivated by the mechanism described above. Thus, the effects of diphtheria, a dreadful disease, can be explained on a molecular level. Incidentally, all eucaryotic cells including yeast are sensitive to the diphtheria toxin in the presence of NAD^+, but procaryotic cells are completely insensitive.

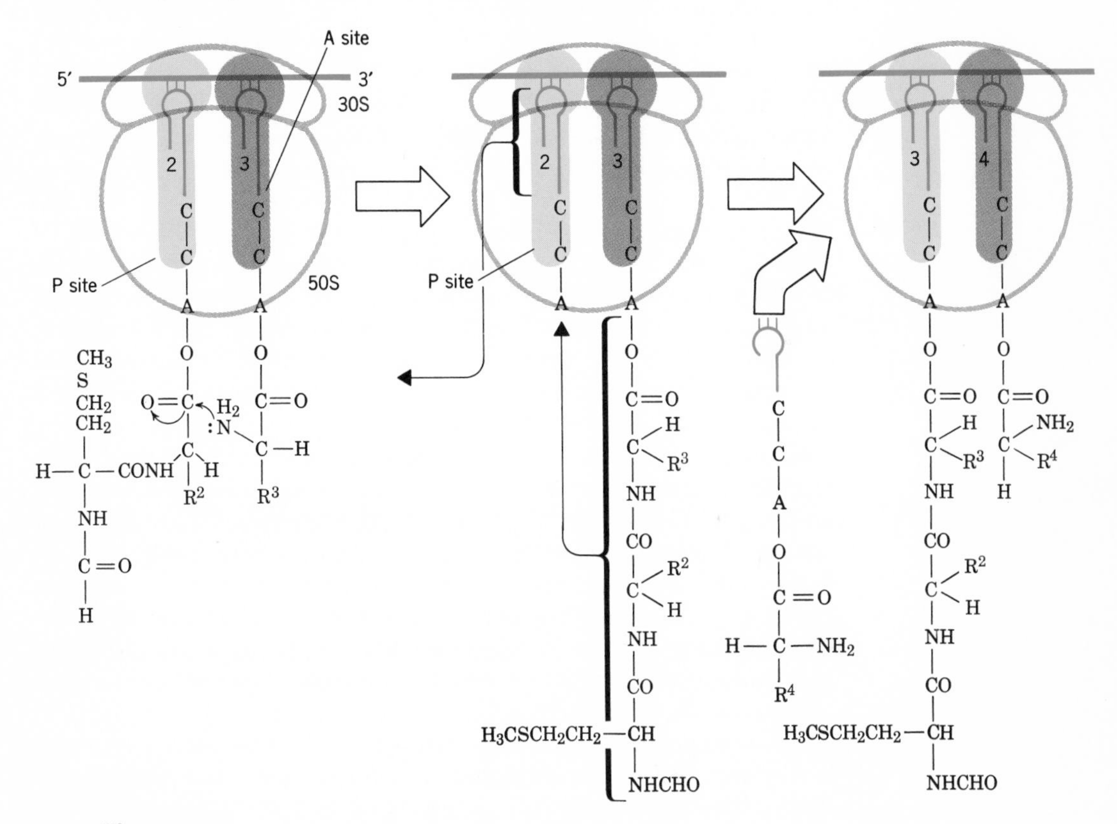

Figure 19-2

Peptide bond formation showing schematic sequence of reactions. Factors Tu, Ts, and G are not included, since their precise roles are not known.

Termination. The termination reaction consists of two events: (*a*) the recognition of a termination signal in the mRNA and (*b*) the hydrolysis of the final peptidyl tRNA ester linkage to release the nascent protein. The termination codons are UAA, UAG, and UGA. Three protein factors are required—R1, R2, and a S factor. The R1 factor is required for the recognition of the codons UAA and UAG, and R2 is required for the recognition of UAA and UGA. The third protein, S, has no release activity, but appears to aid in terminator codon recognition. The picture that is emerging suggests that the termination step can be divided into a terminator codon-dependent R1 or R2 factor binding reaction and a hydrolytic reaction in which either R1 or R2 converts the peptidyl transferase activity at site P into a hydrolytic reaction with the transfer of the peptidyl tRNA to water rather than to another amino acyl–tRNA. A final factor, TR, is involved in discharging the residual tRNA from site P. Once the tRNA is removed, the 70S ribosome dissociates from the

Figure 19-3

Schematic summary of protein synthesis. The semilunar cap (I) represents the free 30S subunit. Initiation of protein synthesis involves attachment of mRNA to a 30S subunit (I) to form complex II; this process requires Mg^{2+} as well as initiation factor F3. Subsequent attachment of fMET–tRNA in response to initiation codon AUG to form complex III requires GTP and initiation factors F1 and F2. Junction of the 50S subunit to complex III produces complex IVa; it is probable that prior to complex IVa an intermediate state IV exists where fMET–tRNA is in the "A" or nonpuromycin reactive site; the transition from IV to IVa possibly occurs on hydrolysis of GTP. Enzymic recognition of internal codons involves factors Tu, Ts, and GTP: the Tu:GTP:Ala–tRNA complex binds to the ribosomes in response to the GCU codon to form complex V. Peptide bond formation occurs by transfer of the fMET (peptidyl) group to form fMET–Ala (VI); peptidyl transfer requires only ribosomes and K^+. Translocation involves several coordinate processes: one codon movement of mRNA and ribosome with respect to each other, precisely positioning the next codon UCU into position for translation; release of deacylated $tRNA_f^{met}$; and coordinate movement of peptidyl tRNA (fMET–Ala) from the A to the P site. By repetition of the codon recognition step Ser–tRNA enters the A site in response to the codon UCU, forming complex VII. Complex VIII represents a peptidyl–tRNA with a polypeptide almost completed. Transpeptidation and translocation produces complex IX, with a completed protein still attached to tRNA and a termination codon UAA in the next recognition site. In response to a release factor, the completed protein is released and perhaps also tRNA after translocation (X). Provided no further cistrons are to be translated, the ribosome may be dissociated into 30S and 50S subunits with release of mRNA (XI). Alternatively, mRNA may be degraded prior to this stage. (Reproduced with permission of S. Pestka and the Annual Reviews, Inc., 1971.)

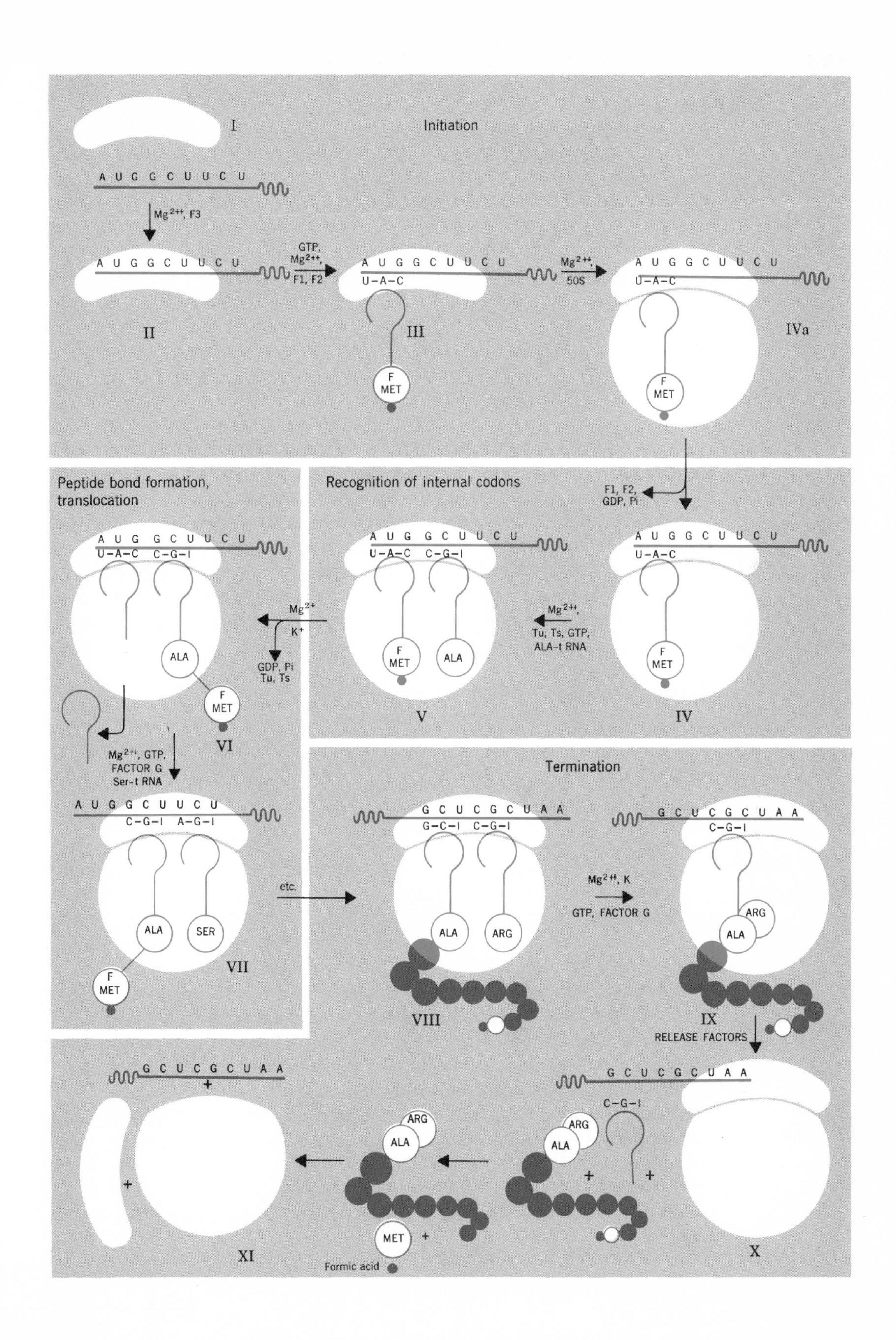

Initiation
I
A U G G C U U C U
Mg 2++, F3
II
GTP, Mg 2++, F1, F2
III
U–A–C
F MET
Mg 2++, 50S
IVa
F1, F2, GDP, Pi
Recognition of internal codons
IV
Mg 2++, Tu, Ts, GTP, ALA–t RNA
V
ALA
Peptide bond formation, translocation
Mg 2+
K+
GDP, Pi
Tu, Ts
VI
C–G–I
Mg 2++, GTP, FACTOR G
Ser–t RNA
VII
A–G–I
SER
etc.
Termination
G C U C G C U A A
G–C–I
VIII
ARG
Mg 2++, K
GTP, FACTOR G
IX
RELEASE FACTORS
X
XI
MET
Formic acid

mRNA into a 30S and a 50S subunit and is ready to reenter the ribosomal cycle for the synthesis of another protein molecule.

The nascent protein presumably has a formyl methionyl NH_2-terminus which must be removed before the protein completes its folding sequence. Two enzymes may participate at this final stage:

(1) a specific deformylase:

$$\text{Formyl methionyl peptide} \longrightarrow \text{Formic acid} + \text{Methionyl peptide}$$

(2) a specific amino peptidase:

$$\text{Methionyl peptide} \longrightarrow \text{Methionine} + \text{peptide}$$

Figure 19-3 summarizes the initiation, elongation, and termination steps.

In vitro Synthesis of Complete Proteins

We have outlined the complex array of steps necessary for the complete synthesis of a protein. Until a few years ago, protein synthesis was observed by counting the incorporation of ^{14}C labeled amino acids into a poorly defined trichloroacetic acid precipitate of denatured proteins. However, with the elucidation of the detailed steps involved in polypeptide biosynthesis, it is now possible to synthesize specific enzymes employing the appropriate template RNA. For example, the gene for β-glucosyl transferase comprises 0.3–1% of the T-4 phage DNA. Thus, the investigator can set up the following scheme:

$$\text{T-4 phage DNA} \xrightarrow{\text{RNA polymerase}} \text{mRNA} \xrightarrow[\text{Complete protein-synthesizing components (Table 19-2)}]{\text{fMET–tRNA}_f^{met}}$$

$$\beta\text{-Glucosyl transferase} + \text{Other proteins}$$

This and other experiments of this type completely confirm the concepts developed to describe the mechanism of protein synthesis.

Biosynthesis of Insulin

It is appropriate to outline the general biosynthetic aspects of the important hormone, insulin, to illustrate the intriguing complexities of eucaryotic protein synthesis. D. F. Steiner has described in a series of elegant experiments the biosynthesis of insulin by the β cells of the islets of Langerhans of the pancreas in a number of species.

The classic work of F. Sanger of England on the precise amino acid sequence of insulin made possible a detailed molecular picture of insulin. Until 1965, insulin was believed to be synthesized as two separate polypeptides which in some manner were oriented to allow the specific formation of disulfide linkages between the two chains to yield insulin.

In 1967, Steiner demonstrated that a protein molecule larger than insulin was formed in pancreatic β cells which exhibited all the properties of a precursor of insulin. Called *proinsulin,* its molecular weight was 9000 (insulin, 6500 mol wt), and it had 81 amino acid residues (insulin, 51). It could be rapidly converted to a fully physiologically active hormone by the proteolytic action of trypsin.

The biosynthesis of insulin that is now emerging is, in its general features, illustrated in Figure 19-4. In the presence of the protein-synthesizing enzymes, factors, and the appropriate mRNA, the ribosomes clustered around the rough endoplasmic reticulum (RER) synthesize proinsulin. The single polypeptide rapidly folds and the disulfide bridges form as the proinsulin is translocated into the cisternal (interior) spaces of the RER, transported via the vesicular tubules of RER to the contiguous Golgi apparatus. The time interval for these events is approximately 10 minutes. An hour later, immature secretory granules are formed by vesiculation from the periphery of the Golgi apparatus. Having a single limiting membrane, they contain proinsulin, proteolytic enzymes, and zinc ions. A rapid conversion to mature granules takes place in the periphery of the β cells, with complete transformation of proinsulin to zinc insulin and the C peptide (see Figure 19-5). At the appropriate signal, the mature granules are secreted by reverse pinocytosis into the blood stream, where the insulin is released. Not only insulin but α-amylase, ribonuclease, etc., are synthesized in the pancreatic exocrine cells by this mechanism.

The question arises as to why the cell first forms proinsulin. Some years ago C. Anfinsen had clearly shown that the amino acid sequence of many proteins is decisive in directing the folding of polypeptide chains into their native conformation. This can be readily demonstrated by unfolding these proteins by reduction in 8*M* urea, allowing them to reform disulfide bonds by exposure to air oxidation, and observing that the reoxidized proteins are

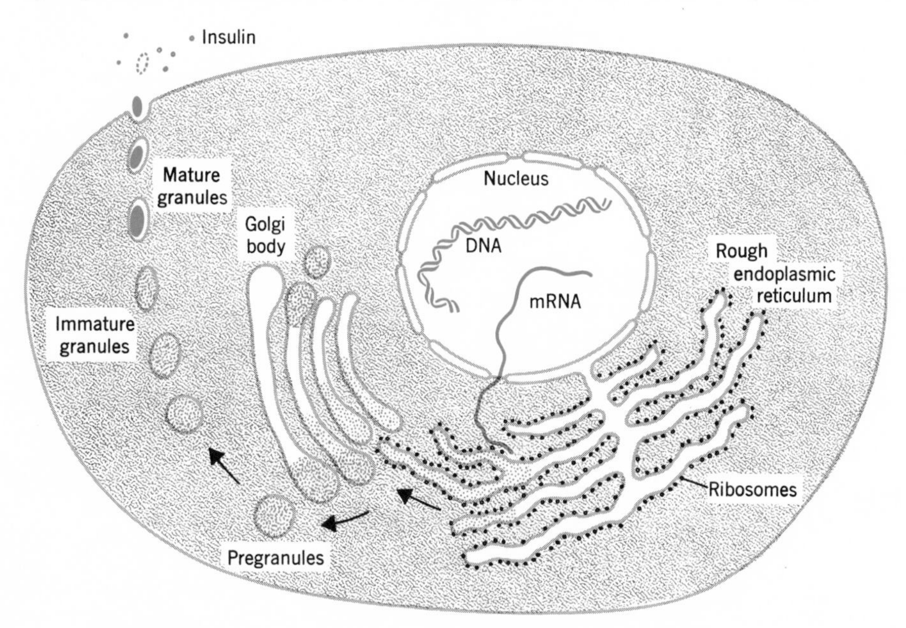

Figure 19-4

Schematic representation of the biosynthesis of insulin in a β cell of the islets of Langerhans in pancreatic tissue. (Modified from a diagram by permission of D. F. Steiner.)

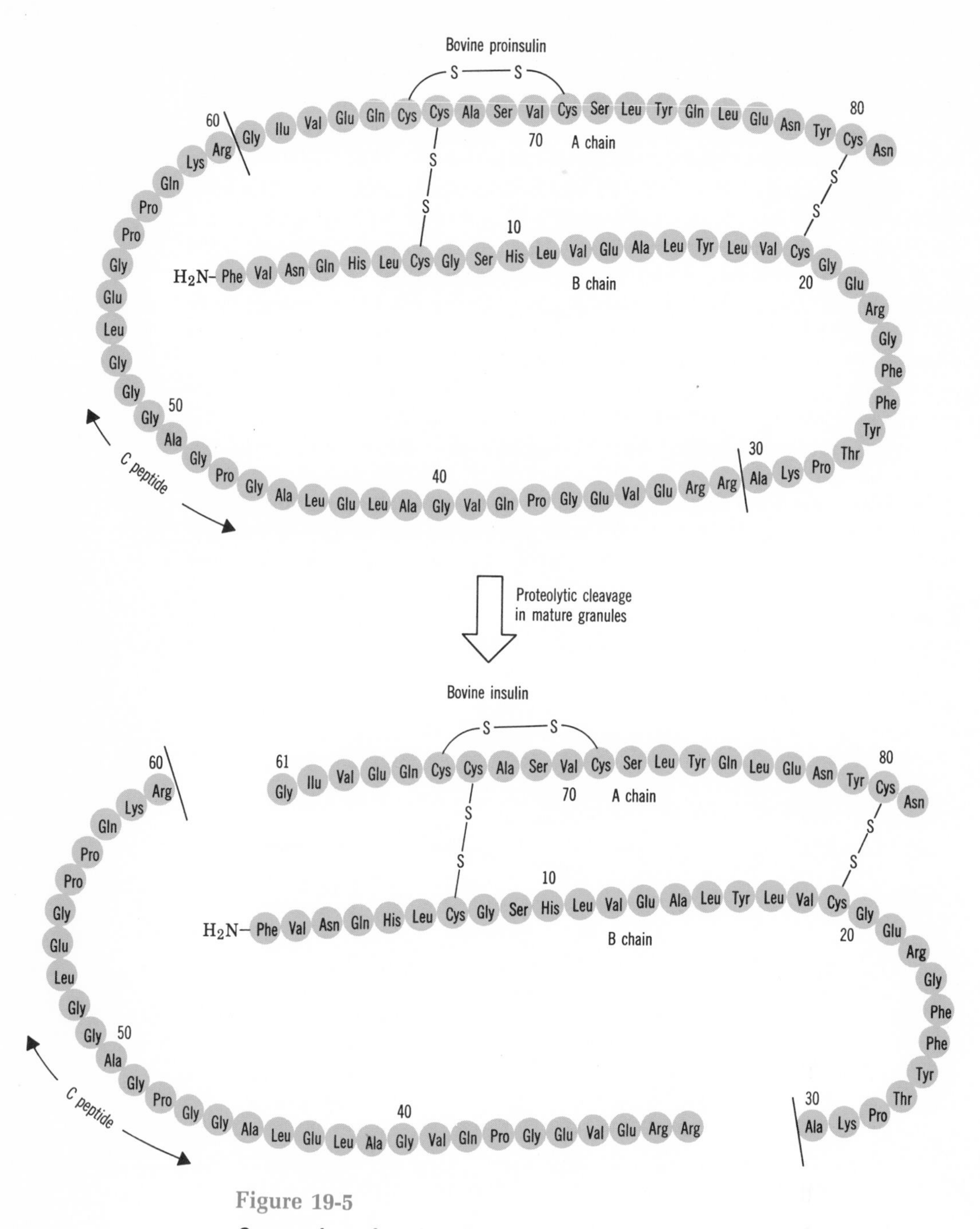

Figure 19-5

Conversion of proinsulin to insulin.

identical to the native ones. This type of experiment is, however, only possible with simple single-polypeptide-chain proteins. Insulin, with its two polypeptide chains, does not readily recombine to form the typical native structure, but rather polymerizes in a random fashion. This observation would suggest that insulin conformation is not thermodynamically highly favored and is dependent upon the integrity of the disulfide bonds.

In sharp contrast, the reduced polypeptide chain of proinsulin, when allowed to stand in dilute alkaline solution under air, is rapidly restored to its native immunological reactivity to levels as high as 80% of the initial values. Under the same conditions, insulin is reconstituted only to about 1%. Thus, one can conclude that a major function of proinsulin in biosynthesis is the facilitation of the proper formation of disulfide bonds under conditions that are thermodynamically highly favored.

Nature of the Genetic Code

Until recently, one of the most intriguing puzzles in modern biology was how to code for twenty amino acids in an unambiguous manner with only four nucleotide residues. Obviously, a definite nucleotide sequence could serve as a code. How long should this sequence be? If each sequence is two residues long, only 4^2 or 16 possible different binary combinations would be available, which is less than the twenty amino acids which must be coded. If each sequence is three residues long, a total of 4^3 or 64 different combinations would be available which would be more than adequate. The solution to this puzzle is one of the most interesting chapters in modern biochemistry.

As we have seen, the sequence of bases on the mRNA directs the precise synthesis of the amino acid sequence of a protein. The codon, the unit that codes for a given amino acid, consists of a group of three adjacent nucleotide residues on mRNA; the next three nucleotide residues on the mRNA code for the next amino acid, etc. The evidence for these conclusions is based on considerable data. In 1961, M. Nirenberg performed a classical experiment. He employed a system from *E. coli* which consisted of a centrifuged supernatant solution and ribosomes supplemented with tRNA. To this system he added a mixture of radioactive amino acids and polyuridylic acid (poly-U) which had been prepared by the action of polynucleotide phosphorylase on UDP. From this complex mixture of amino acids, the only amino acid incorporated into an acid-insoluble fraction consisting of newly synthesized protein was phenylalanine. The product proved to be polyphenylalanine. Nirenberg correctly concluded that the synthetic poly-U was in effect serving as mRNA and was providing the information which specified that only phenylalanyl–tRNA units should become associated with ribosomal nucleoprotein poly-U complex. Randomly mixed polynucleotides were then prepared by the addition of varying amounts of CDP, ADP, or GDP together with UDP to polynucleotide phosphorylase. When the synthetic RNA polymers of different base composition were added to the test system we have described, different amino acids were incorporated into the proteins. A minimum coding ratio of three nucleotide bases for each amino acid was determined, although the precise order of the base in each triplet code was not yet known.

In 1964, Nirenberg devised a simple method whereby trinucleotides of known sequence were employed to decipher the code. A trinucleotide of known sequence was mixed with a labeled amino acyl–tRNA and ribosomes. After a given period of incubation, the suspension was filtered through a fine-porosity filter. Binding of the amino acyl–tRNA to ribosomes depended on the presence of a specific trinucleotide. If no binding occurred, the amino acyl–tRNA would pass through the filter; however, if binding did occur, the amino acyl–tRNA ribosomal complex would remain on the filter and could be easily counted for radioactivity. In effect, the synthetic trinucleotides served as model codons. Employing this trinucleotide binding effect, Nirenberg examined 64 trinucleotides with 20-odd amino acyl–tRNA's.

During this same period K. Khorana, employing organic synthetic techniques as well as enzymatic techniques, prepared synthetic polyribonucleotides with completely defined repeating sequences. Thus, the repeating sequence of CUC UCU CUC . . ., when added to the protein-synthesizing system and radioactive amino acid, yielded a polypeptide which contained only alternating residues of leucine and serine.

From these and other experiments, biochemists were finally able to assign specific amino acids to 61 out of 64 possible codons, with the remaining three recently being designated as terminator codons. Table 19-3 summarizes these results, namely the genetic code.

Table 19-3

The Genetic Code

First position (5′ end)	Second position				Third position (3′ end)
	U	C	A	G	
U	Phe	Ser	Tyr	Cys	U
	Phe	Ser	Tyr	Cys	C
	Leu	Ser	Term[a]	Term	A
	Leu	Ser	Term	Trp	G
C	Leu	Pro	His	Arg	U
	Leu	Pro	His	Arg	C
	Leu	Pro	GluN	Arg	A
	Leu	Pro	GluN	Arg	G
A	Ileu	Thr	AspN	Ser	U
	Ileu	Thr	AspN	Ser	C
	Ileu	Thr	Lys	Arg	A
	Meth (initiation)	Thr	Lys	Arg	G
G	Val	Ala	Asp	Gly	U
	Val	Ala	Asp	Gly	C
	Val	Ala	Glu	Gly	A
	Val	Ala	Glu	Gly	G

[a]Chain-terminating.

We shall now list briefly some generalizations concerning the code:

(1) *The code is universal;* that is, all procaryotic and eucaryotic organisms use the same codons to specify each amino acid.
(2) *The code is degenerate;* that is, more than one arrangement of nucleotide triplets specify the same amino acid. Thus, UUA, UUG, CUU, CUC, CUA, and CUG are codons for leucine. We immediately notice that the first two sets of bases are specific but the third base is flexible. This suggests that a change in the third base by mutation may still allow the correct translation of a given amino acid into protein. Degeneracy often involves only the third base in the codon. For example, the codons for phenylalanine are UUU or UUC. The third base need simply be a pyrimidine. Two of the codons for leucine also begin with UU which in this case must have its third base occupied by a purine. The general pattern of codon assignment suggests that the nucleotide at the 3′ end of the codon may be occupied by either of two bases, a purine or a pyrimidine. This base fit in the third position is termed *wobble*. Therefore, nearly all codons can be represented as xy^{A}_{G} or xy^{U}_{C}. Beside the usual four bases A, U, G,

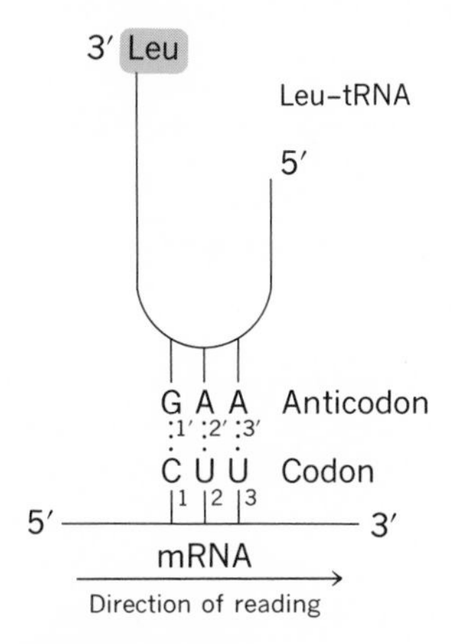

Scheme 19-5

and C, a fifth base I (inosine) is frequently found to form part of the anticodon; I never occurs in codon systems (Scheme 19-5). It, however, occurs as a base in the triplet anticodon of tRNA and invariably complements the third or flexible base of the codon. Thus, the codon CUU can be read by the anticodon, AAG or IAG (read from the 5′ ⟶ 3′ direction). The reason for the flexibility of I is that I can hydrogen bond with U, A, or C.
(3) *The code is nonoverlapping;* that is, adjacent codons do not overlap.
(4) *The code is commaless;* that is, there are no special signals or commas between codons.
(5) Of the 64 possible triplet codons, 61 are employed for encoding amino acid. Three, UAA, UAG, and UGA had been originally called nonsense codons, but now are recognized as specific termination codons.
(6) The codon AUG is of considerable interest, since it is the only codon for

methionine regardless of whether fMET–tRNA$_f^{met}$ or MET–tRNA$_m^{met}$ are employed as methionyl carriers. It serves as the extremely important initiator codon as well as for internal methionyl insertion. Presumably, the initiator factors F1, F2, and F3 or the secondary structure of the mRNA discriminate as to the correct use of AUG. The roles of the initiator and terminator codons are depicted as in Scheme 19-6.

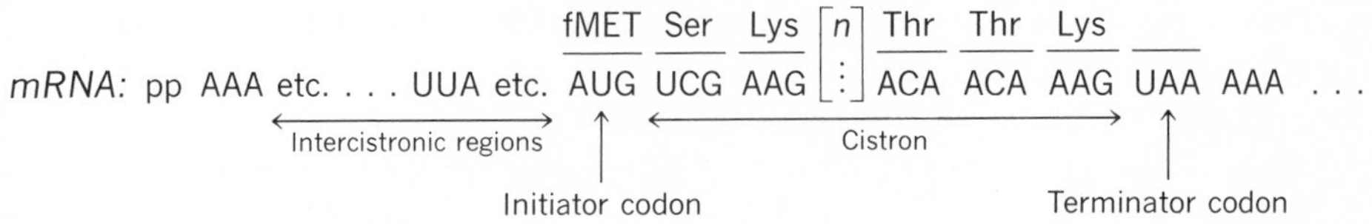

```
                                    fMET Ser  Lys [n] Thr  Thr  Lys  ___
mRNA: pp AAA etc. . . . UUA etc.    AUG  UCG  AAG [:] ACA  ACA  AAG  UAA  AAA . . .
       <-------------------------->  ↑   <--------------------------->  ↑
          Intercistronic regions     |            Cistron               |
                              Initiator codon                   Terminator codon
```

Scheme 19-6

The AUG signal does not directly follow a terminator signal (UAG, UAA, or UGA). There are regions along a polycistronic mRNA that are not translated. The function of such intercistronic regions is not clear.

(7) In general, amino acids with hydrocarbon residues have U or C as the second base; those with branched methyl groups have U as the second base. Basic and acidic amino acid have A or G as the second base.

Genetic Defects in Metabolism

Having examined the mechanisms of replication, transcription, and translation, let us now briefly apply our knowledge to an extremely interesting area of biochemistry.

In the wild-type or normal organism, the total metabolism of the organism is so geared by its array of enzymes that no metabolic intermediates accumulate. However, by a genetic mutation in which a key enzyme is no longer synthesized in an active form, intermediates can either accumulate to significant levels or are excreted. These genetic defects are extremely useful in determining the intermediate steps of a metabolic route under *in vivo* conditions and have been exploited in a large variety of organisms. With humans, these defects lead to the so-called *inborn* errors of metabolism, tragic diseases which in many cases are incurable.

Lower Organisms. The bread mold *Neurospora crassa* has provided excellent material for the biochemical geneticist. Wild strains of *N. crassa* will usually grow well in a simple culture medium composed of sugar, salts, and biotin. When these cultures are exposed to a mutagenic agent such as x rays, we can obtain mutants which only grow when suitable nutritional additions are made to the initial medium. A systematic analysis of the needs of the mutant will frequently indicate a single new nutritional requirement. We do not discuss here the details of the genetic analysis that relates the new nutritional requirement to a position or locus on the chromosomes, but instead we indicate by several examples the great value of this general method in metabolic studies.

Biosynthesis of arginine. Three genetically distinct mutants of *N. crassa* have been observed and thoroughly documented in the metabolism of arginine; these mutants will grow when one or more of three amino acids, namely arginine, citrulline, and ornithine, are added to the minimal medium. Mutant 1 grows only when supplied with arginine but not when given ornithine or citrulline. Mutant 2 can use both citrulline and arginine, but not ornithine, and mutant 3 will grow on any of the three amino acids. These results can be summarized as in our diagram, where the vertical bars indicate a metabolic block in a mutant.

$$\text{Chain of synthesis} \overset{\text{Mutant 3}}{\nrightarrow} \text{Ornithine} \overset{\text{Mutant 2}}{\nrightarrow} \text{Citrulline} \overset{\text{Mutant 1}}{\nrightarrow} \text{Arginine}$$

The nutritional mutant will in general grow on substrates that come after the metabolic block but not on those coming before the block. There may, on some occasions, be an actual accumulation of an intermediate because it is not further metabolized. Thus, in mutant 1 citrulline may accumulate since its further metabolism is blocked by the absence of the enzyme required for its conversion to arginine. By this analysis the biochemist can state that the sequence of synthesis of arginine must follow the order $\longrightarrow$ ornithine $\longrightarrow$ citrulline $\longrightarrow$ arginine.

Biosynthesis of lysine. This method can be applied to organisms other than *N. crassa* to reveal a different or alternate pathway of biosynthesis. Mutants requiring lysine for growth have been found in *N. crassa* and *E. coli,* both of which normally synthesize lysine from sugar and inorganic nitrogen compounds such as nitrate and ammonia. In *N. crassa*, α-amino adipic acid is converted to lysine by some mutants, but these will not use diaminopimelic acid. Some *E. coli* mutants will grow on this acid with ease, however. Diaminopimelic acid and its precursors will also accumulate in different *E. coli* mutants. The mutants that accumulate precursors are deficient in a normally present enzyme which permits utilization of a given precursor. These results are pictured in our diagram.

The value of this type of study is apparent; it reveals new pathways as well as confirms established routes in a variety of organisms. Similar studies have been carried out with mutants from a large number of organisms in the metabolism of amino acids, nucleic acids, vitamins, porphyrins, pigments, and fatty acids. Besides contributing greatly to our knowledge of metabolism, these studies also indicate a direct relation between the enzymatic potential of an organism and its heredity and have led to the hypothesis of *one gene–one polypeptide chain,* which states that a single gene controls the synthesis of a single polypeptide. Separate chains aggregate to yield the active enzyme. Thus, mutant 2 in the arginine pathway no longer has the capacity to synthesize the active critical enzyme protein needed to produce arginine because of the destruction of a specific genetic locus.

Although the one gene–one polypeptide hypothesis is at first glance a simple one, there are at least three ways by which a genetic modification

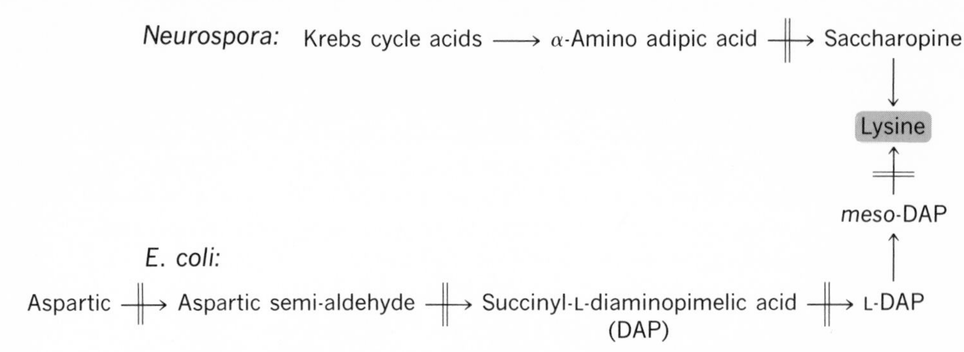

could affect enzyme activity. It could (*a*) cause a change in the molecular structure of the enzyme; (*b*) decrease the concentration of the enzyme and thereby modify the rate of the reaction; or (*c*) exert an indirect effect that involves no change in the enzyme itself. Some aspects of these problems are discussed in Chapter 20.

Inborn Errors of Metabolism in Mammals. In man several diseases are related to a genetic block. These include *alkaptonuria,* in which there is a genetic block in the utilization of homogentisic acid, an intermediate in the oxidation of tyrosine; *phenylketonuria,* in which phenylalanine cannot be converted to tyrosine; and *galactosemia,* in which galactose cannot be directly utilized. The biochemical explanation for galactosemia has been discussed earlier (Chapter 10).

A remarkable inheritable disease is called *sickle cell anaemia*. Human hemoglobin consists mostly of hemoglobin A, which is made up of 2 α-peptide chains and 2 β-peptide chains, $\alpha_2\beta_2$. Patients with sickle cell anaemia inherit the disease in a Mendelian fashion. The heterozygous individual, carrying one normal and one abnormal allele, produces approximately equal amounts of hemoglobin A and S. The homozygous individual, however, produces only abnormal hemoglobin S. Hemoglobin S has a lower solubility and is responsible for the abnormal or sickle shape of the erythrocyte. Since these cells tend to be destroyed by the spleen, severe anaemia will usually result in these patients. The difference between hemoglobin A and S is the substitution of a valine residue for a glutamic residue in the sixth amino acid from the NH_2-terminus of the β-peptide chain:

Normal:

β chain NH_3^+–Val–His–Leu–Thr–Pro–Glu–Glu–Lys . . .

Abnormal:

β chain NH_3^+–Val–His–Leu–Thr–Pro–Val–Glu–Lys . . .

The α-peptide chains in both A and S hemoglobins are the same. Thus, a difference of one amino acid in over 300 residues will cause a drastic and serious change in the physical properties of the pigment concerned primarily with oxygen transport in the body. A mutation such as this, where there is a single amino acid replacement, is called a *point mutation*. Presumably, a single base change has occurred in the portion of the DNA which is involved with the coding of hemoglobin A. It is of interest that an examination of the codons in messenger RNA for glutamic acid shows the following base triplets: GAA and GAG; and for valine GUA and GUG.

One final example of an inborn error in metabolism is the *Tay–Sachs disease,* a fatal cerebral degenerative disorder transmitted in an autosomal recessive manner. The primary defect is the total absence of the hydrolytic enzyme β-D-*N*-acetylhexosamidase A which normally cleaves the terminal *N*-acetylgalactosamine residue from the stored cerebral ganglioside. In its absence massive amounts of this polysaccharide accumulate in the cerebrum, leading to profound mental and motor deterioration and death by age 2–4 years. Fortunately, a precise prenatal assay for this enzyme can be carried

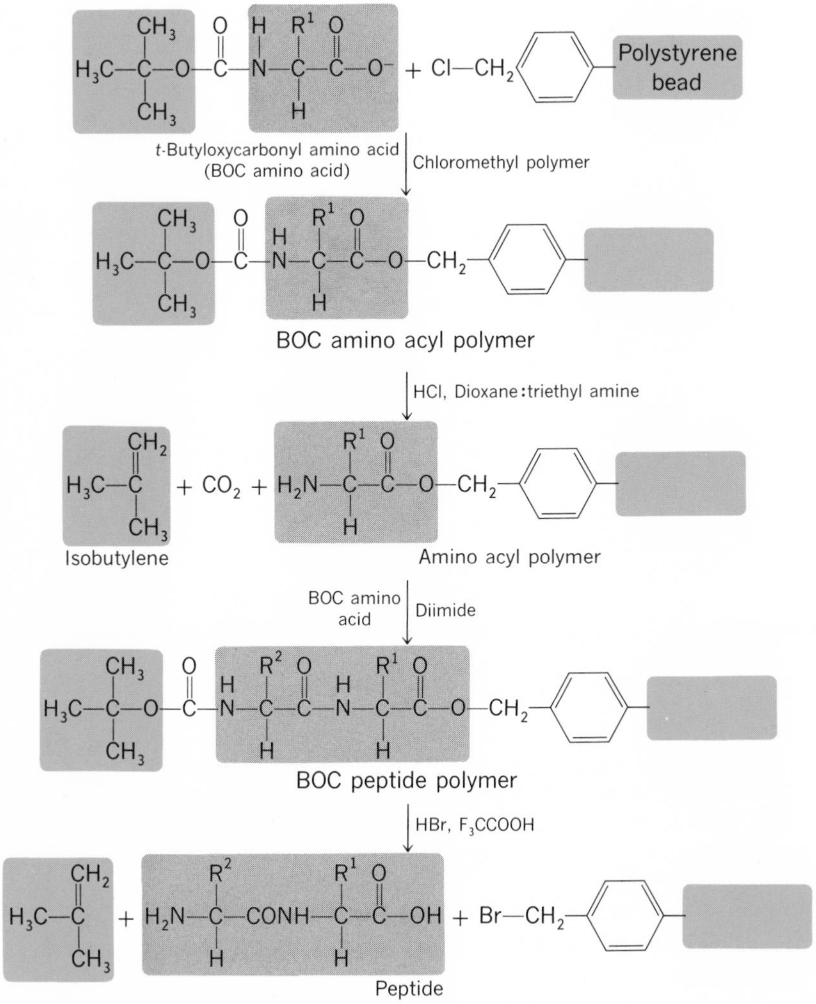

$$(CH_3)_3C{-}O{-}CO{-}NH{-}CHR^1{-}COO^- + Cl{-}CH_2{-}C_6H_4{-}\text{Polystyrene bead}$$

t-Butyloxycarbonyl amino acid (BOC amino acid)

Chloromethyl polymer

$$(CH_3)_3C{-}O{-}CO{-}NH{-}CHR^1{-}CO{-}O{-}CH_2{-}C_6H_4{-}$$

BOC amino acyl polymer

HCl, Dioxane:triethyl amine

$$CH_2{=}C(CH_3){-}CH_3 + CO_2 + H_2N{-}CHR^1{-}CO{-}O{-}CH_2{-}C_6H_4{-}$$

Isobutylene

Amino acyl polymer

BOC amino acid

Diimide

$$(CH_3)_3C{-}O{-}CO{-}NH{-}CHR^2{-}CO{-}NH{-}CHR^1{-}CO{-}O{-}CH_2{-}C_6H_4{-}$$

BOC peptide polymer

HBr, F_3CCOOH

$$CH_2{=}C(CH_3){-}CH_3 + H_2N{-}CHR^2{-}CONH{-}CHR^1{-}COOH + Br{-}CH_2{-}C_6H_4{-}$$

Peptide

Figure 19-6

Chemical reactions employed in solid-phase peptide synthesis.

out on pregnant women suspected of being heterozygous carriers by analyzing for the hydrolase in amniotic fluid or amniotic cells. High correlation of low amounts of the enzyme in Tay–Sachs carriers and fetuses having the disease allows for a recommendation for therapeutic abortion, since the disease is incurable.

Organic Synthesis of Proteins

One of the remaining challenges in the field of chemical synthesis is the total synthesis of biologically active proteins. Since the days of Emil Fischer, who was the first to devise methods to synthesize polypeptides, organic chemists have struggled with this formidable problem. The standard techniques have involved the building up of long chains by repeated condensations of individual amino acids or by combining a series of small peptides to form a single large peptide. Such multisteps are not only very laborious and time-consuming but, more important, the yields are vanishingly small.

In recent years, R. B. Merrifield has devised an ingenious solid-phase peptide synthesis of polypeptides. Essentially, a peptide chain can be synthesized in a stepwise manner from the carboxyl terminal amino acid by having the carboxyl terminus attached by a covalent linkage to an insoluble support. After the synthesis has been completed the peptide chain can be smoothly cleaved from its solid-phase site. Figure 19-6 illustrates the procedure. The technique has now been automated so that six amino acids can be added to a growing peptide every 24 hours.

The procedure is rapid, the yields are high, and no racemization occurs. Simple oligopeptides and complete biologically active proteins can be synthesized. For example, bradykinin (9 residues) can be synthesized in less than a week. Insulin (51 residues), ferredoxin (55), acyl carrier protein (77), and even ribonuclease (129) have been effectively synthesized.

The method opens up vast opportunities for the synthesis of hormones and enzymes, and it also allows the protein chemist to modify these biologically active compounds to determine active sites, etc.

References

1. J. D. Watson, *Molecular Biology of the Gene*. 2nd Ed. New York: Benjamin, 1970.
   An excellent account of the modern concepts of protein synthesis for the interested undergraduate.
2. C. B. Anfinsin, Jr., ed., *Aspects of Protein Biosynthesis*. vol. I (1970), vol. II (1972). New York: Academic Press.
   A thorough account of the whole process at an advanced level.
3. J. Lucas-Lenard and F. Lipmann, *Protein Biosynthesis. Ann. Rev. Biochem.* **40,** 409–441 (1971).
   A recent evaluation of the present status of protein synthesis.

Metabolic Regulation

Introduction

The growth and maintenance of a cell require a highly integrated coordination of anabolic and catabolic processes. Since the functioning unit of the metabolic machinery is the enzyme-catalyzed reaction, the control of this unit becomes the essential feature in metabolic regulation.

Mechanisms of metabolic regulation have been intensively investigated in the past ten years in both procaryotic and eucaryotic organisms. While the subject is complex and still in its infancy, unifying principles are beginning to emerge. Being a multifaceted term, metabolic regulation involves, (*a*) compartmentation of enzymes, (*b*) alternate or separate pathways for catabolism and anabolism of a key substrate, (*c*) kinetic factors involving the interactions of substrates, cofactors and enzymes, and (*d*) the control of enzyme concentration.

Enzyme Compartmentation

In Chapter 6, we considered in some detail the structural components of both procaryotic and eucaryotic cells. The procaryotic cell has for over twenty years been employed by biochemists as a model cell to explore all aspects of cell metabolism. In terms of structure it is a rather simple cell with a plasma membrane onto which an important number of key enzymes are associated (Chapter 6) and a cytoplasmic region in which the principal pathways of metabolism are carried out in an astonishingly orderly manner. The plasma membrane apparently serves as a substitute for organelle enzymes in that the enzymes frequently associated with eucaryotic organelle membranes, such as those which participate in the respiratory chain, and oxidative phosphorylation, as well as phospholipid biosynthesis, are found in the procaryotic plasma membranes. At first glance the cytoplasm of the procaryote may exhibit little structure, but there is increasing evidence that even in this region enzymes may assume a loose, fragile organizational structure. Recent evidence suggests, for example, that acyl carrier protein, a highly soluble protein essential for fatty acid synthesis, rather than being uniformly dispersed in the cytoplasmic region of the *E. coli* cell, is rather loosely associated or layered on the inner surface of the plasma membrane of the cell. It is quite possible that loose, unstable aggregates of enzymes which are involved in sequential metabolic reactions indeed exist in the procaryotic cell, but are immediately disrupted when the cell is subjected to the violent probings of the biochemist.

In the eucaryotic cell, however, an entirely different situation exists. In these cells, compartmentation of metabolic machineries occur for very specific purposes. As in procaryotic organisms, the plasma membrane of eucaryotic organisms is involved in selective transport of important cations, anions, and neutral compounds as well as serving as a barrier from the external milieu. The nucleus is the site for genetic information and for the transcription of this information, i.e., the biosynthesis of mRNA and tRNA in the nucleoplasm and rRNA in the nucleolus and the subsequent modification and transportation of these informational molecules to the cytoplasm for translation into catalytic units, namely, enzymes. The mitochondrion is characterized by its complex of enzymes involved in maintaining the energetics of the entire cell. The endoplasmic reticulum serves as a site of important membrane enzymes. Lysosomes are specific compartments for a host of hydrolytic enzymes which if released into the cytoplasm would cause swift autolysis of that cell. In plants, the chloroplast is the prime organelle for the generation of oxygen, ATP, and reducing power for the plant cell. The Golgi bodies in eucaryotic cells are involved in the formation of secretory bodies (Chapter 6) and also participate in the formation of cell membranes and cell walls.

Another consideration is the spatial separation of multienzyme systems from each other. Thus, in the degradation of glucose to carbon dioxide and water, at least three pathways are involved: glycolysis, the pentose cycle, and the tricarboxylic acid cycle. The glycolytic enzymes and the enzymes of the pentose cycle are found outside the particles, whereas enzymes of the tricarboxylic acid cycle are associated with mitochondria as are the tightly bound particulate enzymes of electron transport and oxidative phosphorylation. A close partnership must exist between the three metabolic sequences, and any interference in that partnership will result in a breakdown or modification of glucose metabolism. Furthermore, any change in the concentration of phosphate and magnesium ions, the ratio of ADP to ATP, $NADP^+$ to NADPH, NAD^+ to NADH, or the tension of oxygen and carbon dioxide would also affect this partnership.

Still another factor in metabolic control and regulation is the ability of mitochondria, for example, to concentrate coenzymes, substrates, and enzymes far above the concentration found outside the particles. By this mechanism the kinetic responses of enzyme-catalyzed reactions in mitochondria are greatly changed.

A final but difficult factor to evaluate is the possible physical compartmentation of enzymic sequences, which would introduce new variables such as permeability barriers toward substrates, enzymes, and cofactors. By compartmentation we mean the actual physical separation and organization of enzymes or substrates into areas such as mitochondria in the cell.

An understanding of the subtle interplay of these organelles in terms of their functions is an extremely important area of investigation.

Alternate Pathways for Metabolism of a Substrate

An important number of biochemical reactions which appear to be reversible are so because of the involvement of two separate enzymes, one catalyzing the forward reaction and the other the backward reaction. These are called *opposing unidirectional reactions:*

$$A \underset{b}{\overset{a}{\rightleftharpoons}} B$$

Typical examples of varying complexity can be cited:

(1)
(*a*) $\text{Glucose} + \text{ATP} \xrightarrow{\text{Hexokinase}} \text{Glucose-6-phosphate} + \text{ADP}$
(*b*) $\text{Glucose-6-phosphate} + H_2O \xrightarrow{\text{Gluco-6-phosphatase}} \text{Glucose} + \text{Pi}$

(2)
(*a*) $\text{Fructose-6-phosphate} + \text{ATP} \xrightarrow{\text{PFK}} \text{Fructose-1,6-diphosphate} + \text{ADP}$
(*b*) $\text{Fructose-1,6-diphosphate} \xrightarrow{\text{Fructo-1,6-phosphatase}} \text{Fructose-6-phosphate} + \text{Pi}$

(3)
(*a*) $\text{Acetate} + \text{ATP} + \text{CoA} \xrightarrow{\text{Thiokinase}} \text{Acetyl–CoA} + \text{AMP} + \text{PP}$
(*b*) $\text{Acetyl–CoA} + H_2O \xrightarrow{\text{Thioesterase}} \text{Acetate} + \text{CoA}$

(4)
(*a*) $\text{Acetyl–CoA} + CO_2 + \text{ATP} \xrightarrow{\text{Acetyl–CoA carboxylase}} \text{Malonyl–CoA} + \text{ADP} + \text{Pi}$
(*b*) $\text{Malonyl–CoA} \xrightarrow{\text{Malonyl–CoA decarboxylase}} \text{Acetyl–CoA} + CO_2$

(5)
(*a*) $\text{Phosphoenol pyruvate} + \text{ADP} \xrightarrow{\text{Pyruvic kinase}} \text{Pyruvate} + \text{ATP}$
(*b*) $\text{Pyruvate} + CO_2 \xrightarrow{\text{ATP}} \text{OAA} \xrightarrow{\text{GTP}} \text{Phosphoenol pyruvate} + CO_2$

(6)
(*a*) $\text{Glucose-1-phosphate} + \text{UTP} \longrightarrow \text{UDPG} \longrightarrow \text{Glycogen}$
(*b*) $\text{Glycogen} + \text{Pi} \xrightarrow{\text{Phosphorylase}} \text{Glucose-1-phosphate}$

(7)
(*a*) $n\ \text{Acetyl–CoA} \xrightarrow{CO_2} n\ \text{Malonyl–CoA} \xrightarrow{\text{Acetyl–CoA}} \text{RCOCoA}$
(*b*) $\text{RCOCoA} \xrightarrow[\text{NAD}^+]{\text{CoA}} \frac{R}{2}\ \text{Acetyl CoA}$
β-Oxidation

In all cases, the forward reaction (*a*) is catalyzed by a specific enzyme, while the back reaction (*b*) is catalyzed by a completely different enzyme, which is usually hydrolytic and thus essentially irreversible. The cell utilizes these reactions involving two completely different sets of enzymes to allow fine regulation of reactions *a* and *b*, since it would be very difficult to control reactions *a* or *b* by employing a single enzyme. However, controls must be imposed in these systems, since otherwise these opposing reactions would couple and lead to futile cyclic activities. Thus, reaction 1, if not coupled to other systems, could lead to a net hydrolysis of ATP to ADP + Pi. In the case of the synthesis and degradation of fatty acids, the cell has imposed further restrictions on these catabolic and anabolic systems in eucaryotic organisms by localizing the β-oxidation enzymes in mitochondria (and in some plants in glyoxysomes) while the synthetase is localized in the cytoplasm. In addition, β-oxidation employs as one of its intermediates the L-β-hydroxyacyl–CoA derivative with CoA as the exclusive thioester component, while the synthetase in all organisms utilizes the D-β-hydroxyacyl thioester with acyl carrier protein as the thioester moiety. In procaryotic organisms, both the degradative and the synthetic systems are soluble, but the β-oxidation system occurs in very low concentrations prior to induction by exposure to fatty acid substrates.

Kinetic Factors

General. The kinetics of a single reaction is governed by the concentration of the enzyme, substrates, coenzymes, cations, and anions; by the temperature and pH; and, where applicable, by the —SS—/—SH ratio, the NADPH/NADP$^+$ and NADH/NAD$^+$ ratios, and the levels of activators or inhibitors.

Other factors include the conversion of proenzymes to the fully active enzymes, as well as synthesis, and of breakdown of enzyme proteins and coenzymes. These factors are of considerable importance when comparing protein (and enzyme) turnover in procaryotic and eucaryotic organisms. For example, in bacteria the total activity of a specific enzyme in a culture increases when its inducer is added to the culture, and this activity remains constant even when the inducer is removed. Total enzyme activity only diminishes when, in the absence of the inducer, the cells continue to grow, with a resulting dilution of the enzyme. In sharp contrast, in animal tissues, the level of the enzyme can be increased by the action of hormones, substrates, or changes in nutrition. However, as soon as the stimulus is removed, the enzyme activity returns to its basal activity. These results are shown in the illustration. There is a continuous synthesis and degradation of proteins in animal cells as documented some forty years ago by R. Schoenheimer. This continual turnover is in sharp contrast to the lack of degradation of protein in exponentially growing bacterial cells. For example, the replacement of protein in the rat liver is rapid, with at least 50% of the protein replaced in 4–5 days. This turnover

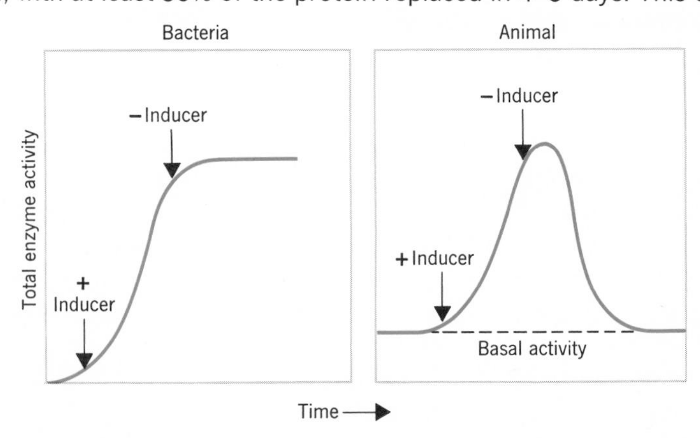

is intracellular; that is, the protein is not excreted but is being synthesized and degraded in the cell. In addition, there is a marked difference in degradation rates of cellular organelle proteins. For example, nuclear protein of rat liver has a half-life of 5.1 days; mitochondrial protein 6.8 days, and endoplasmic reticulum 2.0 days. Enzymes also have a rapid but different turnover rate. For example, glutamic–alanine transaminase has a half-life of 2–3 days, catalase 30 hours, and tyrosine transaminase 2–4 hours.

Nutritional factors play an extremely important role in determining enzyme levels in animal tissues. For example, in a normal balanced diet, fatty acid synthetase will have a basal level of activity which is easily detectable. However, very soon after the diet has been changed to a high-fat, low-carbohydrate diet, the fatty acid synthetase level practically disappears, to reappear within hours if the diet is shifted back to a normal or a low-fat, high-carbohydrate diet. Many examples of this type of variation can be cited to demonstrate the effect of nutrition on enzyme activity in animal tissues. A good understanding of the mechanism of balancing the rate of synthesis and the rate of degradation of enzymes and other proteins in animal cells is at present not available. We shall, however, examine examples of enzyme regulation in animal cells in this chapter which reflect other mechanisms.

One rather limited mechanism involves the conversion of an inactive zymogen to an active enzyme. Most digestive enzymes are formed as zymogens, since they cannot be formed in the active form in ribosomes because this would result in self-destruction of the synthesizing system. Thus, as zymogens, they are inactive but are converted to the active enzymes at the appropriate site of their action, usually in the digestive tract. We have already discussed the conversion of pepsinogen to pepsin, chymotrypsinogen to chymotrypsin, trypsinogen to trypsin, and proinsulin to insulin. This mechanism is a special type of activation which is of limited value in the fine regulation of metabolism.

Enzyme Activation and Inhibition. Mechanisms which modulate enzyme activity by either activating or inhibiting an enzyme will now be considered.

Product inhibition. A rather simple inhibition of a reaction is called product inhibition, where the product of the reaction, by mass action effect, inhibits its own formation. Thus, in the conversion of glucose to glucose-6-phosphate by the enzyme hexokinase, as glucose-6-phosphate begins to accumulate the reaction slows down. It is for this reason that enzyme assays should be carried out at the initial period of the reaction to avoid inhibition by the accumulating product.

Feedback (end-product) inhibition. An even more subtle type of control of enzyme action is designated as feedback inhibition. This is demonstrated most easily by considering the following sequence:

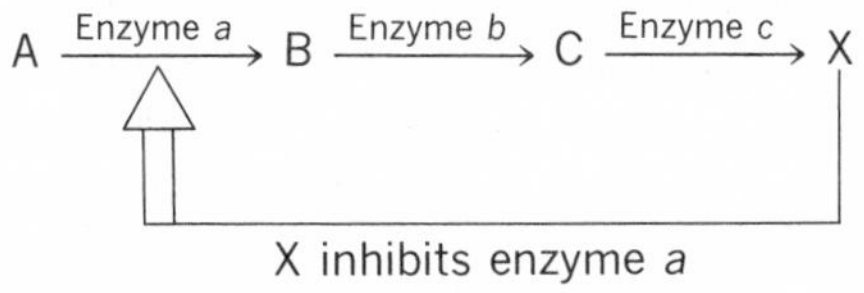

$$A \xrightarrow{\text{Enzyme } a} B \xrightarrow{\text{Enzyme } b} C \xrightarrow{\text{Enzyme } c} X$$

X inhibits enzyme *a*

Here X, the ultimate product of the sequence, serves to prevent the formation of one of its own precursors by inhibiting the action of enzyme *a*. The first enzyme of the sequence, which is also called a monovalent *regulatory* or *allosteric enzyme,* namely enzyme *a*, can also be called the *pacemaker* since the entire sequence is effectively regulated by inhibiting it. An actual example is the formation in *E. coli* of cytidine triphosphate, CTP, from aspartic acid and carbamyl phosphate (Scheme 20-1).

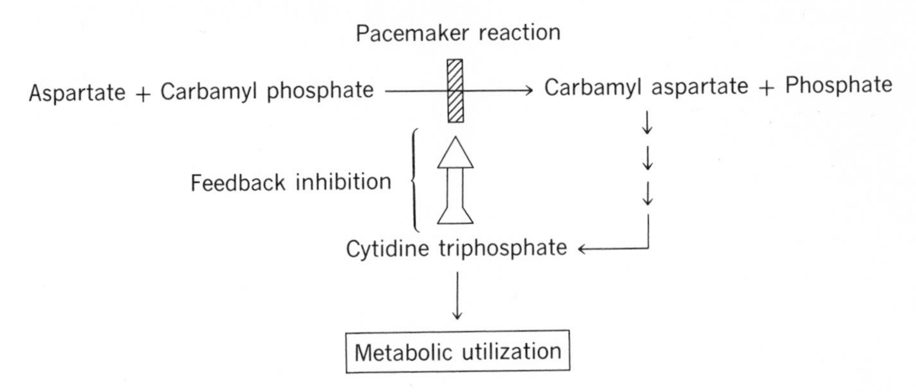

Scheme 20-1

In the diagram, when metabolic utilization is low and CTP concentration is high, feedback inhibition operates. When metabolic utilization is high and CTP concentration is low, feedback inhibition is inoperative.

As a critical concentration of CTP is built up, the triphosphate slows down its own formation by inhibiting the enzyme, aspartate transcarbamylase (ATCase), which catalyzes the pacemaker step for the synthesis of carbamyl aspartate. When the concentration of the triphosphate is sufficiently lowered by metabolic utilization, inhibition is released, and its synthesis renewed (Scheme 20-1).

In all feedback inhibitions, the inhibitor (effector or modulator) usually has no structural similarity to the substrate of the enzyme it is regulating. Thus, CTP in no way resembles aspartic acid, the substrate for aspartic transcarbamylase. Furthermore, all allosteric enzymes so far examined are *oligomeric* enzymes, that is, they have two or more distinct subunits. For example, aspartic transcarbamylase can be readily dissociated into two large subunits, one of which carries the catalytic site and the other the regulatory site. The first subunit, once separated from the second or regulatory subunit, has normal Michaelis–Menten kinetics rather than sigmoidal kinetics and now is no longer affected by CTP. The second subunit has no catalytic activity but binds CTP strongly.

Let us now consider variations of feedback inhibition of metabolic sequences. The regulation of the linear sequence referred to earlier is a straightforward end-product inhibition of the first enzyme in the sequence, a monovalent allosteric enzyme. However, regulation of a branched biosynthetic pathway by X and Y would lead to a situation where an excess of one end product would lead not only to a decrease in the synthesis of X, but also the other end product, not a very good control system. However, a number of mechanisms have been observed which resolve this dilemma.

Monovalent feedback
(by X alone; Y inactive)
Divalent feedback
(by X and Y)

A —a→ B ⟶ C ⟶ D —d→ X; D —d′→ Y

Isofunctional enzymes. In this mechanism the first common step is catalyzed by two different or *isofunctional* enzymes which convert the same substrate to the same product. However, enzyme *a* is under the specific feedback

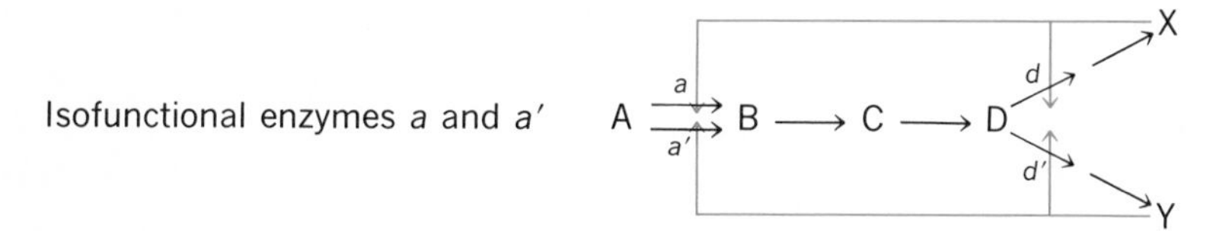

control of X while enzyme *a*′ is insensitive, whereas enzyme *a*′ is under the specific control of Y and is insensitive to X. Since enzyme *a* in the latter instance would still be involved in the synthesis of B, C, and D, a secondary feedback control must be exerted by the two end products; namely X on enzyme *d* and Y on enzyme *d*′. Thus, if an excess of X is formed, it will not only inhibit enzyme *a* and also *d*, but will not interfere with the synthesis of Y. An excellent example of this control mechanism has been described by G. Cohen of France in the biosynthesis of lysine, methionine, threonine and isoleucine from aspartic acid, as depicted in the simplified pathway shown in Scheme 20-2.

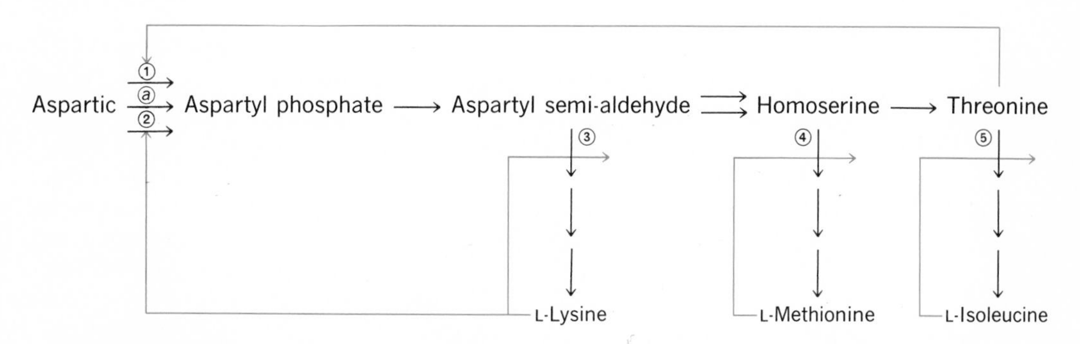

Scheme 20-2

Feedback control of amino acid synthesis: ①, ②, ⓐ—isofunctional aspartic kinases; ①—monovalent feedback inhibited by threonine; ②—monovalent feedback inhibited by lysine; ⓐ—not a regulatory enzyme; ③, ④, ⑤—enzymes under feedback control by lysine, methionine, and isoleucine, respectively.

Sequential feedback control. In this mechanism, enzyme *a* is not regulated by either of the end products of the branched pathway. However, X will inhibit the enzyme that converts the last common substrate D to precursors

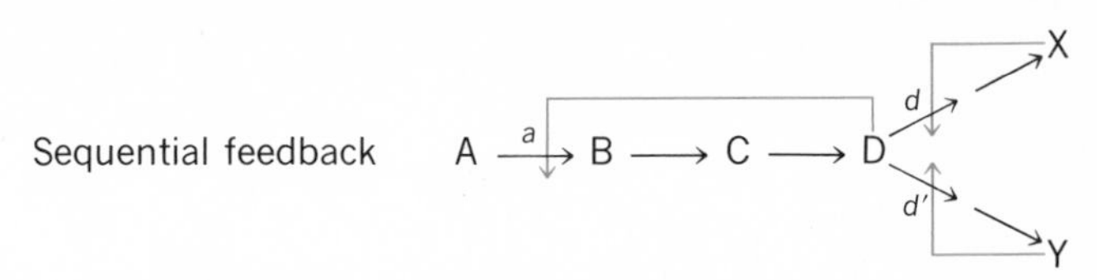

of X, and Y will inhibit the enzyme that will convert D to precursors of Y. Thereby, D will accumulate and inhibit enzyme *a*, which shuts off the entire pathway. An example of this pathway is observed in the biosynthesis of aromatic acids in a number of bacteria and in the regulation of threonine and isoleucine biosynthesis in *Rhodopseudomonas spheroides*.

Concerted feedback inhibition. In this system, enzyme *a* is insensitive to X or Y alone, but when both are present they act in concert to inhibit enzyme *a*. Again, both X and Y exert secondary controls by having X inhibit enzyme

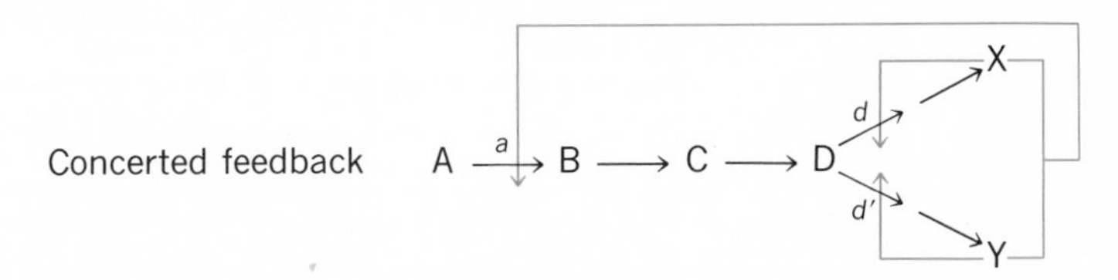

d and Y enzyme *d′*. Thus, if there is an excess of X synthesized, it will only inhibit its own synthesis by controlling the activity of enzyme *d*, allowing Y to be synthesized. As Y accumulates, both X and Y can now, in concert, inhibit enzyme *a*, which is sensitive to inhibition only in the presence of both X and Y. A good example is the inhibition of aspartyl kinase from *Rhodopseudomonas capsulatus* by the combination of both threonine and lysine. Alone these amino acids are ineffective inhibitors.

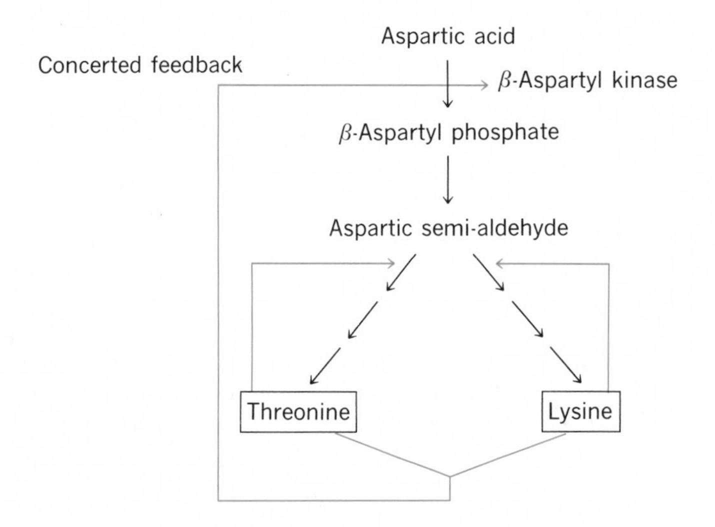

Cumulative feedback inhibition. In this mechanism, X and Y, in saturating concentrations, only cause partial inhibition of enzyme *a*, but when they are both present simultaneously a cumulative effect is observed. Thus, if X at

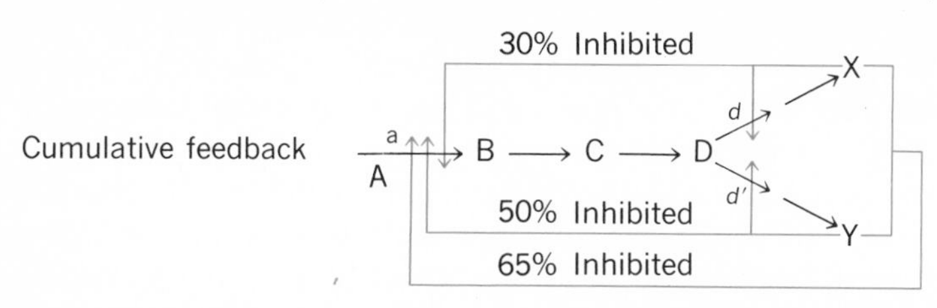

saturating concentration inhibits *a* so that its residual activity is, for example, 70% and Y alone inhibits *a* by 50%, then when X and Y are both present in saturating concentrations the residual activity will be 0.7×0.5 or 35% of the total activity. Inhibition will be 65%. An excellent example of this type of regulation is in the regulation of glutamine synthetase, which will be described in some detail in this chapter.

We have seen that both linear as well as branched biosynthetic pathways can be selectively controlled by different types of feedback control. A number of these regulatory enzymes are controlled by having the effectors induce conformational changes in the protein structures which modify the catalytic site of the enzyme physically. An additional mechanism involves *chemical modification* of the *regulatory enzyme* by covalent attachment of specific groups to the regulated enzyme leading to changes in primary, and thus tertiary, structure of the enzyme. The modifying proteins are specific enzymes which are involved in inserting and removing the specific groups which include phosphoryl and adenylyl components. Table 20-1 lists a number of enzymes which are regulated by chemical modification.

We shall now examine in some detail the regulation of the glutamine synthetase of *E. coli* by a chemical modification mechanism. This enzyme

Table 20-1

Enzymes Regulated by Chemical Modification

Enzyme	Origin	Mechanism of modification	Changes
Glycogen phosphorylase	Mammals; fungi	Phosphorylation/ dephosphorylation	Increased/ decreased
Phosphorylase *b* kinase	Mammals	Phosphorylation/ dephosphorylation	Increased/ decreased
Glycogen synthetase	Mammals; fungi	Phosphorylation/ dephosphorylation	Decreased/ increased
Pyruvate dehydrogenase	Mammals	Phosphorylation/ dephosphorylation	Decreased/ increased
Glutamine synthetase	*E. coli*	Adenylation/ deadenylation	Decreased/ increased

is subject to control by at least four different mechanisms: (*a*) kinetic factors including concentration of ATP and divalent cations, (*b*) repression and derepression of enzyme synthesis in response to the nitrogen source in which the organism grows, (*c*) cumulative feedback inhibition by the multiple end products of glutamic metabolism, and finally (*d*) chemical modification of the synthetase by the attachment and release of adenyl residues by specific enzymes. We shall discuss this last mechanism in some detail.

Detailed investigations by E. Stadtman and H. Holzer of Freiburg, Germany have shown clearly that glutamine synthetase, with a molecular weight of 600,000, consists of twelve identical subunits or protomers, each consisting of a subunit of 50,000 mol wt, with one site for the adenylation by ATP. The acceptor for the adenyl moiety is the hydroxyl group of a tyrosine residue in the polypeptide chain of the monomeric subunit. It is of interest that when ^{14}C-adenyl-labeled glutamine synthetase is digested by proteolytic enzymes, a decapeptide is isolated which, in addition to tyrosine, contains three proline residues. As we have seen earlier (page 87), a high proline content in a critical region of the regulatory site, namely the tyrosine residue, provides a region of minimal secondary structure. Perhaps the reactive tyrosyl residue in this region must be so positioned that the adenyl group can be readily bound or removed. Scheme 20-3 summarizes the present knowledge of this interesting and important enzyme.

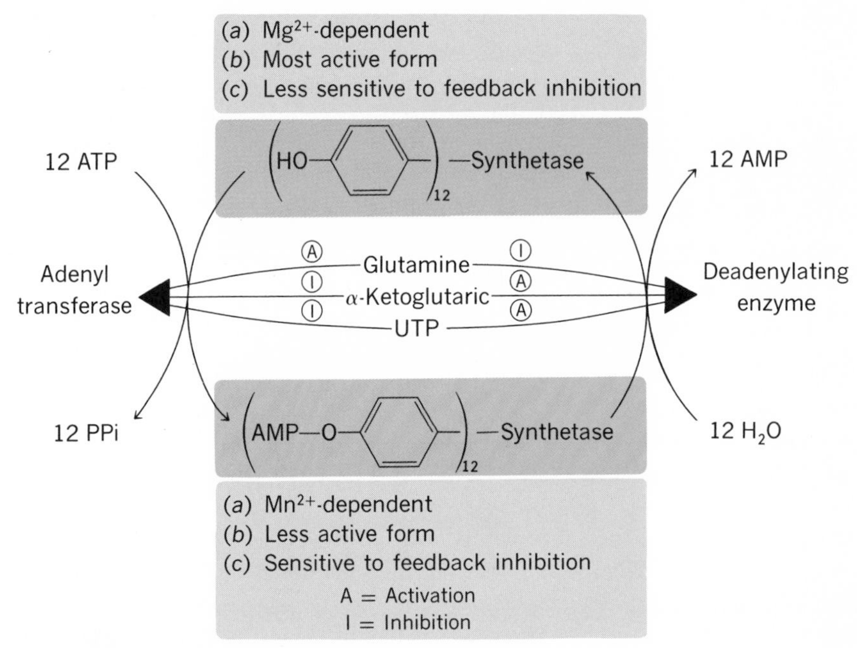

Scheme 20-3

Glutamine synthetase is under *cumulative feedback* control by tryptophan, histidine, CMP, AMP, glucosamine-6-phosphate, and carbamyl phosphate—all end products of glutamine metabolism (see page 476): A, activation; 1, inhibition.

Of unusual interest, glutamine, α-ketoglutaric acid, and UTP are allosteric positive or negative effectors of the two modifying enzymes, namely, adenyl transferase and the deadenylating enzyme.

Whenever one considers the function of regulatory enzymes and their effectors it is always important to relate the molecular events as relevant to *in vivo* conditions of the cell. For example, when the bacterial cell is exposed to conditions of low nitrogen nutrition, there is a relative decrease in nitrogen-containing metabolites in the cell sap which includes glutamic acid and its products, and an increase in α-ketoglutaric acid, the carbon skeleton of glutamine (Scheme 20-4). Under these conditions, the deadenylating enzyme becomes activated, while the adenyl transferase will be inhibited by α-ketoglutaric acid which would rise while glutamine concentration drops. This leads in turn to a deadenylated glutamine synthetase which is more active and less sensitive to feedback inhibition. Ammonia will be rapidly converted to glutamine with high efficiency. But with an excess of nitrogen nutrition, the opposite conditions would hold with a concomitant increase in feedback controls leading to an increase in glutamine and a decrease in its utilization. As a result, the deadenylating enzyme would be inhibited by glutamine, and the adenyl transferase would be activated, leading to the fully adenylated glutamine synthetase which is relatively less active than is the deadenylated synthetase and much more sensitive to cumulative feedback inhibition. Thus, the level of glutamine formation would be lowered.

In summary, we have seen whereby glutamine, a central compound in the nitrogen metabolism of a bacterial cell, has its synthesis under rigid control. The regulatory system involves the three enzymes already referred to which regulate glutamine synthesis, namely (*a*) glutamine synthetase, (*b*) the adenyl transferase, and (*c*) the deadenylating enzyme, the first of which is modified by the second and third enzymes which in turn are under feedback control by glutamine, α-ketoglutaric acid, and UTP. These regulatory events match perfectly the *in vivo* conditions the cell might be exposed to.

Kinetic aspects of allosteric enzymes. The characteristic kinetic factor for most allosteric enzymes is the atypical relationship of activity and substrate concentration. So far, we have considered enzymes that possess independent substrate binding sites, i.e., the binding of one molecule of substrate has no effect on the intrinsic dissociation constants of the vacant sites. Such enzymes yield normal hyperbolic velocity curves. However, if the binding of one substrate (or effector) molecule induces structural or electronic changes that result in altered affinities for the vacant sites, the velocity curve will no longer follow Michaelis–Menten kinetics and the enzyme will be classified as an "allosteric" enzyme. In all likelihood, the multiple substrate (or effector) binding sites of allosteric enzymes reside on different protein subunits. Generally, allosteric enzymes yield sigmoidal velocity curves. The binding of one substrate (or effector) molecule facilitates the binding of the next substrate (or effector) molecule by increasing the affinities of the vacant binding sites. The phenomenon has been called *cooperative binding,* or *positive cooperativity* with respect to substrate binding, or a *positive homotropic*

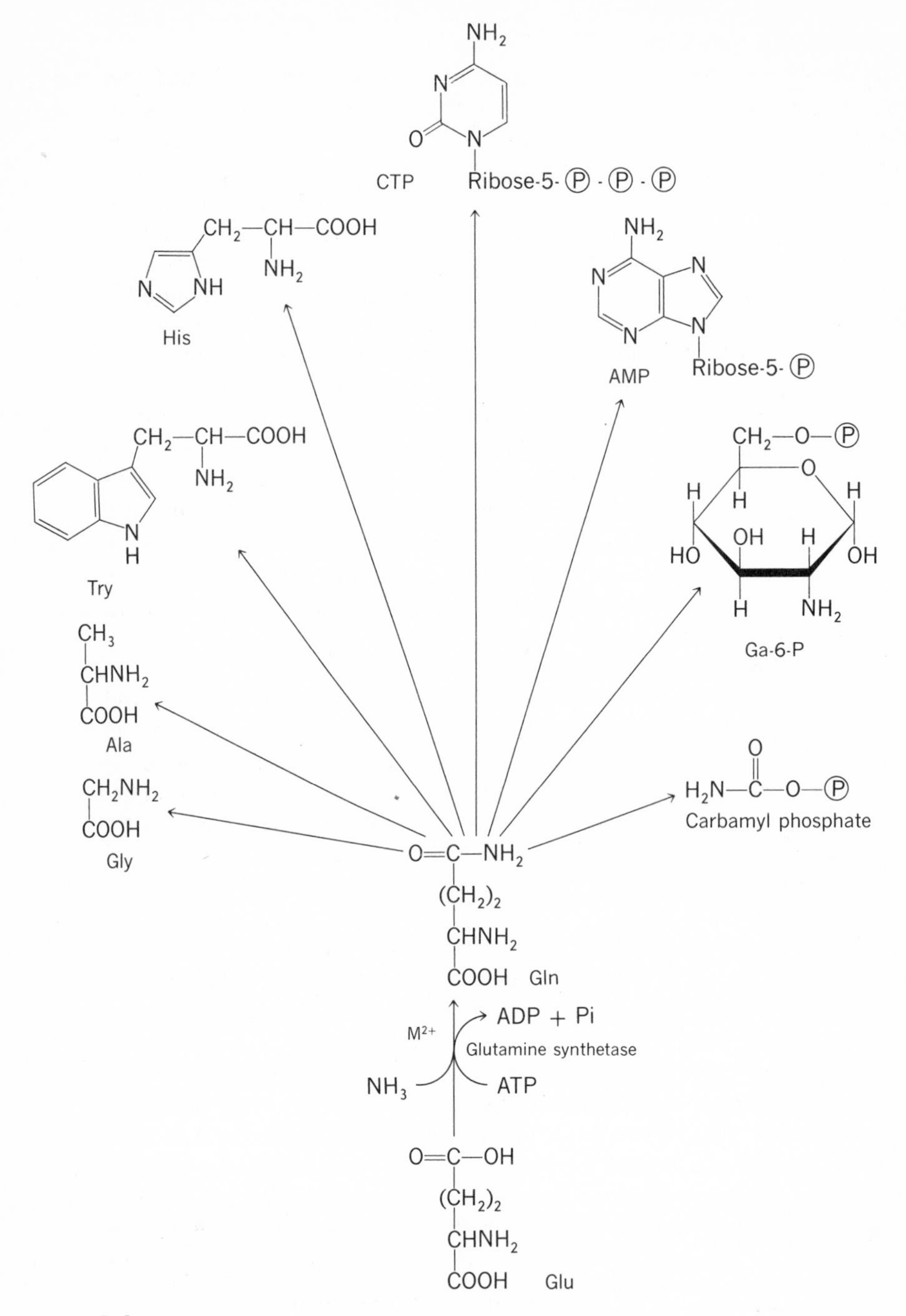

Scheme 20-4

Biosynthesis and metabolic routes of glutamine (Gln).

response. A *positive heterotropic response* signifies that an effector other than the substrate is being bound at a specific regulatory site which increases the affinities of the vacant binding sites.

The potential advantages of a sigmoidal response to varying substrate is illustrated in Figure 20-1. For comparison, a normal hyperbolic velocity with the same $[S]_{0.9}$ is shown. Between $[S] = 0$ and $[S] = 3$, the hyperbolic response curve decelerates, but still rises to 0.75 V_{max}. The sigmoidal curve accelerates exponentially, but only attains 0.10 V_{max} between the same limits

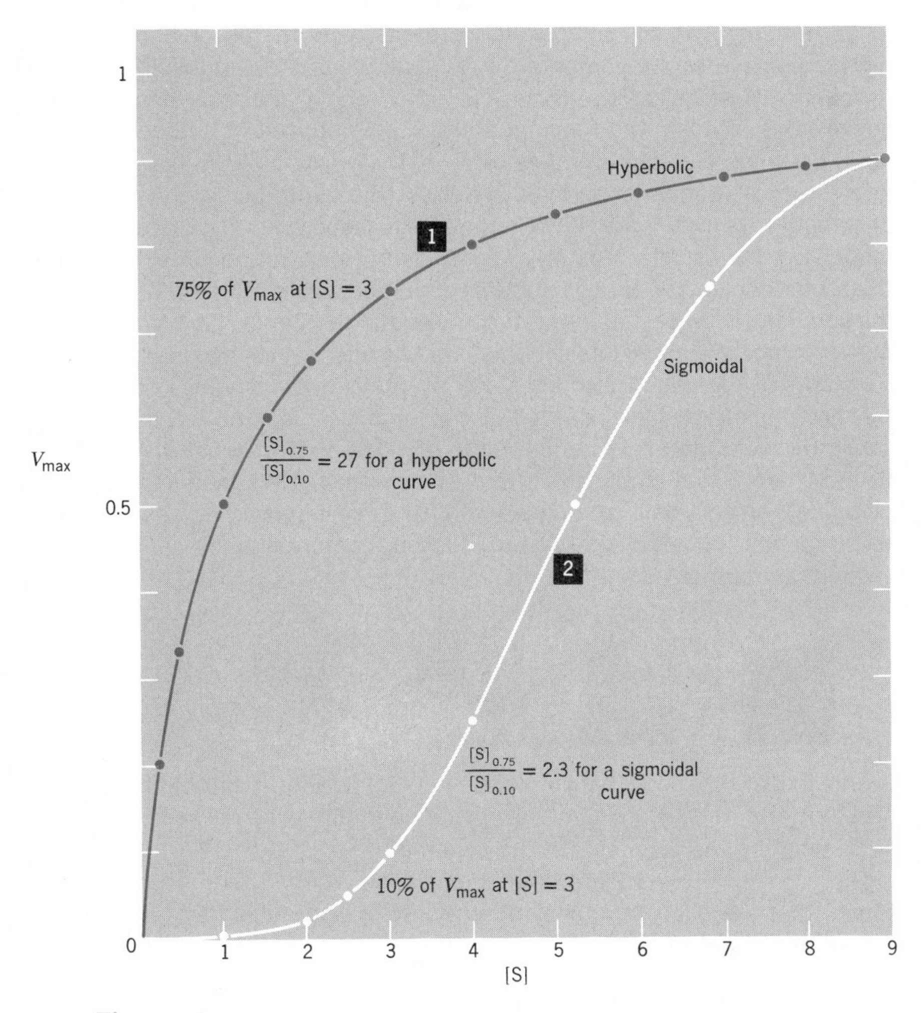

Figure 20-1

The effect of substrate concentrations on the velocity of a reaction catalyzed by [1] a Michaelis–Menten type of enzyme and [2] an allosteric enzyme with sigmoid kinetics. The $[S]_{0.9}/[S]_{0.1}$ ratio for the hyperbolic curve is exactly 81. The $[S]_{0.9}/[S]_{0.1}$ ratio for the sigmoidal response is 9. [S] represents substrate concentration; the subscript indicates the V_{max} at the given [S]. (Reproduced with permission of Irwin H. Segel.)

of [S]. However, the sigmoidal curve increases from 0.10 V_{max} to 0.75 V_{max} with only an additional 2.3-fold increase in [S]. In order to cover the same specific velocity range, the hyperbolic curve requires a 27-fold increase in [S]. Thus, the sigmoidal response acts, in a sense, as an "off–on switch." Also, at moderate specific velocities, the sigmoidal response provides a much more sensitive control of the reaction rate by variations in the substrate concentration.

Two major models for cooperative binding have been proposed. These are the "progressive," or "sequential" interaction model, and the "concerted" or "symmetry" model. Both models are based on the observation that all allosteric enzymes are composed of subunits, i.e., they are oligomers. The "sequential" model of Koshland, Nemethy, and Filmer (based on earlier suggestions of Adair and Pauling, and on the "induced fit" model of Koshland) assumes that the affinities of vacant sites for a given ligand change in a progressive manner as sites are filled (thus introducing the possibility of negative as well as positive homotropic responses). The "sequential" model can be visualized as follows: A ligand (substrate or effector) binds to an unoccupied site on one subunit of an oligomeric enzyme. As a result, the subunit undergoes an induced conformational change. New interactions between subunits are established and this results in a change in the binding constants of the unoccupied sites. For example, if the binding constant for the first substrate molecule is K_B, the binding constant for the second substrate molecule might be altered to iK_B. The second substrate molecule bound changes the binding constant of the vacant sites by another factor, j (to ijK_B), and so on. The sequential change in effective K_B requires that the subunits undergo the ligand-induced conformational change in a sequential manner:

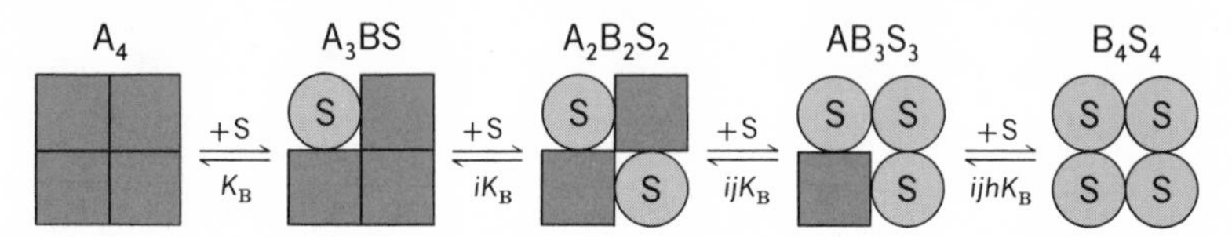

The sequential model can be made very general and applicable to most allosteric enzymes by providing for restricted interactions between subunits, as dictated by the geometry of the oligomers.

The "concerted-symmetry" model of Monod, Wyman, and Changeux assumes that the oligomeric enzyme preexists as an equilibrium mixture of higher- and lower-affinity forms. When a substrate binds preferentially to the

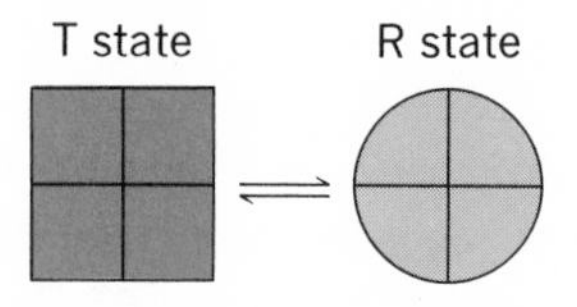

higher-affinity R ("relaxed") state, the equilibrium is displaced in favor of that state:

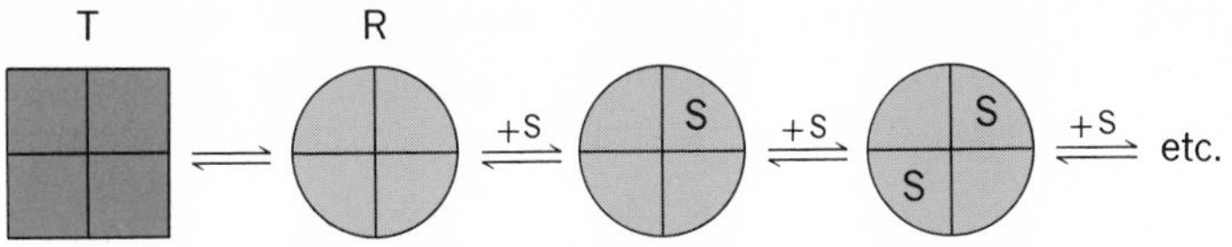

T R

S S S

+S +S +S etc.

The transition between states is concerted, i.e., all the subunits of the oligomers change conformation simultaneously. Because more sites are produced in the transition than were used up by the binding of ligand, the substrate saturation curve is sigmoidal. The concerted-symmetry model does not permit negative homotropic responses. Compounds that bind preferentially to the T ("tight" or "taut") state act as inhibitors; compounds that bind preferentially to the R state act as activators (i.e., they mimic the substrate by promoting the appearance of more higher-affinity substrate sites).

A number of allosteric enzymes exhibit competitive-type kinetics since they involve changes in the apparent K_m of the substrate (i.e., $[S]_{0.5}$) but do not change the V_{max}. They are called *K* systems. Conversely, noncompetitive systems are referred to as *V* systems because they involve changes in the V_{max} but not in the K_m.

Once again the student should note the amplification of stimulation or inhibition in an allosteric enzyme system in contrast to that observed with a nonallosteric system. The feedback mechanism is therefore responsible for continually adjusting the rate of synthesis of metabolic intermediates according to the demands of the cell by positive or negative regulation of allosteric enzymes, the key enzymes in all feedback control mechanisms.

Repression and Induction

Repression and induction—control of enzyme concentration by regulation of transcription or translation. Another important mechanism of the cell is that of *repression*. This mechanism controls the formation of a number of key biosynthetic enzymes. A small molecule—frequently an end product in a biosynthetic path and called a *corepressor*—inhibits the formation of a *repressible* enzyme which acts at an early stage in the biosynthesis of the corepressor. For example, tryptophan synthetase is formed in *E. coli* only when the cells are grown in a tryptophan-free medium. The addition of tryptophan, a corepressor, completely represses the formation of the synthetase. Since the synthesis of enzymes ceases rather rapidly, enzyme already present in the cell continues to catalyze its reaction until it is removed either by degradation or by dilution because of cell growth. Repression is too slow for the continuous adjustment of metabolic sequences we have considered, whereas the mechanism of feedback inhibition is rapid. Repression of the formation of an enzyme by a product of the enzyme's action enables the cell to dispense with making more enzymes than it actually needs for optimal growth. The cell thus releases amino acids for the synthesis of other enzymes which catalyze the formation of compounds in short and needed supply.

Just as *repression* cuts back on the formation of a critical enzyme by a product of that sequence, so *induction* is an important means by which the rate of synthesis of an enzyme can be stimulated several thousandfold. This is accomplished by the addition of the enzyme's substrate to the medium in which the cell is growing. The substrate is called an *inducer* and the enzyme whose synthesis is greatly stimulated by the inducer is called an *inducible* enzyme.

By feedback control, repression, and induction, enzymes in metabolic pathways can be maintained at precise levels so that substrates or intermediates can in turn be kept at physiologically proper concentrations. While inducibility is the rule for *catabolic* enzyme sequences—degradation of exogenous substrates—repressibility is the rule for *anabolic* sequences involved in the synthesis of amino acids and nucleotides. Thus, both repression and induction are highly specific, but inducers are *substrates* of the sequences, whereas corepressors are *products* of the sequences.

J. Monod and F. Jacob of France proposed in 1961 an ingenious mechanism to explain induction and repression. In part for this work and for other major contributions in the field of biochemistry they were awarded the Nobel Prize in Medicine in 1965.

To understand the mechanism of repression, we shall examine the regulation of β-galactosidase in *E. coli* which has been intensely studied and well-characterized. The enzyme catalyzes the reaction

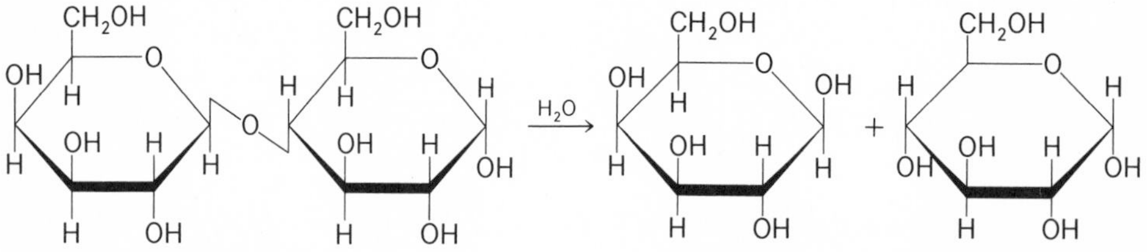

$$\text{Lactose (Nonutilized)} \xrightarrow{H_2O} \text{Glucose (Utilized)} + \text{Galactose (Utilized)}$$

Indirectly, *E. coli* utilizes lactose. First, a galactoside permease which permits entry of lactose into the cell must be induced; and second, β-galactosidase which hydrolyzes the disaccharide to galactose and glucose must be induced. A third enzyme, thiogalactoside transacetylase, is also induced, but its function is unknown. Thus, when *E. coli* is grown in the presence of lactose as the sole carbon source, these three enzymes are induced in large quantities. What is the mechanism of induction? This mechanism is well-understood and serves as a model for the general concept of induction and repression.

The site on the *E. coli* DNA which codes for the enzymes responsible for the utilization of lactose is called the *lactose operon*. The operon consists of four key components: (*a*) a number of structural genes which serve as templates for the mRNA's responsible for the translation of information for the synthesis of enzymes involved in lactose metabolism, namely, β-galactosidase, galactoside permease, and thiogalactoside transacetylase, respectively; (*b*) the operator gene, O, adjacent to the first structural gene; (*c*) the promotor region, P, which is in turn contiguous to the operator gene; and (*d*) the repressor protein which is in turn the product either of a topographically independent or a closely associated regulatory gene. The *lac* operon is thus visualized as shown in Scheme 20-5.

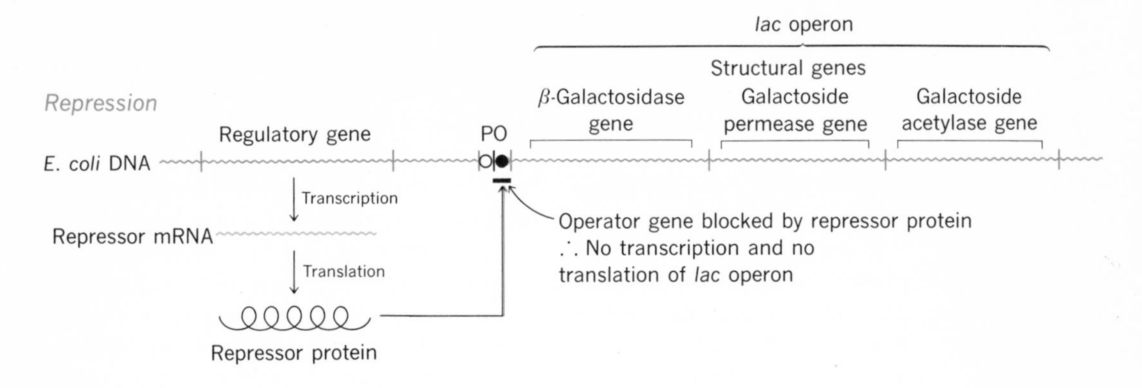

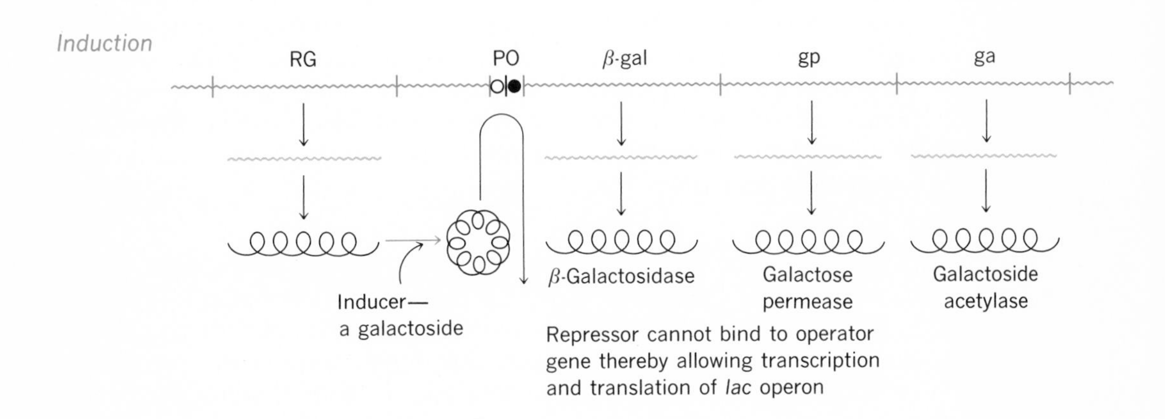

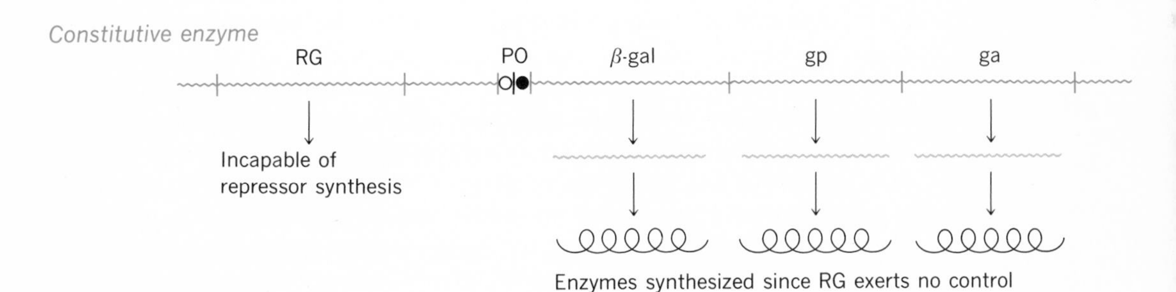

Scheme 20-5

The regulatory gene codes for the transcription of the repressor mRNA which in turn serves as the template for the formation of the protein, the galactosidase repressor. This protein has been recently isolated and purified by Walter Gilbert. It is a tetramer with a molecular weight of 160,000, each monomer consisting of 40,000 mol wt. This protein is unique in that its property to bind specifically to the nucleotide sequence called the operator gene is lost if a specific molecule, the inducer, is present. It is believed that the inducer (like an allosteric effector) modifies the conformation of the repressor protein so that binding will not occur at the operator site. In this case lactose, or its derivatives, is the inducer. If the repressor protein now cannot bind at the operator site, the transcription of the adjacent structural genes for their specific mRNA's is not blocked and RNA polymerase which binds to the promotor site will now initiate the synthesis of the *lac* mRNA's. These mRNA's are synthesized and then translated into the three enzymes for lactose metabolism.

If mutations occur at the regulatory gene sites, nonfunctional repressor proteins may be formed which do not bind at the operator site. These mutants are called *constitutive mutants* and the enzymes synthesized regardless of need are called *constitutive enzymes*. In addition, the structure of the operator gene can undergo mutation so that a repressor cannot bind and constitutive enzymes will be synthesized. *Corepressors*, like inducers, are small-molecular-weight compounds which are able to convert inactive repressor proteins to active proteins fully capable of binding at their specific operator site. These include amino acids such as tryptophan which quickly shuts off the synthesis of the tryptophan synthetase by activating the specific repressor protein for effective binding. The effect of the end product, tryptophan, shutting off the whole series of enzymes responsible for its synthesis is called *end-product repression*.

Catabolite Repression

In general, bacteria only make the enzymes required for the utilization of a specific carbon compound when that compound is present in the medium as the only source of carbon. For example, *E. coli* does not normally metabolize lactose and the enzymes responsible for its metabolism, namely the galactoside permease, and β-galactosidase are both missing. Addition of lactose as the only source of carbon results in the synthesis of large amounts of these enzymes. If, however, glucose is added to the suspension, utilization of lactose is sharply curtailed since the synthesis of the enzymes necessary for its utilization is sharply repressed. This effect is called *catabolite repression*.

An explanation of this phenomena was not forthcoming until it was observed that *E. coli* contained cyclic AMP and that glucose lowered the concentration of this nucleotide very rapidly. Addition of cyclic AMP to cultures in which the lactose-degrading enzymes were repressed by glucose rapidly overcame the repression. Only cyclic AMP and no other adenosine nucleotide proved to be effective. Therefore, repression was somehow related to lowered cyclic AMP concentration. The cAMP effect is general in that the synthesis of all enzymes subject to glucose repression are stimulated by cAMP. These include enzymes involved in transport and metabolism of carbohydrates, amino acid metabolism, and pyrimidine metabolism. Enzymes not subject to glucose repression, such as tryptophan synthetase and alkaline phosphatase, are insensitive to cAMP derepression.

Additional evidence in support of the role of cAMP was supplied by isolating mutants of *E. coli* which were deficient in the enzyme which synthesizes cAMP, namely adenyl cyclase:

$$\text{ATP} \longrightarrow \text{cyclic AMP} + \text{P–Pi}$$

Such mutants contained no cAMP and were unable to grow on lactose and a number of sugars unless cAMP was added, whereupon the normal induction responses to these sugars could be demonstrated. Although it is not clear how glucose regulates the level of cAMP in the cell, there is evidence that glucose facilitates the excretion of the nucleotide from the cell. Whatever the role of glucose might be in regulating cAMP levels in the cell, there is now a reasonable explanation for the role of cAMP itself in the stimulation of inducible enzymes. It is believed that cAMP stimulates the synthesis of *lac* mRNA's by increasing the frequency of initiation of *lac* mRNA chains at the promoter site, P, since P mutants do not respond to cAMP. In very recent *in vitro* studies, I. Pastan and his colleagues have demonstrated that cAMP binds to a specific protein called *cAMP receptor protein* (CRP, 45,000 mol wt), which in turn becomes associated with a specific cAMP–CRP binding site in the P region of *lac* genome. The formation of this complex between cAMP–CRP and *lac* DNA is required for the binding of RNA polymerase to its own specific site in the P region of the *lac* DNA and therefore regulates the rate of initiation of transcription of the mRNA's essential for the formation of the enzymes responsible for the metabolism of lactose. The mechanism of catabolite repression is depicted as shown in Scheme 20-6.

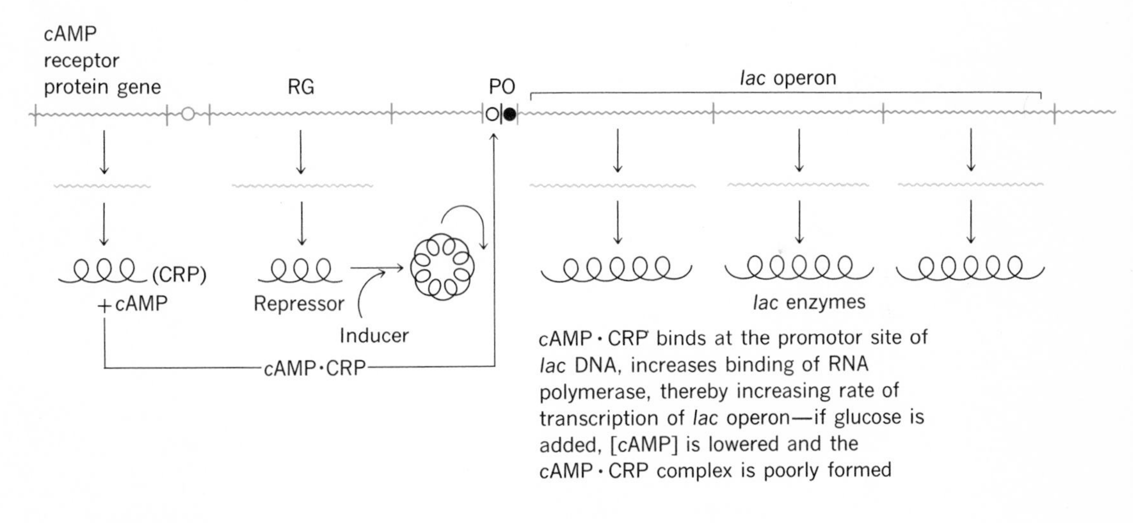

Scheme 20-6

Catabolite repression—a special case of repression.

Effect of Hormones on Metabolic Regulation

The subject of endocrinology is beyond the scope of this book. However, since hormones exert a profound effect on cellular metabolism, it is appropriate to review some new concepts of hormone action.

It appears that, in the higher vertebrates, evolution has developed a complicated set of hormonal controls that regulate cell processes. At all levels of cellular organization, hormones regulate the rate of transport of substrate by altering the permeability of cell membranes, and affect nucleic acid and protein synthesis as well as the metabolism of carbohydrates, fats, and proteins. The hormones include the steroid hormones of the reproductive and adrenal tissues; the peptide hormones of the pituitary; thyroxin, insulin, and glucagon of the pancreas; and hormones of the parathyroid glands. How do these hormones exert their effects on the organism?

In recent years the concept has developed that, rather than having a direct effect on metabolic enzymes, some hormones operate indirectly by way of a two-messenger system. They are considered as the first messengers which travel from the site of synthesis to their specific target cells where they stimulate the formation of a second messenger which at the present time is believed to be cyclic AMP. The level of cAMP is controlled by two important enzymes, adenyl cyclase:

$$\text{ATP} \xrightarrow{Mg^{2+}} 3',5'\text{-}c\text{AMP} + \text{PPi}$$

and a specific cAMP diesterase:

$$c\text{AMP} \xrightarrow{H_2O} 5'\text{-AMP} + \text{Pi}$$

These two enzymes and cAMP are ubiquitous in procaryotic and eucaryotic organisms. Whereas adenyl cyclase is chiefly associated with plasma membranes (and also membranes of mitochondria and endoplasmic reticulum), the diesterase may occur both in the soluble and in the particulate form. The profound metabolic importance of cAMP resides in the fact that adenyl cyclase activity responds to a wide variety of hormones in intact cells. Although the mechanism of activation of the hormones on the enzyme adenyl cyclase is not clear, there is considerable evidence to suggest that the hormones react with specific *hormone receptor sites* (HRS) in the target membrane which are in close proximity to the adenyl cyclase site and activate the enzyme protein by an allosteric interaction. The concept of the first messenger (hormone) and the second messenger (cAMP) system is illustrated in Scheme 20-7.

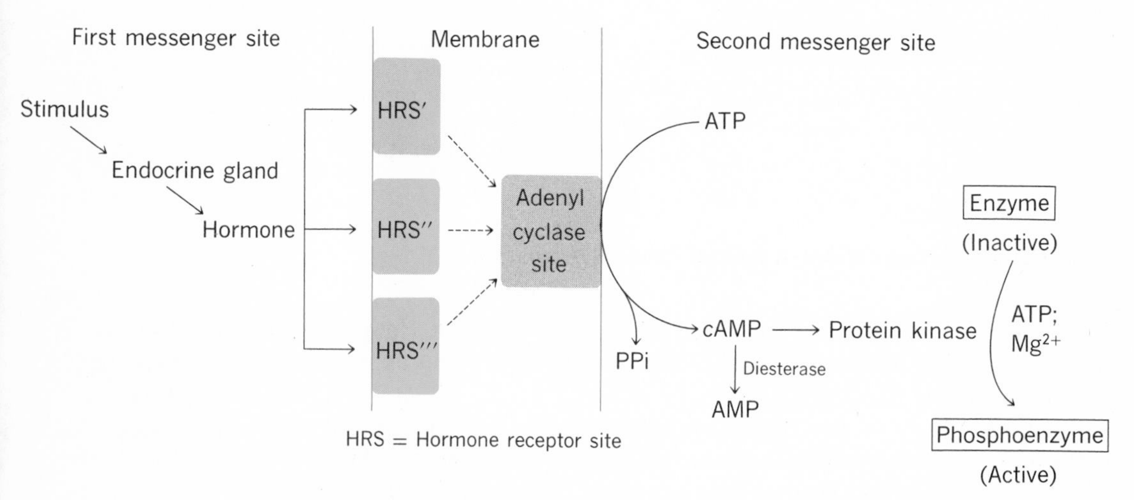

Scheme 20-7

An elegant example of the role of cAMP is in the effect of cAMP on glycogen phosphorylase. The investigations of Earl Sutherland, E. Fischer, and E. Krebs have led to a rather detailed understanding of the regulation of the enzyme glycogen phosphorylase in higher vertebrates. Sutherland received the Nobel Prize in 1971 for his pioneering work in this field.

Induction of glycogenolysis (i.e., breakdown of glycogen) by epinephrine in intact animals or in perfused isolated organs is accompanied by a rapid conversion of glycogen phosphorylase *b* to *a* in heart muscle, and skeletal muscle. Administration of this hormone leads to an increase in cAMP and in the activity of phosphorylase *b* kinase in these tissues. Thus, epinephrine exerts its glycogenolytic effect through the following events: (*a*) activation of adenyl cyclase; (*b*) increased concentration of cAMP; (*c*) activation of protein kinase by cAMP; (*d*) activation of phosphorylase *b* kinase by phosphorylation; and (e) conversion of phosphorylase *b* to *a*. In the meantime, glycogen synthetase *a* is rapidly converted to the *b* form by the same sequence of events as indicated in Scheme 20-8.

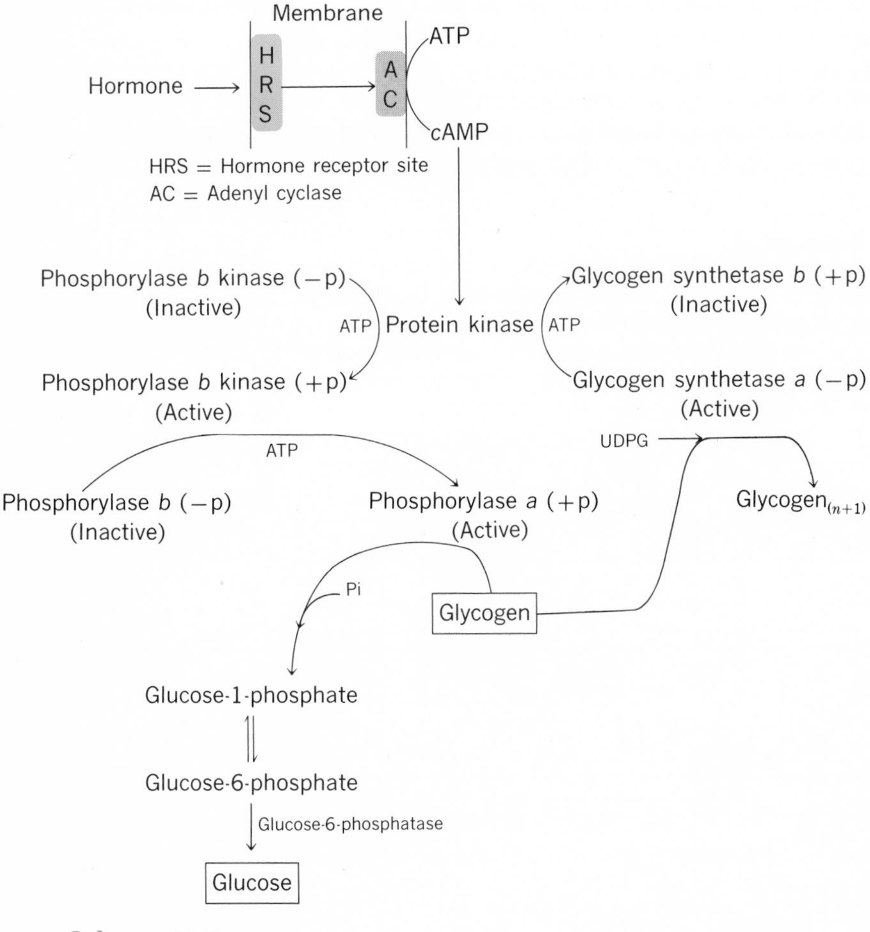

Scheme 20-8

The specific activation effect of cAMP is on a very interesting enzyme, protein kinase. Protein kinase has recently been shown to consist of two subunits:

(*a*) Catalytic unit C (60,000 mol wt)
(*b*) Regulatory unit R (80,000 mol wt)

The unusual feature is that the protein kinase, R·C, is inactive as such, but cAMP binds specifically to the R subunit, thereby favoring dissociation to C, the active form of the enzyme which does not bind cAMP and is not directly affected by cAMP. The following scheme depicts these changes:

$$\underset{\text{Inactive}}{\text{R}\cdot\text{C}} + \text{cAMP} \rightleftharpoons \text{R}\cdot\text{cAMP} + \underset{\text{Active}}{\text{C}}$$

Thus, these results explain in some detail the role of a hormone on the regulation of an important physiological response, namely, the breakdown of glycogen to glucose, and may serve as a model system to explain the action of a hormone on other systems. Certainly the isolation and characterization of the hormone receptor site buried in the plasma membrane will be a landmark discovery in the field of endocrinology.

In conclusion, we have touched on some of the highlights presently known to regulate enzyme activity and enzyme concentration. It is obvious that this area of intensive investigation will in the near future reveal important data relevant to a better understanding of the biochemistry of the cell.

References

1. E. R. Stadtman, in *The Enzymes,* P. D. Boyer, ed. 3rd ed., vol. I. New York: Academic Press, 1970, p. 397.
   An unusually clear account of a difficult subject.
2. G. N. Cohen, *The Regulation of Cell Metabolism*. New York: Holt, Rinehart and Winston, 1968.
   A general account of the mechanisms of control of cell metabolism.

APPENDIX 1

Buffer and pH Problems

Solution of Quadratic Equations

In Chapter 1 the solution of the quadratic equation 1-30 is referred to the appendix. The equation is

$$\frac{x^2}{1-x} = 1.8 \times 10^{-5}$$

This may be rearranged to

$$x^2 = (1.8 \times 10^{-5})(1-x)$$
$$x^2 = 1.8 \times 10^{-5} - (1.8 \times 10^{-5})x$$
$$x^2 + (1.8 \times 10^{-5})x - 1.8 \times 10^{-5} = 0$$

This equation is then in the form: $ax^2 + bx + c = 0$, in which

$$a = 1$$
$$b = 1.8 \times 10^{-5}$$
$$c = -1.8 \times 10^{-5}$$

The solution of a quadratic equation is found as

$$x = \frac{-b \pm \sqrt{b^2 - 4ac}}{2a}$$

Substituting the values for a, b, and c in the quadratic solution,

$$x = \frac{-(1.8 \times 10^{-5}) \pm \sqrt{(1.8 \times 10^{-5})^2 - 4(-1.8 \times 10^{-5})}}{2}$$
$$= \frac{-(1.8 \times 10^{-5}) \pm \sqrt{3.24 \times 10^{-10} + 7.2 \times 10^{-5})}}{2}$$
$$= \frac{-(1.8 \times 10^{-5}) \pm \sqrt{72 \times 10^{-6}}}{2}$$
$$= \frac{-1.8 \times 10^{-5} \pm 8.48 \times 10^{-3}}{2}$$
$$= +4.231 \times 10^{-3} \quad \text{or}$$
$$-4.249 \times 10^{-3}$$

Since in this problem x is the concentration of hydrogen ions $[H^+]$ and can have only positive values, the positive value for x is appropriate. Therefore,

$$[H^+] = 4.23 \times 10^{-3} \text{ mole/liter}$$

Review of Logarithms

There are two systems of logarithms; one is the natural or Naperian system, which employs the base e, and the other is the common system, which has 10 as its base. The logarithm (x or y, respectively) of any number a to the base number e or 10 is the power to which the base e or 10 must be raised to equal a. These may be written

$$x = \log_e a \qquad y = \log_{10} a$$
$$= \ln a$$

The two systems are related by

$$x = \log_e a = 2.303 \log_{10} a = 2.303y$$

In this book logarithms to the base 10 are used almost exclusively. Examples of logarithms to the base 10 are

$$\log 10 = 1$$
$$\log 100 = \log 10^2 = 2$$
$$\log 1000 = \log 10^3 = 3$$
$$\log 0.001 = \log 10^{-3} = -3$$
$$\log 1 = 0$$

For numbers between 1 and 10, tables of logarithms are available or may be read directly from a slide rule. Examples:

$$\log 2 = 0.301$$
$$\log 3 = 0.477$$
$$\log 6 = 0.778$$
$$\log 7 = 0.845$$

The student should be familiar with the operations employed in logarithms. For example, the logarithms are added in multiplication; in division, the logarithms are subtracted. Examples:

$$4 \times 6 = 24$$
$$\log 24 = \log 4 + \log 6$$
$$= 0.602 + 0.778$$
$$= 1.380$$

As a check,

$$\log 24 = \log (10 \times 2.4)$$
$$= \log 10 + \log 2.4$$
$$= 1.0 + 0.380$$
$$= 1.380$$

In pH problems two operations are frequently encountered. As an example, when the $[H^+]$ is given,

$$[H^+] = 3 \times 10^{-4} \text{ mole/liter}$$

calculate the pH:

$$\begin{aligned} pH &= \log \frac{1}{[H^+]} = -\log [H^+] \\ &= -\log (3 \times 10^{-4}) \\ &= -\log 3 - \log 10^{-4} \\ &= -0.477 - (-4) \\ &= 3.523 \end{aligned}$$

The other common operation is to calculate the $[H^+]$ from a given pH. Calculate the $[H^+]$ of a solution whose pH is 9.26:

$$\begin{aligned} pH &= 9.26 \\ [H^+] &= \text{antilog } -9.26 \\ &= \text{antilog } (-10 + 0.74) \\ &= 10^{-10} \times 5.5 \\ &= 5.5 \times 10^{-10} \text{ mole/liter} \end{aligned}$$

A few representative problems are found here and on the following pages. Many additional problems together with their solutions are to be found in the inexpensive paperback *Biochemical Calculations* by I. H. Segel, John Wiley & Sons, New York (1968).

Problems

1. Calculate the pH of

$10^{-4}M[H^+]$	*Answer:*	4.00
$7 \times 10^{-5}M[H^+]$		4.16
$5 \times 10^{-8}M[H^+]$		7.30
$3 \times 10^{-11}M[H^+]$		10.52

2. Calculate the $[H^+]$ of a solution whose pH is given:

pH		
2.73	*Answer:*	$1.86 \times 10^{-3}M[H^+]$
5.29		$5.13 \times 10^{-6}M[H^+]$
8.65		$2.24 \times 10^{-9}M[H^+]$
11.12		$7.59 \times 10^{-12}M[H^+]$

Problems on Chemical Stoichiometry. (*a*) Concentrated H_2SO_4 is 96% H_2SO_4 by weight and has a density of 1.84. Calculate the amount of concentrated acid required to make 750 ml of $1N$ H_2SO_4.

Answer: One liter of concentrated acid weighs 1840 g and contains 1840×0.96 or 1760 g of H_2SO_4. One liter of concentrated H_2SO_4 is therefore 1760/98 or 18 molar (18*M*). Since H_2SO_4 is a diprotic acid producing two protons for 1 mole of H_2SO_4, concentrated H_2SO_4 is 36 normal (36*N*); 750 ml of $1N$ H_2SO_4 contains 0.75 eq or 750 meq. Therefore, 750/36 or 20.8 ml of concentrated H_2SO_4 will contain 750 meq. If 20.8 ml of concentrated H_2SO_4 are diluted to 750 ml with H_2O, the solution will be $1N$.

(*b*) Concentrated HCl is 37.5% HCl by weight and has a density of 1.19. Describe the preparation of 500 ml of $0.2N$ HCl.

Answer: Dilute 8.18 ml of concentrated HCl to 500 ml with H_2O.

(*c*) Glacial CH_3COOH is 100% CH_3COOH by weight and has a density of 1.05. Describe the preparation of 300 ml of 0.5*N* CH_3COOH.

Answer: Dilute 8.6 ml of glacial CH_3COOH to 300 ml with H_2O.

(*d*) Calculate the $[H^+]$ of the final solution with 100 ml of 0.1*N* NaOH is added to 150 ml of 0.2*M* HCl. *Answer:* 0.08*M*

(*e*) Calculate the $[H^+]$ of the final solution when 100 ml of 0.1*N* NaOH is added to 150 ml of 0.2*M* H_2SO_4. *Answer:* 0.2*N*

Buffer Problems. (*a*) Calculate the pH of the final solution when 100 ml of 0.1*M* NaOH is added to 150 ml of 0.2*M* CH_3COOH ($K_a = 1.8 \times 10^{-5}$). 150 ml of 0.2*M* CH_3COOH contains 0.03 mole of CH_3COOH; similarly, 100 ml of 0.1*M* NaOH contains 0.01 mole of NaOH. When these are mixed, 0.01 mole of NaOH will neutralize an equal amount of CH_3COOH to form 0.01 mole of sodium acetate; 0.02 mole of CH_3COOH will remain. Both of these are contained in a volume of 250 ml. The pH may be solved by use of the Henderson–Hasselbalch equation:

$$\text{pH} = \text{p}K_a + \log \frac{[\text{Conjugate Brönsted base}]}{[\text{Brönsted acid}]}$$

Calculate the pK_a first:

$$\begin{aligned} \text{p}K_a &= -\log 1.8 \times 10^{-5} \\ &= -\log 1.8 - \log 10^{-5} \\ &= -0.26 + 5 \\ &= 4.74 \end{aligned}$$

Therefore,

$$\begin{aligned} \text{pH} &= 4.74 + \log \frac{[CH_3COO^-]}{[CH_3COOH]} \\ &= 4.74 + \log \frac{(0.01/250)}{(0.02/250)} \end{aligned}$$

Note, however, that the volume (250 ml) which contains the acetate anion and acetic acid is found in both the numerator and denominator. The last equation simplifies to

$$\begin{aligned} \text{pH} &= 4.74 + \log \tfrac{1}{2} \\ &= 4.74 - \log 2 \\ &= 4.74 - 0.30 \\ &= 4.44 \end{aligned}$$

(*b*) The pK_a's for H_3PO_4 are p$K_{a_1} = 2.1$; p$K_{a_2} = 7.2$; p$K_{a_3} = 12.7$. Describe the preparation of a phosphate buffer, pH 6.7, starting with a 0.1*M* solution of H_3PO_4 and 0.1*M* NaOH.

Answer: The second dissociation of phosphoric acid will be the buffer system.

$$H_2PO_4^- \rightleftharpoons HPO_4^{2-} + H^+ \qquad \text{p}K_{a_2} = 7.2$$

The ratio of conjugate base (HPO_4^{2-}) to the Brönsted acid ($H_2PO_4^-$) may be calculated from the Henderson–Hasselbalch equation:

$$pH = pK_{a2} + \log \frac{[HPO_4^{2-}]}{[H_2PO_4^-]}$$

$$6.7 = 7.2 + \log \frac{[HPO_4^{2-}]}{[H_2PO_4^-]}$$

$$-0.5 = \log \frac{[HPO_4^{2-}]}{[H_2PO_4^-]}$$

$$0.5 = \log \frac{[H_2PO_4^-]}{[HPO_4^{2-}]}$$

$$\text{Ratio } \frac{[H_2PO_4^-]}{[HPO_4^{2-}]} = \text{antilog } 0.5$$

$$\frac{[H_2PO_4^-]}{[HPO_4^{2-}]} = \frac{3.16}{1}$$

In this buffer there will be 316 parts of $H_2PO_4^-$ and 100 parts of HPO_4^{2-} for a total of 416. Since all the phosphate buffer components must come from 0.1*M* H_3PO_4, start by taking 41.6 ml of 0.1*M* H_3PO_4 and add 41.6 ml of 0.1*N* NaOH to neutralize the first proton, which dissociates at $pK_{a1} = 2.1$. Then add 10.0 ml more of alkali to produce 1.0 meq of HPO_4^{2-} and leave 3.16 meq of $H_2PO_4^-$. This would give the desired ratio of $H_2PO_4^-/HPO_4^{2-}$ and consequently a pH of 6.7. The buffer concentration would be equal to the milliequivalents of H_3PO_4 (4.16) divided by the milliliters of the final solution (93.2), or 0.045*M*.

(c) Describe the preparation of 100 ml of 0.1*M* phosphate buffer, pH 6.7, starting with 1*M* H_3PO_4 and 1*M* NaOH.

Answer: The same ratio of $H_2PO_4^-/HPO_4^{2-}$ of 3.16 must be obtained. To prepare 100 ml of 0.1*M* phosphate buffer, take 10 ml of 1*M* H_3PO_4. Then add 10 ml of 1*M* NaOH to neutralize the first proton that dissociates. Then, to obtain the correct ratio, add $10 \times 1/4.16$ or 2.4 ml more of 1*M* NaOH and dilute to final volume of 100 ml.

Problems

1. What would be the pH and concentration of the resulting buffer solution when 3.48 g of K_2HPO_4 and 2.72 g of KH_2PO_4 are dissolved in 250 ml of deionized water? *Answer:* pH = 7.2; the concentration is 0.16*M*

2. A buffer solution contains 0.1*M* CH_3COOH and 0.1*M* sodium acetate (that is, it is a 0.2*M* acetate buffer). Calculate the pH after addition of 4 ml of 0.025*N* HCl to 10 ml of the buffer. The pK_a for acetic acid is 4.74.
Answer: pH = 4.65

3. Describe the preparation of a glutaric acid buffer at pH 4.2 starting with 0.1*M* NaOH and 0.1*M* glutaric acid ($pK_{a1} = 4.32$; $pK_{a2} = 5.54$).

Answer: Add 100 ml NaOH to 232 ml of glutaric acid or any similar ratio of base to acid.

4. Pyridine is a conjugate base which reacts with H^+ to form pyridine hydrochloride. The hydrochloride dissociates to yield H^+ with a pK_a of 5.36. Describe the preparation of a pyridine buffer at pH 5.2 starting with 0.1*M* pyridine and 0.1*M* HCl.

Answer: Add 14.5 ml of 0.1*M* HCl to 24.5 ml of 0.1*M* pyridine.

5. Describe the preparation of 1 liter of a 0.1*M* ammonium chloride buffer, pH 9.0, starting with solid ammonium chloride ($pK_a = 9.26$) and 1*M* NaOH.

Answer: Dissolve 5.35 g NH_4Cl in approximately 500 ml of H_2O, add 35.5 ml of 1*M* NaOH, and dilute to 1.0 liter.

6. Describe the preparation of 1 liter of 0.1*M* ammonium chloride buffer, pH 9.0, starting with 1*M* NH_4OH and 1*M* HCl.

Answer: Add 64.5 ml of 1*M* HCl to 100 ml of 1.0*M* NH_4OH and dilute to 1 liter.

7. What volume of glacial acetic acid and what weight of sodium acetate trihydrate ($CH_3COONa \cdot 3H_2O$) are required to make 100 ml of 0.2*M* buffer at pH 4.5 (pK_a of acetic acid is 4.74)?

Answer: 0.725 ml glacial acetic acid and 0.993 g of sodium acetate trihydrate

8. What weight of sodium carbonate (Na_2CO_3) and sodium bicarbonate ($NaHCO_3$) are required to make 500 ml of 0.2*M* buffer, pH 10.7 (pK_{a_1} of H_2CO_3 is 6.1; $pK_{a_2} = 10.3$)?

Answer: 7.58 g of Na_2CO_3 and 2.40 g of $NaHCO_3$

9. What volume of concentrated HCl and what weight of tris-(hydroxymethyl)amino methane (as the base) are required to make 100 ml of 0.25*M* buffer, pH 8.0 (pK_a of Tris hydrochloride is 8.0)?

Answer: 3.025 g of Tris (as the base) and 1.025 ml of concentrated HCl

10. Describe the preparation of 250 ml of 0.6*M* triethanolamine buffer, pH 7.2, from the free amine and concentrated HCl (pK_a for the amine hydrochloride is 7.8).

Answer: Dissolve 22.4 g of amine in approximately 100 ml of H_2O, add 9.85 ml of HCl, and dilute to 250 ml.

11. (*a*) What weight of glycine ($pK_{a_1} = 2.4$; $pK_{a_2} = 9.6$) and what volume of 1*N* HCl are required to make 100 ml of 0.3*M* buffer, pH 2.4? (*b*) What weight of glycine and what volume of 1*N* NaOH are required to make 100 ml of 0.3*M* buffer, pH 9.3?

Answers: (*a*) 2.25 g of glycine and 15 ml of 1*N* HCl; (*b*) 2.25 g of glycine and 10 ml of 1*N* NaOH

12. An enzyme-catalyzed reaction was carried out in a solution containing 0.2*M* Tris buffer ($pK_a = 8.0$). The pH of the reaction mixture was 7.7 at the start of the experiment. During the reaction 0.033 mole/liter of H^+ were consumed. (Note that the utilization of H^+ ions has the same effect on the buffer as the production of an equivalent amount of OH^- ions.) (*a*) What was the ratio of Tris (free base) to Tris hydrochloride (acid form) at the start of the reaction? (*b*) What was the ratio of Tris/Tris·HCl at the end of the reaction? (*c*) What was the final pH of the reaction mixture?

Answers: (*a*) 0.5; (*b*) 1.0; (*c*) pH 8.0

APPENDIX 2

Methods in Biochemistry

Introduction

Some of the techniques employed in biochemical research have been collected in this appendix, not to serve as a laboratory guide, but rather to acquaint the student with the terms and methods that are the language of the practicing biochemist.

Glass Electrode

The most effective way of accurately measuring the pH value of a biochemical system is to employ a pH meter with a glass electrode. The potential of the glass electrode (E_g) relative to the external reference electrode (E_{ref}) is related to the pH as follows:

$$pH = \frac{E_g - E_{ref}}{0.0591} \quad \text{at } 25°C$$

The typical glass electrode assembly consists of

$$\underbrace{Ag, AgCl(s), HCl(0.1M)}_{\text{Silver–Silver chloride electrode}} \| \text{Glass membrane} \| \text{Solution X} | \underbrace{KCl(Sat), Hg_2Cl_2(s), Hg}_{\text{Calomel half-cell}}$$

When two solutions of different H^+ ion concentrations are separated by a thin glass membrane, a potential difference related to differences in pH of the two solutions is obtained. A typical glass electrode is illustrated in Figure A-2-1.

The potential difference ($E_g - E_{ref}$) is carefully measured either with a potentiometer type of pH meter or with a direct-reading pH meter (line-operated) consisting normally of a simple triode amplifier using the negative feedback principle. Regardless of how the potential difference is measured, the student should note that the results obtained are in terms of *activity* (a_H) rather than concentration of hydrogen ion [H^+]. Unless special glass membranes are used, pH responses are usually adequate between 1 and 11, but above and below these values errors do become evident, and correc-

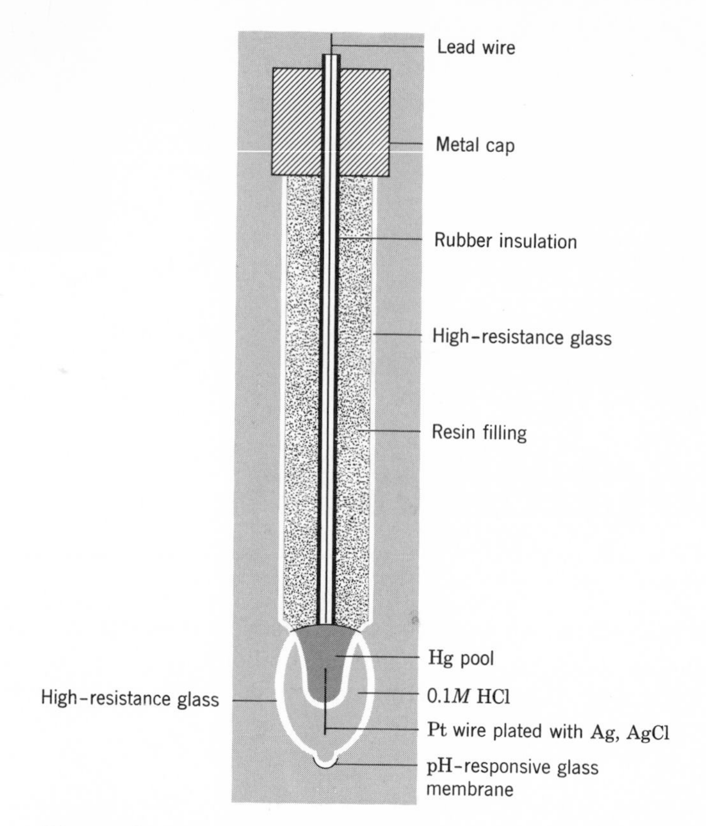

Figure A-2-1

Diagram of a typical glass electrode.

tions must be introduced. The glass electrode should be carefully washed after each pH determination, particularly after dealing with protein solutions, since proteins may absorb on the glass membrane surface, with serious errors resulting. In nonaqueous solutions a partial dehydration of the glass membrane may occur with changes in the potential difference, also leading to errors. Poorly buffered solutions should be thoroughly stirred during measurements, since a thin layer of solution at the glass solution interface may not reflect the true activity of the rest of the solution. It must also be noted that in organic solvents dissociation of acids is decreased and thereby the pH is raised. The student should be aware of these factors. Despite these difficulties the glass electrode pH meter is the preferred tool, since it is an extremely sensitive and stable instrument.

Isotopic Methods

The single most important technique in biochemistry is the critical, careful use of radioisotopes and stable isotopes.

Radioisotopes. From a biochemical standpoint the most useful radioisotopes, ^{14}C, ^{35}S, ^{32}P, and ^{3}H, are β-ray emitters; that is, when the nuclei of these atoms disintegrate, one of the products is an electron which moves with energies characteristic of the disintegrating nucleus. The so-called β rays interact with the molecules through which they traverse, causing dissociation, excitation, or ionization of the molecules. It is the resultant ionization property which is used to measure quantitatively the amount of radioisotope present. See Table A-2-1 for some properties of useful radioisotopes.

Units. A *curie* is the amount of emitter which exhibits 3.7×10^{10} disintegrations/sec (dps). More common units are a millicurie, mc (10^{-3} curie), and a microcurie, μc (10^{-6} curie).

Specific activity. This is defined as disintegrations/minute per unit of substance (mg, μmole, etc.).

Dilution factor. The factor is defined as

$$\frac{\text{Specific activity of precursor fed}}{\text{Specific activity of compound isolated}}$$

This factor is used frequently to express the precursor relation of a compound in the biosynthesis of a second compound. Thus, in the sequence $A \longrightarrow B \longrightarrow C \longrightarrow D$, the dilution factor for $C \longrightarrow D$ would be small, whereas for A it would be large. Therefore, a small dilution factor would indicate that compound C fed to a tissue has a better precursor relationship to the final product than compound A with a large dilution factor.

Percentage of incorporation. This is also used to compare the proximity of a precursor in the biosynthesis of a second compound. If labeled compound A is administered to an experimental system and some of the radioactivity is incorporated into compound D, the percentage of incorporation is expressed as counts/min in D divided by counts/min in $A \times 100$ $[(\text{cpmD}/\text{cpmA}) \times 100]$.

Measurements. Liquid scintillation counting is probably the most popular technique for measuring radioisotopes. The technique is based on the use of a scintillation solution containing fluors and a multiplier phototube. The scintillation solution converts the energy of the radioactive particle into light; the multiplier phototube responds to the light by producing a charge which can be amplified and counted by a scaling circuit.

In liquid scintillation counting, the radioactive substance is usually dis-

Table A-2-1
Some Properties of Useful Radioisotopes

Element	Radiation	Half-life	Energy of radiation (meV)[a]
^{3}H	β^-	12.1 years	0.0185
^{14}C	β^-	5100 years	0.156
^{32}P	β^-	14.3 days	1.71
^{35}S	β^-	87.1 days	0.169

[a] Million electron-volts.

solved in a suitable organic solvent containing the fluor. Alternately, the radioactive sample, which can even consist of filter paper containing the sample, is suspended or immersed in the scintillation fluid. Under these conditions the energy of the radioactive particle is first transferred to the solvent molecule, which may then ionize or become excited. It is the electronic excitation energy of the solvent which is transferred to the fluor (solute). When the excited molecules of the solute return to their ground state, they emit quanta of light that are detected by the phototube.

One problem associated with this technique is the quenching of the light output by colored substances in the sample. In addition, the fluor molecules may be quenched if foreign substances absorb their excitation energy before it is released as light. Methods are available for determining the amount of quenching exhibited by the radioactive sample.

Scintillation counting is particularly useful for determining the weak β particles of tritium (^3H) and carbon-14 (^{14}C). The efficiency of counting these particles can be as high as 50 and 85%, respectively.

Stable Isotopes. Stable isotopes of several of the biologically important elements are available in enriched concentrations and therefore may be used to "tag" or label compounds. As an example, deuterium, the hydrogen atom with mass of 2, is present in most H_2O to the extent of only 0.02%. The remainder of the hydrogen atoms has, of course, a mass of 1. This concentration of 0.02% is known as the normal abundance of deuterium. It is possible to obtain *heavy* water in which 99.9% of the hydrogen atoms are deuterium. The concentration of a heavy isotope is usually measured as *atom % excess;* this is the amount, in percent, by which the isotope exceeds its normal abundance. Thus, the two stable isotopes of nitrogen are $^{14}_{7}N$ and $^{15}_{7}N$, which have a normal abundance of 99.62 and 0.38%, respectively. If a sample of nitrogen gas contains 4.00% $^{15}_{7}N$ (and 96.00% $^{14}_{7}N$), the concentration of $^{15}_{7}N$ in this sample is said to be 3.62 atom % excess. Other stable isotopes that are available in enriched concentrations and therefore may be used as tracers in biochemistry are $^{17}_{8}O$, $^{18}_{8}O$, $^{13}_{6}C$, $^{33}_{16}S$, and $^{34}_{16}S$; the normal abundance of these isotopes can be found in any chemical handbook.

The principles underlying the use of stable isotopes are similar to those employed with radioisotopes. The stable isotopes are measured quantitatively in a mass spectrometer, however. A discussion of the different types of spectrometer available may be found in reference 1 at the end of this appendix. Prior to the development of the spectrometer, methods based on the refractive index, density, and thermal conductivity were available for measuring the concentration of stable isotopes.

Uses of Isotopes. Countless techniques have been developed to study biochemical reaction sequences. Hundreds of commercially available biochemicals labeled with different isotopes at known positions are used in modern research, and this book cites many examples. Some precautions should be pointed out, most important being the isotope effect affecting *rates* of reaction. Because of differences in atomic weight, slight changes in reaction rates will be noted. With tritium (^3_1H) the rate effect is large and may represent a twentieth of the rate of cleavage of a C—$^{1}_{1}H$ bond. With deuterium ($^{2}_{1}H$) the rate effect is about one-sixth. With ^{12}C—^{14}C cleavage the rate effects are small, providing these are the rate-limiting steps.

It is also of note that with both tritium and deuterium such bonds as N—$^{3}_{1}H$ and O—$^{3}_{1}H$ rapidly exchange with water ($^{1}_{1}H_2O$) in the medium and are washed out. The $^{3}_{1}H$ label in acetic acid, CH_3COO^3H, will be immediately washed out because of the great exchange by ionization with normal protons in water. In addition, all compounds that are counted must be carefully purified; another technique is to remove any occluded contaminating radioisotopic substance by "washing out" with the corresponding nonradioactive compound. Thus, CH_3C^*OOH (carboxyl-labeled acetic acid) is readily removed from a desired compound by adding large amounts of normal ($^{12}_{6}C$) acetic acid which could mix with and greatly dilute out the contaminating C*-acetic acid. Another useful criterion is purification to constant specific activity.

Spectrophotometry

This technique is of prime importance in biochemical research. Three different usages are commonly found. (*a*) If the absorbancy index (a_s) at a specific wavelength is known, the concentration of a compound can be readily determined by measuring the optical density at that wavelength. With a large a_s, as in nucleotides, very small quantities of the absorbing material (2–4 μg) can be accurately measured. (*b*) The course of a reaction can be determined by measuring the rate of formation or disappearance of a light-absorbing compound. Thus, NADH absorbs strongly at 340 mμ, whereas the oxidized form (NAD^+) has no absorption at this wavelength. Therefore, reactions involving the production or utilization of NADH (or NADPH) can be assayed by this technique. (*c*) Compounds can frequently be identified by determining their characteristic absorption spectra in the ultraviolet and visible regions of the spectrum.

Two fundamental laws are associated with spectrophotometry; these are Lambert's and Beer's laws. Lambert's law states that the light absorbed is directly proportional to the *thickness* of the solution being analyzed:

$$A = \log_{10}\frac{I_0}{I} = a_s b$$

where I_0 is the incident light intensity, I is the transmitted light intensity, a_s is the absorbancy index characteristic for the solution, b is the length or thickness of the medium, and A is the absorbancy.

Beer's law states that the amount of light absorbed is directly proportional to the *concentration* of solute in solution:

$$\log_{10}\frac{I_0}{I} = a_s c$$

and the combined Beer–Lambert law is $\log_{10} I_0/I = a_s bc$. If b is held constant by employing a standard cell or cuvette, the Beer–Lambert law reduces to

$$A = \log_{10}\frac{I_0}{I} = a_s c$$

The second type of detector device is a *hydrogen flame ionization* detector. It has extreme sensitivity, a wide linear response, and is insensitive to water. In theory, when organic material is burned in a hydrogen flame, electrons and ions are produced. The negative ions and electrons move in a high-voltage field to an anode and produce a very small current, which is changed to a measurable current by appropriate circuitry. The electrical current is directly proportional to the amount of material burned.

The use of gas chromatography has revolutionized the analysis of fat, fatty acids, flavor components, gaseous mixtures, and any compound which can be converted into a volatile material. Recently, great advances have been made in converting quantitatively the nonvolatile amino acids to volatile derivatives, and if this research is successful it will greatly expedite research on protein structure.

Paper Chromatography

Like all simple techniques, this method has revolutionized the separation or detection of reaction products and the determination and identification of compounds. Developed in 1944 by Martin in England, filter paper strips are used to support a stationary water phase while a mobile organic phase moves down the suspended strip of paper in a cylinder, as indicated in Scheme A-2-3. The substances to be separated are spotted near the top of the hanging sheet. Separation is based on a liquid–liquid partition of the compounds.

The ratio of the distance traveled by the compounds to the distance traveled by the solvent front from the original spot at the top of the paper

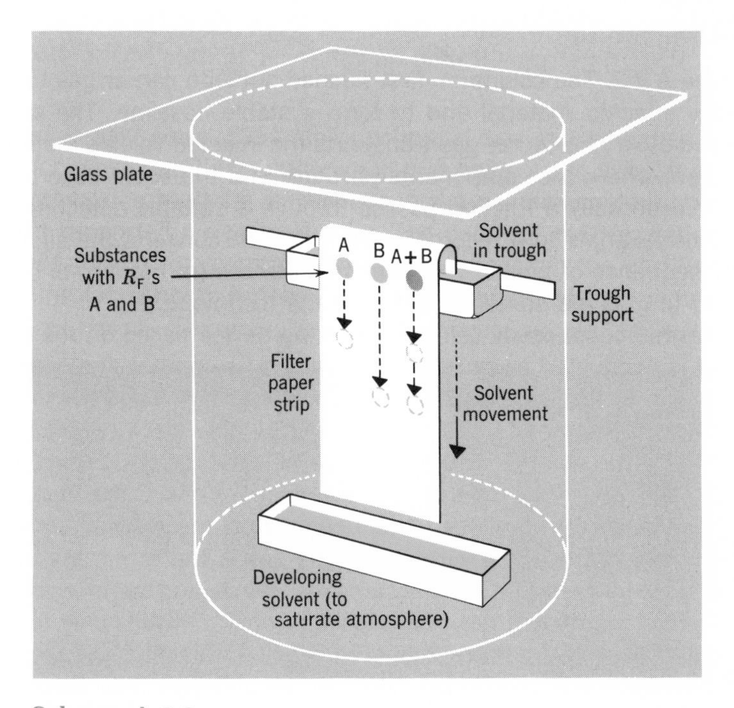

Scheme A-2-3

sheet is called the R_F value of the compound. Under strictly controlled conditions the R_F is an important constant for identification purposes. With a knowledge of how a variety of compounds move in a series of solvents, much can be said about the functional groups of unknown compounds.

The method just described is one-dimensional chromatography. Two-dimensional chromatography is a variant with considerable separatory power, since two different solvents can be employed in sequence to move a single compound (Scheme A-2-4).

A large number of variations on paper chromatography have been developed. They include (*a*) reverse-phase chromatography, wherein the stationary phase is made nonpolar and the mobile phase is polar; and (*b*) a combination of paper chromatography and electrophoresis that involves partition chromatography and electrical mobility of ionic species.

A simple experiment for the student consists of chromatographing writing inks on Whatman No. 1 filter paper. Samples of different inks (try Sheaffer's Permanent Jet Black) are placed as small spots along the shorter edge of a piece of Whatman No. 1 filter paper (20 × 25 cm), 2 cm from the edge. After it has dried, the two longer edges of filter paper are stapled together to form a cylinder with the ink spots on one circumference. That end is placed in a jar containing H_2O to a depth of 1 cm. The water will rapidly rise (in 1 hr) and the different colored components of the inks will migrate in amounts related to their solubility in H_2O and adsorption on cellulose.

Thin-Layer Chromatography

Thin-layer chromatography is adsorption chromatography performed on open layers of adsorbent materials supported on glass plates. A thin uniform film of silica gel containing a binding medium, such as calcium sulfate, is spread onto a glass plate. The thin layer is allowed to dry at room temperature and then is activated by heating in an oven between 100 and 250°C, depending upon the degree of activation which is desired. The activated plate is then placed flat on the laboratory bench and samples spotted carefully on the surface of the thin layer. Material ranging from 0.05 to 50 mg or more are readily spotted with micropipettes. After the solvent has evaporated, the plates are placed vertically in a glass tank which contains a suitable solvent (see Scheme A-2-5). Within 5–30 min a separation is produced by the solvent rising through the thin layer, differentially carrying the components of the

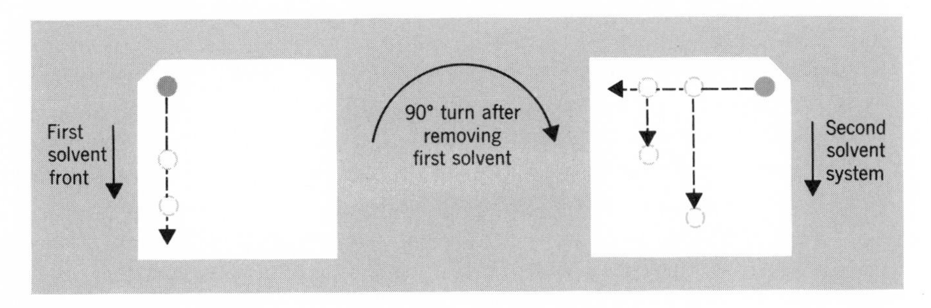

Scheme A-2-4

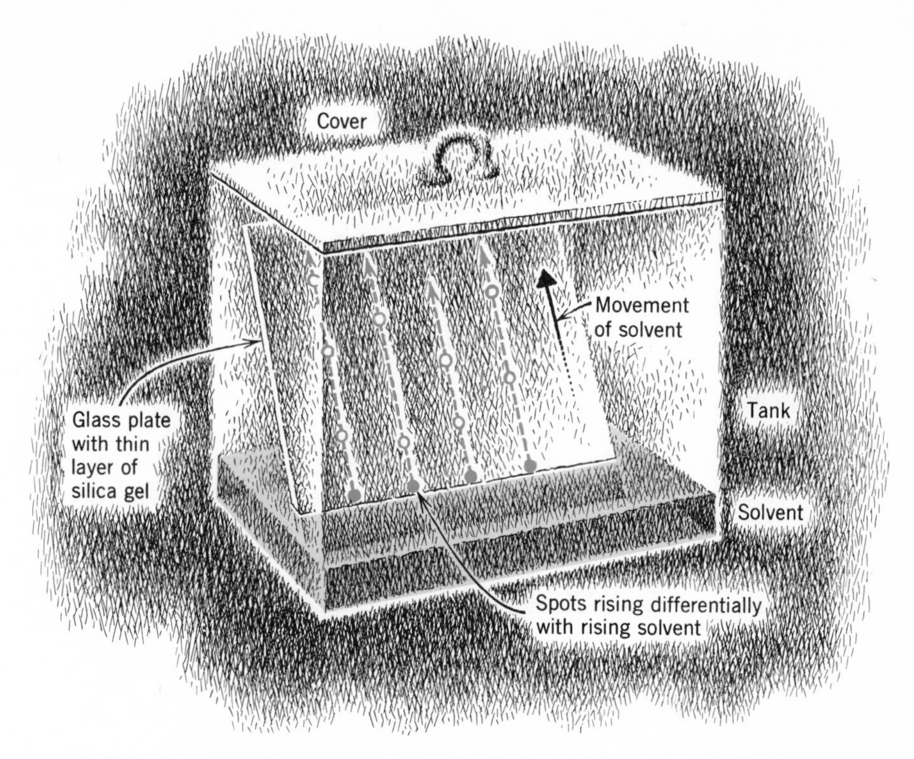

Scheme A-2-5

spots from the origin, depending on adsorption of the components on the silica gel or because of a distribution between the mobile solvent and the water held by the silica gel. The plate is removed from the solvent tank, permitted to dry briefly and then, depending on the type of compounds on the gel, the spots are detected by spraying the plate with a variety of reagents or dyes. Moreover, the thin inorganic layer of adsorbent can be used with reagents of a more corrosive nature. The possibility of using high-temperature techniques such as carbonization in conjunction with a spray of concentrated sulfuric acid, offers a universal means of detection of great sensitivity. Thus, the speed, efficiency, and sensitivity of thin-layer chromatography has made this technique one of the most powerful available to the biochemist.

Ion Exchange Electrostatic attraction of oppositely charged ions on a polyelectrolyte surface forms the basis of ion exchange chromatography. Typical systems include the synthetic resin polymers, such as the strongly acidic cation exchanger Dowex-50, a polystyrene sulfonic acid, and the strongly basic anion exchanger Dowex-1, a polystyrene quaternary ammonium salt. Cellulose derivatives such as carboxymethyl cellulose (CMC) and diethylaminoethyl cellulose (DEAE) exchangers have been very successfully used in protein purification.

The basic principle involves an electrostatic interaction with the exchanging ions and the normal charge on the surface of the resin. These reactions are considered to be equilibrium processes and involve diffusion of a given ion to the resin surface and then to the exchange site, the actual exchange, and finally diffusion away from the resin. The rate of movement of a given ionizable compound down the column is a function of its degree of ionization, the concentration of other ions, and the relative affinities of the various ions present in the solution for charged sites on the resin. By adjusting the pH of the eluting solvent and the ionic strength, the electrostatically held ions are eluted differentially to yield the desired separation.

An example of the use of ion exchange resins in the purification of cytochrome *c* can be cited. Cytochrome c has an isoelectric point of 10.05; that is, at pH 10.05 the number of positive charges will equal the number of negative charges. A column containing a cation exchanger buffered at pH 8.5 is prepared. This column has a full negative charge. Cytochrome *c* at pH 8.5 has a full positive charge. An impure solution of cytochrome *c* at pH 8.5 is placed on the column, and water is passed through the column. The contaminating proteins pass freely through the column (the p*I* of proteins is usually 7.0 or less) but cytochrome *c* is held firmly by electrostatic attraction to the resin beads. If the eluting solvent pH is now raised to about 10, the cytochrome *c* will have a net zero charge and will pass rapidly through as a pure component.

Resin columns are extremely useful in the separation and purification of nucleotides, small-molecular-weight compounds with ionizable groups, amino acids, and peptides. Because of the limited available surface and the lability of proteins, ion resins have not proved too successful in protein purification.

The cellulose derivatives have therefore been developed, since they have high absorptive capacities but still hold proteins rather weakly. This means that by mere adjustments of pH and salt concentration, efficient elution of adsorbed proteins can be made. Two very common derivatives already mentioned are CMC, a cationic exchanger, and DEAE, an anionic exchanger.

In practice, the steps we list here may be taken.

CM—Cellulose:

$$\mathrm{CMC}^{\ominus}_{\mathrm{pH}\,4} + \left.\begin{matrix}\text{Protein}\\ \text{mixture}\end{matrix}\right\}^{\oplus}_{1+2+3} \longrightarrow \mathrm{CMC}^{-} - \begin{bmatrix}\text{Protein}\\ \text{mixture}\end{bmatrix}^{\oplus}_{1+2+3}$$

Adsorption

Elution by increasing pH of eluting solvent, thereby depressing ⊕ charges on proteins or by increasing salt concentration, which displaces proteins at constant pH

Separation
- Protein 1
- Protein 2
- Protein 3

The reverse procedure may be used for DEAE columns, namely placing protein on a DEAE column at pH 8 and eluting by decreasing pH or increasing salt concentration or both.

Gel Filtration

The technique of separating molecules of different size by passage through a gel column is called gel filtration. The polysaccharide, dextran, is carefully cross-linked to give small beads of a hydrophilic, insoluble nature which when placed in water swells considerably to form an insoluble gel. The commercial name for the gel is *Sephadex*. The property of Sephadex to exclude solutes of large molecular size, and to be accessible for diffusion to molecules of small dimension is the basis of the separation method.

The general expression for the appearance of a solute in an effluent is

$$V = V_0 + K_D V_i$$

where V is the elution volume of a substance with a given K_D, V_0 is the void volume or the total volume of the external water (outside the gel grain), V_i is the internal water volume in the gel grain, and K_D is the distribution coefficient for a solute between the water in the gel grain and the surrounding water. A substance with K_D of zero is completely excluded from the gel beads, and substances with K_D values between 0 and 1 are partially excluded. If a sample containing a solute with a $K_D = 1$ and another with $K_D = 0$ is introduced in the column, the latter will appear in the effluent after a volume V_0, and the former will appear after a volume $V_0 + K_D V_i$.

The procedure of dialysis can be readily carried out on a suitable Sephadex column. The column is first equilibrated with the new buffer. The protein solution is introduced to the top of the column and eluted with the new buffer. When the volume V_0 has passed, the protein is eluted in the new buffer medium while the original buffer and small-molecular-weight compounds, etc., are eluted after a volume of $V_0 + K_D V_i$. The process is very rapid and hence is very useful when working with labile proteins. Since K_D varies with proteins of different molecular weights, it is possible to fractionate proteins on gel filtration. The biochemist has a choice of several types of Sephadex beads to prepare columns for this procedure. Thus, Sephadex G-25 excludes compounds of molecular weight of 3500–4500, Sephadex G-50, 8000–10,000, and Sephadex G-75, 40,000–50,000. Sephadex G-100 and G-200 gels may be used for higher-molecular-weight proteins. Figure A-2-2 gives some typical separation results.

An equally useful application of column gel filtration is its use as a method to determine molecular weights even if the protein has not been extensively purified. Depending on the possible molecular weight of the protein, a suitable gel is chosen. Usually, Sephadex G-100 or G-200 is selected, a column is carefully prepared, and the elution volumes of pure proteins with known molecular weights and stabilities are determined to establish a calibration curve. The protein whose molecular weight is to be determined is placed on the same column and its elution volume is determined under conditions identical to those used to elute the known proteins. The results are plotted, K_D vs log mol wt, as depicted in Figure A-2-3.

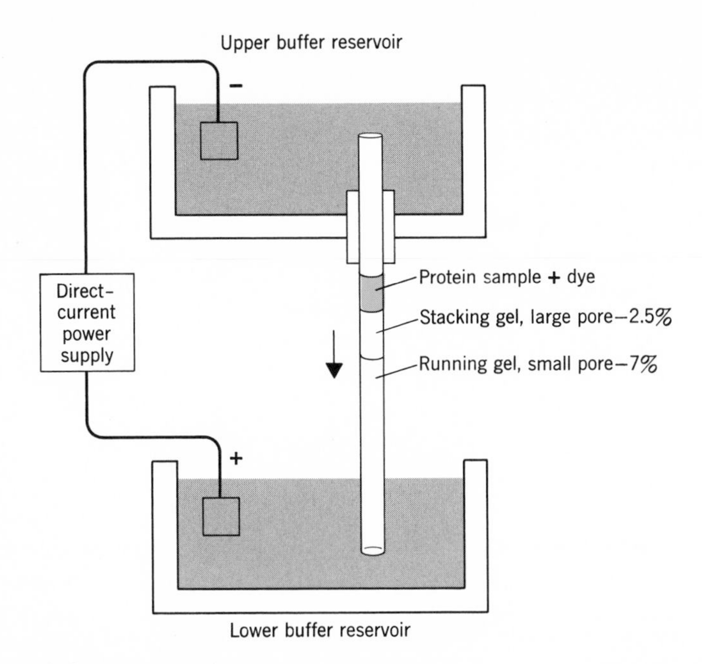

Scheme A-2-8

As depicted in Scheme A-2-8, the apparatus involves a direct-current power supply, and upper and lower reservoir buffer systems which are connected by glass tubes containing in order the protein sample, a stacking gel (2.5% gel), and the running gel (about 6–7%). The gel tubes are prepared by mixing acrylamide with the cross-linking component, methylene-bis-acrylamide and the polymerizing initiator, ammonium persulfate, in the glass tube. With the running gel in place, the stacking gel is prepared above it, the tube is appropriately mounted in the apparatus, the protein solution is added above the stacking gel, and the current is turned on. Frequently, a tracking dye is added with the protein mixture to serve as an indicator of the front of the moving zone as it descends down the tube. When the tracking dye moves to the bottom of the gel column, the current is turned off, and the gel tube is removed, stained with an appropriate dye, and inspected for the number of protein components (Scheme A-2-9).

By a minor modification, namely running an unknown protein and a known protein at different gel concentrations and plotting their log R_{m} against gel concentration and in turn the slopes of each curve against the molecular weight, an accurate molecular weight of the unknown protein can be determined (see Figure A-2-4). Thus, one can ascertain the purity as well as the molecular weight by gel electrophoresis employing microgram quantities of proteins. Modifications involving incorporation of detergents or urea in the gel system allow an estimation of the number of subunits and their molecular weights in a given protein.

Specific Activity/Coenzyme Ratio. If a protein has a coenzyme firmly associated with it, and if, by diverse series of precipitations, a constant ratio of specific

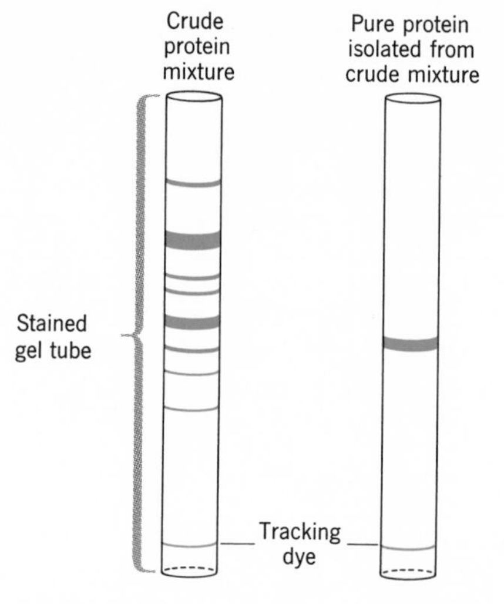

Scheme A-2-9

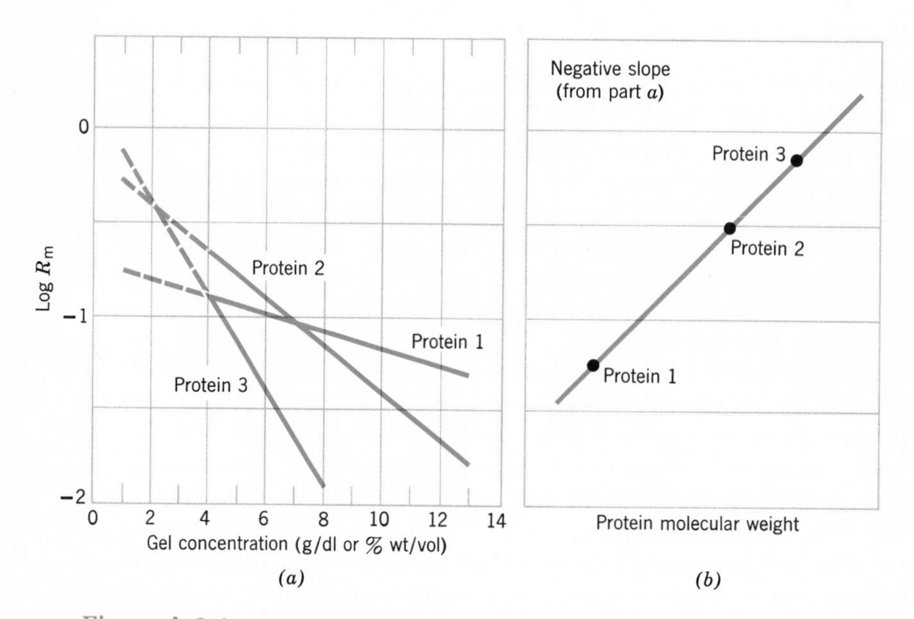

Figure A-2-4

Two graphs showing (*a*) a log R_m plot vs gel concentration where R_m = the ratio of distance which the protein has moved into the running gel divided by the distance that the tracking dye has moved into the running gel; (*b*) plot of negative slope vs molecular weight. (Figures published with permission from *Biochemical Experiments*, by G. Bruening, R. Criddle, J. Preiss, and F. Rudert, Wiley-Interscience, New York, 1970, page 113.)

activity of an enzyme function to the coenzyme concentration is attained, this would be suggestive evidence that a reasonable degree of purification has been achieved.

It might also indicate, however, that the ingenuity of a complex protein system is greater than that of the investigator and that his technique may have achieved little or no separation of contaminating proteins!

Ultracentrifuge. This instrument can measure certain properties of a molecule, such as molecular weight, shape, size, and density, as well as the number of components in a protein solution. The ultracentrifuge subjects a small volume of solution (less than 1 ml contained in a quartz cell) to a carefully controlled centrifugal force, and records, by means of optical and photographic systems, the movement of the macromolecules in the centrifugal field.

A specific method, in which the ultracentrifuge operates at about 55,000 rpm, will be described. As indicated in Figure A-2-5, the solute molecules, which are initially uniformly distributed throughout the solution in the cell, are forced toward the bottom of the cell by the centrifugal field. This migration

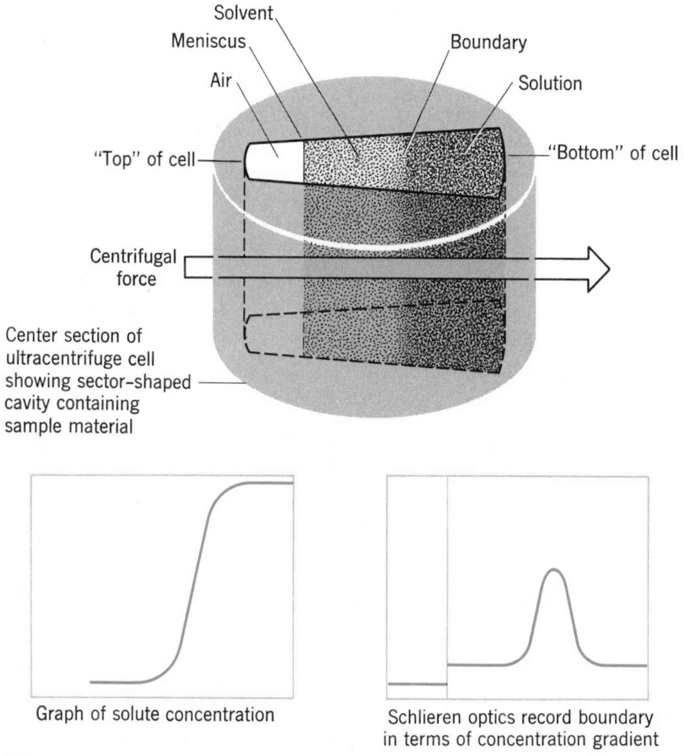

Figure A-2-5

A typical sedimentation velocity study, showing how a boundary formed between solvent and solute molecules can be recorded by a method known as schlieren optics. (Courtesy of Beckman Instruments, Inc.)

leaves a region at the top of the cell that is devoid of solute and contains only solvent molecules. The migration also leaves a region in the cell where the solute concentration is uniform. A boundary is set up in the cell between solvent and solution in which concentration varies with distance from the axis of rotation. The measurement of the boundary's movement, which represents the movement of the protein molecules, is the basis of the analytic method. By the data obtained, namely, the sedimentation rate, the Svedberg unit (S) can be calculated. A Svedberg unit, named in honor of T. Svedberg, the Swedish pioneer in the field, is defined as the velocity of the sedimenting molecule per unit of gravitational field or 1×10^{-13} cm/sec/dyne/g. Typical S values are 4.4 for bovine serum albumin, 1.83 for cytochrome c, and 185 for tobacco mosaic virus. With a knowledge of the diffusion coefficient, molecular weights can be readily calculated. The basic equation relates S and molecular weight:

$$\text{mol wt} = \frac{RTS}{D(1 - V\rho)}$$

where R is the gas constant, T the absolute temperature, S the Svedberg unit, D the diffusion constant, V the partial specific volume, and ρ the density of the solution.

To determine the number of components in a solution, a simple centrifugation can be readily made and the number of boundaries based on concentration gradient peaks can be determined. Diffusion coefficient measurements need not be made.

Methods for Determining Amino Acid Sequences in a Protein

The sequence of amino acids in proteins can be determined by means of three basic analytical procedures: (*a*) identification of the NH_2-terminal amino acid in the protein; (*b*) identification of the COOH-terminal amino acid; and (*c*) partial cleavage of the original polypeptide into smaller polypeptides whose sequence can be determined. In the last procedure, cleavage of the original protein must be carried out in at least two different ways so that the smaller polypeptides produced in one procedure "overlap" those produced in a second procedure and provide an opportunity for identifying the sequence of amino acid in the area of the original chain where the cleavage occurs. The protein whose structure is to be determined must obviously be free of any contaminating amino acids or peptides. Knowing its molecular weight and amino acid composition, the number of times each residue occurs in the protein can be determined. With this information, the determination of sequence can proceed.

Identification of the NH_2-Terminal Amino Acid. When a polypeptide is reacted with 2,4-dinitrofluorobenzene (page 80) the NH_2-terminal group (and the ε-amino group of any lysine that is present in peptide linkage) reacts to form the intensely yellow 2,4-dinitrophenyl derivative of the polypeptide. Subsequent hydrolysis of the peptide with 6*N* HCl hydrolyzes all the peptide bonds, and the yellow derivative of the NH_2-terminal residue (and that of lysine) can be separated by paper chromatography from the free amino acids, compared with known derivatives of the amino acids, and identified. The NH_2-terminal residue can also be identified with the dansyl reagent (page 81).

The reaction of polypeptides with phenylisothiocyanate in dilute alkali (page 81) is the basis for a sequential degradation of a polypeptide that has been devised by P. Edman. In this procedure, the NH_2-terminal group reacts to form a phenylthiocarbamyl derivate. Treatment now with mild acid conditions causes cyclization and cleavage of the NH_2-terminal amino acid as its phenylthiohydatoin. This compound can be separated and compared with the same derivative of known amino acids and thereby identified. The acid conditions utilized to cleave off the phenylthiohydantoin are not sufficiently drastic as to break any other peptide linkages. As a consequence, this method results in the removal and identification of the NH_2-terminal amino acid together with the production of a polypeptide containing *one less* amino acid than the original. This new polypeptide can now be treated with more phenylisothiocyanate in alkali in the same manner and the process repeated many times to degrade the original polypeptide in a stepwise manner.

Phenylisothiocyanate + Polypeptide $\xrightarrow{\text{Alkali}}$ Phenylthiocarbamyl derivative $\xrightarrow{\text{Anhydrous acid}}$ Phenylthiohydantoin derivative of NH_2-terminal amino acid + Polypeptide lacking NH_2-terminal amino acid

Phenylisothiocyanate: $C_6H_5{-}N{=}C{=}S$

Polypeptide: $NH_2{-}CH(R^1){-}C(=O){-}NH{-}CH(R^2){-}C(=O){-}NH{-}CH(R^3){-}C(=O){-}NH{-}CH(R^4){-}C(=O)\cdots$

Phenylthiocarbamyl derivative: $C_6H_5{-}NH{-}C(=S){-}NH{-}CH(R^1){-}C(=O){-}NH{-}CH(R^2){-}C(=O){-}NH{-}CH(R^3){-}C(=O){-}NH{-}CH(R^4){-}C(=O)\cdots$

Phenylthiohydantoin derivative of NH_2-terminal amino acid (ring: $O{=}C{-}N(C_6H_5){-}C({=}S){-}NH{-}C(H)(R^1)$)

Polypeptide lacking NH_2-terminal amino acid: $NH_2{-}CH(R^2){-}C(=O){-}NH{-}CH(R^3){-}C(=O){-}NH{-}CH(R^4){-}C(=O)\cdots$

Identification of the COOH-Terminal Amino Acid. The carboxyl-terminal group of a polypeptide (and the distal carboxyl groups of aspartic and glutamic acid residues in the peptide) can be reduced to the corresponding alcohol with lithium borohydride, $LiBH_4$. It is first necessary to protect the free amino groups by acetylation and to esterify the carboxyl groups. The polypeptide can then be hydrolyzed with acid to produce its constituent amino acids and the amino alcohol corresponding to the COOH-terminal residue. The alcohol can be separated, compared with reference compounds, and identified.

The action of the enzyme carboxypeptidase on polypeptides can also be used to identify the COOH-terminal amino acid, since its action is to hydrolyze that amino acid off the polypeptide. The chief disadvantage is that the enzyme does not act exclusively on the original polypeptide but will also hydrolyze the new COOH-terminal peptide bond as soon as it is formed. Therefore, the investigator must follow the rate of formation of free amino acids to learn which residue represents the terminal in the original polypeptide.

Cleavage of Protein into Smaller Units. Both enzymatic and chemical procedures have been utilized to produce smaller polypeptides that overlap in sequences with the present protein. Partial hydrolysis by dilute acid can be employed. Cyanogen bromide (CNBr) is also used since conditions can be chosen which will cleave only those peptide bonds in which the carbonyl group belongs to a methionine residue. The methionine residue becomes a substituted lactone of homoserine that is bound to one of the two peptides produced in the reaction. This procedure allows one to determine the amino acids in the region of the methionine residues in the original peptide. In addition, knowing the number of methionine residues in the original polypeptide, one can predict the number of smaller polypeptides that will result from treatment with CNBr.

Original polypeptide: ⋮—NH—CH(R^1)—C(=O)—HN—HC(CH_2—CH_2—S—CH_3)—C(=O)—NH—CH(R^2)—C(=O)—⋮ + CNBr ⟶

New polypeptide containing modified methionine residue: ⋮—NH—CH(R^1)—CO—HN—HC—C(=O)—O (lactone ring: HC—CH_2—CH_2—O—C=O)

+ H_3C—S—CN + HBr

+ New polypeptide produced on cleavage: NH_2—CH(R^2)—C(=O)—⋮

Proteolytic enzymes have been extensively used to cleave proteins into smaller polypeptides which can then be analyzed by the procedures described above. Trypsin, for example, is known to hydrolyze those peptide bonds in which the carbonyl group is contributed by either lysine or arginine. As with the cyanogen bromide reaction, one can predict the number of polypeptides that will be formed by the action of trypsin if the number of lysine and arginine residues in the protein is known.

Chymotrypsin will hydrolyze those peptide bonds in which the carbonyl group belongs to phenylalanine, tyrosine, or tryptophan. Pepsin cleaves the peptide bonds in which the amino group is furnished by phenylalanine, tyrosine, tryptophan, lysine, glutamic, and aspartic acids. By utilizing trypsin, whose action is quite specific, and either chymotrypsin or pepsin, the investigator can obtain fragments of the original protein or polypeptide that overlap in sequence. Once the sequence of amino acids in these fragments is known, the process of fitting together the individual fragments can proceed. If, as in the case of insulin (page 454), the original protein can be easily separated into two parts by simple reduction of disulfide bonds, the sequence determination of the two separate chains can proceed.

References

1. S. P. Colowick and N. O. Kaplan, eds., *Methods in Enzymology*. New York: Academic Press, 1955 to date.
   This multivolume work contains general and specific references to nearly all procedures and methods employed in biochemistry. The articles have been authored by the experts in the field.
2. G. Bruening, R. Criddle, J. Preiss, and F. Rudert, *Biochemical Experiments*. New York: Wiley-Interscience, 1970.
   A very useful laboratory manual with good discussions of techniques of general interest to biochemists. The experiments are guaranteed to work!

APPENDIX 3

Biochemical Literature

The biochemical literature is vast in size. As with any rapidly developing field there are several publications exclusively devoted to the rapid communication of new and significant advances. The list of journals that contain full-length research papers dealing with some area of biochemistry presently exceeds 50. Numerous serial publications that appear annually contain review articles that cover the advances since the preceding review paper. For the student who wishes to examine more deeply the biochemical literature, we cite here some of the more familiar sources of information:

Review publications containing articles written by an expert in the field:

Annual Reports of the Chemical Society
Annual Review of Biochemistry
Advances in Enzymology and Related Areas of Molecular Biology
Biochemical Society Symposia
Essays in Biochemistry

Research journals:

A. Journals designed for rapid communications of new advances.
Proceedings of the National Academy of Science (U.S.)
Biochemical and Biophysical Research Communications
Letters of the Federation of European Biochemical Societies (FEBS Letters)
Science
Nature
Naturwissenschaften
(In addition, short communication sections are found in a number of the journals listed below.)

B. Journals containing complete research papers.
Analytical Biochemistry
Archives of Biochemistry and Biophysics
Biochemical Journal
Biochemistry
Biochimica et Biophysica Acta
Canadian Journal of Biochemistry
European Journal of Biochemistry
Hoppe-Seyler's Zeitschrift fur Physiologische Chemie
Journal of Biochemistry (Tokyo)
Journal of Biological Chemistry
Journal of Molecular Biology
Lipids
Plant Physiology
Phytochemistry

Multivolume publications in biochemistry:

The Enzymes, 3rd ed. P. D. Boyer, ed. New York: Academic Press, 1970–present.
The third edition of an early series of enzymology will cover all phases of enzyme chemistry.

Methods in Enzymology. N. O. Kaplan and S. Colowick, eds. New York: Academic Press, 1955–present.
This series covers all phases in the preparation and assays of enzymes, and describes many biochemical techniques.

Comprehensive Biochemistry. M. Florkin and E. H. Stotz, eds. New York: Academic Press.
A multivolume work that began appearing in 1967.

Biology and the Future of Man. P. Handler, ed. New York: Oxford University Press, 1970.
This collection projects the role of biochemistry to the solution of world problems.

The student will also find it worthwhile to browse in bookstores and examine the numerous textbooks that are available. They include both general texts and monographs on specific topics.

Index